REVIEWS IN MINERALOGY   VOLUME 28

HEALTH EFFECTS OF MINERAL DUSTS

George D. Guthrie, Jr.,
Brooke T. Mossman, *Editors*

Proceedings of a short course endorsed by
THE AMERICAN COLLEGE OF CHEST PHYSICIANS
and THE U. S. GEOLOGICAL SURVEY

COVER: Scanning electron micrograph of ferruginous bodies extracted from autopsied human lung. The individual was exposed primarily to chrysotile. The particles consist of asbestos fibers coated by an iron-rich material believed to derive from proteins such as ferritin or hemosiderin. The ferruginous bodies are generally about 5 to 30-µm long; the species of asbestos is not known. Photo courtesy of Lesley S. Smith and Anne F. Sorling (Department of Pathology, Fox Chase Cancer Center, Philadelphia, Pennsylvania).

MINERALOGICAL SOCIETY OF AMERICA
WASHINGTON, D.C.

COPYRIGHT 1993
MINERALOGICAL SOCIETY OF AMERICA
Printed by BookCrafters, Inc., Chelsea, Michigan.

REVIEWS IN MINERALOGY
(Formerly: SHORT COURSE NOTES)
ISSN 0275-0279
Volume 28: *Health Effects of Mineral Dusts*
ISBN 0-939950-33-2

ADDITIONAL COPIES of this volume as well as those listed below may be obtained at moderate cost from the MINERALOGICAL SOCIETY OF AMERICA, 1130 Seventeenth Street, N.W., Suite 330, Washington, D.C. 20036 U.S.A.

Vol.	Year	Pages	Editor(s)	Title
1,3,4,6	*out*	*of*	*print*	
2	1983	362	P.H. Ribbe	FELDSPAR MINERALOGY (2nd edition)
5	1982	450	P.H. Ribbe	ORTHOSILICATES (2nd edition)
7	1980	525	C.T. Prewitt	PYROXENES
8	1981	398	A.C. Lasaga R.J. Kirkpatrick	KINETICS OF GEOCHEMICAL PROCESSES
9A	1981	372	D.R. Veblen	AMPHIBOLES AND OTHER HYDROUS PYRIBOLES—MINERALOGY
9B	1982	390	D.R. Veblen, P.H. Ribbe	AMPHIBOLES: PETROLOGY AND EXPERIMENTAL PHASE RELATIONS
10	1982	397	J.M. Ferry	CHARACTERIZATION OF METAMORPHISM THROUGH MINERAL EQUILIBRIA
11	1983	394	R.J. Reeder	CARBONATES: MINERALOGY AND CHEMISTRY
12	1983	644	E. Roedder	FLUID INCLUSIONS (Monograph)
13	1984	584	S.W. Bailey	MICAS
14	1985	428	S.W. Kieffer A. Navrotsky	MICROSCOPIC TO MACROSCOPIC: ATOMIC ENVIRONMENTS TO MINERAL THERMODYNAMICS
15	1990	406	M.B. Boisen, Jr. G.V. Gibbs	MATHEMATICAL CRYSTALLOGRAPHY (Revised)
16	1986	570	J.W. Valley H.P. Taylor, Jr. J.R. O'Neil	STABLE ISOTOPES IN HIGH TEMPERATURE GEOLOGICAL PROCESSES
17	1987	500	H.P. Eugster I.S.E. Carmichael	THERMODYNAMIC MODELLING OF GEOLOGICAL MATERIALS: MINERALS, FLUIDS, MELTS
18	1988	698	F.C. Hawthorne	SPECTROSCOPIC METHODS IN MINERALOGY AND GEOLOGY
19	1988	698	S.W. Bailey	HYDROUS PHYLLOSILICATES (EXCLUSIVE OF MICAS)
20	1989	369	D.L. Bish, J.E. Post	MODERN POWDER DIFFRACTION
21	1989	348	B.R. Lipin, G.A. McKay	GEOCHEMISTRY AND MINERALOGY OF RARE EARTH ELEMENTS
22	1990	406	D.M. Kerrick	THE Al_2SiO_5 POLYMORPHS (Monograph)
23	1990	603	M.F. Hochella, Jr. A.F. White	MINERAL-WATER INTERFACE GEOCHEMISTRY
24	1990	314	J. Nicholls J.K. Russell	MODERN METHODS OF IGNEOUS PETROLOGY—UNDERSTANDING MAGMATIC PROCESSES
25	1991	509	D.H. Lindsley	OXIDE MINERALS: PETROLOGIC AND MAGNETIC SIGNIFICANCE
26	1991	847	D.M. Kerrick	CONTACT METAMORPHISM
27	1992	508	P.R. Buseck	MINERALS AND REACTIONS AT THE ATOMIC SCALE: TRANSMISSION ELECTRON MICROSCOPY

Reviews in Mineralogy   Volume 28

HEALTH EFFECTS OF MINERAL DUSTS

FOREWORD

The Mineralogical Society of America (MSA) has been sponsoring short courses in conjunction with their annual meetings with the Geological Society of America since 1974. This volume represents the proceedings of a course by the same title held at Harbor House Resort and Conference Center on Nantucket Island off the coast of Massachusetts, October 22-24, 1993. George Guthrie (Los Alamos National Laboratory) and Brooke Mossman (University of Vermont) convened the short course for reasons detailed in the Preface below. They also had primary responsibility for assembling the papers from the many contributors representing mineralogical, biological, and medical disciplines. This is the first truly interdisciplinary effort in the history of MSA short courses, and we have great expectations for it and this book.

I personally thank George Guthrie for heroic efforts to make this volume our technically most uniform product; e.g., he reformatted and retyped all the tables! The attention he and Brooke Mossman gave to assure that the scientific content was both relevant and correct is greatly appreciated. Mineralogy graduate student Brett Macey worked long hours in preparation of the camera-ready copy, and secretary Margie Sentelle contributed in her usual cheerful and skillful manner.

Paul H. Ribbe
Series Editor
Blacksburg, VA
September 15, 1993

PREFACE AND ACKNOWLEDGMENTS

Numerous minerals are known to induce pulmonary diseases. The asbestos minerals (chrysotile and asbestiform amphiboles) are by far the most infamous. However, a number of silica polymorphs, clays, and zeolites have also been studied in great detail, as have several titania polymorphs, hematite, and magnetite (which are often used as negative controls in biological experiments). In fact, the relatively recent attention received by erionite (a fibrous zeolite) has arguably made it the most notorious of the minerals studied thus far.

The processes that lead to the development of disease (or pathogenesis) by minerals very likely occur at or near the mineral–fluid interface (as do many geochemical processes!). Thus the field of "mineral-induced pathogenesis" is a prime candidate for interdisciplinary research, involving mineral scientists, health scientists, petrologists, pathologists, geochemists, biochemists, and surface scientists, to name a few. The success of such an interdisciplinary approach rests on the ability of scientists in very different fields to communicate, and this is hampered by vocabulary barriers and an unfamiliarity with concepts, approaches, and problems. It can be difficult enough for a geoscientist or bioscientist to maintain fluency in the many fields tangential to his or her own field, and this problem is only exacerbated when one investigates problems that are cross-disciplinary. Nevertheless, important advances can be facilitated if these barriers are overcome.

This review volume and the short course upon which it was based are intended to provide some of the necessary tools for the researcher interested in this area of interdisciplinary research. The chapters present several of the important problems, concepts, and approaches from both the geological and biological ends of the spectrum. These two extremes are partially integrated throughout the book by cross-referencing between chapters. Chapter 1 also presents a general introduction into the ways in which these two areas overlap. However, many of the areas ripe for the interdisciplinarian will become obvious after reading the various chapters. The final chapter of this book discusses some of the regulatory aspects of minerals. Ultimately, the regulatory arena is where this type of interdisciplinary approach can make an impact, and hopefully better communication between all parties will accomplish this goal. A glossary is included at the end of this book, because the complexity of scientific terms in the two fields can thwart even the most enthusiastic of individuals.

We thank several organizations for recognizing the importance of an interdisciplinary approach to this problem and, thus, for supporting our efforts. The Division of Engineering and Geosciences (W.C. Luth) of the Office of Basic Energy Sciences and the Health Effects and Life Sciences Section (K. Johnson) of the Office of Health and Environmental Research (both of the U.S. Department of Energy) jointly provided generous support for the short course. Without the support from each of these organizations, this short course would have been difficult to accomplish. The U.S. Department of Energy, through the Los Alamos National Laboratory, provided initial funds for Guthrie's contribution to the editing of the book and very substantial financial support for the first printing of this volume.

In addition to this monetary support, we have received endorsement from the American College of Chest Physicians and the U.S. Geological Survey.

We are grateful for the work done by Susan Myers and the Mineralogical Society of America to organize the logistics of the short course. We thank A. Werner and B. Carey for reviewing some of the manuscripts; and B. Hahn, C. White, A. Garcia, and E. Montoya for help with corrections to manuscripts and the redrafting of some figures. Finally, we are indebted to Paul Ribbe and his staff for making the book a reality. Paul transformed the manuscripts into presentable, camera-ready chapters, and this was no small task. At the Geological Society of America annual meeting following this short course, Paul will receive the Distinguished Public Service Award for his contributions as Series Editor for the *Reviews in Mineralogy* series. This volume is the 28th of a series that began in 1974, and Paul has been Series Editor for almost all of them.

    George Guthrie   Brooke Mossman
    Los Alamos, New Mexico   Burlington, Vermont

9 September 1993

TABLE OF CONTENTS

	Page
Copyright; List of additional volumes of *Reviews in Mineralogy*	ii
Foreword; Preface and Acknowledgments	iii

Chapter 1   G. D. Guthrie, Jr. & B. T. Mossman

MERGING THE GEOLOGICAL AND BIOLOGICAL SCIENCES: AN INTEGRATED APPROACH TO THE STUDY OF MINERAL-INDUCED PULMONARY DISEASES

Introduction	1
References	5

Chapter 2   C. Klein

ROCKS, MINERALS, AND A DUSTY WORLD

Introduction	7
Some Mineralogical Background	9
General references	9
Quantitative characterization of mineral particles in dust	10
Chemical variation in minerals	10
Chemical characterization of dust particles	13
Examples of Potentially Hazardous Minerals in Dust	14
Amphiboles	14
Chrysotile	17
Other layer silicates	19
Kaolinite	20
Vermiculite	21
Montmorillonite	21
Talc	22
Muscovite	22
Chlorite	23
Sepiolite and palygorskite	25
Silica (SiO_2) group	25
Zeolite group	27
Roggianite	29
Mazzite	30
Erionite	30
Mordenite	31
Natural Dusts	32
Generation and migration of natural dusts	32
The hydrologic cycle and weathering processes	32
Erosion rates and source areas of natural dust	35
Wind action	37
Short distance eolian transport	39
Long distance eolian transport	39
Volcanic activity	41
Determination of the background levels of natural dusts	42
Lung particulate burdens	42

        Global background level for fiber counts in the troposphere 44
        Measurements from fluvial sources 46
        Dust from Antarctic ice cores 47
    The geology of two major natural fiber sources 49
        New Idria (Coalinga) chrysotile region, California 49
        Riebeckite and crocidolite in the Hamersley Range of Western
            Australia 52
Concluding Remarks 54
Acknowledgments 56
References 56

Chapter 3   D. R. Veblen & A. G. Wylie

MINERALOGY OF AMPHIBOLES AND 1:1 LAYER SILICATES

Introduction 61
    Existing review literature 61
Crustal Elemental Abundances, Structure Determination, and Representation of
    Silicate Mineral Structures 63
        Crustal abundance of the chemical elements 63
        Determination and refinement of crystal structures 63
        Coordination structure and their representation in two and three
            dimensions 64
        Polymerization in silicates 66
Mineral Habit 67
    Asbestos and the asbestiform habit 69
    Cleavage and parting 71
1:1 Layer Silicates 71
    Basic 1:1 layer structure 71
    Nomenclature 73
    Geological occurrence of 1:1 layer silicates and associated minerals 74
        Serpentine group 74
        Kaolin group 75
        Other 1:1 layer silicates 76
    Crystal structures and chemistry of the serpentine group 76
        Lizardite 77
            Layer conformation 77
            Polytypism and symmetry 77
            Chemical variations 78
        Chrysotile 78
            Layer conformation 78
            Chrysotile polymorphs and polytypes 80
            Chemical variations 81
        Antigorite 81
            Layer conformation 81
            Structural disorder and defects in antigorite 83
            Polytypism and chemical variations 83
        Carlosturanite 84
        Polygonal serpentine 84
        Serpentine intergrowths and other hybrid serpentine structures 84
    Crystal structures and chemistry of the kaolin group 85
        Kaolinite, dickite, and nacrite 85
            Layer conformation 87
            *Layer stacking, octahedral vacancy distribution, and hyd
                rogen bonding* 87
            Structural disorder in kaolinite 88

Chemical variations	90
Halloysite	90
Layer conformation	90
Layer stacking and other structural details	90
Crystal structures of other 1:1 layer silicates	91
Amesite and kellyite	91
Berthierine, brindleyite, and fraipontite	91
Odinite	92
Cronstedtite	92
Nepouite and pecoraite	92
Habit of 1:1 layer silicates	92
Serpentine-group minerals	92
Kaolin-group minerals	94
Other 1:1 layer silicates	94
Surface chemistry, surface charge, and dissolution kinetics of chrysotile	95
Surface chemistry and structure	95
Surface charge	97
Dissolution kinetics	97
Amphiboles	100
Basic structure, chemistry, and nomenclature	100
Geological occurrence of amphiboles	101
Amphibole crystal structure types	102
Clinoamphiboles	103
Polyhedral distortions and rotations	103
Space-group symmetry	104
Orthoamphiboles	105
Protoamphibole	105
The biopyribole polysomatic series and ordered pyriboles related to amphibole	105
Defects and grain boundaries in amphiboles	107
Chain-width errors	107
Twinning	108
Stacking faults	108
Dislocations	108
Exsolution lamellae	109
Grain boundaries	109
Compositional variations in amphibole asbestos	109
Crystallization, mineralogy, and structure of amphibole asbestos and other amphibole habits	110
Asbestiform amphibole	110
Byssolite and nephrite	116
Cleavage and parting fragments of nonasbestiform amphiboles	116
Dimensions of amphibole fibers and cleavage fragments	116
Surface chemistry, surface charge, and dissolution kinetics of amphibole asbestos	117
Surface chemistry and structure	117
Surface charge	119
Dissolution kinetics	119
Implication of 1:1 Layer Silicates, Amphiboles, and Wide-Chain Pyriboles	120
Optical microscopy (OM)	120
Kaolin	120
Serpentine	120
Amphiboles and wide-chain pyriboles	121
X-ray diffraction (XRD)	122
1:1 layer silicates	122
Amphiboles and wide-chain pyriboles	123
Scanning electron microscopy (SEM) and electron microprobe analysis	123

The scanning electron microscope	123
Quantitative and qualitative chemical analysis in the SEM	124
Electron microprobe analysis	125
Identification of 1:1 layer silicates and asbestos with the SEM	125
Transmission electron microscopy (TEM) methods for asbestos identification	126
Chrysotile asbestos	127
Amphibole asbestos	127
Analysis of asbestos in bulk samples	128
Analysis of asbestos in air samples and lung burden	130
Acknowledgments	130
References	131

Chapter 4   D. L. Bish & G. D. Guthrie, Jr.
MINERALOGY OF CLAY AND ZEOLITE DUSTS (EXCLUSIVE OF 1:1 LAYER SILICATES)

Introduction	139
Hydroxides	142
Geological occurrence	142
Crystal chemistry	142
Crystal structures	143
Microstructures and morphologies	145
Surface properties	146
Adsorption characteristics	148
2:1 Layer Silicates	149
Geological occurrence	149
Crystal chemistry	150
Crystal structures	154
Microstructures and morphologies	160
Surface properties (potentially active surface sites)	160
Catalytic properties	161
Chain-structure Layer Silicates	162
Geological occurrence	162
Common associated minerals	162
Crystal chemistry	162
Crystal structures	163
Microstructures and morphologies	165
Surface properties	166
Exchange characteristics	167
Uses	167
Zeolites	168
Geological occurrence	168
Crystal chemistry	169
Crystal structure	172
Morphologies	178
Molecular sieving, exchange, and catalytic properties	179
References	181

Chapter 5
P. J. Heaney & J. A. Banfield
STRUCTURE AND CHEMISTRY OF SILICA, METAL OXIDES, AND PHOSPHATES

Introduction	185
The Silica System	186
Phase equilibria and geological occurrences	186
Crystalline silica	186
Amorphous silica	188
Phase stability	190
Silica: Applications and regulations	190
Uses	190
Restrictions	191
Structures of the silica polymorphs	194
Quartz	194
Microcrystalline quartz	195
Tridymite and cristobalite	197
Coesite and stishovite	200
Amorphous silica	200
Impurity elements	202
Solubilities and surface structure	203
Interactions between organic molecules and silica	205
Carcinogenicity of silica	206
Geochemistry of Iron and Titanium Oxides	206
Introduction	206
The TiO_2 system	208
Occurrence and commonly associated minerals	208
Crystal structures	209
Rutile	209
Brookite and anatase	209
$TiO_2(B)$	209
Trace element chemistry and defect microstructures	211
Particle morphology	212
Potentially active surface sites	212
Solubility	216
Iron oxides and iron-titanium oxides	216
Occurrence and commonly associated minerals	217
Crystal structures	218
Hematite (α-Fe_2O_3) and ilmenite	218
Magnetite and ülvospinel	219
Maghemite and titanomaghemite	220
Wüstite	220
Trace element chemistry and defect microstructures	220
Particle morphology	221
Potentially active surface sites	221
Solubility	222
Phosphates	223
Acknowledgments	224
References	225

Chapter 6   S. J. Chipera, G. D. Guthrie, Jr. & D. L. Bish
PREPARATION AND PURIFICATION OF MINERAL DUSTS

Introduction 235
Sources for Mineral Specimens 236
Preparation of Mineral Dust Specimens 236
    Disaggregation 236
    Purification 237
        Sieving 237
        Magnetic separation 237
        Density separation 238
        Separation based on settling velocity 240
        Combined separation techniques 243
    Size fractionation 244
Summary 245
References 246
Appendix I 246
    Mineral specimens 246
    Equipment/Supplies 247
Appendix II 247
    Principle 247
    Equipment and materials 247
    Methodology 248

Chapter 7   G. D. Guthrie, Jr.
MINERAL CHARACTERIZATION IN BIOLOGICAL STUDIES

Introduction 251
    Mineral species 254
Mineral Content 255
    X-ray diffraction 255
    Transmission electron microscopy 260
Mineral Structures 260
    Deviations from ideal structure 261
Mineral Compositions 261
    Electron probe microanalysis 261
    Analytical electron microscopy 266
    PIXE and SIMS 267
        Proton-induced X-ray emission (PIXE) 267
        Secondary ion mass spectrometry (SIMS) 268
Mineral Surfaces 268
    Surface structure 269
        Scanning electron microscopy 269
        Scanning probe microscopies 269
        Low energy electron diffraction 269
        Transmission electron microscopy 270
    Determination of surface composition 270
Acknowledgments 270
References 270

Chapter 8   M. F. Hochella, Jr.

SURFACE CHEMISTRY, STRUCTURE, AND REACTIVITY OF HAZARDOUS MINERAL DUST

Introduction	275
The General Nature of Mineral Surfaces	276
Surface composition	277
Surface atomic structure	278
Surface microtopography	279
Surface charge	282
Relationship between Mineral Surfaces and Their Biological Activity	284
Evidence for activity dependence on surface composition	284
Evidence for activity dependence on surface atomic structure	286
Evidence for activity dependence on surface microtopography	287
Evidence for activity dependence on surface charge	288
The Surfaces of Chrysotile and Crocodolite	289
Chrysotile	289
General surface description	289
Surface site character and surface charge	290
Surface site reactivity	291
Dissolution behavior	292
Implications for biological activity	292
Crocidolite	293
General surface description	293
Surface site character and surface charge	296
Surface site reactivity	296
Dissolution behavior	297
Implications for biological activity	299
Treatment with deferoxamine	300
Summary and Conclusions	303
Acknowledgments	305
References	305

Chapter 9   R. P. Nolan & A. M. Langer

LIMITATIONS OF THE STANTON HYPOTHESIS

Introduction	309
Early Experimental Animal Studies Relating Fiber Morphology to Carcinogenicity	310
Dosimetry of fibers	316
Conclusions of the implantation studies	319
Limitations of the Stanton hypothesis	321
References	325

Chapter 10   R. F. Giese, Jr. & C. J. van Oss

THE SURFACE THERMODYNAMIC PROPERTIES OF SILICATES AND THEIR INTERACTIONS WITH BIOLOGICAL MATERIALS

Introduction	327
Surface Thermodynamic Theory	328

Surface tension, interfacial tension and surface free energies 328
    The apolar component 329
    The polar component 329
Free energy of adsorption 330
Contact angles and the Young equation 331
Powdered materials and the Washburn equation 333
Hydrophobic and hydrophilic surfaces 334
Electrostatic Interactions 335
    ζ-potential 335
    Total interaction energy between two particles 336
Surface Tension of Minerals and Related Materials 336
    Asbestos 337
    Sepiolite and palygorskite 338
    2:1 layer silicates 339
        Smectite 339
        Muscovite and illite 339
        Talc and pyrophyllite 340
    Quartz 340
    Glass, natural and synthetic 340
Effects of Surface-Sorbed Organic Material 341
Surface Properties of Biological Materials 342
Interactions Between Mineral Surfaces and Biological Materials 342
    Average non-specific repulsion 343
    Discrete site attraction 343
    Low energy surfaces 344
References 344

Chapter 11   A. B. Kane

EPIDEMIOLOGY AND PATHOLOGY OF ASBESTOS-RELATED DISEASES

Historical Introduction 347
Key Epidemiologic Studies of Asbestos-related Diseases 348
    Major diseases associated with asbestos exposure 348
    Studies in occupational epidemiology of asbestos-related diseases 349
        Case series 349
        Case control studies 349
        Cohort studies 350
        Limitations of epidemiologic studies 352
Clinical and Pathologic Features of Asbestos-related Diseases 352
    Pulmonary diseases 352
        Asbestosis 352
        Lung cancer 353
    Pleural reactions 353
    Other diseases associated with asbestos exposure 354
    Pathogenesis of asbestos-related diseases 354
    Implications of epidemiologic studies for prevention of asbestos-related diseases 356
Acknowledgments 358
References 358

Chapter 12 — M. Ross, R. P. Nolan, A. M. Langer & W. C. Cooper

HEALTH EFFECTS OF MINERAL DUSTS OTHER THAN ASBESTOS

Introduction...361
The Silica Minerals and Amorphous Silica.................................362
    Mineralogy of silica...362
    Diseases related to exposure to silica dust........................363
        Silicosis..363
        Silicotuberculosis...364
        Cancer...364
    Epidemiological studies of occupational cohorts exposed to crystalline silica dust...366
        Minnesota iron ore miners (magnetite-bearing rock)..............366
        Iron ore miners (hematite-bearing rock)........................369
        European iron ore miners.......................................370
        Gold miners (South Dakota).....................................371
        Gold miners (Kalgoorlie, Western Australia)....................375
        Granite workers (Vermont)......................................375
        Slateworkers (Vermont, North Wales, and Germany)...............377
        Diatomaceous earth workers.....................................378
        Sandblasting...381
        Cohort studies of certified silicotics.........................381
Coal...383
    Mineralogy...383
    Diseases related to exposure to coal dust..........................383
        Coal workers' pneumoconiosis...................................383
        Coal workers' silicosis.......................................384
    Epidemiological studies of occupational cohorts exposed to coal dust.......384
        Pneumoconiosis mortality and morbidity.........................385
        Cancer mortality...385
The Silicate Minerals (Other than Asbestos)..............................386
    Mineralogy of the silicates..386
    Health effects of selected silicate minerals.......................387
        Talc and pyrophyllite..387
        Kaolinite..388
        Bentonite..388
        Palygorskite/attapulgite and sepiolite.........................388
        Micas..390
        Vermiculite..390
        Wollastonite...394
        Zeolites...394
Environmental Exposures to Fibrous Minerals..............................394
    Environmental exposure to mineral fiber in Turkey, Cyprus, and Greece..394
    Fibrous zeolites and the Karain experience.........................396
    Experimental animal studies corroborate the human studies..........398
References...401

Chapter 13 — A. Churg

ASBESTOS LUNG BURDEN AND DISEASE PATTERNS IN MAN

Introduction...410
Analytical Methods...410

The Relationship between Physical/Chemical Properties of the Asbestos
  Minerals and Biological Behavior..412
Asbestos Fibers in the Lungs of the General Population...............................414
Fiber Concentration in the Lungs of Occupationally-exposed Groups...............416
Effects of Fiber Distribution...421
Effects of Fiber Size..422
Acknowledgments..424
References..424

Chapter 14

B. E. Lehnert

DEFENSE MECHANISMS AGAINST INHALED PARTICLES AND ASSOCIATED PARTICLE-CELL INTERACTIONS

Introduction..427
Deposition of Particles in the Respiratory Tract..427
Particles in the Conducting Airways...431
  Local defensive mechanisms against deposited particles.........................431
  Particle-airway epithelial cell interactions..435
  Solublization of particles in the conducting airways...............................437
  Kinetics of particle clearance from the conducting airways.....................437
Particles in the Pulmonary-alveolar Region..439
  Defense against particles by alveolar phagocytes...................................439
  Detrimental aspects of the phagocytic protective mechanism..................447
  Alveolar macrophage-mediated particle clearance..................................448
  Other mechanisms of particle clearance and retention............................451
  Kinetics of particle clearance from the alveoli...455
  Particle "overloading"..456
Acknowledgments..460
References..460

Chapter 15

J. M. G. Davis

IN VIVO ASSAYS TO EVALUATE THE PATHOGENIC EFFECTS OF MINERALS IN RODENTS

Methods of Exposure...471
Pathogenic Potential of Amphiboles and Serpentines..................................472
  Non-pulmonary pathogenesis..474
  Importance of fiber geometry..474
  Relative pathogenic potential of serpentine and amphibole....................475
Pathogenic Potential Man-made Mineral Fibers...476
Pathogenic Potential of Clay Minerals..477
  Kaolin minerals...477
  Talc..478
  Mica...479
  Sepiolite and palygorskite ("attapulgite")...479
  Vermiculite..480
Pathogenic Potential of Zeolites...480
The Pathogenic Potential of Silica..481
References..483

Chapter 16   K. E. Driscoll

IN VITRO EVALUATION OF MINERAL CYTOTOXICITY AND INFLAMMATORY ACTIVITY

Introduction 489
In Vitro Approaches to Study Mineral Dust Toxicity 490
Cytotoxicity Assays 490
      Erythrocyte hemolysis 490
      Macrophage toxicity 492
      Hemolytic activity and cytotoxicity: Their relationship to *in vivo* toxicity . 495
Macrophage Activation and Inflammatory Activity 496
      Active oxygen species 496
      In vitro effect of minerals on macrophage cytokine production 500
Summary 504
References 506

Chapter 17   B. T. Mossman

CELLULAR AND MOLECULAR MECHANISMS OF DISEASE

Introduction 513
The Disease Process 513
Carcinogenesis and Fibrosis: General Principles 516
Mechanisms of Lung Cancer by Asbestos and Silica 517
Mechanisms of Mesothelioma by Asbestos 518
Mechanisms of Pulmonary Fibrosis (Asbestosis and Silicosis) 519
Acknowledgments 520
References 520

Chapter 18   U. Saffiotti, L. N. Daniel, Y. Mao, A. O. Williams,
                              M. E. Kaighn, N. Ahmed & A. D. Knapton

BIOLOGICAL STUDIES ON THE CARCINOGENIC MECHANISMS OF QUARTZ

Coordinated *In Vivo*, Cellular, and Molecular Approaches 523
Crystalline Silica Used by Biological Studies 524
Animal Models and Species Susceptibility 525
      Cytokines and gene expression in the pathogenesis of silicosis, alveolar t
        ype II hyperplasia and lung carcinogenesis 532
      Cellular Models for Toxicity and Neoplastic Transformation 536
DNA Binding and DNA Damage by Crystalline Silica 539
Conclusions 541
References 542

Chapter 19   V. T. Vu
Regulatory Approaches to Reduce Human Health Risks Associated with Exposures to Mineral Fibers

Introduction..545
Major Regulations and Guidelines on Asbestos..546
    Occupational exposure limits and work practices 546
    Air emissions control and waste disposal..547
    Control of asbestos exposure in schools and buildings......................548
    Health standards for drinking water and effluent guidelines 550
    Restriction or prohibition of the use of asbestos in certain products
        and applications..550
Regulatory Activities on Other Mineral Fibers...551
    Erionite..551
    Refractory ceramic fibers...551
    Glass fiber and mineral wool..552
Conclusions 552
References..553

Glossary 555

Index 577

CHAPTER 1

MERGING THE GEOLOGICAL AND BIOLOGICAL SCIENCES: AN INTEGRATED APPROACH TO THE STUDY OF MINERAL-INDUCED PULMONARY DISEASES

George D. Guthrie, Jr.

Geology and Geochemistry Group, EES–1
Los Alamos National Laboratory
Los Alamos, New Mexico 87545 U.S.A.

Brooke T. Mossman

Department of Pathology
College of Medicine
University of Vermont
Burlington, Vermont 05405 U.S.A.

INTRODUCTION

Most natural solids are minerals.[1] Minerals are ubiquitous on the surface of the Earth: They are the materials from which rocks are made, and they are major constituents of soils. In addition, numerous minerals and other crystalline solids are exploited industrially and residentially for their unique properties. Industrially, minerals (and their synthetic analogs) are commonly used as abrasives, pharmaceuticals, catalysts, ion-exchangers, molecular sieves, additives to impart specific rheological properties, ceramics, fillers, anti-caking agents in spices, building materials, insulation, pigments,... Many of these same properties are utilized in household products. Consequently, each of us is exposed to minerals daily.

Unfortunately, several minerals are known to induce a variety of pulmonary diseases (pneumoconioses and cancers) following inhalation. These mineral-induced diseases include fibrosis (a hardening of the lung associated with excess collagen), lung cancers, and mesothelioma (a rare cancer frequently associated with exposure to fibrous minerals).

Research on the initiation of diseases (pathogenesis) by minerals has progressed rapidly over the last century. From about 1900 through the 1960s, epidemiological evidence amassed implicating asbestos exposure as a cause of fibrosis, lung cancer, and mesothelioma (see chronology in Murray, 1990). Similarly, other minerals were recognized during this period as potential pulmonary toxicants. These observations led to numerous studies during the 1940s, 1950s, and 1960s on the pathologies of diseases associated with exposure

---

[1]   In this book, the term "mineral" is used in its strict sense, *i.e.*, a naturally occurring crystalline solid.

to mineral dusts, and it became clear that a variety of minerals could affect the respiratory system.

The research in this field has proceeded rapidly to the point of addressing the question: What are the mechanisms by which a mineral can induce disease? During the 1980s and 1990s, this question has been a primary focus of numerous studies on mineral-induced pathogenesis. Clearly, being able to answer this question depends on biological, biochemical, and pathological research. However, any model for mineral-induced pathogenesis necessarily entails mineralogical and geochemical factors, because the biochemical processes that lead to disease occur at or near the mineral surface. Factors such as durability, solubility, tensile strength, surface reactivity, surface structure, surface charge, mineral composition, and mineral structure are necessary to explain the large variations in biological responses observed for minerals irrespective of their particle morphologies.

The data compiled by Stanton and coworkers exemplify the need to include mineralogical and geochemical factors into any analysis of the tumorigenic potential of minerals. Figure 1 shows the data from Stanton et al. (1981), which were obtained by applying a 40-mg dose of particles to the pleural surface[2] of rats.[3] The curve shows the familiar relationship derived by Stanton et al.: Fibrous minerals are carcinogenic but non-fibrous minerals are inert. Although the data do suggest a positive correlation between the number of fibers and tumorigenic

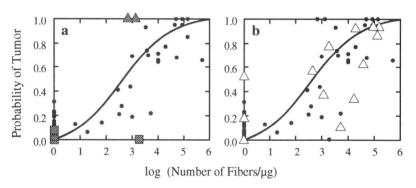

Figure 1. Graph of tumorigenic potential of mineral samples as a function of their fiber content (from Stanton et al., 1981). Fibers were defined as particles with diameters ≤0.25 µm and lengths >8.0 µm. (a) Shaded triangles indicate tremolite, shaded squares indicate talc; (b) White triangles indicate crocidolite (asbestiform riebeckite).

potential, the large degree of scatter exhibited by the data suggests that other factors are needed to explain mineral toxicity fully. Figure 1a highlights the data for tremolite (triangles) and talc (squares) from within the larger data base. Clearly *mineral species strongly affects tumorigenic potential*, because tremolite is more tumorigenic than talc with comparable morphology. Figure 1b highlights

---

[2] The pleura is the membrane lining the cavity containing the lungs. Cells associated with the pleura (mesothelial cells) are involved in at least one type of mineral-induced cancer (mesothelioma).

[3] This study used 30–50 rats for each of the numerous particle types investigated.

the data for the various samples of crocidolite[4] (shown as triangles). The crocidolite data suggest that *other factors also affect tumorigenic potential*, because morphology and mineral species alone are insufficient to explain the scatter exhibited by the data. For example, "other factors" caused relatively non-fibrous samples of crocidolite—e.g., the one at (0, 0.53)—to be more tumorigenic than relatively fibrous samples—e.g., the one at (3.73, 0.10).

The identification of the mineralogical, geochemical, and biochemical mechanisms important in mineral-induced pathogenesis requires an interdisciplinary approach, with bioscientists and geoscientists working together closely. This book presents much of the basic information necessary to facilitate this interaction. The interfingering of the two fields becomes evident throughout this book. For example, the geological processes leading to a natural background level for mineral dusts (including the asbestos minerals) are discussed in Chapter 2, and the potential effect of this background contribution on the mineral burden in human lungs is discussed in Chapter 13. Also noted in Chapter 13 is the interesting observation that chrysotile is underrepresented in lung burdens compared to amphibole asbestos, despite the relative abundances of the two minerals in the dusts to which individuals are exposed. This observation might be explained by the solubility of chrysotile in lung-like fluids, which is discussed in Chapters 3 and 8.

This book begins with chapters covering some of the important geological, mineralogical, and geochemical concepts that are necessary for individuals interested in studying mineral-induced pathogenesis:

- Chapter 2 introduces many of the minerals of interest and presents a discussion of the geological mechanisms that result in the generation of dusts. The natural background exposure to mineral dusts that each of us receives presents an interesting challenge for epidemiologists: the choice of a control population must be made such that the background exposures are comparable between the two groups. Background exposures include a global component that produces a roughly uniform exposure over large regions; however, the background exposure also includes a local component that will change depending on bedrock, soil type, wind direction/speed, and numerous other factors. In addition to being a concern in epidemiological studies, these background exposures should be important considerations in risk assessments and regulations. The potential significance of natural background exposures is well illustrated by the cases of mesotheliomas associated with a natural background exposure to erionite (see Chapter 12).

- Chapters 3, 4, and 5 present detailed accounts of the mineralogy of amphiboles and 1:1 layer silicates (Chapter 3), which include the asbestos minerals; clays and zeolites (Chapter 4), which include a number of minerals that commonly occur as particles of respirable size; and various

---

[4] "Crocidolite" refers to blue asbestos (which is normally riebeckite). Chapter 9 discusses the crocidolite samples used in the experiments by Stanton and coworkers in greater detail.

oxides and phosphates (Chapter 5), which include the silica and titania polymorphs and iron oxides. The mineralogical detail covered in these chapters may initially overwhelm all but the most serious of mineralogists. However, many of the mineralogical factors that potentially affect a mineral's pathogenic potential are covered, so these chapters are important references even for those who are not mineralogists.

- Chapters 6 and 7 discuss some of the mineralogical techniques that are necessary to prepare and to characterize a sample that will be used in a biological assay. Most minerals occur in polymineralic assemblages, so it may be necessary to purify a sample before use. Furthermore, properties of minerals vary among samples, so mineralogical characterization is often required on a sample-by-sample basis.

- Chapters 8, 9, and 10 present some of the factors that relate to the mineralogical mechanisms of mineral-induced pathogenesis. Chapter 8 discusses the nature of mineral surfaces, which often differs from the nature of the bulk minerals. Many processes can occur at mineral surfaces, including sorption, oxidation/reduction, catalysis, and dissolution, and these processes may be important in pathogenesis. Chapter 9 presents a critical review of the Stanton Hypothesis. As alluded to above in Figure 1, the morphology of a particle is an important component of pathogenicity, but it is insufficient alone to explain the biological data. Chapter 10 develops the thermodynamic theory necessary to describe surfaces and surface processes. This theory is applied to the characterization of mineral surfaces and the surfaces of several biological materials.

The second half of this book contains several chapters covering some of the important epidemiological, pathological, biological, and biochemical concepts underlying mineral-induced pathogenesis:

- Chapters 11 and 12 cover the epidemiology in a mineral-species-specific manner. Chapter 11 focuses on the asbestos minerals, and it also describes many of the pathological features associated with exposure to asbestos. Chapter 12 reviews the epidemiological observations made on populations exposed to various minerals, including the silica polymorphs and zeolites.

- Chapters 13 and 14 discuss various aspects of the inhalation of minerals. The exposure each of us receives from airborne mineral dusts results in the deposition of material in the lungs. Chapter 13 describes the mineral contents found in the lungs of different populations and relates these lung burdens to disease patterns. Chapter 14 covers the response of the lung to mineral dusts, including lung defense mechanisms and mineral–cell interactions.

- Chapters 15, 16, 17, and 18 review the evaluation of the pathogenic properties of minerals and present the biological mechanisms by which minerals may induce disease. Chapter 15 presents a critical discussion of the animal models used to assess the pathogenic potential of mineral dusts and reviews the data obtained from these models. Chapter 16 has a similar presentation for cellular and molecular assays used to assess minerals and used to determine the cellular response to mineral exposure. Chapter 17 presents a summary of the mechanisms by which minerals induce various responses (e.g., inflammation, fibrosis and carcinogenesis) as well as an overview of these processes in general. Chapter 18 investigates the whole-animal, cellular, and molecular responses to silica exposure and formulates a model for silica-induced pathogenesis.

The final chapter of this book covers the regulatory aspects of minerals. This arena is unfamiliar to many scientists, and the regulations of minerals affords an interesting peak into the complexities associated with applying science to societal issues and concerns.

Interdisciplinary research is a challenge because of the number of fields with which an individual must be familiar. This book brings together many of the diverse topics that relate to mineral-induced pathogenesis in order to foster the interdisciplinary effort that is needed in this field. In this spirit, a glossary is presented at the end of this book to explain many of the terms specific to each of the individual disciplines. Each of the topics covered in chapters of this book could easily be expanded into a book of its own. Hence, this book is intended as a base upon which a truly interdisciplinary understanding of this topic can be built.

REFERENCES

Murray, R. (1990) Asbestos: a chronology of its origins and health effects. Brit. J. Indus. Med. 47, 361–365.

Stanton, M.F., Layard, M., Tegeris, A., Miller, E., May, M., Morgan, E. and Smith, A. (1981) Relation of particle dimension to carcinogenicity in amphibole asbestoses and other fibrous minerals. J. Nat'l Cancer Inst. 67, 965–975.

Chapter 2

ROCKS, MINERALS, AND A DUSTY WORLD

Cornelis Klein
Department of Earth and Planetary Sciences
University of New Mexico
Albuquerque, New Mexico 87131 U.S.A.

INTRODUCTION

The Earth's troposphere (the lower portion of the atmosphere with an average upper ceiling at about 10–12 km) and hydrosphere contain abundant naturally generated dust. A continuous supply of dust results from weathering and erosion, as part of the hydrologic cycle (see below); an additional, intermittent supply is provided by explosive volcanism; and outer space is also a source of extraterrestrial dust. Although there are additional anthropogenic sources of dust, this paper concerns itself only with dust generated by geological and meteorological processes that act upon the inorganic (mineral) components of the Earth's crust, and dust generated by explosive volcanism.

The ultimate source materials from which the terrestrially produced dust is generated are the various rock types exposed at the Earth's surface. These same rocks are also the source of the thin cover of unconsolidated materials (gravels, sands, soils, etc.) that contribute a large component of naturally produced dust. Natural dust (airborne and/or suspended in water), therefore, is the result of weathering and erosion of several sources: unconsolidated sources, such as deserts and soil cover; and consolidated (lithic) sources, such as high mountain terranes with non-vegetated areas and subarctic and arctic regions with essentially no growth cover and an absence of permanent ice.

Natural dust (in the size range of about 0.1–30 μm) therefore is a composite of (1) lithic, primary mineral grains[†] (particles that are the result of comminution of rocks and little concomitant chemical alteration); (2) mineral grains formed by secondary chemical reactions (those that have formed as a result of complete chemical weathering and alteration of lithic mineral grains and which reside mainly in the unconsolidated Earth cover); (3) volcanic ash and dust; (4) salts from sea sprays; (5) extra-terrestrial dust; and (6) biological materials.

[†] Because dust is defined in terms of very small particulate size (~<0.1–30 μm), most such particles turn out to be monomineralic, that is consisting of a single mineral species. Aggregates of dust particles may consist of several individual mineral grains that are clustered together. This is the result of the fragmentation and chemical alteration of the original generally multimineralic rock from which the particles were derived. Volcanic explosions commonly generate dust particles that are composites of glass and mineral grains, and extraterrestrial dust can also be multi-phase, e.g., consisting of a silicate mineral with fused on Fe-Ni globules.

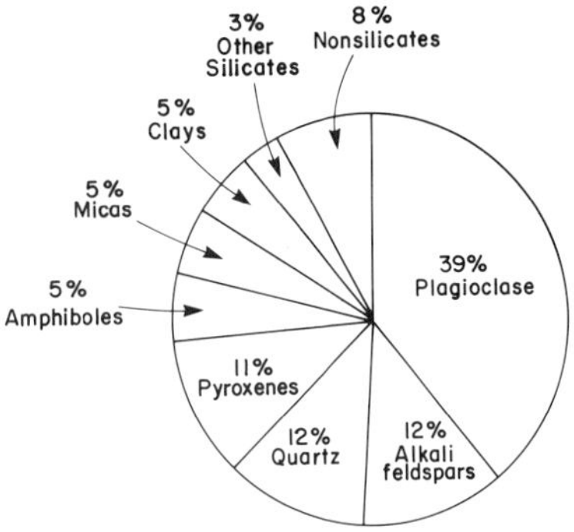

Figure 1. Estimated volume percentages for the common minerals in the Earth's crust, inclusive of continental and oceanic crust: 92% are silicates (from Ronov, and Yaroshevsky, 1969, reprinted with permission from Klein and Hurlbut, 1993, *Manual of Mineralogy*, 21st edition, John Wiley and Sons, New York).

Lithic mineral particles are the result of rapid fragmentation of the original source rock with little chemical alteration. Such mineral particles are the result mainly of physical weathering processes in low-temperature regions (e.g., at high altitudes in mountain terranes) and in arid climates. These mineral particles reflect the composite surface exposure of igneous, sedimentary, and metamorphic rocks. The average mineralogy of all these rock types in the Earth's crust (assuming a continental crustal thickness of approximately 30 km and an oceanic crust thickness of about 12 km) can be estimated and is shown in Figure 1. This shows that about 51 vol % of crustal lithic mineralogy is represented by various members of the feldspar group, about 16% consists of Ca-Mg-Fe silicates such as pyroxenes and amphiboles, 12% is quartz, and 5% is clay (layer silicates that are major constituents of soils and shales); the remaining silicates are other layer silicates such as mica, serpentine, and talc; another 8 vol % is made up of a combination of mainly oxides, carbonates, sulfates and phosphates.

Mineral particles that are part of the unconsolidated materials on the Earth's surface (e.g., soils, deserts, and beaches) consist mainly of various types of layer silicates ("clay") and quartz. Volcanic ash and dust consist mostly of glass and aggregate particles of glass and various common silicates such as feldspar, pyroxene, and amphibole. Extraterrestrial dust is a mixture of silicate minerals and Fe-Ni particles.

Estimates of the average numbers of particles suspended in remote continental air (away from the vicinity of immediate urban influences) are approximately as follows: one m^3 of air contains about 100,000 particles in the 10- to 1-µm range (i.e., ~100 particles/l); about 20 million particles in the 1- to 0.1-µm range (~20,000 particles/l); and 300 million particles in the 0.1- to 0.01 µm range (300,000 particles/l) (Cadle, 1966). The estimated total contribution to the natural aerosol (consisting of solid and liquid particles suspended in air) as released by nature is 2000 x 10^6 tons per year, compared to some 300 x 10^6 tons per year produced by anthropogenic activities (from the SMIC Report quoted in Jaenicke, 1980).

In subsequent parts of this paper the various pathways to the natural generation of dust (via the hydrologic cycle) will be discussed, and two geologically well-known natural dust sources will be described, paying particular attention to quantitative measurements of the dusts from these areas. General dust studies that provide data on possibly global background levels will be presented as well. However, before these subjects are introduced, it may be useful to address a few general aspects of the mineralogical characterization of dust particles, and to discuss briefly some of the mineralogy of several mineral groups that will be part of subsequent presentations. Much of what follows in the next section is discussed in greater detail in subsequent chapters.

SOME MINERALOGICAL BACKGROUND

General references

For researchers dealing with mineral dust, it is useful to have one or several mineralogical texts at hand for reference mainly to the microscopic properties, morphology, general chemical composition, structure type, and general geological occurrence of specific minerals or mineral groups. Such references may include:

> Deer, W.A., Howie, R.A. and Zussman, J. (1992) *An Introduction to the Rock-Forming Minerals*, 2nd edition John Wiley and Sons, New York, 691 p.
>
> Skinner, H.C.W., Ross, M. and Frondel, C. (1988) *Asbestos and Other Fibrous Materials, Mineralogy*, Crystal Chemistry and Health Effects. Oxford University Press, New York, 204 p.
>
> *Definitions for Asbestos and Other Health-related Silicates* (1984) Benjamin Levadie, ed., ASTM Special Technical Publ. 834, Philadelphia, PA., 213 p.

Examples of college-level mineralogy texts that not only deal with a broad overview of the most common minerals but also with principles such as those of crystal chemistry, mineral chemistry, internal structure, space groups, X-ray and optical properties, etc., are:

> Blackburn, W.H. and Dennen, W.H. (1988) *Principles of Mineralogy*. Wm. C. Brown Publishers, Dubuque, Iowa, 413 p.
>
> Klein, C. and Hurlbut, C.S., Jr. (1993) *Manual of Mineralogy*, 21st edition. John Wiley and Sons, New York, 681 p.
>
> Zoltai, T. and Stout, J.H. (1984) *Mineralogy, Concepts and Principles*. Macmillan Publishing, New York, 504 p.

Standard mineralogical reference works (complete references are given in the reference list at the end of this chapter) are:

> *Dana's System of Mineralogy*, as three volumes published in 1944, 1951, and 1962.
>
> *Rock-Forming Minerals*, by W.A. Deer, R.A. Howie, and J. Zussman originally published in 1962 in five volumes and since then supplemented by more detailed volumes (1A, 1B, and 2A, respectively, published in 1982, 1986, and 1978).

An in-depth treatment of topical subjects in mineralogy is provided by the *Reviews in Mineralogy*, volumes 1 to 27, published by the Mineralogical Society of America, Washington, D.C.

Quantitative characterization of mineral particles in dust

The commonly measured aspects of dust particles are grouped as follows:

morphological measurements which include grain size, shape, and habit (e.g., aspect ratio and/or fibrous nature); these properties can be evaluated on a high-magnification optical microscope, on a scanning electron microscope (SEM) or in the scanning and transmission modes of a transmission electron microscope (TEM);

chemical composition, which is generally measured with an energy dispersive spectrometer (EDS) attached to a TEM or SEM;

structural information as produced by transmission electron microscopes as electron diffraction patterns and structure images.

Analytical transmission electron microscopy (ATEM) is one of the most powerful techniques for the semi-quantitative to quantitative characterization of individual dust particles, permitting morphological observations, chemical analysis, and electron diffraction data that generally lead to the unique identification of a mineral (dust) grain with a spatial resolution much smaller than that available with a light optical microscopy, X-ray diffraction, or electron microprobe analysis. For example, using ATEM, a submicrometer mica particle is uniquely distinguishable from a sillimanite or feldspar grain, even though all three contain large amounts of Si and Al as their main chemical constituents. An example of the excellent ATEM characterization of mineral particles in lungs is the study of Paoletti et al. (1987; see **Lung particulate burdens**, below). They identified particles using a combination of morphological, structural and chemical information. Their final results are listed in a mostly unambiguous and mineralogically correct fashion. The particles identified in their study were: micas (muscovite and phlogopite), clays (kaolin and pyrophyllite), gypsum, talc, chrysotile, silicon oxides (probably SiO_2), amphiboles (tremolite and crocidolite), chlorite, vermi-culite, rutile, zeolites, pyroxenes, and feldspar.

Chemical variation in minerals

Although there normally is little or no need for quantitative chemical analysis of a specific dust particle (a list of detected elements and the range of their weight percentages generally suffices), it is helpful to address briefly the main underlying reasons for the highly variable chemistry of most minerals.

All the common rock-forming minerals except quartz, SiO_2 (see Fig. 1), show extensive variation in composition which is generally referred to as *solid solution*. A solid solution *is a mineral structure in which specific atomic site(s) are occupied in variable proportions by two or more different chemical elements (or ionic groups)*. The main factors that determine the extent of solid solution taking place in a crystal structure are:

(1) *The comparative sizes of the ions, atoms or ionic groups that substitute for each other.* Generally a wide range of substitution is possible if size differences between the ions (or atoms) is *less than about 15%*. If the

radii of the two elements differ by 15–30%, substitution is limited or rare. If the radii differ by >30%, little substitution is likely.

(2) *The charges of the ions involved in the substitution.* If the charges are the same, as in Mg^{2+} and Fe^{2+}, the structure in which the ionic replacement occurs will remain electrically neutral. If the charges are not the same, as in the case of Al^{3+} substituting for Si^{4+}, additional ionic substitutions elsewhere in the structure must take place in order to maintain overall electrostatic neutrality.

(3) *The temperature at which the substitution takes place.* There is, in general, a greater tolerance toward atomic substitution at higher temperatures when thermal vibrations (of the overall structure) are greater and the sizes of available atomic sites are larger. Therefore, in a given structure, one expects a greater variability in its composition at higher temperature than at lower temperature.

Although there are three major types of solid solution, referred to as (1) *substitutional*, (2) *interstitial*, and (3) *omission* solid solution, only the most common of these, namely *substitutional solid solution* will be addressed here. The simplest types of ionic substitutions are *simple cationic* or *simple anionic* substitutions. In a compound of the type A^+X^-, A^+ may be partly or wholly replaced by B^+. In this instance there is no valence change. Such a substitution is illustrated by the substitution of Rb^+ in the K^+ position of KCl or biotite. A simple anionic substitution can be represented in an A^+X^- compound in which part or all of X^- can be replaced by Y^-. An example is the incorporation of Br^- in the structure of KCl in place of Cl^-. An example of a *complete binary solid solution series* (meaning substitution of one element by another over the total possible compositional range, as defined by two end member compositions) is provided by olivine $(Mg,Fe)_2SiO_4$, where Mg^{2+} can be replaced in part or completely by Fe^{2+}. The end members of the olivine series between which there is complete solid solution are Mg_2SiO_4 (forsterite) and Fe_2SiO_4 (fayalite). Another example of a complete solid solution series, at elevated temperatures (above about 650°C) is given by the alkali feldspar series. This series ranges from $KAlSi_3O_8$ (K-feldspar) to $NaAlSi_3O_8$ (albite) and exhibits homogeneous solid compositions anywhere between these two end members (above 650°C), due to the simple cationic substitution $K^+ \rightarrow Na^+$ (see Fig. 2a). An example of a complete anionic series between two compounds is given by KCl and KBr. The size of the two anions is within 10% of each other, allowing for complete substitution of Cl^- and Br^-, and vice versa.

If in a general composition of $A^{2+}X^{2-}$ a cation B^{3+} substitutes for some of A^{2+}, electrical neutrality can be maintained if an identical amount of A^{2+} is at the same time replaced by cation C^+. This can be represented by

$$2A^{2+} \rightarrow B^{3+} + C^+ \qquad (1)$$

with identical total electrical charges on both sides of the equation. This type of substitution is known as *coupled substitution*. The plagioclase feldspar series can be represented in terms of two end members, $NaAlSi_3O_8$ (albite) and

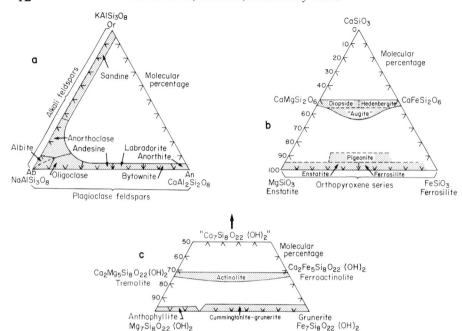

Figure 2. Schematic graphical illustrations of the extent of substitutional solid solution in three major rock-forming silicate groups. Shading represents the approximate compositional fields. (a) Feldspars, with endmember compositions expressed as Or (= orthoclase), Ab (= albite), and An (= anorthite). (b) Pyroxenes, with the upper field in the triangle are commonly known as Ca-clinopyroxenes (including diopside, hedenbergite, and "augite"). (c) Amphiboles with calcic amphiboles shown in the upper compositional range and Ca-poor amphiboles along the lower edge of the triangle (from Klein and Hurlbut, 1993, *Manual of Mineralogy*, John Wiley and Sons, New York, reprinted with permission).

$CaAl_2Si_2O_8$ (anorthite). The complete solid solution between these two end member compositions illustrates coupled substitution:

$$Na^+ \; Si^{4+} \rightarrow Ca^{2+} \; Al^{3+} \tag{2}$$

This means that for each Ca^{2+} that replaces Na^+ in the feldspar structure, one Si^{4+} is replaced by Al^{3+} in the Si-O framework. The equation shows that the total electrical charges on both sides of the equation are identical; as such the structure remains electrically neutral. An example of only limited coupled solid solution is provided by two pyroxenes, diopside ($CaMgSi_2O_6$) and jadeite ($NaAlSi_2O_6$). The coupled replacement can be represented as:

$$Ca^{2+} \; Mg^{2+} \rightarrow Na^+ \; Al^{3+} \tag{3}$$

Although the pyroxene structure is neutral with either type of cation pair, geometrical constraints due to differences in ionic radii do not allow for a complete solid solution of this type.

Atomic sites that in some structures are *unfilled* (or *vacant*) may become partly or wholly occupied due to a substitutional scheme in which a vacancy is involved. For example, a partial coupled substitution in the amphibole tremolite, $\square Ca_2Mg_5Si_8O_{22}(OH)_2$, where $\square$ is the normally vacant A site in the structure, may result from

$$\square + Si^{4+} \rightarrow Na^+ + Al^{3+} \tag{4}$$

In this coupled substitutional scheme Al^{3+} replaces Si^{4+} in the tetrahedral position, and the additional Na^+ is housed in a site that is normally vacant (shown as $\square$). Much of the above discussion of solid solution was taken from Klein and Hurlbut (1993, p. 233–235). Standard graphical illustrations of several common substitutional solid solution series in some rock-forming silicate groups are shown in Figure 2.

The determination of the exact composition of a mineral, in order to establish with precision its possible solid solution parameters, is an important aspect of mineralogical/petrological/geochemical investigations. The exact composition of a mineral as part of a mineral assemblage (in a rock) is commonly diagnostic of its origin (including the cooling history of igneous minerals; the temperature-pressure-composition travel path of a metamorphic mineral; and aspects of the very low temperature of formation, at about 25°C, of some sedimentary minerals during diagenesis). Therefore, the precise composition of a mineral that may commonly display some (or extensive) solid solution may be an important parameter in deciphering the geological history of a specific mineral (or rock) occurrence. Furthermore, because compositional variations affect many mineralogical properties, such variations may also affect a mineral's biologic activity.

Chemical characterization of dust particles

The need for such exact compositional determinations of individual particles in dust is uncommon. After combining the morphological observations on a particular dust particle with data on its general structure (by electron diffraction or by X-ray diffraction if a larger bulk sample is available), the investigator can determine what type of mineral the particle is, e.g., a member of the amphibole family, or a feldspar, or a layer silicate. Additional semi-quantitative chemical information (most commonly obtained by X-ray energy dispersive techniques) should confirm the earlier conclusions and furthermore, will give chemical input with regard to what part of a solid solution series the particle represents. For example (see Fig. 2c), a semi-quantitative analysis may show an amphibole particle to contain large amounts of Ca, Fe, Mg, Si, and only minor Al. This would group the particle as part of the tremolite–ferro-actinolite series. If on the other hand it contained significant Fe, Mg, and Si but no Na and at most minor Ca (<0.5 wt % CaO), it would be identified as part of either the anthophyllite–ferro-anthophyllite series or the magnesio-cummingtonite–grunerite series (see Fig. 2c), depending on its structure. For members of the feldspar group, high Na and K concentrations (with little to no Ca) would identify a particle as an alkali feldspar; high Ca concentrations in addition to Na, Si, and Al as a plagioclase feldspar (see Fig. 2a). Such characterizations are appropriate for most mineral dust studies, and further quantitative chemical data are most often not needed. Certain regulations may require the chemical composition of a suspect mineral to be specific between certain concentration values. For example, they may require a distinction among tremolite [$Ca_2Mg_5Si_8O_{22}(OH)_2$], actinolite [$Ca_2(Mg,Fe)_5Si_8O_{22}(OH)_2$], and hornblende [$(Ca,Na)_{2-3}(Mg,Fe,Al)_5Si_6(Si,Al)_2O_{22}(OH)_2$], all of which are amphiboles and part of an extensive substitutional solid solution series (see Fig. 2c). In such instances, a combination of morphological data, electron or X-ray diffraction data (for knowledge of the structure type of the mineral or mineral group), and semi-quantitative chemical data will generally provide sufficient information to characterize the mineral. However, SEM studies combined with chemical data are not conclusive, because of the lack of information on the mineral structure type.

EXAMPLES OF POTENTIALLY HAZARDOUS MINERALS IN DUST

Naturally produced dust is a direct reflection of the average mineralogy of the Earth's crust (see Fig. 1). Most dust particles are commonly finely granular or equant (this holds for quartz, feldspar and most pyroxenes); some are platy (as exemplified by almost all layer silicates); and some are relatively stubby (as shown by many amphiboles). Only a very small, and relatively rare group of minerals is considered hazardous to human health if present in high concentrations in dust, and most of these are highly fibrous. Several of these minerals are discussed in subsequent chapters; the following is a short synopsis.

Amphiboles

All members of the chemically complex amphibole group of silicates exhibit long prismatic, or acicular (needlelike), crystal habits or morphologies. They also show pronounced elongate (prismatic) cleavage, and some varieties may be asbestiform in unusual geological occurrences. The common morphologically elongate or acicular habit is probably the direct result of the internal structure, which consists of double tetrahedral chains with Si_4O_{11} composition that extend "infinitely" along the c axis (see Fig. 3). The complex chemistry and extensive solid solution in the amphibole group are the result of ionic substitution in the various structural sites (see Fig. 3). A general formula for all amphibole compositions is $W_{0-1}X_2Y_5Z_8O_{22}(OH,F,Cl)_2$, where W represents Na^+ and K^+ in the A site, X denotes Ca^{2+}, Na^+, Mn^{2+}, Fe^{2+}, Mg^{2+}, and Li^+ in the $M4$ sites, Y represents Mn^{2+}, Fe^{2+}, Mg^{2+}, Fe^{3+}, Al^{3+} and Ti^{4+} in the $M1$, $M2$, and $M3$ sites, and Z refers to Si^{4+} and Al^{3+} in the tetrahedral sites of the double chains. Essentially complete ionic substitution may take place between Na and Ca and among Mg, Fe^{2+}, and Mn^{2+}. There is limited substitution between Fe^{3+} and Al, and between Ti and other Y-type cations, and there is partial substitution of Al for Si in the tetrahedral chains. Partial substitution of F, Cl, and O for OH in the hydroxyl sites is also common. Several of the common compositional series are shown in Figure 2c in terms of Ca-, Mg-, and Fe-rich end-member compositions. This figure shows the approximate compositional extent of anthophyllite, cummingtonite–grunerite, and tremolite–actinolite. One of the most common amphiboles, hornblende, is compositionally closely related to actinolite but with additional Na^+ substituting for Ca^{2+}, and Al^{3+} substituting for Y- and Z-type cations.

Another example of extensive solid solution can be found in the Na-rich amphiboles (see Fig. 4). These amphiboles, which include riebeckite and its

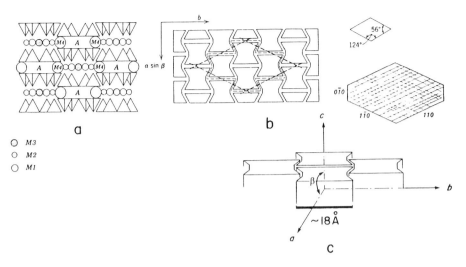

Figure 3. (a) Schematic projection of a monoclinic amphibole crystal structure on a plane perpendicular to the c axis. *M1, M2, M3* and *M4* are cation sites (shown as circles) of which *M1, M2* and *M3* are situated between two facing double chains, and *M4* is in a position that cross-links several double chains. The *A* site is commonly an empty site, although it may house Na in Na-rich amphiboles. The tetrahedral sites (*T*) are not specifically shown but are located in the centers of the tetrahedrons. (b) Control of cleavage angles by breakage through the *A* sites (between the backs of silicate chains) and across the *M4* site, resulting in a prismatic cleavage of two directions (parallel to the **c** crystallographic axis) at ~56° and ~124°. (c) Sketch of the tetrahedral-octahedral-tetrahedral (*t-o-t*) strips parallel to the **c** axis (**a** and **b** from Klein and Hurlbut, 1993, *Manual of Mineralogy*, John Wiley and Sons, reprinted with permission).

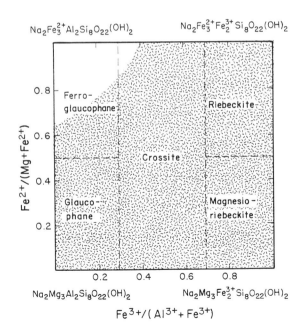

Figure 4. The compositional range and nomenclature of Na-rich amphiboles. Shading shows the extent of the most common compositions.

asbestiform variety crocidolite, are geologically less common than the various Ca-rich amphibole compositions. For a detailed discussion of all substitutional mechanisms of amphiboles and their specific geologic occurrences see Chapter 3, Veblen (1981), and Veblen and Ribbe (1982).

The commercially used asbestiform amphiboles are crocidolite (identical to riebeckite in composition; see Fig. 4, upper right) and amosite, the asbestiform variety of grunerite (Fig. 2c, lower right). This varietal name is derived from the word *Amosa*, an acronym for the company "Asbestos Mines of South Africa." Hand specimens of amosite and crocidolite asbestos are shown in Figure 5. (The

Figure 5. (a) Amosite, an asbestiform variety of the amphibole grunerite (also known as "brown asbestos" in the trade), Penge, Transvaal Province, Republic of South Africa. (b) Crocidolite, an asbestiform variety of riebeckite (also known as "blue asbestos" in the trade), Kuruman, Cape Province, Republic of South Africa.

terms "amosite" and "crocidolite" are not mineral species names but are names applied to brown and blue amphibole asbestos, respectively.) Other amphiboles that may occur in asbestiform varieties are anthophyllite, tremolite, and actinolite (compositional ranges for these minerals are shown in Fig. 2c). Amphiboles are common rock-forming minerals in igneous and metamorphic rocks (see Fig. 1), and as such, some will enter the erosional cycle to produce a natural amphibole component of dust (see discussion of its occurrence in Mount St. Helens ash and in air and water in the Hamersley Range, Western Australia under the heading *Volcanic activity*, below).

Common natural amphibole is *non*-asbestiform but will upon fracture of larger grains (through cleavage) produce abundant grains that have an aspect ratio (length-to-width ratio) of much greater than 3. In the laboratory, brief grinding in a mortar and pestle of massive amphibole specimens will quickly yield fragments with lengths in the range of 10–50 μm and aspect ratios in the range of 3–30. Similar brief grinding of amphibole asbestos samples (e.g., crocidolite, amosite, and tremolite) yields needles with lengths of about 200 μm and aspect ratios on the order of 300 and above. Hence, in dust samples with particulate sizes of about 0.1–10 μm, it is very difficult to determine whether a specific amphibole grain is the result of fragmentation of precursor asbestiform amphibole or of a "common" prismatic and more stubby amphibole. Yet, Wylie (1990) shows how, even at these very small sizes, cleavage fragments of common ("stubby") amphiboles may be clearly distinguished from amphibole asbestos fibers under the optical microscope (e.g, see Chapter 3). Mean aspect ratios of asbestiform amphiboles tend to 20:1 or greater in fibers longer than 5 μm. Wylie (1990) shows that careful optical measurements of aspect ratios and width distributions clearly distinguish between asbestiform and normal amphiboles.

It should be noted that the morphological distinction, in hand specimen, between asbestiform amphibole (e.g., amosite, crocidolite, and tremolite) and the much more common stubby to acicular hornblende and actinolite is immediately obvious. Mean aspect ratios of amphibole asbestos are commonly far greater than 100, in hand specimen, whereas aspect ratios of common (more stubby) amphiboles tend to be small, ranging from 2 to 30 or more.

Chrysotile

Chrysotile is one of three silicate mineral structures (one of three polymorphs) of a fairly constant composition, $Mg_3Si_2O_5(OH)_4$. The other two polymorphs are known as antigorite and lizardite, with all three making up the serpentine mineral group. The essential difference in the structures of the three minerals results from the nature of the 1:1 layers in the structure. (For a discussion of 1:1 layers, or *t-o* layers, see **Other layer silicates** below.) In lizardite these layers are planar (that is, lizardite has a structure typical of the layer silicate group); in chrysotile the 1:1 layers are concentrically coiled to produce hollow tubes parallel to the **a** axis; and in antigorite, 1:1 layers are corrugated. Figures 6 and 7, respectively, illustrate the internal structure in a TEM-image and the hand specimen appearance of chrysotile (known as "white asbestos" in the trade).

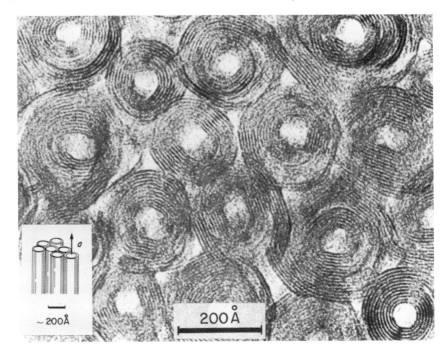

Figure 6. Transmission electron micrograph of the cross-sections (perpendicular to the fiber axis) of chrysotile fibers. Specimen from the Transvaal Province, South Africa (photograph reproduced with permission from Plate 46 in Sudo et al., 1981, *Electron Micrographs of Clay Minerals*, Elsevier, Amsterdam). Insert is a sketch of the packing of chrysotile fibrils parallel to the **a** crystallographic axis. Compare with Figures 8, 9, 12, and 13 in the chapter by Veblin and Wylie, this volume.

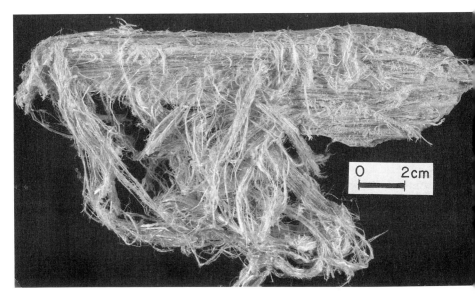

Figure 7. Hand specimen of chrysotile asbestos, Thetford Mines, Quebec, Canada.

Chapter 3 presents a detailed discussion of the structures of the 1:1 layer silicates, including the serpentine minerals. Various chemical, crystallographic, and geological aspects of chrysotile are also presented by Ledoux (1979), Bailey (1988), Hanley (1987), Hanley et al. (1989) and Otten (1993).

Approximately 95% of all asbestos produced is chrysotile (Schreier, 1989). This represents mineable, ore-grade material produced in a very few localities around the world. However, non-commercial occurrences of chrysotile are much more widespread as parts of extensively altered bodies of serpentinite rocks, which result from the hydrothermal alteration of ultrabasic rocks (including dunites, peridotites and pyroxenites). These serpentinite bodies are fairly common on a world-wide basis, and Schreier (1989) has made a compilation of such occurrences (his Fig. 4). In North America, serpentinites are most common in the Appalachians, in the Rocky Mountains, and in California. McGuire et al. (1982) compiled a map of all chrysotile-asbestos and serpentine-and-peridotite regions (with associated chrysotile) in California, including the large serpentinite region of New Idria, near Coalinga; see below.

Other layer silicates

As a result of the processes of chemical weathering of primary (mainly igneous and metamorphic) minerals, hydrous minerals are formed that are stable at the low-temperature, water-rich, and generally oxidizing conditions of the Earth's crust. Most of these are members of the layer silicate group. Although this group is large (see Bailey, 1984, 1988), this discussion will be restricted to those that are most common in soil environments (kaolinite, montmorillonite, and vermiculite) and in low-grade metamorphic rocks such as talc schists, muscovite (sericite) schists, and chlorite schists. Mica (e.g., muscovite), which may be a major constituent of some igneous rocks, and two layer silicates with distinct fibrous habits, sepiolite and palygorskite, will also be discussed.

Most layer silicates may occur in very fine grain sizes (on the µm level) and as such are part of naturally generated dust. Furthermore, most members of this silicate group have a platy or flaky habit due to one prominent cleavage, a characteristic which arises from the dominance in the structure of two types of indefinitely extended sheets. A *tetrahedral sheet* consists of SiO_4 tetrahedra polymerized in two dimensions to form a sheet with an overall composition of Si_2O_5. An *octahedral sheet* consists of $M(OH)_6$ octahedra polymerized in two dimensions to form a sheet with an overall composition of $M_2(OH)_6$ for a *dioctahedral sheet* (where M is usually a trivalent cation such as Al^{3+}) or $M_3(OH)_6$ for a *trioctahedral sheet* (where M is usually a divalent cation, such as Mg^{2+} or Fe^{2+}). A tetrahedral sheet can bond to either side of an octahedral sheet by replacing two of the six hydroxyl groups on the octahedral sheet with its apical oxygens. (The apical oxygens are those oxygens that are not part of the SiO_4 polymerization.) When a tetrahedral sheet is bonded to one side of an octahedral sheet, the resulting struture is a *t-o* unit (or *1:1 layer*), with an overall composition of $[Si_2O_5][M_2(OH)_4]$ or $[Si_2O_5][M_3(OH)_4]$. These formulae are generally written as $M_2Si_2O_5(OH)_4$ and $M_2Si_2O_5(OH)_4$. The minerals chrysotile and kaolinite (Fig. 8a) are examples of 1:1 layer silicates. When tetrahedral sheets

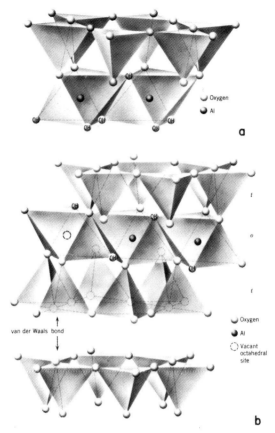

Figure 8. (a) Diagrammatic sketch of the structure of kaolinite in which a tetrahedral sheet is bonded on one side of an octahedral layer. (b) Diagrammatic sketch of the structure of pyrophyllite which is made up of a *t-o-t* sequence. Talc has the same structure but with Mg^{2+} in the octahedral sheet instead of Al^{3+} as in pyrophyllite (from Klein and Hurlbut, 1993, *Manual of Mineralogy*, 21st ed., John Wiley and Sons, reprinted with permission).

are bonded to both sides of an octahedral sheet, the resulting struture is a *t-o-t* unit (or *2:1 layer*), with an overall composition of $M_2Si_4O_{10}(OH)_2$ or $M_2Si_4O_{10}(OH)_2$. The stacking of 1:1 layers or 2:1 layers and the possible incorporation of interlayer cations such as K^+ (mainly the micas) leads to various structural types as shown in Figure 9.

Kaolinite. Kaolinite, $Al_2Si_2O_5(OH)_4$, shows little compositional variation. It consists of stacked *t-o* layers (see Figs. 8a and 9); hence, kaolinite is a 1:1 layer silicate. It has a characteristic platy morphology (see Fig. 10) and the fine particle size that typifies clay minerals. In addition to its presence in soils (Dixon and Weed, 1977), kaolinite is a major component of kaolin (or "China clay") deposits that are mined world-wide. Detailed mineralogical aspects are discussed in Chapter 3.

Vermiculite. Vermiculite occurs as a dioctahedral and trioctahedral layer silicate with a structural arrangement similar to that of talc (see Figs. 8b and 9) but with interlayer water molecules between the stacked *t-o-t* units (or 2:1 layers). Its ideal composition is $(Mg,Ca)_{0.3-0.5}(Mg,Fe^{2+},Al)_3(Al,Si)_4O_{10}(OH)_2 \cdot 4H_2O$. Mg and Ca occur as hydrated interlayer cations, coordinated by water molecules (i.e., the "$\cdot 4H_2O$"). The dioctahedral variety is most common in soils. Although vermiculite is an important component of soils (Dixon and Weed, 1977), it also occurs as coarser-grained rock masses resembling the micas, as in the mineable deposits of Libby, Lincoln County, Montana. An Al-rich vermiculite from red soils is illustrated in Figure 11. Detailed mineralogical aspects are presented in Chapter 4.

Montmorillonite. As a member of the smectite group, montmorillonite is an important clay mineral in soils (Dixon and Weed, 1977) and sediments. It has a generalized composition of $(Na,Ca)_{0.3}(Al,Mg)_2Si_4O_{10}(OH)_2 \cdot nH_2O$, and its structure is very similar to that of dioctahedral mica (see muscovite in Fig. 9) with

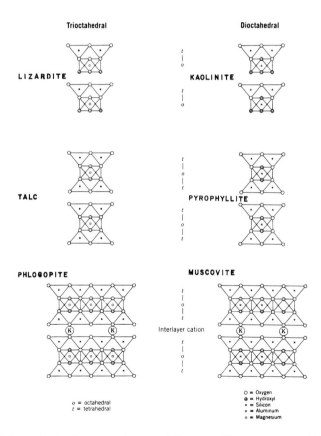

Figure 9. Schematic development of some of the layer silicate structures (from Klein, and Hurlbut, 1993, *Manual of Mineralogy*, 21st ed., John Wiley and Sons, New York, reprinted with permission).

smaller amounts of interlayer cations (Na and/or Ca) and extra water. In addition to its common presence in soils, montmorillonite is also found as in *bentonite*, a rock formed by the chemical alteration of volcanic ash in prehistoric lakes and estuaries. Figure 12 illustrates a montmorillonite occurrence from a bentonite deposit in Japan. Mineralogical aspects of smectites, including montmorillonite, are presented in Chapter 4.

Talc. Talc, with composition $Mg_3Si_4O_{10}(OH)_2$, consists of 2:1 layers that are stacked on top of each other without interlayer cations (see Figs. 8b and 9). Most talc shows little compositional variation, however, significant amounts of Fe (up to 1.22 of the 3 Mg sites) have been reported (Evans and Guggenheim, 1988). Other compositions with similar structures also occur in nature (e.g., willemseite, in which Ni replaces some Mg, and pyrophyllite, $Al_2Si_4O_{10}(OH)_2$, a dioctahedral analog of talc). Talc often occurs in large, relatively pure deposits which are mineable, commonly in dolomite-rich rocks that have been altered by metamorphism and in ultramafic rocks that have been serpentinized. Further mineralogical aspects of talc and pyrophyllite are presented in Chapter 4.

Muscovite. A member of the mica group, muscovite has composition $KAl_2(AlSi_3O_{10})(OH)_2$ and consists of stacked 2:1 layers with K^+ as the interlayer cation (see Fig. 9). The perfect cleavage parallel to the layers allows the mineral to be split into very thin sheets. Muscovite is a widespread and common rock-forming mineral in igneous as well as metamorphic rocks. Muscovite and other members of the mica group—e.g., biotite $K(Mg,Fe)(Si_3Al)O_{10}(OH)_2$—are also abundant in soils.

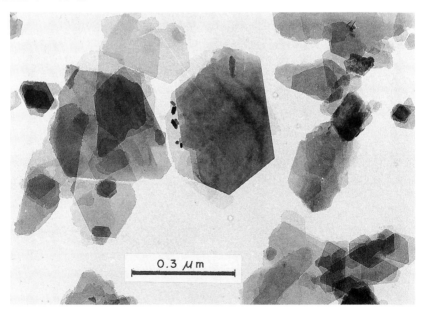

Figure 10. Electron micrograph of pseudohexagonal plates of kaolinite, Kampaku mine, Tochigi Prefecture, Japan (photograph reproduced with permission from Plate 1 in Sudo et al., 1981, *Electron Micrographs of Clay Minerals*, Kodansha Scientific Ltd., Tokyo).

Chlorite. Chlorite consists of alternating *t-o-t* units and *o* units, and it has a highly generalized composition of $(Mg,Fe)_3(Si,Al)_4O_{10}(OH)_2\cdot(Mg,Al,Fe)_3(OH)_6$ (see Fig. 13). By analogy to the 2:1 layer silicates like muscovite, the octahedral sheet in chlorite occupies the interlayer region (i.e., that region occupied by K^+ in muscovite). Because chlorite shows extensive solid solution (e.g., substitution of

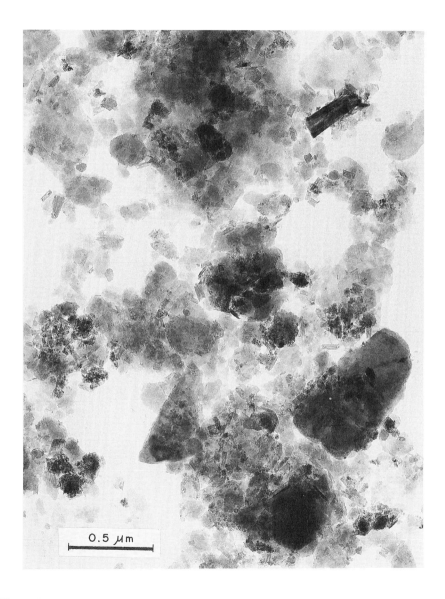

Figure 11. Electron micrograph of an Al-rich vermiculite showing an aggregate of platy particles with angular borders. From the Sumitomo Cement Gifu mine, Gifu Prefecture, Japan (photograph reproduced with permission from Plate 76 in Sudo et al., 1981, *Electron Micrographs of Clay Minerals*, Kodansha Scientific Ltd., Tokyo).

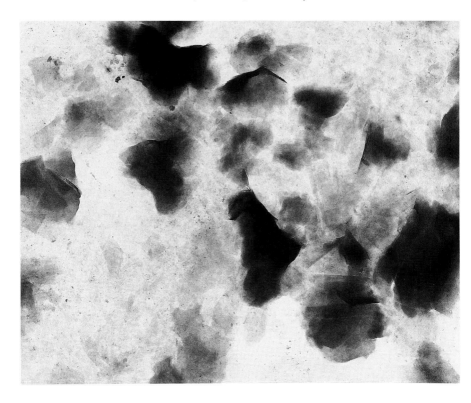

Figure 12. Electron micrograph of montmorillonite from a bentonite occurrence. It shows very thin particles with angular edges. From the Kunimine Aterazawa mine, Japan (photograph reproduced with permission from Plate 77 in Sudo et al., 1981, *Electron Micrographs of Clay Minerals*, Kodansha Scientific Ltd., Tokyo).

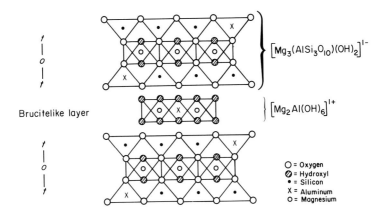

Figure 13. Schematic illustration of the structure of chlorite, consisting of *t-o-t* layers with an octahedral interlayer of approximate $Mg_2Al(OH)_6$ composition (from Klein and Hurlbut, 1993, *Manual of Mineralogy*, 21st ed., John Wiley and Sons, reprinted with permission).

Mg^{2+} by Fe^{2+}, Fe^{3+}, and Al^{3+}; and extensive replacement of Si by Al) many endmembers species names have been proposed (Bailey, 1988) on the basis of chemical composition. Although chlorite is common in many low- to medium-grade metamorphic rocks, it is also found in igneous rocks as an alteration product of silicates such as pyroxenes and amphiboles. Figure 14 illustrates chlorite from a metamorphic rock occurrence.

Sepiolite and palygorskite. These minerals (also known as "attapulgite" clays) are less common in nature, but they have been used for centuries because of their diverse and unique properties. For example, sepiolite ("meerschaum") has been used for making tobacco pipes. Sepiolite [with a generalized formula of $Mg_4Si_6O_{15}(OH)_2 \cdot 6H_2O$] and palygorskite [with an approximate formula of $(Mg,Al)_2Si_4O_{10}(OH) \cdot 4H_2O$] are both classified as layer silicates because they contain continuous tetrahedral sheets of Si_2O_5 composition. In these structures, however, the octahedral sheets are discontinuous between the tetra-hedral sheets and occur as ribbons (Jones and Galan, 1988). Both minerals commonly have a characteristically fibrous habit (Figs. 15 and 16). Mineralogical aspects of sepiolite and palygorskite are presented in Chapter 4.

Silica (SiO_2) group

Of the various members of the silica group, quartz and its microcrystalline varieties (e.g., chert) are most common in nature. Both are of essentially constant composition, SiO_2, with a structure in which all SiO_4 tetrahedra are linked into a continuous framework arrangement (see Chapter 5 for an extensive discussion of the silica minerals). Although quartz makes up about 12 vol % (see Fig. 1) of primary (mainly igneous and metamorphic) rocks, it becomes much more abundant in the unconsolidated sedimentary cover of the Earth's crust as a result of the weathering cycle. Because quartz is highly resistant to both chemical weathering and fragmentation (it lacks pronounced cleavage), it tends to survive the weathering cycle as a major constituent of soils, sands, and gravels. (See **The hydrologic cycle and weathering processes**, below, for a discussion of fragmentation.)

Quartz is a major component of soils, comprising as much as 90–95% of all sand and silt fractions in a soil. It is a major constituent also of beach sands, dunes and deserts. Because of its ubiquitous presence it is a major constituent of airborne dust and fluvial systems. Quartz-laden dust storms are well known during the spring in the arid regions of the southwestern USA (see also discussion of eolian transport, below, and Fig. 30).

Whereas quartz is a major constituent of many major rock types (igneous, metamorphic, and sedimentary), chert is most common in rocks that have formed as chemical precipitates. Banded iron-formation, with abundant chert (or jasper) is such a chemical sedimentary rock. Other fine-grained examples of silica include diatomaceous Earth, which is a very fine-grained, chalky appearing deposit consisting of siliceous diatom tests that accumulated on the seafloor. An example

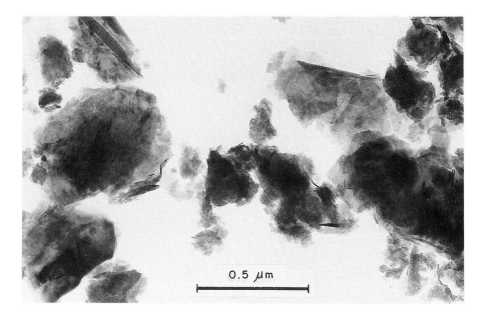

Figure 14. Electron micrograph of chlorite from a metamorphic rock consisting of calcite-garnet-hematite-epidote. From Ichinokoski, Toyama Prefecture, Japan (photograph reproduced with permission from Plate 10 in Sudo et al., 1981, *Electron Micrographs of Clay Minerals*, Kodansha Scientific Ltd., Tokyo).

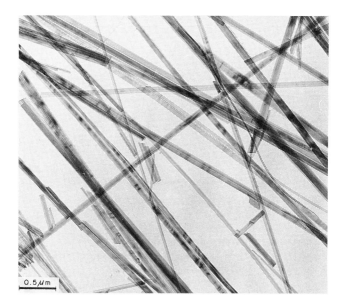

Figure 15. Electron micrograph of sepiolite consisting of elongate fibers. From the Karasawa mine, Tochigi Prefecture, Japan (photograph reproduced with permission from Plate 109 in Sudo et al., 1981, *Electron Micrographs of Clay Minerals*, Kodansha Scientific Ltd., Tokyo).

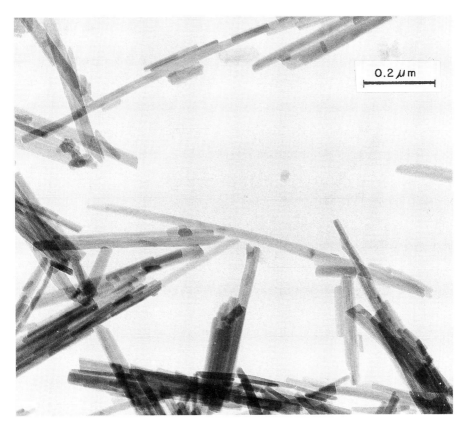

Figure 16. Electron micrograph of palygorskite consisting of elongate fiber particles. From the Karasawa mine, Tochigi Prefecture, Japan (photograph reproduced with permission from Plate 113 in Sudo et al., 1981, *Electron Micrographs of Clay Minerals*, Kodansha Scientific Ltd., Tokyo).

of very fine-grained but well-crystallized authigenic quartz is shown in Figure 17 (authigenic meaning "formed in place" after deposition of the original sediment).

Zeolite group

The zeolite group of minerals consists of about fifty species (Gottardi and Galli, 1985), but only a few can be considered as common rock-forming minerals, namely natrolite, chabazite, heulandite, and stilbite (cf. Klein and Hurlbut, 1993). The zeolites are hydrated aluminosilicates of alkali and alkaline earth cations, arranged in three-dimensional, framework ("tectosilicate") structures that are similar to those of quartz and feldspar but with much larger void space. Their ability to lose water fairly continuously over an elevated temperature range (150 to 400°C) without collapse of the framework structure and to resorb it from the atmosphere at room temperature is one of the unique properties that qualify these minerals as zeolites. Furthermore, the unique properties of zeolites are exploited extensively in processes requiring catalysis, ion-exchange, and molecular sieving.

Figure 17. Electron micrograph of authigenic quartz showing typical hexagonal prism terminated by combination of two unequally developed rhombohedra (reproduced, with permission, from Plate 122 in Sudo et al., 1981, *Electron Micrographs of Clay Minerals*, Kodansha Scientific Ltd., Tokyo).

Figure 18. (a and b) Roggianites from Pizzo Marcio, Italy. Both photos from Gottardi and Galli (1985) *Natural Zeolites*, Springer-Verlag, New York, reprinted with permission.

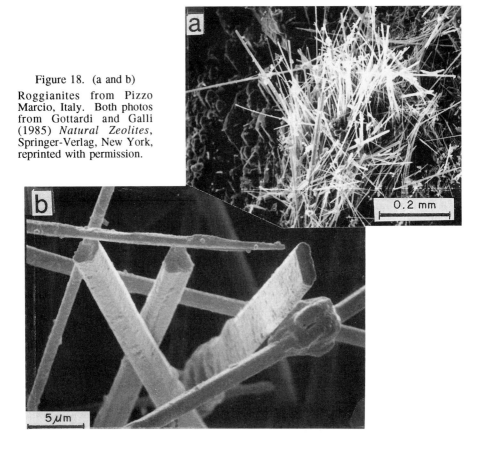

Many zeolites occur in well-developed crystal form, and their overall morphological habits can be classified as (1) platy, (2) equant, or (3) fibrous. Only a small number of some uncommon zeolites have finely fibrous habits such that their particulates may be health hazards under specific circumstances. Four of these zeolites will be briefly discussed here. They are: roggianite, mazzite, erionite, and mordenite. For complete descriptions of these and other zeolites the reader is referred to Chapter 4 and Gottardi and Galli (1985).

Roggianite. The roggianite structure is based on a framework of Si, Al and Be tetrahedra with an ideal composition of $Ca_{15}(Si,Al,Be)_{48}O_{90}(OH)_{16} \cdot 34H_2O$. Roggianite has tetragonal symmetry, and it commonly occurs as aggregates of thin fibers, less than 1 μm in diameter (Fig. 18a). The mineral appears to be of late hydrothermal origin in association with albitite dikes (albitite is a coarse-grained rock composed chiefly of albite). Morphologically it exhibits several parallel

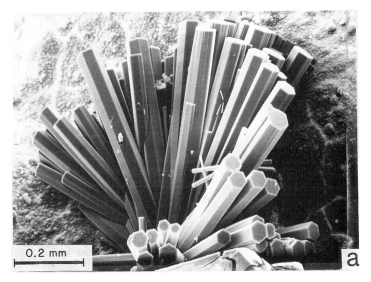

Figure 19. (a and b) Mazzites from Mt. Semiol, France. Both photos from Gottardi and Galli (1985) *Natural Zeolites*, Springer-Verlag, New York, reprinted with permission.

elongate prisms without termination (Fig. 18b). Gottardi and Galli (1985) report its occurrence from only two localities in Italy.

Mazzite. The mazzite structure is based on a framework of Si and Al tetrahedra having an ideal composition of $K_2CaMg_2(Si,Al)_{36}O_{72} \cdot 28H_2O$. It has hexagonal symmetry, as can be seen by its morphology (Fig. 19a). It has been found only at Mont Semiol, Montbrison, Loire, France where it occurs as a hydrothermal deposit in vugs of olivine-rich basalts.

Erionite. With a composition of $K_2NaMgCa_{1.5}(Al_8Si_{28})O_{72} \cdot 28H_2O$, the erionite structure is based on a hexagonal framework of Si and Al tetrahedra containing 6-fold rings (see Fig. 20b). Erionite forms as a result of zeolitic

Figure 20. (a) Fibrous erionite from Cape Lookout, Oregon. (b) Sedimentary erionite from Durkee, Oregon. Both photographs from Gottardi and Galli (1985) *Natural Zeolites*, Springer-Verlag, New York, reprinted with permission.

alteration of tuffaceous rhyolitic rocks that are of volcanic origin (tuffaceous is applied to a sedimentary rock that contains broken volcanic fragments; rhyolite is a volcanic rock which is equivalent in bulk composition to that of intrusive granite). It also occurs as a result of hydrothermal deposition in vugs of lava flows. It has been identified in several localities world-wide and was found to be a serious health hazard in the Cappadocian region in Turkey (Baris et al., 1989).

Mordenite. The mordenite structure is based on a framework of Al and Si tetrahedra with pronounced development of hexagonal sheets. Its ideal chemical composition is $K_{2.8}Na_{1.5}Ca_2(Al_9Si_{39})O_{96} \cdot 29H_2O$; its symmetry is orthorhombic. See Figure 21. It is a fairly common zeolite and is of sedimentary as well as

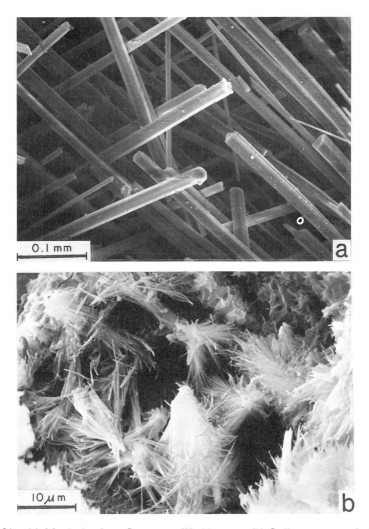

Figure 21. (a) Mordenite from Stevenson, Washington. (b) Sedimentary mordenite from Ponza, Italy. Both photographs from Gottardi and Galli (1985) *Natural Zeolites*, Springer-Verlag, New York, reprinted with permission.

hydrothermal origin. The transformation of glassy components in volcanic tuffs to mordenite is considered sedimentary in origin, whereas the occurrence of mordenite as a vein-filling in volcanic rocks is considered hydrothermal. Two samples of mordenite have been shown to be much less active than erionite in a variety of *in vitro* and *in vivo* assays (e.g., Hansen and Mossman, 1987; Palekar et al., 1988; Suzuki and Kohyama, 1984); however, these samples contained ~50 to 65% impurities (Guthrie and Bish, unpublished data reported in Guthrie, 1992), so the biological results for pure mordenite remain to be determined.

NATURAL DUSTS

Generation and migration of natural dusts

The hydrologic cycle and weathering processes. It is well known that planet Earth is continually subjected to dynamic forces, the often violent expressions among which are earthquakes and volcanic activity. In our daily, lives we are also well aware of yet other dynamic forces that are the result of motion of surface fluids, water, and air. Photographs taken from outer space clearly show that the outer surface of the Earth consists mainly of water, and that the tropospheric envelope around it is in constant motion. Figure 22 shows an Earth view, taken by the crew of Apollo 17 on December 7, 1992, with brilliant white swirling cloud masses that are the expression of fluid motion. The complex cycle by which water moves from the oceans to the troposphere, to the land, and back to the oceans is known as the *hydrologic cycle*. The energy source that drives this cycle is heat from the sun. It does so mainly by evaporating water from the surface of warm oceans in the tropics. This is transported by winds, which themselves are driven primarily by the temperature differences between hot and cold parts of the globe. The water in the troposphere condenses to clouds and eventually returns to the surface as precipitation. Some of the major elements of the hydrologic cycle are illustrated in Figures 23, 28 and 35. Much of the precipitation soaks into the ground by infiltration to form groundwater, whereas much of the remainder collects as runoff, which eventually returns to the oceans via river systems.

Throughout the hydrologic cycle, water is a very powerful long-distance transporting agent of fine particles in the range ~0.1 to 30 μm, as part of the suspended load. (This range is also most common in atmospheric dust.) The world's rivers carry to the ocean each year about 7 billion tons of suspended fine sediment (clays and silts) and 1–2 billion tons of coarser bed-load sediment (mainly as sands moved by saltation; Press and Siever, 1986, p. 208).

Water also plays a very important role in the lowering of mountainous areas by erosion, producing a wide range of sedimentary materials that are the result of two main types of weathering: (1) *mechanical weathering*, or *comminution*, and (2) *chemical weathering*, which results in the chemical breakdown of originally fresh rock and mineral materials. Comminution (fragmentation) is the physical break-up of rock into boulders, pebbles, sand, silt-and-clay-sized particles. (In geology, clay-sized particles are <4 μm; silt-sized particles are in the range 4 to 62 μm; and sand-sized particles are in the range 62–2000 μm.) In the first stages of

Figure 22. Earth as photographed by the crew of Apollo 17 on December 7, 1972. The major cloud masses and storms are centered over the Antarctic continent. The African continent, with Madagascar to its east, is clearly visible. The top right shows the Red Sea, separating the Arabian peninsula from northern Africa. NASA photo courtesy of Astronaut Harrison H. Schmitt, Apollo 17.

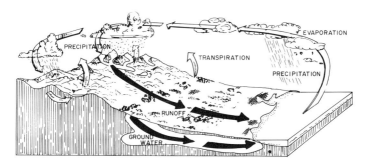

Figure 23. The hydrologic cycle. Evaporation of ocean water causes cloud formation, which in turn causes the precipitation of rain or snow. Much of the water that falls on land returns to the oceans by run-off and groundwater seepage (adapted from Hamblin, 1989, *The Earth's Dynamic Systems*, Macmillan Publishing Co., New York).

fragmentation, water, through freeze-thaw cycles, is a major (mechanical) weathering agent, and chemical weathering is of lesser importance. In regions where there is little liquid water and limited bacterial and plant life (e.g., polar-and-arctic regions and mountains at high altitude), mechanical weathering is the most important agent. Such is also the case in arid climates (see Fig. 24). Fragmentation of rocks results from fractures in the rocks that can be occupied by water or ice, and also by plant roots, and bacteria; all of these are instrumental in enlarging and wedging the fractures apart. A very important force in regions where the temperatures range about the freezing point is *frost wedging*. The freezing of water in fractures and cracks, with the concomitant volume increase of about 9% over that of liquid water, results in stresses that tend to fracture further and to disrupt the original rock. The primary result of rapid mechanical

Figure 24. (a) High altitude peaks in the Mackenzie Mountains, Northwest Territories, Canada. In subarctic climates, mechanical weathering is the main agent producing the fragmented slope debris. (b) Side view of the broken rock fragments (talus) in the same area as a. (c) Coarse, rounded, and fractured bolders at the bottom of a cliff in the arid climate of the San Juan Basin, northwest New Mexico. (d) Accumulations of rock fragments of variable size in the same region as c. Photographs by C. Klein.

weathering is the production of gravel and sand deposits that have a mineralogical composition very much like the original (parent) rock. When chemical weathering comes into play, as in streams and rivers, most of the original minerals will be altered mainly to layer silicates, such as clays.

It is useful to compare the degradation of rocks and minerals on the Earth to the analogous process on the moon, as elucidated by petrological studies of lunar rocks. Although the rocks exposed on the moon range in age from about 4200 to 3000 million years, they are completely fresh, that is they are devoid of any chemical alteration. Because the moon lacks an atmosphere with oxygen, carbon dioxide and water, the weathering processes on the moon involves fragmentation only. This fragmentation is the result mainly of the surface bombardment by meteorites and micrometeorites; meteorite impacts, producing the larger craters can be seen clearly on the surface of the moon; fine dust in the lunar regolith ("soil") is the result of a combination of large impacts and bombardment by small meteorites (see Fig. 25). The dust, therefore, is composed of the same minerals as found in the parent rocks.

On Earth, physical and chemical weathering tend to reinforce each other in most locations, except those that are particularly cold or arid. The smaller the fragment that is produced by mechanical action, the greater the surface area that can be attacked by water and, hence, the faster the rate of its chemical alteration and decay. Chemical weathering is the response of minerals (mainly silicates, oxides, and to a lesser extent carbonates) that originally formed at elevated temperatures and pressures at lower levels in the Earth's crust (as part of igneous or metamorphic processes) to the hydrous, oxidizing and low-temperature environment at the Earth's surface. Under these conditions such minerals tend to become unstable. This results in a complex set of chemical reactions that involves (1) the leaching of minerals (such as the feldspars) and (2) the complete alteration of the original phases (which are stable at high-temperature) to phases stable at lower temperatures, such as a large variety of clays and other layer silicates. Chemical weathering is most rapid and pervasive in humid, tropical climates, and it is less rapid in temperate regions. One common mineral, quartz (SiO_2), is able to survive the combined attack of physical and chemical weathering as exemplified by the abundance of quartz sand beaches, dunes and quartz-rich sedimentary rocks, known as sandstones. The majority of other silicate minerals that are subjected to long periods of chemical attack will finally dissolve, or they will react to form layer silicates in the size range of clays (<4 µm) or silts (4–62 µm).

Erosion rates and source areas of natural dust. Erosion of the Earth's surface involves the mechanical destruction of the exposed materials and the removal thereof by running water (including rainfall), wind, moving ice, waves and currents. The rates of erosion are highly variable for different parts of the globe. Figure 26 shows a compilation by Gregory and Walling (1973) of various data sources for rates of erosion. This shows that major mountain chains, such as the Rocky Mountains, the Alps, and the Himalayas, erode relatively fast when compared with the erosion rates of relatively low relief continental regions worldwide. Although the erosion rates of relatively low-lying continental regions are

Figure 25. Earth-rise over the lunar surface, photographed by the crew of Apollo 17 on December 15, 1972. NASA photograph courtesy of Harrison H. Schmitt, Apollo 17.

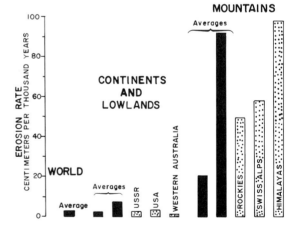

Figure 26. Some rates of erosion. This illustrates the amounts of material eroded as a function of physiography (from Gregory and Walling, 1973, who list the sources for the various data shown).

small, the areal extent of such regions is very large, as compared with the global extent of mountain chains. Such low-lying continental regions include very large areas of exposed Precambrian rocks (see Fig. 27) ranging in age from about 3900 to 570 million years old; Precambrian rocks are also present in parts of elevated mountainous regions. The high relief mountains and the low relief Precambrian regions are probably the major global source areas of primary mineral particles that have undergone little to no chemical alteration.

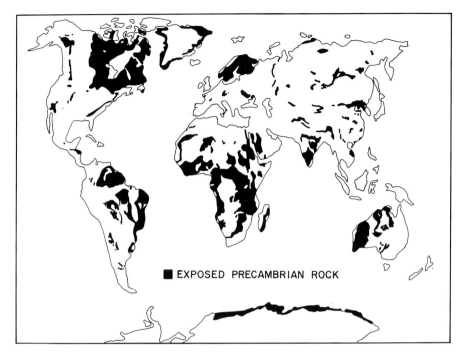

Figure 27. Distribution of exposed Precambrian rocks on modern continents (adapted from Goodwin, 1976, in *The Early History of the Earth*, Windley, ed., John Wiley and Sons, New York, reprinted with permission).

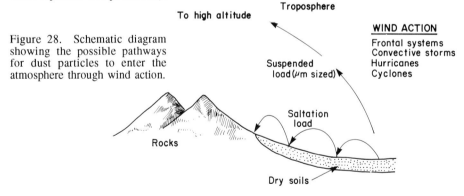

Figure 28. Schematic diagram showing the possible pathways for dust particles to enter the atmosphere through wind action.

Another major dust source is soil. The major mineral components of such soil-generated dust are various clay minerals and quartz. Yet another source of dust is the exposed bedrock world-wide, especially where there is little to no vegetation cover (such as high mountain terranes and the subartic-to-artic regions where there is no permanent ice cover).

Wind action. Wind action is an erosional agent that contributes much naturally occurring fine particulate matter (dust) to the troposphere, and therefore to the hydrologic cycle. Wind can erode, transport and deposit materials, and the laws of fluid motion that govern liquids apply to gases as well. However, because

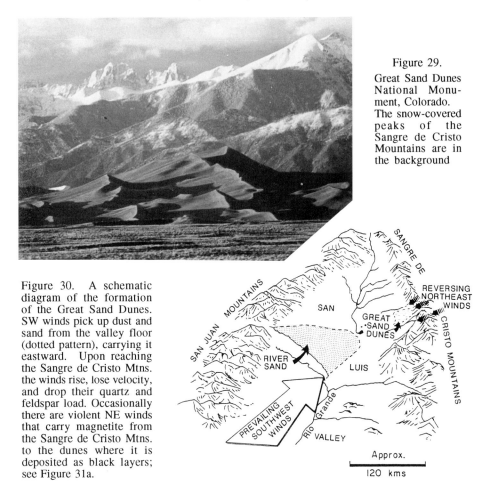

Figure 29. Great Sand Dunes National Monument, Colorado. The snow-covered peaks of the Sangre de Cristo Mountains are in the background

Figure 30. A schematic diagram of the formation of the Great Sand Dunes. SW winds pick up dust and sand from the valley floor (dotted pattern), carrying it eastward. Upon reaching the Sangre de Cristo Mtns. the winds rise, lose velocity, and drop their quartz and feldspar load. Occasionally there are violent NE winds that carry magnetite from the Sangre de Cristo Mtns. to the dunes where it is deposited as black layers; see Figure 31a.

(Sketched from a wall illustration in the Great Sand Dunes National Monument).

(1) the density of air is much less than that of water and (2) the airflow is mainly unconfined (not restricted to channels as in the case of water) the materials that can be carried by wind are generally much smaller than those carried in e.g., a fast-flowing stream, and the extent of wind-related (eolian) particle travel and deposition is large and high, reaching into the stratosphere.

Wind can transport particulate matter by *saltation* and *suspension* (see Fig. 28, above). Saltation is the jumping mode by which larger particles can be moved over limited distances by strong winds (or strong river currents). Particles of sizes greater than about 100 μm (Gillette, 1981) tend to settle back quickly to the surface when turbulence associated with strong winds subsides. Suspension is the wind-related transport of particles that are smaller in size than 10 μm. Material that is transported in this manner over very long distances in the troposphere is on average less than 2 μm in size (Péwé, 1981). Wind action in general is most effective in dry climatic periods and/or in arid climatic zones.

Short distance eolian transport. A very informative and geologically well documented example of the effectiveness of saltation is provided by the Great Sand Dunes National Monument in the San Luis Valley of the Southern Rockies of Colorado (see Fig. 29). The quartz (and lesser feldspar) sand in the dunes (with an average grain size of about 300 μm; see Fig. 31) was originally derived from rocks of the San Juan Mountains about 80 km to the west (Johnson, 1967; see Fig. 30). Stream erosion of the San Juan Mountains deposited the weathered products (mainly quartz sand with lesser feldspar) onto the valley floor where the prevailing southwest winds pick up the sand by saltation. When the winds rise on the eastern side of the Sangre de Cristo Mountains, they lose velocity and drop their sand load (of mainly quartz and lesser feldspar) in the dune area. However, the dunes also contain thin black bands of magnetite with a somewhat smaller average grain size than that of the quartz and feldspar (see Fig. 31). Magnetite (Fe_3O_4) is not transported as easily because its specific gravity (5.2) is much greater than that of quartz or feldspar (both ~2.6). Interestingly enough, the magnetite does not originate from the San Juan Mountains to the west, but results from weathering of Precambrian rocks in the Sangre de Cristo Mountains immediately to the east. Although westerly is the prevailing wind direction, there are violent winds in the early part of the year that come from the northeast, carrying the magnetite (Harris and Kiver, 1985).

Much smaller particles (in the range of 5 to 50 μm; Péwé, 1981), as part of common wind blown dust and loess storms, may travel up to 100 km.

Long distance eolian transport. The particle size of the aerosol that remains suspended in the air until it is brought down by rainfall onto water or land surfaces ranges generally between 1 and 10 μm (Péwé, 1981), with most of the aerosol dust <5 μm (Peterson and Junge, 1971). One source of this dust is the deserts, and it is estimated that fine airborne dust amounts to 500×10^6 tons/yr worldwide (Peterson and Junge, 1971), which accounts for about one quarter of the total naturally generated dust (2000×10^6 tons/yr). Arid and semiarid lands cover about 30% of the Earth's land surface (Meigs, 1953). Dust storms carry desert dust particles over thousands of km (Péwé, 1981) and Figure 32 shows some of the major directions and distances of dust transport.

Examples of major desert source areas are the Sahara, the southern coast of the Mediterranean, the northeast Sudan, the Arabian Peninsula, northern and western China, central Australia, southwest United States, and the Kalahari of southern Africa. For example, powerful dust storms that are common in the Sahara, transport dust to both North and South America (Péwé, 1981). Desert dust is the result of *in-situ* weathering of bedrock and the abrasion of bedrock by wind-blown sand. A major mineral component of such dust is quartz.

Betzer et al. (1988) report the transport of what they call "giant mineral aerosol particles" (>75 μm in size) over 10,000 km from their source region. These particles were generated by a major dust outbreak between March 15 and 17, 1986, in China. This dust plume was sampled in the North Pacific at Midway, Oahu, and Adios Station. About 95% of the mass influx of particles >75 μm was found to be quartz. Other particles (over a range of 10–75 μm) were found to be

Figure 31. (a) Fine banding of light colored sand (made up of quartz and feldspar) and black magnetite sand. Small scale divisions are centimeters. Taken in the field on a dune surface of the Great Sand Dunes National Monument, Colorado. (b) A magnified view of the sand under a microscope. The black grains are magnetite. Photographs by C. Klein.

a mixture of Ca-rich particles (possibly calcite), Al-Si-Ca-rich (probably plagioclase feldspar), clays, ferro-magnesian silicates (possibly pyroxene), Al-Si-K-Fe-rich (possibly mica) and several Fe-Ti-rich materials (oxides). The authors conclude that "the appearance of the giant particles more than 10,000 km from their source cannot be explained using current acknowledged atmospheric transport mechanisms."

Volcanic activity. Not only does dust travel over very large distances around the globe, it can also reach very high altitudes in the stratosphere. One of the greatest volcanic explosions ever witnessed, the 1883 explosion of Krakatoa, in the strait between Sumatra and Java, Indonesia, introduced enormous amounts of fine dust into the stratosphere. This dust drifted around the Earth for several years reducing the Earth's annual mean temperature by a few degrees (as a result of partial blocking of solar radiation) and producing brilliant red sunsets.

Other very large eruptions took place on March 28 and April 3 and 4, 1982, in the El Chichón volcano in Mexico, introducing massive amounts of dust into the stratosphere. Mackinnon et al. (1984) report on the mineralogy of this volcanic dust as collected between altitudes of 16.8 and 19.2 km. A significant proportion of the dust particles studied ranged between 2 and 40 µm in size. Their identifications, based on data obtained from optical microscopy, SEM, and ATEM, are shown graphically, as a modal analysis, in Figure 33. Over 80 vol % of the collected particles consisted of volcanic glass (with a composition rich in Al and Si), with the remainder consisting of feldspars (of various compositions), pyroxenes, amphiboles, SiO_2 (an unidentified silica mineral), magnetite, and calcium sulfate. The calcium sulfate is considered to represent a condensation product having formed in the explosive plume during eruption.

The Mount St. Helens explosion in the state of Washington on May 18, 1980 also introduced a large volume of volcanic glass and minerals into the stratosphere. The mineralogical results of a sample (with a general range of particle sizes of ~0.1–30 µm) are given in Table 1, below. A study by Rietmeijer (1993) of volcanic dust from El Chichón collected in May, 1985, at an altitude between 34 and 36 km reports no glass, but instead indicates all Si-rich materials to be crystalline silica phases (either cristobalite or keatite; see Table 1). Each particle in this study was identified by a combination of TEM imaging, selected area electron diffraction (SAED) for identification of crystallinity and crystallographic (structural) properties, and energy dispersive (EDS) analysis.

More recently, a series of spectacular eruptions of the Mount Pinatubo volcano in the Philippines, beginning June 12, 1991, has produced ash clouds reaching well into the stratosphere, representing the greatest ash clouds that have been observed since the beginning of the satellite era (Minnis et al., 1993). These authors describe the strong cooling effect due to the volcanic aerosol, with continued cooling through September, 1992.

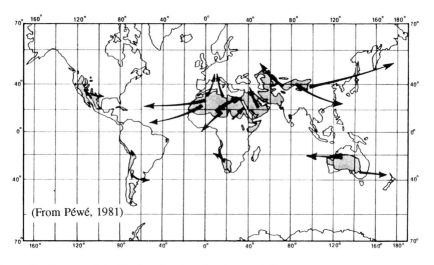

Figure 32. Deserts of the Earth and major directions and distances of dust transport

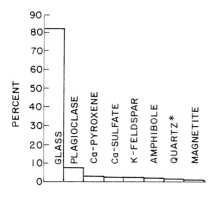

Figure 33. Mineralogy of the dust particles from the El Chichón stratospheric clouds, collected in May, 1992, between 16.8 and 19.2 km with a high-altitude flight. Particles identified as "quartz" are actually grains of an unidentified silica mineral (from Mackinnon et al., 1984).

Determination of the background levels of natural dusts

Lung particulate burdens. There is a considerable volume of literature (e.g., Stovin and Partridge, 1982; Churg, 1983; Case et al., 1988; Case and Sébastien, 1987, 1988, and 1989; Tuomi and Tammilehto, 1989) that deals with the characterization of fiber types in lung burdens of occupationally-exposed individuals, with special emphasis on the identification of chrysotile and/or various types of amphibole (e.g., tremolite, crocidolite, amosite, or anthophyllite; cf. Chapter 13). Much less is known about the general particulate burden of subjects who have not been exposed occupationally to sources of fibers, as generated by mining and manufacturing activities involving e.g., asbestos. Examples of some studies that address the question of the general mineralogical make-up of the lung burden are, e.g., Churg and Wiggs (1987), Stettler et al. (1991), and Paoletti et al. (1987); see also Chapter 13. The study by Churg and Wiggs (1987) is based on autopsy material from 10 male cigarette smokers living in an urban setting. The particles were identified by a combination of energy

Table 1

Mineralogical composition of Mount St. Helens ash and El Chichón ash, expressed as volume percent

Mount St. Helens ash[a]		El Chichón ash [b]	
Glass	37	Silica (cristobalite, keatite)	35
Plagioclase	16	Feldspar	36
Hornblende	21	Orthopyroxene and olivine	15
Pyroxene	13	Barite ($BaSO_4$)	8
Other: SiO_2, possibly cristobalite		Plattnerite (PbO_2)	3
		Maghemite (Fe_2O_3)	2
		Magnetite (Fe_3O_4)	1

Sources:
[a] Farlow et al. (1981); collected at 17-km altitude.
[b] Rietmeijer (1993); collected at 34–36-km altitude.

Table 2

Mineral particles identified in lung samples of residents of Rome, Italy

Mineral type			No. of cases
mica	{	muscovite phlogopite	10
clay	{	kaolin pyrophyllite	10
gypsum			3
talc			10
chrysotile			1
silicon dioxide			10
amphibole	{	tremolite crocidolite	6
chlorite	{	chlorite vermiculte	6
rutile			2
others	{	plagioclase zeolite pyroxene feldspar	9
unidentified			8

Source: Paoletti et al. (1987).

dispersive X-ray analysis, supplemented by electron optical morphological determinations and electron diffraction. Only particles greater than 0.1 μm in diameter, with a mean particle size of 0.8 ± 0.1 μm, and a range of < 0.25 to >2.26 μm, were studied. The particles identified and their respective volume percentages are: kaolinite (28%), silica (20%), mica (16%), feldspars (9%), talc (7%), titanium-rich (6%), aluminum-rich (4%), miscellaneous (10%).

The study by Stettler et al. (1991) deals with the particulate lung burden of 91 subjects from the Cincinnati, Ohio urban area. The particles ranging in size from 0.37 to 1.02 μm were identified by a combination of SEM and X-ray dispersive X-ray analysis. Such a combination of techniques can process a large number of particles but it provides only very qualitative mineralogical results. Stettler et al. report an average of 38.1% by volume of Al-silicate (this includes feldspar of various types, kaolinite, and possibly some mica), 21.0% silica (probably mainly quartz), with the remainder a mixture of rutile-like (10.2%), Al-rich (4.4%), Si-rich (3.1%), Ti-rich (3.0%), and Mg-silicate (2.5%) compositions.

The Paoletti et al. (1987) study involved the characterization of mineral particles by AEM technique in the lung samples of 10 subjects who had lived in the Rome area but who had not been occupationally exposed to mineral dust. All particles larger than 0.1 μm were identified; approximately 70% had diameters between 1 and 5 μm, and no particles larger than 30 μm in size were encountered. The results of this study are given in Table 2. The authors note that micas, clays, talc, and silicon dioxide (probably mainly quartz) make up approximately 90% of all the mineral particles observed. Paoletti et al. (1987) do not speculate about the source of this particulate material in the urban environment of Rome. They note, however, that tremolite occurs more commonly than chrysotile, and they suggest that the higher percentage of tremolite-like amphibole is the result of contamination in talc dusts. A geological (or mineralogical) perspective can lead to a different interpretation. The layer silicates (including micas, clays, and chlorite), the SiO_2 (probably mainly quartz), the amphiboles, and others (including plagioclase, zeolites, pyroxenes, and other feldspar) may all represent dust particles produced by natural weathering processes of the Earth's crust (compare the findings in Table 2 with the mineral group listings in Fig. 1). Particles that result most likely from human activity are: gypsum, from plaster of Paris or plaster board in the drywall construction of houses and buildings; talc, used as a filler in paints, plastics, paper, and rubber and as talcum powder and face powder; rutile as the white pigment base, TiO_2, used in paint, plastics, rubber, and paper; and vermiculite as used extensively in building aggregate and insulation. The Paoletti et al. study probably represents the results of filtering (by the lungs) of urban air which in this instance, because of the relative proximity of Rome to the Appenines and the Alps, may contain a large percentage of the primary lithic mineral particles that likely result from the weathering of the various rock types (and their partial loose cover of fragmentary material and soils) in the close-by mountain chains.

Global background level for fiber counts in the troposphere. There are very few studies that have attempted to characterize quantitatively the background particle content of continental or oceanic air (for a summary, see

Health Effects Institute, 1991, Table 4-8). Kohyama (1989), as part of a study of airborne asbestos fibers in various non-occupational environments in Japan, collected a total of 16 air samples over the remote island of Chichijima (for location with respect to Japan, see Fig. 34) and three additional samples over the Pacific ocean between Japan and the island. The fibers were studied by AEM (using a combination of morphology, energy dispersive X-ray analysis, and electron diffraction patterns) and were <5 µm in length. All the fibers were identified as chrysotile and their concentrations for the 19 samples are given in Table 3. The geometric mean of the concentration for these samples is 9.7 fibers/l, and the average is 14.1 fibers/l, with a total range of 4 to 48 fibers/l. As can be seen from Figure 34, Chichijima Island (part of the Ogasawara Island chain) is about 1000-km south of Tokyo. The population on the island is about 15,000, with some automobiles but the effect of these on the airborne chrysotile level was considered to be extremely small (Kohyama, pers. comm.). The bedrock of the island is mainly volcanic with lesser limestone. There is no asbestos mine and no asbestos manufacturing on the island. The samples were taken in October 1982

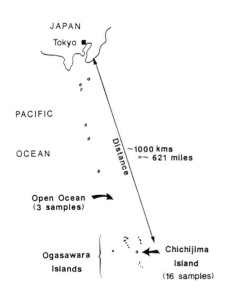

Figure 34. Location of air samples collected over the Pacific Ocean and on Chichijima Island (from Kohyama, 1989).

Table 3

Airborne chrysotile levels on Chichijima Island and over the ocean between Tokyo and the island

Location	Number of samples	Asbestos concentration (fibers/l)
on the sea	3	11.7 30.6 12.8
meteorological observatory	4	<7.7 8.0 23.5 <4.0
Ogamiyama Park	4	47.6 15.0 <7.4 4.7
subtropical agriculture center	4	15.7 6.2 <7.3 36.7
aerospace tracking station	4	<8.1 8.0 8.6 <4.0
geometric mean		9.7
average		14.1

Source: Kohyama (1989); expanded data base listed here from Kohyama (pers. comm.).

and 1983 when high atmospheric pressure overlies the island such that there is no air stream from the Japanese islands to the Ogasawara Islands (Kohyama, pers. comm.). Kohyama writes "Therefore I considered that the air over the Chichijima Island is not affected by man-produced fiber and not affected by air blowing from the Japanese islands slightly polluted by asbestos. Consequently, we selected the Chichijima Island as a sampling point that would show a background level of airborne asbestos on the Earth." These data, therefore, may be representative of a global (chrysotile) fiber background with a geometric mean of 9.7 fibers/l and an average (arithmetic mean) of 14.1 fibers/l. These are not only very important baseline data for the possible background chrysotile content of global air, but they also raise the important question of why there is this level of chrysotile with no reported amphibole. This question will be addressed further below under New Idria (Coalinga) chrysotile region, California. Finally, it should be noted that Kohyama's findings are at the high end of findings from the limited number of other studies in rural environments, as summarized by the Health Effects Institute (1991). However, all of the studies reported non-zero values for the background levels of dusts.

Measurements from fluvial sources. Fragments and dust particles that result from the erosion of bedrock and the loose sand and soil cover on the Earth are very effectively moved by flowing water, be it runoff or groundwater. Dust particles suspended in the troposphere are returned to the Earth's surface (see Fig. 35) and added to the hydrosphere, mainly by a combination of dry (gravity) deposition and precipitation scavenging (e.g., Barlow and Latham, 1983; Rodhe, 1983). Dry deposition is the result of the influences of gravity and airflow on dust particles, in the absence of precipitation (Battan, 1966). Dust incorporated in cloud water (precipitation scavenging) and returned to the Earth in precipitation may be many thousand to million of times more concentrated than dust in the air (Pruppacher et al., 1983). The dust that is returned to the Earth's surface may become available for resuspension unless it is incorporated in wet soils, or in lakes, streams and oceans. Gonzales and Murr (1977) report on the identification of particles scavenged by individual raindrops in the Socorro, New Mexico, area.

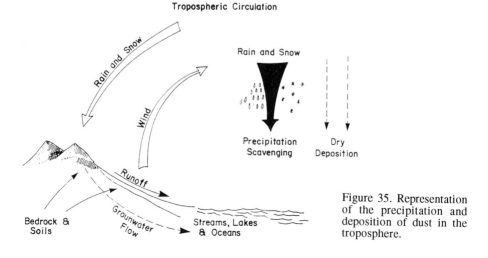

Figure 35. Representation of the precipitation and deposition of dust in the troposphere.

Table 4

Comparison of asbestos fiber contents (in fibers/l) of some natural waters

Source	Natural water unaffected by asbestos-rich geology	Natural water affected by asbestos-rich geology
lakes	$<10^6$	10^6-10^8
		10^7-10^9
rivers	$<10^6$	10^6-10^{13}
		10^7-10^8
		10^6-10^9
		10^8-10^{10}
		10^7-10^{11}
water supplies	$<10^6$	10^7-10^8
	$<10^6$	10^6-10^8

Source: Schreier (1989) Table 11, which lists 21 references as the original sources for these values and ranges.

The particulates were observed to range in size from less than 0.01 to 3 µm, with a mean particle size of 0.1 µm. Essentially all the particulates were found to be crystalline single grains or polycrystalline aggregates composed mainly of mica, kaolinite, and other layer silicates.

Although little is known, even qualitatively, about the mineralogic make-up of natural dust that is carried in suspension in fluvial system (see e.g., McDowell-Boyer et al., 1986; Murr and Kloska, 1976), there has been a multitude of studies since early 1970 (see Schreier, 1989, for a compilation of sources) on the fiber (asbestos) content of municipal drinking water systems and of rivers and lakes. Table 4 gives a synopsis of the data in Schreier (1989, his Table 11) for natural waters unaffected and affected by asbestos-rich geology of the bedrock. These data show that asbestos fibers in natural surface waters in the range of 10^5-10^7 fibers/l are commonplace. What is especially noteworthy about these fiber data is that the majority of such water studies lists chrysotile asbestos fiber as the main or only fiber type. From general geological and weathering considerations, one would expect amphibole fibers to occur as well. However, amphibole fibers are identified in fluvial systems only in a small number of reports. Nelson (1986) reports considerable amphibole fiber concentrations in surface waters from Beaver Bay and Duluth, Minnesota (Lake Superior) and Everett, Washington. Both of these regions are rich in amphibole-containing bedrock. Stewart (1976) reports the presence of amphibole fibers in the water supplies of Boston, San Francisco, Philadelphia, Denver, and Seattle in addition to the everpresent chrysotile component. The question of the origin of chrysotile fiber in almost all waters tested in North America will be discussed further under the heading of New Idria (Coalinga) chrysotile region.

Dust from Antarctic ice cores. The ice sheet of East Antarctica may represent a depositional record of the chemical and physical constituents accumulated in the snow from the troposphere and stratosphere that spans more than 500,000 years (Mosley-Thompson, 1980). The study of particle records in deep ice cores allows for the detection of variations in atmospheric particle loading as a function of time. Mosley-Thompson (1980) reports on a 911-year record of microparticle deposition at the south pole. In this study approximately 6000 samples obtained from a 101-m ice core were studied by a combination of light microscopy, scanning electron microscopy, and X-ray energy dispersive spectroscopy. The minimum size of the particles studied was 0.5 µm. Figure 36 is from Mosley-Thompson (1980). The two spheres at the upper-left of the figure could be cosmic "silicate" spheres; the particulates in the two upper-right photos are most likely of terrestrial volcanic origin, although the uppermost figure has a morphology suggestive of that of chrysotile; and the shard-like particles in the two photographs at the bottom could be either of terrestrial volcanic or extra-terrestrial origin. As is the case for dust particles in fluvial systems, the overall mineralogical make-up of representative suites of dust from ice cores is only very partially known.

Kohyama (1989) reports on the chrysotile asbestos content of ice core samples from Antarctica for which the ages of the original snow falls are known. The results expressed as fibers/l of water are given in Table 5. The concentration values for these samples are very similar to those reported in Table 4, in the first

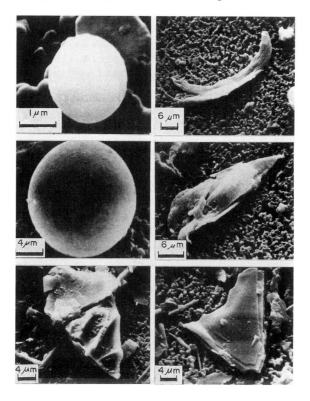

Figure 36. Scanning electron microscope photographs of selected particles in an ice core (101 m long) from Antartica (reproduced from Plate IV of Mosley-Thompson, 1980).

Table 5

Chrysotile asbestos concentrations in
snow and ice samples, Antarctica

Sample location	Age of snow	Concentration (fibers/l)
Mizuho base 70–100-cm depth (collected in 1977)	1970–1973	2.25×10^5
Mizuho base 5-m depth	ca. 1930	1.09×10^5
Yamato Mountains	>10,000 years	5.83×10^5

Source: Kohyama (1989) Table 2.

column entitled "Natural water unaffected by asbestos-rich geology." The similarity in the fiber concentrations of modern water samples and old ice core samples from the Yamato Mountains (see Table 5; dated from long before the industrial revolution) suggests that there may be a relatively constant ambient chrysotile asbestos concentration that reflects a global background value (see also New Idria-Coalinga-chrysotile region, below).

The geology of two major natural fiber sources

Here follows a brief geologic synopsis of two very large natural sources of asbestos fiber. In the northern hemisphere, the biggest natural chrysotile fiber source appears to be the New Idria (Coalinga) area of California. In the southern hemisphere there is the enormous source of riebeckite amphibole (and associated crocidolite asbestos) in the Hamersley Range of Western Australia.

New Idria (Coalinga) chrysotile region, California. The very fine-grained chrysotile in this region is hosted in a serpentinite body which outcrops over a 50-sq-mi area in Fresno and San Benito Counties, 35 miles northwest of Coalinga, California (Mumpton and Thompson, 1975; see Fig. 37 for location). The age of the serpentinite and the mechanism of alteration to chrysotile are not well known. However, the association with Cretaceous and Tertiary rock types (Coleman, 1986), suggests that the serpentinite is at least late Cretaceous in age, or ~65 million years old. The chrysotile occurrence in this region is unique for two reasons: (1) the ore parts of the serpentinite contain more than 50% recoverable chrysotile (as compared with other chrysotile ores that contain 5–10% as cross- or slip-fiber veins) and (2) the ore occurs as soft, powdery agglomerates of very finely matted chrysotile fibers of very small size. The electron micrograph in Figure 38 shows that the diameters of the chrysotile fibers are < 0.02 μm and their lengths are on the order of 5–15 μm. The extremely small sizes of these fibers are the result of intensive shearing, brecciation, and chemical alteration of the serpentinite, producing large amounts of material of extremely small particle size (Mumpton and Thompson, 1975). These fine chrysotile grains are present not

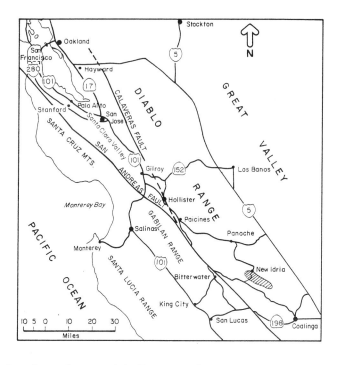

Figure 37. Location and extent of the New Idria (Coalinga) serpentine mass (shown with a dashed pattern) which is the source rock of the very fine chrysolite illustrated in Figure 38 (from Coleman, 1986).

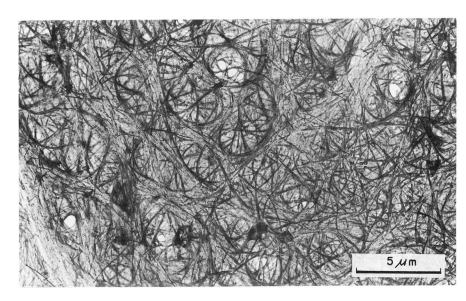

Figure 38. Electron micrograph of typical chrysotile fibers from the New Idria, California region (from Woolery and Cohan, 1968, p. 9).

only in the major ore zones of the serpentinite (mined from pit mines) but occur throughout the ~50 square-mile region of the serpentinite, in bedrock exposures as well as the soil cover. Most of the slopes in the region are fairly barren but richly strewn with blocks and fragments of chrysotile-rich serpentinite. The general vegetation in the region is relatively sparse, probably directly related to the underlying chrysotile-rich serpentine which provides an unfavorable environment for plant growth (Schreier, 1989). The sparseness of vegetation is also a result of the generally arid climate of the region. Most of the annual precipitation occurs from November through April, and the average rainfall from May through October is generally less than a few centimeters (Woolery and Cohan, 1968). Mumpton and Thompson (1975) quote that "in the late 1950s, J.H. Bright, a Union Carbide Corporation geologist, reported that the friable ore is not merely a surface weathering phenomenon, but that it extends several tens of feet deep and is relatively uniform over a 15 square-mile area."

The combination of the aridity of the climate, the relative sparseness of the vegetation cover, and the extremely fine grain size of the pervasive chrysotile fibers make this a uniquely erodable natural source of fibers. The chrysotile fibers are available at the Earth's surface in such a state that they need no further fragmentation before they become part of the hydrosphere or troposphere. Indeed, the 50-sq-mi area underlain by serpentinite and chrysotile-rich bedrock and soil cover is a ready-made source of fiber that is efficiently transported by air or by water. Cooper et al. (1979) report that dustfalls along roads and trails in a recreational area located within the perimeter of the New Idria serpentinite, consist of ≥90% chrysotile asbestos, with a maximum airborne fiber count of 5300 chrysotile fibers/l (in air) for fibers longer than 5 μm (as measured with dust collectors at various locations close to roads and motorcycle trails in the Clear Creek recreational area). McGuire et al. (1982) report values of 5600 million chrysotile fibers/l (in water) in an aqueduct near Coalinga, California, and 15,000 million chrysotile fibers/l south thereof at the Kern River intertie. Hayward (1984) reports similarly high chrysotile concentrations in various sources of drinking water in many parts of California, with some of the concentrations the highest of any in the world, ranging from 14 to 260,000 million fibers/l (with a median value of 1000 million fibers/l).

The New Idria (Coalinga) serpentinite mass and its intimately associated very fine-grained chrysotile fiber is the source of chrysotile-rich sediments in streams and rivers, as a result of runoff. When these sediments collect in playas and basins they become a source of wind-blown and air-suspended chrysotile fiber during dry periods. Similarly, this region is a primary source of air-suspended chrysotile fiber, mainly during the arid summer and fall months of the year.

It appears very likely that the 50-sq-mi New Idria region of chrysotile-laden serpentinite, with a common natural fiber length ranging from several to several tens of μm, is a major natural supply source of the global ambient chrysotile values measured by Kohyama (1989) over the Pacific Ocean southeast of Japan (see Table 3) and in the old ice cores of Antarctica (see Table 5). The age of the serpentinite and other surrounding rocks is probably at least Late Cretaceous (or about 65 million years in age) which makes it highly likely that the serpentinite rich in chrysotile has been an erosional source of global fine chrysotile for a very long time. At least over the past 5000 years this region has had similar climatic and relief conditions (Les McFadden, pers. comm.).

Riebeckite and crocidolite in the Hamersley Range of Western Australia. The Hamersley Range of Western Australia is geologically best known for the vast deposits of iron-rich chemical sedimentary rocks, known as banded iron-formations, which are about 2,500 million years in age. These Precambrian banded iron-formations occur in a well-banded, essentially horizontally layered sedimentary sequence. Their total thickness is about 3900 feet (which equals about 1200 m; Trendall and Blockley, 1970). This sequence of rocks is exposed in a range of low mountains with the top of the range at about 3500 feet above sea level. The outcrop area of the Hamersley Group which contains all the major iron-formations is about 60,000 km^2 (Trendall, 1983). The region identified in Figure 38 with the two approximate distances of 500 and 160 kms, respectively, includes the Hamersley Group as well as parts of the underlying Fortescue Group exposures; this calculates to be approximately 80,000 km^2, or about the size of the State of Indiana. Intimately interbedded with these vast iron-formations are bands rich in the Na-amphibole, riebeckite, and/or the asbestiform variety crocidolite. Trendall and Blockley (1970) estimate the inferred crocidolite reserves (of the mineable long-fiber variety) to be just over 400,000 tons, with the largest known deposit containing 225,000 tons. The last crocidolite mine in the area was closed in December, 1966 (Nevill and Rogers, 1992). In addition to the restricted ore zones of crocidolite, however, non-asbestiform riebeckite is present in intimate association with iron-rich minerals (e.g., magnetite, Fe_3O_4, and hematite, Fe_2O_3) and quartz throughout the thick iron-formation sequence. As such the present Hamersley Range is a major natural source of Na-rich amphibole fiber, be it crocidolite or riebeckite.

It appears geologically very likely (Trendall, 1983) that the original extent of the Hamersley Group (which contains all the iron-formations rich in Na-amphibole) was about three times larger than it is now. This is shown by the broad dashed line around the "egg-shaped" basin in Figure 39. The aerial extent of this egg-shaped basinal region is about the size of the State of Oregon. This means that over geologic time about two thirds of the total egg-shaped region has had its cover of Hamersley Group rocks removed by erosional processes. Morris (1989) presents a chronology for the Hamersley Basin dating back from its origin of about 2500 million years ago to the present. He concludes that the Hamersley Group has been continually exposed at the Earth's surface since about 150 million years ago. This means that an area about the size of the State of Oregon (the egg-shaped region outlined by the dashed circumference in Fig. 39) has been subject to erosion since 150 million years ago. If one were to assume an erosion rate of about 1 cm per thousand years for Western Australia (see Fig. 26), then one can simply calculate (assuming a steady erosion rate, which is probably incorrect) that in 150 million years about 1500 meters of Hamersley Group sequence has been likely removed by erosional processes. This volume is equivalent to a 1500-m high (or 5000-ft high = 1-mile high) pile of rocks on an area of the State of Oregon. If this rock pile contained on average ~5 vol % riebeckite (and/or crocidolite), then it is clear that the Hamersley region of

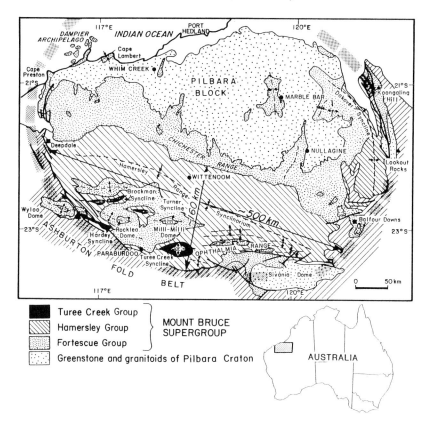

Figure 39. Generalized geologic map of the Hamersley Range, Western Australia. The region of present-day Hamersley Group exposures (which includes banded iron-formations and associated riebeckite and/or crocidolite) is shown to be about 500 km by 160 km in size. The prior extent of the Hamersley Group sequence is shown by the much larger egg-shaped region outlined by the broad dashed lines (from Trendall, 1983).

Western Australia has been a very large and unique natural (erosional) source of Na-rich amphiboles in the southern hemisphere for an extremely long time.

Monitoring of amphibole fibers in the air and in drinking water supplies in areas close to Wittenoom, Western Australia (for location see Fig. 39), reveals the following information. A Western Australian Department of Conservation and Environmental (1986) report gives mean amphibole fiber concentrations in air measured in one of the steep gorges near Wittenoom (removed from mining tailings and road cover rich in amphibole) of 0.2–0.6 fibers/l, with a maximum of 1.3 fibers/l. These air fiber counts refer to amphibole only. A report by the Western Australian Government Chemical Laboratories (1982) gives amphibole fiber counts for four boreholes (for drinking water) in the Wittenoom region ranging from 0 to 100,000 fibers/l. As is to be expected, natural erosional processes are the cause for these low-level (background) values for sodium amphibole fibers in northwestern Western Australia.

CONCLUDING REMARKS

Geologic weathering and erosional processes are responsible for the generation of large amounts of natural dust that enter the hydrosphere and troposphere. Although the processes for the generation of natural dust are well understood, little is known of the quantitative mineralogical make-up of such dust. At this time there are no complete and quantitative studies of natural dust in various remote (away from anthropogenic sources of dust) regions of the globe. A program of dust collection in the southwestern United States that could provide some important answers about dust influx rates and mineral content has been on-going at the U.S. Geological Survey in Denver, Colorado, since about nine years under the direction of Marthis C. Reheis. A manuscript in progress by Reheis will summarize five years of annual data for 60 sites and relates the amount and composition of dust to climate, source area, soils, and human disturbance (see also M.C. Reheis, in review).

Knowledge of the quantity of such natural dust in the troposphere and hydrosphere, combined with its mineralogic characterization, would allow for the recognition of global background levels of natural dust. Knowledge of such background levels is needed before regulatory agencies can set standards of acceptable environmental values. Furthermore, this information is an essential component of any epidemiological study aimed at understanding pulmonary diseases such as fibrosis, lung cancer, and mesothelioma. Even in cases of smokers with no occupational exposure to mineral dusts it should be realized that each of us has had at least a background exposure to many mineral dusts. Consequently several mineralogical questions concerning dust should be addressed, including: What is the global background of chrysotile fibers in the troposphere? What are the average background levels of quartz particulates in the troposphere? What are the maxima for quartz particulates during dust storms in arid climates (as in many urbanized parts of the southwestern U.S.)? These values should be known so that regulations can be set a levels consistent with the natural flux of global dust. At this time, only a limited number of studies have investigate the background levels of dust, e.g., the study in which Kohyama (1989) attempted to establish an ambient (background) level for natural and globally distributed chrysotile fibers in the troposphere (see Table 3). This study reports an average concentration of 14.1 fibers/l and a geometric mean of 9.7 fibers/l. Other, similarly quantitative studies of chrysotile fibers, in other remote areas of the globe are necessary to provide a stronger data base for these natural background averages.

Regarding quartz, the *Wall Street Journal* of March 22, 1993, has an extensive discussion of "How sand on the beach came to be defined as a human carcinogen." It reports that the Environmental Protection Agency "may take action at limiting public exposure to silica." Before regulations are promulgated a regulating agency should establish the range of quartz dust values that people in arid and desert climates are, and have been, exposed to for hundreds of years, without any apparent indication of disease. It should also take into account the discovery of long-range transport (over more than 10,000 km from their source) of giant (>75 μm) quartz particles (Betzer et al. 1988); smaller respirable particles should be transported even longer distances. "Appropriate" dust levels of all kinds of minerals can be regulated only after the fluxes and characterization of natural mineral dusts have been well established.

Modern instrumentation (such as the analytical transmission electron microscope) is routinely capable of providing unique identification of dust particles as small as 0.1 μm in diameter through a combination of morphological study, structure analysis and chemical composition. And theoretically TEM can be used to identify dust particles as small as a few unit cells (i.e., on the order of 10^{-3} μm); however, characterization of particles this small can be very tedious. Additional techniques are available for further evaluation of provenance of a single particulate or a bulk dust sample. The minor and trace element abundances and isotopic make-up of single dust particles can be determined using a secondary ion mass spectrometer (SIMS). The limiting condition on a SIMS is the 10-μm ion beam. As such, SIMS analyses on particles larger than 10 μm can provide quantitative chemical results for all elements and isotopes, at levels of parts per million to parts per billion. Synchrotron X-ray fluorescence analysis has been used by Sutton and Flynn (1988) for the determination of the trace element composition of stratospheric particles and by Flynn and Sutton (1990) to determine quantitatively the minor and trace element abundance of cosmic dust particles, ranging in size from 7 to 30 μm. Flynn and Sutton (1990) report that element sensitivities below 10 ppm were achieved for 20 μm particles. Minor-element, trace-element, and isotopic signatures of dust particulates would allow comparison with similar data for rock types that would be suspected to be their possible source.

Age determinations on single grains larger than 25 μm in diameter are now possible by isotopic measurements using the ion microprobe SHRIMP, as available at the Australian National University in Canberra, Australia. This instrument is a high resolution ion mass spectrometer. The dating of individual dust particles allows for the evaluation of their provenance through the knowledge of the ages of various rock types that are suspected to represent a possible source.

The oxygen isotopic ratios of bulk samples of quartz particulates from present-day atmospheric aerosols were compared by Jackson (1981) with those of sediments and soils in the northern and southern hemisphere to deduce possible differences in provenance of the airborne dust. Differences in the $\delta^{18}O$[†] values of northern and southern hemisphere dust are attributed to: (1) the abundant presence of microcrystalline quartz in the form of chert, which is of low-temperature sedimentary origin, in the northern hemisphere and (2) the

---

[†] Oxygen isotope measurements are typically reported as $\delta^{18}O$, which is defined as:

$$\delta^{18}O = \left[\frac{(^{18}O/^{16}O)_{sample} - (^{18}O/^{16}O)_{SMOW}}{(^{18}O/^{16}O)_{SMOW}}\right] \times 10^3 ‰,$$

where $(^{18}O/^{16}O)_{SMOW}$ is the ratio of the two oxygen isotopes in standard mean ocean water (Faure, 1977, p. 325). The ratio of the oxygen isotopes in a specific sample reflects a partitioning coefficient that varies as a function of conditions; hence, this ratio can be used to fingerprint the last environment with which a rock or mineral equilibrated.

contrasting erosion of high-temperature quartz in igneous and metamorphic rocks in the southern hemisphere (Jackson, 1981).

Soil scientists who have known for years that airborne dust is an important component of soils are now developing techniques using $^{87}Sr/^{86}Sr$ and Ca/Sr ratios of dust, bedrock and soil carbonate to evaluate the contribution of bedrock and dust to soil carbonates (Quade et al., in review).

Finally, there are now a range of microanalytical techniques available that will allow for studies of the provenance of dust particles, providing a better understanding of the processes responsible for the commonly high background levels of mineral dusts from natural sources.

ACKNOWLEDGMENTS

I am grateful to a departmental colleague, Frans J.M. Rietmeijer, for very informative and inspiring discussion sessions, over several years, on natural dust. I thank Jacques Dunigan, of the Asbestos Institute, Sherbrooke, Quebec, Canada, for assisting in the location of several pertinent references. Richard Clarke, Chemistry Centre, Perth, Western Australia provided me with several environmental reports that I would have been unable to locate otherwise. Ermanno Galli, University of Modena, Italy and Saburo Aita, JEOL Nihon Denshi Co., Ltd., Tokyo, Japan kindly provided me with some excellent photography of zeolites and layer silicates, respectively. Thoughtful reviews which considerably improved the final version of this manuscript were provided by Frans J.M. Rietmeijer, Malcolm Ross, Les D. McFadden, and George D. Guthrie, Jr. I thank Mabel Chavez for the word processing and Dag Lopez for preparing the majority of the illustrations.

REFERENCES

Bailey, S.W., ed. (1984) Micas. Rev. Mineral. 13, 584 p.
Bailey, S.W., ed. (1988) Hydrous Phyllosilicates (Exclusive of Micas). Rev. Mineral. 19, 725 p.
Baris, I., Simonato, L., Artvinli, M., Pooley, F., Saracci, R., Skidmore, J. and Wagner, C. (1987) Epidimiological and environmental evidence of the health effects of exposure to erionite fibers: A four-year study in the Cappadocian region of Turkey. Int'l J. Cancer, 39, 10–17.
Barlow, A.K. and Lathan, J. (1983) A laboratory study of the scavenging of sub-micron aerosol-charged raindrops, In Precipitation Scavenging, Dry Deposition and Resuspension, vol. 1, Precipitation Scavenging, H.R. Pruppacher, R.G. Seminon and W.G.N. Slinn, eds., Elsevier, New York, 551–558.
Battan, L.J. (1966) The Unclean Sky, a Meteorologist Look at Air Pollution. Doubleday, Garden City, New York, 141 p.
Betzer, P.R., Carder, K.L., Duce, R.A., Merrill, J.R., Tindale, N.W., Uematsu, M., Costello, D.K., Young, R.W., Feely, R.A., Breland, J.A., Bernstein, R.E. and Greco, A.M. (1988) Long-distance transport of giant mineral aerosol grains. Nature 336, 568–571.
Blackburn, W.H. and Dennen, W.H. (1988) Principles of Mineralogy. Wm. C. Brown Publs., Dubuque, Iowa, 413 p.
Cadle, R.D. (1966) Particles in the Atmosphere and Space. Reinhold, New York, 226 p.
Case, B.W. and Sébastien, P. (1987) Environmental and occupational exposure to chrysotile asbestos: a comparative microanalytic study. Archives of Environmental Health 42, 185–191.
Case, B.W. and Sébastien, P. (1988) Biological estimation of environmental and occupational exposure to asbestos. Annals Occupational Hygiene, 32, 181–186.

Case, B.W. and Sébastien, P. (1989) Fiber levels in lung and correlation with air samples. In Nonoccupational exposure to Mineral Fibers, J. Bignon, J. Peto, and R. Saracci, eds., IARC Publication no. 90, 207–218.
Case, B.W., Sébastien, P. and McDonald, J.C. (1988) Lung fiber analysis in accident victims: a biological assessment of general environmental exposures. Archives of Environmental Health 43, 178–179.
Churg, A. (1983) Nonasbestos pulmonary mineral fibers in the general population. Environ. Res. 31, 189–200.
Churg, A. and Wiggs, B. (1987) Types, numbers, sizes and distribution of mineral particles in the lungs of urban male cigarette smokers. Environ. Res. 42, 121–129.
Coleman, R.G. (1986) Field trip guide book to the New Idria area, California. Int'l Mineral. Assoc. Meetings, Stanford University, Stanford, CA.
Cooper, W.C., Murchio, J., Popendorf, W. and Wenk, H.R. (1979) Chrysotile asbestos in a California recreational area. Science, 206, 685–688.
Dana, J.D., A System of Mineralogy, 7th ed., vol. 1, 1944; vol. 2, 1951; vol. 3, 1962. John Wiley & Sons, New York, Rewritten by C. Palache, H. Berman and C. Frondel.
Deer, W.A., Howie, R.A. and Zussman, J. (1962) Rock-Forming Minerals, 5 vols., John Wiley & Sons, New York. Three completely revised volumes (1A, 1B, and 2A) were published in 1982, 1986, and 1978, respectively.
Deer, W.A., Howie, R.A. and Zussman, J. (1992) An Introduction to the Rock Forming Minerals. John Wiley & Sons, New York, 691 p.
Department of Conservation and Environment (1986) Wittenoom airborne asbestos study. Technical Series 7, Perth, Western Australia, 35 p.
Dixon, J.B. and Weed, S.B., eds. (1977) Minerals in Soil Environments. Soil Science Soc. Amer., Madison, Wisconsin, 948 p.
Evans, B.W. and Guggenheim, S. (1988) Talc, pyrophyllite, and related minerals In Hydrous Phyllosilicates (exclusive of micas), S.W. Bailey, ed., Rev. Mineral. 19, 225–294.
Farlow, N.H., Oberbeck, V.R., Snetsinger, K.G., Ferry, G.V., Polkowski, G. and Hayes, D.M. (1981) Size distribution and mineralogy of ash particles in the stratosphere from eruptions of Mount St. Helens. Science, 211, 832–834.
Faure, G. (1977) Principles of Isotope Geology. John Wiley & Sons, New York, 464 p.
Flynn, G.J. and Sutton, S.R. (1990) Synchrotron X-ray fluorescence analyses of stratospheric cosmic dust: new results for chondritic and low-nickel particles. Proc. 20th Lunar Planet. Sci. Conf., 335–342.
Gillette, D.A. (1981) Production of dust that may be carried great distances. In Desert Dust: Origin, Characteristics, and Effect on Man, T.R. Péwé, ed., Geol. Soc. Amer. Spec. Paper 186, 11–26.
Goodwin, A.M. (1976) Giant impacting and the development of continental crust. In The Early History of the Earth, Windley, B.F., ed., John Wiley & Sons, New York, 77–95.
Gonzales, T.W. and Murr, L.E. (1977) An electron microscopy study of particulates present in individual raindrops. J. Geophysical Res. 82, 3161–3166.
Gottardi, G. and Galli, E. (1985) Natural Zeolites. Springer-Verlag, New York, 409 p.
Government Chemical Labs. (1982) The Significance of Asbestos Fibers in Drinking Water. Internal Report, Perth, Western Australia.
Gregory, K.J. and Walling, D.E. (1973) Drainage Basin Form and Process, a Geomorphological Approach. John Wiley & Sons, New York, 456 p.
Guthrie, G.D., Jr. (1992) Biological effects of inhaled minerals. Amer. Mineral. 77, 225–243.
Hamblin, W.K. (1989) The Earth's Dynamic Systems. Macmillan Publishing Co., New York, 576 p.
Hanley, D.S. (1987) The origin of chrysotile asbestos veins in southeastern Quebec. Canadian J. Earth Sci. 24, 1–9.
Hanley, D.S., Chernosky, J.V., Jr., and Wicks, F.J. (1989) The stability of lizardite and chrysotile. Canadian Mineral. 27, 483–493.
Hansen, K., and Mossman, B.T. (1987) Generation of superoxide (O_2^-) from alveolar macrophages exposed to asbestiform and nonfibrous particles. Cancer Research, 47, 1681–1686.
Harris, D.V. and Kiver, E.P. (1985) The Geologic Story of the National Parks and Monuments, 4th ed. John Wiley & Sons, New York, 464 p.
Hayward, S.B. (1984) Field monitoring of chrysotile asbestos in California waters. J. Amer. Water Works Assoc. 76, 66–73.
Health Effects Institute (1991) Asbestos in Public and Commercial Buildings: A Literature Review and Synthesis of Current Knowledge. Health Effects Institute—Asbestos Research, Cambridge, 366 p.
Jackson, M.L. (1981) Oxygen isotopic ratios in quartz as an indicator of provenance of dust. In Desert Dust: Origin, Characteristics, and Effect on Man, T.L. Péwé, ed., Geol. Soc. Amer. Spec. Paper 186, 27–36.
Jaenicke, R. (1980) Natural Aerosols. Annals New York Acad. Sciences, 338, 317–329.

Johnson, R.B., (1967), The Great Sand Dunes of Southern Colorado. U.S.Geol. Surv. Prof. Paper 575-C, C177–C183.
Jones, B.F. and Galan, E. (1988) Sepiolite and palygorskite, In Hydrous Phyllosilicates (Exclusive of Micas), S.W. Bailey, ed., Rev. Mineral. 19, 631–674.
Klein C. and Hurlbut, C.S., Jr. (1993) Manual of Mineralogy, 21st ed. John Wiley and Sons, New York, 681 p.
Kohyama, N. (1989) Airbone asbestos levels in non-occupational environments in Japan, In Non-occupational Exposure to Mineral Fibres, J. Bignon, J. Peto, R. Sacarri, eds., IARC Publication no. 90, 262–276.
Ledoux, R.L., ed. (1979) Short Course in the Mineralogical Techniques of Asbestos Determination. Mineral. Assoc. Canada, Royal Ontario Museum, Toronto, 279 p.
Levadie, B., ed. (1984) Definitions for Asbestos and Other Health-related Silicates. ASTM Special Publ. 834, Philadelphia, PA, 213 p.
Mackinnon, I.D.R., Gooding, J.L., McKay, D.S. and Clanton, U.S. (1984) The El Chichón stratospheric cloud: solid particles and settling rates. J. Volcanology Geothermal Res. 23, 125–146.
McDowell-Boyer, L.M., Hunt, J.R. and Sitar, N. (1986) Particle transport through porous media. Water Resources Res. 22, 1901–1921.
McGuire, M.J., Bowers, A.E. and Bowers, D.A. (1982) Asbestos analysis case history: surface water supplies in southern California. J. Amer. Water Assoc. 74, 471–478.
Meigs, P. (1953) World distribution of arid and semiarid climates. In Review of Research on Arid Zone Hydrology, UNESCO, Pans. Arid Zone Programs 1, 203–210.
Minnis, P., Harrison, E.F., Stowe, L.L., Gibson, G.G., Denn, F.M., Doelling, D.R. and Smith, W.L. (1993) Radiative climate forcing by the Mount Pinatubo eruption. Science 259, 1411–1415.
Morris, R.C. (1989) The iron ores of the Hamersley Province. Exploration Research News, CSIRO, Wembley, WA, 6–7.
Mosley-Thompson, E. (1980) 911 years of microparticle deposition at the south pole: a climatic interpretation. Institute of Polar Studies, Report 73, Ohio State Univ., Columbus, OH, 130 p.
Mumpton, F.A. and Thompson, C.S. (1975) Mineralogy and origin of the Coalinga asbestos deposit. Clays and Clay Minerals 23, 131–143.
Murr, L.E. and Kloska, K. (1976) The detection and analysis of particulates in municipal water supplies by transmission electron microscopy. Water Resources 10, 469–477.
Nelson, H. (1986) Estimated national occurrence and exposure to asbestos in the public drinking water supplies. Environmental Protection Agency Report, Contract no. 68-01-7166, 121 p.
Nevill, M. and Rogers, A. (1992) Inquiry into asbestos issues at Wittenoom. Australian Government Report, 55 p.
Otten, M.T. (1993) High-resolution transmission electron microscopy of polysomatism and stacking defects in antigorite. Amer. Mineral. 78, 75–84.
Palekar, L.D., Most, B.M., and Coffin, D.L. (1988) Significance of mass and number of fibers in the correlation of V79 cytotoxicity with tumorigenic potential of mineral fibers. Environmental Res. 46, 142–152.
Paoletti, L., Batisti, D. Caiazza, S., Petrelli, M.G., Taggi, F., De Zorzi, L., Dina, M.A. and Donelli, G. (1987) Mineral Particles in the lungs of subjects resident in the Rome area and not occupationally exposed to mineral dust. Environmental Res. 44, 18–28.
Peterson, S.T. and Junge, C.E. (1971) Sources of particulate matter in the atmosphere. In: Study of Man's Impact on Climate, W.H. Mathews, W.W. Kellop and G.D. Robinson, eds., MIT Press, Cambridge, MA, 310–320.
Péwé, T.R. (1981) Desert dust: an overview. In Desert Dust: Origin, Characteristics, and Effect on Man, T.R. Péwé, ed., Geol. Soc. Amer. Spec. Paper 186, 1–10.
Press, F. and Siever, R. (1986) Earth. 4th ed. W.H. Freeman and Co., New York, 656 p.
Pruppacher, H.R., Semonin, R.G. and Slinn, W.G.N. (1983) Precipitation Scavenging, Dry Deposition and Resuspension, vols. 1 and 2. Elsevier, New York, 1462 p.
Quade, J., Chivas, A. and McCulloch, M.T. Strontium and carbon isotopic tracers and the origins of soil carbonate in South Australia and Victoria. Paleogr. Palaeoecol. (1991 SLEADS Conf. vol.), in rev.
Reheis, M.C., Dust deposition and its effect on soils. Chapter for U.S. Geol. Survey Bull., in rev.
Rietmeijer, F.J.M. (1993) Volcanic dust in the stratosphere between 34 and 36 km altitude during May, 1985. J. Volcanology and Geothermal Res. 55, 69–83.
Rodhe, H. (1983) Precipitation scavenging and tropospheric mixing, In Precipitation Scavenging, Dry Deposition and Resuspension, vol. 1, Precipitation Scavenging, H.R. Pruppacher, R.G. Seminon and W.G.N. Slinn, eds., Elsevier, New York, 719–729.
Ronov, A.B. and Yaroshevsky, A.A. (1969) Chemical composition of the Earth's crust. Amer. Geophys. Union Monograph no. 13, 50.
Schreier, H. (1989) Asbestos in the Natural Environment. Elsevier, New York, 159 p.

Skinner, H.C.W., Ross, M. and Frondel, C. (1980) Asbestos and Other Fibrous Materials: Mineralogy, Crystal Chemistry and Health Effects. Oxford Univ. Press, New York, 204 p.

SMIC (1971) Inadvertent Climate Modification. Report of the Study of Man's Impact on Climate. MIT Press, Cambridge, MA.

Stettler, L.E., Platek, S.F., Riley, R.D., Mastin, J.P. and Simon, S.D. (1991) Lung particulate burdens of subjects from the Cincinnati, Ohio urban area. Scanning Microscopy 5, 85–94.

Stewart, I.M. (1976) Asbestos in the water supplies of the ten regional cities. Environmental Protection Agency Report, Contract no. 68-01-2690.

Stovin, P.G.I. and Partridge, P. (1982) Pulmonary asbestos and dust content in East Anglia. Thorax 37, 185–192.

Sudo, T., Shimoda, S., Yotsumoto, H. and Aita, S. (1981) Electron Micrographs of Clay Minerals. Elsevier, Amsterdam, 203 p.

Sutton, S.R. and Flynn, G.J. (1988) Stratospheric particles: Synchrotron X-ray fluorescence determination of trace element contents. Proc. 18th Lunar Planet. Sci. Conf., 607–614.

Suzuki, Y., and Kohyama, N. (1984) Malignant mesothelioma induced by asbestos and zeolite in the mouse peritoneal cavity. Environmental Res. 35, 277 – 292.

Trendall, A.F. and Blockley, J.G. (1970) The iron-formations of the Precambrian Hamersley Group, Western Australia. Geol. Survey Western Australia, Bull. 119, 366 p.

Trendall, A.F. (1983) The Hamersley Basin, In Iron-Formation: Facts and Problems. A.F. Trendall and R.C. Morris, eds., Elsevier, New York, 69–129.

Tuomi, T. and Tammilehto, L. (1989) Mineral fiber concentration in lung tissue of mesothelioma patients in Finland. Amer. J. Industrial Medicine 16, 247–254.

Veblen, D.R., ed. (1981) Amphiboles and Other Hydrous Pyriboles—Mineralogy. Rev. Mineral. 9A, 372 p.

Veblen, D.R. and Ribbe, P.H., eds. (1982) Amphiboles: Petrology and Experimental Phase Relations. Rev. Mineral. 9B, 390 p.

Woolery, R.G. and Cohan, W.T. (1968) New Idria chrysotile; an unusual ore yields new products. Union Carbide Corp., private report, Niagara Falls, N.Y., 30 p.

Wylie, A.G. (1990) Discriminating amphibole cleavage fragments from asbestos: rationale and methodology. Proc VIIth Int'l Pneumoconioses Conf., Dept. Health and Human Resources (NIOSH) Publ. no. 90-108, part II, 1065–1069.

Zoltai, T. and Stout, J.H. (1984) Mineralogy, Concepts and Principles. MacMillan, New York, 504 p.

CHAPTER 3

MINERALOGY OF AMPHIBOLES AND 1:1 LAYER SILICATES

David R. Veblen

Department of Earth and Planetary Sciences
The Johns Hopkins University
Baltimore, Maryland 21218 U.S.A.

Ann G. Wylie

Laboratory for Mineral Deposits Research
Department of Geology
University of Maryland
College Park, Maryland 20742 U.S.A.

INTRODUCTION

Although amphiboles and 1:1 layer silicates most commonly are not asbestiform, these mineral groups do contain the two primary types of asbestos minerals, specifically amphibole asbestos and serpentine (or chrysotile) asbestos. Thus, because amphibole and serpentine asbestos are the most infamous of mineral pathogens, the crystal chemistry and geological occurrences of amphiboles and 1:1 layer silicates are of preeminent importance when considering the health effects of mineral dusts. Furthermore, nonasbestiform amphibole can enter the environment in various ways, and other 1:1 layer silicates, notably kaolinite, are widely used in industry.

This chapter is therefore devoted to reviewing briefly the ways in which mineralogists and crystal chemists represent complex silicate structures; the basic nomenclature for amphiboles and 1:1 layer silicates; the geological occurrences of these minerals; their crystal structures and defect structures; the various morphologies, or habits, of amphibole and 1:1 layer silicate crystals; and the potentially active surface sites and dissolution kinetics of such particles. We end the chapter with a discussion of how 1:1 layer silicates, amphiboles, and other chain silicates related to amphiboles are identified in the laboratory. Some of the specific analytical techniques necessary for identification are, however, described in detail in Chapter 7.

Existing review literature

This review is intended to be only an introduction to those basic aspects of 1:1 layer silicates and amphiboles that are of most importance to the health professional. Intended to be read in an hour or two,[†] it cannot be comprehensive or detailed. There are, however, many excellent reviews of these structural

---

[†] The interested reader is advised to enroll in a speed reading course.—*Eds.*

groups already in the literature, and these can be consulted for additional detail on crystal chemistry and occurrence. Indeed, we have drawn heavily on these review sources in writing the present treatment, and we wish to express our gratitude to their authors for assembling literally thousands of references to the primary literature on crystal chemistry, petrology, physical properties, and geological occurrence. Among the best sources are earlier volumes of the MSA *Reviews in Mineralogy* series, especially volumes 9A (Veblen, 1981) and 9B (Veblen and Ribbe, 1982) on amphiboles and volume 19 (Bailey, 1988a) on 1:1 and some other layer silicates.

The following is a partial annotated list of references:

I. Amphiboles
   A. Crystal chemistry and properties
      1. Hawthorne (1981a)—nomenclature, crystal structure, crystal chemistry
      2. Hawthorne (1983)—more detail than Hawthorne (1981a), plus spectroscopy
      3. Hawthorne (1981b)—spectroscopy (Mössbauer, vibrational, electronic absorption)
      4. Cameron and Papike (1979)—crystal structures and crystal chemistry
      5. Papike (1988)—crystal structures and chemical variations
      6. Ghose (1981)—cation distribution and exsolution
   B. Relationships among biopyriboles
      1. Thompson (1981)—symmetry and chemographic relationships
      2. Veblen (1981)—observed structures and polysomatic reactions
   C. Petrology and geology
      1. Robinson et al. (1982)—natural metamorphic amphiboles
      2. Gilbert et al. (1982)—experimental petrology
      3. Wones and Gilbert (1982)—natural igneous amphiboles
      4. Ernst (1968)—overview of petrology and crystal chemistry
   D. Amphibole asbestos *per se*
      1. Zoltai (1981)—mineralogy of amphibole asbestos
      2. Whittaker (1979)—crystal chemistry of amphibole asbestos
      3. Ross (1981)—geology of amphibole and serpentine asbestos
      4. Skinner et al. (1988)—monograph on fibrous minerals
      5. Chisholm (1983)—electron microscopy and diffraction of asbestos
      6. Hodgson (1979)—properties and chemistry of asbestos

II. 1:1 layer silicates
   A. Crystal chemistry
      1. Bailey (1988a,b)—structure and polytypism of 1:1 layer silicates
      2. Giese (1988)—structures and stability of kaolin minerals
      3. Wicks and O'Hanley (1988)—structures of serpentine minerals
      4. Bailey (1988c)—structures and chemistry of other 1:1 minerals
   B. Petrology and geology
      1. Murray (1988)—formation and occurrence of kaolin minerals
      2. Wicks and O'Hanley (1988)—petrology of serpentine minerals
      3. Chernosky et al. (1988)—stability and thermodynamics of serpentine
   C. Serpentine asbestos *per se*
      1. Wicks (1979)—mineralogy and crystal chemistry
      2. Chisholm (1983)—electron microscopy and diffraction of asbestos
      3. Ross (1981)—geology of amphibole and serpentine asbestos
      4. Skinner et al. (1988)—monograph on fibrous minerals
      5. Hodgson (1979)—properties and chemistry of asbestos

CRUSTAL ELEMENTAL ABUNDANCES, STRUCTURE DETERMINATION, AND REPRESENTATION OF SILICATE MINERAL STRUCTURES

Crustal abundance of the chemical elements

The minerals we find at or near the earth's surface are largely controlled by the abundances of the chemical elements in the crust of the earth. Oxygen is by far the most abundant element, making up approximately 45.5 wt % of the continental crust (Faure, 1991). The next eight elements by weight are the metals Si (26.8%), Al (8.40%), Fe (7.06%), Ca (5.3%), Mg (3.2%), Na (2.3%), K (0.90%), and Ti (0.5%). Together, these nine elements comprise over 99% of the earth's crust. Because the anion oxygen forms chemical bonds with at least partial ionic character with these metallic cations, the common rock-forming minerals are oxygen-based compounds such as oxides, carbonates, sulfates, phosphates, and, by far the most abundant, silicates and silica minerals. (Silicates are compounds containing oxygen, silicon, and at least one other cation, whereas the silica minerals have the nominal formula SiO_2.) Also geochemically notable is hydrogen, a highly mobile and pervasive element in the crust. Although present only at the 0.14% level by weight (Mason and Moore, 1982), this light element is important on an atomic % basis, as a catalyst in many geologically important reactions, as a chemical transport medium when in the form of water, and as a fundamental constituent of naturally occurring 1:1 layer silicates and amphiboles.

Determination and refinement of crystal structures

Much of what we know about crystal structure has been derived from observations of the way in which crystals scatter various types of radiation, most notably the scattering of X-rays. Discovered in 1912, X-ray diffraction quickly led to the determination of numerous mineral structures (for a brief historical sketch of the early days of X-ray diffraction, see von Laue, 1969). Although crude by modern standards, these early structure determinations allowed important generalizations to be made about the structures of crystalline materials (see the next section).

By definition, minerals are crystalline,[1] and one property of crystalline materials is translational periodicity. This property allows the entire structure of a crystalline material to be described by a small subset of the atoms present in a crystal. Whereas typical individual crystals of even µm-sized grains consist of $>>10^9$ atoms, the structure of such crystals can be described by a unit cell containing typically $<<10^2$ atoms. In such a description, the shape of the unit cell—i.e., the atom coordinate system described by three axes (**a**, **b**, **c**)[2] and their angular relationships ($\alpha = \mathbf{b}\angle\mathbf{c}$, $\beta = \mathbf{a}\angle\mathbf{c}$, $\gamma = \mathbf{a}\angle\mathbf{b}$)—and the positions of the atoms within the unit cell are sufficient to describe the entire crystal structure. The

---

[1] Minerals are generally defined as naturally occurring, inorganic, crystalline substances. Some materials satisfy the first of these two criteria but are noncrystalline or amorphous. These types of substances (e.g., amorphous silica) are referred to as mineraloids.

[2] Some crystallographers use **X, Y, Z** to denote crystallographic axes. Here we follow the convention used by the *American Mineralogist*. We use **a, b,** and **c** to denote the crystallographic axes (i.e., vectors); *a, b,* and *c* to denote the magnitudes of these axes (i.e., the lattice parameters); and **X, Y,** and **Z** to denote the principle optical directions.

remainder of the structure can be obtained by simply translating the unit cell along each of its axes. The description can be further simplified by utilizing point-group symmetry operations (e.g., mirror planes, rotational axes) and Bravais lattices, and the crystallographically allowed combinations for these symmetry operations and Bravais lattices are known as *space groups*. A more detailed discussion of space groups, symmetry operations, and unit cells can be found in Klein and Hurlbut (1993), and a compilation of the possible space groups and their symmetry characteristics can be found in the *International Tables for Crystallography* (Hahn, 1983).

Today, the determination of a mineral's basic crystal structure is routine, assuming that relatively defect-free crystals of reasonable size (e.g., 100 μm) can be obtained. Typically, diffracted X-ray intensities are measured with a computer-controlled, single-crystal diffractometer. Once the basic structure is known, the exact positions for the atoms within one unit cell are derived by a least-squares *crystal structure refinement*, which involves adjusting the model structure in order to minimize the differences between the observed diffracted intensities (or, more technically, structure factors) and those calculated from the model structure. In a very well-refined structure, the atomic positions may be known to an accuracy of approximately 0.001 Å (in this paper, we use the ångström unit for length measurements; 1 Å = 0.1 nm).

Unfortunately, the 1:1 layer silicates of the serpentine and kaolin groups typically are very fine-grained, structurally disordered, and/or deformed, and the fibers of amphibole asbestos inherently are too thin to obtain good single crystals suitable for crystal structure refinement. What we know, therefore, about the detailed structures of these layer silicates comes primarily from only a few single-crystal refinements based on fortuitously well-crystallized material or refinements based on powder diffraction data. In addition, the asbestiform amphiboles can occur with macroscopic nonasbestiform habits, and structural data obtained from this type of material can be used as a starting point for structural studies of amphibole asbestos. Structural data from X-ray diffraction investigations can be sup-plemented with information from electron diffraction, high-resolution transmission electron microscopy (HRTEM) (Buseck, 1992), and spectroscopy experiments (Hawthorne, 1988). It should be kept in mind, however, that the very properties that make asbestos and clays technologically interesting also preclude us from characterizing their crystal chemistry with the same detail that is possible with other, better-crystallized minerals.

Coordination structures and their representation in two and three dimensions

Several of the most important generalizations about the crystal structures of ionically bonded materials were formalized in the five famous statements known as Pauling's Rules (Pauling, 1929). The first of these rules begins as follows: "Around every cation, a co-ordinated polyhedron of anions forms...." This statement means that in stable crystal structures the cations are typically surrounded by anions to which they are bonded. If one connects the centers of these anions, the resulting network of lines defines a polyhedron, which is called the *coordination polyhedron* of the cation. Structures of this type can be called *coordination compounds*, and all of the rock-forming silicates, silica minerals, oxides, fluorides, etc. crystallize with structures of this type. The number of anions bonded to a cation is called the *coordination number* for that cation.

Cations are commonly coordinated to (or bonded to) three, four, or six oxygen anions, to form anionic groups such as CO_3^{2-} (carbonate), SiO_4^{4-} (silicate), or MgO_6^{10-}. Coordination numbers greater than six also occur, as is commonly the case for Na, K, and Ca in silicate structures (e.g., in the M4 site of many amphiboles). Coordination structures can be represented in several different ways. In three-dimensional models, spheres of different sizes and colors to represent different elements can be glued together. Most commonly, large spheres that touch each other are used to represent oxygen anions, with smaller balls representing cations stuffed into the small interstices between oxygens. Such models are called packing models. Though useful for simple crystal structures, packing models are of less use for showing the structures of complex materials (e.g., silicates), because it is not easy to see very far into the interior of the model, even if transparent balls are used to represent oxygen. We will not, therefore, use packing models in this review.

A more common type of model for silicate structures consists of spheres with different colors representing different atomic species. These spheres are connected by metal rods, which represent the chemical bonds between anions (oxygen) and cations. Such models are usually called *ball-and-stick models*, and two-dimensional drawings made with circular atoms connected by lines representing the bonds are also common. Figure 1a shows coordinations of four and six drawn in this fashion.

A third type of model can be constructed from polyhedra that represent the actual coordination polyhedra in the structure, and in two dimensions, these coordination polyhedra can be drawn as projections. The polyhedra for coordination numbers four (commonly called tetrahedral coordination) and six (octahedral coordination) are shown in Figure 1b. The vertices of the polyhedra indicate the positions of oxygen atoms, and the cations (not shown) lie at the centers of the polyhedra.

Ball-and-stick models and polyhedral models both have strengths and weaknesses. Because they can be constructed with rods of variable length and with variable angles between rods, ball-and-stick models are excellent for visualizing the detailed distortions of coordination polyhedra (or variations in

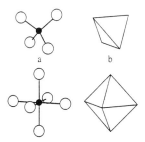

Figure 1. Representations of cation coordination numbers four (tetrahedral) and six (octahedral). (a) Ball and stick representations; open circles are oxygen atoms, and closed circles are cations. (b) Polyhedral representations; the vertices of the polyhedra are the oxygen atoms, and the cation is in the center of the polyhedron.

bond distances and angles) that occur in most minerals. For very complex structures, however, this representation can be confusing (it can be hard to see the forest for the trees). Because a polyhedron represents a number of bonds and atoms in a single geometrical object, the polyhedral representation is usually excellent for quickly grasping the essentials of a structure. In three dimensions, however, polyhedral models are usually constructed from mass-produced, perfectly regular polyhedra. Therefore, although they show the basic structure, they do not show structural subtleties such as distortions of the coordination polyhedra. In two-dimensional drawings, however, polyhedral distortions can be shown, though the exact positions of the coordinated cations inside the polyhedra generally are not indicated. In many structures, the coordination polyhedra for cations having coordination numbers greater than six are very distorted or irregular. In this case, it is common to mix the two representations, showing the regular coordination polyhedra as actual polyhedra, but showing the cations with irregular coordination as circles connected by lines (bonds) to their coordinating oxygens.

Polymerization in silicates

At relatively low pressures such as those found in the earth's crust, Si is invariably coordinated by four O atoms, thus forming a tetrahedron as the coordination polyhedron. In most silicates and in silica minerals, some or all of the oxygen anions are bonded not to one, but to two Si atoms, i.e., oxygens are shared between two Si tetrahedra. Through such oxygen sharing, the tetrahedra are thus connected, or *polymerized*, to form larger anionic groups that are sometimes called silicate polyanions. The groups thus produced can consist of connected pairs of tetrahedra, closed rings of tetrahedra, linear chains, two-dimensional sheets, or three-dimensionally polymerized frameworks. The geometry of these polyanions forms the basis for the structural classification of silicate structures (Liebau, 1980; 1985).

Amphiboles are double-chain silicates characterized by units such as that shown in both ball-and-stick and polyhedral representations in Figure 2. Some of the oxygen atoms ("bridging oxygens") are shared by two tetrahedra, but others ("nonbridging oxygens") are bonded to only one Si atom, and these may connect the tetrahedral chain to other coordination polyhedra, by sharing these oxygens with other cations, such as Na, Mg, Al, Ca, and Fe. As a further chemical complication, Al atoms may substitute for some of the Si atoms in the double chain.

The layer silicates (also called sheet silicates or phyllosilicates) are characterized by two-dimensional sheets ("tetrahedral sheets") of polymerized tetrahedra, each of which shares three of its four oxygen atoms with other tetrahedra. Such a silicate sheet from the serpentine mineral lizardite is shown in Figure 3. As in the chain silicates, the nonbridging oxygen atoms (those that are bonded to only one Si) can form bonds to other cations. In the 1:1 layer silicates, these other cations most commonly are six-coordinated Al, Mg, and Fe that form a sheet of edge-sharing octahedra. By sharing of oxygens, such an octahedral sheet can be connected to a parallel tetrahedral sheet, and it is the 1:1 ratio of tetrahedral sheets to octahedral sheets that gives this group of layer silicates its name.

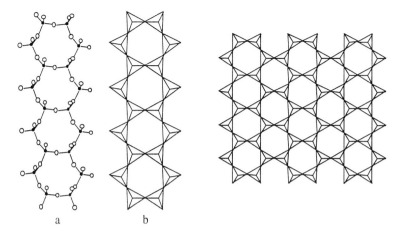

Figure 2. [left] (a) Ball-and-stick representation of a double silicate chain characteristic of amphiboles. This chain is from fluor-riebeckite (Hawthorne, 1978). Open circles represent oxygen atoms, smaller filled circles represent Si, and lines represent Si–O bonds. The viewing direction is close to **a***; the chain has been rotated slightly, so that some of the O atoms do not block the Si atoms. (b) Polyhedral representation of the same chain, projected on to (100). The vertices of the tetrahedra are the oxygen positions, and the Si atoms (not shown) lie in the centers of the tetrahedra.

Figure 3. [right] The silicate sheet from lizardite-1T (Mellini, 1982), projected onto the (001) plane.

In this review, we will use primarily the polyhedral representation to introduce the crystal structures of 1:1 layer silicates and amphiboles, and we will use the structures to rationalize the complicated chemistry of these minerals.

MINERAL HABIT

The habit of a mineral is the shape or morphology that its crystals or aggregates of crystals adopt during crystallization (Klein and Hurlbut, 1993). Many minerals, such as the amphiboles, can crystallize in a wide variety of habits. Figure 4 illustrates some of the habits found in single crystals and crystal aggregates and the terms that are generally used to describe these habits. The nomenclature for describing habits is not rigorous, several terms overlap, and some terms may include others (e.g., asbestiform minerals are fibrous, but not all fibrous minerals are asbestiform). Despite these shortcomings, habit nomenclature is widely used in describing mineral specimens, and there is consensus among mineralogists on the meanings of most terms.

Habit is controlled by the environment in which crystallization occurs. Many variables affect crystal growth, and several of these probably play a role in the development of a *fibrous habit*. A mineral is said to be fibrous if it "gives the appearance of being composed of fibers, whether the mineral actually contains separable fibers or not" (Zoltai, 1981). Fibrous minerals generally grow either from an aqueous fluid in open spaces such veins or cavities, or during retrograde metamorphism and alteration of preexisting minerals in the presence of an aqueous fluid. In the latter case, the fibrous minerals commonly occupy a smaller

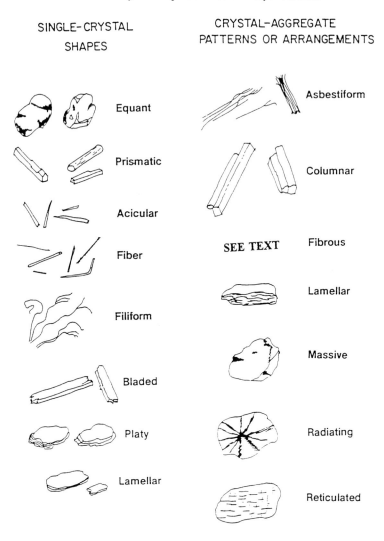

Figure 4. Examples of different mineral habits found in single crystals and crystal aggregates.

volume than the minerals they replace, allowing for interfiber porosity. Many investigators have shown that mineral habit can be controlled by impurities in the aqueous fluid that interfere with the growth of certain crystal faces. Stanton (1972) discusses this phenomenon and reviews the literature on the subject. The interfering phase or ion may form an intimate intergrowth on specific crystallographic surfaces, limiting further development of these faces. If the interfering material actually forms a crystalline layer on the surface of its host, such a relationship is referred to as epitaxial growth. In other cases, the interference is not due to true epitaxy but simply to an adsorbed layer or substitution of an ion along certain faces that limits further development of the crystal on that surface. Growth of fibrous materials is discussed in detail by Zoltai (1981).

Asbestos and the asbestiform habit

Asbestos is defined as a group of highly fibrous silicate minerals that readily separate into long, thin, strong fibers that have sufficient flexibility to be woven, are heat resistant and chemically inert, are electrical insulators, and therefore are suitable for uses where incombustible, nonconducting, or chemically resistant material is required (Gary et al., 1974). As a result of these useful properties, asbestos has been used in literally thousands of applications, including a wide variety of building materials such as insulation, textured paints, vinyl tile, spackling compounds, wall board, roofing tile, and concrete pipe. Asbestos may be laminated as paper, woven for cloth, mixed with binders to form friction products (e.g., brake pads), and used as filters. A recent summary of the products likely to contain asbestos was published by the Health Effects Institute (1991).

The asbestiform habit has a number of characteristics that differentiate it from other habits, but the main differences are the fibrillar structure and the fiber flexibility and strength. Fibrils are single, elementary fibers that have very small widths. In amphibole asbestos, the fibrils are single crystals (sometimes twinned and commonly with defects) having irregular cross sections and being separated from the other fibrils by grain boundaries. In chrysotile (serpentine) asbestos, the fibrils are individual rolls of silicate layers or tubes consisting of concentric cylindrical silicate layers; most chrysotile fibrils have a circular cross section or nearly so. Fibrils form in bundles and share a common axis of elongation, which is the **c** crystallographic axis in amphiboles and usually the **a**-axis in chrysotile. Amphibole fibrils tend to be misoriented with respect to the other two crystallographic directions. The widths of the fibrils vary significantly among asbestos minerals, and even asbestos of a single mineral species may have significantly different fibril dimensions in different geological occurrences. Separated (or milled) asbestos fibers may consist of either single fibrils or bundles containing a number of fibrils. (For an example, see the image of an amphibole asbestos fiber, Fig. 27, in the SEM section of this chapter.) Table 1 gives the mean widths of asbestos fiber populations from some of the major commercial asbestos deposits.

Unfortunately, there is much confusion concerning the terms *asbestos* and *asbestiform*. In part, this confusion stems from the fact that the regulatory and mineralogical definitions of asbestos differ. In this paper, we will use *asbestos* for the serpentine mineral chrysotile and amphiboles that crystallize with the properties described in the last two paragraphs. For the term *asbestiform*, we will follow Zoltai's (1981) usage, which states that any mineral resembling asbestos is asbestiform. Skinner et al. (1988) suggest that only chrysotile and the five amphiboles that most commonly form asbestos deposits (anthophyllite, tremolite, actinolite, grunerite, riebeckite) should be called asbestiform (i.e., they wish the term asbestiform to connote not only habit, but also a specific and very restrictive set of mineral species). We believe this usage is too restrictive. Even the amphibole in some extensively mined asbestos deposits is not one of the usual five. For example, the chemical analysis of Bolivian crocidolite given by Hodgson (1979) shows it to be the amphibole magnesio-riebeckite, not riebeckite (Leake, 1978). As another example, chain silicate that has been mined from the "anthophyllite" asbestos deposit at Orijärvi, Finland, is actually a mixture of anthophyllite, jimthompsonite, and highly disordered material to which a well-defined mineral name cannot be applied (Schumacher and Czank, 1987). The predominant amphibole in a recently discovered asbestos deposit in Texas is potassian winchite, again not one of the usual five asbestiform amphiboles (Wylie and Huggins, 1980). Given these and similar examples, we will follow Zoltai (1981) and use *asbestiform* to indicate any mineral having the same habit as asbestos.

Table 1

Average Width of Asbestos Populations as Established by TEM

Mineral species *Sample type*	Location	*Dimensions*[†] (μm)		*Reference*
Chrysotile				
bulk	Calidria	mean width	0.07	(1)
		median width	0.03	(3)
bulk	Quebec	mean width	0.17	(2)
		median width	0.15	(3)
airborne	Quebec	median width	0.05, 0.06	(4)
	Rhodesia	mean width	0.16	(2)
Anthophyllite				
bulk	Finland	median width	0.61	(3)
airborne	Finland	median width	0.44, 0.52, 0.70	(5)
Actinolite				
	Minnesota	mean width	0.41	(6)
		median width	0.24	(6)
	Germany	median width	0.17	(3)
Tremolite				
bulk	Greece	median width	<0.5	(7)
bulk	Montana	median width	<0.4	(7)
Grunerite				
bulk	Transvaal Province (South Africa)	mean width	0.35	(8)
		mean width	0.29	(6)
airborne	Transvaal Province (South Africa)	median width	0.20–0.26	(4)
Riebeckite				
bulk	Cape Province (South Africa)	mean width	0.23	(2)
		median width	0.20	(3)
		mean width	0.12	(8)
		mean width	0.09, 0.06	(9)
		for l > 2 μm:		
		mean width	0.11, 0.10	
airborne	Cape Province (South Africa)	median width	0.07–0.09	(4)
bulk	Australia	mean width	0.09, 0.08	(9)
		for l > 2 μm:		
		mean width	0.14, 0.11	
bulk	Transvaal Province (South Africa)	mean width	0.12, 0.13, 0.20, 0.13, 0.12	(9)
		for l > 2 μm:		
		mean width	0.23, 0.25, 0.36, 0.47, 0.60	
bulk	Bolivia	mean width	0.18	(9)
		for l > 2 μm:		
		mean width	0.4	

[†] All lengths (l) are included unless otherwise stated.

Sources:
1–Cambell et al. (1980); 2–Spurney et al. (1980); 3–Pott et al. (1987); 4–Gibbs and Hwang (1980); 5–Timbrell (1989); 6–Cook et al. (1982); 7–Langer et al. (1990); 8–Wylie et al. (1982); 9–Shedd (1985).

Asbestos can form in cross-fiber veins, where the fiber axis is perpendicular to the vein walls, in slip-fiber veins, where the fiber axis is at an angle to the vein walls, and in mass-fiber deposits, where the fiber bundles are intertwined and randomly oriented (Ross, 1981). In some circumstances, usually as a result of weathering, the latter may be physically resilient and referred to as "mountain leather" or "mountain cork" (such materials can form from a number of different fibrous minerals). Mountain leather has been described as looking like wetted paper, and it commonly forms a very tough material, hence the reference to leather. In the older literature, amphibole asbestos, especially actinolite asbestos, was sometimes referred to as amianthus. A thorough review of the asbestiform habit and related morphologies is given by Zoltai (1981).

Cleavage and parting. Because they affect the morphology of mineral fragments, it is necessary to define cleavage and parting. Cleavage refers to breakage of a mineral on an approximately planar surface that is controlled by its crystal structure. All samples of a given mineral should show similar cleavage properties. Parting refers to approximately planar breakage along planes that are not cleavage planes. Typically, parting surfaces are produced by planar defects, such as twin planes, or exsolution lamellae (approximately planar lamellae of another mineral that have precipitated out of the host mineral in the solid state). Because the microstructures of minerals vary from occurrence to occurrence, not all specimens of a given mineral will show the same parting properties. Cleavage is described in terms of its quality by terms such as perfect (e.g., micas), good, fair, and poor (Klein and Hurlbut, 1993, p. 252–253).

1:1 LAYER SILICATES

Basic 1:1 layer structure

As indicated above, all 1:1 layer silicates are characterized by two-dimensional units consisting of one silicate sheet connected to one octahedral sheet. As in other layer-silicate groups, such as the micas (Bailey, 1984), there are two different end-member types of octahedral sheets that can occur in the 1:1 structures. In the first of these, called a *trioctahedral sheet*, all of the octahedrally coordinated sites are occupied by metal atoms (Fig. 5a). In the other type, called a *dioctahedral sheet*, two-thirds of the octahedral sites are filled by metal atoms, whereas the remaining sites are vacant (Fig. 5b). In both trioctahedral and dioctahedral 1:1 layer silicates, the silicate sheet is connected to the octahedral sheet by sharing of the apical (nonbridging) oxygens of the tetrahedral sheet, as shown for the lizardite structure in Figure 6 and the kaolinite structure in Figure 7. Both the tetrahedral and octahedral sheets can carry a net electrostatic charge, but the entire 1:1 layer must be electrostatically neutral (i.e., if charged, the two sheets must have equal but opposite charges). Unlike the micas, 1:1 layer silicates do not have interlayer cation sites between the layers. In the hydrated form of halloysite, however, there are water molecules between the layers.

Within the basic 1:1-layer-silicate category, a number of important structural and chemical variations can occur: (1) differences in the conformation of the layers [e.g., planar, curled, or corrugated]; (2) variations, either ordered or disordered, in the way the layers are stacked together [polytypism]; (3) variations,

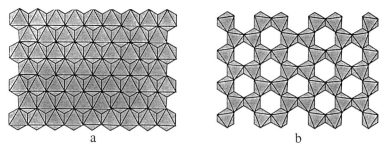

Figure 5. The two end-member types of octahedral sheets found in layer silicates. (a) The trioctahedral sheet in lizardite-1T (Mellini, 1982). (b) The dioctahedral sheet in kaolinite (Bish and Von Dreele, 1989). Note that shared octahedral edges are shortened relative to unshared edges.

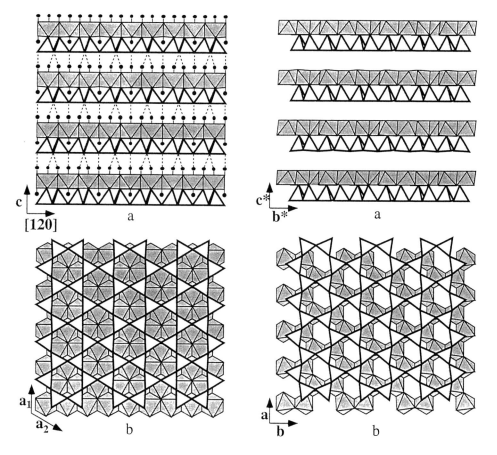

Figure 6. [left] The crystal structure of lizardite-1T (Mellini, 1982). (a) [100] projection. Small black circles are hydrogen atoms, and dotted lines are hydrogen bonds. (b) [001] projection. Note the very slight tetrahedral rotation.

Figure 7. [right] The crystal structure of kaolinite (Bish and Von Dreele, 1989). The H positions are not indicated. (a) [100] projection. (b) Projection onto (001). Note the tetrahedral rotations and resulting ditrigonal distortion of the hexagonal rings that make up the silicate sheet.

generally minor, in the fraction of octahedral sites that are vacant ["di–tri substitutions"]; and (4) variations in the detailed chemistry of the cations that occupy both the tetrahedral and octahedral sites [isomorphous, or homologous, substitutions]. Layer stacking disorder (#2 above) deserves special emphasis, because it is particularly common in layer silicates and can considerably complicate their structural investigation. Stacking disorder typically is manifested either as fine-scale twinning on (001), the plane parallel to the layers, or as defects called stacking faults. A structure formed by ordered stacking of essentially identical layers in a specific sequence is called a *polytype*. In addition to structural variations, different 1:1 layer silicates also occur under different geological conditions. In order to discuss these important aspects of this structural group, we first must briefly address the nomenclature used to denote the different chemical variants of the 1:1 minerals.

Nomenclature

The mineral names assigned to 1:1 layer silicates depend on the chemical compositions of the octahedral and tetrahedral sheets. The usual formula unit used for these structures contains five oxygen anions, four hydroxyl groups (OH), three octahedrally coordinated metal sites, and two tetrahedrally coordinated metal sites. Up to one octahedral site per formula unit (p.f.u.) may be vacant (i.e., the dioctahedral case), so that a generalized structural formula can be written $M_{2-3}[T_2O_5](OH)_4$, where M indicates the 6-coordinated cations and T represents those that are 4-coordinated. The silicate polyanion (i.e., the silicate sheet), can be emphasized by enclosing it in square brackets, as shown.

Mineral names and end-member chemical formulae for 1:1 layer silicates are given in Table 2. In natural specimens, there is generally at least some homolo-

Table 2. Mineral names and end-member or observed chemical formulae of 1:1 layer silicates

Dioctahedral species (kaolinite group)	
kaolinite, dickite, nacrite	$Al_2[Si_2O_5](OH)_4$
halloysite	$Al_2[Si_2O_5](OH)_4 \cdot 2H_2O$
Trioctahedral species	
serpentine: lizardite and chrysotile	$Mg_3[Si_2O_5](OH)_4$
serpentine: antigorite	$Mg_{3-x}[Si_2O_5](OH)_{4+2x}$
amesite	$(MgAl)[SiAlO_5](OH)_4$
kellyite	$(Mn_{1.8}Mg_{0.2}Fe^{3+}_{0.1}Al_{0.9})[Si_{1.0}Al_{1.0}](OH)_4$
berthierine	$(R^{2+}_a R^{3+}_b \square_c)_{\Sigma 3.0}[Si_{2-x}Al_xO_5](OH)_4$
brindleyite	$(Ni_{1.75}Al_{1.0}\square_{0.25})[Si_{1.5}Al_{0.5}O_5](OH)_4$
fraipontite	$(Zn_{2.35}Al_{0.65})[Si_{1.35}Al_{0.65}O_5](OH)_4$
odinite	$(R^{2+}_{1.35}R^{3+}_{1.05}\square_{0.60})[Si_{1.85}Al_{0.15}O_5](OH)_4$
cronstedtite	$(Fe^{2+}_2Fe^{3+})[SiFe^{3+}O_5](OH)_4$
nepouite, pecoraite	$(Ni_3)[Si_2O_5](OH)_4$

gous substitution, for example, of Mg→Fe and Mg+Si→2Al, and there can be major departures from the end-member compositions.

Although Table 2 lists quite a number of 1:1 layer silicate minerals, some of them are quite rare, and only two are commonly exploited in relatively pure form by industry: kaolinite and the serpentine mineral chrysotile. Massive serpentine rock (serpentinite), which commonly is a mixture of two or even three serpentine minerals (Wicks and Plant, 1979), also is exploited, for example, as a building stone and as crushed stone used for a variety of applications. In the following discussions, we therefore emphasize the occurrence and crystal chemistry of the serpentine and kaolin-group minerals.

Geological occurrence of 1:1 layer silicates and associated minerals

The geological occurrences and petrology of 1:1 layer silicates are discussed in detail by Wicks and O'Hanley (1988), Murray (1988), and Bailey (1988c).

Serpentine group. The serpentine group of minerals is very common in hydrothermally altered, Mg-rich rocks such as altered basalt, peridotite, and dunite. Many such rocks have been almost completely altered to serpentine and are referred to as serpentinites. Lizardite is the most common form of serpentine. Retrograde metamorphism (change in a rock occurring with decreasing temperature and pressure) and pseudomorphic replacement of preexisting minerals favor the formation of lizardite over that of antigorite, the second most abundant form of serpentine. Antigorite formation is favored by prograde metamorphism (increasing T and P). Chrysotile, the most widely known but least abundant of the serpentine minerals, commonly occurs in veins or masses in serpentinite, where it typically forms during late-stage hydrothermal activity. It has been commercially exploited as asbestos in Quebec, Ontario, Vermont, California, Rhodesia, the former USSR, South Africa, Australia, and elsewhere (Ross, 1981). Chrysotile may also occur intimately intergrown with lizardite. Polygonal serpentine also occurs commonly in serpentinites. The serpentine minerals are usually associated with magnetite and with a variety of other Mg-rich minerals, which usually occur in small amounts. These accessory minerals include tremolite, chlorite, dolomite, magnesite, diopside, grossular, talc, and brucite. If alteration is incomplete, remnant olivine and/or pyroxene may be present in serpentinites. Serpentinites composed predominantly of antigorite or lizardite commonly have been exploited as crushed stone and are widely dispersed along the east and west coasts of the United States.

As observed in hand samples and petrographic thin sections, the textures of serpentine minerals can be very complex. They depend not only upon the *intensive* thermodynamic parameters during growth (e.g., temperature and pressure), but also on the mineral that is being replaced by serpentine (e.g., olivine or pyroxene) and on whether the growth is during a prograde or retrograde phase of metamorphism. Complicating matters further is the common occurrence of polymetamorphic assemblages (i.e., more than one period of metamorphism). The modern analysis of serpentine textures in terms of the serpentine mineralogy, the mineral replaced, prograde *versus* retrograde conditions, deformation conditions, and the P and T of formation has largely been pioneered by Wicks and his coworkers (Wicks, 1984a,b,c; 1986; Wicks and Plant, 1979; Wicks and Whittaker, 1977; Wicks et al., 1977; Wicks and Zussman, 1975). Based on textural criteria developed in these references, it may be possible to make a good guess as to the precise serpentine mineral(s) present in a given specimen. Despite the richness of textural variation, however, rigorous identification still requires diffraction methods, except for one case: cross-fiber asbestos veins in serpentinites can be confidently identified as chrysotile (Wicks and O'Hanley, 1988).

Altered carbonate rocks also may serve as hosts to all forms of serpentine. In this association, calcite, dolomite, talc, chlorite, phlogopite, tremolite–actinolite, other amphiboles (such as richterite and winchite), and occasionally quartz and alkali feldspar may occur with serpentine. Chrysotile is not so common in this association as lizardite and antigorite, but there are commercial chrysotile asbestos deposits in altered carbonates in Arizona and South Africa. Lizardite and/or antigorite are common in many commercial talc deposits, whereas chrysotile is rare, though it has been reported.

Kaolin group. The kaolin group[3] of minerals has a wide variety of commercial applications. Kaolinite is used as a filler in plastics, paint, rubber, and paper. It is a major raw material in the manufacture of brick and ceramics, and it may be used as a petroleum-cracking catalyst. Dickite and nacrite are used as refractory materials, and halloysite is used in ceramics and as a catalyst. China clay, ball clay, and fire clay refer to clay materials dominated by kaolinite. Clays classified as miscellaneous by the United States Bureau of Mines usually contain kaolin and are used in a wide variety of applications, including an abrasive termed "rottenstone," the "clay dummies" used to pack dynamite in blasting holes, and fillers in paint and asphalt (Hosterman, 1973).

Geological settings in which the kaolin minerals usually form are described by Murray (1988). In general, these clay minerals replace other silicate minerals, especially feldspar and muscovite, as a result of weathering, the action of groundwater, or hydrothermal alteration. Kaolinite is an extremely common mineral in soils and hence in natural aerosols. Halloysite may also occur in soils, though less commonly than kaolinite, and it appears to be favored by alternating wet and dry conditions as opposed to weathering below the water table (Giese, 1988). Kaolinite, halloysite, dickite, and nacrite may form by hydrothermal alteration of other minerals and hence are often found as gangue minerals in sulfide ore deposits. Nacrite is rare and is restricted to hydrothermally altered rocks. Kaolinite and dickite are also known to form as authigenic sedimentary minerals, forming by alteration of detrital feldspar or muscovite by groundwater.

Commercial kaolin deposits formed by weathering are often referred to as residual deposits, since it is the kaolin minerals that are left behind while most other components of the original rock are dissolved and removed. Necessary to form residual deposits of kaolin are (1) source rocks that have high aluminum and

---

[3] The word *kaolin* is used for the group of minerals indicated in Table 2; "kaolinite group" is also used sometimes. In addition, the noun kaolin is commonly used to denote mixtures of, or containing, kaolin-group minerals (Murray, 1988).

silicon but low iron content; (2) high rainfall to promote weathering, rapid drainage, and sufficient groundwater movement to remove soluble cations; and (3) a low water table, so that the deposit may have sufficient depth for commercial exploitation. Murray (1988) summarizes the geology of the major commercial deposits throughout the world. Kaolinite is the most common kaolin mineral in commercial residual deposits, although it is frequently accompanied by small amounts of halloysite. Common in residual deposits as gangue minerals are quartz, illite, mica, smectite minerals, and occasionally feldspar.

Commercial deposits of kaolin minerals formed by hydrothermal fluids usually contain a wider variety of minerals than residual deposits. It is also in these deposits that dickite and nacrite may occur, indicating high temperatures of formation (Hanson et al., 1981), probably between 100° and 400°C. Associated with the kaolin-group minerals in hydrothermal deposits are quartz, alunite, pyrite, sericite, chalcedony, allophane, opal, cristobalite, sulfates, and residual feldspar.

Some commercial deposits of kaolin minerals are classified as sedimentary, because the majority of the kaolin was formed by weathering elsewhere; it later was eroded, transported, and deposited by water to form the deposit. In some such deposits, authigenic kaolinite (i.e., kaolinite formed in place) may be an important component. In the sedimentary association, quartz, small amounts of smectite, illite, and partially altered feldspar, and occasionally bauxite are found with kaolinite. The extensive deposits of kaolinite in Georgia and South Carolina also contain small amounts of biotite, magnetite, ilmenite, tourmaline, zircon, goethite, and other heavy minerals.

Other 1:1 layer silicates. Amesite typically occurs as white, green, or pink hexagonal prisms, but it has been identified in only four localities (Bailey, 1988c) as an alteration product of other silicates. Berthierine, formerly sometimes referred to by the chlorite name chamosite, is a common mineral in unmetamorphosed sedimentary iron formations and as a precursor to chlorite in very low-grade metamorphic rocks. A Ti-rich variety has been found associated with serpentine (Arima et al., 1985), a Ni-rich berthierine analog named brindleyite occurs in bauxite deposits in Greece (Maksimovic and Bish, 1978), and a Zn-rich berthierine is found in the weathering zone of some Zn deposits (Fransolet and Bourguignon, 1975). Odinite, a green mineral resembling glauconite, occurs in continental shelf regions and reef lagoons close to the mouths of many rivers in tropical climates. Cronstedtite occurs as black crystals in low-temperature sulfide veins. Ni-rich green clays known as garnierite are found associated with lateritic Ni deposits derived from the weathering of serpentinite. Garnierite contains several layer silicates, including nepouite and pecoraite.

Crystal structures and chemistry of the serpentine group

The idealized crystal structures, structural variations, and chemical substitutions found in minerals of the serpentine group are reviewed in detail by Wicks and O'Hanley (1988), and the reader is directed to their paper for an in-depth treatment.

An important point made by Wicks and O'Hanley and by numerous other authors before them is that a free-standing, pure-Mg, trioctahedral sheet as found in brucite, $Mg_3(OH)_6$, for example, should have a **b**-axis length of $b \approx 9.43$ Å. However, a free, pure-Si tetrahedral sheet would have $b \approx 9.1$ Å. Thus, there is an inherent misfit between these two sheets. In the 2:1 layer silicates, such as talc, the silicate sheets occur symmetrically on both sides of the octahedral sheet and thereby balance each other, so that the 2:1 layers tend to remain planar. In serpentine, however, there is a tetrahedral sheet on only one side of the octahedral sheet, and therefore the stress due to misfit is not balanced and can cause the layers to curl, much as a bimetallic strip in a thermostat curls in response to changes in temperature or a lipid bilayer curls. Interestingly, this tendency to curl was first suggested on purely theoretical grounds by Linus Pauling (1930), although his suggestion seems to have been largely ignored until the 1950s (Wicks and O'Hanley, 1988, p. 102). Of the three basic serpentine structures, lizardite has planar 1:1 layers, whereas chrysotile and antigorite possess layers that are at least locally curled. In turn, the layer curvature or lack thereof seems to control the degree of chemical substitutions in these three structures, for example, the amount of Al that can substitute for Mg and Si.

Lizardite. The detailed crystal structures of the serpentine mineral lizardite are known primarily from three-dimensional crystal structure refinements of two different structural variants (polytypes—see below) known as lizardite-$1T$ (Fig. 6) and lizardite-$6H_1$ (Mellini, 1982; Mellini and Zanazzi, 1987). Indeed, because no equivalent, high-quality data exist for chrysotile or antigorite, the lizardite structures refined by Mellini provide the primary experimental insight into the detailed crystal chemistry of the serpentine group (e.g., the subtleties of polyhedral distortions and hydrogen bonding).

Layer conformation. As shown by the projection of the lizardite structure in Figure 6a, its 1:1 layers are quite planar, with little or no buckling in individual oxygen sheets.

Polytypism and symmetry. Figure 6b shows that the individual layers of lizardite have trigonal (or 3-fold rotational) symmetry. The trigonal symmetry is preserved by the $1T$ stacking of the layers, in which each layer is perfectly superimposed over the others, with no shifts or rotations. As noted above, a specific way of stacking the layers together is called a *polytype*, so that we can refer to this structure as the $1T$ polytype of lizardite, or, more compactly, as lizardite-$1T$. The numeral in the symbol $1T$ indicates that this is a one-layer polytype (one layer per unit cell, so that the structure repeats after one layer), and the T indicates that it has trigonal symmetry (one of the trigonal space groups, $P31m$). In lizardite-$2H_1$, however, the layers are staggered so that the octahedra of one layer do not directly overlie those of the adjacent layers. The 2 indicates that there are two layers per unit cell, H implies that this polytype has hexagonal symmetry (space group $P6_3cm$), and the subscript 1 is an arbitrary sequence number indicating that this is the first of several two-layer hexagonal polytypes. Theoretically, there are an infinite number of possible polytypes, although only a few have been observed to occur in nature. Lizardite polytypes and their space-group symmetries are discussed in detail by Wicks and Whittaker (1975),

Hall et al. (1976), and Wicks and O'Hanley (1988, p. 93). One important aspect of the different lizardite polytypes is that the layers generally are stacked in such a way that some of the H atoms of one layer are involved in hydrogen bonding with an adjacent layer (Fig. 6a).

Chemical variations. Chemical variations in lizardite have been the subject of several studies and are summarized by Wicks and O'Hanley (1988). Dioctahedral substitutions (i.e., vacancies in octahedral sites) are very limited or absent in lizardite. The primary chemical variations involve substitution of Al in both the octahedral and tetrahedral sites (an amesite-type substitution); the substitution of Fe^{3+} for Si in the tetrahedral sites; and the substitution of both Fe^{2+} and Fe^{3+} in the octahedral sheet. With the exception of Fe^{2+} in the octahedral sites, these substitutions tend to reduce the misfit between the octahedral and tetrahedral sheets (they make the tetrahedral sheet larger and the octahedral sheet smaller). Hence, these substitutions reduce the tendency for the 1:1 layers to curl. Lizardite can be quite aluminous, and there may well be complete solid solution between the lizardite end member and amesite. Lizardite analyses also commonly show excess H_2O (Whittaker and Wicks, 1970), the reason for which is not rigorously understood.

Other chemical substitutions in lizardite are largely controlled by the bulk rock chemistry. For example, most ultramafic rocks contain at least some Ni, and hence lizardite produced during serpentinization of such rocks generally contains trace to minor amounts of Ni. Similarly, minor Co, Mn, Cr, and Zn may be incorporated in lizardite forming in appropriate geological environments.

Chrysotile. Although chrysotile is the least abundant of the three traditional serpentine minerals, it accounts for some 95% of world asbestos production (Ross, 1981) and hence is of key importance when considering the health effects of serpentine dusts.

Layer conformation. Unlike the planar layers of lizardite, the 1:1 layers of chrysotile are tightly curled, and there are no reversals in the sense of curvature (as there are in antigorite—see below). The curvature of the chrysotile layers is one way of partially compensating for the inherent misfit between the Mg octahedral sheet and the Si tetrahedral sheet. The chrysotile structure is commonly described in terms of a cylindrical lattice. The rather complicated theory of cylindrical lattices and the details of their diffraction effects are beyond the scope of this review, but the papers by Whittaker and by Toman and Frueh cited by Wicks and O'Hanley (1988) cover this topic thoroughly.

Much of what we know about the detailed conformations of the layers in chrysotile comes from the pioneering high-resolution transmission electron microscopy (HRTEM) studies of Yada (1967; 1971; 1979). As seen in the HRTEM images of Figure 8a,b, the 1:1 layers of chrysotile can take the form of concentric tubes, increasing in diameter away from the fibril axis and creating a fibril that has cylindrical shape (as indicated above, each discrete particle consisting of a set of concentric cylinders in chrysotile is defined as a separate fibril). The cross sections of the cylinders can either be circular, or they can be deformed into

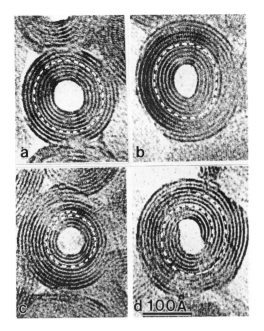

Figure 8. HRTEM images of chrysotile fibrils, viewed parallel to the fiber direction (modified from Yada, 1971).

(a) Fibril with cylindrical layers and nearly circular cross section.
(b) Fibril with deformed cylindrical layers.
(c) Fibril with single-layer spiral structure (like a single carpet rolled up).
(d) Fibril with multi-layer spiral structure (like several carpets rolled up together).

ellipses or less-regular shapes (Fig. 8a,b). Another common type of fibril consists of one or more 1:1 layers rolled up like a carpet (Fig. 8c,d). Although most naturally occurring chrysotile fibrils apparently are approximately cylindrical in shape, other, more complicated shapes also occur (Chisholm, 1983; Yada, 1971). In chrysotile produced in the laboratory, cone-in-cone and spiral cone shapes also have been observed (Yada and Iishi, 1977).

Whittaker (1957) calculated that the ideal, strain-free diameter for a cylindrical, pure-Mg chrysotile layer would be about 176 Å. Because the 1:1 layers energetically cannot withstand too tight a curvature, chrysotile fibrils generally possess inner tubes (i.e., hollow cores), as can be seen in Figure 8. Furthermore, even when the cylindrical fibrils are arranged in a closest-packed array (Fig. 9; Baronnet, 1992a), there is pore space between them. Numerous papers have been written on the question of whether or not the axial tubes in chrysotile fibrils are truly hollow and on the nature of the possible filling materials. The possibility also exists that the spaces between fibrils may be partially or completely filled by some material. Data bearing on these questions (density variations, surface area measurements, electron microscopy) have been reviewed elegantly by Chisholm (1983), who concludes that some, but not all, chrysotile specimens have some sort of filling material between and inside of the individual fibrils. However, the exact identity of such filling materials (e.g., amorphous silica or curved strips of 1:1 layers) is still not known with certainty.

In addition to the minimum diameter exhibited by the innermost 1:1 layer cylinder, chrysotile fibrils from a given occurrence also tend to show fairly restricted variations in the outer diameters of the fibrils. Observed ranges in inner and outer diameters are discussed in the section on particle morphology, below.

Chrysotile polymorphs and polytypes. The variations in layer orientation, stacking types, and the resulting structural subtleties in chrysotile are detailed by Wicks and O'Hanley (1988). To discuss the two major subdivisions of the chrysotile structures, it is necessary to define **a** and **b** axes for a single 1:1 layer, as shown in Figure 10. The layers of most chrysotile specimens are oriented so that **a** is parallel to the fibril axis. Some samples, however, have **b** parallel to the cylinder axis, and these are known as parachrysotile (Whittaker, 1956c).

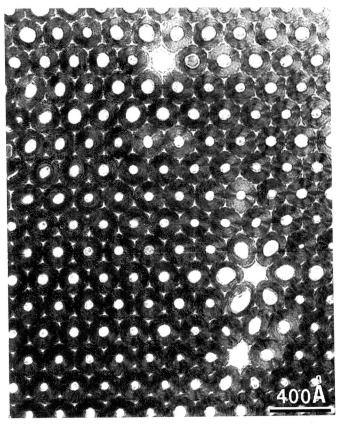

Figure 9. Packing of chrysotile fibrils in chrysotile asbestos (from Baronnet, 1992a). In most areas, the cylindrical fibrils are closest-packed. Not all chrysotile samples show packing that is this regular.

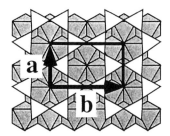

Figure 10. Definition of the **a**- and **b**-axes for a 1:1 layer, as found in chrysotile.

Parachrysotile is designated by Wicks and O'Hanley (1988) as a *polymorph*[4] of the more normal **a**-parallel types. Parachrysotile typically occurs in relatively small amounts intergrown with other chrysotile structures, so that its structural details are only poorly known.

More normal chrysotile, with **a** parallel to the fibril axis, occurs as at least several different stacking variants, or polytypes. These include a two-layer clinochrysotile (Whittaker, 1956a) designated as chrysotile-$2M_{c1}$ (2 for two 1:1 layers in the unit cell, M for monoclinic, c to indicate a cylindrical lattice, and 1 to indicate that this is the first of the possible monoclinic two-layer structures); a two-layer orthochrysotile designated chrysotile-$2Or_{c1}$ (Whittaker, 1956b); and a one-layer clinochrysotile (Zvyagin, 1967) known as chrysotile-$1M_{c1}$. Polytypes in chrysotile are more complicated than those described for normal, noncylindrical crystals. Because each successive layer going out from the cylinder axis must be larger, it is a geometrical necessity that a mismatch develop between layers, similar to the situation in what are called vernier structures or noncommensurate, misfit layer structures (Makovicky and Hyde, 1981). Thus, hydrogen bonding and other relationships between layers presumably are variable from place to place along the **b** direction, though the details are not experimentally known. The two-dimensional refinements of Whittaker (1956a,b) also show that there are subtle shifts parallel to the fibril axis in some oxygen positions, further complicating the description of chrysotile layer stacking and increasing the number of possible polytypes (Wicks and O'Hanley, 1988). It goes without saying that the description of symmetry in the various chrysotile types is a very complicated matter, due to curvature of the lattice. In addition, some chrysotile samples show disordered layer stacking, which can be indicated by the polytype designation chrysotile-D_c.

Chemical variations. As in lizardite, the most common substituents in chrysotile are Al and Fe (Wicks and O'Hanley, 1988). The amounts of these substitutions, however, tend to be much more limited than those observed in lizardite, apparently due to the octahedral-tetrahedral mismatch required for tight layer curvature. Whereas complete solid solution may exist between lizardite and amesite (see Wicks and O'Hanley, 1988, Fig. 4), substitutions of Al for Si and Al+Fe for Mg in chrysotile seldom, if ever, exceed 10%. It also has been observed that there typically is less Al in chrysotile than in coexisting lizardite (Wicks and Plant, 1979). Like lizardite, chrysotile commonly shows excess H_2O, beyond that required by the ideal chrysotile stoichiometry (Whittaker and Wicks, 1970). Hodgson (1979) and Ross (1981) present chemical analyses for commercial chrysotile asbestos from several major mining districts.

Antigorite. Intermediate in abundance between lizardite and chrysotile is antigorite, which has variable stoichiometry that differs from the $Mg_3[Si_2O_5](OH)_4$ structural formula of the other serpentine minerals (Table 2).

Layer conformation. The corrugated structure of antigorite represents a novel adaptation to the inherent misfit between the Mg-rich octahedral and

---

[4] A polymorph is a term applied to materials with the same composition but with different structures. Polytype (defined above under "Basic 1:1 layer structure") is a subset of polymorph.

Si-rich tetrahedral sheets that are common to the serpentine group. The antigorite structure also is among the most poorly known of the common rock-forming minerals, due to the lack of relatively large, well-ordered crystals necessary for high-quality X-ray diffraction (XRD) studies. Antigorite possesses a long-period, wavelike modulation parallel to the **a**-axis, giving a unit-cell parameter that varies considerably around a value of $a \approx 43.3$ Å, compared to the $a \approx 5.3$ Å value common to lizardite and chrysotile.

According to the structural model developed in a series of papers by Kunze (1956; 1958; 1959; 1961), the Mg-rich octahedral sheet is similar to that found in lizardite, though it differs in having a wavelike deformation (Fig. 11). The tetrahedral sheet, on the other hand, is quite different, in that the direction of the apical oxygen atoms of the Si tetrahedra flips from $+\mathbf{c}$ to $-\mathbf{c}$ where the curvature of the sheet reverses (Fig. 11). Tetrahedral sheets with apical oxygens pointing in different directions are an important characteristic of the group of structures known as modulated layer silicates, which are reviewed by Guggenheim and

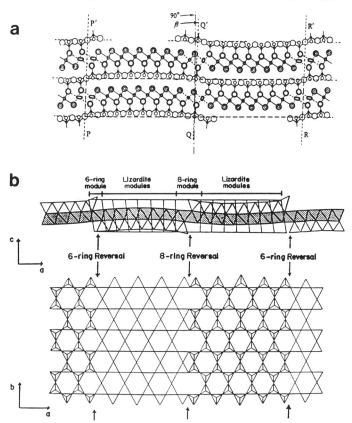

Figure 11. (a) The Kunze (1956) structural model for antigorite. (b) Polysomatic representation of the Kunze model for the antigorite structure. Serpentine (lizardite-like?) slabs are combined with two types of curvature-reversal slabs, one with 4- and 8-membered rings, and the other with 6-membered rings (Livi and Veblen, 1987, after Spinnler, 1985).

Eggleton (1987; 1988). Recent improvements to Kunze's basic structural model for antigorite have been suggested by Spinnler (1985) and by Uehara and Shirozu (1985).

Spinnler (1985) introduced a polysomatic representation of the antigorite structure (Ferraris et al., 1986; Livi and Veblen, 1987; Thompson, 1978; Veblen, 1991), as shown in Figure 11b. In this view, the antigorite structure can be thought of as consisting of a series of (100) lizardite-like slabs separated alternately by slabs having six-fold tetrahedral rings or four- and eight-fold rings. Wicks and O'Hanley (1988) suggested that the lizardite-like slabs should more properly be called idealized serpentine slabs, since details of the lizardite structure (e.g., exact configuration of hydrogen bonding) may not be present in the structure of antigorite, which is more poorly known.

Structural disorder and defects in antigorite. Most of our information about structural variations in antigorite comes from HRTEM and selected-area electron diffraction (SAED) studies (Spinnler, 1985; Spinnler et al., 1983; Yada, 1979; Mellini et al., 1987; Otten, 1993). Although variations in the *a* lattice parameter were first noted in XRD studies, electron microscopy has shown that there can be considerable variation in the modulation wavelength, even within what appears to be a single crystal (Spinnler, 1985; Mellini et al., 1987; Otten, 1993; Veblen, 1991, his Fig. 12; Veblen, 1992, his Fig. 6-25). This variation in antigorite periodicity can be understood most easily with reference to Figure 11b: simply by inserting or removing serpentine modules, the wavelength can be varied. It is very important to note that the serpentine modules and the 8- and 6-ring modules do not have the same stoichiometry. Therefore, variations in the wavelength, and hence in the ratios of the different types of slabs, also imply variations in stoichiometry. By varying the numbers of lizardite-like slabs in a nonperiodic way, the stoichiometry of the antigorite can be adjusted almost smoothly within a certain range. The generalized formula for antigorite is $Mg_{3-x}[Si_2O_5](OH)_{4+2x}$. Other defects in antigorite include the termination of modulations at superstructure dislocations, stacking faults, twins, and small shifts between adjacent layers that can cause the superstructure reflections in SAED patterns to be irrationally oriented with respect to the basic serpentine substructure (Spinnler, 1985). These defects are particularly well-illustrated by the very high-resolution TEM images of Otten (1993). In the only systematic study of antigorite microstructures as a function of metamorphic grade (i.e., increasing temperature), Mellini et al. (1987) showed that all of these types of disorder occur primarily in low-temperature antigorite samples, and much of the disorder is annealed out at higher grades.

Polytypism and chemical variations. Whereas most antigorite appears to consist of a 1-layer structure, 2-layer polytypes have been noted from HRTEM studies (Veblen, 1980; Yada, 1979). In addition, most antigorite samples possess at least some stacking disorder, the degree of which appears to vary with metamorphic grade (Mellini et al., 1987).

As with chrysotile, the occurrence of the antigorite structure, with its alternating layer curvature, requires the inherent misfit that exists between the

Mg-rich octahedral sheet and the Si-rich tetrahedral sheet. Thus, like chrysotile, the substitution of Al and Fe in the octahedral sheet is limited, although not so stringently as in chrysotile (Wicks and O'Hanley, 1988). Only minor Al and, rarely, Fe^{3+}, substitute for Si. Unlike lizardite and chrysotile, antigorite does not appear to contain excess H_2O.

Carlosturanite. This recently discovered mineral has, in the past, apparently been mistaken for antigorite. The proposed structure is closely related to a planar serpentine structure, but ordered Si vacancies break the silicate sheets into chains (Mellini et al., 1985). Thus, carlosturanite is not a layer silicate, but a triple-chain silicate having a different type of triple chain from that in jimthompsonite (Veblen and Burnham, 1978b). We will not, therefore, discuss carlosturanite further, other than to note that at least some material formerly called picrolite (a supposed serpentine material with splintery habit) is actually carlosturanite (Mellini and Zussman, 1986), and it appears that this long-overlooked mineral may occur commonly in serpentinites. Carlosturanite is discussed further by Wicks and O'Hanley (1988) and by Guggenheim and Eggleton (1988).

Polygonal serpentine. This unusual material has also been called polygonal chrysotile and Povlen chrysotile. As initially suggested by Middleton and Whittaker (1976) and reviewed by Wicks and O'Hanley (1988), polygonal serpentine fibers commonly have an ordinary, cylindrical chrysotile core (for example, chrysotile-$2M_{c1}$). The polygonal serpentine overgrows and/or replaces the chrysotile core. The exact structural character of the polygonal material appears to be variable and is not yet fully understood, variously being reported as planar chrysotile-$2M_{c1}$ (the meaning of "planar chrysotile" is far from clear), as lizardite-$1H_1$, or as multilayer lizardite. A variety of structural variations have been reported by Cressey and Zussman (1976) and by Mellini (1986).

One fascinating observation is that polygonal serpentine fibers typically contain either 15 or 30 sectors (Yada and Wei, 1987). The number of sectors has recently been explained on structural grounds by Chisholm (1992) and by Baronnet (1992b). Though derived independently, the structural models of Chisholm and Baronnet are similar. Both recognized that each successive layer of rolled serpentine (either cylindrical or polygonal) must add five lattice spacings parallel to **b**, compared to the preceding layer, due to the fact that it has a larger diameter. Chisholm noted that if one octahedron (equivalent to **b**/3) is added at each sector boundary, then a 15-sector fiber will result. If one octahedron is added only every other layer, then a 30-sector fiber occurs. Baronnet's structural model describes the added octahedral spacings as partial dislocations, with 30-sector fibers resulting from staggering the dislocation positions from layer to layer. Baronnet also noted that the polytype of the lizardite-like planar serpentine must change from sector to sector in a periodic way, thus reducing the apparent 15- or 30-fold symmetry of the fibers to 5-fold rotational symmetry. In a sense, polygonal serpentine can then be thought of as a mineralogical example of a one-dimensional quasicrystal with 5-fold symmetry.

Serpentine intergrowths and other hybrid serpentine structures. As noted above, XRD studies have shown that serpentinites commonly contain not one, but two or three serpentine minerals. HRTEM studies have further clarified the scale of these intergrowths. In many cases, different types of serpentine are finely intergrown, as shown in Figure 12 for antigorite, chrysotile, and polygonal serpentine (Spinnler, 1985).

In other cases, however, the serpentine minerals are so finely intergrown that they may be considered to form hybrid, or mixed structures. One example is polygonal serpentine, described above, which commonly consists of planar serpentine (lizardite-like?) sectors surrounding a chrysotile core. A different type of intimate intergrowth is illustrated in Figure 13, where planar serpentine layers pass without interruption into serpentine with a tightly curled lattice, as is generally associated with chrysotile (Veblen and Buseck, 1979a; Wicks, 1986). Such intimate intergrowths also occur between the lizardite and antigorite structures. Livi and Veblen (1987) observed lamellae of planar serpentine that were occasionally interrupted by pairs of antigorite offsets; this serpentine was intergrown with phlogopite. Again, individual layers passed without break from lizardite structure to antigorite structure, and back again.

The tripartite classification of serpentine structures into lizardite, chrysotile, and antigorite was initially proposed by Whittaker and Zussman (1956). As reiterated by Wicks and Whittaker (1975), this classification divides "the serpentine minerals into three structural groups based on cylindrical layers (chrysotile), corrugated layers (antigorite), and flat layers (lizardite)." This is a convenient and flexible classification, because at least a tentative identification can be based either on diffraction methods or on HRTEM images, simply by observing the shape of the layers, which implies the shape of the lattice. (X-ray diffraction still is the preferred method of identification, however, because antigorite viewed down the **a**-axis might easily be mistaken for lizardite.) Regardless of identification method, it is essential to recognize that extremely fine-scale intergrowths and hybrids of these three species do occur, and even individual, unbroken 1:1 layers may in one place conform to one species and in another place fit the definition of a different serpentine mineral. As pointed out by Wicks and O'Hanley (1988), perhaps the most difficult question of nomenclature arises where lizardite passes continuously into slightly curved serpentine (i.e., not so tightly curled as normal chrysotile). Is this less-curled material lizardite or chrysotile? Questions of this sort have yet to be resolved in a systematic way.

Crystal structures and chemistry of the kaolin group

As noted above, clay minerals of the kaolin group are dioctahedral 1:1 layer silicates having Al as the predominant octahedrally coordinated cation. Their crystal chemistry and thermodynamic stabilities have been reviewed by Giese (1988), and Bailey (1988b) provides a clear explanation of the layer stacking in kaolin minerals.

Kaolinite, dickite, and nacrite. These three polymorphs of $Al_2[Si_2O_5](OH)_4$ are structurally similar. Dickite and nacrite are uncommon, typically form hydrothermally, and can crystallize in relatively well-ordered crystals large

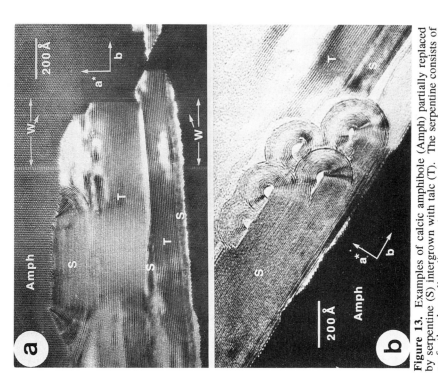

Figure 13. Examples of calcic amphibole (Amph) partially replaced by serpentine (S) intergrown with talc (T). The serpentine consists of perfectly planar lizardite passing continuously into the tightly curled layers characteristic of the chrysotile structure. In (a), W refers to chain-width errors in the amphibole (Veblen and Buseck, 1979a).

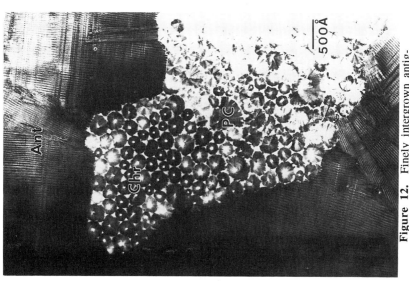

Figure 12. Finely intergrown antigorite (Ant), chrysotile (C), and polygonal serpentine (PC) (modified from Spinnler, 1985).

enough for single-crystal XRD studies (e.g., hundreds of μm on a side). Kaolinite is much more common but tends to possess a high degree of structural disorder and to crystallize in clay-sized crystals that are unsuitable for single-crystal XRD studies.

Layer conformation. The dioctahedral 1:1 layers of kaolinite (Fig. 7), dickite, and nacrite are ideally planar. Given the small size and inherent weakness of typical kaolinite crystals, however, it is common for them to be deformed from a strictly planar shape. It is also common for clay-sized flakes of kaolinite to be at least slightly curled at the edges. However, the degree of layer curling in kaolinite, dickite, and nacrite is negligible compared to that observed in chrysotile or tubular halloysite (see below).

Layer stacking, octahedral vacancy distribution, and hydrogen bonding. Bailey (1963) introduced twelve standard polytypes for 1:1 layer silicates based on simple stacking sequences involving either no shift or shifts of successive layers parallel to the **a**- or **b**-axis, and these are found in the three planar kaolin minerals (Bailey, 1988b; Giese, 1988). The layers of both kaolinite and dickite are arranged according to the $1M$ stacking type, in which each successive layer is shifted by the same amount parallel to **a**. For both structures, however, the arrangement of vacancies violates the maximum monoclinic symmetry possible with a one-layer unit cell. Bailey defined three octahedral sites, B and C being related by the mirror plane in an ideal one-layer monoclinic structure, and A being a third, symmetrically unrelated site (see Bailey, 1988b, Fig. 2; Giese, 1988, Fig. 2). The vacancies in kaolinite are on either the B or C sites, thus violating the mirror and reducing the symmetry to triclinic. Placing the vacancies on either of these sites results in a one-layer structure with space group $C1$ or $P1$, designated $1Tc$. Bookin et al. (1989) have suggested that the vacancies actually have a preference for the B site. Bish and Von Dreele (1989) present convincing evidence that kaolinite is C-centered, not primitive (assuming the usual type of layer-silicate unit cell). In contrast to the vacancy arrangement in kaolinite, the vacant octahedral site in dickite alternates between the B and C sites in successive layers, producing a monoclinic two-layer structure with space group Cc, designated $2M$.

The structure of nacrite is based on the standard $6R$ polytype, which in ideal form has space group $R3c$. Ordering of the octahedral vacancies violates the 3-fold axis, however, degrading the symmetry from rhombohedral to monoclinic. Bailey (1963) showed that a two-layer unit cell could be chosen, rather than the six-layer cell for the ideal $6R$ polytype. Like dickite, nacrite has Cc space-group symmetry, but this two-layer unit-cell setting for nacrite has its **a**- and **b**-axes reversed compared to those of most other layer silicates.

Structures for the three planar kaolin minerals are illustrated in Figures 7 and 14 (Bish and Von Dreele, 1989; Giese, 1988). All three exhibit structural distortions typical of those in other layer silicates, such as tetrahedral rotations and shortening of shared octahedral edges. The diagrams for kaolinite and dickite in Figure 14 show the experimentally determined H positions from Young and Hewat (1988) and Joswig and Drits (1986) respectively, and the nacrite

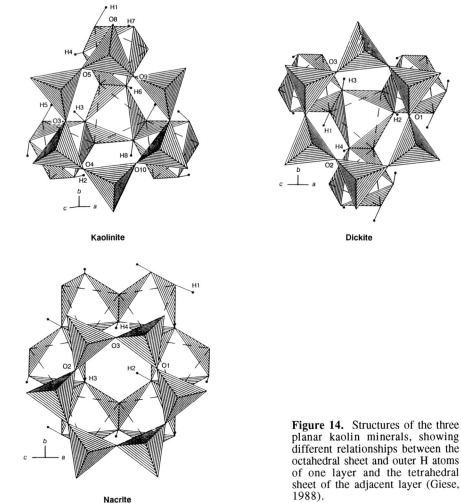

Figure 14. Structures of the three planar kaolin minerals, showing different relationships between the octahedral sheet and outer H atoms of one layer and the tetrahedral sheet of the adjacent layer (Giese, 1988).

projection includes H positions estimated from electrostatic energy calculations by Giese and Datta (1973). Although the details differ, the H atoms of the *outer surface* hydroxyl groups are oriented toward tetrahedral oxygens of the adjacent 1:1 layer in all three structures. (The outer surface hydroxyls are those that lie in the surface of the octahedral sheet that is not shared with the tetrahedral sheet, i.e., adjacent to the interlayer region.) These orientations are consistent with hydrogen bonding between the surface OH groups and the bridging tetrahedral oxygen atoms.

Structural disorder in kaolinite. The observation that kaolinite is highly variable in defect state is illustrated by Figure 15. The powder XRD pattern for the Keokuk sample, a hydrothermal kaolinite, shows sharp, discrete peaks, indicative of reasonably good structural perfection. In contrast, sample IV-L, a sedimentary kaolinite from Georgia that initially formed by weathering, produces

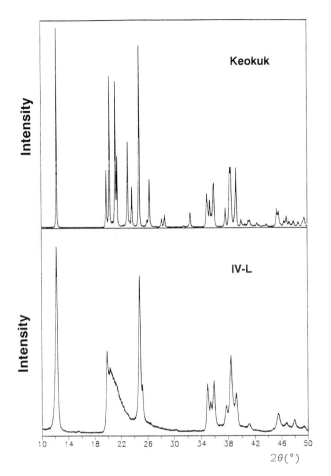

Figure 15. Powder XRD diffraction patterns for an unusually well-crystallized kaolinite sample (Keokuk) and one showing substantial structural disorder (IV-L). (Giese, 1988).

much higher diffuse background, and many of the peaks are smeared into continuous bands of intensity. This sort of diffraction pattern is characteristic of crystalline materials with a high degree of structural disorder. The Keokuk sample is an unusual kaolinite: most kaolinites exhibit a high degree of disorder in XRD patterns.

Ideally, transmission electron microscopy would be used to identify the types of defects present in kaolinite and their density. Unfortunately, kaolinite suffers from extremely rapid electron beam damage (amorphization) in the electron microscope, and direct characterization of kaolinite defect states has not proved feasible. As thoroughly described by Giese (1988), various techniques (e.g., ^{27}Al-NMR and electrostatic energy calculations) have been used to study disorder in kaolinite, but most of the work has involved the comparison of experimental powder XRD patterns with calculated patterns based on possible defect models. Suggested defect types have included stacking faults (layer shifts of $\pm n\mathbf{b}/3$ or other values), twins (rotations of $\pm 2\pi/3$ or other values), and disorder involving displacements of the vacant octahedral sites (Plançon and Tschoubar, 1977a,b). The initial work of Plançon and Tschoubar (1977a,b) suggested that octahedral vacancy displacements are the dominant defect. More recent studies, however, indicate that layer shifts are more important (Bookin et al., 1989; Plançon et al., 1989). In addition, there are indications that some kaolinite samples may consist of at least two different types of kaolinite having different defect densities (Giese, 1988).

Chemical variations. Compared to other clays, the planar kaolin-group minerals show only minor deviations from the ideal $Al_2[Si_2O_5](OH)_4$ composition. There is some evidence for deviations from the ideal Si/Al ratio and for incorporation of the minor elements Fe, Ti, and K, which may be present as structural substituents, surface coatings, or intergrown non-kaolin layers.

Halloysite. This member of the kaolin group is reviewed in detail by Giese (1988). As suggested by its structural formula (Table 2), the common weathering product halloysite consists of 1:1 layers similar to those of kaolinite, but with water molecules located in the interlayer region between these layers. The interlayer spacing of such hydrated halloysite is approximately 10.1 Å. Under room-temperature conditions, however, halloysite dehydrates irreversibly to a form having an interlayer spacing of about 7.3 Å (close to that of kaolinite), in some cases dehydrating even if it is kept in contact with water. During dehydration, halloysite apparently passes through transitional states having intermediate values of interlayer spacing.

Layer conformation. The layers of halloysite are commonly tightly curled and adopt numerous shapes, forming cylindrical tubes reminiscent of chrysotile, polygonal tubes reminiscent of polygonal serpentine, spheres, plates, oblate spheroids, stubby cylinders, and irregular shapes (Dixon and McKee, 1974; Giese, 1988). Typical sizes of the resulting particles are given below in the section on habit of the kaolin-group minerals. Whereas serpentine fibrils typically have the **a**-axis parallel to the cylinder axis, tubular halloysite most commonly has its **b**-axis parallel to the particle axis (as in parachrysotile).

Layer stacking and other structural details. Halloysite typically produces poor XRD patterns, with only a few relatively broad peaks and high diffuse background. Electron diffraction patterns of dehydrated halloysite(7 Å) indicate a two-layer structure (Giese, 1988), as do patterns of halloysite(10 Å) obtained with an environmental cell (Kohyama et al., 1978). The space group Cc is suggested by Kohyama et al., but little else is known of the structural details.

Due to the high degree of disorder in natural halloysite and the resulting difficulty of obtaining structural information, Costanzo et al. (1980) prepared hydrated kaolinite(10 Å) to study the positions of the water molecules characteristic of halloysite(10 Å). These better-ordered materials allowed the identification of two structurally distinct types of water; the detailed arrangement and associated hydrogen bonding are discussed by Giese (1988).

Crystal structures of other 1:1 layer silicates

Because other 1:1 layer silicates generally are not used in industry and are much less abundant than minerals of the serpentine and kaolin groups, their structures will be described only briefly. However, each is naturally occurring, so that an environmental or background exposure is at least possible. Also, some of these materials could be used effectively in bioassays to elucidate the relationships between biological responses and mineralogical properties (e.g., structure, composition, trace metal content, defect state, etc.). For example, one could compare biological responses over the lizardite–amesite series to assess the role of Al in the octahedral sheet. These data could then provide insight into the role of octahedral Al in the biological activity of chrysotile. Hence, the following summary may aid researchers in their design of future experiments.

The chemical compositions and structures of these other 1:1 layer silicates are ably reviewed by Bailey (1988c), from whom the following brief notes are largely derived. End-member or observed compositions for these minerals are listed in Table 2.

Amesite and kellyite. The highly aluminous, serpentine-like mineral amesite may be one end member of a solid solution series with lizardite (Wicks and O'Hanley, 1988). It crystallizes with several polytypes, including $2H_2$, $2H_1$, $6R_1$, and a $6R_2$-type stacking with triclinic distortion. The different polytypes can occur intergrown with each other on a fine scale, disordered stacking has been reported, and amesite also is commonly twinned. Kellyite, a Mn- and Al-rich 1:1 structure, exhibits disordered stacking, as well as polytype $6R_2$ and a polytype with a distorted $2H_2$ structure having monoclinic or triclinic symmetry.

Berthierine, brindleyite, and fraipontite. Berthierine is a relatively common, Fe-rich 1:1 layer silicate found in unmetamorphosed sedimentary iron formations and other Fe-bearing, diagenetic or low-grade metasediments. Berthierine apparently crystallizes primarily in two one-layer structures, a $1T$ form having the same simple stacking type (no interlayer shifts or rotations) found in lizardite $1T$, and a monoclinic $1M$ polytype. Intergrowths of the two structures are common (Bailey, 1988c), and mixed-layer intergrowths with other layer structures also have been reported (Banfield et al., 1989). Given the difficulty of recognizing 1:1 layer silicates that are finely intergrown with chlorite (Reynolds, 1988), it may well be that berthierine is much more widespread in low-grade metamorphic rocks than previously thought. For the berthierine formula of Table 2 (Bailey, 1988c), the tetrahedral Al contents range from x = 0.45–0.90 and are generally exceeded by the trivalent octahedral cations, R^{3+} (mainly Al, 0.37–1.03 atoms p.f.u., and Fe^{3+}, 0.01–0.27 p.f.u.), the excess of which are charge balanced by octahedral vacancies, which are designated with the symbol □ (i.e., b − x = 2c). Divalent octahedral cations R^{2+} are dominantly Fe^{2+}, 1.33–1.84 p.f.u., with some Mg substitution, 0.04–0.66 p.f.u. Related species are a Ni-rich form known as brindleyite (Maksimovic and Bish, 1978) and a Zn-rich form called fraipontite (Fransolet and Bourguignon, 1975) (Table 2).

Odinite. This recently described mineral occurs in young, relatively shallow, continental-shelf sediments in tropical regions (Bailey, 1988d). Its intermediate dioctahedral-trioctahedral structure contains substantially more Fe^{3+} than berthierine and more octahedral vacancies to balance the charge. Odinite apparently transforms easily to chlorite and hence has not been recognized in older rocks.

Cronstedtite. This 1:1 layer silicate mineral forms in hydrothermal sulfide veins. Cronstedtite shows much variation in its layer stacking sequences, and at least eight polytypes have been recognized with single-crystal XRD methods. Because it tends to form relatively good crystals, there are several crystal structure refinements available, as described by Bailey (1988c).

Nepouite and pecoraite. These Ni-rich minerals typically occur with other minerals in very fine-grained masses that are given the group name garnierite (see below). Nepouite is a platy species, apparently analogous to lizardite, whereas pecoraite forms tubes and rods and can be thought of as analogous to chrysotile. Because of their small crystal size and poor crystallinity, these Ni-rich analogs of the serpentine structures have not been characterized crystallographically with nearly the same detail as lizardite and chrysotile (Bailey, 1988c).

Habit of 1:1 layer silicates

As defined in a previous section, habit is a term that describes the shape of mineral crystals or crystal aggregates. For example, elongated crystals shaped like knife blades possess bladed habit, whereas groups of individual crystals with columnlike morphology can be called columnar (Klein and Hurlbut, 1993, p. 250).

Serpentine-group minerals. Lizardite is usually green but may be quite pale to white. Its most common habit is fine-grained and massive (with individual crystals having a platy habit), but vein lizardite may be platy to fibrous (but not asbestiform). It is often associated with chrysotile as a separate phase, but it also occurs as a polygonal overgrowth over a cylindrical chrysotile core, forming an intimate hybrid structure as described above (Cressey, 1979; Cressey and Zussman, 1976). As crushed fragments, lizardite tends to lie on the (001) cleavage plane, producing a selected area electron diffraction (SAED) pattern made up of a hexagonal array of spots (Whittaker and Zussman, 1971; Zussman et al., 1957).

Antigorite is most commonly massive, but it also occurs in a foliated or splintery habit in slip veins or on shear surfaces. It may also occur as fracture-filling veins in a splintery habit known as picrolite; as noted above, however, at least some "picrolite" has been shown to be carlosturanite (Mellini and Zussman, 1986), and it is not yet clear how much of this splintery material is actually antigorite. This habit probably forms from chrysotile veins during prograde metamorphism.

Chrysotile is the least abundant of the three serpentine minerals, although it is widely known as the most important commercial asbestos mineral. As asbestos, it is exploited from cross-fiber or slip-fiber veins cutting a massive serpentinite host and occasionally also from mass-fiber deposits. Table 3, modified from Wicks and O'Hanley (1988), illustrates the commercial types of chrysotile deposits as a function of the type of protolith and the occurrence of chrysotile. Chrysotile also may occur as a minor component in lizardite- and antigorite-dominated massive serpentinites and as a component of some altered dolomite-bearing carbonate rocks.

Table 3. Classification of chrysotile asbestos deposits (after Wicks & O'Hanley, 1988)

Lithology		Dominant Fiber Type	Example
Alpine-type peridotites, Phanerozoic	tectonized peridotites	cross-fiber slip- & mass-fiber mass-fiber	Cassiar, B.C. Coalinga, CA Asbestos, Quebec
Differentiated Precambrian	sills komatiitic flows	cross-fiber cross-fiber	Great Dike, Zimbabwe Abitibi Belt, Ontario
Serpentinized limestones, both Phanerozoic & Precambrian	limestone	cross-fiber	Globe, Arizona

The asbestiform habit of chrysotile is characterized by bundles of individual, elementary fibers that share a common axis of elongation. As defined above, these elementary fibers are called fibrils, and each chrysotile "roll," such as those in Figure 8, is a fibril. Fibrils tend to be about 250 Å in diameter with a central tube of about 70 Å, but there is variation in these dimensions. Chisholm (1983) summarizes fibril dimensions obtained with TEM in his Table 4.1. Composite fibers with lengths up to a meter have been reported, although individual fibrils do not attain such lengths in their native states (processed or milled chrysotle exhibits different fiber lengths). Most chrysotile asbestos fibers have lengths on the order of a centimeter or less. The lengths and widths of populations of chrysotile fibers differ somewhat among deposits and among commercial products from a single mine. Table 1 summarizes dimensional data for a number of commercial asbestos samples, including chrysotile. Only dimensional data gathered by transmission electron microscopy are included, since most chrysotile fibrils and the smaller composite fibers are not visible by optical or scanning electron microscopy.

Chrysotile also may occur in a massive, nonfibrous form. Massive chrysotile is found as a component of cross- or slip-fiber veins with its asbestiform variety or as coatings on shear surfaces (Wicks and O'Hanley, 1988). It may also occur in a banded, splintery, and pseudofibrous form commonly associated with lizardite (Wicks and Whittaker, 1977), or in the form of polygonal serpentine (Cressey and Zussman, 1976; Middleton and Whittaker, 1976). Although it is common practice in regulatory matters to assume that all chrysotile is asbestiform, this is a false assumption, and analytical methods that rely solely on bulk measurements, such as XRD, may give incorrect estimates of the amount of asbestiform chrysotile. Likewise, "fiber counting" methods that do not also utilize rigorous mineral identification techniques (e.g., electron diffraction) do not unambiguously estimate the chrysotile component of fibrous material.

Kaolin-group minerals. Kaolinite occurs as euhedral, platy, pseudohexagonal particles; as vermicular stacks or books of plates; and, rarely, as fibers. The individual particles range from about 0.25 μm to 3 mm, with most less than 2 μm in maximum dimension. The particle size may reflect the origin of the deposit. For example, in Georgia and South Carolina, there are both Tertiary and Cretaceous sedimentary deposits dominated by kaolinite. The Cretaceous kaolinites are fairly coarse, with 50–70% of the particles smaller than 2 μm, whereas 80–90% of the Tertiary kaolinite particles are smaller than 2 μm (Murray, 1988). Dombrowski and Murray (1984) have demonstrated that the thorium content is higher in the Cretaceous than in the Tertiary kaolinites. Murray and Jansen (1984) have shown through the application of oxygen isotopes that the large vermicular kaolinite crystals found in the Cretaceous deposits were probably derived from feldspar alteration *in situ* (authigenic kaolinite). Murray (1988) believes that the Tertiary kaolins are reworked Cretaceous kaolins, which would account for their smaller particle size, the absence of vermicular books, and their lower Th contents.

Halloysite is found in a variety of morphologies, including tubes, spheres, plates, oblate spheroids, stubby cylinders, and irregular shapes (Bates et al., 1950; Dixon and McKee, 1974; Kirkman, 1981). The most common form is tubular, with the long axis of the tube parallel to the **b**-axis, although some specimens are elongated parallel to the **a**-axis (Honjo et al., 1954). Bates et al. (1950) report that the inside diameters of the tubes range from 200 to 1000 Å, with a median of 400 Å, and the outside diameters range from 400 to 1900 Å, with a median of 700 Å. Most halloysite tubes are about 1 μm in length, although Bates et al. describe a sample from North Carolina with tube lengths greater than 10 μm. Platy, irregular particles of halloysite are probably split and/or unrolled tubes. Kirkman (1981) reports "squat ellipsoidal" halloysite in tephra and soils derived from volcanic ash. These particles have diameters generally less than 0.5 μm. Dixon and McKee (1974) describe spheroidal halloysite, which has a platy exterior with shapes characteristic of kaolinite. The spheres have diameters of about 1500 Å and cores of about 180 Å; the cores do not scatter electrons so strongly as the exterior parts of the particles. Diamond and Bloor (1970) describe globular clusters of quasi-tubular halloysite radiating from common centers. The overall diameters of the clusters are 10 to 20 μm. Individual particles are about 3000 Å in diameter and several micrometers long. It is the general consensus that the morphology of halloysite is a primary feature formed during growth and does not result from hydration of a kaolinite 7-Å layer (Giese, 1988). Merino et al. (1989) have suggested that it is Al substitution for Si that causes the sheets to bend and form the tubes of halloysite. They also suggest that this substitution is controlled by pH in the environment of formation.

Other 1:1 layer silicates. Most other 1:1 layer silicates have platy morphologies. However, those found in garnierite may be fibrous. Garnierite has been shown to contain a mixture of minerals, including talc, chlorite, smectite, sepiolite, and a 1:1 layer silicate (Faust, 1966) for which Maksimovic (1973) proposed the name nepouite. Brindley and Hang (1973) and Uyeda et al. (1973) describe tube- and rod-shaped particles of nepouite, and Uyeda et al. found similarly shaped particles in many garnierite samples, which they identified as a Ni-rich form of chrysotile. Faust et al. (1969; 1973) gave the name pecoraite to this mineral. Brindley's (1980) suggestion that the name pecoraite be used for all 1:1 layer silicates that are fibrous and contain Ni as the dominant octahedral cation is generally followed.

Surface chemistry, surface charge, and dissolution kinetics of chrysotile

Because chrysotile is the most important asbestos mineral, its health effects have been the topic of numerous studies. Here we briefly discuss three important health-related topics: the structure and chemistry of the chrysotile surface (the part of the fibers that interacts directly with cells and fluid in the lung or other organs), the surface charge of chrysotile fibers, and the dissolution kinetics of chrysotile (which control, at least in part, the residence time of fibers in the lung). For a much more thorough discussion of the field of surface chemistry as it relates to mineral health effects, see Chapter 8.

Surface chemistry and structure. Because the periodicity of a free-standing silicate sheet is inherently smaller than that of a Mg-hydroxide sheet, the 1:1 layers of chrysotile are curled with the silicate sheet directed inward, toward the axis of the fibril, and the Mg-hydroxide surface on the outside. The atomic layer of an ideal chrysotile fiber that is directly in contact with its environment is thus a centered hexagonal array of hydroxyl groups, similar to the surface of brucite, $Mg(OH)_2$. The analogous surface hydroxide sheet, the underlying sheet of Mg atoms, and the silicate surface have been imaged directly in the planar serpentine mineral, lizardite, with atomic force microscopy by Wicks et al. (1992) (Fig. 16).

The basic structure of this array of surface hydroxyl groups and underlying Mg atoms is known from the crystal structure refinement of $1T$ lizardite (Mellini, 1982) (Fig. 5a). It is also true, however, that the structure of a free surface of a crystal can reorganize and in general will not be exactly the same as that of the same plane deep inside the bulk crystal. In an aqueous fluid, it can be expected that the H atoms of the chrysotile surface will form at least transient hydrogen bonds with a relatively organized surface layer of water molecules.

There are additional, important complications when considering the surface structure and chemical behavior of the chrysotile surface. One important consequence of the detailed structure of chrysotile fibrils, as shown clearly by the HRTEM images of Figure 8, is the fact that all chrysotile fibrils should not be structurally identical. Fibrils with cylindrical layers (Fig. 8a,b) should present a continuous Mg-hydroxide surface to their environment. Fibrils in which the layers are in a spiral configuration (i.e., rolled up like a carpet; Fig. 8c,d), however, will present a permanent ledge to their surroundings, as shown diagrammatically in Figure 17. The possible influences of these surface structural differences on dis-solution and other reactions are discussed briefly below.

Another complication is that chrysotile fibrils are not ideal, solid cylinders, but instead they possess a central pore. As discussed above and in detail by

Chisholm (1983), experimental evidence suggests that these central pores can be at least partially filled by some material other than serpentine, and the spaces between chrysotile fibrils also can be filled. When the chrysotile fibrils are disaggregated (e.g., during milling), these interfibril and intrafibril materials presumably remain on the fibril surface and in the central pore respectively. The initial surface presented by chrysotile to a fluid therefore may be at least in part a coating material, rather than the ideal Mg-hydroxide sheet. Similarly, a fluid will be in contact with a material that fills the fibril pores at the ends of the fiber.

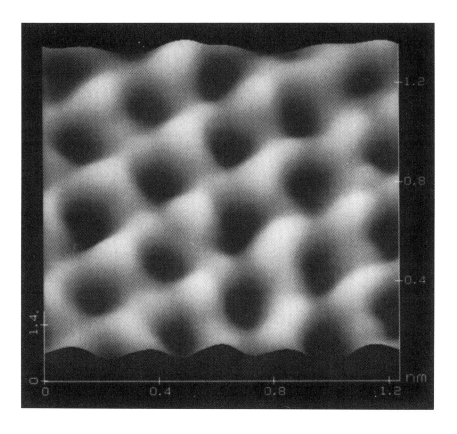

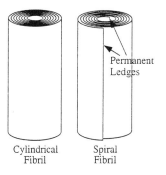

Figure 16. [above] Oblique view of a filtered atomic-force microscopy image of the OH plane of the octahedral sheet of lizardite (from Wicks et al., 1992). Scale is in nm, and the field of view is 1.2 × 1.2 nm.

Figure 17. [left] Schematic diagrams of the surface structures of two different types of chrysotile fibrils. (a) Fibril with cylindrical layers has a smooth cylinder surface. (b) Fibril with spiral layers has permanent surface steps on both the outer and inner surfaces, which may be preferred sites for dissolution and other processes.

Finally, when not filled, the pores of chrysotile are large enough to allow access to aqueous fluids and even many macromolecules. The interior pore surfaces in an ideal, unfilled fibril will be the silicate sheet of the serpentine structure. Although the interactions of an aqueous fluid with the pore surfaces may be minor compared to interactions with the fibril surface, there are no experimental data addressing this question. It may well be that tubular chrysotile fibrils present not one, but two structurally and chemically different surfaces to their environment. Furthermore, the interior surfaces in zeolites are important in many catalytic reactions, where the macromolecules enter the zeolite, react with the interior surface, and then exit the crystal. It is conceivable that the central pores in chrysotile could behave in an analogous fashion.

Surface charge. The surface charge of particles is normally reported as zeta (ζ) potential, which is a measure of the potential energy difference between a bulk solution and the boundary between the free solvent and the solvent adhering to the fiber surface. Zeta potential is proportional to the total surface potential. To determine zeta potential, particles are dispersed in a solution controlled for pH, and an electric field is applied. The time required for the particles to migrate set distances is reported as the electrophoretic mobility, which can then be converted to zeta potential after accounting for the viscosity and the dielectric constant of the suspending liquid.

Apparently, the octahedral sheet of chrysotile possesses a net positive charge in aqueous solutions, and, since this sheet forms the outer layer of the chrysotile fibrils, their exterior surfaces are positively charged. The silicate sheet forming the surface of the interior tube may impart a negative charge to this surface, but surface charge measurements always show a significant net positive charge on chrysotile fibers (Light and Wei, 1977). The net positive charge may be the result of the larger size of the exterior surface, the presence of material inside the tube insulating it from the fluid medium in which fibers are suspended for surface charge measurements, or simply a lack of permeability.

Light and Wei (1977) report the zeta potential of two samples of chrysotile to be +40.5 and +52.5 mV in distilled water with pH = 7.4 at room temperature. The positive sign reflects the positive surface charge. Light and Wei show that the surface charge of chrysotile is readily reduced by acid leaching or by the addition of a surfactant to the suspending fluid. It is probable that surface charge with a spiral or cone-in-cone structure will be less than that of fibers with a simple cylindrical form, but this would be difficult to demonstrate experimentally.

Dissolution kinetics. There have been a number of studies that bear on the kinetics of chrysotile dissolution (Choi and Smith, 1972; Chowdhury, 1975; Gronow, 1987; Jaurand et al., 1977; Luce et al., 1972; Morgan et al., 1971; Thomassin et al., 1977). Of particular relevance is the recent experimental work by Hume and Rimstidt (1992) and Hume (1991), who provide discussion of the earlier work. Using existing data, Hume and Rimstidt showed that human lung fluids are very undersaturated with respect to both Mg and Si, so that chrysotile should dissolve in the lung. In their experiments, they monitored the release of Mg and Si to aqueous fluids at 37°C (approximately human body temperature) as

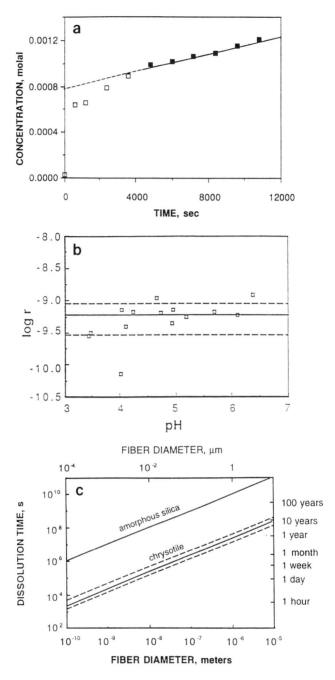

Figure 18. Results from chrysotile dissolution experiments in aqueous solutions at 37°C. (a) Increase of Si content of the fluid as a function of time (Hume and Rimstidt, 1992). (b) Log of the rate constant for dissolution as a function of pH (Hume, 1991). (c) Dissolution times for chrysotile and amorphous silica as a function of fiber diameter, calculated from dissolution rate constants (Hume and Rimstidt, 1992).

a function of time. Consistent with earlier work, they found that Mg is initially leached preferentially from the chrysotile (i.e., the dissolution is incongruent), with release of Si apparently limiting the dissolution reaction. Therefore, in their experiments they focused on Si in the fluid, observing Si concentration profiles such as that shown in Figure 18a.

From their concentration data, Hume and Rimstidt (1992) extracted rate constants for chrysotile dissolution. They found that within the acidity range (pH = 4 to 7) and ionic strength range of fluids found in the lung, dissolution rate is constant within the errors of their analysis (Fig. 18b), and the Si dissolution kinetics are zero-order. The average rate determined by Hume and Rimstidt is $k = 5.9$ (± 3.0) $\times 10^{-10}$ mol m^{-2} sec^{-1}. Using a simple shrinking solid-cylinder model for dissolution of a chrysotile fiber, they calculated that a fiber with a diameter of 1 µm should dissolve completely in 9 (± 4.5) months (the error given is 1σ). This dissolution time is of the same order as the time (20 months) indicated by the rate model of Parry (1985). Using a rate constant for dissolution of amorphous silica from Rimstidt and Barnes (1980), Hume and Rimstidt (1992) calculated that it would take a silica-glass fiber with the same diameter (1 µm) approximately two orders of magnitude longer to dissolve (438 years) than the chrysotile fiber (Fig. 18c). If fibers are cleared from the lung by dissolution, then, it is expected that the residence time for chrysotile is relatively short and much shorter than the residence time of silica-glass fibers. These experimental dissolution data provide one possible explanation of the lung burden data reviewed in Chapter 13: The reason chrysotile is underrepresented in mineral populations in the lung may be that it dissolves rapidly. The details of the calculations summarized by Hume and Rimstidt (1992) are found in Hume (1991).

There are several structural issues that have not yet been addressed in the literature on chrysotile dissolution. First, fibrils having a spiral structure possess a permanent ledge (Fig. 17), similar to the ledge produced by the emergence of a screw dislocation at the surface of a crystal. Such ledges are absent for fibrils with cylindrical layers. It is known from both theory and experiment that surface ledges affect both crystal growth and dissolution kinetics (Lasaga, 1980), and it is probable that the ledges on spiral chrysotile fibrils enhance their dissolution rates compared to those of cylindrical fibrils. It is conceivable, then, that early in dissolution experiments (and perhaps in organisms) the kinetics are controlled by spiral fibrils, with cylindrical fibrils dissolving more slowly and controlling the rates after complete dissolution of the spiral ones. Unfortunately, experimental testing of this speculation would be very difficult.

The dissolution literature also has not resolved the problems of open versus closed axial pores and surface coatings, discussed above. Fibrils with filled pores may well show different dissolution kinetics, compared to open fibrils. Assuming that the filling material is not serpentine, even the chemistry of early dissolution may be different for chrysotile having pore fillings and coatings due to materials between the fibrils of the original serpentine asbestos rock. Given the range of fibril structures and pore fillings that apparently exists, it may well be that different chrysotile samples will eventually be shown to dissolve at different rates and possibly with different degrees of incongruence.

AMPHIBOLES

Basic structure, chemistry, and nomenclature

As discussed above, amphiboles are double-chain silicates, with silicate polyanions such as that shown in Figure 2. Such a chain has a Si:O ratio (or Si+Al:O, if there is tetrahedral Al) of 4:11, and the oxygen atoms of the chains are coordinated not only to Si, but to a variety of other cation sites, yielding the following simplified generic formula:

$$A_{0-1}B_2C_5T_8O_{22}(OH, F, Cl, O)_2$$

(Hawthorne, 1981a). In this formula, T refers to the tetrahedrally coordinated sites within the silicate chain, C to fairly regular octahedral cation sites, B to less-regular octahedral or eight-coordinated cation sites, and A to relatively large, irregular cation sites having coordination numbers of six to 12. Although other substitutions can and do occur, especially in rocks with exotic compositions, the primary occupants of the four types of cation sites in rock-forming amphiboles are as follows:

A = Na, K   C = Mg, Fe^{2+}, Mn, Al, Fe^{3+}, Ti
B = Na, Li, Ca, Mn, Fe^{2+}, Mg   T = Si, Al

The presence of these four very different types of cation sites accounts for the notoriously wide variations in chemical composition observed for amphiboles. Because of this great chemical compliance, amphibole has been variously characterized as a "mineralogical garbage can" and as a "mineralogical shark in a sea of unsuspecting elements" (Robinson et al., 1982). Nomenclature for amphiboles is based on variations in their crystal structures and on their chemical compositions, and hence it is complicated. However, strict adherence to the correct nomenclature for amphibole species means that a mineral name conveys much useful information on structure and composition. This is in marked contrast to terms such as amosite and crocidolite, which provide only information such as the color of the asbestos (i.e., brown or blue).

The amphibole nomenclature was simplified and made more rational by the Amphibole Subcommittee of the International Mineralogical Association Commission on New Minerals and Mineral Names. Their recommended nomenclature is presented by Leake (1978) and was summarized by Hawthorne (1981a; 1983).

Amphiboles are classified into four major chemical groups, based on the occupancy of the B sites of the above general formula:

$(Ca + Na)_B < 1.34$	iron–magnesium–manganese group ("ferromagnesian" amphiboles)
$(Ca + Na)_B \geq 1.34$ $Na_B < 0.67$	calcic amphibole group
$(Ca + Na)_B \geq 1.34$ $0.67 \leq Na_B < 1.34$	sodic-calcic amphibole group
$Na_B \geq 1.34$	alkali (or sodic) amphibole group

Within each of these four groups, there are numerous basic amphibole names, and these names can be modified by using prefixes and adjectival modifiers denoting certain chemical variations. All in all, there are literally hundreds of possible amphibole names, and it is not, therefore, within the scope of this paper to present a detailed discussion of amphibole nomenclature.

Table 4. Mineral and asbestos names of amphiboles that have been mined as asbestos and their idealized chemical formulae

riebeckite (crocidolite)	$Na_2Fe^{3+}{}_2(Fe^{2+},Mg)_3[Si_8O_{22}](OH)_2$
grunerite (amosite)	$(Fe^{2+},Mg)_7[Si_8O_{22}](OH)_2$
anthophyllite (anthophyllite asbestos)	$(Mg,Fe^{2+})_7[Si_8O_{22}](OH)_2$
tremolite (tremolite asbestos)	$Ca_2Mg_5[Si_8O_{22}](OH)_2$
actinolite (actinolite asbestos)	$Ca_2(Mg,Fe^{2+})_5[Si_8O_{22}](OH)_2$

Although the number of naturally occurring amphiboles is large, only a few of them occur in the form of asbestos and in large enough amounts to have been mined as such. For these amphiboles, idealized compositions are listed in Table 4. For information on the exact compositional ranges covered by these names, see Hawthorne (1981a; 1983) or Leake (1978). Note that *amosite* (not to be confused with the 1:1 layer silicate amesite) and *crocidolite* have long been used for the asbestiform varieties of the amphiboles grunerite and riebeckite respectively. These varietal names for asbestos predate the mineralogical identification of these materials. However, the terms amosite and crocidolite are still useful, especially now that we know that some, if not all, asbestiform amphiboles contain intergrown minerals other than amphibole, especially layer silicates (see "Grain boundaries" below). If amphibole asbestos is truly a mixture, it is then technically a rock, not a mineral, and the varietal names could be considered *rock names*.[†] (In fact, even if amphibole asbestos were pure amphibole, it should perhaps be considered a monomineralic *rock* with very unusual texture, since rock is defined as an aggregate consisting of crystals of one or more minerals.) Also note that other amphiboles can occur with the asbestiform habit, though typically in noncommercial amounts, and that magnesio-riebeckite has been mined as asbestos, in addition to those amphiboles listed in Table 4.

Geological occurrence of amphiboles

The amphiboles are widely distributed, common minerals. They are important rock-forming minerals in a variety of igneous rocks that range in composition from granite to gabbro. They are common in many metamorphic rocks, especially in those derived from Fe- and Mg-rich igneous rocks and carbonates, and they may be found as a minor component in sandstones and

---

† As implied by the designation *rock* for the terms crocidolite and amosite, one should bear in mind that any sample of asbestos or other rock/mineral may contain other minerals as major, minor, or trace contaminants. The purchase of a mineral specimen is not analogous to the purchase of a reagent-grade chemical, regardless of whether the material is obtained from a commercial supplier or a museum. All samples of a rock/mineral should be analyzed to confirm mineral content, mineral composition and structure, surface properties, etc.—*Eds.*

other coarse-grained sedimentary rocks. Fibrous amphiboles, including amphibole asbestos, by and large are confined to metamorphic rock. Small amounts of fibrous amphiboles also may be found as veins and seams formed by hydrothermal fluids in some ore deposits and in igneous rocks. In commerce, amphiboles commonly are found in materials used as crushed stone, in industrial talc, in sands produced from crushed stone or sand deposits in recently glaciated terrane, and in dimension stone. Tremolitic talc has important uses in ceramics and as a filler in paint. Nephrite, a massive variety of fibrous actinolite, is highly prized as one of the forms of jade. In the asbestiform habit, amphiboles have been used in thousands of different applications, as noted previously.

Amphibole crystal structure types

The structures of all amphiboles share a basic unit consisting of a strip of octahedrally coordinated cations sandwiched between two double silicate chains (Fig. 19). These units are sometimes referred to as I-beams, in reference to their idealized shape being similar to beams of structural steel. The C cations of the general formula given above are found in the crystallographic sites of the octahedral strip, known as M1, M2, and M3. The B-group cations are situated in the M4 site, which tends to be a comparatively irregular coordination environment, situated at the flanks of the strip of regular octahedral sites (Fig. 19).

In order to describe the amphibole structure types, it is necessary to have some way of describing the orientation of an I-beam. This is commonly done using the duck convention of Thompson (1981), as shown in Figure 20. Clearly, the center duck in this figure is swimming to the right. (Thompson's paper came out during the first year of the Reagan presidency.) The left-hand and right-hand ducks (and hence octahedra) of Figure 20 are swimming toward you and away from you respectively, and these orientations can be designated + and −.

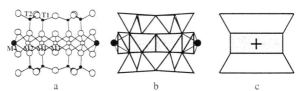

Figure 19. The "I-beam" from the amphibole structure of fluor-riebeckite (Hawthorne, 1978). (a) Ball-and-stick representation, showing nomenclature for M and T metal sites. (b) Polyhedral representation. (c) An idealized I-beam module, as used for simplified diagrams; the + indicates that the ducks are swimming toward you (see Fig. 20).

Figure 20. The duck convention (or convention of ducks?) of Thompson (1981). The center duck is a right-wing republican. The left-hand duck is swimming toward you, and the right-hand duck is swimming away.

Among the structurally ordered amphiboles, there are three basic, topologically distinct ways of stacking the I-beams, which can be described in terms of the directions in which the ducks are swimming as one moves through successive octahedral layers in the direction of the **a**-axis. The stacking types are as follows:

1. ... + + + + + + + ... or simply (+) clinoamphiboles (monoclinic)
2. ... + + − − + + − − ... or simply (+ + − −) orthoamphiboles (orthorhombic)
3. ... + − + − + − + − ... or simply (+ −) protoamphiboles (orthorhombic)

Because they differ primarily in the way the idealized, identical structural layers are stacked, these three structures may be referred to as ideal polytypes. Within each of these basic structure types, there are important structural variations (e.g., changes in space group, polyhedral distortions, and polyhedral rotations) that occur as functions of composition, temperature, and pressure. Each of these structure types is discussed below with reference to the asbestiform amphibole varieties.

In the usual unit-cell settings for all three basic structure types, the **c**-axis is defined as parallel to the silicate chains, the **a**-axis is in the stacking direction, and the **b**-axis is normal to **a** and **c**, pointing across the width of the chains. For clinoamphiboles, two unit-cell settings have been used. Most commonly, the **a**-axis is chosen so that the lattice is C-centered (e.g., the standard space group $C2/m$), but an alternative choice of **a**, analogous to the usual choice of **a** in clinopyroxenes, yields an I-centered (body-centered) clinoamphibole lattice (e.g., space group $I2/m$). This business of alternative unit cells can be very tricky, even for those with much experience in crystallography; the two types of unit cells are described clearly by Thompson (1978) and Hawthorne (1981a).

Clinoamphiboles. All calcic, alkali, and sodic-calcic amphiboles possess the (+) type of stacking, as do some ferromagnesian amphiboles. Of the asbestiform varieties (Table 4), riebeckite (alkali), grunerite (ferromagnesian), and tremolite and actinolite (calcic) fall in this structure type, which is illustrated for fluor-riebeckite in Figure 21 (Hawthorne, 1978). [The O(3) site of the amphibole crystal refined by Hawthorne contained approximately 45% OH and 55% F; hence the fluor- prefix.] Viewing the structure parallel to the **c**-axis (Fig. 21a), it can be seen that the octahedral orientation is + in all of the octahedral layers. (Obviously, if one turned this structure around so as to be looking at the ducks' tails, all octahedra would be in the − orientation; this is simply the twin of the orientation shown.)

Polyhedral distortions and rotations. Subtle rotations and distortions of the coordination polyhedra from their ideal shapes occur as a function of the precise composition (i.e., the chemical occupancies of the various crystallographic sites), temperature, and pressure. Indeed, these variable distortions and the presence of the four very different types of cation sites (A, B, C, T) allow the clinoamphibole structure to accommodate the large variations in composition observed in natural and synthetic samples. Distortions typical of all amphiboles are octahedral flattening due to shortening of shared octahedral edges; minor rotation of the backs of the silicate chains out of the (100) plane (Fig. 21a); and rotation of the tetrahedra, so that the chains are not perfectly straight (Fig. 21b).

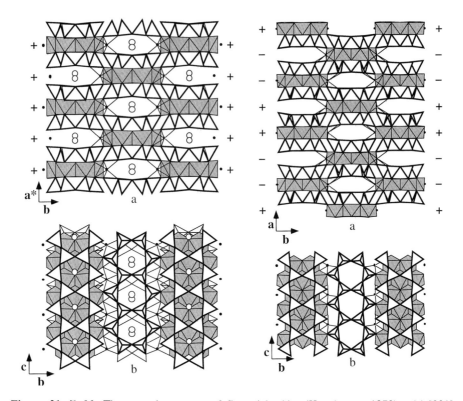

Figure 21. [left] The crystal structure of fluor-riebeckite (Hawthorne, 1978). (a) [001] projection showing the (+) stacking sequence characteristic of clinoamphiboles. The M(1), M(2), and M(3) sites are shown as shaded octahedra, M(4) is represented by black circles, and the A-sites, which are approximately one-third full in this sample, are shown as larger open circles. (b) Projection onto (100).

Figure 22. [right] The crystal structure of anthophyllite (Finger, 1970). (a) [001] projection, showing the (++−−) stacking sequence characteristic of orthoamphiboles. (b) [100] projection.

Polyhedral distortions, rotations, and their rationalization in terms of amphibole crystal chemistry are elaborately described by Hawthorne (1981a; 1983).

Space-group symmetry. Monoclinic amphiboles occur in at least three different space groups, depending on their chemical composition and temperature. Hawthorne (1981a; 1983) describes how the details of the amphibole structure vary for these three space groups. Also note that Maresch et al. (1991) observed a synthetic amphibole that underwent a phase transition from monoclinic ($C2/m$) to triclinic ($C\bar{1}$) symmetry with falling temperature.

1. $C2/m$ (or $I2/m$, if using the clinopyroxene-like unit-cell setting): This is the only space group that has been reported for calcic, alkali, and sodic-calcic amphiboles, and it also occurs for many Mg-Fe-Mn amphibole compositions. Also, primitive cummingtonite inverts to this space group with increasing temperature. All silicate chains are symmetrically

identical in the $C2/m$ structure. Amphiboles with this space group can accommodate relatively large cations in the M(4) site (e.g., Ca, Na).

2. $P2_1/m$: This space group is apparently restricted to very Mg-rich cummingtonite compositions. There are two symmetrically distinct silicate chains.

3. $P2/a$: The only amphibole reported with this space group is joesmithite, a rare, Be-rich form with Pb and Ca in the A-sites.

Orthoamphiboles. Amphiboles with the stacking sequence $(++--)$ are orthorhombic and crystallize in space group *Pnma*. This structure occurs for the ferromagnesian amphiboles anthophyllite and gedrite, as well as the Li-rich species holmquistite. The crystal structure of the orthoamphibole anthophyllite (Finger, 1970), one of the amphiboles that has been mined as asbestos, is illustrated in Figure 22. The *Pnma* structure has a relatively small M4 site that cannot accommodate much Na or Ca, and this structure therefore does not occur for calcic, alkali, or sodic-calcic amphiboles. Anthophyllite asbestos has been mined in the past, most notably in the Paakkila region of Finland, where 350,000 tons of asbestos were produced (Ross, 1981).

Protoamphibole. Amphibole with the stacking sequence $(+-)$ is orthorhombic, space group *Pnmn*, and has been called protoamphibole. This structure is known from a synthetic Li-Mg silicate (Gibbs, 1969).

The biopyribole polysomatic series and ordered pyriboles related to amphibole

Thompson (1978) showed that the amphibole structure can be thought of as consisting of two structurally and chemically different types of slabs, which he called M and P (Fig. 23a). These M and P slabs have structures that are topologically identical to the structures of micas and pyroxenes, so that a mica (or talc) could be constructed purely of M slabs, ...MMMMMM..., or simply (M). Similarly, a crystal made purely from P slabs, ...PPPPPP..., or (P), would have the pyroxene structure and stoichiometry, and the amphibole structure is ...MPMPMP..., or (MP). Modular structures of this type can be called polysomes, and a series of structures made from the same pair of slabs is called a polysomatic series. Thompson (1978) called the pyroxene–amphibole–mica polysomatic series of minerals the *biopyriboles*, from the names biotite, pyroxene, and amphibole. Thus, any structure that can be constructed by assembling M and P slabs is a biopyribole, *including pyroxenes, amphiboles, and micas*. Similarly, any of the chain-silicate biopyriboles (i.e., biopyriboles excluding the layer silicates) may be called a *pyribole*. Another polysomatic mineral with which we are already acquainted is antigorite, which can be thought of as planar serpentine (lizardite-like) slabs interspersed with two other types of structural slab (Fig. 11b). In addition to Thompson's original paper, the concept and applications of polysomatism have been reviewed by Veblen (1991; 1992).

In addition to the rigorously alternating (MP) amphibole structure, there is no structural reason the M and P slabs cannot be assembled in other sequences, such

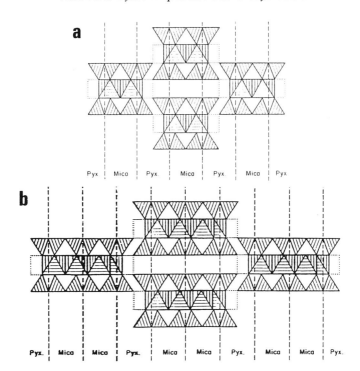

Figure 23. The polysomatic representation of biopyribole structures, which consist of Mica (M) and Pyroxene (P) slabs (Veblen, 1981, after Thompson, 1978). (a) Amphibole, with double chains, has the slab sequence (MP). (b) Clinojimthompsonite, with triple chains, has the slab sequence (MMP).

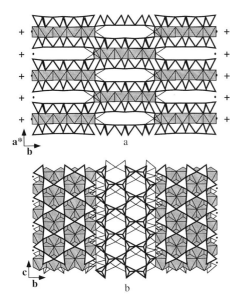

Figure 24. Crystal structure of the triple-chain silicate clinojimthompsonite (Veblen and Burnham, 1978b). (a) [001] projection, showing the (+) stacking sequence as in other clinopyriboles (compare with Fig. 21). (b) Projection onto (001).

as ...MMPMMPMMP..., or (MMP), as shown in Figure 23b. Such a structure would have not double chains like amphibole, but triple silicate chains. Thompson actually predicted that triple-chain silicates of this sort might occur as minerals, though he did not publish his prediction until after such minerals were discovered. In 1975, the triple-chain silicates now known as jimthompsonite and clinojimthompsonite were discovered coexisting with the fibrous, ferromagnesian amphibole anthophyllite (Veblen and Burnham, 1975; 1978a,b; Veblen et al., 1977). Also discovered was the first known mixed-chain silicate, the polysome (MPMMM), which has both double and triple chains and was named chesterite. The triple-chain crystal structure of clinojimthompsonite (Veblen and Burnham, 1978b) is shown in Figure 24, above. Like the monoclinic pyroxenes and amphiboles, clinojimthompsonite has the stacking sequence (+). Similarly, the minerals jimthompsonite and chesterite share the orthorhombic stacking sequence (+ + – –) with the orthopyroxenes and orthoamphiboles.

There are a number of important consequences of the polysomatic model. For example, because they are constructed of two different types of structural slabs, all members of a polysomatic series are stoichiometrically collinear. Thus, the stoichiometry of the amphibole anthophyllite is simply the sum of the formulae of the pyroxene enstatite and the layer silicate talc:

$$Mg_7Si_8O_{22}(OH)_2 = Mg_4Si_4O_{12} + Mg_3Si_4O_{10}(OH)_2 \quad (1)$$
anthophyllite   enstatite   talc

Another very important point about polysomatic structures is that there is no structural reason that the modules must be assembled in an ordered, periodic fashion. As a result, it is structurally permissible to have defective slab sequences such as ...MPMPMPMMPMPMPM.... The extra M slab inserted into the middle of this otherwise normal amphibole structure (MP) is a crystal *defect*, or a mistake, in the amphibole structure. The structure of this particular defect is that of a single slab of triple chains inserted into the normally double-chain amphibole structure. Defects of this type do occur in natural amphiboles, and, along with other types of defects, constitute a very important aspect of amphibole structure, and especially the structure of amphibole asbestos.

Defects and grain boundaries in amphiboles

Chain-width errors. Defects of the type just described involve aperiodicity in the sequence of silicate chains of a pyribole crystal. Chisholm (1973), using the TEM, was the first to observe this type of defect in a chain silicate (amphibole asbestos, in fact). He called them *Wadsley defects*, a term used by solid-state chemists for analogous defects found in synthetic oxides and also called *crystallographic shear planes*. Veblen et al. (1977) used the simple descriptive term *chain-width error* or *chain-width defect*, because these defects involve aperiodicity in the widths of the silicate chains (e.g., single, double, triple, etc.). In keeping with the Liebau (1980; 1985) classification of silicate structures, Czank and Liebau (1980) preferred the term *chain-multiplicity fault*. All of these terms are acceptable and are currently in use by various workers in the field. Generically, faults such as these also can be called *polysomatic defects*, since they involve errors in the otherwise periodic sequence of slabs in a polysomatic crystal structure (Veblen, 1991).

Many different types of chain-width errors have been observed in amphiboles by many different electron microscopists; for reviews of the published literature, see Veblen (1981; 1992). Also, Maresch and Czank (1983) provide an extensive catalog of the defects observed in synthetic amphiboles. Common defects in naturally occurring amphiboles include those consisting of a single (010) slab with single, triple, quadruple, quintuple, sextuple, or wider silicate chains. Combination chain-width errors also occur, consisting of two or more slabs with anomalous chain width. It is also important to recognize that regions of pyribole can occur with random or chaotic sequences of chains having two or more different widths. Such regions cannot be assigned to any particular mineral group such as pyroxene, amphibole, or triple-chain silicate. They can be called chain-width-disordered pyribole, or simply structurally disordered pyribole. For detailed discussion and many fine images of these defects, we suggest browsing some of the original TEM studies on amphibole asbestos (Ahn and Buseck, 1991; Alario Franco et al., 1977; Champness et al., 1976; Chisholm, 1973, 1983; Crawford, 1980; Cressey et al., 1982; Dorling and Zussman, 1987; Hutchison et al., 1975; Pooley and Clark, 1979; Veblen and Buseck, 1979b, 1980; Veblen et al., 1977; Whittaker et al., 1981a,b).

Twinning. Twin planes are another type of crystal defect observed in monoclinic amphiboles, and they seem to be particularly abundant in amphibole asbestos (see below). As noted above, an ideal clinoamphibole crystal has all of its ducks swimming in one direction, either ... + + + + + + + + + + + ... or ... – – – – – – – – – – ... . In a twinned crystal, however, the ducks reverse direction, leading to sequences such as ... + + + + + + – – – – – – ... . The plane where ducks change direction is parallel to (100); it is called the twin plane. The regions of + and – orientation are said to be in a twin relationship to one another.

Stacking faults. A stacking fault is another type of defect, which can occur either in clinoamphiboles or in orthoamphiboles. As with twinning, stacking faults involve those nutty ducks swimming in the wrong direction. In this case, however, the ducks quickly correct their error, so that commonly only one or two layers are affected. One possible stacking-fault structure is
... + + + + + – + + + + + ... ,   with only one layer in anomalous orientation. In pyroxenes (Livi and Veblen, 1989) and amphiboles (Smelik and Veblen, 1993a), however, it appears that stacking faults involving an even number of layers are favored over this type of one-layer stacking fault. Faults of this type would include the inclusion of two layers with the wrong sign in clinoamphibole,
... + + + + – – + + + + ... ,   and likewise in orthoamphibole,
... + + – – + + – – + + + + + – – + + – – + + ...

Dislocations. The above types of imperfections are called planar, because they are two-dimensional (some "planar" defects are actually not planar, but are surfaces with more complex morphology—so much for exactitude in nomenclature!). Dislocations, on the other hand, are linear defects that typically form during deformation, crystal growth, or as a result of a solid-state reaction or phase transformation. Dislocations have been observed in amphibole asbestos in several studies (Ahn and Buseck, 1991; Whittaker et al., 1981a; Cressey et al., 1982). There has been much interest in the possibility that amphibole asbestos fibrils might contain axial screw dislocations, since such dislocations occur in and account for the elongated morphology of many other types of inorganic whiskers (Zoltai, 1981; Walker and Zoltai, 1979), including some minerals (Veblen and Post, 1983). Appropriate amplitude-contrast TEM experiments suggest, however, that such screw dislocations are not present, at least in most amphibole asbestos fibrils (Veblen, 1980). On a tentative basis, then, we conclude that spiral growth around axial screw dislocations is probably not the operative mechanism during the growth of amphibole asbestos. The possibility exists, of course, that dislocations present during crystal growth have annealed out of the amphibole, or that amphibole asbestos from occurrences other than those explored with TEM will turn out to contain axial dislocations. Furthermore, it is quite possible that other fibrous types of amphibole, such as byssolite, form by such a growth mechanism.

Exsolution lamellae. Exsolution lamellae are also common, nonperiodic features in amphiboles. Such lamellae are composed of an amphibole having a different chemical composition from that of the host amphibole crystal, and they form by solid-state precipitation (exsolution) during slow cooling of an amphibole solid solution. Many different types of exsolution have been observed with the petrographic microscope and electron microprobe (for reviews, see Ghose, 1981; Robinson et al., 1982). There also have been many recent reports using TEM of amphibole exsolution occurring at scales beyond the resolution of the light microscope (Shau et al., 1993; Smelik and Veblen, 1991; 1992a,b; 1993a,b; Smelik et al., 1991). However, exsolution lamellae have yet to be reported from any high-quality amphibole asbestos sample, despite the fact that exsolution might be expected for some amphibole asbestos compositions.

Grain boundaries. In that they mark the position where the periodic structure of a crystal stops, grain boundaries also can be considered as a type of crystal defect. The structure of grain boundaries in anthophyllite asbestos, amosite, and crocidolite are discussed below. Perhaps the most important observation is that sheet silicates occur at the boundaries between the fibrils in all amphibole asbestos samples that have been investigated with appropriate TEM experiments. This observation indicates that amphibole asbestos really should be considered a rock, not a mineral. Also, as discussed below, the surfaces of amphibole asbestos may not be equivalent to a more normal amphibole surface prepared from material with a different habit.

Compositional variations in amphibole asbestos

Because of their great structural compliance and the resulting diversity of their compositions, a great deal has been written about amphibole chemical variations and the crystallographic factors that control them. For general information, the reader should consult volumes 9A and 9B of the Reviews in Mineralogy series, Hawthorne (1983), and other sources listed in the Introduction.

Here we restrict ourselves to a few remarks about the chemical variations exhibited by amphibole asbestos. Whereas the composition of chrysotile asbestos is typically close its ideal, end-member formula, $Mg_3Si_2O_5(OH)_4$, with fairly minor substitutions of Al and Fe, amphibole asbestos can show major departures from the idealized formulae given in Table 4. For example, as noted above, although crocidolite asbestos is usually stated to be the amphibole riebeckite, crocidolite can be so far from the end-member formula $Na_2Fe_2^{3+}Fe^{2+}_3Si_8O_{22}(OH)_2$ that it does not even fall within the rather large compositional field devoted to the name riebeckite. Given the possibility that health effects vary with chemical composition of the mineral, such variations should be a sobering thought for those who might wish to associate exact biological activities with specific mineral names or asbestos types.

The major substitutions in amphibole asbestos are those typical of other biopyribole minerals. Fe^{2+} and Mg mix freely in the octahedral sites; some Fe^{3+} occurs in amosite and anthophyllite; there is commonly some substitution of Al for Si, as well as some octahedral Al; although the amphibole asbestos minerals all nominally have empty A-sites, some K, Na, and Ca can be observed in these relatively large cavities (but some of this may be only apparent, resulting from intermixed layer silicates); limited solid solution toward the calcic amphibole series occurs in crocidolite, amosite, and anthophyllite; and Mn substitution can range from zero up to a healthy (or maybe not so healthy!) 0.234 atoms p.f.u. for one amosite reported by Hodgson (1979). The reader should consult Hodgson (1979) or Ross (1981) for amphibole asbestos analytical data.

A final complicating factor in the interpretation of amphibole asbestos is that it seldom, if ever, is pure amphibole. All electron microscopy studies performed with the amphibole c-axis parallel to the electron beam have shown there to be layer silicates intergrown with the amphibole. Any bulk analysis, therefore, is contaminated by these additional minerals. Given the great advances in analytical electron microscopy during the past decade, there is now much room for observations on the both the amphibole and the other intergrown minerals in "amphibole" asbestos.

Crystallization, mineralogy, and structure of amphibole asbestos and other amphibole habits

Asbestiform amphibole. The aqueous fluid involved in the crystallization of amphibole asbestos is likely to be highly supersaturated with respect to the precipitating amphibole. Apparently, once crystallization begins, growth takes place rapidly from many nuclei. Although amphibole fibers are formed by surfaces parallel to the c-axis, the fastest-growing faces are those that are not parallel to c. For monoclinic amphiboles, these may be faces with Miller indices (011) or ($\bar{1}$01). As noted above in the section on defects, it has been proposed that amphibole asbestos fibers grow by growth layers normal to the c-axis in a spiraling pattern around screw dislocations, but experimental evidence for this is lacking, despite attempts to observe such dislocations with TEM.

Many investigators have observed that asbestiform amphiboles are characterized by defects such as twin planes and chain-width errors, or intergrowths with sheet silicates and pyriboles. These may be manifestations of the type of "impurity" that limits growth and controls habit. Alternatively, some of the defects and the layer silicates may form during subsequent alteration of the amphibole by an aqueous fluid. This topic will be discussed in more detail in the following description of asbestos.

The fibrillar structure of amphibole asbestos is probably a factor in controlling a number of the properties for which asbestos is known, including high tensile strength. The subject is discussed in detail by Skinner et al. (1988). The high tensile strength is quite remarkable, exceeding that of the massive varieties by tenfold or more (Walker and Zoltai, 1979). The tensile strength and durability of grunerite asbestos were aptly demonstrated when it abraded a hole through a stainless steel mill during air jet milling (Campbell et al., 1980). Walker and Zoltai attribute the increase in tensile strength of asbestos not only to the very small diameters of the fibrils, since it is known that strength increases as diameter decreases, but also to the small number of flaws on the fibril surface. It may also be that the planar defects and subgrain boundaries parallel to the fiber axes enhance the tensile strength by mitigating the propagation of cracks and by providing sites for interplanar slip (Ahn and Buseck, 1991; Whittaker, 1979).

Although pristine amphibole asbestos is very strong, the tensile strength decreases when amphibole fibers are heated to between 200° and 500°C, before major chemical or structural changes are known to take place (Hodgson, 1965), as shown in Table 5 (from Carr and Herz, 1989, p. 22). In part because of the slight loss of water accompanying heating, Hodgson attributes the loss of tensile

Table 5. Physical properties and acid dissolution of asbestos[1]

	Chrysotile	Crocidolite	Amosite
Tensile strength[2], M N m^{-2}	3100	3500	2500
Elastic modulus[2], G N m^{-2}	160	190	160
Specific gravity	2.55	3.43	3.37
Magnetic susceptibility, mean at 10k Oe	5.3×10^{-6}	78.7×10^{-6}	60.9×10^{-6}
% loss in tensile strength, at			
300°C	0	13	37
400°C	[2.7][3]	63	61
500°C	[13.5][3]	78	80
600°C	84	83	96
% wt loss, boiling refluxed 4M HCl, for			
0.5 h	60	6	8
2 h		7	15
4 h		7.5	22
8 h		8.5	30

[1] Data from Carr and Herz (1989).
[2] Measured on micro-fibers 4 mm long between jaws.
[3] Gain in tensile strength, rather than loss.

strength to the breaking of hydrogen bonds between fibrils, bonds which, in this view, must therefore contribute to the overall tensile strength of composite fibers. The thermal breakdown of amphiboles, including crocidolite and amosite, is reviewed by Ghose (1981), and Gilbert et al. (1982) provide extensive information on the thermal stability limits of many different amphiboles. Available evidence supports Hodgson's view that the amphibole component of amphibole asbestos does not begin to dehydroxylate below 500°C. It seems likely, however, that the low-temperature loss of water, and perhaps also the reduction in tensile strength, are related to changes in the interfibrilar layer silicates that appear to be a universal component of amphibole asbestos (see discussion below, this section); the upper thermal stabilities of many layer silicates tend to be lower than those for chemically related amphiboles.

Flexibility and anomalous optical properties are also characteristics of the asbestiform amphiboles. The flexibility of fibers is enhanced by the small widths and extreme aspect ratios of the fibrils and by the fact that fibrils in bundles may slip past one another. Among commercial amphibole asbestos types, the most flexible is riebeckite asbestos (crocidolite), and the least flexible is anthophyllite asbestos. Since fibril width is greatest for anthophyllite and least for crocidolite, this suggests that flexibility is inversely proportional to fibril width. Most amphibole asbestos samples consist of monoclinic amphibole. However, in cross-polarized light, fiber bundles display parallel extinction, instead of the inclined extinction characteristic of other habits of the same minerals, a property predictable from the fibrillar structure (Wylie, 1979). Riebeckite asbestos (crocidolite) and grunerite asbestos (amosite) were first thought to be orthorhombic because of this property. In noncommercial samples of tremolite and actinolite asbestos, oblique extinction can frequently be seen in the larger fibers that are not bundles. Oblique extinction is never observed in commercial crocidolite or grunerite asbestos, due to their very small fibril diameters.

The question of the role of wide-chain pyriboles (those with chain widths of three or greater) in the development of fibrous amphiboles is not yet fully understood. Chisholm (1973) suggested that the fibrous habit might result from structural faults along (010). Hutchison et al. (1975) confirmed that triple (or multiple) chains parallel to (010) were present in samples of grunerite asbestos and fibrous tremolite. In fibrous anthophyllite, the triple chains were segregated into domains. They noted abundant (100) twinning in grunerite asbestos and riebeckite asbestos. Ahn and Buseck (1991) point out that chain-width defects propagate into fibers from grain boundaries in crocidolite, as in anthophyllite asbestos (Veblen, 1980). Chain-width errors might be important in fiber formation, since they could lead to partings along (010), and, in combination with partings on (100) due to twinning or stacking faults and (110) cleavage, they might lead to the development of fibers (Veblen, 1980; Veblen et al., 1977). However, chain-width errors and stacking faults may simply preclude the lateral growth of the fibers initially. They would have little impact on the (001), (011), or ($\bar{1}01$) faces, which are the most likely surfaces experiencing rapid growth during crystallization. However, in all these studies, only a very limited number of samples were studied. Therefore, Dorling and Zussman (1987) designed a study to see if they could generalize any relationships between the various habits and defects observed in tremolite and actinolite. They examined 29 prismatic and massive, 8 byssolite, and 16 asbestiform calcic amphiboles with optical microscopy, scanning electron microscopy, and TEM. Table 6 summarizes the results of their investigation. Although on average asbestiform samples and nephrite (amphibole jade) contain more chain-width defects than other habits, Dorling and Zussman did not find a clear correlation between chain-width errors and the fibrous habit. In fact, the fibrous byssolites lacked these defects almost entirely. They did find that an abundance of (100) parting surfaces due to twinning is restricted to asbestiform varieties, consistent with the earlier work of Hutchison et al. (1975) on grunerite asbestos. It is clear, therefore, that although chain-width errors are common in asbestiform amphibole, alone they do not produce the asbestiform habit.

The interfaces between asbestos fibrils have been investigated by Alario Franco et al. (1977), Crawford (1980), Veblen (1980), Cressey et al. (1982), and Ahn and Buseck (1991). Although the fibrils have their c-axes (i.e., the fiber axis) nearly, but not perfectly, parallel, their orientations are more disordered perpendicular to the c-axis. The grain boundaries are irregular and tend to be bounded by (010) and (110) surfaces for only short distances. Cressey et al. (1982) present evidence that most of the boundaries in amosite were formed during growth, but some were produced by shear stress following fibril formation. Cressey et al. also report that while sheet silicates may be present at fibril interfaces in amosite, they are so abundant in anthophyllite asbestos that they never observed anthophyllite fibrils in contact with each other. Veblen (1980) did, however, observe amphibole-amphibole contacts in another sample of anthophyllite asbestos (Fig 25a). Both studies found talc, chlorite, and serpentine (lizardite, chrysotile, and antigorite) in the grain boundaries (Fig. 25b,c), although confirmation of compositions by analytical electron microscopy is still needed. Veblen (1980) showed that the crystallographic orientation of an individual anthophyllite fibril controls the orientation of the interfibril sheet silicates, but orientation of the sheet silicate with the adjoining fibril is irregular (Fig. 25b). Cressey et al. (1982) report that the regularity between amphibole and sheet silicate orientation is greatest for talc but may be less well-developed for chlorite and serpentine. Ahn and Buseck (1991) found a mica similar to Na-biotite between fibrils of crocidolite, with the a-axis of the mica parallel to the c-axis of riebeckite. Ahn and Buseck (1991), Crawford (1980), and Veblen (1980) found that when the rotational misalignment of adjacent fibrils is very slight, the fibril boundaries may be bridged by wedgelike wide-chain pyriboles (Fig. 25a). Based on available evidence, and comparing asbestiform amphibole with fragments produced by cleavage and parting (see below), it seems likely that the primary mechanism for fibril formation in amphibole asbestos is separation of the fibers along the grain boundaries between fibrils. Because these grain boundaries contain layer silicates, it is likely that the amphibole asbestos surface with which a biological system will interact in many cases may not actually be amphibole. The likelihood of a layer-silicate interface is variable, depending on the sample, and contact with a number of different layer silicates is possible. Because of the potential biological ramifications of the interfibril layer silicates in amphibole asbestos, their analysis with state-of-the-art AEM methods is much needed.

Table 6

Morphology and properties of tremolite–actinolite: a general comparison

Morphology	Crystal units	Multiple twinning	Well developed crystal faces	Multiple chain (010) defects	Dislocation networks and sub-grain boundaries	Exsolution	Chemical substitutions		
							Na in A Site	Na for Ca	Al for Si
prismatic	large crystals	none	common	abundant in one specimen; few in others	common	some specimens	varied	varied	varied
acicular byssolite	slender needles or laths	none	common	rare	none	none	appreciable	very little	appreciable byssolite
massive nephrite	submicroscopic short laths	none	rare	abundant	sub-grain boundaries common	none	†	†	†
asbestos	submicroscopic fibrils	common	rare	abundant in tremolite; few in actinolite	none	none	varied	little	almost none

Source: Dorling and Zussman (1987)
† Insufficient samples for generalization

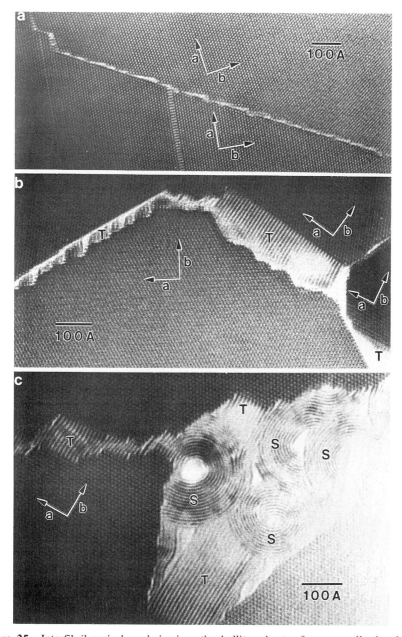

Figure 25. Interfibril grain boundaries in anthophyllite asbestos from a small, abondoned mine at Pelham, Massachusetts (Veblen, 1980). (a) Two adjacent fibrils showing amphibole-amphibole contact; where the boundary changes orientation at the upper left, it assumes a wedgelike, wide-chain pyribole structure. (b) Three adjacent fibrils with the layer silicate talc (T) in their grain boundaries. (c) A very low-angle grain boundary (~1° rotation between the fibrils). The amphibole at the top is one fibril, and the patches in the lower left and lower right both belong to the other fibril. The boundary contains talc (T) and serpentine mostly of the chrysotile persuasion (S), though a few planar serpentine (lizardite) layers are also present (distinguished from talc by smaller spacing between layers).

It is not known whether the interfibril sheet silicates and intergrown wide-chain pyriboles formed as an alteration of preexisting amphibole fibers, or simply grew epitaxially on the fiber surfaces during amphibole growth in the direction of the c-axis. If the latter is the case, it may be that the sheet silicates or wide-chain pyriboles were the interfering phase that limited lateral growth of amphibole and resulted in the formation of the very fine fibril diameters characteristic of asbestos. However, many of the complex textures observed in asbestos amphibole are similar to those produced by postcrystallization reactions in other, nonasbestiform occurrences.

Byssolite and nephrite. Dana applied the term byssolite to the stiff, fibrous variety of actinolite (Ford and Dana, 1932, p. 574). Byssolite grows as single or twinned, very fine fibers, usually in a matted mass, and the fibers do not occur in bundles as in asbestos. Individual fibers are on the order of 1-2 μm in width, are quite brittle, and readily crush to shorter particles. They lack the tensile strength and fibrillar structure of amphibole asbestos. Wylie (1979) and Dorling and Zussman (1987) also have used the term byssolite to distinguish these fibers from asbestos. Nephrite, a form of jade, is composed of actinolite fibers in a mass-fiber habit, although the fibers in nephrite are not separable, nor is nephrite asbestiform.

Cleavage and parting fragments of nonasbestiform amphiboles. In examining many crushed amphibole specimens by optical microscopy, we have found that among amphiboles that are not asbestiform, those that form the more elongated cleavage fragments commonly have well-developed (100) parting surfaces. Although less common, (010) parting surfaces are frequently present as well. Elongation of amphiboles during comminution is probably a function not only of (110) cleavage, but also the degree of development of (010) chain-width errors, (100) stacking faults and twinning, and possibly other structural defects that promote parting. Whereas cleavage and parting apparently cannot alone produce high-quality amphibole asbestos, they can produce highly elongated cleavage fragments. In fact, upon crushing, mixtures of anthophyllite, chesterite, and jimthompsonite with many chain-width errors produce mostly very elongated fragments, some of which show extreme elongation and a high degree of flexibility (Veblen and Burnham, 1978a; Veblen et al., 1977). This material does not, however, possess the fibrilar microstructure associated with good asbestos. It thus appears that narrow, flexible fibers can be produced as the result of structural defects and cleavage. However, the very easy formation of extremely narrow fibers associated with amphibole asbestos probably occurs primarily as a result of the material separating along the grain boundaries that occur between the fibrils, rather than by separation along defects.

Dimensions of amphibole fibers and cleavage fragments. There are many data sets describing the distributions of length and width of asbestos fibers in both bulk samples and airborne populations. There are also data available on populations of crushed massive to prismatic amphiboles and airborne amphibole cleavage fragments, and some data are available on the distribution of amphiboles found in lung tissue. The widths of cleavage fragments of nonasbestiform amphiboles increase as their lengths increase, and for particles longer than 5 μm, widths less than 1.0 μm are uncommon (Siegrist and Wylie, 1980). On the other hand, more than 90% of amphibole asbestos fibers have widths less than 1.0 μm, and most are less than 0.5 μm. Furthermore, the width of an asbestos fiber is independent of its length. Almost all amphiboles identified in lung tissue are also less than 1 μm in width, and although it may be impossible to ascertain the growth habit of individual particles, the dimensions of most are entirely consistent with those of asbestos. Wylie et al. (1993) summarize the data on the distributions of amphibole widths in populations of asbestos, cleavage fragments, and particles comprising lung burden. Table 1 provides width data on asbestos fiber populations.

Surface chemistry, surface charge, and dissolution kinetics of amphibole asbestos

For general comments on surface chemistry, surface charge measurements, and dissolution kinetics, see the analogous section on chrysotile, above.

Surface chemistry and structure. As with chrysotile, we can speculate on the likely surface chemistry and atomic-scale surface structure by noting that the amphibole I-beams are rather strongly bonded units, but the I-beams themselves are connected with weaker bonds. Probably hundreds of publications, from the earliest research papers on chain silicate structure to introductory mineralogy texts (Klein and Hurlbut, 1993, p. 477, 490, for example) have assumed that the cleavage surface in pyriboles leaves the I-beams intact, winding through the regions of weaker structure between them. Figure 26 shows a likely possibility for the resulting cleavage surface structure. Although the positions of oxygen atoms that remain or have been removed are quite arbitrary, this surface structure does address the more significant question of which metal-ion coordination polyhedra are exposed on the surface. The cation sites to which a solution would have easy access are M(4), the tetrahedral sites, and the A-sites (not shown in Fig. 26, because most amphibole asbestos has a nominally vacant A-site, but see Fig. 21). There may be more limited access to the M(2) site from the surface, but the other cation sites, M(1) and M(3), will not immediately come in contact with the environment but will interact once dissolution has begun; in fact, if the I-beams remain intact, this will be the case for any amphibole surface parallel to the c-axis, regardless of the orientation of the surface within the [001] zone.

In terms of the cations occupying the near-surface sites, M(4) is nominally occupied by the B-group cations indicated in Table 4. T(1) and T(2) are occupied primarily by Si, with limited substitution by Al. The A-site, when partially occupied, will contain limited amounts of Na and K.

If pure, untwinned amphibole were to separate by parting along (010) or (100), these alternative surface structures might appear as in Figure 26. Again, the solution would have primary access to M(4), the tetrahedral sites, and the A–sites, if occupied. It is believed by most workers, however, that these partings occur along defects, so that the local structures would differ at least somewhat from those shown. Nonetheless, if our ideas about the defect structures for (100)

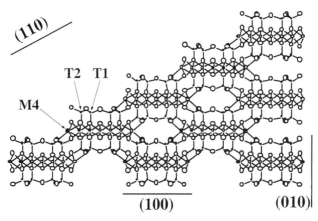

Figure 26. [001] projection of a possible structure for a (110) cleavage surface on the clinoamphibole fluor-riebeckite (Hawthorne, 1978). This structure assumes that the amphibole I-beams remain essentially intact on either side of the corrugated cleavage surface. Although the details of bonds to oxygen and the dangling bonds to removed oxygen ions obviously are not known, it appears likely that the cations with which a medium in contact with a fresh cleavage surface will interact most directly are those occupying the M(4) sites, the tetrahedral sites T(1) and T(2), and, if occupied, the A-sites (not shown; see Fig. 21a). Lines labeled with Miller indices (100) and (010) are the traces of these planes.

twinning and (010) chain-width errors are correct, the parting surfaces would still be very similar to those of pure amphibole parting with the I-beams intact. It is also likely on structural grounds, however, that the (100) partings will cut through the octahedral strips of half the I-beams, leaving a smoother surface with all of the M sites exposed. The gross structures of these cleavage and parting surfaces may well be resolvable with scanning probe microscopy methods (see Chapter 9).

Although we can speculate on the details of these idealized surface states, there are many observations, discussed above in the section on the asbestiform habit in amphiboles, that suggest that there are other, very important factors that may largely control the surface structure and chemistry of amphibole asbestos fibers. First, although TEM studies of the fibrils of amphibole asbestos do show surfaces parallel to the amphibole cleavage planes, other surfaces also occur, and the surface is commonly quite irregular. Second, all amphibole asbestos that has been observed by TEM with the electron beam parallel to the c-axis has contained layer silicates intergrown between the amphibole fibrils. The amounts and identities of these layer silicates vary considerably from sample to sample, and in some cases the boundaries contain complex assemblages of several layer-silicate species (e.g., Fig. 25c).

If the primary mechanism for fiber formation in amphibole asbestos is separation along the grain boundaries between fibrils (as argued above and by many previous workers), then both the noncleavage amphibole surfaces and the interfibril layer silicates will significantly affect the overall surface structure and chemistry that is exposed to a fluid or to lung tissue. Perhaps the most important generalization we can make about the surfaces of amphibole asbestos fibers is that no generalization is possible. Within any single sample there are likely to be *many* different types of surfaces that vary in structure, in orientation, and in chemistry. Additionally, from the TEM studies cited previously, it is clear that the exposed surfaces in different samples can be very different with respect to the same factors. Unlike the homogeneous, single-phase, gemlike crystals one tends to see in museum mineral collections, amphibole asbestos is a very, very complex mixture that varies substantially from locality to locality, and perhaps even within a single mine.

Surface charge. In contrast to chrysotile, the net surface charge is negative for those amphiboles that have been investigated. It is also observed that the surface charge on amphiboles, as measured by zeta potential, increases as the particle aspect ratio increases, so that asbestiform amphiboles have a higher surface charge than cleavage fragments. Schiller et al. (1980) suggest that the ends of amphibole cleavage fragments are positively charged, whereas the surfaces parallel to the **c**-axis have negative charge. As the aspect ratio increases, the relative amount of surface parallel to **c** also increases, so that more elongate fragments of fibers have a larger net negative surface charge.

Walker (1981) further proposes that the presence of surface defects may affect surface charge, since certain defects may expose cations to the solution, and that amphibole asbestos has fewer surface defects than other varieties. These exposed cations would have the effect of reducing the net negative surface charge. Walker's (1981) preliminary observation of the density of etch pits formed on amosite and grunerite cleavage fragments support his hypothesis about defect density. However, the observations are limited and qualitative, so that the importance of defects to surface charge in amphibole asbestos must be considered an unresolved question.

Light and Wei (1977) report the room-temperature zeta potential in distilled water with pH = 7.4 for crocidolite, amosite, and anthophyllite asbestos as -50.5, -58.5, and -54.0 mV respectively. These magnitudes are approximately equal to their measurements of zeta potential for chrysotile, although the sign is opposite. Nolan et al. (1991) report the zeta potentials of seven samples of tremolite asbestos and two samples of tremolite cleavage fragments. For the asbestiform fibers, mean zeta potentials range from -32.9 to -56.3 mV. Mean zeta potentials of the cleavage fragments are -13.3 and -13.5 mV. The differences between cleavage fragments and asbestos fibers are consistent with the explanation of Schiller et al. (1980) for variation of surface charge with aspect ratio.

Dissolution kinetics. It is generally believed that dissolution kinetics for amphibole asbestos are considerably slower than those for chrysotile. For example, Table 5 shows much slower weight loss for amphibole asbestos in boiling, refluxed 4M hydrochloric acid, as well as considerably faster dissolution for amosite versus crocidolite (Carr and Herz, 1989). Boiling 4M HCl is, however, a considerably more corrosive environment than the human lung. Unfortunately, there appear to be no existing data on dissolution rates that can be used to calculate the time necessary to dissolve amphibole fibers in the human lung (Hume and Rimstidt, 1992). It would appear that experimental geochemists who specialize in mineral dissolution kinetics could make valuable contributions to our understanding of dissolution rates of amphiboles and amphibole asbestos; rate constants for amphibole dissolution under appropriate conditions are necessary for an understanding of amphibole residence time in the lung.

There have been several observations on amphibole-asbestos dissolution that are related to its mineralogy. Crawford (1980) and Crawford and Miller (1981) note TEM evidence indicating that crocidolite fibers are eroded during *in vitro* exposure to normal human blood serum. Erosion is most notable where (100) planar defects intersect the fiber surface, although general surface dissolution is also noted. Ralston and Kitchener (1975) detected a small amount of surface dissolution under mild leaching of amphibole asbestos. They also observed strong adsorption of certain cationic surfactants, indicating specificity of surface reactions occurring on amphibole asbestos (Skinner et al., 1988). Cook et al. (1982) report that the number of fibers of ferroactinolite asbestos and grunerite asbestos increases with residence time in the lung of rats, perhaps indicating dissolution or disaggregation along fibril boundaries. On mineralogical grounds, it seems very unlikely to us that non-asbestiform amphibole would disintegrate into asbestos-like fibers under these conditions.

IDENTIFICATION OF 1:1 LAYER SILICATES, AMPHIBOLES, AND WIDE-CHAIN PYRIBOLES

Many analytical methods have been employed for the identification of 1:1 layer silicates and amphiboles. Here we discuss identification by the most common methods: optical microscopy, XRD, SEM, and the various TEM methods (imaging, electron diffraction, and microanalysis). For general discussions of these methods, see Chapter 8 by Guthrie, this volume.

Optical microscopy (OM)

Kaolin. Minerals of the kaolin group are frequently so fine-grained that optical properties other than mean index of refraction are quite difficult to measure. However, the literature suggests that the kaolin polymorphs can be distinguished based on the optical properties (Fleischer et al., 1984). Halloysite has the lowest average index of refraction, with ß = 1.556–1.555. Both isotropic and biaxial forms (negative, 2V = 90°) have been reported. Dickite is optically positive with ß = 1.563 and 2V = 55–80°. Nacrite has ß = 1.562, is optically negative, and has 2V = 40–90°. Kaolinite is also optically negative, with ß = 1.565, but its 2V value is smaller than that of nacrite (2V = 24–50°).

Serpentine. The refractive indices of the serpentine group of minerals show variations that can be associated with differences in Fe, Ni, and Al content and, to some degree, with structure. Chrysotile has the lowest indices of refraction, with ß as low as 1.546 (up to 1.569). The indices of refraction of chrysotile are known to increase with heat treatment, so that the indices of refraction of the chrysotile found in some building products, such as furnace insulation, may be higher. The index of refraction ß of antigorite has been reported to range from 1.566 to 1.603, while the magnitude of ß for lizardite ranges from 1.555 to 1.569. In general, as the indices of refraction increase, so does the birefringence.

Asbestiform chrysotile has much lower indices of refraction than all types of amphibole asbestos. It has been confused with fibrous talc, but the birefringence of talc is much higher than that of serpentine. Most chrysotile is elongate along the **a**-axis, parallel to the Z optic direction (i.e., the maximum index of refraction is parallel to the direction of fiber elongation). However, fibers of parachrysotile are elongate parallel to the **b**-axis and the Y optic direction (i.e., ß is parallel to elongation). In both cases, the fibers are length-slow and exhibit parallel extinction.

Amphiboles and wide-chain pyriboles. In most bulk samples, amphiboles can be identified by optical microscopy. However, to determine exactly which amphibole is present, optical properties are not always sufficient. For example, the optical properties of actinolite may overlap those of hornblende, and tremolite may overlap richterite optically. The International Mineralogical Association recommends that a specific amphibole name be assigned only when a chemical analysis is available (Leake, 1978). However, the optical properties are frequently a guide to the general grouping of amphiboles. There are many good sources for the optical properties of the amphiboles, including Phillips and Griffen (1981), Nesse (1991), and Fleischer et al. (1984).

Asbestiform amphiboles may present some special problems in identification. While the optical properties of the major commercial amphibole asbestos minerals are well-known and are adequate for identification, if the amphibole asbestos is from another source, for example, some local deposit, the optical properties may not indicate the type of amphibole. This is especially true when the fibrils have widths less than about 0.5 µm and/or the fibrils are characterized by polysynthetic (100) twinning. In such cases, monoclinic amphiboles may possess orthorhombic optical properties (Wylie, 1979). Furthermore, when fibril widths are small, it is difficult or impossible to observe interference figures, which are quite important for identification.

The optics of amphiboles that have altered partially to sheet silicates or that contain significant amounts of other intergrown pyriboles may vary from those given in the literature for the unaltered amphibole. Generally, the indices of refraction and 2V can be expected to decrease and the birefringence to increase as alteration progresses. This is illustrated by the sequence anthophyllite, chesterite, jimthompsonite, fibrous talc, talc (Veblen and Burnham, 1978a). Anthophyllite is the "pure" amphibole and so has only double silicate chains, chesterite has alternating double and triple chains, jimthompsonite is a pure triple-chain silicate, fibrous talc is typically an intimate intergrowth of talc, anthophyllite, and other, disordered chain silicates, and talc is the pure, end-member layer silicate. The value of 2V varies from close to 90° in anthophyllite, to 71° for chesterite, 62° for jimthompsonite, to 0–30° for talc. The birefringence in the same order is 0.013–0.022 for anthophyllite, 0.023 for chesterite, 0.028 for jimthompsonite, 0.030–0.045 for fibrous talc, and 0.04–0.05 for most talc. The indices of refraction in fibrous talc decrease as the proportion of anthophyllite decreases (as determined by XRD). It is more difficult to generalize about indices of refraction in the anthophyllite, chesterite, jimthompsonite series because of the confounding variable of the iron content, which can have a significant effect on the optical properties. However, specimens of chesterite and jimthompsonite, each with about 13% iron, have ß = 1.632 and 1.626 respectively. In thin section, amphiboles also can be differentiated from the wide-chain minerals by the characteristic angles between their cleavage planes (Veblen and Burnham, 1978a). These angles are approximately 56°/124° for amphiboles, 44.7°/135.3° for chesterite (and its unnamed monoclinic analog), and 37.8°/142.2° for jimthompsonite and clinojimthompsonite.

X-ray diffraction (XRD)

Ideally, any mineral that is recognized as a valid species would produce a unique XRD pattern. Therefore, it should be possible to identify any mineral with XRD. In practice, however, this is not always a simple task. In the first place, samples in the real world frequently contain mixtures of several minerals, the diffraction patterns of which may overlap and thereby obscure each other. Particle size and structural disorder (e.g., stacking disorder in layer silicates and chain-width disorder in pyriboles) may alter the widths and shapes of reflections (see the comparison of two kaolinite patterns in Fig. 15). Furthermore, both preferred orientation and structural disorder can significantly affect the intensities of diffraction peaks. Therefore, while it is usually possible to differentiate between major mineral groups using powder XRD, for example serpentine from kaolin, it may be very difficult to differentiate species within a mineral group. It can, for example, be difficult to tell chrysotile from lizardite, or to distinguish one monoclinic amphibole from another. In this section, we will examine what is normally possible with powder XRD.

1:1 layer silicates. Bailey (1969; 1988b) has shown that the planar 1:1 layer silicates can be divided structurally into 12 standard planar polytypes. Further, these 12 polytypes can be categorized into four groups of three polytypes defined by the direction of interlayer shift and the occupancy pattern of octahedral cation sets in the structure. These four groups are readily distinguished in monomineralic samples, based on the positions of the most intense reflections. Further differentiation within these groups requires the weaker X-ray reflections, which may be difficult or impossible to observe in many cases. Bailey (1969) gives calculated XRD patterns for all 12 standard polytypes for a Mg-rich composition. Bailey (1988b,e) further describes the XRD identification of the polytypes of serpentine. Natural examples of all four polytype groups and of 10 of the 12 polytypes occur in the Mg-rich 1:1 layer silicates. In the Al-rich 1:1 layer silicates, only two ordered polytype groups are represented, with dickite and kaolinite belonging to the same group. Powder XRD patterns for various polytypes of amesite, cronstedtite, kellyite, and odinite are tabulated by Bailey (1988c).

Since the discovery of platy halloysite (Noro, 1986), fibrous kaolinites (Souza Santos et al., 1965), and spherical kaolinites (Tomura et al., 1985), it has been recognized that morphology alone is insufficient to distinguish halloysite from the other kaolin minerals. The diagnostic criterion for halloysite is the presence of interlayer water (Churchman and Carr, 1975). Whereas hydrated halloysite has an interlayer spacing of approximately 10 Å, the dehydrated form halloysite(7 Å) can be confused with kaolinite. Churchman et al. (1984) have shown that even when halloysite is largely dehydrated, it intercalates formamide and N-methylformamide at a faster rate that kaolinite, and the two can be distinguished by this property using XRD. Additional data are presented by Churchman (1990), along with a good discussion on differentiating these kaolin minerals by different methods.

Whittaker and Zussman (1956) describe the XRD criteria for distinguishing antigorite, lizardite, and chrysotile. The criteria for lizardite polytypes are available in Bailey (1969) and Bailey (1988b,e). Polygonal serpentine cannot be conclusively identified by XRD. Antigorite is most easily distinguished from the other two forms of serpentine by its unique (330) reflection at 1.56–1.57 Å, and the (11l) reflections of lizardite are not present in chrysotile. When serpentine is mixed with other minerals or when preferred orientation occurs, these polymorphs usually cannot be distinguished. Wicks and Zussman (1975) describe a microbeam XRD system for analyzing small areas of serpentine in thin section to partly overcome these problems. Wicks and O'Hanley (1988) list XRD data for two lizardite polytypes, two chrysotile polytypes, and four antigorite specimens.

Amphiboles and wide-chain pyriboles. Amphiboles can be identified by XRD as belonging to the amphibole group, provided a few of the major reflections can be obtained. However, though differentiation of orthorhombic from monoclinic amphiboles is usually possible, it is usually not possible to differentiate among the individual members of these large groups with confidence. Furthermore, the various habits of the amphiboles cannot be distinguished by XRD (i.e., it is not possible to distinguish between prismatic cleavage fragments and amphibole asbestos). XRD data for amphiboles can be found in the Powder Diffraction File (International Centre for Diffraction Data, 1993). Structural data, such as unit-cell parameters and atomic positions, from which powder XRD patterns can be calculated, are given by Hawthorne (1983) for all published amphibole structure refinements up to 1983. Calculated powder XRD patterns for the wide-chain biopyriboles jimthompsonite, clinojimthompsonite, and chesterite are illustrated and tabulated by Veblen and Burnham (1978a).

Scanning electron microscopy (SEM) and electron microprobe analysis

The scanning electron microscope. Imaging and chemical analysis with the scanning electron microscope are covered in detail by the classic text of Goldstein et al. (1981). The SEM produces a very narrow electron beam with energies up to a few tens of keV. By way of electrostatic coils, the beam is rastered across the specimen, and simultaneously the signal from an electron detector displays the intensity of electron scattering on a cathode-ray tube operating in synchronization with the beam raster. Most commonly, a secondary-electron detector, which measures low-energy electrons emitted from near the sample surface, is employed, yielding a fairly high-resolution image showing the morphology of the specimen. An example of this type of image is given in Figure 27, showing that the SEM can be used to produce useful images of fibrous amphiboles. In addition, for thick, flat, polished samples, a backscattered electron detector can be used, producing images that can be interpreted as high-resolution maps of average atomic number in the sample. Because fine-grained 1:1 layer silicates and asbestos seldom produce good samples of this sort, it is less common to study them with this type of detector. It is probably worth noting that, manufacturers claims of superb resolution to the contrary, the ultimate resolution is seldom achieved in the real world with real samples. Though an improvement in resolution over the light microscope, much of what is seen with SEM also can be seen by a careful observer using OM.

Quantitative and qualitative chemical analysis in the SEM. A chemical analysis can be considered to be quantitative if it indicates not only what components are in the sample, but also how much of each component. In addition, for an analysis to be considered quantitative, the analyst must have evaluated the likely errors in the quantities, which is generally done by using various statistical methods, coupled with multiple analyses of homogeneous standards. (It means little to say that a certain soap is 99.9% pure, if one does not know whether the statistical and systematic errors used to derive that figure are 0.1% or 100%!) If the various components in the sample are known, but the amounts present or errors are not, then the analysis should be considered qualitative.

Figure 27. A secondary electron image of a fiber of asbestiform potassian winchite, a sodic-calcic amphibole, obtained with a SEM (Wylie and Huggins, 1980). The image shows that the fibers consist of smaller fibrils, a characteristic typical of asbestiform minerals.

Chemical analyses in the SEM are usually acquired using an energy-dispersive X-ray detection system (commonly called *EDS*, for energy-dispersive spectrometry), which measures the numbers of X-ray photons as a function of energy. Since a given element produces X-rays with specific energies, an element present in the sample will produce peaks in the X-ray spectrum. The energies of these peaks allow identification of the elements present in the sample, i.e., a qualititative analysis. When analyzing thick, flat samples in an SEM, it is possible to use very well-constrained correction procedures to obtain quantitative analyses, the errors for which vary with the element being analyzed, the amount present in the sample, and the exact analytical conditions. Similarly, it is possible, though much more difficult, to obtain quantitative analyses of particles or fibers by accounting for their shapes (Armstrong, 1991). The present authors believe that if these procedures are not used, the analysis should not be considered quantitative, and it should not be reported as such. Most chemical analyses obtained from powders or fibers of 1:1 layer silicates and asbestos are qualitative, and variations other than extreme ones in peak intensities should not be assumed to simply reflect variations in sample composition; exact sample geometry and other factors also can have profound effects on intensities.

Electron microprobe analysis. Electron microprobes are rather sophisticated electron-beam instruments that are largely designed for quantitative chemical analysis. Early "probes" were dedicated instruments with extraordinarily poor imaging capabilities (typically only OM of the most disastrous sort). Today's probes are based on SEM columns and can have excellent scanning electron as well as optical microscopy capabilities. Many microprobes now have ancillary energy-dispersive X-ray detectors, but, most important, they have high-resolution, wavelength-dispersive spectrometers, which use a crystal as a Bragg's-Law filter to select out the energy of a specific X-ray emission line. The X-rays are then counted with very linear detectors that are capable of very high throughput. Counting statistics are thus greatly improved, and if the sample geometry is carefully controlled and attention is paid to correction procedures, excellent quantitative analyses can result (even for many minor elements, if one is willing to count X-rays for a long enough time).

Unfortunately, even the best electron microprobe will not make up for poor analytical procedures. Analyses from particles and fibers currently cannot achieve the accuracy or precision possible with thick, flat, polished samples. It is essential that any analyst setting out to use an electron microprobe for particle or fiber analysis first characterize carefully the analytical errors peculiar to this type of analysis (Armstrong, 1991), rather than believing accuracy claims appropriate to polished bulk samples.

Identification of 1:1 layer silicates and asbestos with the SEM. The primary information used for identification of fine-grained silicates with an SEM instrument are an image showing the morphology, plus a qualitative analysis indicating what major elements are present. The kaolinite-group minerals all will produce analyses showing Al and Si. In general, the images will show clay-sized particles. Kaolinite, dickite, and nacrite all occur in plates, which sometimes show good pseudohexagonal morphology, and hence are not differentiable; however, unless the occurrence is hydrothermal, kaolinite is the likely mineral. Halloysite occurs in many morphologies, including tubes and spheres, and hence will commonly look different from kaolinite. Their morphologies do, however, overlap. For a good discussion of kaolinite–halloysite, see Churchman (1990). Other Al-Si minerals (e.g., pyrophyllite) can also be fine-grained and platy. Analyses of the serpentine minerals show major Mg and Si, and commonly minor Fe, and, like the kaolin minerals, lizardite and antigorite cannot be distinguished by SEM alone.

Chrysotile asbestos also contains primarily Mg and Si, but then so do anthophyllite asbestos, sepiolite, and palygorskite. Whereas all of these can be differentiated with diffraction methods, morphology and qualitative analysis alone are not sufficient. The same can be said of the other forms of amphibole asbestos. There are simply too many silicate minerals containing (in addition to Si) the major elements Na, Mg, Ca, and Fe to make a rigorous identification as to mineral species. If, however, the problem is fairly specific, for example to examine a known asbestos sample and determine whether it is crocidolite or amosite, or to determine aspect ratios for particles of a known mineral, the SEM may be able to provide a useful answer. The ability to make a specific mineral identification also can be improved by using quantitative analysis procedures, rather than relying on qualitative analysis.

Transmission electron microscopy (TEM) methods for asbestos identification

Transmission electron microscopy and chemical analysis in the transmission microscope are described in detail in Hirsch et al. (1977), Buseck et al. (1988), Buseck (1992), and Joy et al. (1986). TEM methods have been used extensively to study the serpentine minerals and have provided much of the structural information we have about chrysotile and antigorite (see the section on serpentine structures for numerous references). Similarly, there have been some TEM investigations of kaolin-group minerals, but these are hampered by rapid electron beam damage during observation. In this section, we focus instead on the specific question of asbestos identification. Because it can provide the critical combination of imaging, diffraction, and compositional information, TEM is ideally suited for this task. However, the transmission electron microscope is a complex instrument, the understanding of which requires considerable theoretical background. This is especially so in the case of defective crystalline materials like asbestos, for which a thorough grounding in the subtleties of electron diffraction is essential. Incorrect identification can easily result from operator ignorance or incompetence.

Identification of asbestos with TEM is discussed in two excellent reviews by Champness et al. (1976) and by Chisholm (1983), and the reader is guided to those papers for details. In short, with TEM methods, one can (1) quickly recognize with imaging that a particle is a fiber, i.e., that it has a high aspect ratio such as 20:1 or greater; (2) determine with electron diffraction (SAED) the general crystal structure type to which the mineral belongs, or perhaps even determine the exact species; and (3) with an EDS analytical system obtain a qualitative or even quantitative chemical analysis, which should isolate a mineral within its structure group to the exact species. (The errors associated with quantitative analyses in the TEM are quite large, compared to those for an electron microprobe, but they can be assessed.) We will briefly discuss these types of information in turn for chrysotile and amphibole asbestos.

Chrysotile asbestos. Images of chrysotile asbestos commonly show individual fibrils with their typical diameters of about 250 Å. Images of such fibrils may even show the central pore, though caution is advised, since images of fibers without pores taken at large defocus values may also show apparent pores due to displacement of Fresnel fringes to the fiber axis (see Chisholm, 1983, for a discussion of pore imaging). Electron diffraction patterns of chrysotile are unlikely to be confused with those from other minerals, due to the well-ordered cylindrical or spiral layer lattice of this form of asbestos. Specifically, many of the diffraction spots for chrysotile are in the form of comets (i.e., a well-defined head with a diffuse tail), and the arrangement of these comets and the sharp reflections should serve to identify chrysotile, even without a chemical analysis. Diffraction from chrysotile is discussed by Whittaker (1956a,b,c; 1957), Zussman et al. (1957), Whittaker and Zussman (1971), Wicks and O'Hanley (1988), and Chisholm (1983). An EDS analysis obtained in the TEM should show Mg and Si as major components and commonly will also show Fe, Al, and other less abundant constituents. If the analysis is quantitative, Mg:Si should be close to 3:2.

Amphibole asbestos. Obviously, TEM images of amphibole asbestos fibers show elongate particles. Amphibole asbestos commonly appears to be lathlike, compared to chrysotile. If imaging conditions are appropriate, defects such as twinning and chain-width errors may be present, though they are not of much interest if the goal is simply identification. However, they may be of great interest if the goal is to characterize those aspects of a sample that may affect its properties (e.g., biological activity).

Electron diffraction patterns are essential for identification of amphibole asbestos; it is not too strong a statement to say that without appropriate SAED patterns, asbestos cannot be identified rigorously with the TEM. Amphibole asbestos fibers almost invariably are deposited parallel to the surface of the TEM support film, which means that the **c**-axis is normal to the beam. This also means that the reciprocal axis **c*** will be within about 15° of proper recording orientation, and it is a simple matter for the microscopist to bring one of many possible diffraction patterns containing **c*** into orientation. Such patterns will show prominent layer lines with the ~5.2 Å value of d_{001} characteristic of amphiboles. Unfortunately, such a value is also characteristic of *all* pyriboles, including pyroxenes, *all* layer silicates, and a lot of other silicate minerals. To identify a fiber as amphibole based on this layer-line spacing alone is naive. It is necessary at least to index the other primary direction in the electron diffraction pattern (i.e., to do a full two-dimensional indexing of the pattern), and even then ambiguities will exist between amphibole and other pyriboles for some orientations.

Furthermore, even two-dimensional indexing of an electron diffraction pattern may not preclude error. Any electron microscopist with much experience will know that it is not uncommon to misidentify the orientation of a SAED pattern from a known material or to misidentify the material based on a single SAED pattern, especially for relatively low-symmetry materials. This is the nature of the game and results from the instrumental errors inherent to the formation of SAED patterns in electron microscopes and from the very large number of crystalline compounds that are possible candidates for identification. Chances for an incorrect identification are greatly reduced if two, or even more, electron diffraction patterns are acquired and indexed.

Amphibole asbestos SAED patterns commonly contain diffuse streaks and/or doubled spots that result from defects and can complicate indexing. Streaks parallel to b^* typically result from chain-width errors (most commonly triple chains) intergrown with normal amphibole structure on (010). Streaks parallel to a^* and twinned diffraction patterns are characteristic of twinning on (100), which is more common in amphibole asbestos than in other amphibole habits. In fact, these fine structures in SAED patterns may help, rather than hinder, the microscopist in identifying the material. For example, Hutchison and Whittaker (1979) showed that amosite and crocidolite can be differentiated by observing the exact positions of the spots in SAED patterns from twinned fibrils. Chisholm (1983) provides detailed discussion of these and other subtleties in electron diffraction patterns from amphibole asbestos.

Even if SAED patterns serve to identify a fiber as amphibole, they usually will not permit identification as to species. For this, a qualitative EDS chemical analysis is generally sufficient to differentiate among the five types of amphibole asbestos. Note, however, that amphibole fibers of other compositions can and do occur. A quantitative analysis will provide enough information to confidently give the amphibole a species name and to place the amphibole within the compositional space characteristic of the name.

Transmission electron microscopy (imaging, diffraction, chemical analysis) is currently the best method for rigorously identifying individual asbestos particles in a dispersed sample (i.e., a sample that is not bulk material directly from a mine). However, for a truly ironclad identification of amphibole asbestos with the TEM, one must have an image showing the particle, certainly one but preferably two or more well-indexed SAED patterns, and a chemical analysis. Although EDS systems were not yet common on TEM instruments, Chisholm's (1983) comments still put this in a useful perspective: "...the measurements involved are time-consuming, and the cost high in terms of both capital equipment and skilled manpower. [These methods are not] practical for large-scale monitoring of air- or water-borne particles, but electron microscopy is essential if such particles are to be studied in any detail. A compromise may perhaps be necessary in which a less rigorous standard of identification is accepted in exchange for speed and cheapness." The exact nature of this compromise is still an unresolved issue.

Analysis of asbestos in bulk samples

Determination of the amount of an asbestiform component in a bulk sample is particularly difficult when the same mineral in a nonasbestiform habit is also present, because under these circumstances, bulk analytical techniques such as XRD cannot be used, and some type of microscopy must be employed. The regulation of asbestos in commerce frequently specifies the amount of asbestos as measured by weight percent. For example, the Environmental Protection Agency requires that building materials cannot contain more that 1 wt % asbestos, and the Hazardous Communication Act put forth by the Occupational Safety and Health Administration requires any product that contains more than 0.1 wt % asbestos to be labeled as containing asbestos. The variation in the lengths and widths of asbestos fibers determines how the mass of asbestos is distributed as a function of particle size, which in turn will determine how an analysis for asbestos should be performed. Wylie (1993) has shown that the lengths of asbestos fibers follow a distribution of the form

$$\log N_l = D \log l + C \qquad (2)$$

where N_l is the number of fibers with lengths greater than l, D is the fractal dimension of the length, and C is a constant. Siegrist and Wylie (1980) showed that the width (w) of asbestos fibers can be modeled as a function of length by the following:

$$\log w = F \log l + C \qquad (3)$$

where F is a "fibrosity index" and C is a constant. Assuming a cross-sectional shape (cylindrical for chrysotile, rectangular for amphibole: Wylie et al., 1982), it is possible to calculate the relationship

$$\log N_m = D \log m + C \qquad (4)$$

where N_m is the number of fibers with mass greater than m, D is the fractal dimension of the mass distribution, and C is a constant. By using these relationships, Wylie (1993) has shown that the mass distribution for chrysotile from the Calidria deposit in California and from a long-fiber sample from the Jeffrey Mine, Quebec, are distinctly different, as are the distributions of mass for crocidolite, amosite, and tremolite asbestos. These data are shown in Figure 28. For the Calidria chrysotile and crocidolite, the smaller fibers are so numerous that they contain a substantial portion of the mass. The small fibers are less abundant, however, in the chrysotile from Jeffrey Mine, in tremolite asbestos, and in amosite, so that the mass of the population is concentrated in the less-abundant larger fibers and fiber bundles. In terms of their mass distribution, the Jeffrey chrysotile, amosite, and tremolite asbestos are similar to most crushed rock, i.e., the larger the particles the greater the proportion of the mass they contain (Turcotte, 1986).

These differences in the mass distributions as a function of fiber dimensions for the asbestos minerals means that an analysis of a sample performed by TEM may provide very different results from those of an analysis by SEM or OM, which overlook the smallest fibers. The error will depend largely on the distribution of mass as a function of particle size for the matrix minerals.

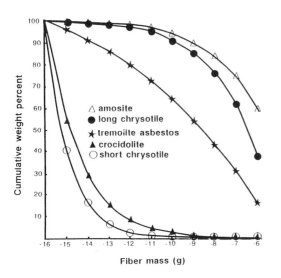

Figure 28. Calculated wt % asbestos in samples of crocidolite, amosite (grunerite asbestos), Calidria chrysotile, and tremolite asbestos as a function of individual fiber mass. The distributions assume that the smallest fiber in the population has a mass of 10^{-16} g and that the largest fiber bundle has a mass of 10^{-6} g.

Analysis of asbestos in air samples and lung burden

In air samples and lung burden studies, the analyst is faced with very small individual particles. X-ray diffraction techniques using silver membrane filters have been employed successfully in analyzing extremely small quantities of material from both of these sources. Most commonly employed, however, is the scanning or transmission electron microscope. In most published SEM studies, mineral identification is based on qualitative chemical composition as determined by EDS and morphology. In TEM studies, asbestos is identified by EDS qualitative chemical analysis, morphology, and recognition of an electron diffraction pattern. For chrysotile, the tubular morphology is often quite apparent in TEM images, and the electron diffraction pattern is quite characteristic. For the amphiboles, however, the major criterion generally employed is the presence of layer lines separated by an interval of approximately 5.3 Å. When the analyst is dealing with lung burdens from those occupationally exposed to amphibole asbestos or with air samples from an environment where asbestos is known to occur, these criteria may be indicative of amphibole asbestos, and the analyst can have some confidence that the identification is correct. However, any pyribole (e.g., pyroxene, jimthompsonite) can produce a layer-line spacing of 5.3 Å, and all major pyribole groups contain the same major metallic elements, such as Si, Al, Mg, Fe, Ca, and Na. Therefore, layer-line spacing and qualitative chemical analysis are not specific to amphibole, and significant error may result in identification by assuming that they are. As discussed above, the combination of quantitative chemical analysis and full indexing of several SAED patterns from the same particle should allow accurate identification.

ACKNOWLEDGMENTS

We thank George Guthrie and Paul Ribbe for editing most of the tasteless jokes out of this manuscript. During the production of this review, DRV was supported by NSF grant EAR8903630 and DOE grant DE-FG02-89ER14074; he thanks his diminutive helpers, Bonnie, Krista, and Annie, for their assistance in the final stages of manuscript preparation. AGW appreciates support of the Department of Geology, University of Maryland. Crystal structure projections presented in Figures 1, 2, 3, 5, 6, 7, 19, 21, 22, 24, and 26 were prepared with the computer program ATOMS for the Macintosh (v. 1.2, © 1992, Shape Software, 521 Hidden Valley Rd., Kingsport, TN 37663) and embellished with Canvas 3.0.

REFERENCES

Ahn, J.H., and Buseck, P.R. (1991) Microstructures and fiber formation mechanisms of crocidolite asbestos. Am. Mineral., 76, 1467–1478.
Alario Franco, M., Hutchison, J.L., Jefferson, D.A., and Thomas, J.M. (1977) Structural imperfection and morphology of crocidolite (blue asbestos). Nature, 266, 520–521.
Arima, M., Fleet, M.E., and Barnett, R.L. (1985) Titanian berthierine: a Ti-rich serpentine group mineral from Picton ultramafic dyke, Ontario. Can. Mineral., 23, 213–220.
Armstrong, J. T. (1991) Quantitative elemental analysis of individual microparticles with electron beam instruments. In K. F. J. Heinrich and D. E. Newbury, Eds., Electron Probe Quantitation, pp. 261–316. Plenum Press, New York.
Bailey, S.W. (1963) Polymorphism of the kaolin minerals. Am. Mineral., 49, 1196–1209.
Bailey, S.W. (1969) Polytypism of trioctahedral 1:1 layer silicates. Clays & Clay Minerals, 17, 355–371.
Bailey, S.W., Ed. (1984) Micas. Rev. Mineral., v. 13. Mineralogical Society of America, Washington, D.C.
Bailey, S.W. (1988a) Introduction. In S. W. Bailey, Ed., Hydrous Phyllosilicates, Rev. Mineral., 19, 1–8.
Bailey, S.W. (1988b) Polytypism of 1:1 layer silicates. In S. W. Bailey, Ed., Hydrous Phyllosilicates, Rev. Mineral., 19, 9–27.
Bailey, S.W. (1988c) Structures and compositions of other trioctahedral 1:1 phyllosilicates. In S. W. Bailey, Ed., Hydrous Phyllosilicates, Rev. Mineral., 19, 169–188.
Bailey, S.W. (1988d) Odinite, a new dioctahedral-trioctahedral Fe^{3+}-rich 1:1 clay mineral. Clay Minerals, 23, 237–247.
Bailey, S.W. (1988e) X-ray diffraction identification of the polytypes of mica, serpentine, and chlorite. Clays & Clay Minerals, 36, 193–213.
Banfield, J.F., Karabinos, P., and Veblen, D.R. (1989) Transmission electron microscopy of chloritoid: Intergrowth with sheet silicates and reactions in metapelites. Am. Mineral., 74, 549–564.
Baronnet, A. (1992a) Polytypism and stacking disorder. In P. R. Buseck, Ed., Minerals and Reactions at the Atomic Scale: Transmission Electron Microscopy, Rev. Mineral., 27, 231–288.
Baronnet, A. (1992b) Polygonized serpentine as the first mineral with fivefold symmetry (abstr.). Abstracts, v. 3, p. 682, 29th International Geological Congress, Kyoto, Japan.
Bates, T.F., Hildebrand, F.A., and Swineford, A. (1950) Morphology and structure of endellite and halloysite. Am. Mineral., 35, 463–484.
Bish, D.L., and Von Dreele, R.B. (1989) Rietveld refinement of non-hydrogen atomic positions in kaolinite. Clays & Clay Minerals, 37, 289–296.
Bookin, A.S., Drits, V.A., Plançon, A., and Tchoubar, C. (1989) Stacking faults in kaolin-group minerals in the light of real structural features. Clays & Clay Minerals, 37, 297–307.
Brindley, G.W. (1980) The structure and chemistry of hydrous nickel-containing silicates and nickel aluminum hydroxy minerals. Bull. Minéral., 103, 161–169.
Brindley, G.W., and Hang, P.T. (1973) The nature of garnierites—I. Structures, chemical compositions and color characteristics. Clays & Clay Minerals, 21, 25–37.
Buseck, P.R., Ed. (1992) Minerals and Reactions at the Atomic Scale: Transmission Electron Microscopy. Rev. Mineral., v. 27. Mineralogical Society of America, Washington, D. C.
Buseck, P.R., Cowley, J.M., and Eyring, L., Eds. (1988) High-Resolution Transmission Electron Microscopy. Oxford University Press, New York.
Cameron, M., and Papike, J.J. (1979) Amphibole crystal chemistry: A review. Fortschr. Mineral., 57, 28–67.
Campbell, W.J., Huggins, C.W., and Wylie, A.G. (1980) Chemical and physical characterization of amosite, chrysotile, crocidolite, and nonfibrous tremolite for oral ingestion studies by the National Institute of Environmental Health Sciences. U.S. Bureau of Mines Report of Investigations No. 8452.
Carr, D.D., and Herz, N., Eds. (1989) Concise Encyclopedia of Mineral Resources. MIT Press, Cambridge, Massachusetts.
Champness, P.E., Lorimer, G.W., and Cliff, G. (1976) The identification of asbestos. J. Microscopy, 108, 231–249.

Chisholm, J.E. (1983) Transmission electron microscopy of asbestos. In S.S. Chissick and R. Derricott, Eds., Asbestos, p. 85–167. John Wiley & Sons, New York.
Chisholm, J.E. (1992) The number of sectors in polygonal serpentine. Can. Mineral. 30, 355–365.
Choi, U. and Smith, R.W. (1972) Kinetic study of dissolution of asbestos fibers in water. J. Coll. Interf. Sci. 40, 253–261.
Chowdhury, A. (1975) Kinetics of leaching of asbestos minerals at body temperature. J. Appl. Chem. Biotechnol. 25, 347–353.
Churchman, G.J. (1990) Relevance of different intercalation tests for distinguishing halloysite from kaolinite in soils. Clays & Clay Minerals 38, 591–599.
Churchman, G.J. and Carr, R.M. (1975) The definition and nomenclature of halloysites. Clays & Clay Minerals 23, 382–388.
Churchman, G.J., Whitton, J.S., Claridge, G.C.C. and Theng, B.K.G. (1984) Intercalation method using formamide for differentiating halloysite from kaolinite. Clays & Clay Minerals 32, 241–248.
Cook, P.M., Palekar, L.D. and Coffin, D.L. (1982) Interpretation of the carcinogenicity of amosite asbestos and ferroactinolite on the basis of retained fiber dose and characteristics in vivo. Toxicol. Lett. 13, 151–158.
Costanzo, P.M., Clemency, C.V. and Giese, R.F. (1980) Low-temperature synthesis of a 10-Å hydrate of kaolinite using dimethylsulfoxide and ammonium fluoride. Clays & Clay Minerals 28, 155–166.
Crawford, D. (1980) Electron microscopy applied to studies of the biological significance of defects in crocidolite asbestos. J. Microscopy 120, 181–192.
Crawford, D. and Miller, K. (1981) Structure and structural changes in crocidolite asbestos associated with biological systems. Micron, 12, 25–28.
Cressey, B.A. (1979) Electron microscope studies of serpentine textures. Can. Mineral. 17, 741–756.
Cressey, B.A., Whittaker, E.J.W. and Hutchison, J.L. (1982) Morphology and alteration of asbestiform grunerite and anthophyllite. Mineral. Mag. 46, 77–87.
Cressey, B.A. and Zussman, J. (1976) Electron microscopic studies of serpentinites. Can. Mineral. 14, 307–313.
Czank, M. and Liebau, F. (1980) Periodicity faults in chain silicates: A new type of planar lattice fault observed with high resolution electron microscopy. Phys. Chem. Minerals 6, 85–93.
Diamond, S. and Bloor, J.W. (1970) Globular cluster microstructure of endellite (hydrated halloysite) from Bedford, Indiana. Can. Mineral. 18, 309–312.
Dixon, J.B. and McKee, T.R. (1974) Internal and external morphology of tubular and spheroidal halloysite particles. Clays & Clay Minerals 22, 127–137.
Dombrowski, T. and Murray, H.H. (1984) Thorium—a key element in differentiating Cretaceous and Tertiary kaolins in Georgia and South Carolina. Proc. 27th Int. Geol. Cong. 15, 305–317.
Dorling, M. and Zussman, J. (1987) Characteristics of asbestiform and non-asbestiform calcic amphiboles. Lithos, 20, 469–489.
Ernst, W.G. (1968) Amphiboles. Springer-Verlag, New York.
Faure, G. (1991) Principles and Applications of Inorganic Geochemistry. Macmillan, New York.
Faust, G.T. (1966) The hydrous nickel-magnesium silicates—the garnierite group. Amer. Mineral. 51, 279–298.
Faust, G.T., Fahey, J.J., Mason, B. and Dwornik, E.J. (1969) Pecoraite, $Ni_6Si_4O_{10}(OH)_8$, nickel analog of clinochrysotile, formed in the Wolf Creek meteorite. Science 165, 59–60.
Faust, G.T., Fahey, J.J., Mason, B. and Dwornik, E.J. (1973) The disintegration of the Wolf Creek meteorite and the formation of pecoraite, the nickel analog of clinochrysotile. U. S. Geol. Surv. Prof. Paper 384-C, 107–135.
Ferraris, G., Mellini, M. and Merlino, S. (1986) Polysomatism and the classification of minerals. Rend. Soc. Italiana Mineral. Petrol. 41, 181–192.
Finger, L.W. (1970) Refinement of the crystal structure of an anthophyllite. Carnegie Inst. Wash. Year Book, 68, 283–288.
Fleischer, M., Wilcox, R.E. and Matzko, J.J. (1984) Microscopic determination of the nonopaque minerals. U. S. Geol. Surv. Bull. 1627, 1–453.
Ford, W.E. and Dana, E.S. (1932) A Textbook of Mineralogy, 4th Ed. John Wiley & Sons, New York.
Fransolet, A.M. and Bourguignon, P. (1975) Donnees nouvelles sur la fraipontite de Moresnet (Belgique). Bull. Soc. Fr. Mineral. Crystallogr. 98, 135–244.
Gary, M., McAffee, R. and Wolf, C., Ed. (1974) Glossary of Geology. American Geological Institute, Washington, D. C.
Ghose, S. (1981) Subsolidus reactions and microstructures in amphiboles. In D. R. Veblen, Ed., Amphiboles and Other Hydrous Pyriboles—Mineralogy, Rev. Mineral. 9A, 325–372.
Gibbs, G.V. (1969) The crystal structure of protoamphibole. Mineral. Soc. Amer. Spec. Paper 2, 101–110.
Gibbs, G.W. and Hwang, C.Y. (1980) Dimensions of airborne asbestos fibres. In J. C. Wagner and W. Davis, Eds., Biologic Effects of Mineral Fibres, p. 69–78. IARC, Lyon, France.
Giese, R.F., Jr. (1988) Kaolin minerals: Structures and stabilities. In S.W. Bailey, Ed., Hydrous Phyllosilicates, Rev. Mineral. 19, 29–66.
Giese, R.F. and Datta, P. (1973) Hydroxyl orientations in kaolinite, dickite, and nacrite. Amer. Mineral. 58, 471–479.

Gilbert, M.C., Helz, R.T., Popp, R.K. and Spear, F.S. (1982) Experimental studies of amphibole stability. In D. R. Veblen and P. H. Ribbe, Eds., Amphiboles: Petrology and Experimental Phase Relations, Rev. Mineral. 9B, 229–353.
Goldstein, J.I., Newbury, D.E., Echlin, P., Joy, D.C., Fiori, C.E. and Lifshin, E. (1981) Scanning Electron Microscopy and Xray Microanalysis. Plenum Press, New York.
Gronow, J.R. (1987) The dissolution of asbestos fibres in water. Clay Minerals 22, 21–35.
Guggenheim, S. and Eggleton, R.A. (1987) Modulated 2:1 layer silicates: Review, systematics, and predictions. Amer. Mineral. 72, 724–738.
Guggenheim, S. and Eggleton, R.A. (1988) Crystal chemistry, classification, and identification of modulated layer silicates. In S. W. Bailey, Ed., Hydrous Phyllosilicates, Rev. Mineral. 19, 675–725.
Hall, S.H., Guggenheim, S., Moore, P. and Bailey, S.W. (1976) The structure of Unst-type 6-layer serpentines. Can. Mineral. 14, 314-321.
Hahn, Theo (1983) International Tables for Crystallography, Vol. A: Space-Group Symmetry. D. Reidel Publishing Co., Dordrecht, Holland, 854 p.
Hanson, R.F., Zamora, R. and Keller, W.D. (1981) Nacrite, dickite, and kaolinite in one deposit in Nayarit, Mexico. Clays & Clay Minerals 29, 452–453.
Hawthorne, F.C. (1978) The crystal chemistry of the amphiboles. VIII. The crystal structure and site chemistry of fluor-riebeckite. Can. Mineral. 16, 187–194.
Hawthorne, F.C. (1981a) Crystal chemistry of the amphiboles. In D.R. Veblen, Ed., Amphiboles and Other Hydrous Pyriboles—Mineralogy, Rev. Mineral. 9A, 1–102.
Hawthorne, F.C. (1981b) Amphibole spectroscopy. In D.R. Veblen, Ed., Amphiboles and Other Hydrous Pyriboles—Mineralogy, Rev. Mineral. 9A, 103–139.
Hawthorne, F.C. (1983) The crystal chemistry of the amphiboles. Can. Mineral. 21, 173–480.
Hawthorne, F.C., Ed. (1988) Spectroscopic Methods in Mineralogy and Geology. Rev. Mineral. 18, 698 p.
Health Effects Institute (1991) Asbestos in Public and Commercial Buildings: A Literature Review and Synthesis of Current Knowledge. Health Effects Institute, 141 Portland St., Cambridge, Massachusetts, USA.
Hirsch, P., Howie, A., Nicholson, R.B., Pashley, D.W. and Whelan, M.J. (1977) Electron Microscopy of Thin Crystals, 2nd ed. Krieger Publishing, Malabar, Florida, USA.
Hodgson, A.A. (1965) Fibrous Silicates. Royal Institute of Chemistry, London.
Hodgson, A.A. (1979) Chemistry and physics of asbestos. In L. Michaels and S.S. Chissick, Eds., Asbestos, p. 67–114. John Wiley & Sons, New York.
Honjo, F., Kitamura, N. and Mihama, K. (1954) A study of clay minerals by means of single-crystal electron diffraction diagrams—The structure of tubular kaolin. Clay Minerals Bull. 2, 133–141.
Hosterman, J.W. (1973) Clays. In D. Brobst and W. P. Pratt, Eds., United States Mineral Resources, U.S. Geol. Surv. Prof. Paper 820, 123–132.
Hume, L.A. (1991) The Dissolution Rate of Chrysotile. M.S. thesis, Virginia Polytechnic Institute and State University, Blacksburg, Virginia.
Hume, L.A. and Rimstidt, J.D. (1992) The biodurability of chrysotile asbestos. Amer. Mineral. 77, 1125–1128.
Hutchison, J.L., Irusteta, M.C. and Whittaker, E.J.W. (1975) High-resolution electron microscopy and diffraction studies of fibrous amphiboles. Acta Crystallogr. A31, 794–801.
Hutchison, J.L. and Whittaker, E.J.W. (1979) The nature of electron diffraction patterns of amphibole asbestos and their use in identification. Environ. Res. 20, 445–449.
International Centre for Diffraction Data (1993) Powder Diffraction File, Sets 1–42. ICDD, Newtown Square Corporate Campus, 12 Campus Blvd., Newtown Square, Pennsylvania.
Jaurand, M.C., Bignon, J., Sebastien, P. and Goni, J. (1977) Leaching of chrysotile asbestos in human lungs. Environ. Res. 14, 245–254.
Joswig, W. and Drits, V.A. (1986) The orientation of the hydroxyl groups in dickite by X-ray diffraction. N. Jb. Mineral. Mh. H1, 19–22.
Joy, D.C., Romig, A.D., Jr., Goldstein, J.I., Eds. (1986) Principles of Analytical Electron Microscopy. Plenum Press, New York.
Kirkman, J.H. (1981) Morphology and structure of halloysite in New Zealand tephras. Clays & Clay Minerals 29, 1–9.
Klein, C. and Hurlbut, C.S., Jr. (1993) Manual of Mineralogy (after J.D. Dana), 21st ed. John Wiley & Sons, New York.
Kohyama, N., Fukushima, K. and Fukami, A. (1978) Observation of the hydrated form of tubular halloysite by an electron microscope equipped with an environmental cell. Clays & Clay Minerals 26, 25–40.
Kunze, G. (1956) Die gewellte Struktur des Antigorits, I. Z. Kristallogr. 108, 82–107.
Kunze, G. (1958) Die gewellte Struktur des Antigorits, II. Z. Kristallogr. 110, 282–320.
Kunze, G. (1959) Fehlordnungen des Antigorits. Z. Kristallogr. 111, 190–212.
Kunze, G. (1961) Antigorit. Strukturtheoretische Grundlagen und ihre praktische Bedeutung für die weiters Serpentin-Forschung. Fortschr. Mineral. 39, 206–324.
Langer, A.M., Nolan, R.P., Addison, J. and Wagner, J.C. (1990) Critique of the "Health Effects of Tremolite," The official statement of the American Thoracic Society adopted by the ATS Board of Directors, June 1990, in OSHA Docket H-033d, December 1990.

Lasaga, A.C. (1980) The atomistic basis of kinetics: Defects in minerals. In A.C. Lasaga and R.J. Kirkpatrick, Eds., Kinetics of Geochemical Processes, Rev. Mineral. 8, 261–319.
Leake, B.E. (1978) Nomenclature of amphiboles. Amer. Mineral. 63, 1023–1053.
Liebau, F. (1980) Classification of silicates. In P.H. Ribbe, Ed., Orthosilicates, Rev. Mineral. 5, 1–24.
Liebau, F. (1985) Structural Chemistry of Silicates. Springer-Verlag, Berlin.
Light, W.G. and Wei, E.T. (1977) Surface charge and hemolytic activity of asbestos. Environ. Res. 13, 135–145.
Livi, K.J.T. and Veblen, D.R. (1987) "Eastonite" from Easton, Pennsylvania: A mixture of phlogopite and a new form of serpentine. Amer. Mineral. 72, 113–125.
Livi, K.J.T. and Veblen, D.R. (1989) Transmission electron microscopy of interfaces and defects in intergrown pyroxenes. Amer. Mineral. 74, 1070–1083.
Luce, R.W., Bartlett, R.W. and Parks, G.A. (1972) Dissolution kinetics of magnesium silicates. Geochim. Cosmochim. Acta, 36, 35–50.
Makovicky, E. and Hyde, B.G. (1981) Non-commensurate (misfit) layer structures. Struc. Bond. 46, 101–168.
Maksimovic, Z. (1973) The isomorphous series lizardite-nepouite. Zap. vses. mineral. Obshch. 102, 143–149. Quoted by Bailey (1988c).
Maksimovic, Z. and Bish, D.L. (1978) Brindleyite, a nickelrich aluminous serpentine mineral analogous to berthierine. Amer. Mineral. 63, 484–489.
Maresch, W.V. and Czank, M. (1983) Problems of compositional and structural uncertainty in synthetic hydroxyl-amphiboles; with an annotated atlas of the realbau. Per. Mineral.—Roma, 52, 463–542.
Maresch, W.V., Miehe, G., Czank, M., Fuess, H. and Schreyer, W. (1991) Triclinic amphibole. Eur. J. Mineral. 3, 899–903.
Mason, B. and Moore, C.B. (1982) Principles of Geochemistry. John Wiley & Sons, New York.
Mellini, M. (1982) The crystal structure of lizardite 1T: Hydrogen bonds and polytypism. Amer. Mineral. 67, 587–598.
Mellini, M. (1986) Chrysotile and polygonal serpentine from the Balangero serpentinite. Mineral. Mag. 50, 301–306.
Mellini, M., Ferraris, G. and Compagnone, R. (1985) Carlosturanite: HRTEM evidence of a polysomatic series including serpentine. Amer. Mineral. 70, 773–781.
Mellini, M., Trommsdorff, V. and Compagnoni, R. (1987) Antigorite polysomatism: Behavior during progressive metamorphism. Contrib. Mineral. Petrol. 97, 147–155.
Mellini, M. and Zanazzi, P.F. (1987) Crystal structures of lizardite-1T and lizardite-$2H_1$ from Coli, Italy. Amer. Mineral. 72, 943–948.
Mellini, M. and Zussman, J. (1986) Carlosturanite (not 'picrolite') from Taberg, Sweden. Mineral. Mag. 50, 675–679.
Merino, E., Harvey, C. and Murray, H.H. (1989) Aqueous chemical control of the tetrahedral-aluminum content of quartz, halloysite, and other low temperature silicates. Clays & Clay Minerals 37, 135–142.
Middleton, A.P. and Whittaker, E.J.W. (1976) The structure of Povlen-type chrysotile. Can. Mineral. 14, 301–306.
Morgan, A., Holmes, A. and Gold, C. (1971) Studies of the solubility of constituents of chrysotile asbestos in vivo using radioactive tracer techniques. Environ. Res. 4, 558–570.
Murray, H.H. (1988) Kaolin minerals: Their genesis and occurrences. In S.W. Bailey, Ed., Hydrous Phyllosilicates, Rev. Mineral. 19, 67–89.
Murray, H.H. and Jansen, J. (1984) Oxygen isotopes—indicators of kaolin genesis? Proc. 27th Int'l Geol. Cong, 15, 287–303.
Nesse, W.P. (1991) Introduction to Optical Mineralogy, 2nd ed. Oxford University Press, New York.
Nolan, R.P., Langer, A.M., Oechsle, G.W., Addison, J. and Colflesh, D.E. (1991) Association of tremolite habit with biological potential: Preliminary report. In R.C. Brown, Ed., Mechanisms in Fibre Carcinogenesis, p. 231–251. Plenum Press, New York.
Noro, H. (1986) Hexagonal platy halloysite in an altered tuff bed, Kamaki City, Aichi Prefecture, central Japan. Clay Minerals 21, 401–415.
Otten, M. (1993) High-resolution transmission electron microscopy of polysomatism and stacking defects in antigorite. Amer. Mineral. 78, 75–84.
Papike, J.J. (1988) Chemistry of the rock-forming silicates: Multiple-chain, sheet, and framework structures. Rev. Geophys. 26, 407–444.
Parry, W.T. (1985) Calculated solubility of chrysotile asbestos in physiological systems. Environ. Res. 37, 410–418.
Pauling, L. (1929) The principles determining the structure of complex ionic crystals. J. Amer. Chem. Soc. 51, 1010–1026.
Pauling, L. (1930) The structure of the chlorites. Proc. Nat. Acad. Sci. 16, 578–582.
Phillips, W.R. and Griffen, D.T. (1981) Optical Mineralogy: The Nonopaque Minerals. W.H. Freeman and Co., San Francisco.
Plançon, A., Giese, R.F., Jr., Snyder, R., Drits, V.A. and Bookin, A.S. (1989) Stacking faults in the kaolin-group minerals: Defect structures of kaolinite. Clays & Clay Minerals 37, 203–210.
Plançon, A. and Tschoubar, C. (1977a) Determination of structural defects in phyllosilicates by X-ray diffraction—I. Principle of calculation of the diffraction phenomenon. Clays & Clay Minerals 25, 430–435.

Plançon, A. and Tschoubar, C. (1977b) Determination of structural defects in phyllosilicates by X-ray diffraction—II. Nature and proportion of defects in natural kaolinites. Clays & Clay Minerals 25, 436–450.

Pooley, F.D. and Clark, N. (1979) Fiber dimensions and aspect ratio of crocidolite, chrysotile, and amosite particles detected in lung tissue specimens. Ann. New York Acad. Sci. 330, 711–716.

Pott, F., Zien, U., Reiffer, F., Huth, F., Ernst, H. and Mohn, U. (1987) Carcinogenicity studies on fibers, metal compounds and some other dusts in rats. Exper. Path. 32, 129–152.

Ralston, J. and Kitchener, J.A. (1975) The surface chemistry of amosite asbestos, an amphibole silicate. J. Colloid Interf. Sci. 50, 242–249.

Reynolds, R.C., Jr. (1988) Mixed layer chlorite minerals. In S.W. Bailey, Ed., Hydrous Phyllosilicates, Rev. Mineral. 19, 601–629.

Rimstidt, J.D. and Barnes, H.L. (1980) The kinetics of silica-water reactions. Geochim. Cosmochim. Acta, 44, 1683–1699.

Robinson, P., Spear, F.S., Schumacher, J.C., Laird, J., Klein, C., Evans, B.W. and Doolan, B.L. (1982) Phase relations of metamorphic amphiboles: Natural occurrence and theory. In D.R. Veblen and P.H. Ribbe, Eds., Amphiboles—Petrology and Experimental Phase Relations, Rev. Mineral. 9B, 1–227.

Ross, M. (1981) The geologic occurrences and health hazards of amphibole and serpentine asbestos. In D.R. Veblen, Ed., Amphiboles and Other Hydrous Pyriboles—Mineralogy, Rev. Mineral. 9A, 279–323.

Schiller, J.E., Payn, S.L. and Khalafalla, S.E. (1980) Surface charge heterogeneity in amphibole cleavage fragments and amphibole asbestos fibers. Science 209, 1530–1532.

Schumacher, J.C. and Czank, M. (1987) Mineralogy of triple- and double-chain pyriboles from Orijärvi, southwest Finland. Amer. Mineral. 72, 345–352.

Shau, Y.-H., Peacor, D.R., Ghose, S. and Phakey, P.P. (1993) Submicroscopic exsolution in Mn-bearing alkali amphiboles from Tirodi, Maharashtra, India. Amer. Mineral. 78, 96–106.

Shedd, K.B. (1985) Fiber dimensions of crocidolite from Western Australia, Bolivia, and the Cape and Transvaal provinces of South Africa. U.S. Bureau of Mines Report of Investigations No. 8998.

Siegrist, H.G. and Wylie, A.G. (1980) Characterizing and discriminating the shape of asbestos fibers. Environ. Res. 23, 248–361.

Skinner, H.C.W., Ross, M. and Frondel, C. (1988) Asbestos and Other Fibrous Materials. Oxford University Press, New York.

Smelik, E.A., Nyman, M.W. and Veblen, D.R. (1991) Exsolution within the calcic amphibole series: Natural evidence for a miscibility gap between actinolite and hornblende. Amer. Mineral. 76, 1184–1204.

Smelik, E.A. and Veblen, D.R. (1991) Exsolution of cummingtonite from glaucophane: A new orientation for exsolution lamellae in clinoamphibole. Amer. Mineral. 76, 971–984.

Smelik, E.A. and Veblen, D.R. (1992a) Exsolution of Ca-amphibole from glaucophane and the miscibility gap between sodic and calcic amphiboles. Contrib. Mineral. Petrol. 112, 178–195.

Smelik, E.A. and Veblen, D.R. (1992b) Exsolution of hornblende and the solubility limits of Ca in orthoamphibole. Science 257, 1669–1672.

Smelik, E.A. and Veblen, D.R. (1993a) A transmission and analytical electron microscope study of exsolution microstructures and mechanisms in the orthoamphiboles anthophyllite and gedrite. Amer. Mineral. 78, 511–532.

Smelik, E.A. and Veblen, D.R. (1993b) Complex exsolution in glaucophane from Tillotson Peak, north-central Vermont. Can. Mineral. 31, in press.

Souza Santos, P. de, Brindley, G.W. and Souza Santos, H. de (1965) Mineralogical studies of kaolinite-halloysite clays: Part III. A fibrous kaolin mineral from Piedade, Sao Paulo, Brazil. Amer. Mineral. 50, 619–628.

Spinnler, G.E. (1985) HRTEM study of antigorite, pyroxene–serpentine reactions, and chlorite. Ph.D. dissertation, Arizona State Univ., Tempe, Arizona.

Spinnler, G.E., Veblen, D.R. and Buseck, P.R. (1983) Microstructure and defects of antigorite. Proc. Elec. Micros. Soc. Amer. 41, 190–191.

Spurney, K.R., Stober, W., Opiela, H. and Weiss, G. (1980) On the problem of milling and ultrasonic treatment of asbestos and glass fibers in biological and analytical applications. J. Amer. Indust. Hygiene Assoc. 41, 198–203.

Stanton, R.L. (1972) Growth and growth structures in open space. In Ore Petrology, p. 199–226. McGraw-Hill Book Co., New York.

Thomassin, H.H., Goni, J., Baillif, P., Touray, J.C. and Jaurand, J.C. (1977) An XPS study of the dissolution kinetics of chrysotile in 0.1 N oxalic acid at different temperatures. Phys. Chem. Minerals 1, 385–398.

Thompson, J.B., Jr. (1978) Biopyriboles and polysomatic series. Amer. Mineral. 63, 239–249.

Thompson, J.B., Jr. (1981) An introduction to the mineralogy and petrology of the biopyriboles. In D.R. Veblen, Ed., Amphiboles and Other Hydrous Pyriboles—Mineralogy, Rev. Mineral. 9A, 141–188.

Timbrell, V. (1989) Review of the significance of fibre size in fibre-related lung disease: A centrifuge cell for preparing accurate microscope-evaluation specimens from slurries used in inoculation studies. Ann. Occup. Hygiene, 33, 483–505.

Tomura, S., Shibasak, Y., Mizuta, H. and Kitamura, M. (1985) Growth conditions and the genesis of spherical and platy kaolinite. Clays & Clay Minerals 33, 237–239.

Turcotte, D.L. (1986) Fractals and fragmentation. J. Geophys. Res. 91, 1921–1926.

Uehara, S. and Shirozu, H. (1985) Variations in chemical composition and structural properties of antigorites. Mineral. J. 12, 299–318.
Uyeda, N., Hang, P.T. and Brindley, G.W. (1973) The nature of garnierites II. Electron optical study. Clays & Clay Minerals 21, 41–50.
Veblen, D.R. (1980) Anthophyllite asbestos: Microstructures, intergrown sheet silicates, and mechanisms of fiber formation. Amer. Mineral. 65, 1075–1086.
Veblen, D.R. (1981) Non-classical pyriboles and polysomatic reactions in biopyriboles. In D.R. Veblen, Ed., Amphiboles and Other Hydrous Pyriboles—Mineralogy, Rev. Mineral. 9A, 189–236.
Veblen, D.R. (1991) Polysomatism and polysomatic series: A review and applications. Amer. Mineral. 76, 801–826.
Veblen, D.R. (1992) Electron microscopy applied to nonstoichiometry, polysomatism, and replacement reactions in minerals. In P.R. Buseck, Ed., Minerals and Reactions at the Atomic Scale: Transmission Electron Microscopy, Rev. Mineral. 27, 181–229.
Veblen, D.R. and Burnham, C.W. (1975) Triple-chain biopyriboles: Newly discovered intermediate products of the retrograde anthophyllite-talc transformation, Chester, VT (abstr.). Trans. Amer. Geophys. Union (EOS), 56, 1076.
Veblen, D.R. and Burnham, C.W. (1978a) New biopyriboles from Chester, Vermont: I. Descriptive mineralogy. Amer. Mineral. 63, 1000–1009.
Veblen, D.R. and Burnham, C.W. (1978b) New biopyriboles from Chester, Vermont: II. The crystal chemistry of jimthompsonite, clinojimthompsonite, and chesterite, and the amphibole-mica reaction. Amer. Mineral. 63, 1053–1073.
Veblen, D.R. and Buseck, P.R. (1979a) Serpentine minerals: Intergrowths and new combination structures. Science 206, 1398–1400.
Veblen, D.R. and Buseck, P.R. (1979b) Chain width order and disorder in biopyriboles. Amer. Mineral. 64, 687–700.
Veblen, D.R. and Buseck, P.R. (1980) Microstructures and reaction mechanisms in biopyriboles. Amer. Mineral. 65, 599–623.
Veblen, D.R., Buseck, P.R. and Burnham, C.W. (1977) Asbestiform chain silicates: New minerals and structural groups. Science, 198, 359–365.
Veblen, D.R. and Post, J.E. (1983) A TEM study of fibrous cuprite (chalcotrichite): Microstructures and growth mechanisms. Amer. Mineral. 68, 790–803.
von Laue, M. (1969) Historical introduction. In N.F.M. Henry and K. Lonsdale, Eds., International Tables for X-ray Crystallography, 3rd ed., v. 1, p. 1–5. Kynoch Press, Birmingham, England.
Walker, J.S. (1981) Asbestos and the asbestiform habit. M.S. thesis, Univ. of Minnesota, Minneapolis.
Walker, J.S. and Zoltai, T. (1979) A comparison of asbestos fibers with synthetic crystals known as "whiskers." Ann. New York Acad. Sci. 330, 687–704.
Whittaker, E.J.W. (1956a) The structure of chrysotile. II. Clinochrysotile. Acta Crystallogr. 9, 855–862.
Whittaker, E.J.W. (1956b) The structure of chrysotile. III. Orthochrysotile. Acta Crystallogr. 9, 862–864.
Whittaker, E.J.W. (1956c) The structure of chrysotile. IV. Parachrysotile. Acta Crystallogr. 9, 865–867.
Whittaker, E.J.W. (1957) The structure of chrysotile. V. Diffuse reflections and fiber texture. Acta Crystallogr. 10, 149–156.
Whittaker, E.J.W. (1979) Mineralogy, chemistry and crystallography of amphibole asbestos. In R.L. Ledoux, Ed., Mineralogical Techniques of Asbestos Determination, p. 1–34. Mineral. Assoc. Canada, Montreal.
Whittaker, E.J.W., Cressey, B.A. and Hutchison, J.L. (1981a) Edge dislocations in fibrous grunerite. Mineral. Mag. 44, 287–291.
Whittaker, E.J.W., Cressey, B.A. and Hutchison, J.L. (1981b) Terminations of multiple chain lamellae in grunerite asbestos. Mineral. Mag. 44, 27–35.
Whittaker, E.J.W. and Wicks, F.J. (1970) Chemical differences among the serpentine "polymorphs." Amer. Mineral. 55, 1025–1047.
Whittaker, E.J.W. and Zussman, J. (1956) The characterization of serpentine minerals by X-ray diffraction. Mineral. Mag. 31, 107–126.
Whittaker, E.J.W. and Zussman, J. (1971) The serpentine minerals. In J. A. Gard, Ed., Electron-Optical Investigation of Clays, Ch. 5. Mineral. Soc. Great Britain, London.
Wicks, F.J. (1979) Mineralogy, chemistry and crystallography of chrysotile asbestos. In R.L. Ledoux, Ed., Mineralogical Techniques of Asbestos Determination, p. 35–78. Mineral. Assoc. Canada, Montreal, Quebec.
Wicks, F.J. (1984a) Deformation histories as recorded by serpentinites. I. Deformation prior to serpentinization. Can. Mineral. 22, 185–195.
Wicks, F.J. (1984b) Deformation histories as recorded by serpentinites. II. Deformation during and after serpentinization. Can. Mineral. 22, 197–204.
Wicks, F.J. (1984c) Deformation histories as recorded by serpentinites. III. Fracture patterns developed prior to serpentinization. Can. Mineral. 22, 205–209.
Wicks, F.J. (1986) Lizardite and its parent enstatite: A study by X-ray diffraction and transmission electron microscopy. Can. Mineral. 24, 775–788.
Wicks, F.J., Kjoller, K. and Henderson, G.S. (1992) Imaging the hydroxyl surface of lizardite at atomic resolution with the atomic force microscope. Can. Mineral. 30, 83–91.

Wicks, F.J. and O'Hanley, D.S. (1988) Serpentine minerals: Structures and petrology. In S.W. Bailey, Ed., Hydrous Phyllosilicates, Rev. Mineral. 19, 91–167.
Wicks, F.J. and Plant, A.G. (1979) Electron-microprobe and X-ray-microbeam studies of serpentine textures. Can. Mineral. 17, 785–830.
Wicks, F.J. and Whittaker, E.J.W. (1975) A reappraisal of the structures of the serpentine minerals. Can. Mineral. 13, 227–243.
Wicks, F.J. and Whittaker, E.J.W. (1977) Serpentine textures and serpentinization. Can. Mineral. 15, 459–488.
Wicks, F.J., Whittaker, E.J.W. and Zussman, J. (1977) An idealized model for serpentine textures after olivine. Can. Mineral. 15, 446–458.
Wicks, F.J. and Zussman, J. (1975) Microbeam X-ray diffraction patterns of the serpentine minerals. Can. Mineral. 13, 244–258.
Wones, D.R. and Gilbert, M.C. (1982) Amphiboles in the igneous environment. In D. R. Veblen and P. H. Ribbe, Eds., Amphiboles: Petrology and Experimental Phase Relations, Rev. Mineral. 9B, 355–390.
Wylie, A.G. (1979) The optical properties of the fibrous amphiboles. Ann. New York Acad. Sci. 330, 611–619.
Wylie, A.G. (1993) Modeling asbestos populations: A fractal approach. Can. Mineral. 31, in press.
Wylie, A.G., Bailey, K.F., Kelse, J.W. and Lee, R.J. (1993) The importance of width in asbestos fiber carcinogenicity and it implications for public policy. J. Amer. Indust. Hygiene Assoc., in press.
Wylie, A.G. and Huggins, C.W. (1980) Characteristics of a potassian winchite-asbestos from the Alamore talc district, Texas. Can. Mineral. 18, 101–107.
Wylie, A.G., Shedd, K.B. and Taylor, M.E. (1982) Measurement of the thickness of amphibole asbestos fibers with the scanning electron microscope and the transmission electron microscope. In K.F. Heinrich, Ed., Microbeam Analysis, p. 181–187. San Francisco Press, San Francisco.
Yada, K. (1967) Study of chrysotile asbestos by a high resolution electron microscope. Acta Crystallogr. 23, 704–707.
Yada, K. (1971) Study of microstructure of chrysotile asbestos by high resolution electron microscopy. Acta Crystallogr. A27, 659–664.
Yada, K. (1979) Microstructures of chrysotile and antigorite by high-resolution electron microscopy. Can. Mineral. 17, 679–691.
Yada, K. and Iishi, K. (1977) Growth and microstructure of synthetic chrysotile. Amer. Mineral. 62, 958–965.
Yada, K. and Wei, L. (1987) Polygonal microstructures of Povlen chrysotile observed by high resolution electron microscopy. (Abstr.). Euroclay 87, Sevilla, Spain,
Young, R.A. and Hewat, A.W. (1988) Verification of the triclinic crystal structure of kaolinite. Clays & Clay Minerals 36, 225–232.
Zoltai, T. (1981) Amphibole asbestos mineralogy. In D.R. Veblen, Ed., Amphiboles and Other Hydrous Pyriboles—Mineralogy, Rev. Mineral. 9A, 237–278.
Zussman, J., Brindley, G.W. and Comer, J.J. (1957) Electron diffraction studies of serpentine minerals. Amer. Mineral. 42, 133–153.
Zvyagin, B.B. (1967) Electron-Diffraction Analysis of Clay Mineral Structures. Plenum Press, New York.

CHAPTER 4

MINERALOGY OF CLAY AND ZEOLITE DUSTS (EXCLUSIVE OF 1:1 LAYER SILICATES)

David L. Bish and George D. Guthrie, Jr.

Geology and Geochemistry Group, EES–1
Los Alamos National Laboratory
Los Alamos, New Mexico 87545 U.S.A.

INTRODUCTION

Clays and zeolites are among the most important of natural dusts by virtue of their occurrence throughout the world on the earth's surface and their important industrial uses. Mankind thus has ample opportunity for exposure to these materials, and at least some of these have been shown to be biologically active. Consequently, this paper will present information on the properties of several common clays and zeolites. A variety of hydroxide minerals, including gibbsite, boehmite, diaspore, brucite, goethite, and lepidocrocite, are common minerals in soil environments and in altered rock masses, and these minerals are included in this chapter. Many of the so-called 2:1 layer silicates (talc, pyrophyllite, smectite, vermiculite, illite, biotite, and muscovite) are also included, as are the chain-structure clays, sepiolite and palygorskite. Finally, because of their industrial importance, their occurrence near the surface in many areas, and their use as building stone in many parts of the world, the natural zeolites erionite, mordenite, clinoptilolite, phillipsite, and chabazite are also discussed. The 1:1 layer silicates, including the serpentine and kaolin minerals, will not be addressed in this chapter as they are discussed in Chapter 3.

The chemistry, structures, and properties of all of these minerals are discussed in greater detail in numerous publications. For example, the Mineralogical Society of America recently published a 584-page *Reviews in Mineralogy* volume (Bailey, 1984b) discussing the micas (a subset of 2:1 layer silicates). In contrast, this chapter is not meant to be an exhaustive summary of the available information on these minerals. Instead, this chapter is meant to provide the reader with basic information on a variety of important aspects of each mineral, including crystal structure diagrams of each mineral and references to more detailed discussions. The following is a partial annotated list of references for any reader interested in a more in-depth presentation of topics in this chapter:

I. Clays

1. Bailey (1984b)—A *Reviews in Mineralogy* volume that presents a detailed treatment of the structures, chemistries, properties, and occurrences of many of the 2:1 layer silicates, including the micas (e.g., muscovite, biotite, phlogopite) and the clay mineral illite.

2. Bailey (1988)—A *Reviews in Mineralogy* volume that presents a detailed treatment of the structures, chemistries, properties, and occurrences of many layer silicates, including the 1:1 layer silicates (e.g., the kaolin group minerals, the serpentine group minerals), talc, palygorskite (attapulgite), sepiolite, smectite, chlorite, and vermiculite.
3. Brindley and Brown (1980)—A detailed treatment of the structures and chemistries of many clay minerals, including the 1:1 layer silicates (discussed in Chapter 3 of this book), 2:1 layer silicates, chain-layer silicates (also referred to as modulated layer silicates), oxides, and hydroxides.
4. Newman (1987)—Similar to Brindley and Brown (1980).
5. Brown (1961)—The use of X-ray methods to identify and to characterize clay minerals
6. Moore and Reynolds (1989)—A detailed discussion of the preparation, characterization, and analysis of clay minerals. Includes a discussion of methods for the purification of clay-bearing material and particle size separation.
7. Cairns-Smith and Hartman (1986)—Discusses clay minerals as possible templates for the origin of life. Although this process may appear unrelated to (or, better, the antithesis of) mineral-induced pathogenesis, it entails the interactions between clay minerals and biologically important organic molecules; hence, it may have a direct bearing on clay-mineral-induced pathogenesis.
8. Grim (1968)—An introduction to many aspects of clay minerals.
9. *Clays and Clay Minerals* and *Clay Minerals*—Two scientific journals that publish research papers dealing with clay minerals.

II. Zeolites

1. Mumpton (1977)—A *Reviews in Mineralogy* volume that presents the structures, occurrences, and uses of the natural zeolites.
2. Barrer (1978)—A detailed presentation of many of the important properties of zeolites, including their sorptive, sieving, and catalytic capabilities.
3. Gottardi and Galli (1985)—A detailed treatment of structural and compositional aspects of natural zeolites. Contains more detail than Mumpton (1981). Also discusses the synthesis conditions for many zeolites.
4. Kalló and Sherry (1988)—A collection of numerous papers on properties, uses, occurrence, applications, and synthesis of zeolites.
5. *Zeolites*—A scientific journal that publishes research papers on zeolites.

a) b)

Figure 1. Polyhedral representations of octahedrally (a) and tetrahedrally (b) coordinated cations. Two projections of each are shown.

The hydroxide and silicate minerals discussed here have structures containing both octahedrally and tetrahedrally coordinated cations. An octahedrally coordinated cation has six nearest-neighbor anions (typically O^{2-}, OH^-, F^-, or neutral H_2O) and derives its name from the shape of the polyhedron formed by the coordinating anions (i.e., an octahedron with anions at the six apices) (Fig. 1a). Cations that typically occur in octahedrally coordinated sites include Al^{3+}, Mg^{2+}, Fe^{2+}, Fe^{3+}, and other cations whose ionic radii are similar to these. Tetrahedrally-coordinated cations have four nearest-neighbor anions, usually oxygen, which ideally form a polyhedron in the shape of a tetrahedron (Fig. 1b). The tetrahedral cations in the minerals discussed here are almost exclusively Al^{3+} and Si^{4+}.

In general, the hydroxide minerals are made up of octahedral units linked together in specific ways, often in the form of sheets, and the 2:1 layer silicates are made up of sheets of both octahedral and tetrahedral cations. Octahedra are linked together, via shared edges, to form continuous sheets. If all possible octahedra in a sheet are occupied by cations, the sheet is termed trioctahedral (Fig. 2a), and if only two out of three octahedra are occupied, the sheet is termed dioctahedral (Fig. 2b). A so-called 1:1 layer (structures discussed in Chapter 3) derives its name from the fact that each layer is composed of one tetrahedral sheet bonded to one octahedral sheet (Fig. 2c). A so-called 2:1 layer derives its name from the fact that each layer is composed of two tetrahedral sheets bonded to one octahedral sheet (Fig. 2d). The tetrahedral-octahedral-tetrahedral (t-o-t) units are not always continuous, as they are in the 2:1 layer silicates, and they may form chain structures if the t-o-t units are linked together in an alternate manner, such as is found in the chain-structure minerals sepiolite and palygorskite. Zeolites do not have layer structures but instead are comprised of tetrahedra linked to form a three-dimensional framework structure. Zeolites have no specific octahedrally-coordinated sites.

Because all of these minerals can accommodate a wide range of cations and anions, it is common for chemical substitutions to cause polyhedra to be locally electrically unbalanced. For example, the substitution of Al^{3+} for Si^{4+} in a tetrahedral site gives rise to a net negative charge on a particular structural unit. Substitution of Al^{3+} for Mg^{2+} in an octahedral site gives rise to a net positive charge. Such substitutions are common in the 2:1 layer silicates, and they can be balanced by the incorporation of cations between the individual 2:1 layers. If no simple structural mechanism exists for balancing such substitutions, i.e., there is no

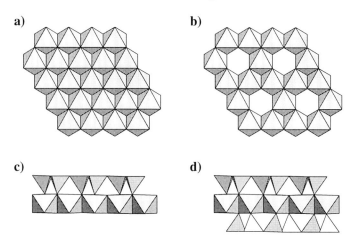

Figure 2. Polyhedral representations of a trioctahedral (a) and dioctahedral (b) sheet viewed perpendicular to the sheets and a 1:1 (c) and 2:1 (d) layer viewed parallel to the layers.

structural site for a balancing ion, they typically do not occur on a large scale. Substitution of Al^{3+} for Si^{4+} in zeolites likewise gives rise to a net negative charge on the tetrahedral framework. This negative charge is balanced by the incorporation of cations in the interstices or structural cavities in the frameworks or by the formation of a hydroxyl group.

HYDROXIDES

Geological occurrence

The hydroxide minerals occur in a variety of geological environments, and many of the minerals discussed here are important in soils. Brucite most commonly occurs as an alteration product of periclase (MgO) in metamorphosed dolomites and limestones, but it also occurs as a low-temperature hydrothermal vein mineral in serpentinites, in chloritic schists, and in boiler scale (Deer et al., 1962, p. 92). It is not a common soil component and is, in general, fairly rare. Important associated contaminants include the serpentine minerals and low-temperature alteration products such as hydromagnesite and calcite.

The most important occurrences of gibbsite, diaspore, and boehmite are in bauxites and laterites, often as major sources of Al. Both deposits result from severe weathering and leaching of Al-silicate rocks (e.g., granites). Although gibbsite is the most common Al-hydroxide, all three can also be important constituents of tropical soils, along with several other Fe-oxide and hydroxide minerals and kaolin minerals. Common contaminants associated with these three Al-hydroxides include goethite, hematite, and kaolinite; associated silica minerals are rare due to the occurrence of these Al-hydroxides in highly leached soils.

Goethite is the most common Fe-oxide soil mineral and occurs in virtually every type of soil. It is the stable soil iron oxide mineral in humid climates, and it and lepidocrocite are very common in soils formed from iron-bearing rocks weathered under oxidizing conditions. They also form in bogs, springs, and lakes as direct precipitates. In many parts of the world they make up the primary Fe-bearing mineral assemblage in economic deposits (e.g., in bog iron ores), in association with hematite, Mn-minerals, calcite, and quartz. Lepidocrocite is less common than goethite and hematite, and it occurs in reducing, anaerobic soils with associated Fe^{2+} ions (Newman, 1987).

Crystal chemistry

The compositions (Table 1) and structures of these minerals are comparatively simple. Brucite [$Mg(OH)_2$] accepts small amounts of other divalent cations similar in size to Mg^{2+}, such as Ni, Fe, Zn, and Mn. An Fe-rich variety, ferrobrucite, has been identified, but admixture with magnetite may be the source of Fe. Gibbsite [$Al(OH)_3$] is typically close to end-member composition, with only very minor substitutions by Fe, Ti, and Mg. Likewise, diaspore [α-AlOOH] and boehmite [γ-AlOOH] show little chemical substitution other than small amounts of Fe^{3+} (up to mol 5%) and Mn. Goethite [α-FeOOH], which is isostructural with diaspore, shows considerable isomorphous substitution in nature, particularly in

Table 1

Classification of common 2:1 layer silicates
based on layer charge per $O_{10}(OH)_2$

Charge	Mineral group	Sub-groups	Species
~0	talc-pyrophyllite	dioctahedral	pyrophyllite
		trioctahedral	talc
0.2–0.6	smectite	dioctahedral	montmorillonite, beidellite, nontronite
		trioctahedral	saponite, hectorite
0.6–0.9	vermiculite	dioctahedral	no specific species
		trioctahedral	no specific species
~1.0	mica	dioctahedral	muscovite, paragonite
		trioctahedral	phlogopite, biotite, lepidolite
~2.0	brittle mica	dioctahedral	margarite
		trioctahedral	clintonite, anandite

soils. The Al-for-Fe substitution is important in soil goethites (e.g., Schulze, 1984), and up to ~33 mole % Al can substitute for Fe in goethite. The substitution of numerous other cations, including Mn, Cr, Cd, Co, Cu, Ni, Pb, Zn, Ti, and V, also has been documented in numerous publications (e.g., Schwertmann et al., 1989), and soil and lateritic goethites can contain several percent of Al_2O_3, Cr_2O_3, ZnO, and NiO. Lepidocrocite [γ-FeOOH] is isostructural with boehmite, and the presence of impurities in the soil system appears to inhibit the crystallization of lepidocrocite. Although there are little or no detailed chemical data on lepidocrocite due to its common occurrence in mixtures of very fine-grained soil minerals, its formation is inhibited by the presence of Al, Si, or Cu^{2+}, and its formation is encouraged by the presence of Co^{2+} or Mn^{2+} (Newman, 1987). The chemistry of these minerals has been further described by Taylor (1987).

Crystal structures

In spite of the similarity in compositions of these hydroxide minerals, their structures are significantly different. Brucite has a simple layer structure (Zigan and Rothbauer, 1967), often used as an example of the building block for the octahedral sheet of trioctahedral layer silicates. The hydroxyl groups in brucite are in hexagonal close packing in the sequence ...ABABA..., and each elementary layer consists of two sheets of hydroxyls parallel to the basal plane (the plane of the layers), with a sheet of Mg atoms between them (Fig. 3). The Mg atoms in each sheet are each surrounded by six hydroxyl groups forming a polyhedron resembling an octahedron, and the Mg atoms are thus said to be in octahedral (6-fold) coordination. Thus the individual layers are composed of octahedra sharing edges. The individual layers are weakly held together by van der Waals forces, and individual Mg–OH layers are stacked as closely as possible, with one hydroxyl of one layer neighboring three hydroxyls of the next layer.

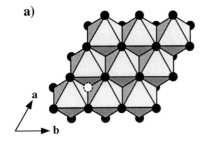
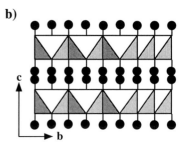
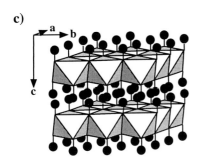

Figure 3. The brucite structure: (a) projected down the c-axis; (b) projected down the a^*-axis (the a^*-axis is normal to the b- and c-axes); (c) oblique projection nearly down the a^*-axis. The octahedra contain a magnesium atom at the center and oxygen atoms at each apex; the black circles represent hydrogen atoms. The white circle with a dashed outline (Fig. 3a) shows the positions occupied by hydrogen atoms from the next hydroxyl sheet up; the hydrogen atoms from adjacent layers are not juxtaposed as the a^*-axis projection suggests.

The gibbsite structure (Saalfeld and Wedde, 1974) is related to that of brucite, and, like brucite, its structure is used as an example of the building block for the octahedral sheet of dioctahedral layer silicates (Fig. 4). As in brucite, the hydroxyl sheets in gibbsite are approximately close-packed, with two hydroxyl sheets in each elementary layer. The Al atoms in gibbsite are each surrounded by six hydroxyl groups, but only two out of three possible positions are occupied. Again, individual layers are made up of octahedra sharing edges; in unusual fashion, the shared edges are longer than unshared edges. In contrast to the layer stacking in brucite, each hydroxyl in one layer of gibbsite is situated directly opposite to a hydroxyl in the next layer rather than fitting as closely as possible. The hydroxyl layers are in the sequence ...ABBAABBA... The interlayer attractions are weak and are dominated by hydrogen bonding.

Diaspore and goethite have the same structure (Busing and Levy, 1958), with hexagonal close-packed oxygens. Aluminum atoms lie between pairs of three oxygen atoms forming strips two octahedra wide; each oxygen atom is bonded to three Al atoms (Fig. 5). The hydrogen atom is associated with only one of two crystallographically independent oxygen atoms. The octahedral site in diaspore is relatively uniform, with three Al–O distances of ~1.85 Å and three at ~1.98 Å. Similarly, goethite has three Fe–O distances of ~1.9 Å and three of ~2.1 Å, reflecting the larger ionic radius of Fe^{3+}. Interestingly, substitution of Mn^{3+} for Fe in goethite causes a significant distortion of the octahedral site, with the two apical metal–O bonds lengthening and the equatorial metal–O bonds shortening significantly with increasing Mn substitution. The changes on substitution of Mn for Fe are due to Jahn-Teller distortions in the Mn-containing octahedra. There is

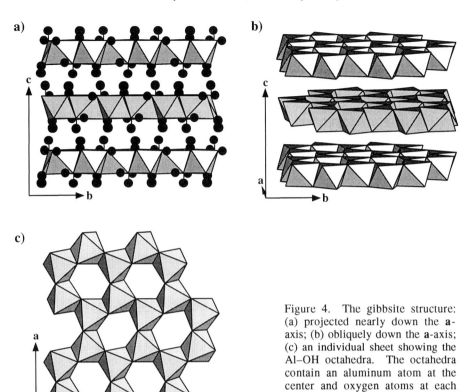

Figure 4. The gibbsite structure: (a) projected nearly down the **a**-axis; (b) obliquely down the **a**-axis; (c) an individual sheet showing the Al–OH octahedra. The octahedra contain an aluminum atom at the center and oxygen atoms at each apex; the black circles represent hydrogen atoms. (Hydrogen atoms were removed from b and c.)

evidence that Mn-substituted goethites are significantly strained due to the presence, in individual crystallites, of unit cells of significantly different size and geometry. This strain undoubtedly has implications for the stability and solubility of such goethites.

Boehmite and lepidocrocite are isostructural and have a structure made up of double sheets of Al octahedra (Fig. 6). The sheets are composed of chains of octahedra which share edges along the **a**-axis and corners along the **c**-axis. The double sheets are held together by hydrogen bonding between similar hydroxyls in adjacent sheets (Hill, 1981).

Microstructures and morphologies

Macrocrystalline brucite has platy cleavage as a result of its layer structure and weak interlayer bonding. However, brucite can also form as fibers, after chrysotile, and a fibrous form of brucite has been called nemalite in the literature. Gibbsite, like brucite, has perfect basal cleavage and often occurs as small tabular crystals or as platelets. Diaspore has perfect cleavage and often occurs as

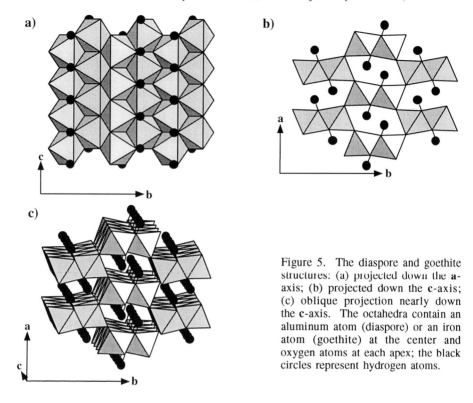

Figure 5. The diaspore and goethite structures: (a) projected down the a-axis; (b) projected down the c-axis; (c) oblique projection nearly down the c-axis. The octahedra contain an aluminum atom (diaspore) or an iron atom (goethite) at the center and oxygen atoms at each apex; the black circles represent hydrogen atoms.

elongated thin plates; it can sometimes be acicular. Boehmite is usually submicroscopic, usually in tabular crystals or laths, much like lepidocrocite. As a result of the hydrogen bonding between octahedral sheets, they both have good {010} cleavage. Goethite often occurs in fibrous masses, particularly in economic deposits. However, when any of these hydroxide minerals occurs in soils, most or all are very finely crystalline and often submicroscopic.

Surface properties

Because most occurrences of these minerals are as very finely crystalline (and high surface area) material, their surface properties are quite important. In contrast to the 2:1 layer silicates, most hydroxides have pH-dependent surface charges due to O and OH associated with broken bonds on these high-surface area minerals. Surface charge associated with isomorphous substitutions is much less important with hydroxides than with 2:1 layer silicates. The surface charge on these minerals has been reviewed in detail by Taylor (1987), who also includes a number of general references on the subject. Much of the following discussion is from Taylor (1987).

According to Taylor (1987), virtually any hydroxide mineral that has been in contact with water will have hydroxyl surfaces, both from structural hydroxyls and from interactions with H^+ and OH^- in water. Even the structural O atoms (e.g., in boehmite or goethite) will likely be hydroxylated as a result of their

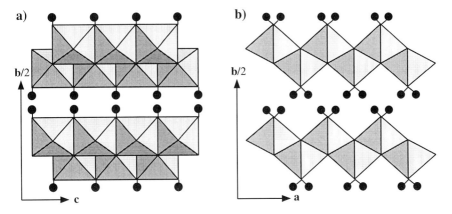

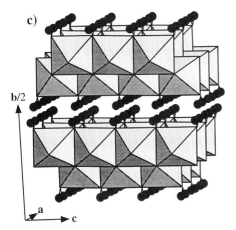

Figure 6. The boehmite and lepidocrocite structures: (a) projected down the **a**-axis; (b) projected down the **c**-axis; (c) oblique projection nearly down the **a**-axis. The octahedra contain an aluminum atom (boehmite) or an iron atom (lepidocrocite) at the center and oxygen atoms at each apex; the black circles represent hydrogen atoms.

residual negative surface charge. These surface OH groups are chemisorbed, but surface H_2O layers are generally physisorbed via hydrogen bonding to these OH groups. Through reversible adsorption of H^+ or OH^- ions (amphoteric dissociation of the surface OH groups), a charge develops on the surfaces, which act as Brønsted acids. At the pH of point-of-zero-charge (PZC), the net surface charge is zero and the surface anion and cation exchange capacities are equal. Similarly, the isoelectric point (IEPS) is the pH at which the net surface charge is zero, in the absence of any specifically adsorbable ions. PZC values for the hydroxide minerals include 5.0 for gibbsite, 6.9 for goethite, and 8.2 for boehmite (Stumm and Morgan, 1981). The surface of a mineral will be positively charged when solution pH is lower than the PZC and the mineral will act as an anion adsorber. Conversely, at pHs higher than the PZC, surfaces will be negatively charged. As Bowden et al. (1973) showed, the energetics of ion adsorption at surfaces are determined by an electrostatic contribution (function of ionic charge), a "covalent" contribution (function of ionic charge, size, and polarizability), and a "chemical" contribution due to a specific chemical (favorable free energy) interactions. Specific adsorption refers to a free-energy driven interaction, regardless of surface charge or activity in solution. It is these interactions at the surfaces of these minerals that will largely determine how they interact with biologic materials. Such specific adsorption processes can significantly affect the surface properties of the mineral because they can change the effective surface charge. For example, surface interaction with a positive metal cation will cause the fixed surface charge to increase. Specific adsorption is often associated with anions of weakly dissociated acids and results from coordination of a surface metal cation with oxy-anions or replacement of surface OH groups by anions coordinating the surface metal cation (Taylor, 1987).

Adsorption characteristics. Although there are few studies on specific cation adsorption on the crystalline Al-hydroxides, there are considerably more data on specific adsorption of anions and molecular species due to the importance of Al-hydroxides in agriculture (Taylor, 1987). Al-hydroxides appear to be particularly important in the adsorption of phosphate, and Parfitt et al. (1977a) showed that ligand exchange occurs at low concentrations between phosphate and singly coordinated OH and H_2O groups on the edge faces of gibbsite. Taylor (1987) suggested that the same adsorption mechanism probably occurs with other similar hydroxides. Al-hydroxides also specifically adsorb sulfate, with adsorption increasing with decreasing pH. As these hydroxides are selective for phosphate over sulfate, phosphate can replace all surface sulfate, but simple anions such as chloride do not compete as they are not specifically adsorbed (Harward and Reisenauer, 1966). According to Sims and Bingham (1968) and Jacobs et al. (1970), Al-hydroxides, and to a lesser extent, Fe-oxides and hydroxides, are important in controlling both arsenate and B retention in soils. Al-hydroxides also adsorb silicic acid as a function of the surface area of the minerals, with lesser adsorption for the Fe-hydroxides (Jones and Handreck, 1963; McKeague and Cline, 1963).

Similar to the mechanism for phosphate adsorption on gibbsite, fulvic acid is adsorbed by a ligand-exchange process with singly coordinated OH groups on the (100) edge face, although fulvic acid adsorption is less than phosphate adsorption due to steric limitations. Na-humate and Na-fulvate are also specifically adsorbed, with resultant release of OH into solution due to ligand exchange (Parfitt et al., 1977b). Parfitt et al. (1977c) showed that oxalate and benzoate are also strongly adsorbed on the (100) edge face of gibbsite as bidentate complexes.

In contrast to the Al-hydroxides, there is a considerable amount of literature on the specific adsorption of cationic species on surfaces of Fe-hydroxides (and oxides). According to Taylor (1987), cations are assumed to adsorb as bidentate metal–OH complexes, involving the replacement of two surface H atoms by one metal–OH complex. A variety of studies have shown that goethite is selective for Cu, Zn, and Pb but less selective for Co, Cd, and Mn, and adsorption appears also to be a function of the anion in solution, adsorption being greater in sulfate-containing solutions than in chloride or nitrate solutions. Data also show that Fe-hydroxides interact with complex cationic and anionic species, such as Pu(IV) and Pu(V) (Sanchez et al., 1985).

The specific anion adsorption behavior of Fe-hydroxides is similar to that for Al-hydroxides, and in many cases the postulated mechanisms are the same. For example, Atkinson et al. (1972) suggested that phosphate formed a bridging ligand structure of binuclear complexes on (100) goethite faces, with phosphate replacing two singly coordinated surface OH groups. Two of the phosphate O atoms are thus coordinated to the goethite Fe^{3+}. However, due to the fact that surface OH in goethite may be bonded to one, two, or three Fe atoms, it is possible for the adsorption energy to be insufficient to exchange more strongly bonded OH (Taylor, 1987). Due to steric restrictions, ligand exchange occurs at only one site per two unit cells, and there appears to be a relationship between phosphate adsorption and goethite surface area. According to Taylor (1987), such singly coordinated OH groups also occur on some crystal faces of lepidocrocite, facilitating the formation of similar surface complexes. Parfitt and Russell (1977), Parfitt and Smart (1977), and Parfitt and Smart (1978) showed that sulfate may be adsorbed in a similar fashion to goethite and lepidocrocite. Taylor (1987) provided a list of other ions that can be specifically adsorbed onto goethite, including silicate, molybdate, chloride, oxalate, benzoate, selenite, and fulvic and humic acids. Rochester and Topham (1979) showed that goethite adsorbs pyridine at Lewis acid sites and via H-bonding with surface OH groups, and CO_2 and NO are adsorbed as bicarbonate-carbonate and nitrite–nitrate groups, respectively. It is important to note (Taylor, 1987) that competition between different species can affect the relative amounts of the species adsorbed and that specific adsorption is a function of solution pH, both due to the nature of the surface and the speciation of the adsorbate.

2:1 LAYER SILICATES

Geological occurrence

The 2:1 layer silicates talc, pyrophyllite, smectite, vermiculite, illite, biotite, and muscovite occur in a variety of geological environments around the world. Talc and pyrophyllite typically occur in metamorphic rocks and are uncommon in weathering environments. Talc forms through the hydrothermal alteration of ultrabasic rocks, often containing serpentine, tremolite, and/or chlorite, and it often occurs in veins and/or faults and shear zones. Poorly crystalline talc-like minerals form through the low-temperature alteration of ultramafic rocks under humid, tropical conditions giving rise to laterites. Biotite and/or muscovite occur in virtually every rock type and are common as primary igneous minerals, as metamorphic minerals, and in sedimentary environments. They are also very common in weathering environments and in soils, often partially weathered to smectite or vermiculite. Illite is a common mineral in sedimentary rocks and in low-grade metamorphic rocks, and interstratifications between illite and smectite occur in a variety of sedimentary rocks and other rock types that have been subjected to low-temperature alteration processes. Montmorillonite (smectite) is commonly the major constituent of bentonites (altered volcanic ash), and smectites are very common components of soils. Vermiculite is common as an alteration product of biotite and can occur either as large crystals, pseudomorphous after the parent biotite, or as a fine-grained soil mineral. Expanded vermiculite is commonly used as a packing material or soil additive, although much of the material that is marketed as vermiculite is actually hydrobiotite, an interstratification of biotite and vermiculite. Macrocrystalline varieties of these materials expand upon rapid heating when the two molecular layers of water in the interlayers turn to steam and abruptly drive the individual layers apart. The geological occurrence of these minerals is the subject of numerous publications, and the interested reader is referred to Bailey (1984a, 1988) and Newman (1987) and the chapters contained therein.

Crystal chemistry

The structures of the 2:1 layer silicates are more complex than those of the common hydroxide minerals, and, because of this complexity, their compositions (Table 2) show significant variability. Considerable chemical substitution is possible in both the octahedral and tetrahedral sheets, and it is these octahedral and tetrahedral substitutions that give rise to the different mineral species listed in Table 1. Talc and pyrophyllite represent the "parent" trioctahedral and dioctahedral species, respectively, with no significant octahedral or tetrahedral substitutions that give rise to charge imbalances. Talc commonly exhibits substitution of Fe^{2+} for Mg and can contain up to 4% Al_2O_3 and a small amount of fluorine substituting for OH. Alkali and alkaline-earth cation contents are low, reflecting the lack of significant charge on the layers and the lack of significant

Table 2

Average chemical compositions of the several 2:1 layer silicates

talc	$Mg_3Si_4O_{10}(OH)_2$
pyrophyllite	$Al_2Si_4O_{10}(OH)_2$
smectites	
montmorillonite	$Z^+_{2x}(Al_{1-x}Mg_x)_2(Si_4)O_{10}(OH)_2 \cdot nH_2O$
beidellite	$Z^+_{4y}(Al)_2(Si_{1-y}Al_y)_4O_{10}(OH)_2 \cdot nH_2O$
nontronite	$Z^+_{4y}(Fe^{3+})_2(Si_{1-y}Al_y)_4O_{10}(OH)_2 \cdot nH_2O$
hectorite	$Z^+_{3x}(Mg_{1-x}Li_x)_3(Si)_4O_{10}(OH)_2 \cdot nH_2O$
saponite	$Z^+_{4y}(MgFe^{2+})_3(Si_{1-y}Al_y)_4O_{10}(OH)_2 \cdot nH_2O$
vermiculite	$Z^+_x(MgFe^{2+}Al)_3(SiAl)_4O_{10}(OH)_2 \cdot 4H_2O$ (where "x" is total layer charge)
illite	$Z^+_{2x+4y}[Al_{1-x}(MgFe^{2+})_x]_2(Si_{1-y}Al_y)_4O_{10}(OH)_2$
biotite	$K(MgFe^{2+})_3(Si_3Al)O_{10}(OH)_2$
muscovite	$K(Al)_2(Si_3Al)O_{10}(OH)_2$
margarite	$Ca(Al)_2(Si_2Al_2)O_{10}(OH)_2$

interlayer cations. Under unusual circumstances, e.g., the alteration or weathering of ultramafic rocks, Ni-rich talcs can form by low-temperature processes in a lateritic environment. Such deposits are actively mined throughout the world and comprise important reserves of Ni. Talc can also contain significant amounts of Mn, Fe, and Al, with associated Na, Ca, and K, but such talcs are unusual and are often complex submicroscopic intergrowths of several different phases. Pyrophyllite shows a limited amount of variation from the ideal composition, with minor substitution of Al for tetrahedral Si (up to ~0.4 Al atoms out of eight tetrahedral positions) and minor replacement of octahedral Al by Fe^{3+} and Mg (up to ~0.1 atoms out of four octahedral sites). Interlayer or exchangeable cations (including Ca, Na, and K) are minor and nonessential, and in fact the ideal pyrophyllite structure cannot accommodate interlayer cations without forming local mica-like regions. There is some evidence in synthetic materials for a relationship between tetrahedral trivalent cations and excess OH (e.g., Rosenberg, 1974), but natural samples do not exhibit evidence for this substitution.

The presence of significant isomorphous substitutions in either the tetrahedral or octahedral sheets of the 2:1 layer silicates gives rise to 2:1 layers that are negatively charged. However, the basic structure of the 2:1 layer remains the same. Such isomorphous substitutions commonly include Al^{3+} (or rarely Fe^{3+}) for Si^{4+} in the tetrahedral sheets and Mg^{2+} or Fe^{2+} for octahedral Al^{3+} in dioctahedral layer silicates. These substitutions produce a 2:1 layer with a net negative charge that is usually balanced by the presence of alkali or alkaline-earth cations in the interlayer region. Of course, substitution of Al^{3+} for octahedral Mg^{2+} or Fe^{2+} in trioctahedral 2:1 layers silicates locally produces a net positive charge, but this is offset by a greater tetrahedral negative charge. The possible variations in isomorphous substitution give rise to layer silicates with a great range in properties, from low-charge minerals such as smectites to high-charge micas such as muscovite and biotite. Such materials differ from talc and pyrophyllite in numerous respects, including the manner of layer stacking, the presence of interlayer cations, the possible presence of interlayer water, and the possibility of expanding in water and organic liquids. This group of isomorphously substituted 2:1 layer silicates comprises the most geologically and industrially important layer silicates, and the applications of these minerals are too numerous to list.

Traditionally, the 2:1 layer silicates have been classified primarily on the basis of their layer charge and secondarily on the basis of their composition (e.g., Bailey, 1980) as in Table 1. This classification has proven useful in that is differentiates minerals with significantly different properties. For example, the swelling capabilities, cation-exchange capacities, and hardnesses differ greatly from talc to phlogopite, from pyrophyllite to muscovite. Margarite, often known as a brittle mica, is a 2:1 layer silicate in which tetrahedral Al-for-Si substitution exists to such an extent that the layer charge per $O_{10}(OH)_2$ is twice that of a typical muscovite or biotite. As a result, the electrostatic attraction between negatively charged layers and the Ca^{2+} interlayer cations is greater and margarite sheets are much less flexible than muscovite (thus the name "brittle" mica). Margarite, and the related species clintonite and anandite, are not common micas, and their properties are not discussed further in this chapter. Interested readers are referred to Guggenheim and Bailey (1975) and Newman and Brown (1987).

As the layer charge increases from zero to ~0.2 per $O_{10}(OH)_2$, the population of cations in the interlayer increases to balance the negative charge on the layers. In such materials, known as smectites (Table 2), both dioctahedral and trioctahedral species exist, with the dioctahedral smectites being much more common. Table 2 shows the common dioctahedral and trioctahedral smectites and their compositional variations. Tetrahedral cations usually include Si and Al, with minor Fe^{3+}, and octahedral cations include Al, Mg, Ni, Fe^{2+}, Fe^{3+}, Mn, and Li. Minor or trace amounts of many other cations similar in size and charge to these can easily substitute in the octahedral and tetrahedral sites. The interlayer cations balancing the isomorphous substitutions in the 2:1 layers are usually Ca^{2+}, Na^+, or K^+, (and occasionally Mg) and, depending on humidity and their hydration energy, these cations may or may not be hydrated in the interlayer (e.g., see Gillery, 1959; Suquet et al., 1975). As is obvious from Table 2, the negative charge in smectites can originate either in the octahedral or tetrahedral sheets, for example by substitution of Mg^{2+} for Al^{3+} or Al^{3+} for Si^{4+}, respectively. In addition, a charged layer may also originate through octahedral vacancies and deprotonation of hydroxyls (Guven, 1988). Smectites typically span the layer-charge range from 0.2 to ~0.6, although as Guven pointed out, high-charge smectites begin to overlap in composition and properties with vermiculites and then are usually distinguished from each other based on crystallite size and paragenesis. A general formula for a dioctahedral smectite can be expressed as:

$$Z^+_{x+y}(Al^{3+}_{2-y}Mg^{2+}_y)(Si_{4-x}Al_x)O_{10}(OH)_2 \cdot nH_2O,$$

where x and y represent the tetrahedral and octahedral substitutions, respectively, and Z^+ represents the interlayer cation (Guven, 1988). As is common in mineralogical nomenclature, the break between species is defined as the midpoint between two compositions. Thus, dioctahedral smectites with y > x are defined as montmorillonites and those with x>y are called beidellites. Dioctahedral smectites with significant substitutions of Fe^{3+} for Al are called nontronites.

Increasing isomorphous substitution, producing layer charges between 0.6 and 0.9, yields the mineral vermiculite. Vermiculites appear to be much more limited in compositional range than smectites, and they are commonly trioctahedral, although Newman and Brown (1987) suggested that the apparent rarity of dioctahedral vermiculites is due to their occurrence as mixtures, precluding accurate chemical analysis. Since vermiculites commonly form through alteration of parent trioctahedral micas or chlorites, their compositions are similar to the phlogopite–biotite series, the primary differences being the lower layer charge, the presence of interlayer water, and the oxidation of structural Fe (Newman and Brown, 1987). Vermiculites typically have more than one in four tetrahedral cation positions occupied by Al, and the primary octahedral cation is Mg, with often more than 0.5 Fe^{3+}, resulting in a small positive charge on the octahedral sheet. As with smectites, a wide range in composition of the 2:1 layers is possible, and both Fe- and Ni-rich vermiculites occur. There appears to be a correlation between the minor-element content of vermiculites and their geological association, with samples associated with ultramafic bodies enriched in Ti, Ni, Cr, and Zn (de la Calle and Suquet, 1988). According to Newman and Brown, the presence of a nominal positive (+0.15 to +0.61) octahedral charge coupled with the large negative tetrahedral charge (>1) distinguishes vermiculites from high-charge saponites. The interlayer cation in vermiculites is usually Mg, with minor Ca and Na possible; due to the high hydration energy of Mg^{2+}, vermiculites commonly exist with two layers of water in the interlayer providing an octahedral coordination complex around each Mg ion.

The mineral illite and related or identical species, such as sericite and glauconite, have layer charges less than one, with fairly restricted compositions. However, their occurrence as mixtures gives rise to considerable variety in apparent chemical composition. Historically, the definition of illite has varied and has been based on composition, structure, paragenesis, and crystallite size. Srodon and Eberl (1988) defined it as a "nonexpanding, dioctahedral, aluminous, potassium mica-like mineral which occurs in the clay-size (<4 µm) fraction." In general, illites have less K and Al and more H_2O and Si than muscovite. Octahedral and tetrahedral substitutions in illite are similar to those in other 2:1 layer silicates, and the interlayer cation is predominantly K, with lesser amounts of Na, Ca, and NH_4^+. Unlike the interlayer cations in smectites and vermiculites, the K in illite is not easily exchangeable. Glauconite is similar to illite but is an Fe^{3+}-rich species, typically with greater octahedral trivalent cation substitutions (Odom, 1988). Due to its great importance in sedimentary rocks, the literature on illite is voluminous; see Hower and Mowat (1966) and the recent summary by Srodon and Eberl (1988) for complete information on illite.

The common micas listed here all have total layer charges ~1, with the charge arising primarily from tetrahedral Al-for-Si substitutions. The micas, consisting of the dioctahedral species muscovite, paragonite, phengite, and celadonite and the trioctahedral species, biotite and phlogopite exhibit compositional variations similar to those found for the other 2:1 layer silicates, namely substitution of Al and Fe^{3+} for tetrahedral Si and substitution of Al, Mg, Fe^{2+}, Fe^{3+}, and, less commonly, Li in the octahedral sites. A variety of other less common micas exist, including the Li-rich micas which occur primarily in pegmatites. There are several lengthy discussions of the crystal chemistry of these less common micas (e.g., Deer et al., 1962; Bailey, 1984a; Newman and Brown, 1987) and the reader is referred to these publications for more information. A variety of much less common substituents exist, primarily in the octahedral sheets, including Ti, Mn, Ni, Cr, V, Co, Cu, and Zn, with B and Be substituting in the tetrahedral sites. There is essentially a continuous gradation between muscovite, $KAl_2(Si_3Al)O_{10}(OH)_2$, and phengite, $K[(Al,Fe^{3+})_{1.5}(Mg,Fe^{2+})_{0.5}](Si_{3.5}Al_{0.5})O_{10}(OH)_2$, and other substitutions of Fe^{3+} and Mg for Al produce celadonite, $KFe^{3+}MgSi_4O_{10}(OH)_2$ (e.g., Newman and Brown, 1987). Interlayer cations in all of these micas are usually K, Na, and Ca, although minor Ba, Sr, Rb, Cs, and NH_4^+ may be present. The layer charge in all of these micas is sufficient to overcome the hydration energy of the interlayer cations, and none of these micas has molecular water in its interlayer, although several authors have postulated the occurrence of H_3O^+ as an interlayer species. Although a number of studies have searched for evidence of tetrahedral Al–Si ordering in the micas, only the brittle micas, margarite in particular, show appreciable tetrahedral ordering.

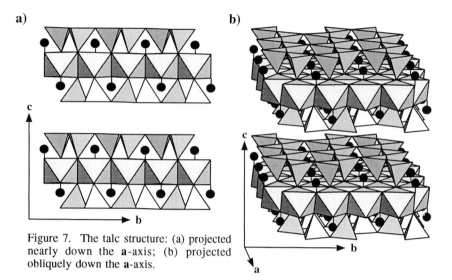

Figure 7. The talc structure: (a) projected nearly down the **a**-axis; (b) projected obliquely down the **a**-axis.

Crystal structures

The crystal structures of these minerals are all based on a fundamental 2:1 unit, composed of a *t-o-t* sandwich. Both talc and pyrophyllite have neutral layers, due to the lack of significant chemical substitutions in either octahedral or tetrahedral sheets. Talc is composed of individual 2:1 layers which have almost monoclinic symmetry (Fig. 7). Two tetrahedral sheets with opposite orientation provide octahedral coordination polyhedra around the Mg atoms. The two crystallographically distinct octahedral sites are nearly identical in talc, one M1 site, usually vacant in dioctahedral 2:1 layer silicates, and two M2 sites. The two hydroxyls occupy a *trans* configuration in M1 and a *cis* configuration in M2. Although the octahedral sheet is thinned slightly, increasing its lateral dimensions, most of the mismatch between octahedral and tetrahedral sheets appears to be accommodated by the rotation of individual tetrahedra of the tetrahedral sheet from their ideal hexagonal geometry, decreasing the lateral dimensions of the sheet. The distortion in the octahedral "brucite" sheet of talc is comparable to that found in the brucite structure (Guven, 1988). Octahedral thinning and tetrahedral rotation are common mechanisms operating in layer silicate structures to accommodate the misfit between octahedral and tetrahedral sheets.

The stacking in both talc and pyrophyllite is not determined by interlayer cations, as in the micas, or by optimum hydrogen bonding, as in the 1:1 layer silicates and chlorites. Instead, it appears to be determined by minimization of Si^{4+}–Si^{4+} repulsions (Evans and Guggenheim, 1988). Individual 2:1 layers are displaced, causing oxygen atoms on the interlayer surfaces to "pack" together and decreasing the repeat distance; if the hexagonal oxygen rings of the interlayer surfaces were in exact opposition, as they are in micas, the layers would be 0.27 Å farther apart (Rayner and Brown, 1973; Perdikatsis and Burzlaff, 1981). The manner of stacking the 2:1 layers, in which adjacent layers are not held in

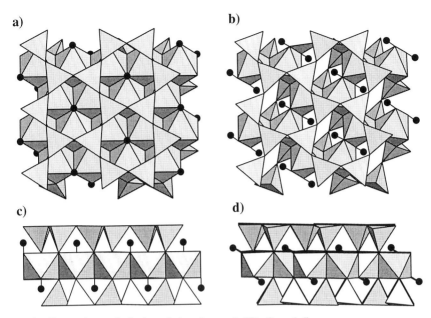

Figure 8. Comparison of talc (a and c) and pyrophyllite (b and d).

registry by interlayer cations or hydrogen bonding, causes the structure to be triclinic. Thus, in an ideal talc, there is no potentially reactive interlayer region containing exchangeable cations as there is in other, expanded, layer silicates. In addition, the structure does not readily expand in water or other polar organic liquids. The layer–layer interactions in talc and pyrophyllite appear to include both van der Waals and a small electrostatic attraction (Giese, 1975). It appears that earlier reports of a two-layer monoclinic polytype of talc were incorrect, possibly due to the occurrence of twinning, and talc appears to form only a one-layer triclinic polytype. These layer silicates are therefore not simply structurally analogous to micas containing no interlayer cations.

Unlike talc, pyrophyllite occurs in both two-layer monoclinic and one-layer triclinic polytypes, *2M* and *1Tc*, respectively, although both polytypes appear to have similar properties. In pyrophyllite, only two (crystallographically identical) sites (M2, the *cis* site) out of three possible octahedral sites are occupied, by Al, leaving a third and larger unoccupied site (Fig. 8). Shared Al octahedral edges in pyrophyllite are significantly shortened to minimize Al–Al repulsion, producing occupied octahedral sites that are significantly more distorted than in trioctahedral minerals such as talc. Apart from these significant differences, pyrophyllite is similar to talc in most respects. The tetrahedral-octahedral sheet misfit in pyrophyllite is minimized by: (1) rotation of the tetrahedral sheet which reduces its lateral dimensions; (2) tilting of the tetrahedra, creating a thickening of the tetrahedral sheet and corrugation of the basal plane oxygens; and (3) flattening of the octahedral sheet with consequent enlargement of its lateral dimensions. The layer repeat in pyrophyllite is ~9.19 Å (Newman and Brown, 1987), and there is insufficient space in the interlayer for cations or water

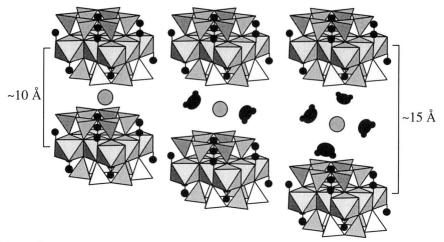

Figure 9. The smectite structure viewed oblique to the 2:1 layers. From left to right, structures show schematically a collapsed smectite, a 1-water-layer smectite, and a 2-water-layer smectite.

molecules. Like talc, pyrophyllite does not undergo interlayer expansion, although Lee and Guggenheim (1981) noted that large single crystals were sensitive to solvents such as acetone, resulting in splitting along cleavage planes. Considering the weak interlayer bonding compared with the relatively strong intralayer bonding and the excellent (001) cleavage in both pyrophyllite and talc, it is likely that interlayer siloxane surfaces make up the majority of potentially reactive surfaces. Such surfaces probably consist primarily of Si–OH groups.

Although the basic structure of the 2:1 layers in smectites is similar to the talc–pyrophyllite structures, there are important differences in their overall structures due to their low to intermediate layer charges and consequent weak interlayer bonding (see Guven, 1988, for a detailed summary of smectite structures and compositions) (Fig. 9). The weak interlayer bonding in smectites yields crystallites that are usually very poorly arranged, in a crystallographic sense. Large smectite crystals do not occur and even the very small crystallites that do form exhibit, in most cases, almost a total lack of three-dimensional order. Only a few samples of beidellite and saponite exhibit regular stacking, with some three-dimensional order (Méring, 1975). In most cases, the only significant crystallographic order existing in smectites is within individual layers, and crystallites or stacks of individual layers resemble a deck of cards thrown on the floor, each layer rotationally disordered with respect to the next one. This type of stacking, with no relationship between adjacent layers, is termed turbostratic. Thus, because of their disordered nature, detailed three-dimensional crystallographic information on smectites is largely lacking, and many details are inferred by comparison with related better-ordered minerals. Guven (1988) summarized recent structural data for smectites, primarily the results of Tsipursky and Drits (1984), who concluded that most of the dioctahedral smectites examined by them had monoclinic ($C2/m$) symmetry, with variations in the details of the vacant octahedral site. However, it appears that most well-ordered dioctahedral smectites follow the normal octahedral ordering pattern, with the M1 *trans* site vacant. It is clear that smectites in which the origin of the layer charge is in the tetrahedral sheets have a greater degree of layer stacking order due to the localization of charge near the interlayer (de la Calle and Suquet, 1988).

In addition to the prevalence of random stacking sequences in smectites, there are data suggesting that the individual 2:1 smectite layers are compositionally heterogeneous, as one might expect from the considerable compositional variations possible. For example, many smectites have compositions intermediate between montmorillonite and beidellite, with the negative layer charge arising from substitutions in both the tetrahedral (beidellitic layers) and octahedral (montmorillonitic layers) sheets. Such smectites must be very heterogeneous on an individual layer scale (e.g., see Newman and Brown, 1987). Furthermore, this heterogeneity in the 2:1 layers gives rise to heterogeneities in the interlayer, with interlayer cations concentrated in regions of higher layer charge. Guven (1988) showed how the site of the negative layer charge (tetrahedral or octahedral) can influence the preferred site of the interlayer cation. A detailed description of such heterogeneities is beyond the scope of this chapter, but this brief discussion serves to illustrate how non-ideal smectites are. In fact, the interlayer region in smectites is considerably different from that in most other 2:1 layer silicates, and this region is responsible for most of the interesting and technologically useful properties of smectites (Guven, 1988). Smectite interlayers commonly contain alkali and/or alkaline-earth cations sufficient to balance the negative charge on the 2:1 layers, together with variable amounts of water that depend on the type of cation and on the activity of water (e.g., relative humidity). Small and/or highly charged interlayer cations commonly are surrounded by one or two hydration shells of water, depending on the humidity. At low relative humidities (<20%) Ca-smectites have mainly one layer of water molecules between the 2:1 layers, whereas they have two water layers at higher relative humidities (e.g., Suquet et al., 1975). Na- smectites transform from a zero-water layer structure at very low humidities to one water layer at intermediate humidities (10 to 50%) and to a two-water layer structure at high humidities. K-smectites, because of the small hydration energy of K^+, exist in a zero-water layer structure up to ~40% relative humidity, transforming to a one-water layer structure at higher humidities. In addition, it is important to note that many smectites undergo osmotic swelling in aqueous suspensions, often expanding to many times their dried state (e.g., Norrish, 1972). Such osmotic swelling behavior is quite variable and depends on many factors, including the layer charge, the site of the layer charge, the interlayer cation composition, and the water composition. In any case, a smectite particle interacting in an aqueous suspension is likely to be expanded and the clay–water mixture behaves as a colloidal suspension (see van Olphen, 1977).

Naturally, the nature of the interlayer cation also affects the manner in which water is lost from the clay. Large univalent cations such as K^+ with relatively small hydration energies are much easier to dehydrate in the interlayer than smaller divalent cations such as Ca^{2+}, with a higher hydration energy. It is primarily the balance between the hydration energy of the interlayer cation, the activity of water, and the energetics of layer-cation attractions that determines whether or not smectites will expand. In smectites, the cation hydration energy is sufficient to overcome the layer-cation-layer attraction and the minerals expand. The lack of strong interlayer bonding in smectites gives rise to numerous interesting and valuable properties. For example, the expanded interlayer facilitates interaction of the clay with a variety of polar organic molecules, and numerous different chemical reactions can be carried out in the interlayer, including precipitation of large interlayer complexes (see Rupert et al., 1987). The comparatively weak bonding of the interlayer cations also allows ready exchange of these cations with other cationic species. Finally, the weak layer–layer attraction allows smectites to expand and contract reversibly as a function of activity of water or exchange with organic liquids. The interlamellar expansion and cation-exchange properties of smectites are subjects of numerous papers and books, and the interested reader is referred to them (MacEwan and Wilson, 1980; Fripiat et al., 1984; Lagaly, 1984; Thomas, 1984; Newman, 1987; Rausell-Colom and Serratosa, 1987; Theng, 1974). The intracrystalline swelling behavior of smectites has not been modeled quantitatively, although it is generally agreed that several factors contribute to this behavior (see Giese, 1978; Bleam, 1993). Although talc and pyrophyllite have neutral 2:1 layers, the individual layers are meshed closely together so that organic liquids that could potentially interact with the layer surfaces cannot easily gain access to the interlayer region. Furthermore, there are no interlayer species that can interact with organic liquids. Thus, the presence of interlayer cations and the expanded nature of the interlayer region in the 2:1 layer silicates with charged layers contributes to their ability to expand and contract.

Vermiculites have a higher layer charge and consequently usually possess a higher degree of three-dimensional order than smectites. They can occur in large single crystals, often pseudomorphous after the parent mineral (often biotite or chlorite), although they also often occur as fine crystallites in soils. Only the general structural features of vermiculites will be presented here, and the reader is referred to de la Calle and Suquet (1988) for further details. A number of studies of the crystal structure of vermiculite have been published, and more than ten different modes of stacking have been identified. However, crystals with ordered stacking sequences are rare and are limited to cation-exchanged vermiculites. Natural Mg-vermiculites are only partially ordered, i.e., they have semi-random stacking (de la Calle and Suquet, 1988). Clearly, as some cation-exchanged (e.g., Na or Ca) vermiculites exhibit ordered stacking sequences, the cation-exchange process causes not only a change in interlayer cation but can cause a significant change in the way the layers interact with one another (de la Calle and Suquet, 1988). Semi-random stacking refers to a crystallite in which adjacent layers are randomly translated or shifted in definite crystallographic directions by discrete amounts. The structure of the individual 2:1 layer of vermiculite is closely related to that of its parent materials, typically biotite, and differences arise primarily from oxidation of Fe^{2+} and partial loss of hydroxyl H ions. Detailed studies show distortions in the 2:1 layer similar to those found in the better-ordered 2:1 layer silicates, e.g., tetrahedral rotations. The interlayer of vermiculite typically contains Mg^{2+} ions, surrounded by a double layer of H_2O molecules, providing approximately octahedral coordination around the Mg. Numerous authors have shown that both interlayer cations and water occupy definite sites in the interlayer (Bailey, 1980), and detailed analyses (de la Calle and Suquet, 1988) show that diffuse or non-Bragg X-ray scattering is partially due to ordering of the interlayer cations.

As with many of the well-ordered and complex mineral systems, there is considerable published crystallographic information on the di- and trioctahedral micas. Thus, a detailed discussion here of mica structures is not warranted, and the interested reader is referred to Bailey (1984a) and the references contained therein. The mica structure is similar to that of the 2:1 layer silicates described above, with negatively charged 2:1 layers and univalent interlayer cations. The brittle micas, with divalent interlayer cations and a layer charge approximately twice that of the true micas, are relatively rare and will not be discussed here. Unlike the 2:1 layer silicates with low layer charges, the micas have a layer charge sufficient to cause the individual layers to be keyed together by the large interlayer cations, usually K or Na. The layer charge is ideally -1.0 per $O_{10}(OH)_2$, significantly higher than the ~-0.3 value for smectites and ~-0.6 value for vermiculites. Like these other 2:1 layer silicates, however, the layer charge can arise through any combination of (1) substitution of a trivalent cation for tetrahedral Si^{4+}; (2) substitution of a univalent or divalent cation for a divalent or trivalent cation, respectively, in the octahedral sheet; or (3) octahedral vacancies. The magnitude of the layer charge and the keying effect of the interlayer cation causes adjacent 2:1 layers to superimpose so that the basal-tetrahedra hexagonal rings line up and provide approximately octahedral (ideally 12-coordinated) coordination for the interlayer cations. The presence of the interlayer cation and the approximate matching of opposing layers gives rise to an interlayer separation in micas ~3.3 Å compared with a separation in pyrophyllite and talc of 2.8 to 2.9 Å. It is apparently the interlocking of adjacent layers by the interlayer cation that limits the variability in stacking of micas to modifications occurring at the octahedral sheet (Bailey, 1980). A detailed discussion of the numerous possible different modes of stacking individual layers (polytypes) is probably not of primary concern when attempting to understand the potential reactivity of micas. Suffice it to say that a rich diversity of polytypes of both di- and trioctahedral micas has been identified and is discussed by many authors.

Although the micas are probably partially ordered on a short-range scale, their long-range symmetry, in most cases, is consistent with both octahedral and tetrahedral disorder. For example, the space groups of common trioctahedral micas allow only one independent tetrahedral site, requiring that Al and Si be disordered. Any tetrahedral cation ordering would cause a reduction in symmetry which has not been observed. As with the other common 2:1 layer silicates, the common trioctahedral micas have two independent octahedral sites, a single *trans*-OH site and two *cis*-OH sites. However, in spite of the resultant possibility of octahedral cation ordering, it is seldom observed. The common polytype of the dioctahedral micas possesses long-range symmetry that allows tetrahedral cation ordering, but significant tetrahedral ordering has not been found. The octahedral cation sites are ordered in the sense that the vacant octahedron is always the single *trans*-OH site, which is larger than the two equivalent (and occupied) *cis*-OH octahedral sites (Bailey, 1984a). A variety of other less common polytypes and micas exist, some exhibiting interesting and unusual cation-ordering patterns, and these are summarized by Bailey (1980; 1984a).

Microstructures and morphologies

All of the 2:1 layer silicates have, as a result of their structure, platy cleavage, that is they break into sheets. Their platy cleavage thus controls, to a large extent, the nature of their exposed surfaces. The microstructure and morphology of these minerals is usually studied in the dry state by scanning electron microscopy, transmission electron microscopy (TEM), or optical microscopy, but the microstructure of smectite in aqueous pastes has also been studied using small-angle X-ray and neutron scattering and TEM. Such studies of pastes show that Ca-smectites often form a network of quasi-crystals containing pores about 1 μm in size and made up of face-to-face bonded smectite layers. Na-smectite pastes do not contain such pores and instead consist of only a few layers with water located between the layers (e.g., Ben Rhaiem et al., 1987).

Virtually all 2:1 layer silicates occur as plates, but several sometimes occur as laths or even fibers, due to particular growth mechanisms or, in a few cases, due to complex intergrowths with other minerals along certain crystallographic directions. Illite in particular is noted for its occurrence in a fibrous form, particularly in pores in sandstones. Although such material is typically termed hairy or filamentous illite, it is actually lath-like on a very fine scale, as are most "fibrous" layer silicates. The laths, often displaying perfect (110) terminations, are typically up to 30 mm in length, 0.1 to 0.3-mm wide, and from 20 to 200-Å thick (Srodon and Eberl, 1984). According to the review of these materials by Srodon and Eberl (1984), most hairy illites appear to be "normal" 1M illites with compositions typical of illites formed from smectites, i.e., there is nothing unusual about them other than their morphology. Rutstein (1979) described similar fibrous muscovite intergrown with chlorite from a shear zone in a phyllite, possibly elongate along (100). The fibrous nature of the material was attributed to crystal growth in an opening shear zone. Fibrous vermiculite has been reported by Weiss and Hofmann (1952).

Surface properties (potentially active surface sites)

A tremendous amount of literature exists on the adsorptive properties of the 2:1 layer silicates, particularly the smectites and vermiculites. These minerals interact with cations and, to a lesser extent, anions in solution at interlayer sites, siloxane surface sites, and at the sites of broken bonds. The minerals that have little or no layer charge interact only at the sites of broken bonds on edges and interlayer surfaces. For example, Yvon et al (1987) showed that the cation-exchange capacity of ground talc arises primarily from the dissociation of silanol groups on lateral surfaces. Rouxhet and Brindley (1966a,b) showed that wet-grinding muscovite produces surfaces that appear to be covered by strongly held contaminating organic material, in addition to an increased amount of strongly held water molecules. Thus, these mineral surfaces can interact strongly with species as simple as H_2O (Rouxhet and Brindley, 1966a,b) or complex organic molecules such as nuclein bases (Samii and Lagaly, 1987). In general, the amount of species adsorbed is a function of the layer charge and usually of the surface area of the mineral. There are also numerous examples of co-adsorption phenomena, i.e., cases in which adsorption of one compound is increased in the presence of a different compound, and adsorption of organic compounds is also often affected by the presence of inorganic salts, possibly affecting the structure of water on or near the clay surface (Samii and Lagaly, 1987).

Probably the most important adsorption phenomenon associated with the 2:1 layer silicates is cation exchange, which occurs most often in these minerals via an exchange of interlayer cations. Again, there are numerous publications discussing this phenomenon (e.g., Barrer, 1978). Bolt (1987) recently briefly reviewed the process of cation adsorption in clay systems and discussed some of the recent theories for describing ion binding at surfaces. This paper provides a brief introduction to the field of cation adsorption; readers desiring additional information and theory on this subject are directed to the recent book on mineral–water interface geochemistry (Hochella and White, 1990).

Not only can simple inorganic cations and organic species adsorb to the surfaces and interlayers of these minerals, but large hydroxy-complexes can be precipitated in the interlayers of both smectites and vermiculites. This process creates materials similar to chlorites in many respects, but the interlayer hydroxide sheet is incomplete, yielding materials with very high surface areas (300 to 500 m^2/g) and large interlayer pore volumes (e.g., Pinnavaia, 1983; Occelli, 1987). Such materials are useful in a variety of hydrocarbon cracking applications.

Catalytic properties

Exploitation of the catalytic properties of layer silicates has a long history described in several texts (e.g., Grim, 1962; Theng, 1982; Pinnavaia and Mortland, 1986; Rupert et al., 1987). Clays and layer silicates have been used in numerous catalytic applications, including cracking, isomerization, polymerization, desulphurization, hydrolysis, hydrogenation, and oxidation reactions. Although the use of clays and layer silicates in many of these applications was supplanted first by silica-alumina gel catalysts and later by zeolites, the catalytic properties of clays and layer silicates are certainly very important aspects of these minerals. It is also likely that their catalytic properties are very important in affecting or even determining the nature of many of the biological effects of minerals. Much of the catalytic behavior of these minerals is based on the presence of high surface areas and surface acidity, and many authors have demonstrated that Al^{3+} in these minerals is the source of surface acidity (Rupert et al., 1987). Preparation of a natural clay or layer silicate for use as a catalyst usually involves treatment of the material with sulfuric or hydrochloric acid at elevated temperatures, washing, drying, and calcining at 500 to 600°C. As a detailed discussion of the catalytic behavior of clays and layer silicates is beyond the scope of this chapter, the interested reader is referred to Grim (1962) and Rupert et al. (1987) for further information and a list of appropriate references. However, it cannot be overemphasized that these properties of clays (and zeolites, as discussed below) are

CHAIN-STRUCTURE LAYER SILICATES

Geological occurrence

The geological occurrences of sepiolite and palygorskite have been discussed in detail recently by Jones and Galan (1988), and the interested reader is referred to their paper for an excellent summary of the literature. Both palygorskite and sepiolite occur around the world, but in limited amounts; indeed, Jones and Galan (1988) stated that they are relatively rare. Both minerals have been reported in deep oceanic sediments, although palygorskite appears to be much more common than sepiolite in sediments. Both minerals occur in soils from arid to semi-arid environments and are particularly important in paleosols and calcretes. Recent information suggests that both minerals can form via a dissolution–precipitation reaction or by direct precipitation from soil waters in the near-surface soil environment in arid and semi-arid climates (Vaniman et al., 1992). Singer (1984) reviewed the pedogenic (formed in the soil environment) occurrences of palygorskite and stated that the mineral is common in soils of arid environments; he favors a formation mechanism involving precipitation from solution rather than alteration of precursor minerals. Although these minerals are often found in fault zones (e.g., Riddle, Oregon), there appears to be no documented occurrence that they ever form by a hydrothermal mechanism (Jones and Galan, 1988). According to Jones and Galan, the best documented pedogenic (soil-related) deposits of these minerals occur in caliche, calcrete paleosols, calcareous lacustrine deposits, or surficial carbonates.

Common associated minerals. Other clay minerals are often absent from deposits containing sepiolite and/or palygorskite, but such deposits commonly occur with smectites, amorphous silica, or chert, and sometimes with kaolinite, serpentine minerals, alkali zeolites, carbonates, sulfates, and other salts (Jones and Galan, 1988). Silica minerals, particularly quartz, can occur with either palygorskite or sepiolite.

Crystal chemistry

Both palygorskite and sepiolite are essentially hydrated magnesium silicates with minor amounts of substituting elements. It is important to note that attapulgite is structurally and chemically identical to palygorskite, and the term "attapulgite" is not accepted by the International Mineralogical Association. The argument that attapulgite has a significantly different morphology than palygorskite is irrelevant, insofar as morphology is not considered when defining mineral names. Therefore, the name attapulgite should not be used as a substitute for palygorskite. The approximate structural formula (Jones and Galan, 1988) for palygorskite is

$$(Mg_{5-y-z}R^{3+}_y\square_z)(Si_{8-x}R^{3+}_x)O_{20}(OH)_2(H_2O)_4\ R^{2+}_{(x-y+2z)/2} \cdot 4H_2O$$

Tetrahedral cations ($Si_{8-x}R^{3+}_x$) are predominantly Si and R^{3+} is usually Al, ranging from 0.01 to ~0.69 out of eight tetrahedral positions (Weaver and Pollard, 1973); no other tetrahedral cations usually occur in significant amounts. However, recent magic-angle spinning nuclear magnetic resonance ^{27}Al spectra of palygorskite samples from Florida strongly suggest that Al^{3+} occurs only in octahedral coordination (Guven et al., 1992). Mg is the most common octahedral cation ($Mg_{5-y-z}R^{3+}_y\square_z$) but others can include Al, Fe^{2+}, Fe^{3+}. The total number of octahedral cations ranges from 3.76 to 4.64, averaging 4.0, suggesting that the mineral is dioctahedral. Vacancies ($\square$) appear to be concentrated at the edges of each octahedral sheet (Jones and Galan, 1988). Apparently approximately four H_2O molecules are situated in the channels and four H_2O molecules are coordinated to octahedral cations at the edges of the sheets, although the exact positions of all water molecules are hard to quantify and the amounts of water can vary significantly. Exchangeable cations ($R^{2+}_{(x-y+2z)/2}$) are usually Ca, Na, and K and range up to ~0.7 (Singer and Galan (1984).

The sepiolite structural formula, based on the structure of Brauner and Preisinger (1956), is

$$(Mg_{8-y-z}R^{3+}_y\square_z)(Si_{12-x}R^{3+}_x)O_{30}(OH)_4(H_2O)_4\, R^{2+}_{(x-y+2z)/2} \cdot 8H_2O\ .$$

Octahedral cations ($Mg_{8-y-z}R^{3+}_y\square_z$) are predominantly Mg, with minor amounts of Mn, Fe, Al, or Ni possible; the amount of vacancies, $\square_z$, is usually small but almost always less than one out of eight octahedral cations. Tetrahedral cations ($Si_{12-x}R^{3+}_x$) are predominantly Si, with very minor amounts of Al and possibly Fe^{3+}. Exchangeable cations ($R^{2+}_{(x-y+2z)/2}$) include small amounts of Ca and Na, with lesser amounts of K and occasional NH_4. Sepiolite contains less strongly bonded water than palygorskite but more loosely held water in the larger structural channels. The cation-exchange capacity of both sepiolite and palygorskite typically ranges from 20 to 30 meq/100 g.

Crystal structures

The crystal structures of both palygorskite and sepiolite are closely related, consisting of 2:1 layer silicate ribbons, continuous along the **c**-axis, linked to adjacent ribbons by inversion of SiO_4 tetrahedra at the ribbon boundary. The width of the ribbons in the **b**-axis is two pyroxene-like (or talc-like) single chains in palygorskite (Fig. 10) and three chains in sepiolite (Fig. 11). This width corresponds to five octahedral sites, in the typical dioctahedral sequence Mg-Al-$\square$-Al-Mg, in palygorskite (Drits and Aleksandrova, 1966) and to eight octahedral sites, all occupied by Mg, in sepiolite. The tetrahedral sheet is continuous from ribbon to ribbon (along the **b**-axis), apart from the inversion of silicate tetrahedra at each ribbon boundary. However, the octahedral portion of each ribbon is discontinuous, ending at the edge of each ribbon. The network thus formed by these ribbons creates pseudo-rectangular channels running parallel to the **c**-axis between neighboring ribbons. The channel dimensions in palygorskite are ~3.7 × 6.4 Å^2 and in sepiolite ~3.7 × 10.6 Å^2. The theoretical total specific surface, including the interior area, is ~900 Å^2. These large channels in palygorskite and sepiolite contain loosely held (i.e., "zeolitic" water) water

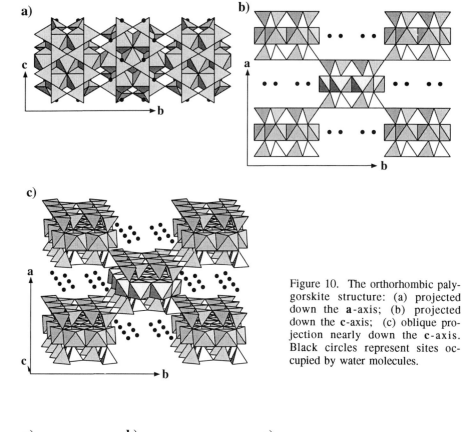

Figure 10. The orthorhombic palygorskite structure: (a) projected down the **a**-axis; (b) projected down the **c**-axis; (c) oblique projection nearly down the **c**-axis. Black circles represent sites occupied by water molecules.

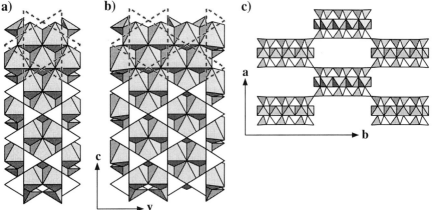

Figure 11. Comparison of the palygorskite (a) and sepiolite (b) structures down the **a**-axis. Each shows a single "I-beam" from which the entire structures can be assembled. Portion of upper tetrahedral strips dashed and unfilled to show the octahedral strips beneath. (c) Sepiolite down the **c**-axis (cf. Fig. 10b). The water molecules within the tunnels are not shown (cf. Fig. 10).

molecules, and exchangeable cations and more strongly held water molecules are coordinated to the edges of the octahedral sheets. Some edge oxygens are apparently also coordinated to H atoms, forming a hydroxyl group.

Variations in the **a**/3 intralayer shifts at the octahedral interface can theoretically give rise to different polytypes, but this appears to occur only for palygorskite, which occurs in one orthorhombic and three different monoclinic structures. Sepiolite apparently occurs only in an orthorhombic structure (Bailey, 1980). Based on unusual X-ray diffraction patterns intermediate between palygorskite and sepiolite, several authors (e.g., Martin-Vivaldi and Linares-Gonzalez, 1962) have suggested the presence of random intergrowths of both structures (Fig. 12), yielding both sepiolite-like and palygorskite-like domains in an individual crystal. Considering the similarities between the two structures, it is likely that such intergrowths are common, at least to a limited extent, in most samples.

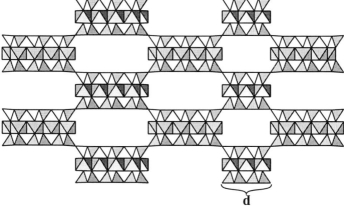

Figure 12. Example intergrowth of a palygorskite lamella (p) in sepiolite. Viewed down the c-axis. The water molecules within the tunnels are not shown (cf. Fig. 10).

Microstructures and morphologies

Both sepiolite and palygorskite are lath-like on a microscopic scale, with the laths clumping together to form frayed bundles on a macroscopic scale (Newman, 1987). Although a few palygorskite and sepiolite samples have asbestiform macroscopic laths up to several cm in length, most are 0.2 to 5 mm long, with widths between 100 and 300 Å and thicknesses between 50 and 100 Å (Martin-Vivaldi and Robertson, 1971). As shown by high-resolution TEM (HRTEM) (Tchoubar et al., 1973), the external surfaces of sepiolite consist primarily of (110) faces, with fewer (100) faces. Serna and Van Scoyoc (1979) used TEM data and information on average fiber sizes to estimate that the average external surface area for both minerals is 300 to 400 m^2/g and the internal (channel) surface area averages 500 to 600 m^2/g.

The industries exploiting palygorskite often describe the mineral product used as "attapulgite." This practice derives from the fact that palygorskite from Attapulgus, Georgia, USA, was first referred to by De Lapparent (1936) as attapulgite. Much of the palygorskite used in this country comes from Miocene deposits in northern Florida and southern Georgia, the deposits containing the original attapulgite. Attapulgite from Attapulgus was only later known to be palygorskite. Palygorskite exhibits a range in lath length, and the Miocene material from the southeastern U.S. is the short-lath (~1-mm long) type (see Fig. 41 in Grim, 1953). These palygorskite laths are often matted together, and palygorskite–smectite intergrowths increase going north. Thus, the palygorskite industry has maintained the name attapulgite (much as a trade name) to define the material that does not consist of long fibers. (W. Moll, personal communication). Since attapulgite is not an accepted mineral name, it should be used only as a trade name.

Surface properties

Serratosa (1979) reviewed the surface properties of sepiolite and palygorskite, and the interested reader is referred to that publication for further information. Many applications of these two minerals arise from the very large external surface areas of 200 to 400 m^2/g for both minerals, and there is a large amount of literature on the catalytic and adsorption behavior of both minerals. Due to their large size, the channels in sepiolite and palygorskite are an important aspect of the surface properties of sepiolite and palygorskite. In spite of their size, only small polar molecules appear to enter after low-temperature outgassing *in vacuo*, and the channels appear to be partially blocked. In fact, even re-adsorption of water into the channels of partially dehydrated palygorskite is hindered, although outer-surface adsorption is rapid. NH_3 can enter the channels of outgassed material through repeated flushing, leading to complete replacement of channel water. Sorption of species other than water and NH_3 appears to be limited to the external surfaces, and the minerals are not true molecular sieves. Higher-temperature outgassing (>300°C in vacuo) removes at least some of the bound water, and alternate ribbons rotate positively and negatively to close the channels and form what are termed folded structures. Newman (1987) showed that there is considerably less than a monolayer of water in the channels of sepiolite and palygorskite, and the hindered sorption of water into the channels must also affect intracrystalline cation exchange. It is thus apparent that the "external" surfaces are most important in determining the adsorption behavior of sepiolite and palygorskite. Raussell-Colom and Serratosa (1987) summarized the adsorption behavior of sepiolite and palygorskite, and much of the following discussion is taken from their chapter. They showed that primarily three kinds of sorption centers exist in both minerals: (1) Oxygen atoms in the tetrahedral sheets; (2) octahedral Mg^{2+} ions at the edges of the octahedral ribbons; and (3) SiOH groups on terminal tetrahedra at the external surfaces. Because few isomorphous substitutions exist in the tetrahedral sheets (i.e., substitutions of Al for Si), the tetrahedral oxygen atoms are weak electron donors and interactions with sorbed species are weak. Edge octahedral Mg ions can participate in sorption either via replacement of the two molecules of coordinated water per cation or by hydrogen bonding to the water. Unlike smectites and vermiculites (and zeolites), the Mg is not easily exchangeable and the coordinated water does not have acidic properties as in smectites. In addition, because the Mg is not easily exchanged, adsorption via cation exchange is of minor importance in sepiolite and palygorskite. SiOH groups are important by virtue of the very large external surface area and consequently the large number of broken Si–O–Si bonds exposed on the surface. The Si–OH groups can interact with both polar and non-polar molecules at the surface and are particularly important in interactions with a variety of organic compounds.

Intracrystalline adsorption is limited in sepiolite and palygorskite due to the sizes of the channels and the non-expanding nature of the clays. Therefore, only small and highly polar molecules interact with the "inner" surfaces, and non-polar organic molecules sorb primarily to external surfaces. Polar organic molecules can penetrate into the channels, but preliminary outgassing of the material is usually necessary to remove "zeolitic" water. For example, short-chain alcohols can penetrate into the channels after outgassing. Raussell-Colom and Serratosa (1987) also showed that sepiolite and palygorskite are specific for organic molecules having sizes and shapes similar to the size of exposed open channels. The differing chromatographic behavior of both minerals for long-chain *n*-paraffins and analogous branched-chain paraffins dissolved in pentane can be explained by the different channel widths in the two minerals. Nederbragt and de Jong (1946) found that the *n*-paraffins were preferentially retained by palygorskite. In addition, both minerals are highly specific for a variety of organic dyes, including acridine orange, methylene blue, and natural indigo. Barrer (1978) also summarized the results of a variety of chromatographic and selectivity experiments with palygorskite.

Exchange characteristics

The cation-exchange capacity of palygorskite usually ranges between 10 and 40 meq/100 g, and higher reported values are probably due to smectite contamination (Jones and Galan, 1988). Exchangeable cations are predominantly Ca, with minor amounts of Na and K. The cation-exchange capacity of sepiolite ranges from 20 to 45 meq/100 g, and exchangeable cations include Ca, Mg, Na, K, NH_4^+, and Cu. As noted above, in spite of the large channel sizes in palygorskite and sepiolite, even migration of intracrystalline water molecules is hindered, and intracrystalline cation exchange is not significant. Most cation exchange appears to occur on the surfaces of the very fine laths.

Uses

The industrial applications of both palygorskite and sepiolite arise from their sorptive properties, their rheological properties, and their catalytic properties, most of which arise from the large external surface area. Alvarez (1984) and Galan (1987) outlined the numerous uses of sepiolite, most of which overlap with those of palygorskite, and the interested reader is directed to these papers. As the following list attests, these minerals are very important industrial commodities, and a full description of their uses is beyond the scope of this chapter. Both minerals are used as decoloring agents for mineral and vegetable oils, paraffin, and greases, as absorbent granules, as cat litter and other odor absorbents, as carriers for insecticides and herbicides, as catalyst carriers, as dispersants, as fillers for rubber and plastic (preventing sagging and pigment settling), as asbestos substitutes, in asphalt coatings, in paints, in "no-carbon-required" paper, in detergents, in the pharmaceutical industry, in the cosmetic industry, in cigarette filters, and in agriculture and animal nutrition.

The rheological behavior (e.g., viscosity) of suspensions of these minerals is related to the type of medium (organic or aqueous), the type and concentration of electrolyte, the pH, and shear stress and pregelification. Both minerals can yield stable suspensions of high viscosity, facilitating their use as drilling muds. Suspensions appear to be composed of agglomerates of fiber bundles and they exhibit non-Newtonian behavior up to pH≈9. Unlike bentonite (smectite) drilling muds, suspensions of palygorskite and sepiolite are stable over a wide range in pH, are relatively insensitive to the presence of electrolytes, and can thus be used as drilling muds for formations containing saline waters. According to Alvarez (1984), sepiolite's sorptive capacity is greater than any other clay, giving rise to numerous applications in water and oil adsorption, organic vapor adsorption, odor adsorption, and NH_3 adsorption. These minerals are also useful as catalyst carriers due to their very high surface area and mechanical and thermal stability, and Alvarez (1984) summarized some of the literature in this area. Palygorskite's high water adsorption properties facilitate its use as an additive in grouting mortar for the prevention of slaking. Both minerals are used, particularly in eastern Europe and the former Soviet Union countries, in purifying hydrocarbons (e.g., Ovcharenko and Kukovsky, 1984), including removing moisture and organic contaminants from natural gas and in purification of benzol.

ZEOLITES

Geological occurrence

The natural zeolites erionite, mordenite, clinoptilolite, phillipsite, and chabazite occur in greatest abundance in a variety of sedimentary deposits near the earth's surface, although most of these minerals may also occur in minor amounts as alteration products in metamorphic and igneous rocks. These zeolites are common in (1) saline, alkaline-lake deposits formed through essentially closed-system alteration of volcanic ash; (2) vertical sedimentary sequences formed through open-system alteration and/or burial diagenesis of volcanic ash; (3) hydrothermally altered volcanic or sedimentary rocks; and (4) deep-sea sediments. Several of these zeolites, notably clinoptilolite, chabazite, mordenite, and phillipsite, also occur in soils in some parts of the world. Salt-affected soils developed on volcanic rocks reportedly contain chabazite and phillipsite in addition to several other zeolites not discussed here (see Ming and Mumpton, 1987, for a summary of the importance of zeolites in soils), apparently forming under conditions similar to those in saline, alkaline lakes. Ming and Mumpton also described the occurrence of clinoptilolite in several soils in the western U.S. and in the former Soviet Union, apparently contributed via aeolian (wind) and/or fluvial processes. These zeolites, particularly clinoptilolite, are also common in soils developed on zeolite-bearing rocks. Important associated minerals in zeolite deposits include a variety of other natural zeolites, often at the trace level, quartz, other crystalline and non-crystalline silica phases, smectite, alkali feldspars, evaporite minerals such as gypsum and halite, and unaltered volcanic glass.

Crystal chemistry

Because the natural zeolites discussed here consist of three-dimensionally linked tetrahedral Si,Al–O frameworks, the major possible tetrahedral substitutions are the same as those for the 2:1 layer silicates, primarily Al^{3+} and Fe^{3+} for Si^{4+}. The tetrahedral Si:(Si+Al) ratio is one of the major discriminating factors between many different zeolites, with some zeolites tending towards the siliceous end and others typically being more aluminous. The charge-balancing extraframework cations can be quite variable in these natural zeolites but are dominantly Ca, Na, K, and Mg. However, the exchangeable nature of the extraframework cations and the large sizes of the sites in which they reside give rise to a wide array of possible extraframework cations. These can include not only the alkali and alkaline earth cations but a variety of other cations. Ideal formulae for these zeolites are listed in Table 3.

Table 3

Chemical compositions of selected natural zeolites

erionite	$NaK_2MgCa_{1.5}(Al_8Si_{28})O_{72} \cdot 28H_2O$
mordenite	$Na_3KCa_2(Al_8Si_{40})O_{96} \cdot 8H_2O$
clinoptilolite	$(Na,K)_6(Al_6Si_{30})O_{72} \cdot 20H_2O$
phillipsite	$K_2(Ca_{0.5},Na)_4(Al_6Si_{10})O_{32} \cdot 12H_2O$
chabazite	$Ca_2(Al_4Si_8)O_{24} \cdot 12H_2O$

The tetrahedral Si:Al ratio has a major effect on the properties of zeolites. Obviously, the Al-for-Si (or Fe^{3+}-for-Si) exchange alters the chemistry of the framework structure, and as discussed above, the chemistry of the extraframework cations varies in response to the framework charge. In other words, the coupled reaction responsible for this exchange is $Si + \square \rightarrow Al(or\ Fe^{3+}) + M^+$, where $\square$ represents a cage-site vacancy and M^+ represents a univalent or divalent cation. Although this exchange reaction is the primary substitution found in natural zeolites, other framework substitutions can occur. One such industrially important substitution is $2Si \rightarrow P + Al$ to form a framework with a composition of $AlPO_4$. Of course, implicit in these substitutions is the possible alteration of framework charge, and this is one of the important properties to consider when relating a biological response to a zeolite with a specific composition. Another important consequence of framework substitutions is that the catalytic properties of a zeolite can be altered. The $Si \rightarrow Al + H^+$ exchange is exploited extensively in the catalyst industry, because the Al-tetrahedra can function as proton donors/acceptors (shown schematically in Fig. 13; see below under "Molecular sieving, exchange, and catalytic properties"). In these types of materials, the catalytic ability is directly related to the framework charge, such that more negative frameworks should have a greater ability to function as catalysts. This general method of catalysis is exploited to break long-chain hydrocarbons into shorter chains Whether the catalytic properties of a zeolite relate to biological

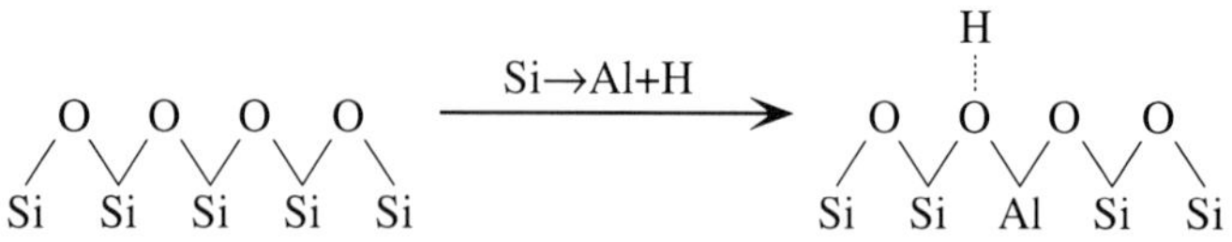

```
              Si→Al+H              H
                                   :
O   O   O   O             O   O   O   O
 \ / \ / \ / \             \ / \ / \ / \
Si  Si  Si  Si  Si        Si  Si  Al  Si  Si
```

Figure 13. Schematic diagram showing the effect of the substitution Si → Al + R^+, where R^+ is a monovalent cation. In this case, the cation is a proton, and this site can function as a catalytic proton donor/acceptor site.

activity has not been explored sufficiently, but it should be an extremely important consideration in future studies of zeolite-pathogenicity. For example, numerous studies have shown that erionite is highly pathogenic, whereas a limited number of studies indicate that mordenite is at most mildly pathogenic. If this observation is found to be generally true for erionites and mordenites, one possible explanation is that zeolite-pathogenicity is related to framework charge, because (as discussed below) erionite typically has a higher framework charge than mordenite. The typical range for Si:Al ratio (along with several other properties) of the five zeolites discussed in this chapter is presented in Table 4.

As with any mineral that exhibits compositional variation, zeolite composition clearly affects mineralogical properties. Consequently, we summarize below the various compositional ranges that are observed for several natural zeolites. These variations include those related to the framework substitutions, as discussed above. It should be remembered, however, that the composition of the cage sites is very easily altered. For example, the cations present in the cation sites can be easily exchanged with cations in solution, so the types of the cation present in a specific sample will actually reflect the last fluid with which the zeolite equilibrated. This exchange can further occur during the biological experiment, i.e., the zeolite will exchange cations with the medium in which it is placed. This property can, of course, be exploited in a biological experiment by "loading" a zeolite sample with a specific cation, in order to assess its role in pathogenesis. In other words, the zeolite can be used to provide a source of a metal *in vitro* or *in vivo*. Finally, as with the cage-site cations, the cage-site water molecules can easily be removed or replaced, and, in fact, the amount and locations of these water molecules will change in response to the type of cage-site cation and to the local humidity.

The tetrahedral Si:(Si+Al) ratio in erionite typically varies from about 75 to 80%, with sedimentary erionites often higher than hydrothermal samples (Gottardi and Galli, 1985). K is typically the dominant extraframework cation, although in some sedimentary samples either Ca or Na can be dominant. Mg is usually subordinate to these three cations and Fe may be either in tetrahedral or extraframework sites, although a detailed study of the site of Fe has apparently not been done. The amount of water varies with the nature of the exchangeable cations and is usually between 26 and 30 water molecules per 72 oxygens.

Mordenite has a tetrahedral Si:(Si+Al) ratio ranging from 80 to 85%, with sedimentary samples again having the highest ratio (i.e., being the most siliceous).

Table 4

Several properties of selected natural zeolites

Mineral	Lattice parameters[a]	Si:(Si+Al)[b]	Void space[a,d]	Free aperture size
erionite	a = 13.26 Å c = 15.12 Å	0.75–0.78	0.36	3.6 Å × 5.2 Å[a] (8-rings ⊥ [001])
mordenite	a = 18.13 Å b = 20.49 Å c = 7.52 Å	0.80–0.85	0.26	6.7 Å × 7.0 Å[a] (12-rings ‖ [001]) 2.9 Å × 5.7 Å[a] (8-rings ‖ [010])
clinoptilolite	a = 17.72 Å b = 17.90 Å c = 7.40 Å β = 116.06°	0.74–0.84	0.34	3.5 Å × 7.9 Å[c] (10-rings ‖ [001]) 3.0 Å × 4.4 Å[c] (8-rings)
phillipsite	a = 9.87 Å b = 14.30 Å c = 8.67 Å β = 124.08°	0.54–0.75	0.30	4.2 Å × 4.4 Å[a,c] (8-rings ‖ [100]) 2.8 Å × 4.8 Å[a,c] (8-rings ‖ [010]) 3.3 Å diameter[a] (8-rings ‖ [001])
chabazite	a = 9.42 Å α = 94.08°	0.59–0.80	0.48	3.7 Å × 4.1 Å[c] (8-rings)

[a] Barrer (1978).
[b] Gottardi and Galli (1985).
[c] Vaughan (1978).
[d] Void space is reported as intracrystalline pore volume, which Barrer (1978) lists as cm³ liquid H_2O per cm³ of crystal.

Roque-Malherbe et al. (1990) suggested that most of the Fe^{3+} in mordenite is contained in tetrahedral sites, although their samples were mixtures of several zeolites. There is little variation in extraframework cation composition, with Na and Ca dominant and K and Mg subordinate (Passaglia, 1975). The water content of mordenite varies from 26 to 30 molecules of H_2O per 96 oxygens.

Clinoptilolite is probably the most common zeolite on the earth's surface, and numerous chemical data are available for this mineral and the structurally related species heulandite. The tetrahedral Si:(Si+Al) ratio of both minerals ranges from 74 to 84% (~27 to 30 Si atoms per 72 oxygens). Clinoptilolite typically has >28 Si atoms per 72 oxygens whereas heulandite usually has fewer than 28 Si atoms, although isolated exceptions to this occur (Gottardi and Galli, 1985). The remaining tetrahedral cations are Al, with minor Fe^{3+} (Roque-Malherbe et al., 1990). The variation in the extraframework cation composition in these minerals is very wide, and samples with significant Ca, Na, K, Mg, Sr, and Ba have been documented. Typically, Ca, Na, and K are the dominant extraframework cations, ranging up to ~4 cations per 72 oxygens. Although not definitive, heulandite often is enriched in Ca. Mumpton (1960) suggested that heulandite could be distinguished from clinoptilolite on the basis of the higher Al content of heulandite, but Mason and Sand (1960) suggested that heulandite has Ca dominant whereas Na and K are dominant in clinoptilolite. Although the thermal behavior of heulandite is distinct from that of clinoptilolite, for many purposes, the distinction between the two species is artificial and can be ignored. The water content of clinoptilolite and heulandite is a function of both the extraframework cation composition and the partial pressure of H_2O (relative humidity) in equilibrium with the zeolite. Samples rich in divalent extraframework cations, such as Ca or Mg, typically have higher water contents than those containing dominantly univalent cations such as Na or K. Heulandites typically have 25 to 27 H_2O molecules per 72 oxygens, whereas clinoptilolites contain 20 to 24 H_2O molecules.

Gottardi and Galli (1985) remarked that phillipsite exhibits perhaps the largest compositional variation of all natural zeolites, and hydrothermal phillipsites can be significantly different from sedimentary samples. Phillipsite is usually more aluminous than clinoptilolite, with 3 to 6.5 Al atoms per 32 oxygens and a wide variation in Si:(Si+Al), ranging from 54 to 77% (e.g., Gottardi and Galli, 1985). Sedimentary phillipsites, particularly those from the alteration of rhyolitic vitric tuff, occupy the siliceous end of this compositional series and are often Na and K rich (e.g., Sheppard and Fitzpatrick, 1989). In general, phillipsites can have a wide range in extraframework cation composition, with up to four Na or K cations per 32 oxygens and up to two Ca cations. Amounts of Fe and Mg are usually very low. Although phillipsite seldom has significant amounts of Sr, it can often contain large amounts of Ba. There is little variation in amount of H_2O, with usually 11 to 13 molecules per 32 oxygens.

As with phillipsite, the highest Si contents in chabazite samples occur for those with sedimentary origins, in which up to 80% of the tetrahedra can be occupied by Si. However, chabazite also shows a wide compositional variation, with Si:(Si+Al) ratios as low as 59% possible. Calcium is the most common extraframework cation, with up to 2 Ca atoms per 24 oxygens, but significant amounts of Na (up to 3.1) and K (up to 2.1) are possible. Sr-rich varieties can occur, but Ba contents are usually very low (Gottardi and Galli, 1985). Significant amounts of both Fe and Mg can occur in chabazite, although, considering the heterogeneous nature of most sedimentary occurrences of all of these zeolites, these cations may be due to impurities.

Crystal structure

Numerous studies of the crystal structures of these five zeolites have been published illustrating that they have framework structures composed of (Si,Al)–O tetrahedral frameworks, with most tetrahedral oxygen atoms shared between two tetrahedra. Gottardi and Galli (1985) provided a valuable summary of zeolite structures in general, the common tetrahedral building blocks making up their structures, and the detailed structures of the minerals discussed here. Similar to

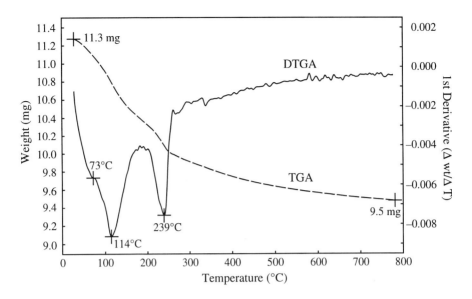

Figure 14. Thermogravimetric analysis of heulandite (see discussion on clinoptilolite in text). Plot shows the weight change (TGA) and derivative of the weight change (DTGA) as functions of temperature. Inflections in the TGA curve are emphasized by the minima in the DTGA curve. Note that at least three distinct weight loss events occur: ~73°C, ~114°C, and ~239°C. Each weight-loss event corresponds to the evolution of water molecules that were located in specific sites within the framework structure.

the layer silicates, the zeolites have a negatively charged framework due to the substitution of Al^{3+} (and rarely Fe^{3+}) for Si^{4+}. The negative framework charge is balanced by the presence of cations (commonly Ca, Na, and K; rarely Li, Mg, Rb, and Sr) in the openings or channels, and much of the remaining "free" space in the structural cavities is occupied by water molecules. The water in zeolites does not passively occupy the structural channels but instead interacts both with the framework and with the extraframework cations. The water molecules are held with a range of energies so that water can evolve over a large temperature range (Fig. 14). The cavities in these zeolites are all large enough so that the extra-framework cations can be easily exchanged with a different cation, and the water molecules can be reversibly removed from the cavities with typically only minor resultant structural effects. Once the water is removed from these structural cavities, the zeolite is often said to be "activated" and many of them readsorb not only water but also a variety of gases and molecules.

The erionite structure (Figs. 15 and 16) is based on an ...AABAAC... stacking sequence of 6-rings of tetrahedra, yielding a framework containing alternating cancrinite cages and hexagonal prisms that are cross-linked with single rings to form large cavities (so-called erionite cages). The cancrinite cage (see Gottardi and Galli, 1985) is created by stacking two 6-rings of tetrahedra, with alternating double 4-rings and 6-rings around the periphery between the two stacked

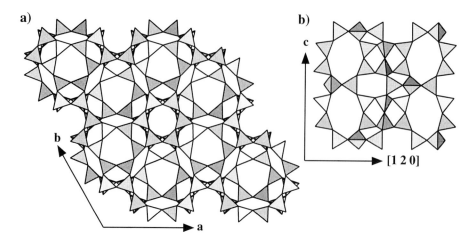

Figure 15. The erionite structure viewed down the c-axis (a), which is the axis along the fiber length, and the a-axis (b).

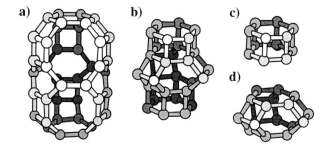

Figure 16. Polyhedral cages that can be used to construct the erionite structure. Spheres correspond to Si, and "bonds" connect Si in adjacent tetrahedra. (a) the erionite cage; (b) composite cage consisting of a double-6-ring cage (c) and a cancrinite cage (d).

6-rings. The K in erionite is located in the cancrinite cage with Ca, Na, and Mg located in the large erionite cage partially coordinated by water molecules. Although the K in the cancrinite cages cannot be removed from erionite via cation exchange at room temperature, upon dehydration Ca displaces K from its position in the cancrinite cage. K moves to the center of the 8-rings of the erionite cage (Schlenker et al., 1977).

The structure of mordenite was first solved by Meier (1961), and numerous refinements of mordenites of different composition have been conducted since. The structure can be viewed as consisting of pseudohexagonal sheets of tetrahedra, pointing both up and down, with additional tetrahedra attached to both the top and bottom of the sheet. The structure is then completed by attaching these sheets through mirror planes (see Gottardi and Galli, 1985, p. 21 to 24). Alternatively the structure may be viewed as being composed of columns of

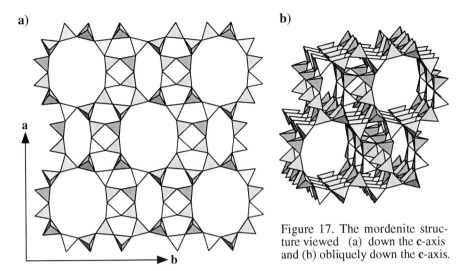

Figure 17. The mordenite structure viewed (a) down the **c**-axis and (b) obliquely down the **c**-axis.

linked 5- and 4-membered rings (rings composed of 5 and 4 tetrahedra, respectively) that are cross linked to produce large, nearly cylindrical channels bounded by continuous 12-membered rings (Fig. 17). Mordenite also contains continuous 8-ring channels parallel to and between the 12-membered rings (Fig. 17) and a zig-zag series of 8-membered rings perpendicular to the 12- and 8-membered rings. For reasons that are presently not universally accepted, most natural zeolites exhibit characteristics consistent with the presence of only 8-membered rings (small-pore type), whereas many synthetic mordenites have sorption properties consistent with the presence of 12-membered rings (large-pore type). Stacking faults are apparently partially responsible for the small-pore behavior (Meier, 1961). This difference in pore size can affect the types of molecules that have access to the internal framework sites, such as the catalytic proton sites. There is partial Si/Al ordering, with the 4-member ring tetrahedra enriched in Al, and the extraframework cations can reduce the space group symmetry to a subgroup of Cmcm (Alberti et al., 1986). In natural specimens, Ca ions usually occur in an 8-coordinated site in the 8-membered rings, bonded to six framework oxygens associated with the Al- enriched tetrahedra and two water molecules. Potassium and other large cations occur in 6-coordinated sites, bonded to five framework oxygens and one water molecule. Mortier et al. (1976) also found Ca^{2+} ions in the centers of the large channels as fully hydrated complexes. Mordenite fibers or needles are characterized by elongation along the **c** axis, i.e., along the 12-membered-ring channels.

The structure of clinoptilolite and the isostructural mineral, heulandite, was first solved and refined by Merkle and Slaughter (1968). Their work and numerous refinements that followed showed that the clinoptilolite/heulandite structure is made up of layers of linked 4- and 5-membered rings lying in the **a**–**c** plane. These layers are linked to yield two major structural channels parallel to the **c**-axis, one composed of 10-membered rings (so-called A rings) and one composed of 8-membered rings (B rings) (Fig. 18b). An additional set of 8-membered rings (C rings) runs parallel to the **a**-axis (Fig. 18a). No structural

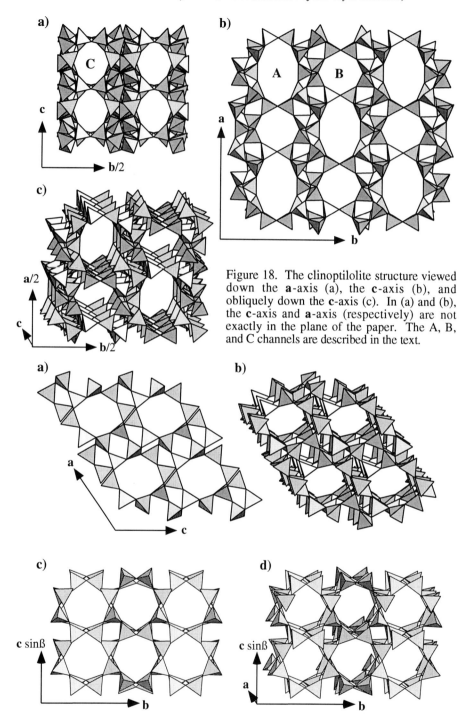

Figure 18. The clinoptilolite structure viewed down the **a**-axis (a), the **c**-axis (b), and obliquely down the **c**-axis (c). In (a) and (b), the **c**-axis and **a**-axis (respectively) are not exactly in the plane of the paper. The A, B, and C channels are described in the text.

Figure 19. The phillipsite structure viewed down the **b**-axis (a); obliquely down the **b**-axis (b); down the **a**-axis (c); and obliquely down the **a**-axis (d).

channels run perpendicular to the **a**–**c** layers. The excellent platy (010) cleavage in these minerals occurs across this network of channels, between the 4- and 5-membered-ring layers (Fig. 18b); to the authors' knowledge, clinoptilolite occurs only in a platy morphology and fibrous varieties are not known. Aluminum ions are always concentrated in the T2 tetrahedra, located between the A and B rings, and this preferential occupancy undoubtedly affects the details of the extraframework cation distribution, particularly Ca^{2+}. Typically four distinct extraframework sites are identified within this network of channels. Most Na cations are 7-coordinated by two framework oxygens and five water molecules, within the A rings. Calcium ions are located primarily within the B rings, coordinated by three framework oxygen atoms and five water molecules. Potassium cations occur in the center of the type-C 8-membered ring, between the A and B rings. It is coordinated to six framework oxygen atoms and three or four water molecules. Magnesium ions are usually placed in the center of the A ring, coordinated by six water molecules (Koyama and Takéuchi, 1977; Armbruster and Gunter, 1991). The exact details of the extraframework sites in clinoptilolite/heulandite minerals vary considerably from sample to sample due to the large variation in chemical composition. In addition, the importance of water molecules in completing the coordination polyhedra of the extraframework cations means that any change in degree of hydration will result in a significant change in the details of the extraframework cation locations. The channels in these minerals obviously constitute a very dynamic system.

The structure of phillipsite was first solved and refined by Steinfink (1962). This work and subsequent refinements showed that the structure consists of layers of 4- and 8-membered tetrahedral rings approximately perpendicular to the **a**-axis (Fig. 19, above). These layers are linked by 4-membered rings forming "crankshafts" with the 4-membered rings of the layers (Rinaldi et al., 1974; see Gottardi and Galli, 1985, their Figs. 0.2K and 0.2L). These linkages produce two major sets of channels, one parallel to the **a**-axis and a smaller channel parallel to the **b**-axis formed by the 8-membered rings resulting from the connection between the crankshafts of the 4-membered rings. The intersections of these two channels form large cages. No Al–Si order has been identified in phillipsite. Two distinct extraframework cation sites exist in phillipsite, a Ca-Na site near the intersection of the two channels and a distinct K site along the edge of the larger channels. The Ca-Na site is only partially occupied, coordinated by four water molecules and three framework oxygens, and the K site is fully occupied, most closely coordinated by four framework oxygens and four water molecules. All water molecules but one [W(3)] occur near the centers of the channels or cages, with W(3) occurring on the circumference of the large channels. In addition, all water molecules bond to the Ca-Na site except for W(2), which is associated with the K site (Rinaldi et al., 1974). Apparently the K site controls substitutions of large cations such as Ba^{2+} and Sr^{2+}, with the maximum of these cations set by a fully occupied K site. The Ca-Na site varies from partially to almost fully occupied, depending on whether the zeolite is Ca or Na rich, respectively.

The structure of chabazite was first determined by Dent and Smith (1958) who showed that the mineral is made up of an AABBCCAA... stacking of 6-rings of tetrahedra forming large cages (Fig. 20; Gottardi and Galli, 1985, Fig. 0.3Ai).

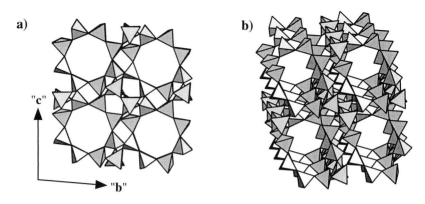

Figure 20. The chabazite structure viewed down the rhombohedral axis (a) and obliquely down the rhombohedral axis (b).

Alternatively, the structure may be envisioned as being built of double 6-rings of tetrahedra linked by tilted 4-membered rings (Calligaris et al., 1982). This stacking of 6-membered rings gives rise to a large cage, often described as the chabazite cage, which is entered by six 8-membered rings or two double 6-rings. Although most refinements have been done in space group $R\bar{3}m$, requiring tetrahedral Al-Si disorder, recently it has been shown that some chabazite crystals possess $P\bar{1}$ symmetry, allowing Al-Si ordering. Four distinct extraframework cation sites have been identified which are variably occupied, depending on the exchangeable-cation composition. These include site C1 on the 3-fold symmetry axis in the center of the double 6-ring which is partially occupied in Ca and Sr forms. This site has nearly octahedral coordination, completed by six framework oxygen atoms. Site C2, also on the 3-fold axis, is just outside the double 6-ring, i.e., on the periphery of the large cage; this site is the most highly occupied extraframework site and has highly distorted octahedral coordination. Site C3 is also on the 3-fold axis near the center of the large cage, usually with low occupancy, and coordinated by water molecules. Site C4 is nearly at the center of the 8-ring port to the large cage, also usually with low occupancy, coordinated to both framework oxygens and water molecules (Alberti et al, 1982). Significant difficulty has been encountered locating all of the water molecules in the chabazite structure; the most highly occupied and well-ordered water site is exactly at the center of the 8-ring window to the large chabazite cage.

Morphologies

Most natural zeolites occur with either fibrous, equant, tabular, or prismatic morphologies. A specific zeolite species can exhibit different morphologies, depending upon its growth history. In general, however, natural samples of erionite and mordenite exhibit a fibrous morphology; phillipsite exhibits a prismatic morphology; clinoptilolite and heulandite exhibit a tabular morphology; and chabazite exhibits equant to tabular morphologies (Gottardi and Galli, 1985).

Molecular sieving, exchange, and catalytic properties

The open framework structures of the natural zeolites give rise to a variety of interesting and technologically very important properties, including molecular sieving, cation exchange, and catalytic behavior. Most of the uses of zeolites, both natural and synthetic, are based on these properties, although rocks containing natural zeolites are also widely used as building stones throughout the world (see Flanigen and Mumpton, 1977; Mumpton, 1977). The large cavities and channels in the zeolites yield diverse molecular sieving properties that vary from zeolite to zeolite by virtue of the differences in their detailed structures and the sizes of the structural cavities and channels (e.g., Table 4). Typically, a zeolite is useful as a molecular sieve only after it has been completely dehydrated, thereby removing much of the occupants of the channels and cavities, i.e., water molecules. Thus, only those zeolites whose structures remain largely intact after complete dehydration are useful as molecular sieves. Of those zeolites discussed here, mordenite, erionite, and chabazite are good examples of molecular sieves; Breck (1974) and Vaughan (1978) discussed the sieving and gas adsorption properties of these species. An example of molecular sieving behavior is the separation of hydrocarbons such as propane, n-butane, n-pentane, and/or n-heptane from branched-chain hydrocarbons, such as isobutane and isooctane by chabazite. The former are rapidly adsorbed by chabazite, whereas the latter are totally excluded. A similar *cation*-sieving effect can influence the manner in which cations are adsorbed by a zeolite, either by excluding larger cations from the smaller channels and cavities or by retarding adsorption of a hydrated cationic species due to size constraints (Breck, 1974).

Likewise, the large channels and cavities, with their associated extra-framework cations that are comparatively weakly bound to the framework, facilitate the very important phenomenon known as cation exchange. Natural zeolites have cation-exchange capacities usually between 200 to 400 meq/100g, superior to most other inorganic cation exchangers. The bulk of the extra-framework cations in these natural zeolites can be readily removed and replaced by other cations, although the details of the exchange process and selectivity sequences depend on the nature of the individual structures and the particular cations present in the zeolite and solution. Certain trends are obvious when comparing different mineral species, and typical selectivity sequences for an open, siliceous zeolite are $Cs \approx Rb \approx K > Na > Li$ and $Ba > Ca \approx Na \approx Sr$. Thus, for example, a typical natural zeolite with an average (Na,K,Ca) extraframework cation composition would be highly selective for Cs over the existing Ca and Na (i.e., it would strongly favor adsorption of Cs). The details of the selectivity sequence depend not only on the nature of the structural channels and cavities, but they depend also on the particular Si:Al ratio (e.g., Table 4) and distribution in the framework. Therefore, the observed overall selectivity sequence will be a function of the individual selectivity sequences exhibited by each of the discrete, different extraframework cation sites in the zeolite structure. Vaughan (1978) briefly summarized the cation-exchange properties of several natural zeolites. In general, those natural zeolites that are open and siliceous are highly selective for large univalent cations such as Cs^+ or NH_4^+ and much less selective for smaller cations such as Li^+. The literature on cation exchange by zeolites is voluminous, and a review of this literature is beyond the scope of this chapter (see summaries in Sherry, 1969; 1971; Breck, 1974; Sand and Mumpton, 1978; Gottardi and Galli, 1985). However, one should be constantly aware that zeolites, like many clays and clay minerals, are active participants in any mineral–fluid interaction and can modify and be modified by their environment.

A third important property of these zeolites is their ability to catalyze certain reactions. Similar to the clays and layer silicates, the catalytic properties of zeolites are also largely a consequence of their very large surface areas (internal, in the channels and cavities) and their Al–Si frameworks. As with many other catalysts, these zeolites, particularly in their hydrogen forms, act as solid acids, making them active in acid-catalyzed reactions (e.g., Fig. 13). According to Breck (1974), cracking catalysts containing zeolites were first used in 1962, and today the vast majority of petroleum cracking employs synthetic zeolites. Although the natural zeolites are typically inferior to synthetic varieties in catalytic applications, the natural varieties nevertheless have high catalytic activities. Vaughan (1978) briefly summarized the catalytic applications of a variety of natural zeolites, and this discussion is taken from his summary. In general, a zeolite is modified either by acid treatment or cation exchange to enhance its catalytic properties. For example, ammonium chabazite is useful for cracking n-hexane, and acid- or caustic-treated clinoptilolite is an active hydrogenation, dehydrogenation, and dealkylation catalyst. H-erionite and Ni-erionite are effective shape-selective hydrocarbon-conversion catalysts, selectively cracking $>C_{12}$ n-paraffins but not $C_{10}-C_{12}$ n-paraffins. Interestingly, both Ca-erionite and Na-erionite are not active catalysts. In 1978, Vaughan stated that erionite was the only natural zeolite then being used commercially as a catalyst, so the natural species have been greatly overshadowed by synthetic varieties. The catalytic behavior of modified mordenites (hydrogen or dealuminated varieties) has also been studied extensively. There are experimental data illustrating the catalytic properties of unmodified natural zeolites (e.g., Ciambelli et al., 1988; Kalló, 1988). Thus, although maximum catalytic activity requires pretreatment of the zeolite and usually involves the use of synthetic species, it is clear that natural, unmodified zeolites exhibit measurable catalytic activity. Such activity could conceivably be of great importance in determining the biological effects of these minerals.

Each of the above three characteristics of zeolites may affect the behavior of a zeolite in a biological fluid. For example, the sieving characteristics will determine which molecules in the biological fluid have access to the internal framework sites. This might be exploited in a biological assay by choosing a series of zeolites with various pore sizes, thereby choosing the types of molecules that can enter the framework. The cation-exchange properties can be important in that the zeolite can alter the composition of the fluid with which it is in contact. A zeolite can release cations to the fluid as well as scavenging cations from the fluid. Hence, a zeolite will modify (or buffer) the equilibrium composition of the fluid. Catalytic properties may also be important in biological systems, because zeolites may be able to catalyze biochemical reactions important in pathogenesis. The variety of zeolite species and large compositional ranges within each species offer the potential to test the importance of each of these properties in mineral-induced diseases. This is a critical step in determining the mineralogical mechanisms of pathogenesis.

REFERENCES

Alberti, A., Galli, E., Vezzalini, G., Passaglia, E. and Zanazzi, P.F. (1982) Position of cations and water molecules in hydrated chabazite. Natural and Na-, Ca-, Sr-, and K-exchanged chabazites. Zeolites 2, 303-309.
Alberti, A., Davoli, P. and Vezzalini, G. (1986) The crystal structure refinement of a natural mordenite. Zeit. Kristallogr. 175, 249-256.
Armbruster, T. and Gunter, M.E. (1991) Stepwise dehydration of heulandite-clinoptilolite from Succor Creek, Oregon, U.S.A.: A single-crystal X-ray study at 100 K. Amer. Mineral. 76, 1872-1883.
Atkinson, R.J., Posner, A.M. and Quirk, J.P. (1972) Kinetics of isotopic exchange of phosphate at the $\alpha FeOOH$–aqueous solution interface. J. Inorg. Nucl. Chem. 34, 2201–2211.
Bailey, S.W. (1980) Structures of layer silicates. In: Crystal Structures of Clay Minerals and Their X-ray Identification, Brindley, G.W. and Brown, G., Eds., Mineralogical Society, London.
Bailey, S.W. (1984a) Classification and Structures of the Micas. Rev. Mineral. 13, 1-12.
Bailey, S.W., Ed. (1984b) Micas. Reviews in Mineralogy 13. Mineralogical Society of America, Washington, D.C., 584 p.
Bailey, S.W., Ed. (1988) Hydrous Phyllosilicates (Exclusive of Micas). Reviews in Mineralogy 19. Mineralogical Society of America, Washington, D.C., 725 p.
Barrer, R.M. (1978) Zeolites and Clay Minerals as Sorbents and Molecular Sieves. Academic Press, London, 497 p.
Ben Rhaiem, H., Pons, C.H. and Tessier, D. (1987) Factors affecting the microstructure of smectites: Role of cation and history of applied stress. Proc. Int'l Clay Conference, Denver, Schultz, L.G., van Olphen, H. and Mumpton, F.A., Eds., 292–297.
Bleam, W.F. (1993) Atomic theories of phyllosilicates: quantum chemistry, statistical mechanics, electrostatic theory, and crystal chemistry. Rev. Geophys. 31, 51–73.
Bolt, G.H. (1987) Cation adsorption in aqueous clay systems: An introductory review. Proc. Int'l Clay Conference, Denver, Schultz, L.G., van Olphen, H. and Mumpton, F.A., Eds., 363–369.
Breck, D.W. (1974) Zeolite Molecular Sieves. Structure, Chemistry, and Use. John Wiley & Sons, New York, 771 p.
Brindley, G.W. and Brown, G., Eds. (1980) Crystal Structures of Clay Minerals and Their X-Ray Identification. Mineralogical Society, London, 495 p.
Brown, G. (1961) The X-Ray Identification and Crystal Structures of Clay Minerals. Mineralogical Society, London, 544 p.
Busing, W.R. and Levy, H. A. (1958) A single crystal neutron diffraction study of diaspore, AlO(OH). Acta Crystallogr. 11, 798–803.
Cairns-Smith, A.G. and Hartman, H. (1986) Clay Minerals and the Origin of Life. Cambridge University Press, Cambridge, 193 p.
Calle, C. de la and Suquet, H. (1988) Vermiculite. Rev. Mineral. 19, 455-496.
Calligaris, M., Nardin, G. and Randaccio, L. (1982) Cation-site location in a natural chabazite. Acta Crystallogr. B38, 602-605.
Ciambelli, P., Bagnasco, G., Czaran, E. and Papp, J. (1988) Changes in the acidity and catalytic activity of natural mordenites after various pretreatments. In: Occurrence, Properties and Utilization of Natural Zeolites, D. Kalló and H. S. Sherry, Eds., Akadémiai Kiadó, Budapest, 625–632.
Deer, W.A., Howie, R.A. and Zussman, J. (1962) Rock-Forming Minerals, Vol. 3, Sheet Silicates. Longmans, London, 270 p.
De Lapparent, J. (1936) The relation of the sepiolite–attapulgite series of phyllitic silicates of the mica type. Compt. Rend. 203, 482–484.
Dent, L.S. and Smith, J.V. (1958) Crystal structure of chabazite, a molecular sieve. Nature 181, 1794-1796.
Flanigen, E.M. and Mumpton, F.A. (1977) Commercial properties of natural zeolites. Rev. Mineral. 4, 165–175.
Galan, E. (1987) Industrial applications of sepiolite from Vallacas-Vicálvaro, Spain: A review. Proc. Int'l Clay Conference, Denver, Schultz, L.G., van Olphen, H. and Mumpton, F.A., Eds., 400–404.
Giese, R.F. (1975) Interlayer bonding in talc and pyrophyllite. Clays and Clay Minerals 23, 165–166.
Giese, R.F. (1978) The electrostatic interlayer forces of layer structure minerals. Clays and Clay Minerals 26, 51–57.
Gottardi, G. and Galli, E. (1985) Natural Zeolites. Springer–Verlag, Berlin, 409 p.

Grim, R.E. (1953) Clay Mineralogy. 1st edition, McGraw-Hill, New York, 384 p.
Grim, R.E. (1962) Applied Clay Mineralogy. McGraw-Hill, New York, NY, 422 p.
Grim, R.E. (1968) Clay Mineralogy. 2nd edition, McGraw-Hill, New York, 596 p.
Guggenheim, S. and Bailey, S.W. (1975) Refinement of the margarite structure in subgroup symmetry. Amer. Mineral. 60, 1023–1029.
Güven, N. (1988) Smectites. Rev. Mineral. 19, 497-560.
Güven, N., de la Caillerie, J-B.d'E. and Fripiat, J.J. (1992) The coordination of aluminum ions in the palygorskite structure. Clays and Clay Minerals 40, 457-461.
Harward, M.E. and Reisenauer, H.M. (1966) Reactions and movement of inorganic soil sulphur. Soil Science 101, 326–335.
Hill, R.J. (1981) Hydrogen atoms in boehmite: A single-crystal X-ray diffraction and molecular orbital study. Clays and Clay Minerals 29, 435–445.
Hochella, M.F., Jr. and White, A.F., Eds. (1990) Mineral–Water Interface Geochemistry. Reviews in Mineralogy 23. Mineralogical Society of America, Washington, D.C., 603 p.
Jacobs, L.W., Syers, J.K. and Keeney, D.R. (1970) Arsenic sorption by soils. Soil Sci. Soc. Amer. Proc. 44, 750–754.
Jones, B.F. and Galan, E. (1988) Sepiolite and Palygorskite. Rev. Mineral. 19, 631-674.
Jones, L.H.P. and Handreck, K.A. (1963) Effects of iron and aluminium oxides on silica in solution in soils. Nature 198, 852–853.
Kalló, D. (1988) Catalysts from Hungarian natural clinoptilolite and mordenite. In: Occurrence, Properties and Utilization of Natural Zeolites, Kalló,D. and Sherry, H.S., Eds., Akadémiai Kiadó, Budapest, 601–624.
Kalló, D. and Sherry, H.S., Eds. (1988) Occurrence, Properties and Utilization of Natural Zeolites, Akadémiai Kiadó, Budapest, 857 p.
Koyama, K. and Y. Takéuchi (1977) Clinoptilolite: the distribution of potassium atoms and its role in thermal stability. Zeit. Kristallogr. 145, 216- 239.
Lee, J.H. and Guggenheim, S. (1981) Single crystal X-ray refinement of pyrophyllite-1Tc. Amer. Mineral. 66, 350–357.
Martin-Vivaldi, J.L. and Linares-Gonzalez, J. (1962) A random intergrowth of sepiolite and attapulgite. Clays and Clay Minerals 9, 592–602.
Mason, B. and Sand, L.B. (1960) Clinoptilolite from Patagonia: The relationship between clinoptilolite and heulandite. Amer. Mineral. 45, 341–350.
McKeague, J.A. and Cline, M.G. (1963) The adsorption of monosilicic acid by soil and other substances. Can. J. Soil Sci. 43, 83–96.
Meier, W.M. (1961) The crystal structure of mordenite (ptilolite). Zeit. Kristallogr. 115, 439-450.
Méring, J. (1975) Smectites in Soil Components 2, Inorganic Components, J.E. Gieseking, Ed., Springer-Verlag, New York, 98–120.
Merkle, A.B. and M. Slaughter (1968) Determination and refinement of the structure of heulandite. Amer. Mineral. 53, 1120-1138.
Moore, D.M. and Reynolds, R.C., Jr. (1989) X-Ray Diffraction and the Identification and Analysis of Clay Minerals. Oxford University Press, Oxford, 332 p.
Mortier, W.J., Pluth, J.J. and J.V. Smith (1976) Positions of cations and molecules in zeolites with the mordenite-type framework. III. Rehydrated Ca-exchanged ptilolite. Mat. Res. Bull. 11, 15-22.
Mumpton, F.A. (1960) Clinoptilolite redefined. Amer. Mineral. 45, 351–369.
Mumpton, F.A., Ed. (1977) Mineralogy and Geology of Natural Zeolites. Reviews in Mineralogy 4, Mineralogical Society of America, Washington, D.C., 225 p.
Mumpton, F. A. (1977) Utilization of natural zeolites. Rev. Mineral. 4, 177–204.
Newman, A.C.D., Ed. (1987) Chemistry of Clays and Clay Minerals. Monograph No. 6, Mineralogical Society, London. Wiley, New York, 480 p.
Occelli, M.L. (1987) Surface and catalytic properties of some pillared clays. Proc. Int'l Clay Conference, Denver, Schultz, L.G., van Olphen, H. and Mumpton, F.A., Eds., 319–323.
Parfitt, R.L., Frazer, A.R., Russell, J.D. and Farmer, V.C. (1977a) Adsorption on hydrous oxides. II. Oxalate, benzoate and phosphate on gibbsite. J. Soil Sci. 28, 40–47.
Parfitt, R.L., Frazer, A.R. and Farmer, V.C. (1977b) Adsorption on hydrous oxides. II. Fulvic acid and humic acid on goethite, gibbsite and imogolite. J. Soil Sci. 28, 289–296.
Parfitt, R.L. and Russell, J.D. (1977) Adsorption on hydrous oxides. IV. Mechanisms of adsorption of various ions on goethite. J. Soil Sci. 28, 297–305.
Parfitt, R.L. and Smart, R.St.C. (1977) Infrared spectra from binuclear bridging complexes of sulphate absorbed on goethite (α-FeOOH). J. Chem. Soc. Faraday Trans. 73, 796–802.
Parfitt, R.L. and Smart, R.St.C. (1978) The mechanism of sulphate adsorption on iron oxides. Soil Sci. Soc. Amer. J. 42, 48–50.
Passaglia, E. (1975) The crystal chemistry of mordenites. Contrib. Mineral. Petrol. 50, 65–77.

Perdikatsis, B. and Burzlaff, H. (1981) Strukturverfeinerung am Talk $Mg_3[(OH)_2Si_4O_{10}]$. Zeit. Kristallogr. 156, 177–186.
Pinnavaia, T.J. (1984) Preparation and properties of pillared and delaminated clay catalysts. In: Heterogeneous Catalysis, Shapiro, B.L., Ed., Texas A&M Univ. Press, College Station, Texas, 142–164.
Pinnavaia, T.J. and Mortland, M.M. (1986) Aspects of clay catalysis. In: Clay Minerals and the Origin of Life, Cairns-Smith, A.G. and Hartman, H., Eds., Cambridge University Press, Cambridge, 131–134.
Raussell-Colom, J.A. and Serratosa, J.M. (1987) Reactions of clays with organic substances, Ch. 8 in Chemistry of Clays and Clay Minerals, A.C.D. Newman, Ed. Monograph No. 6, Mineralogical Soc., London.
Rayner, J.H. and Brown, G. (1973) The crystal structure of talc. Clays & Clay Minerals 21, 103–114.
Rinaldi, R., Pluth, J.J. and Smith, J.V. (1974) Zeolites of the phillipsite family. Refinement of the crystal structures of phillipsite and harmotome. Acta Crystallogr. B30, 2426-2433.
Rochester, C.H. and Topham, S.A. (1979) Infrared studies of the adsorption of probe molecules onto the surface of goethite. J. Chem. Soc. Faraday Trans. 75, 872–882.
Roque-Malherbe, R., Díaz-Aguila, C., Reguera-Ruíz, E., Fundora-Lliteras, J., López-Colado, L. and Hernández-Vélez, M. (1990) The state of iron in natural zeolites: A Mössbauer study. Zeolites 10, 685–689.
Rosenberg, P.E. (1974) Pyrophyllite solid solutions in the system $Al_2O_3-SiO_2-H_2O$. Amer. Mineral. 59, 254–260.
Rouxhet, P.G. and Brindley, G.W. (1966) Experimental studies of fine-grained micas. I. Organic contamination on the surface of wet-ground muscovite. Clay Minerals 6, 211–218.
Rouxhet, P.G. and Brindley, G.W. (1966) Experimental studies of fine-grained micas. II. The water content of wet-ground micas. Clay Minerals 6, 219–228.
Rupert, J.P., Granquist, W.T. and Pinnavaia, T.J. (1987) Catalytic properties of clay minerals, Ch. 6 in Chemistry of Clays and Clay Minerals. A.C.D. Newman, Ed., Monograph No. 6, Mineralogical Soc., London.
Rutstein, M.S. (1979) Fibrous intergrowths of cross muscovite and cross chlorite from shear zones of Pennsylvanian carbonaceous rocks in Rhode Island. Amer. Mineral. 64, 151–155.
Saalfeld, H. and Wedde, M. (1974) Refinement of the crystal structure of gibbsite, $Al(OH)_3$. Zeit. Kristallogr. 139, 129–135.
Samii, A.M. and Lagaly, G. (1987) Adsorption of nuclein bases on smectites. Proc. Int'l Clay Conference, Denver, Schultz, L.G., van Olphen, H. and Mumpton, F.A., Eds., 363–369.
Sanchez, A.L., Murray, J.W. and Sibley, T.H. (1985) The adsorption of plutonium IV and V on goethite. Geochim. Cosmochim. Acta 49, 2297–2307.
Sand, L.B. and Mumpton, F.A., Eds. (1978) Natural Zeolites: Occurrence, Properties, Use. Pergamon Press, Oxford, 546 p.
Schlenker, J.L., Pluth, J.J. and Smith, J.V. (1977) Dehydrated natural erionite with stacking faults of the offretite type. Acta Crystallogr. B33, 3265–3268.
Sheppard, R.A. and Fitzpatrick, J.J. (1989) Phillipsite from silicic tuffs in saline, alkaline-lake deposits. Clays and Clay Minerals 37, 243–247.
Sherry, H.W. (1969) The ion-exchange properties of zeolites. In: Advances in Ion Exchange, Vol. 2, J.A. Marinsky, Ed., Marcel Dekker, New York, 89–133.
Sherry, H.W. (1971) Cation exchange on zeolites. Advances in Chemistry Series, 101, Molecular Sieve Zeolites—I, 350–379.
Sims, J.R. and Bingham, F.T. (1968) Retention of boron by layer silicates, sesquioxides and soil materials. III. Iron- and aluminium-coated layer silicates and soil materials. Soil Sci. Soc. Amer. Proc. 32, 369–373.
Singer, A. (1984) Pedogenic palygorskite in the arid environment. In: Palygorskite-Sepiolite. Occurrences, Genesis and Uses. A. Singer and E. Galan, Eds., Elsevier, p. 169–175.
Srodon, J. and Eberl, D.D. (1984) Illite. Rev. Mineral. 13, 495-544.
Stumm, W. and Morgan, J.J. (1981) Aquatic Chemistry. An Introduction Emphasizing Chemical Equilibria in Natural Waters. John Wiley & Sons, New York, 780 p.
Taylor, R.M. (1987) Non-silicate oxides and hydroxides. In: Chemistry of Clays and Clay Minerals. A.C.D. Newman, Ed., Monograph No. 6, Mineralogical Soc., London.
Theng, B.K.G. (1974) The Chemistry of Clay-Organic Reactions. Hilger, London, 343 p.
Theng, B.K.G. (1982) Clay-activated organic reactions. In: Int'l Clay Conference 1981, H. van Olphen and F. Veniale, Eds., Developments in Sedimentology 35, Elsevier, Amsterdam, 197–238.
Vaniman, D.T., Ebinger, M.H., Bish, D.L. and Chipera, S.J. (1992) Precipitation of calcite, dolomite, sepiolite and silica from evaporated carbonate and tuffaceous waters of southern Nevada, U.S.A. In: Water-Rock Interaction, Y.K. Kharaka and A.S. Maest, Eds. A.A. Balkema, Rotterdam, 687–690.

Vaughan, D.E.W. (1978) Properties of natural zeolites. In: Natural Zeolites. Occurrence, Properties, Use, L.B. Sand and F.A. Mumpton, Eds., Pergamon Press, New York, 353–371.

Weiss, A. and Hofmann, U. (1952) Faserig Vermikulit von Kropfmuhl bei Passau. Acta Albertina Ratisbonensia 20, p. 53 (quoted in Deer et al., 1962, Vol. 3).

Yvon, J., Cases, J.M., Mercier, R. and Delon, J.F. (1987) Effect of comminution on the cation-exchange capacity of talc and chlorite from Trimouns, France. Proc. Int'l Clay Conference, Denver, Schultz, L.G., van Olphen, H. and Mumpton, F.A., Eds., 257–260.

Zigan, F. and Rothbauer, R. (1967) Neutronenbeugungsmessungen am Brucit. N. Jahrb. Mineral. Monat., 137–143.

CHAPTER 5

STRUCTURE AND CHEMISTRY OF SILICA, METAL OXIDES, AND PHOSPHATES

Peter J. Heaney

Department of Geological and Geophysical Sciences
Princeton University
Princeton, New Jersey 08544 U.S.A.

Jillian A. Banfield

Department of Geology and Geophysics
University of Wisconsin–Madison
Madison, Wisconsin 53706 U.S.A.

INTRODUCTION

This chapter will address the structures and chemical properties of the rock-forming oxides and phosphates. In one sense, these materials are less complex than the chain and sheet silicates discussed in previous chapters. In general, their unit cells are smaller and their formulas are shorter. Moreover, the baroque chemical variation that makes amphiboles such a joy for mineralogists (and a nightmare for health scientists) is virtually absent within some oxides, such as the large group of silica (SiO_2) minerals. Instead, the oxides express their individuality through structural permutation, and thereby they illustrate the important concept of *polymorphism*.

Polymorphs are materials that have the same chemical composition but different atomic architectures. Diamond and graphite perhaps are the most famous polymorphic pair. Both are elemental C, but in diamond each carbon atom is covalently bonded to four neighboring carbon atoms, whereas graphite contains weakly bonded sheets in which the atoms are arranged in hexagonal rings (Fig. 1). Despite their chemical identity, diamond and graphite exhibit dramatic differences in their physical properties as a consequence of their structural disparity. Most strikingly, diamond is the hardest mineral known, and graphite is one of the softest. In addition, diamond and graphite are stable under different conditions of temperature and pressure; diamond is stable only at pressures above ~20 kbar, and graphite is stable at low pressures and temperatures.

Similarly, many of the minerals considered within this chapter occur as polymorphs that have different physical properties and different fields of thermodynamic stability. By extension, these polymorphs might be expected to

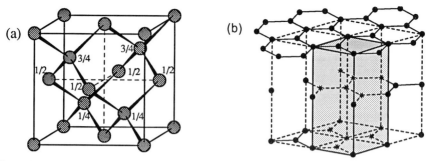

Figure 1. (a) Structure of diamond. (b) Structure of graphite. Courtesy of E.A. Smelik.

display very different behaviors as pathogens. Whereas most biological studies of the sheet and chain silicates must tease apart the combined effects of compositional and structural variation, experiments that explore the toxicity of polymorphs within a single oxide system can pinpoint the role that crystal structure alone plays in the initiation of disease. When glasses of these compositions are included, then the importance of crystallinity itself in the pathogenic process may be judged.

THE SILICA SYSTEM

Phase equilibria and geological occurrences

Crystalline silica. Oxygen and silicon are the two most abundant elements on the surface of the earth; together they account for nearly 85 atom % of the crust (Mason and Moore, 1982). Therefore, it is not surprising that most minerals are built upon a polymerized framework of $(SiO_4)^{4-}$ tetrahedral units. When these structures incorporate additional cations, such as Mg^{2+}, Al^{3+}, or Fe^{2+}, in a balanced, stoichiometric fashion, the resulting minerals are known as *silicates*. The amphiboles, micas, and zeolites discussed in previous chapters are all examples of silicate minerals. If a mineral composition conforms purely to SiO_2, then the substance is considered a *silica* phase. The silica phases are the most common oxides in the earth's crust, and, after the feldspars, they are the most abundant minerals. Because of the geological importance of the silica phases, several excellent surveys exist, including Frondel (1962), Sosman (1965), and Klein and Hurlbut (1993).

The phase diagram reproduced in Figure 2 shows the stability relations among silica polymorphs as a function of temperature and pressure. It will be observed that α–quartz (or low quartz) is stable over most of the temperatures and pressures that characterize the earth's crust. Consequently, α-quartz is by far the most common silica polymorph on the earth's surface and alone comprises 12 wt % of the crust. In addition, quartz has a hardness of 7 out of 10 on the Mohs scale, and it is sparingly soluble in water. Therefore, quartz is highly resistant to mechanical and chemical weathering and persists through geologic time. In fact, even defect microstructures that formed ~2.2 billion years ago are

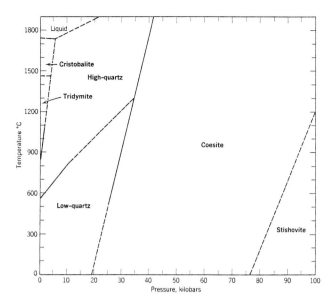

Figure 2. Phase diagram for the silica system. From Klein and Hurlbut (1993, *Manual of Mineralogy*, 21st edition, p. 527; © John Wiley & Sons, Inc. Reprinted with permission.).

preserved in quartz grains from Precambrian banded iron formations (Heaney and Veblen, 1991a).

Quartz occurs within all three of the major rock types. In igneous rocks, quartz is common within granites, granodiorites, and rhyolites, and it is an important constituent of metamorphic phyllites, mica shists, migmatites, and quartzites (Best, 1982; Shelley, 1993). Because of its low susceptibility to alteration, quartz becomes relatively more abundant within beach sands as they mature, and sedimentary sandstones almost always are quartz-rich (Pettijohn, 1975). In addition, quartz is the final product in the crystallization sequence of siliceous oozes on the ocean floor (Hein, 1987), and it is the primary gangue (or matrix) mineral in the metalliferous veins of ore deposits. In microcrystalline fibrous form, quartz (or chalcedony) occurs as a secondary infilling of seams and cavities within rocks.

Although α-quartz is the only silica phase that is thermodynamically stable throughout much of the crust, other silica polymorphs do exist metastably at the earth's surface. As can be seen in Figure 2, cristobalite and tridymite are stable only at high temperatures and low pressures. Nevertheless, they are not uncommon crustal minerals, and they can form in a variety of ways. As silica liquid cools, cristobalite is the first silica solid to crystallize. If cristobalite crystals are quenched by a sudden exposure to atmospheric temperatures (as may happen during a volcanic eruption), then the inversion from cristobalite to quartz is inhibited. Tridymite may be quenched in a similar fashion. In addition, cristobalite and tridymite form from the devitrification of volcanic glasses, such as obsidian.

Lastly, the siliceous oozes on the seafloor that derive from the skeletons of diatoms, radiolarians, and sponges undergo a process known as *diagenesis*—a transformation during burial that results from heating to low temperatures (<300°C) and squeezing at low pressures (<2 kbar). The diagenetic sequence of these siliceous gels is well-established (see reviews by Williams et al., 1985, and Williams and Crerar, 1985). Colloidal silica transforms first to an amorphous solid (opal-A), which in turn alters to a disordered mixture of cristobalite and tridymite (opal-CT). With further heating, opal-CT changes to microcrystalline quartz (chert) and eventually to coarse-grained quartz. As a result, even though diatomaceous earth begins as noncrystalline skeletal silica, X-ray powder diffraction patterns of diagenetic diatomite reveals traces of cristobalite and quartz (Fig. 3).

Just as cristobalite and tridymite occur as metastable polymorphs, the minerals coesite and stishovite have been found in rocks that equilibrated in short-lived high pressure environments, such as meteoritic impact craters. These silica phases were synthesized before they were discovered in nature—coesite in 1953 (Coes, 1953) and stishovite in 1961 (Chao et al., 1962). Coesite also has been observed in high-pressure eclogites from South Africa and China.

Yet another metastable silica polymorph that is receiving increased attention is called moganite, which is found at low temperatures and pressures and is usually intergrown with microcrystalline quartz (Flörke et al., 1984; Heaney and Post, 1992). Drop-solution calorimetry experiments indicate that moganite is less stable than quartz at room temperatures and pressures (Petrovic et al., 1993a), and it seems probable that moganite has no field of true stability.

Amorphous silica. Natural occurrences of amorphous silica also are widespread. As noted above, siliceous oozes on the marine floor solidify to form poorly crystalline opal, or porcellanite. In the absence of diagenesis, these opaline deposits can become quite extensive; in the Miocene Monterey Formation of California, diatomite sequences measure hundreds of meters in thickness (Garrison et al., 1981). In addition, an amorphous siliceous sinter, known as geyserite, precipitates from geyser fluids that contain high concentrations of dissolved silica.

Extrusive magmas may quench to form volcanic glasses upon sudden exposure to air or water. The structure of volcanic glass depends upon the composition of the starting liquid; basaltic glass (or tachylite) devitrifies fairly rapidly, but more silicic glasses (obsidian) may persist for millions of years in large deposits (O'Keefe, 1984). Silica glass also is found within tektites, which are spherical or teardrop-shaped silicate glass bodies associated with impact craters. Tektites usually are compositionally messy, but they may contain particles of pure silica glass known as lechatelierite (Glass, 1984). Silica glass also forms when lightening strikes unconsolidated sand or soil. The resulting fulgarite adopts the shape of the lightening bolt and makes a popular museum display.

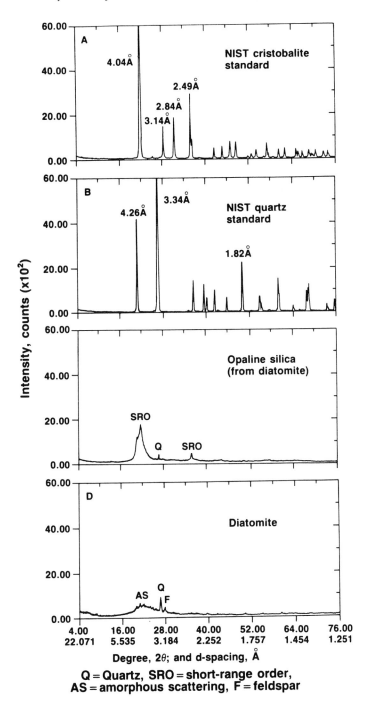

Figure 3. X-ray diffraction patterns from crystalline and amorphous forms of silica. From Ampian and Virta (1992)

Though sometimes ignored by geologists, amorphous silica also is cycled through the environment via biological activity. Internal silicification of plant tissues promotes structural integrity and affords protection against plant pathogens and insects (Chen and Lewin, 1969). Some plants, such as members of the genus *Equisetum*, release exudates through the plant root that depolymerize silica in surrounding soil in order to facilitate silica uptake (Weiss and Herzog, 1977; Sangster and Hodson, 1986). Silica accumulation can account for ~20% of the dry weight of rushes, rice, and sugarcane, and the ash produced by burning the hulls of rice seeds may contain ~95 wt % silica (Kaufman et al., 1981). Likewise, trees can deposit nodules of amorphous silica as phytoliths within their leaves, and, after death and decomposition, this silica is returned to the soil. Geis (pers. comm. cited by Sangster and Hodson, 1986) estimates that the foliage and wood of sugar maples yield 90.1 kg/ha of particulate silica, whereas red pine yields 42.3 kg/ha. Interestingly, no uncontested evidence supports the direct precipitation of crystalline silica within organisms (Mann and Perry, 1986).

Phase stability. The natural profusion of metastable polymorphs within the silica system may be partially explained by examination of the thermodynamic properties of the different silica phases relative to quartz. Even though α-quartz is the stable form of silica at 298 K and 1 bar, calorimetric analysis reveals that most of the observed polymorphs are only slightly less stable than quartz. Specifically, work by Petrovic et al. (1993b) indicates that the enthalpies of the silica polymorphs (including silica glass) at 298 K lie within 15 kJ/mole of the enthalpy of quartz (Fig. 4). Thus, quartz is not strongly favored energetically over its polymorphic relatives. The only exception to this rule is stishovite, whose structure differs considerably from that of the other silica phases (see below).

Silica: Applications and regulations

Uses. The applications for silica are manifold. As a major constituent of sand and gravel, quartz is found in mortar and concrete, and when crushed to fine particle sizes it is used in fluxes, abrasives, porcelains, and paints (Branch of Industrial Minerals, 1992). Because quartz has an extremely high quality factor, since World War II it has been synthesized for use as an electronic oscillator (Frondel, 1945). Diatomaceous earth is an effective filter as a result of its high porosity. Silica glass (sometimes incorrectly called quartz glass) is essential to applications ranging from fiber optics to semiconductors to fiberglass insulation (Fanderlik, 1991). Crystalline silica phases that have been synthesized but not observed in nature would include keatite[†] (Shropshire et al., 1959) and numerous silica-rich zeolites. Industrial demand for these zeolitic polymorphs has steadily increased as a result of their tailor-made catalytic characteristics and their organophilic sorptive properties (Jacobs and Martens, 1987; Cheetham and Day, 1992).

[†] Keatite has recently been reported as a phase in naturally occurring, high altititude, atmospheric dusts (Rietmeijer, 1993, J. Volcan. Geotherm. Res., 55, 69–83). These dusts are believed to originate from volcanic sources.

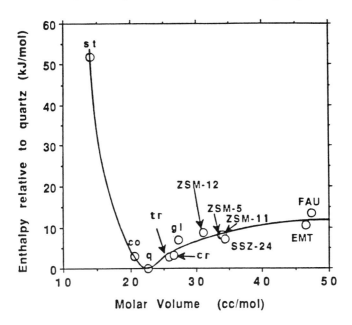

Figure 4. Enthalpies of natural and synthetic silica phases relative to the enthalpy of quartz as a function of molar volume. Polymorphs include stishovite (st), coesite (co), quartz (q), tridymite (tr), cristobalite (cr), glass (gl), and zeolitic silica (ZSM-12, ZSM-5, ZSM-11, SSZ-24, EMT, and FAU). From Petrovic et al. (1993b)

Restrictions. Silica is geologically ubiquitous and technologically indispensible. Although this combination is in many ways ideal, it becomes problematic in light of evidence that silica is a pathogen. Physicians have long recognized the general hazards posed by silicate dusts to miners and other stoneworkers, and the specific causal relationship between crystalline silica and silicosis was well established early in this century (see Rosner and Markowitz, 1991).

More recent studies, however, also have implicated crystalline silica as a carcinogen (for reviews see Bendz and Lindqvist, 1977; Dunnom, 1981; Evered and O'Connor, 1986; and especially Goldsmith et al., 1986). Consequently, the International Agency for Research on Cancer (IARC) has classified crystalline silica as a Group 2A carcinogen (IARC, 1987); this designation indicates that crystalline silica is perceived as a probable carcinogen on the basis of limited evidence in humans and sufficient data involving laboratory animals. On the other hand, the IARC has ruled that the carcinogenicity of amorphous silica has not been demonstrated adequately. Therefore, crystalline silica is subject to the 0.1% Hazard Communication Standard (HCS) of the Occupational Safety and Health Administration (OSHA), whereas amorphous silica is exempt from the HCS. This directive may be all-encompassing for those industries regulated directly or indirectly by OSHA. Even operations targeted at non-silica resources will be affected, since quartz is so widely distributed (Table 1).

Table 1

Silica as an accessory phase in common commodities

Commodity	Type of Silica	Major Applications
antimony	quartz	flame retardants, batteries, ceramics, glass, alloys
bauxite	quartz	Al production, refractories, abrasives
beryllium	quartz	electronics
cadmium	quartz, jasper, opal, chalcedony	batteries, platings, pigments, plastics, alloys
concrete	quartz, chert	aggregate
clay	quartz, opal	paper, ceramics, paint, refractories
copper	quartz	electrical conduction, plumbing, machinery
crushed stone	quartz	construction
diatomite	quartz, opal	filtration
dimension stone	quartz	building facings
feldspar	quartz	glass, ceramics, filler material
fluorite	quartz	acids, steelmaking flux, glass, enamel, weld rod coatings
garnet	quartz	abrasives, filtration
germanium	quartz, opal, chalcedony	infrared optics, fiber optics, semiconductors
gold	quartz, chert	jewelry, dental, industrial, monetary
gypsum	quartz	gypsum board, building plaster
industrial sand	quartz	glass, foundry sand
iron ore	chert, quartz	iron and steel industry
iron oxide pigment	chert, quartz, opal	construction materials, paint, coatings
lithium	quartz	ceramics, glass, aluminum production
magnesite	quartz	refractories
mercury	quartz	chlorine and caustic soda manufacture, batteries
mica	quartz	joint cement, paint, roofing
perlite	opal, quartz	building construction
phosphate rock	quartz, chert	fertilizers
pumice	obsidian	concrete aggregate, building block
pyrophyllite	quartz	ceramics, refractories
sand and gravel	quartz	construction
selenium	quartz	photocopiers, glass manufacturing, pigments
silicon	quartz	silicon and ferrosilicon for steel industry, computers, photoelectric cells
silver	quartz, chert	photographic material, electrical and electronic products
talc	quartz	ceramics, paint, plastics, paper
tellerium	quartz	steel and copper alloys, rubber compounding, electronics
thallium	quartz, chert, chalcedony, opal	electronics, superconductors, glass alloys
titanium	quartz	pigments for paint, paper, plastics, metal for aircraft, chemical processing
tungsten	quartz	cemented carbides for metal machining and wear-resistant components
vanadium	quartz, opal	alloying element in iron, steel, and titanium
zinc	quartz, chert, chalcedony, opal	galvanizing, zinc-based alloys, chemicals, agriculture
zircon	quartz	ceramics, refractories, zirconia production

Just as amorphous and crystalline silica provoke different pathogenic responses, the metastable polymorphs of silica apparently exhibit different levels of toxicity (Hemenway et al., 1986; Saffiotti, 1986). These disparities presumably relate to differences in the intrinsic and surface structures of the silica polymorphs, and an understanding of the pathogenic mechanisms may require an intimate knowledge of their crystal chemistry.

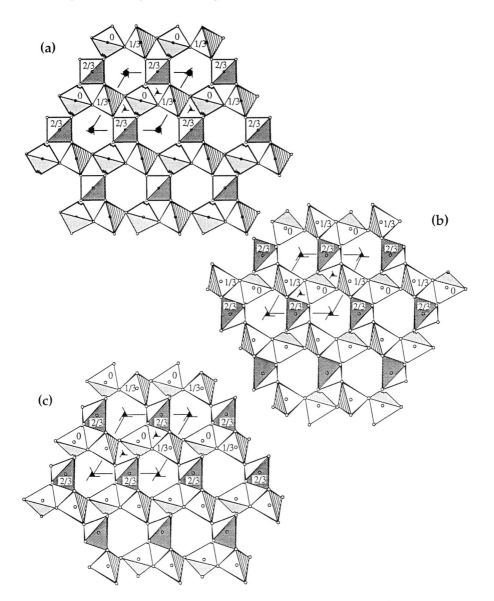

Figure 5. (a) Projection of the structure of β quartz along the **c** axis. (b) and (c) Projections of the structure of the two Dauphiné twin orientations of α-quartz along the **c** axis.

Structures of the silica polymorphs

Quartz. In high-temperature β-quartz, paired helical chains of silica tetrahedra spiral in the same sense about 6_4 or 6_2 screw axes parallel to the **c** axis (Fig. 5a, above). The two chains within each helix are symmetrically related to each other by the twofold axis contained within the sixfold screw axes, and the space groups for β-quartz are *P6$_4$22* and *P6$_2$22* (Bragg and Gibbs, 1925; Wright and Lehmann, 1981). When β-quartz is cooled below 573°C, however, it transforms to α-quartz through a *displacive* transition, characterized by the bending but not the breaking of chemical bonds (Buerger, 1951).

Displacive transitions are common within the framework minerals (tectosilicates), in which the apical oxygens of every tetrahedral unit are shared between two silicon cations. Typically, the transition involves the collapse of an expanded, low-density structure to a higher density polymorph when the mineral is cooled below a specific critical temperature. In the case of quartz, the density increases from 2.53 g/cm^3 (β-quartz) to 2.65 g/cm^3 (α-quartz). This structural contraction involves no activation energy. Consequently, displacive transitions are virtually instantaneous, and the high-temperature β phase will not persist metastably below the critical temperature. Therefore, the existence of β-quartz is of biological importance only to the extent that the transition from β- to α-quartz affects the microstructure of the low temperature polymorph.

The primary defects induced by the α- to β-quartz transition are the so-called Dauphiné twins (Heaney and Veblen, 1991b). When the silica framework of β-quartz collapses at the transition temperature, the individual silica tetrahedra can tilt in one of two possible directions, giving rise to two distortional variants of α-quartz (Fig. 5b,c). The tilting of the tetrahedra lowers the screw symmetry from sixfold to threefold, and projections of the open tunnels parallel to the **c** axis change their geometry from hexagonal to ditrigonal. As can be seen from the ditrigonal tunnels in Figures 5b and 5c, one distortional variant can be transformed to the other by a 180° rotation. Thus, these Dauphiné twin orientations of α-quartz are related by the twofold symmetry lost during the transformation. The space groups for α-quartz are *P3$_1$21* and *P3$_2$21*.

The Dauphiné twin domains can be imaged with the transmission electron microscope (Fig. 6). The presence of these twin domains can alter the physical properties of α-quartz. For instance, the piezoelectric charge associated with one twin orientation is opposite that of the other twin orientation; therefore, if the volumes occupied by the two twin orientations are equal, the net electric charge is zero. In addition, it seems likely that the structure of the Dauphiné twin boundary approximates that of β-quartz (Liebau and Böhm, 1982). Consequently, quartz crystals with high densities of twin domains are structurally heterogeneous. Moreover, some experiments suggest that impurity cations are preferentially partitioned into these β-quartz-like twin boundaries (Heaney and Veblen, 1991c).

In addition to the Dauphiné twins, quartz crystals frequently contain Brazil twin domains that form during crystal growth. These twins describe regions of

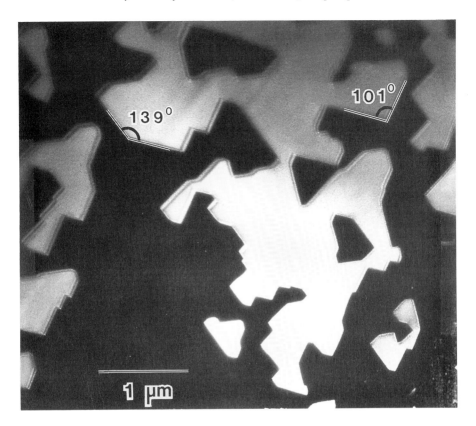

Figure 6. The Dauphiné twin domains as imaged using dark-field TEM ($g = 301$). The black and white areas represent the two twin orientations that correspond to the structures depicted in Figures 5b and 5c.

opposite handedness (i.e., one twin corresponds to space group $P3_121$ and the other to $P3_221$). In Figure 7a, the helical chains of silica tetrahedra spiral downward into the page counterclockwise; these chains are left-handed. Alternatively, the tetrahedral chains can spiral downward in a clockwise fashion, describing right-handed helices (Fig. 7b). These two configurations of α-quartz are called enantiomorphs. The Brazil twins are related by mirror symmetry, and the structure of the Brazil twin boundary has been ascertained through dark-field TEM techniques (McLaren and Phakey, 1966; McLaren and Pitkethly, 1982). Amethyst crystals tend to contain high concentrations of Brazil twin domains, suggesting some kind of causal relationship between substitutional Fe^{3+} and the formation of the twins (Schlössin and Lang, 1965).

Microcrystalline quartz. α-quartz occurs as fine-grained (<1 μm) varieties that are structurally distinct from prismatic quartz. These varieties may be classified into fibrous and non-fibrous species. Fibrous microcrystalline silica is

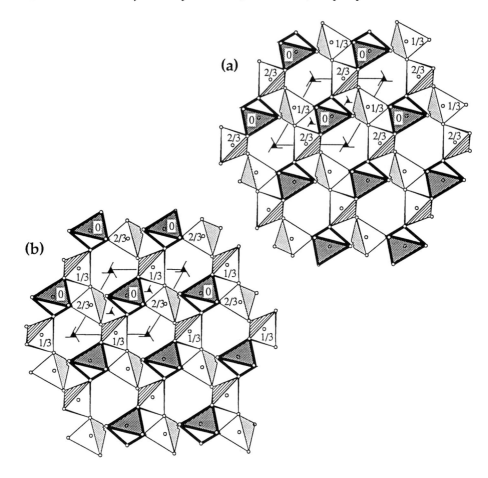

Figure 7. The enantiomorphic Brazil twins of α-quartz.

known as chalcedony and includes agate as a sub-variety. Equigranular microcrystalline silica is called chert and includes flint, jasper, and onyx as subvarieties. Full classification systems are offered by Frondel (1962) and Flörke et al. (1991). Recent research has demonstrated that the crystal behavior of microcrystalline silica departs from that of coarse-grained quartz in a number of ways. Whereas prismatic quartz crystals grow parallel to the **c** axis, chalcedony fibers are elongate parallel to [110], and the fibers are twisted about the fiber axis (Frondel, 1978). Both chalcedony and chert contain extraordinarily high concentrations of Brazil twin boundaries (Miehe et al., 1984), creating enantiomorphic twin lamellae that are ~100 Å thick. In addition, chalcedony and quartz are intimately intergrown with moganite, whose structure consists of slabs of right-handed quartz that alternate with slabs of left-handed quartz at the unit-cell scale (Miehe and Graetsch, 1992). The amount of intergrown moganite is quite variable, but generally moganite accounts for 5 to 20 wt % of a microcrystalline silica specimen (Heaney and Post, 1992).

Figure 8. Chalcedony fibers etched with HF acid reveal marked fibrosity in secondary electron images.

The fibrous habit of chalcedony (Fig. 8) might be worrisome in light of reports suggesting that a fibrous particle shape *per se* plays a role in the induction of cancer (Selikoff, 1977; Stanton et al., 1981; O'Neill et al., 1986). Individual chalcedony fibers typically range from 0.01 to 1 μm in thickness, and their maximum fiber length may be on the order of 10 μms to 1 cm. Thus, the aspect ratio for chalcedony can be extremely high. On the other hand, chalcedony differs markedly from chrysotile and amphibole asbestos in that chalcedony fractures across fiber boundaries. Bonding between two fibers in chalcedony apparently is stronger than bonding within a single fiber, presumably as a result of the pervasive structural defects within individual fibers. Therefore, grinding chalcedony to dust does not produce fibrous particles.

Tridymite and cristobalite. The idealized structures of tridymite and cristobalite are closely related in that they are based upon different stacking sequences of the same module. This fundamental building module is a sheet containing hexagonal rings of silica tetrahedra (Fig. 9), and the tetrahedra within this sheet alternately point above and below the plane defined by the basal oxygen atoms. In tridymite, these sheets are stacked such that the hexagonal rings directly overlie one another, creating continuous tunnels (~4.5 Å across) normal to the sheets (Fig. 10). These tunnels account for the extremely low density of tridymite (2.26 g/cm^3), and they may provide sites for catalytic activity in the fashion of zeolites. Bonding of the sheets in tridymite is achieved by 180° rotation of a given sheet relative to the sheets above and below, and the stacking sequence is denoted by a doubled periodicity of the fundamental module: ...ABABAB... along the stacking direction, where A refers to one type of sheet and B refers to its rotated equivalent.

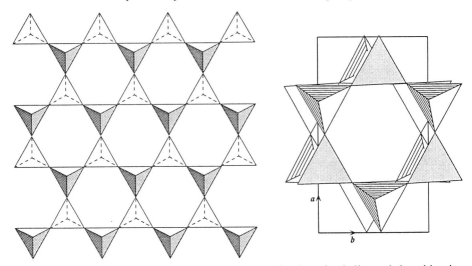

Figure 9. (left) The fundamental stacking module for the cristobalite and the tridymite structures consists of a sheet of silica tetrahedra arranged in hexagonal rings.

Figure 10. (right) In tridymite, the hexagonal rings superimpose so as to create extended tunnels normal to the sheets along **c**. Courtesy of E.A. Smelik.

The stacking sequence in cristobalite involves three tetrahedral sheets that are translated relative to one another in a repetitive fashion: ...ABCABC... As a result of this stacking sequence, the hexagonal rings do not superimpose in cristobalite (Fig. 11), and cristobalite lacks the continuous tunnels normal to the layers that are seen in tridymite. However, tunnel structures can be found parallel to the layers, and the density of cristobalite is only slightly higher than that of tridymite (2.32 g/cm^3). Because tridymite and cristobalite are built from the same stacking module, these minerals may be intimately intergrown with each other (e.g., ...ABABABCAB...), giving rise to high densities of stacking faults (Fig. 12a) (Thompson and Wennemer, 1979; Carpenter and Wennemer, 1985).

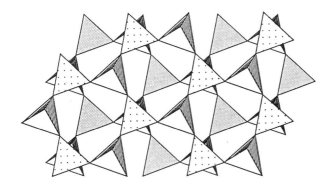

Figure 11. Part of the cristobalite structure normal to the stacking direction along [111]. Courtesy of E.A. Smelik.

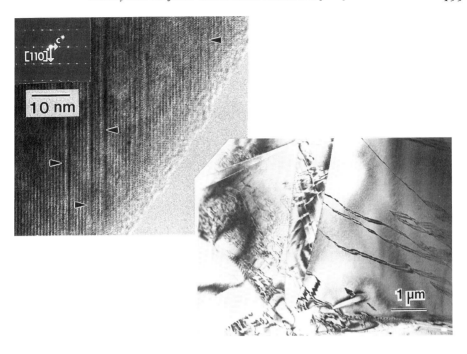

Figure 12. (a) High resolution TEM image of tridymite reveals numerous stacking faults (arrowed) along the (001) planes, giving rise to streaks along c^* in selected area electron diffraction patterns (inset). Courtesy of E.A. Smelik. (b) Dark-field TEM images of cristobalite exhibit twin and antiphase domains. Courtesy of Gordon L. Nord, Jr.

Cristobalite is stable at low pressure between 1728°C and 1470°C, and tridymite is stable between 1470°C and 870°C. As noted above, both of these polymorphs exist metastably at room temperature. The failure of these phases to invert to quartz is related to the dramatic architectural changes required by the transformation. Unlike the displacive transition between β- and α-quartz, the transitions from cristobalite and tridymite to α-quartz necessitate the breaking and reassembly of chemical bonds. These *reconstructive* transitions are characterized by high activation energies, and they are extremely sluggish at low temperatures.

On the other hand, both cristobalite and tridymite experience metastable displacive transitions analogous to the α- to β-quartz inversion. Idealized high tridymite is hexagonal, with space group $P6_3/mmc$ (Gibbs, 1926). However, at temperatures below ~300°C, tridymite undergoes a series of transitions to lower symmetry polymorphs. The nature of these low-temperature structures remains one of the more hotly debated topics in mineralogy today, and reviews of the voluminous literature on this subject can be found in Carpenter and Wennemer (1985) and Smelik and Reeber (1990). In short, it seems clear that no two tridymite crystals behave identically as a function of temperature; the number of distinct tridymite polymorphs reported below 300°C has ranged from 2 (e.g. Buerger and Lukesh, 1942) through 5 (e.g., Wennemer and Thompson, 1984).

Moreover, the critical temperatures at which one metastable variant transforms to another are notoriously unreproducible among different specimens. This crystallographic eccentricity may be attributed to the different ways in which the hexagonal rings can collapse as the temperature is lowered, so that framework distortion may vary markedly over short length scales (Dollase and Baur, 1976). Though not well documented, the transition from hexagonal to orthorhombic or monoclinic symmetry in tridymite must be accompanied by the formation of twin domains.

The transition from high to low cristobalite is better behaved than that of tridymite, but it is not without its own charm (Fig. 12b). Well-crystallized cristobalite transforms sharply from a cubic ($Fd3m$) to a tetragonal ($P4_12_12$) polymorph at ~270°C (Dollase, 1965; Leadbetter and Wright, 1976). This transition involves the formation of 12 distinct twin and antiphase domains (Hatch and Ghose, 1991; Nord, 1992). Antiphase domains are similar to twins, but they are related to each other through translation rather than rotation, reflection, or inversion. Using TEM, Lally et al. (1978) have imaged these domains, but full characterization of the many microstructural defects within tetragonal cristobalite has not yet been achieved.

Coesite and stishovite. The structures of the high-pressure polymorphs of silica are discussed briefly here. Their rare occurrence alone precludes a serious health risk, and epidemiological studies indicate that they are biologically inert (e.g., Wiessner et al., 1988). The tetrahedral framework of coesite is topologically similar to that of the feldspar group (Fig. 13). Rings of four silica tetrahedra are joined to form chains parallel to [001] and [110], and the space group is $C2/c$ (Zoltai and Buerger, 1959; Levien and Prewitt, 1981). The density of coesite is 3.01 g/cm^3, and TEM studies of natural coesites have revealed remarkably few dislocations, or deviations from the ideal crystal structure (Ingrin and Gillet, 1986).

The polymorph with the highest density is stishovite (4.35 g/cm^3), which is remarkable among silica phases for hosting Si in octahedral rather than tetrahedral coordination. Stishovite is isostructural with rutile (Fig. 14). It has space group $P4_2/mnm$, and it consists of chains of edge-sharing octahedra that form tunnels parallel to **c** (Sinclair and Ringwood, 1978; Hill et al., 1983; Spackman et al., 1987). Although stishovite is a rare crustal phase, it may be a more common constituent of the mantle (Navrotsky et al., 1992), and most stuctural studies of stishovite have focused on crystal changes at high pressures and temperatures (Bassett and Barnett, 1970; Endo et al., 1986; Ross et al., 1990).

Amorphous silica. Silica glasses differ from the crystalline polymorphs in that they lack long-range atomic order. Included in this category are opal, diatomaceous earth, Si-rich fiberglass, quartz glass, fused quartz, and silica glass. Since it is not possible with glasses to define specific atomic positions within a unit cell, determinations of glass structures attempt to describe atomic arrangements on a local scale. The kinds of structural characteristics that may be quantified include interatomic bond angles, average distances among atoms, and ring topologies. Experimental methods that exploit the translational symmetries

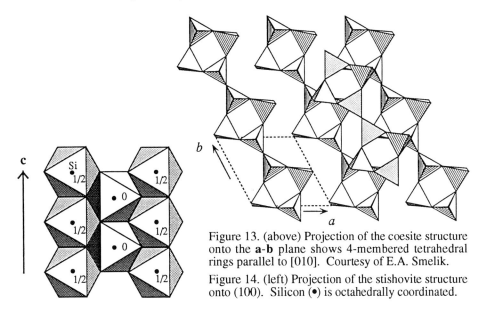

Figure 13. (above) Projection of the coesite structure onto the **a-b** plane shows 4-membered tetrahedral rings parallel to [010]. Courtesy of E.A. Smelik.

Figure 14. (left) Projection of the stishovite structure onto (100). Silicon (•) is octahedrally coordinated.

present in crystals are of limited usefulness, so that diffraction techniques must be supplemented by spectroscopies that are sensitive to near-atomic environments.

Raman spectroscopy of natural glasses reveals little evidence for non-bridging oxygens; in tektites and obsidians virtually all oxygen anions are bonded to two tetrahedral cations, and the framework is fully polymerized (White and Minser, 1984). As with the crystalline polymorphs, the coordination of Si in glasses increases from 4 to 6 with increasing pressure; however, the existence of five-fold coordinated Si at atmospheric pressure is suggested by molecular dynamical (MD) and nuclear magnetic resonance (NMR) studies of silicate melts (Woodcock et al., 1976; Kubicki and Lasaga, 1988; Stebbins and McMillan, 1989; Stebbins, 1991). On the other hand, an NMR investigation of synthetic and natural opals reveals no evidence for Si in five- or six-fold coordination (De Jong et al., 1987).

Neutron diffraction patterns of obsidian, tektite, lechatelierite, and fulgurite are only slightly different from the pattern produced by synthetic silica glass (Wright et al., 1984). The diffraction peaks in these patterns arise from commonly occurring interatomic periodicities (sometimes called radial distribution functions), and they agree with those expected from MD calculations at ~1.60 Å (Si–O), 2.52 Å (O–O), 3.2 Å (Si–Si), and 4.3 Å (Si–O[2]) (Kubicki and Lasaga, 1988). X-ray diffraction of opals reveals a broad hump corresponding to a d-spacing of ~4.10 Å; this interplanar distance corresponds to the thickness of the tetrahedral sheets (Fig. 9) that comprise the structures of cristobalite and tridymite (Jones and Segnit, 1971).

Raman (Dowty, 1987) and NMR (De Jong et al., 1987) spectra of amorphous silica most closely resemble the tridymite spectrum among the crystalline silica polymorphs. Tridymite consists of fairly regular 6-membered rings, and it seems likely that 6-membered rings predominate in at least some types of amorphous silica (Zoltai and Buerger, 1960; Dowty, 1987; Kubicki and Lasaga, 1988); by extension, the distribution of Si–O–Si bond angles in opaline silica also may be most similar to that in tridymite (De Jong et al., 1987). This structural resemblance between amorphous silica and tridymite explains the diagenetic transformation of opal-A to metastable opal-CT rather than to quartz. As suggested by Ostwald's Step Rule, silica diagenesis follows the route that is kinetically rather than thermodynamically most favorable (Jones and Segnit, 1972; Morse and Casey, 1988).

The structure of opaline silica differs from that of other silica glasses on the mesoscopic scale. As revealed by Darragh et al. (1976), opal often consists of closest-packed spheres of amorphous silica that typically measure 150 to 300 nm in diameter. The interstices between the spheres can contain water vapor or liquid, and when the spheres are uniformly sized and strongly cemented together, the opals will diffract visible light, thereby giving rise to the flash of color that is characteristic of gem opal. Upon diagenesis to opal-CT, the spheres are replaced by bladed crystals in rosette-like aggregates (Flörke et al., 1976). As with chalcedony fibers, cleavage of opal is not controlled by particle morphology; fracture planes in opal pass through the spheres rather than around them.

Impurity elements

The crystalline polymorphs of silica do not occur in complete solid solution with any other natural systems at low temperatures and pressures. Trace element analyses suggest that the major impurity species include Al, Fe, Ti, Li, Na, K, and Ca (Frondel, 1962). These impurity cations generally substitute in low concentrations, and they may reveal themselves by imparting characteristic colors to their host. For instance, rose quartz contains Ti^{3+}, amethyst Fe^{3+}, and smoky quartz Al^{3+} (Cohen and Makar, 1985; Cohen, 1985). The concentrations of these cations vary substantially from specimen to specimen, but they rarely exceed 1.0 wt % oxide.[2] Indeed, analyses of 24 microcrystalline silica specimens (flint, chalcedony, moganite, and opal) reveal no impurity cations in excess of 0.22 wt % oxide (Flörke et al., 1982; Graetsch et al., 1987), and most impurity elements are present at levels below 0.05 wt % oxide.

The crystallographic sites and the oxidation states for these cations remain somewhat ambiguous. However, Graetsch et al. (1987) observe that the monovalent cation concentration in microcrystalline quartz generally equals that of the trivalent cations; this pattern suggests that trivalent Al and Fe substitute for tetrahedral Si, and monovalent Na and K occupy interstitial sites to maintain charge balance. By analogy with the rare mineral β-eucryptite, the impurities in quartz probably occupy the tunnels parallel to the **c** axis. β-eucryptite ($LiAlSiO_4$) is a *stuffed derivative* of quartz (Buerger, 1954). The silicate framework is isostructural with that of quartz, but Li and Al exchange for Si and vacancies. In

---

[2] Wt % oxide is a typical way to report compositions of minerals (because most minerals consist of cations coordinated by oxygen). A quartz containing Al at a level of 1 wt % oxide would have a composition of SiO_2 99% and Al_2O_3 1%.

β-eucryptite, Al^{3+} is tetrahedrally coordinated, and the Li^+ cations reside within the tunnels (Schulz and Tscherry, 1972). These channel cations move freely along the [001] direction, and eucryptites are considered one-dimensional superionic conductors (Nagel and Böhm, 1982). Similarly, ionic conductivity in quartz is higher parallel to **c** than normal to **c**.

Perhaps the most significant impurity that occurs within the low-pressure silica polymorphs is hydrogen, present either as a hydroxyl group within silanol (SiOH) or as molecular water (H_2O). The speciation of hydrogen in quartz is of particular importance to structural geologists and materials scientists. In a phenomenon known as hydrolytic weakening, low stresses induce plastic deformation in hydrogenated quartz crystals, whereas hydrogen-free quartz is brittle and strong (Griggs and Blacic, 1965). Aines et al. (1984) suggest that most of the interstitial H in synthetic quartz (<1 wt % H_2O) occurs as small groups of water molecules. By contrast, microcrystalline quartz can contain 1 to 2 wt % H_2O, which is equally distributed between molecular water and silanol (Frondel, 1982; Flörke et al., 1982; Graetsch et al., 1985). The molecular water probably is present in interstices at high-angle grain boundaries, and the hydroxyl groups occur as isolated defects within the structure or perhaps along the Brazil twin boundaries. Opaline silica, often represented as $SiO_2 \cdot nH_2O$,[3] may contain up to 20 wt % water, but water contents between 3 and 10 wt % are more typical. In opal, the H_2O occurs both as liquid water in micropores and as individual molecules trapped within the structure (Langer and Flörke, 1974).

Solubilities and surface structure

The dissolution behavior for the silica polymorphs follows the sequence of silica diagenesis; solubility decreases from amorphous silica to cristobalite/tridymite to chalcedony to quartz (Fournier, 1977; Walther and Helgeson, 1977). The aqueous fluids that are in equilibrium with the different polymorphs are fairly well constrained with respect to silica concentration: quartz with 6 to 10 ppm SiO_2; chalcedony with ~30 ppm SiO_2; cristobalite/tridymite with ~100 ppm SiO_2; and amorphous silica with ~200 ppm SiO_2. These fluids may contain bimodal distributions of silica monomer with silica polymer, and cross-bonding within the polymer population appears to increase with increasing silica concentration (Crerar et al., 1981; Heaney, 1993). As expected, silica solubility also increases with smaller particle size and larger surface area (Gislason et al., 1993).

At low pH (pH = 1–6), the dissolution process produces silicic acid: $SiO_2(s) + 2 H_2O(l) = Si(OH)_4(aq)$. At higher pH, other aqueous species, such as $[SiO(OH)_3]^-$ and $[SiO_2(OH)_2]^{2-}$, are more abundant (Brady and Walther, 1990), and the solvated silica grows increasingly polymerized (Iler, 1979; Schwartzentruber et al., 1987). Moreover, up to a pH of 6, little change occurs in the dissolution rate, but with higher alkalinity the kinetics of dissolution increase by

---

[3] The "· nH_2O" notation signifies that the H_2O exists as molecular water that is not integral to the framework structure. An opal with 10 wt % H_2O would have a formula of $SiO_2 \cdot 0.37\ H_2O$.

more than 2 orders of magnitude both in vitreous silica (Wirth and Gieskes, 1979) and in quartz (Knauss and Wolery, 1988).

The hydrolysis mechanism involving solutions at near-neutral pH appears to follow a two-step process. The first step involves the saturation of the silica surface with silicic acid by the following reaction:

$$-\text{Si}-\text{O}-\text{Si}-\text{OH} \,(s) + \text{H}_2\text{O} = -\text{Si}-\text{O}-\text{Si}(\text{OH})_3 \,(s) \tag{1}$$

The second dissolution step requires severance of the Si–O bond below the saturated surface and produces free monomeric silicic acid:

$$-\text{Si}-\text{O}-\text{Si}(\text{OH})_3 \,(s) + \text{H}_2\text{O} = -\text{Si}-\text{O}-\text{Si}-\text{OH} \,(s) + (\text{H}_4\text{SiO}_4)^0 \,(aq) \tag{2}$$

This second reaction is the rate-determining step in silica dissolution (Fleming, 1986). Rimstidt and Barnes (1980) calculate activation energies for this step of ~70 kJ/mol for quartz, ~66 kJ/mol for cristobalite, and ~62 kJ/mol for amorphous silica. By contrast, these authors observe that when simple diffusion from the surface is the rate-limiting step in dissolution, activation energies range from 17 to 25 kJ/mol, thereby explaining the sparing solubility of silica in aqueous systems.

Solubility of silica is influenced not only by variations in pH but also by the presence of ions in the solution and in the solid phase. For instance, the solubility of silica glass generally decreases as the ionic strength of the electrolyte solution increases. Marshall and Warakomski (1980) observe that this "salting out" behavior is dependent upon the types of cation species in the solution. Specifically, they note that the following cations (in decreasing order of effectiveness) act to suppress silica solubility: $Mg^{2+} > Ca^{2+} > Li^+ > Na^+ > K^+$. Those cations that require the greatest number of hydration anions appear to decrease solubility the most, suggesting that removal of "free" water from solution inhibits Reactions 1 and 2 by decreasing the activity (or thermodynamic concentration) of water.

Surface charge may exert a complementary control on solubility behavior. Both the point-of-zero-charge (pzc) and the isoelectric point for quartz occur below a pH of 2, whereas the pzc for rutile is found at pH = 6 (Li and De Bruyn, 1966). Therefore, unless the ambient solution is highly acidic, the quartz surface is negatively charged from the dissociation of the surface silanol groups: $SiOH + H_2O = SiO^- + H_3O^+$. However, the surface potential of silica at high pH may be depressed in solutions with high ionic strength due to the adsorption of cations to the silica surface, and this adsorption in turn may decrease solubility rates. For instance, for solutions with pH greater than 9, Mg^{2+} strongly adsorbs to the surface of silica glass, and the dissolution rate decreases from that at pH = 8 by 30% (Wirth and Gieskes, 1979). On the other hand, high concentrations of anions in solution will maintain strongly negative surface charge, and these species serve to catalyze dissolution. Wirth and Gieskes (1979) note that Cl⁻

anions enhance dissolution rates of silica glass over the entire pH range tested (pH = 5 to 11).

Solubility of silica in electrolytic solutions is not merely a function of masking and unmasking the surface charge. Even though certain cations may lower solubility rates by adsorption to the surfaces of specific polymorphs, with other silica phases the cations may form complexes that promote dissolution. These complexes may increase silica solubility by the creation of stable silica-ligand species that lower the activity of silicic acid in the solution, and the complexes may increase solubility rates by providing a mechanism for the breakage of Si–O bonds with low activation energies. As an example, quartz exhibits higher solubilities (Fournier, 1983) and higher solubility rates (Dove and Crerar, 1990) in NaCl solutions than in pure aqueous solutions. By contrast, silica glass is less soluble in fluids with high ionic strength. The increase in the solubility rate for quartz is much more dramatic than the enhancement in solubility, and Dove and Crerar (1990) speculate that Na^+ complexes with surface silanol in such a way that the remaining Si–OH bonds are more accessible to hydrolysis.

Evidence also supports complexation of silicic acid with sulfate to produce a species with composition $[Si(OH)_4 \cdot SO_4]^{2-}$ (Marshall and Chen, 1982). Likewise, Reardon (1979) observes enhanced solubility of amorphous silica in low pH solutions with varying concentrations of ferric iron, and he presents evidence for silica–iron complexing according to the following reaction: $[SiO(OH)_3]^- + Fe^{3+} = [FeSiO(OH)_3]^{2+}$. On the other hand, the solubility of silica is depressed when Al contaminates the surface of quartz (Beckwith and Reeve, 1969), perhaps due to the formation of a resistant aluminosilicate at the reaction site.

Interactions between organic molecules and silica

In addition, a number of studies have demonstrated complexing of SiO_2 with organic molecules. Following work of Seifert (1966), Hartwig and Hench (1972) demonstrated epitaxial growth of poly-L-alanine on the (100) and the (101) faces of left-handed quartz. Poly-D-alanine would not grow in an oriented fashion on left-handed crystals of quartz, and neither poly-L-alanine nor poly-D-alanine grew epitaxially on to a ceramic glass. Thus, quartz selectively bonds with amino acids on the basis of a shared chirality. In addition, several researchers have observed altered rates of quartz dissolution in the presence of organics, including soils (Crook, 1968), Na-adenosine triphosphate or ATP (Evans, 1964), and pyrocatechol (Jørgenson, 1967; Iler, 1979).

The assays of Bennett et al. (1988) suggest that multi-protic and multi-functional organic acids tend to enhance quartz solubility; in order of decreasing reactivity, citrate, oxalate, salicylate, and acetate all accelerated quartz dissolution. On the basis of ultraviolet spectroscopy, Bennett et al. (1988) cite evidence for the formation of a silica–organic complex. Similarly, Gaffney et al. (1989) use Raman spectroscopy to demonstrate the complexation of silicic acid with oxalic, citric, succinic, malonic, maleic, tartaric, and phthalic acids. These results would seem to support the possibility of direct C–O–Si bonding between cellular structures and particulate silica. In fact, coating silica surfaces with organic polymers such as polyvinylpyridine-N-oxide (PVPNO) (Nash et al., 1966; Nolan et al., 1981) and (–CN), (–CH$_3$), (`NH$_3$), and –(N(CH$_3$)$_3^+$) (Wiessner et al., 1990) may alter the biological activity of the material.

Carcinogenicity of silica

Studies of the pathogenicity of silica suggest that tridymite and cristobalite are more carcinogenic than quartz, whereas amorphous silica is comparatively unreactive (Saffiotti, pers. comm.). From a purely structural perspective, the enhanced carcinogenicity of tridymite is puzzling, since tridymite more than any other silica polymorph resembles silica glass. The Raman and NMR spectra of tridymite reveal a close kinship with amorphous silica over short-range scales, and the values for solubility and free energy of tridymite are close to those for silica glass. Moreover, the high defect densities observed in tridymite impart an aperiodic character, which is more characteristic of amorphous silica (i.e., silica glass) than crystalline silica (e.g., quartz).

It would seem difficult to reconcile the structural similarities between tridymite and glass with their different levels of toxicity unless silica plays an indirect role in cellular transformation. A number of studies have suggested that silica serves as a catalyst for reactions that produce free radicals and other agents that damage DNA. In contrast to silica glasses and, to a lesser extent, quartz, the tunnels of tridymite and cristobalite offer reactive sites that may promote the formation of peroxides and free radicals. These tunnels are continuous and periodic, and they offer larger internal surface areas with charged sites to enhance reactivity. Future experiments involving high-silica zeolites, such as ZSM-5 and ZSM-11, should clarify the role of silica catalysis.

GEOCHEMISTRY OF IRON AND TITANIUM OXIDES

Introduction

Although Fe-, Ti-, and Fe–Ti-oxide minerals rarely constitute more than a few percent of any single rock (exceptions include Fe-ore and anatase deposits), they are common minor phases in unconsolidated surficial materials and sedimentary, metamorphic, and igneous rocks. Their structures are simple compared with those of amphibole, zeolites, and sheet silicates. Like the silica phases, however, they adopt multiple polymorphs, and their metastable and stable phase relations are complex. The Fe- and Fe–Ti-oxides are distinguished by their relatively high conductivity (compared to most minerals) and by their magnetic properties (see Banerjee, 1991, for an excellent overview of the magnetic properties of Fe–Ti oxides).

Fe- and Ti-oxides are commonly employed in the manufacture of everyday materials. For example, Ti-oxides are widely used in paint, toothpaste, pill capsules, soap, sunscreen, paper, and a variety of foods, and Fe-oxides are used in magnetic recording media. Ti-oxides also are exploited for their surface catalytic properties and are of industrial interest as photocatalysts. Technologically useful catalytic materials generally have exceedingly small particle sizes (orders of magnitude smaller than particles used in paints, for example, but in the size range sometimes encountered in natural materials). All surface-related properties of finely crystalline materials vary dramatically from those of more coarsely crystalline materials. Thus, in assessing the surface activity it may be important to distinguish between tests that involve nanocrystalline and micron-sized materials. In fact, recent studies on the differences in biological activity for very-fine-grained (e.g., <0.01 µm) and larger-grained (e.g., 1 µm) examples of the same material suggest that the very-fine-grained material does, indeed, exhibit a higher biological activity (e.g., Driscoll and Maurer, 1991; Driscoll, pers. comm.). Nevertheless, some studies have reported a low activity for very-fine-grained material (e.g., Ferin and Oberdörster, 1985).

The minerals described in this section have compositions that fall on the TiO_2–FeO–Fe_2O_3 compositional diagram (Fig. 15). Structural data are summarized in Table 2. Pseudobrookite ($Fe^{3+}_2TiO_5$) and ferropseudobrookite

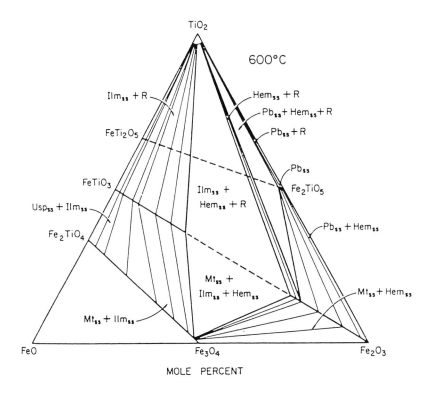

Figure 15. Triangular diagram from Haggerty (1976) showing relationships among minerals in the Fe-Ti-O system at 600°C. Minerals include wüstite (FeO), rutile, anatase, brookite, and $TiO_2(B)$ (TiO_2), magnetite (Fe_3O_4), hematite and maghemite (Fe_2O_3), ulvöspinel (Fe_2TiO_4), and ilmenite ($FeTiO_3$). Tie lines indicate approximate compositional limits for ulvöspinel-magnetite and ilmenite-hematite.

Table 2

Structural details for the four TiO_2 minerals, hematite, ilmenite, magnetite, ulvöspinel, maghemite, and titanomaghemite

Structure	Space group	Density (g/cm^{-3})	Unit cell data (nm)	Reference
Rutile	$P4_2/mnm$	4.13	$a=9{,}4584, c=9{,}2858$	Gonschorek and Feld (1982)
Anatase	$4_1/amd$	3.79	$a=0.3784, c=0.9515$	Horn et al. (1972)
Brookite	P bca	3.99	$a=0.917, b=0.546, c=0.514$	Meagher and Lager (1979)
TiO_2 (B)	C 2/m	3.64	$a=1.216, b=0.374, c=0.651, \beta=107.29°$	Banfield et al. (1991)
Hematite	$R\bar{3}$	5.26	$a=0.5035, c=1.4086$	Blake et al. (1966)
Ilmenite	$R\bar{3}$	4.70	$a=0.5088, c=1.4086$	Weschler and Prewitt (1984)
Magnetite	$Fd\bar{3}$	5.18	$a=0.8396$	Weschler et al. (1984)
Ulvöspinel	$Fd\bar{3}$	4.78	$a=0.8535$	Weschler et al. (1989)
Maghemite	$P3_22_12$	4.88	$a=0.8340, c=2.496$	Greaves (1983)
Titanomaghemite	$P4_132$	Var.	$a=0.8341$	Collyer et al. (1988)

($FeTi_2O_5$) are relatively uncommon high-temperature phases not considered here. The discussion below will focus on the Ti-oxides because these materials have been well studied and because they illustrate an approach that might be brought to bear in toxicity assessment studies of more complex minerals.

The TiO_2 system

Occurrence and commonly associated minerals. TiO_2 adopts at least 7 distinct structures, all of which are shared by many other minerals and inorganic compounds. Of the titania polymorphs, only rutile, anatase, brookite, and TiO_2(B) are known to occur in nature. Anatase, brookite, and TiO_2(B) develop in hydrothermal veins and replace Ti-rich minerals and glass during weathering (e.g., Banfield et al., 1991a, 1991b; Banfield and Veblen, 1992; Mazer et al., 1992). Thus, these minerals are common minor constituents of sediments and soils, and they coexist with a variety of clay minerals and their accessories, such as Fe- and Al-oxides and oxyhydroxides, quartz, feldspar, and zeolite.

Brookite is less common than anatase, and its infrequent development is consistent with its higher structural metastability (Post and Burnham, 1986). Brookite is difficult to synthesize; most attempts have produced anatase or mixtures of anatase and brookite (e.g., Bischoff, 1992). Brookite is most commonly reported as a replacement phase associated with other retrograde reaction products, including chlorite, titanite, etc.

Anatase, brookite, and TiO_2(B) convert to rutile during prograde metamorphism. Rutile is a common accessory mineral in a wide variety of crustal and mantle-derived rocks (e.g., eclogites, kimberlites; see Haggerty, 1991, for review), and it is relatively resistant to weathering. As a result, rutile may be inherited by metasediments.

Crystal structures. All of the TiO_2 polymorphs contain Ti cations that are octahedrally coordinated by oxygen. These octahedra usually are edge- and corner-linked, although pairs of face-sharing octahedra are found in reduced rutile derivatives (TiO_{2-x}) and in Ti_2O_3. Examination of the actual (as opposed to the idealized) structures reveals highly distorted Ti coordination environments. Although Ti–O distances are relatively constant (e.g., 0.1945 and 0.198 nm in rutile), the distances between next-nearest neighbors vary considerably. For instance, the Ti–Ti bond in rutile ranges from 0.296 to 0.357 nm, and shared and unshared edges are characterized by dramatic differences in O–O bond lengths. (Sources for structure refinements are listed in Table 2.) Because of these structural variations, the densities of the four polymorphs are quite different. The four minerals also display very different surface chemistries. In Figure 16, simplified cation arrangements are depicted within the surfaces that are parallel to oxygen planes. It is likely that adsorption of metals or organics to such surfaces and surface catalyzed reactions will strongly depend on the specific cation configuration.

Rutile. The structure of the most commonly occurring TiO_2 phase, rutile, is shared among a number of minerals including pyrolusite (MnO_2), cassiterite (SnO_2), stishovite (SiO_2), and in a distorted form, marcasite (FeS_2). The structure can be described as based on a distorted closest packed array of oxygens with half the octahedral sites filled. In detail (see Fig. 17), the oxygen planes are puckered, resulting in an anion array halfway between hexagonal and cubic closest packed (Hyde et al., 1974; Hyde and Andersson, 1989). Figure 17d presents a projection of the rutile structure along the c-axis, revealing parallel chains of edge-sharing octahedra.

Brookite and anatase. The structures of rutile and brookite are distinguished from those of TiO_2(B) and anatase by the presence of octahedral chains in two orientations rather than in one. The structure of brookite is more complex than that of the other TiO_2 polymorphs (Meagher and Lager, 1979; Waychunas, 1991). It has been described in terms of doubled (Lindsley, 1976) or mixed (Hyde and Andersson, 1989) hexagonal closest packing of anions. However, brookite is structurally very similar to anatase, and both can be constructed by stacking the same layer (Fig. 18a). A conventional representation of anatase is shown in Figure 18b. The (101)-bounded anatase slabs (Fig. 18b) are stacked in a + + − − sequence (+ refers to the octahedral tilt) to form brookite (Fig. 19a,b). Note that the high-pressure TiO_2 polymorph called TiO_2(II) can be constructed from the same slab (+ − + −), as can rutile if the chains are straight rather than kinked (Fig. 19b).

TiO_2(B). TiO_2(B) has a structure directly related to that of anatase. Figure 19a illustrates the (103)-bounded slabs in anatase that, when reconnected, produce TiO_2(B) (Fig. 20).

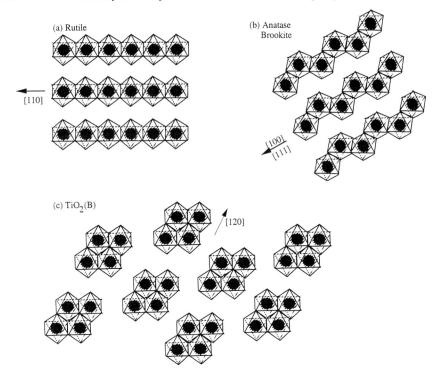

Figure 16. Diagram illustrating arrangements of Ti cations (black spots) at surfaces parallel to oxygen planes in (a) rutile; (b) anatase and brookite; (c) TiO$_2$(B).

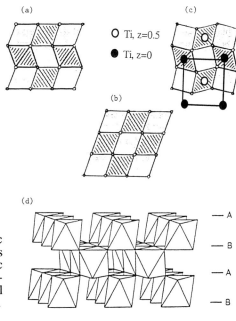

Figure 17. (a), (b), and (c) Schematic anion arrays demonstrate that rutile is intermediate between hexagonal and cubic closest packed. From Hyde and Andersson (1989). (d) Edge-sharing octahedral chains in rutile. From Waychunas (1991).

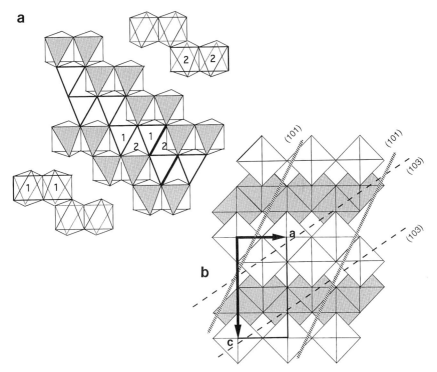

Figure 18. (a) Chains of octahedrally-coordinated Ti cations are arranged in layers (dark stipple) found in both anatase and rutile. Adjacent layers are stacked either in position "1" (same octahedral tilt) or in position "2" (layers rotated by 180°). Different layer stacking sequences distinguish anatase and rutile. (b) Schematic diagram illustrating the anatase structure (viewed down [100]). The (103)-bounded slices of the anatase structure are also found in $TiO_2(B)$, and (101)-bounded slices are present in brookite. Modified from Banfield and Veblen (1992).

Trace element chemistry and defect microstructures. TiO_2 minerals are commonly the predominant host for Ti as well as a variety of minor elements (especially the "high field strength" elements). All polymorphs are white (or colorless) in pure form. The typical red-brown (or occasionally blue) color reflects the presence of impurities, such as Fe. Other common impurities are Nb, Ta, V, Cr, Hf, Zr, Al, and occasionally, Ti^{3+}. Mantle-derived rutiles are rarely stoichiometric, typically containing several wt % oxide (or more) Fe, Cr, Nb, and other impurities (Haggerty, 1991). Rutiles from crustal rocks tend to be more pure, although substitutional Fe is common.

The substitution mechanisms for natural impurities vary. Data for metamorphic rutiles suggest that Fe is accommodated by extended defects that apparently form within homogeneous samples on cooling. These defects are essentially slices of hematite (Fe_2O_3) lamellae intergrown with rutile. Hematite-like arrangements of pairs of face-sharing tetrahedra are systematically related to defects known as crystallographic shear planes (Banfield and Veblen, 1991) that have been reported in synthetic, reduced rutile (e.g., Bursill and Hyde, 1972) and, in rare cases, in natural rutile (Miser and Buseck, 1988). Additional impurity substitution mechanisms involve hydrogen (Vlassopoulos et al., 1990) and coupled trivalent and pentavalent substitutions (Vlassopoulos et al., 1990; Banfield et al., 1993). When such substitutional cations are ordered, superperiodic structures may result. For instance, in trirutile (Ti_2FeO_6) the c-axis is three times that of rutile.

The only other common microstructure in rutile (other than point defects and dislocations) is twinning. The distance between twin composition planes ranges from millimeters to nanometers in highly defective samples. Although {011} is the most common orientation, other twin orientations for the twin composition plane have been reported.

The minor element chemistries of anatase and brookite are similar to that of rutile. Little is known about impurities in $TiO_2(B)$. Extended defects have not been reported in impure anatase and brookite, despite high impurity concentrations.

Particle morphology. TiO_2 minerals produced during low-temperature reactions are typically very finely crystalline. In some cases, crystals may be no more than a few nanometers in diameter (Banfield et al., 1991). Conversely, crystals of rutile, anatase, and brookite found in hydrothermal veins and in metamorphic and igneous rocks may be centimeter-sized.

Several-micrometer-wide aggregates of nanometer-sized, randomly oriented, diamond-shaped anatase crystals are common in metasediments (Banfield et al., 1991). Although rutiles in the submicrometer-size range are common, it is not clear that crystals in the sub-0.01-µm-size range are ever produced, either in nature or in the laboratory. Rutile crystals are typically euhedral or subhedral. Micrometer-sized needle-shaped rutile crystals are common in some metasediments (Banfield and Veblen, 1991) and occur as inclusions within other minerals (e.g., rutilated quartz and garnet).

Equidimensional, centimeter-sized brookite crystals occur in some localities, such as Magnet Cove, Arkansas. Brookite may also develop a fibrous habit (Fig. 21). $TiO_2(B)$ has only been reported from three localities to date (Banfield et al., 1991; Banfield and Veblen, 1992; Mazer et al., 1992), and its morphology has not been documented.

Potentially active surface sites. Much of what is known about the reactivity of Ti-oxides, particularly finely crystalline anatase, comes from the ceramics and materials science literature. Anatase is used as a catalyst and catalyst support (Anderson et al. 1988). Materials for gas separation reactions require controlled pore size distributions and particles in the nanocrystalline range (Bischoff, 1992). Other applications include hydrocarbon catalysis and photodegradation of aromatic and aliphatic organics (Barbeni et al., 1987, Hashimoto et al., 1984; Tunesi and Anderson, 1991; Sabate et al., 1990, 1992;

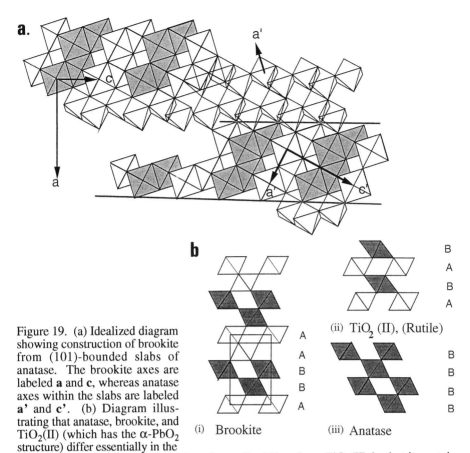

Figure 19. (a) Idealized diagram showing construction of brookite from (101)-bounded slabs of anatase. The brookite axes are labeled **a** and **c**, whereas anatase axes within the slabs are labeled **a'** and **c'**. (b) Diagram illustrating that anatase, brookite, and TiO$_2$(II) (which has the α-PbO$_2$ structure) differ essentially in the stacking sequence of this slab. Note that rutile differs from TiO$_2$(II) in that it contains kinked rather than straight chains. Modified from Banfield and Veblen (1992).

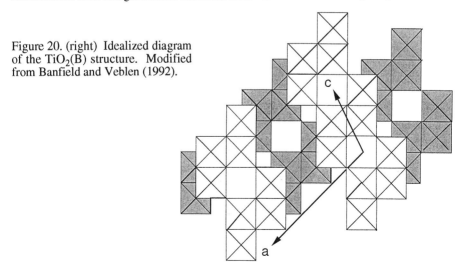

Figure 20. (right) Idealized diagram of the TiO$_2$(B) structure. Modified from Banfield and Veblen (1992).

Figure 21. Selected area electron diffraction pattern (a) and TEM image (b) of fibrous brookite from a sediment from Utah. Note the high aspect ratio of the crystals, which are elongate parallel to their **c** axes.

Yamazaki-Nishida et al., 1993). Although photodegradation generally involves ultraviolet irradiation, impurities (especially Cr) within anatase (and rutile) change the band characteristics, making photocatalysis within the visible region possible.

It is likely that surfaces of a TiO_2 polymorph, particularly small particles with complex surface geometries, offer a heterogeneous array of sites for chemisorption. Mechanisms by which organic and inorganic compounds adsorb to the surface of TiO_2 depend upon the detailed coordination environment of surface atoms. Sites vary with crystallographic orientation of the surface, location on the surface, particle size, and particle morphology.

Infrared studies of adsorption of carbon monoxide on anatase confirm the existence of heterogeneous arrays of adsorption sites (Garrone et al., 1989). Tunesi and Anderson (1991) suggest that the sites most active in controlling adsorption are Ti atoms exposed at the oxide surface that are four-coordinated to oxygen and have two unfilled orbitals. Infrared studies by Carrisoza et al. (1977) indicate that these sites have the highest Lewis acidity of all surface sites, because cations can accept two pairs from electron donors to achieve complete octahedral coordination. Carrisoza et al. (1977) calculated that 1.9 four-coordinated Ti octahedral sites were present per square nanometer of surface area (total surface area 25 m^2/g). Catalytic decomposition of alcohol in the gas phase is also attributed to this type of surface site (Anpo, 1989). Molecules that cannot bind to the surface Ti cations to form a ring appear to be less prone to adsorb onto anatase surfaces (Tunesi and Anderson, 1991). Similar models have been invoked to explain surface complexes formed between salicylate and goethite.

Two different mechanisms, both involving charge transfer (Gutierrez and Salvador, 1986), have been proposed for halide (e.g., iodine) oxidation at TiO_2 surfaces under irradiation. Proposed semiconductor-mediated photocatalytic mechanisms include direct charge transfer from the semiconductor to the dissolved benzene derivative (Hashimoto et al., 1984) and generation of radicals from water decomposition which then attack the aromatic ring. Detailed photo-oxidation mechanisms are discussed by Tunesi and Anderson (1991). These authors also infer that reaction pathways vary with acidity and with concentration of organic molecules in solution. They suggest that electron transfer predominates at low pH and indirect hydroxyl attack at high pH.

It is generally believed that, for materials applications, surface properties are polymorph specific. For example, formic acid reactions are believed to be promoted by anatase but not by rutile (Anderson, pers. comm.). The fundamental differences between anatase and rutile surface structures and surface energies can be established experimentally and predicted based on likely surface topologies (see Fig. 16). For example, Figure 22 compares the attenuated total reference (ATR) spectra of water adsorbed to anatase and rutile. Different spectral characteristics clearly illustrate distinctive arrays of adsorption sites on the two minerals.

The relative toxicities of anatase and rutile have received some attention in the medical literature. The theoretical probability of different biological effects of different TiO_2 polymorphs has been discussed (e.g., Maata and Arstila, 1975; Rode et al., 1981). Zitting and Skytta (1979) support the view that rutile is a nuisance dust (American Conference of Governmental Hygienists, 1984). However, these authors reported that anatase induced significant *in vitro* hemolytic activity and noted that pulmonary clearance of anatase was much slower than that of rutile. Ferin and Oberdörster (1985) exposed rats to aerosols of both anatase and rutile to further evaluate anatase versus rutile particle clearance from the lung. Ferin and Oberdörster's results contradict those of Zitting and Skytta; they conclude that there is no evidence that the crystal structure of TiO_2 alters the biological effects of TiO_2 particles.

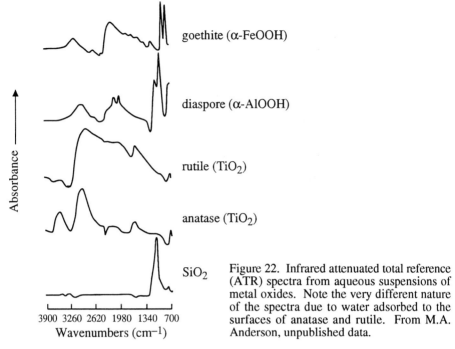

Figure 22. Infrared attenuated total reference (ATR) spectra from aqueous suspensions of metal oxides. Note the very different nature of the spectra due to water adsorbed to the surfaces of anatase and rutile. From M.A. Anderson, unpublished data.

Solubility. A number of studies indicate that it is extremely difficult to solubilize Ti (i.e., to dissolve TiO_2 minerals) under natural conditions (Wedepohl, 1969; Orians, et al., 1990; Van Buren and Burnham, 1992). Under Eh and pH conditions where water is stable and over a considerable range of temperatures, Ti occurs as a very insoluble oxide species (Lee, 1981).

Despite this, ample evidence exists for Ti mobility. For example, anatase cemented sandstones (e.g., Yau et al., 1987), rutile-rich skarns (e.g., Nakano et al., 1989) and pegmatites (Willis et al., 1989), rutile-dominated veins in low-grade metamorphic rocks, and mantle-derived rutile nodules (Haggerty, 1991) are not uncommon. Selverstone (1990) reports rutile vein assemblages and rutile daughter minerals in fluid inclusions in subduction zones. Numerous workers have studied titanium oxide solubility and hydrolysis of Ti^{4+} in aqueous solutions (e.g., Nabivanets and Lukachina, 1964; Ayers and Watson, 1991). Anatase and rutile solubility in 1M NaCl solutions at 200°C and 300°C have also been investigated (Schilling and Vink, 1967). Various aqueous species advocated to explain Ti mobilization include CO_3^{2-}, Cl^-, F^-, PO_4^{3-}, K-titanates, and organic complexes (Kraynov, 1971, Hynes, 1980; Giere, 1986; Murphy and Hynes, 1986). Considerable experimental and theoretical work has also been undertaken to clarify this issue (e.g., Willis and Shock, 1991; Willis, 1992).

Iron oxides and iron–titanium oxides

Oxides of Fe are found in igneous, metamorphic, and sedimentary rocks. Minerals consisting only of oxygen and iron include magnetite (Fe_3O_4), maghemite (γ-Fe_2O_3), hematite (α-Fe_2O_3), and wüstite (FeO). [See Chapter 4 for a discussion of hydroxide minerals.] Ilmenite ($FeTiO_3$) and ulvöspinel (Fe_2TiO_4) represent the Fe–Ti end-members whose structures are derived from those of hematite and magnetite respectively. At elevated temperatures, ilmenite displays a complete solid solution series with hematite and magnetite with ulvöspinel. Maghemite and titanomaghemite are oxides formed by oxidation of magnetite and ulvöspinel–magnetite solid solutions. All of these minerals have relatively simple structures that are well described by reference to cation sites within either hexagonally or cubic closest packed oxygen arrays.

Occurrence and commonly associated minerals. Hematite and magnetite are among the most ubiquitous accessory minerals in igneous and metamorphic rocks. In metamorphic rocks, the occurrence of magnetite and hematite is determined by the rock composition, temperature, pressure, and fugacity of oxygen and sulfur (Frost, 1991). Amphibole-grade rocks may contain both hematite and magnetite after hematite; however, graphitic metasediments are generally free of these minerals (Frost, 1991). Hematite in igneous rocks usually is produced by oxidation-exsolution of magnetite or retrograde replacement of Fe-bearing minerals. Hematite is a common weathering product in soils. The red coloration in many sediments and soils typically indicates its presence. Magnetite persists as a resistant phase and so is also common in near-surface detrital materials and metasediments. Exceptionally Fe-oxide-rich sediments and metasediments are mined as iron ore. Because of extensive dust production during mining activity, hematite has received particular attention with regard to its toxicity.

Because of the tremendous variation in hematite and magnetite paragenesis, it is not useful to itemize the many minerals that are commonly associated with the iron oxides. However, in the more restricted case of Fe-ores, it is possible to describe a typical mineralogic assemblage. In these deposits, minerals include hematite and magnetite in roughly equal proportions, carbonates (siderite > dolomite > calcite), and silicate minerals (greenalite and chert) (Stanton, 1972). Assemblages dominated by hematite, greenalite, and quartz (± siderite) are common at lower grades. Metamorphism and reduction will drive the following reaction:

$$3\ Fe_2O_3 + Fe_3Si_2O_5(OH)_4 \rightarrow 3\ Fe_3O_4 + SiO_2 + H_2O \tag{3}$$

resulting in magnetite-dominated assemblages (Frost, 1991). Metamorphosed equivalents may contain other Fe-rich silicate minerals, including fayalite, laihunite, grunerite, almandine, hedenbergite, and Fe-rich orthopyroxene.

Ilmenite is a common accessory mineral in a wide variety of igneous, metamorphic, and sedimentary rocks. Because of its resistance to weathering, it frequently is concentrated in beach sands. Ilmenite also occurs as lamellar intergrowths in oxidized magnetite–ulvöspinel (Buddington and Lindsley, 1964). Ulvöspinel with the end-member composition does not normally occur in nature. However, near end-member ulvöspinel is found as exsolution lamellae in magnetite.

Maghemite (γ-Fe_2O_3) is one of the more poorly understood Fe-oxide minerals. Maghemite is metastable with respect to hematite and generally converts to hematite at elevated temperatures (300 to 400°C) or at high pressures (Kushiro, 1960). In nature, maghemite forms by oxidation of magnetite at relatively low temperatures (< 300 to 400°C), and in the laboratory maghemites can be synthesized at room temperature. Smith (1979) reported the development of maghemite along cracks in magnetite. Most data suggest that various degrees of magnetite oxidation results in products that display a complete range of solid solution between magnetite and maghemite. The reaction proceeds by loss of Fe rather than by addition of oxygen (Furata and Otsuki, 1985).

The term titanomaghemite refers to nonstoichiometric spinels displaced from the magnetite–ulvöspinel join due to the presence of excess ferric iron and vacancies (Lindsley, 1976). At temperatures below 1000°C, titanomaghemites generally form by oxidation of ulvöspinel–magnetite solid solutions.

Crystal structures. The structures of the iron oxides and iron-titanium oxides are described in the following section.

Hematite (α-Fe_2O_3) and ilmenite. Oxygen in hematite is approximately hexagonally closest packed. Fe cations occupy $2/3$ of the octahedral sites, producing a rhombohedral structure with space group $R\bar{3}c$ and hexagonal cell parameters: **a** = 0.5034 nm, **c** = 1.3750 nm (Blake et al., 1966). This occupancy pattern is apparent when the structure is viewed normal to the closest-packed planes down [001] (hexagonal axes), as seen in Figure 23a. The (001) surfaces are probably most relevant to studies of surface reactivity.

Hematite contains pairs of octahedrally coordinated face-sharing Fe cations that are separated from face-sharing pairs above and below by vacant octahedra (Fig. 23b). Modular strips that consist of vacant sites sandwiched by face-sharing octahedra run parallel to the **c**-axis, and they are cross-linked to form the three-dimensional structure through shared octahedral edges and corners (stippled pairs in Fig. 23b). Distortions of the octahedra in hematite are manifested by a puckering of the closest packed oxygen sheets, and cations are displaced within the octahedra toward the vacant sites. Fe^{3+} pairs that are separated by oxygen planes are antiferromagnetically coupled (i.e., alternate planes are magnetized in opposing directions) via superexchange through the intervening oxygens (Li, 1956). However, the coupling is imperfect insofar as the spins are not exactly antiparallel, and the resulting structure displays weak "parasitic" magnetism (Dzyaloshinsky, 1958; Moriya, 1960).

Hematite can accommodate very substantial quantities of Ti. Ilmenite ($FeTiO_3$), the Ti-rich end member with the same basic structure as hematite, contains equal quantities of Fe^{2+} and Ti. At high temperatures (above ~700°C) there is complete solid solution between ilmenite and hematite (e.g., Burton, 1991). Fe–Ti order in ilmenite produces face-sharing pairs of Fe and Ti octahedra aligned along **c** and destroys the **c**-glide. The symmetry is lowered to $R\bar{3}$. Spins on alternate Fe sites along the **c** axis are opposed, so ilmenite is antiferromagnetic at absolute zero. The magnetic properties for minerals intermediate in composition between ilmenite and hematite vary in a complex fashion, particularly those that are magnetic above room temperature (Ishikawa and Akimoto, 1957; Ishikawa and Syono, 1963; Lawson and Nord, 1984).

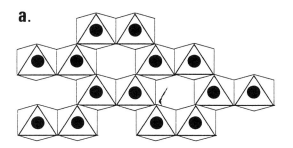
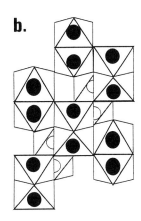

Figure 23. Schematic representations of the hematite structure showing: (a) layers of octahedrally coordinated cations in which $2/3$ of the octahedral sites are occupied (black spots); (b) the stacking sequence of layers shown in (a). Note that pairs of face-sharing octahedra are apparent in (b).

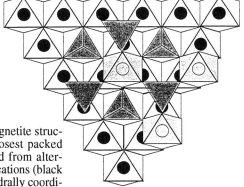

Figure 24. Schematic diagram of the magnetite structure as viewed down [111] on to the closest packed oxygen planes. Magnetite is constructed from alternating layers of octahedrally coordinated cations (black spots) and mixed octahedrally and tetrahedrally coordinated cations (stippled polyhedra).

Magnetite and ulvöspinel. Magnetite (Fe_3O_4) has the inverse spinel structure (Verway and deBoer, 1936). In contrast to normal spinels, the divalent iron cations in magnetite are octahedrally coordinated, and half of the trivalent iron cations are in tetrahedral sites. Thus, the formula for magnetite may be written as $(Fe^{3+})^{IV}(Fe^{2+}Fe^{3+})^{VI}O_4$. Planes of oxygen atoms are stacked along [111] in an almost cubic closest packed array. When viewed along [111], layers containing only octahedral sites alternate with those containing both octahedrally and tetrahedrally coordinated cations (Fig. 24). The [111] view is probably the most useful from the perspective of surface properties because of the prevalence of {111} faces on magnetite crystals.

Ulvöspinel (Fe_2TiO_4), like magnetite, has the inverse spinel structure. Parallel to [111], the sequence of planes is $O-(Fe,Ti)^{VI}-O-Fe^{IV}-(Fe,Ti)^{VI}-Fe^{IV}-O$ (Lindsley, 1976). Fe^{2+} and Ti^{4+} are randomly distributed over the octahedral sites and Fe^{2+} occupies the tetrahedral sites. A number of models have been proposed to describe the cation distributions in members of the ulvöspinel–magnetite solid solution. A strong temperature dependence of ordering patterns was reported by Trestmann-Matts et al. (1983).

Maghemite and titanomaghemite. Maghemite has a defect spinel structure (e.g., $Fe_2\square_1O_3\square_1$), with vacancies ($\square$) generally distributed over octahedral rather than tetrahedral sites. These vacancies reduce the cell edge from 0.8395 nm (magnetite) to 0.832 nm (maghemite). The vacancies may be ordered, but their distribution patterns vary. Consequently, maghemite may adopt a number of lower symmetry structures, some of which are characterized by enantiomorphic space groups, such as *$P4_1$* and *$P4_3$* (Schrader and Büttner, 1963) or *$P4_132$* and *$P4_332$* (Smith, 1979). Like magnetite, maghemite is ferrimagnetic. Much of what is known about this phase results from research motivated by its importance as a magnetic recording material. Evolution of thinking about this phase is summarized by Lindsley (1976) and Waychunas (1991).

Titanomaghemites are distinguished from maghemites by the presence of Ti. Because they are also formed by oxidation of spinels, their structures are analogous to those of maghemite. Studies such as those by Schmidbauer (1987) and Allan et al. (1989) reveal evidence for vacancies on tetrahedral sites and for variations in the amount of tetrahedral Ti.

Wüstite. Wustite (FeO) is rarely encountered at the earth's surface, as it is unstable below 570°C. However, it may be an important constituent of the mantle and core (Hazen and Jeanloz, 1984). Wüstite consists of interpenetrating cubic closest packed arrays of Fe and O, and it is isostructural with NaCl. It always contains some octahedral ferric iron, which is charge balanced by vacancies (McCammon and Liu, 1984).

Trace element chemistry and defect microstructures. The most common impurities found in the magnetite–ulvöspinel and hematite–ilmenite minerals are Mg, Al, Cr and Mn. These impurities generally reflect various degrees of solid solution with end-member spinel group and rhombohedral oxide minerals. Spinels include magnesioferrite ($MgFe_2^{3+}O_4$), jacobsite ($MnFe_2^{3+}O_4$), trevorite ($NiFe_2^{3+}O_4$), franklinite ($ZnFe_2^{3+}O_4$), hercynite ($Fe^{2+}Al_2O_4$), and chromite ($Fe^{2+}Cr_2O_4$). Oxides that are isostructural with hematite include corundum (Al_2O_3), eskolaite (Cr_2O_3), geikielite ($MgTiO_3$), and pyrophanite ($MnTiO_3$).

Mn and Mg substitution in ilmenite is common, particularly in mantle-derived minerals. Substitution of Cr tends to be more restricted when Mn and Mg are present. Many magnetite-ilmenite pairs, especially from basaltic rocks, contain appreciable Mg; for instance, lunar rocks tend to be rich in Mg (Lindsley, 1991). Although corundum shares the hematite structure, Al substitution in hematite is quite limited at moderate temperatures (Turnock and Eugster, 1962).

Metastable aluminous hematites, found in soils and sediments, may form by dehydration of hydrous Fe–Al-oxide precursors (Fysh and Clark, 1982).

Upon cooling, minerals with compositions intermediate between hematite and ilmenite and magnetite and ulvöspinel exsolve, producing fine-scale intergrowths of these minerals (e.g., Nickel, 1958; Price, 1979, 1981). Oxyexsolution of ulvöspinel–magnetite results in the development of ilmenite plates in magnetite (e.g., Haggerty, 1991). Twins are common in spinels along {111} and in the rhombohedral oxides on (001) and {101}.

Particle morphology. The morphology of hematite crystals varies with their origin. Sub-micron-sized crystals occur in cements and are produced during alteration reactions near the Earth's surface. If replacing Fe-silicates, hematite tends to have a plate-like morphology with large extensive surfaces parallel to (001). However, octahedral morphology is typical when hematite replaces magnetite. Rosettes of platy hematite crystals and botryoidal hematite are quite common, particularly in Fe-ores. Ilmenite crystals exhibit morphologies that are similar to those of hematite. Although ilmenite crystals may be more equidimensional, (001) surfaces remain prominent.

Magnetite and ulvöspinel frequently develop as octahedral crystals bound by {111}. The magnetic properties of volcanic glasses may be explained by inclusions of nanometer-scale magnetite crystals (Schlinger et al., 1986). Perhaps the most remarkable magnetite assemblages are those produced within or external to (depending on species) the bodies of bacteria, giving rise to "magnetofossils" (Kirschvink and Chang, 1984). Both fresh water and marine bacteria manufacture uniformly-sized magnetite crystals that are several nanometers in size. Equant octahedra, elongate prisms, and bullet-shaped crystals also have been reported (Vali and Jirschvink, 1989). Mann et al. (1984) describe magnetosomes made up of hexagonal prisms of {011} faces of magnetite. Frankel and Blakemore (1980) suggest that strings of magnetite crystals manufactured within the bacteria serve as a biomagnetic compass.

Potentially active surface sites. The representations of the structures of hematite and magnetite (and, by Ti substitution, ilmenite and ulvöspinel; and by Al substitution in hematite, corundum) in Figures 22 and 24 emphasize the topology of surfaces (and thus cation and anion arrays) most likely to be exposed to solutions. The surface chemistry of oxide minerals and hydroxide minerals is discussed in detail by Schindler and Stumm (1987).

Of the Fe–Ti-oxides, hematite and magnetite raise the most concern with regard to their potential toxicity, because of the scale of mining operations involving ores of these minerals. Hematite is generally considered to be a nuisance dust and is used as a negative control in toxicity tests. Pott and Friedrichs (1972) and Pott et al. (1974) compared the effects on rats of intraperitoneal injections of hematite (0% fibrosis or tumors), chrysotile (40% incidence of tumors), and silica glass (55% incidence of tumors). Similar negative results were obtained for hematite dust inhalation (Vorwald and Karr, 1938) and implantation (Mossman and Craighead, 1982). *In vitro* experiments indicate that hematite is noncytotoxic and nongenotoxic (Dubes and Mack, 1988). Reports on the effects of hematite dusts (and compounding factors) on the health of mine workers were summarized by Guthrie (1992).

Spinels appear to be relatively inert in biological systems. Davis (1972) found that both chromite and magnetite induce the weakest cellular response following a 10-mg intrapleural injection in Balb/C mice, whereas chrysotile induces a strong effect. *In vitro* assays by Davies (1983) support the view that magnetite is relatively inactive, and Davis (1972) found that magnetite is not cytotoxic to mouse peritoneal macrophages. Pezerat et al. (1989) reported that the magnetite surface is capable of reducing dissolved oxygen to form oxygen radicals; however, magnetite was ~40 to 60 times less effective than chrysotile.

Solubility. Solubility of Fe-oxides (and Fe–Ti-oxides) is strongly dependent upon the chemistry of the solution, dissolved ligand concentrations, redox, and pH. The four general dissolution pathways are (i) proton-assisted, (ii) ligand-promoted acid, (iii) reductive, and (iv) ligand promoted reductive (Afonso et al., 1990). The importance of reductive dissolution pathways has been investigated intensively (e.g., Stone and Morgan, 1988; Borghi et al., 1989). There have been dozens of reports on hematite and magnetite dissolution in inorganic and organic solutions. Studies include those with solutions of bicarbonate (hematite: Bruno et al., 1992), citric acid (hematite: Johnsson et al., 1992), hydrogen sulfide (Afonso and Stumm, 1990), ascorbate (magnetite, hematite: Afonso et al., 1990), carboxylic acid (maghemite: Litter et al., 1991), other chelating agents (e.g., hematite: Torres et al., 1990), and thioglycolic acid (magnetite: Baumgartner et al., 1993). Fe-oxide surfaces generally have a net positive charge in acidic solutions, favoring adsorption of anionic ligands to mineral surfaces. Sidhu et al. (1981) suggested that Cl^- increases dissolution rates for Fe-oxides by forming Fe–Cl surface complexes. Biota can also greatly influence dissolution rates by providing protons and complexing ligands.

Dissolution rates of magnetite, maghemite, hematite (as well as goethite, akaganeite, and lepidocrocite) in 0.5 N HCl and $HClO_4$ at 25°C were reported by Sidhu et al. (1981). Rates increased in the sequence goethite–hematite–maghemite–akaganeite–magnetite–lepidocrocite. Constant values for Fe^{2+}/Fe^{tot} suggest that magnetite undergoes congruent dissolution in HCl. The faster dissolution rate of magnetite compared with maghemite may result from the greater ease in rupturing Fe^{2+}–OH bonds than Fe^{3+}–OH bonds; this effect presumably outweighs the heightened solubility imposed by vacancies in maghemite. Reductive dissolution pathways are enhanced for this reason. The more rapid dissolution of maghemite compared to hematite (despite identical composition) may result from the lower density of maghemite and the presence of tetrahedrally coordinated Fe. Magnetite, maghemite, and hematite dissolution rates at 0.5N HCl and 25 °C are summarized in Table 3. Other data for reactions of metal oxides with aqueous solutions are reported by Blesa et al. (1988) and in a forthcoming text by Blesa et al. (1993).

PHOSPHATES

Just as the $(SiO_4)^{4-}$ tetrahedral group is the fundamental unit that underlies the silicate structure, tetrahedral $(PO_4)^{3-}$ anionic groups serve as the basis for the phosphate minerals. Although phosphates commonly are involved in pathological tissue mineralization (Massry et al., 1980; LeGeros and LeGeros, 1984), little epidemiological evidence suggests that phosphate dusts represent a serious health hazard, and OSHA has not targeted phosphates for regulation. Therefore, this treatment of phosphates will be brief.

More than 300 different phosphates occur naturally, representing ~60 distinct structure types (Moore, 1984). Compositional variation within the phosphates is extensive. In addition to cationic solid solution series within these minerals, the $(PO_4)^{3-}$ tetrahedral unit itself can be replaced by an arsenate $(AsO_4)^{3-}$ and a vanadate $(VO_4)^{3-}$ anionic group. For instance, pyromorphite $[Pb_5(PO_4)_3Cl]$ is isostructural with mimetite $[Pb_5(AsO_4)_3Cl]$ and vanadinite $[Pb_5(VO_4)_3Cl]$, and solid solutions exist between each pair of end members.

Despite the large number of different phosphate minerals, only apatite $[Ca_5(PO_4)_3(F,OH,Cl)]$ is sufficiently abundant to qualify as a major rock-forming mineral. As determined by Naray-Szabo in 1930, apatite is hexagonal with space group $P6_3/m$. Within the apatite structure, the phosphate groups are isolated from each other and share no bridging oxygen anions (Fig. 25). Ca cations occupy two structurally distinct sites; Ca(I) is coordinated by 9 oxygen anions from 6 PO_4 groups, and Ca(II) is surrounded by 8 anions—7 oxygens from 5 PO_4 groups and 1 F, OH, or Cl anion. The monovalent ions lie upon the sixfold axes, but the precise positions depend upon the ionic species present. McConnell (1973) and Kanazawa (1989) provide detailed structural reviews.

Although apatites occur as accessory minerals within igneous and metamorphic rocks (Nash, 1984), 95% of the world's phosphate reserves are found in phosphorites, which are sedimentary phosphate deposits (Cook, 1984). The apatite within these deposits is a carbonate-rich fluorapatite called francolite.

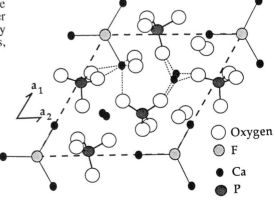

Figure 25. The fluorapatite structure projected on to the (001) plane. After Klein and Hurlbut (1993; used by permission of John Wiley & Sons, New York).

Typically, sedimentary francolite occurs in microcrystalline form (collophane) within spherical nodules. These pisolites sometimes exhibit a concentric banding similar to that seen in carbonate oolites, and the sizes of the pisolites may range from ~10 μm to several cm in diameter (Slansky, 1986). The petrogenesis of the phosphatic nodules found in phosphorites remains incompletely resolved. Most scientists favor a general mechanism that involves upwelling of cold, nutrient-rich waters into a shallow marine environment, thereby allowing the proliferation of a biota that ultimately contributes phosphate to the sediment (Cook and McElhinny, 1979). Today, these extensive phosphorite deposits are heavily mined for use in fertilizers, detergents, and pharmaceuticals.

Apatite is of particular interest to biochemists because it is the mineral that constitutes such calcified tissues as tooth enamel, dentine, and bone. The variety of apatite found in humans is a carbonate-hydroxyapatite known as dahllite $[Ca_5(PO_4,CO_3)_3(OH,F)]$. As with most apatites found in nature, dahllite is non-stoichiometric; the molar Ca:P ratio usually is lower than the ideal value of 1.67, ranging from 1.54 to 1.75 (LeGeros and LeGeros, 1984). Moreover, the way in which carbonate substitutes into the structure has provoked controversy for fifty years. Suggested substitution schemes include:

$CO_3^{2-} + H_3O^+ \leftrightarrow PO_4^{3-} + Ca^{2+}$ (Simpson, 1965)

$4\ CO_3^{2-} + H_3O^+ \leftrightarrow 3\ PO_4^{3-} + Ca^{2+}$ (McConnell, 1973)

$CO_3^{2-} \cdot F^- \leftrightarrow PO_4^{3-}$ (Binder and Troll, 1989)

In addition, debate surrounds the possibility of non-apatitic precursors during the formation of bone, such as amorphous calcium phosphate (ACP) and octacalcium phosphate $[Ca_4H_6(PO_4)_3]$ (Nelson and Barry, 1989). Relative stabilities of these precursor phases is strongly dependent upon pH, such that precipitation of ACP is favored with higher alkalinity at high supersaturation (Nancollas et al., 1989).

Natural apatite is much more reactive than stoichiometric apatite as a result of compositional and structural defects (Posner et al., 1984). For example, increased carbonate content increases the solubility of apatite, and the equilibrium phosphate concentration in solution with stoichiometric fluorapatite may be two orders of magnitude less than the phosphate concentration in solution with natural francolite (Jahnke, 1984). By contrast, increased F content in apatite promotes crystallinity and decreases solubility, thereby accounting for the beneficial effects of fluoridation on tooth enamel (LeGeros and LeGeros, 1984). In high enough concentrations, Mg^{2+} will inhibit the growth of apatite, as will the pyrophosphate ions and their relatives, such as ADP and ATP (Posner et al., 1984).

ACKNOWLEDGMENTS

The authors thank the following people for contributing figures: Eugene A. Smelik, Gordon Nord, Jr., Ivan Petrovic, Cornelis Klein, and Bob Virta. P.J.H. acknowledges the support of NSF Grant EAR-9206031. J.F.B. acknowledges the support of NSF Grant EAR-9207041 and an unrestricted research grant from Du Pont.

REFERENCES

Afonso, M.D.S., Morando, P.J., Blesa, M.A., Banwart, S. and Stumm, W. (1990) Reductive dissolution of iron oxides by ascorbate. The role of carboxylate anions in accelerating reductive dissolution. J. Colloidal Interface Sci. 138, 74–82.

Afonso, M.D.S. and Stumm, W. (1992) Reductive dissolution of iron (III) (hydr)oxides by hydrogen-sulfide. Langmuir 8, 1671–1675.

Aines, R.D., Kirby, S.H. and Rossman, G.R. (1984) Hydrogen speciation in synthetic quartz. Phys. Chem. Minerals 11, 204–212.

Allan, J.E.M., Coey, J.M.D., Sanders, I.S., Schwertmann, U., Friedrich, G. and Wiechowski, A. (1989) An occurrence of a fully-oxidized natural titanomaghemite in basalt. Mineral. Mag. 53, 299–304.

American Conference of Governmental and Industrial Hygenists (1984) ACGIH: Threshold limit values for chemical substances in the work environment adopted by the ACGIH conference for 1984–1985. ACGIH, Cincinnati, Ohio.

Ampian, S.G. and Virta, R.L. (1992) Crystalline silica overview: Occurrence and analysis. U.S. Dept. Interior, Bureau of Mines, Information Circ. 9317.

Anderson, M.A. Gieselmann, M.J. and Xu, Q. (1988) Titania and alumina ceramic membranes. J. Membrane Sci. 39, 243–258.

Anpo, M. (1989) Photocatalysis on small particle TiO_2 catalysts–Reaction intermediates and reaction mechanisms. Res. Chem. Intermediates 11, 67–106.

Ayers, J.C. and Watson, E.B. (1991) Rutile in high-pressure hydrothermal environments. EOS, Trans. Amer. Geophys. Union 72, 141–142.

Banerjee, S.K. (1991) Magnetic properties of Fe-Ti oxides. Rev. Mineral. 25, 107–127.

Banfield, J.F. and Veblen, D.R. (1991) The structure and origin of Fe-bearing defects in metamorphic rutile. Amer. Mineral. 76, 113–127.

Banfield, J.F., Jones, B.J. and Veblen, D.R. (1991) An AEM—TEM study of weathering and diagenesis, Abert Lake, Oregon. (I) Weathering reactions in the volcanics. Geochim. Cosmochim. Acta 55, 2781–2793.

Banfield, J.F., Jones, B.J. and Veblen, D.R. (1991) An AEM—TEM study of weathering and diagenesis, Abert Lake, Oregon. (II) Diagenetic modification of the sedimentary assemblage. Geochim. Cosmochim. Acta 55, 2795–2810.

Banfield, J.F., Veblen, D R. and Smith, D.J. (1991) The identification of naturally occurring $TiO_2(B)$ by structure determination using high–resolution electron microscopy, image simulation and distance–least–squares refinement. Amer. Mineral. 76, 343–353.

Banfield, J.F. and Veblen, D.R. (1992) A TEM study of the conversion of perovskite to anatase and $TiO_2(B)$ and the use of fundamental building blocks to understand the relationships among the TiO_2 polymorphs. Amer. Mineral. 77, 545–557.

Banfield, J.F., Bischoff, B.L. and Anderson, M.A. (1993) TiO_2 accessory minerals: coarsening and transformation kinetics in pure and doped synthetic nanocrystalline materials. Chem. Geol. 106 (in press).

Barbeni, M., Minero, C., Pellizzetti, E., Borgarello, E. and Serpone, N. (1987) Chemical degredation of chlorophenols with fento reagent (Fe-$^{2+}$H$_2$O$_2$). Chemosphere 16, 2225–2237.

Bassett, W.A. and Barnett, J.D. (1970) Isothermal compression of stishovite and coesite up to 85 kilobars at room temperature by X-ray diffraction. Phys. Earth Planetary Interiors 3, 54–60.

Baumgartner, E., Romagnolo, J. and Litter, M.I. (1993) Effect of anionic polyelectrolytes on the dissolution of magnetite in thioglycolic acid solutions. J. Chem. Soc. Faraday Trans. 89, 1049–1055.

Beckwith, R.S. and Reeve, R. (1969) Dissolution and deposition of monosilicic acid in suspensions of ground quartz. Geochim. Cosmochim. Acta 33, 745–750.

Bendz, G. and Lindqvist, I., Eds. (1977) Biochemistry of Silicon and Related Problems. Plenum Press, New York, 591 p.

Bennett, P.C., Melcer, M.E., Siegel, D.I. and Hassett, J.P. (1988) The dissolution of quartz in dilute aqueous solutions of organic acids at 25°C. Geochim. Cosmochim. Acta 52, 1521–1530.

Best, M.G. (1982) Igneous and metamorphic petrology. 630 p. W.H. Freeman and Company, New York.

Binder, G. and Troll, G. (1989) Coupled anion substitution in natural carbon-bearing apatites. Contrib. Mineral. Petrol. 101, 394–401.

Bischoff, B. L. 1992. Thermal stabilization of anatase (TiO_2) membranes. University of Wisconsin, Ph.D. dissertation, 165 p.

Blake, R.L., Hessevick, R.E., Zoltai, T. and Finger, L.W. (1966) Refinement of the hematite structure. Amer. Mineral. 51, 123–129.

Blesa, M.A. (1988) Reactions of metal oxides with aqueous solutions. Materials Science Forum 29, 31–98.

Blesa, M.A., Morando, P.J. and Regazzoni, A.E. (1993) Chemical dissolution of metal oxides. CRC Press, Boca Raton, Florida (in press).

Borghi, E.B., Blesa, M.A., Maroto, A.J.G. and Regazzoni, A.E. (1989) Reductive dissolution of magnetite by solutions containing EDTA and Fe-II. J. Colloid. Interface Sci. 130, 299–310.

Brady, P.V. and Walther, J.V. (1990) Kinetics of quartz dissolution at low temperatures. Chemical Geology, 82, 253–264.

Bragg, W.H. and Gibbs, R.E. (1925) The structure of α- and β-quartz. Proc. Royal Soc. London A109, 405–427.

Branch of Industrial Minerals (1992) Crystalline Silica Primer. U.S. Bureau of Mines, Special Publ. 49 p.

Buddington, A.F. and Lindsley, D.H. (1964) Iron-titanium oxide minerals and synthetic equivalents. J. Petrology 5, 310–357.

Buerger, M.J. (1951) Crystallographic aspects of phase transformations. In: R. Smoluchowski, ed., Phase Transformations in Solids. Wiley, New York, p. 183–211.

Buerger, M.J. (1954) The stuffed derivatives of the silica structures. Amer. Mineral. 39, 600–614.

Buerger, M.J. and Lukesh, J. (1942) The tridymite problem. Science 95, 20–21.

Bursill, L.A. and Hyde, B.G. 1972. Crystallographic shear in the higher Ti oxides: Structure, texture, mechanisms and thermodynamics. In: Progress in Solid State Chemistry, H. Reiss and J.O. McCaldin, Eds., Pergamon Press, Oxford, 177–253.

Burton, B.P. (1991) Interplay of chemical and magnetic ordering. Rev. Mineral. 25, 303–319.

Carpenter, M.A. and Wennemer, M. (1985) Characterization of synthetic tridymites by transmission electron microscopy. Amer. Mineral. 70, 517–528.

Carrisoza, I., Munuera, G. and Castañar, S. (1977) Study of interaction of aliphatic alcohols with TiO_2: III. Formation of alkyl-titanium species during methanol decomposition. J. Catalysis, 49, 265–277.

Chao, E.C.T., Fahey, J.J., Littler, J. and Milton, D.J. (1962) Stishovite, SiO_2, a very high pressure new mineral from Meteor Crater, Arizona. J. Geophys. Res. 67, 419–421.

Cheetham, A.K. and Day, P., Eds. (1992) Solid State Chemistry: Compounds. University Press, Oxford, 304 p.

Chen, C.H. and Lewin, J. (1969) Silicon as a nutrient element for *Equisetum arvense*. Canad. J. Botany 47, 125–131.

Coes, L. (1953) A new dense crystalline silica. Science 118, 131–132.

Cohen, A.J. (1985) Amethyst color in quartz, the result of radiation protection involving iron. Amer. Mineral. 70, 1180–1185.

Cohen, A.J. and Makar, L.N. (1985) Dynamic biaxial absorption spectra of Ti^{3+} and Fe^{2+} in a natural rose quartz crystal. Mineral. Mag. 49, 709–715.

Cook, P.J. (1984) Spatial and temporal controls on the formation of phosphate deposits—A review. In: J.O. Nriagu and P.B. Moore, Eds., Phosphate Minerals. Springer-Verlag, Berlin, p. 242–274.

Cook, P.J. and McElhinny, M.W. (1979) A reevaluation of the spatial and temporal distribution of sedimentary phosphate deposits in the light of plate tectonics. Econ. Geol. 74, 315–330.

Crerar, D.A., Axtmann, E.V. and Axtmann, R.C. (1981) Growth and ripening of silica polymers in aqueous solutions. Geochim. Cosmochim. Acta 45, 1259–1266.

Crook, K.A.W. (1968) Weathering and roundness of quartz sand grains. Sedimentology 11, 171–182.

Darragh, P.J., Gaskin, A.J. and Sanders, J.V. (1976) Opals. Scientific Amer. 234, 84–95.

Davies, R. (1983) Factors involved in the cytotoxicity of kaolinite towards macrophages in vitro. Environ. Health Perspec. 51, 249–252.

Davis, J.M.G. (1972) The fibrogenic effects of mineral dusts injected into the pleural cavity of mice. British J. Exper. Pathol. 53, 190–200.

De Jong, B.H.W.S., van Hoek, J., Veeman, W.S. and Manson, D.V. (1987) X-ray diffraction and ^{29}Si magic-angle-spinning NMR of opals: Incoherent long- and short-range order in opal-CT. Amer. Mineral. 72, 1195–1203.

Dollase, W.A. (1965) Reinvestigation of the structure of cristobalite. Zeits. Kristallogr. 121, 369–377.

Dollase, W.A. and Baur, W.H. (1976) The superstructure of meteoritic low tridymite solved by computer simulation. Amer. Mineral. 61, 971–978.

Dove, P.M. and Crerar, D.A. (1990) Kinetics of quartz dissolution in electrolyte solutions using a hydrothermal mixed flow reactor. Geochim. Cosmochim. Acta 54, 955–969.

Dowty, E. (1987) Vibrational interactions of tetrahedra in silicate glasses and crystals: II. Calculations on melilites, pyroxenes, silica polymorphs and feldspars. Phys. Chem. Minerals 14, 122–138.

Driscoll, K. E. and Maurer, J. K. (1991) Cytokine and growth factor release alveolar macrophages: potential biomarkers of pulmonary toxicity. Toxicol Pathol. 19, 398-405.

Dubes, G.R. and Mack. L.R. (1988) Asbestos-mediated transfection of mammalian cell cultures. *In Vitro* Cell. Develop. Biol. 24, 175–182.

Dunnom, D.D., Ed. (1981) Health effects of synthetic silica particulates. Amer. Soc. Testing and Materials, Philadelphia, 226 p.

Dzyaloshinsky, I. (1958) A thermodynamic theory of "weak" ferromagnetism of antiferromagnetics. J. Phys. Chem. Solids 4, 241–255.

Endo, S., Akai, T., Akahama, Y., Wakatsuki, M., Nakamura, T., Tomii, Y., Koto, K., Ito, Y. and Tokonami, M. (1986) High temperature X-ray study of single crystal stishovite synthesized with Li_2WO_4 as flux. Phys. Chem. Minerals 13, 146–151.

Evans, W.D. (1964) The organic solubilization of minerals in sediments. In: U. Colombo and G.D. Hobson, Eds., Advances in Organic Geochemistry. Macmillan, New York, p. 263–270.

Evered, D. and O'Connor, M., Eds. (1986) Silicon Biochemistry. John Wiley & Sons, Chichester, England, 264 p.

Fanderlik, I. Ed. (1991) Silica glass and its application. Elsevier, Amsterdam, 304 p.

Ferin, J. and Oberdörster, G. (1985) Biological effects and toxicity assessment of titanium dioxides: anatase and rutile. Amer. Indus. Hygiene Assoc. J. 46, 69–72.

Fleming, B.A. (1986) Kinetics of reaction between silicic acid and amorphous silica surfaces in NaCl solutions. J. Colloid Interface Sci. 110, 40–64.

Flörke, O.W., Flörke, U. and Giese, U. (1984) Moganite: A new microcrystalline silica-mineral. N. Jb. Mineral. Abh. 149, 325–336.

Flörke, O.W., Graetsch, H., Martin, B., Röller, K. and Wirth, R. (1991) Nomenclature of micro- and non-crystalline silica minerals, based on structure and microstructure. N. Jb. Mineral. Abh. 163, 19–42.

Flörke, O.W., Hollmann, R., von Rad, U. and Rösch, H. (1976) Intergrowth and twinning in opal-CT lepispheres. Contrib. Mineral. Petrol. 58, 235–242.

Flörke, O.W., Köhler-Herbertz, B., Langer, K., Tönges, I. (1982) Water in microcrystalline quartz of volcanic origin: Agates. Contrib. Mineral. Petrol. 80, 324–333.

Fournier, R.O. (1977) Chemical geothermometers and mixing models for geothermal systems. Geothermics 5, 41–50.

Fournier, R.O. (1983) A method of calculating quartz solubilities in aqueous sodium chloride solutions. Geochim. Cosmochim. Acta 47, 579–586.

Frankel, R.B. and Blakemore, R.P. (1980) Navigational compas in magnetic bacteria. J. Magetism Magnetic Mat. 15–18, 1562–1564.

Frondel, C. (1945) History of the quartz oscillator-plate industry, 1941–1944. Amer. Mineral. 30, 205–213.

Frondel, C. (1962) The System of Mineralogy (7th Ed.). John Wiley & Sons, New York, 334 p.

Frondel, C. (1978) Characters of quartz fibers. Amer. Mineral. 63, 17–27.

Frondel, C. (1982) Structural hydroxyl in chalcedony (Type B quartz). Amer. Mineral. 67, 1248–1257.

Frost, B.R. (1991) Magnetic petrology: Factors that control the occurrence of magnetite in crustal rocks. Rev. Mineral. 25, 489–506.

Furata, T. and Otsuki, M. (1985) Quantitative electron microprobe analysis of oxygen in titanomagnetites with implications for oxidation processes. J. Geophys. Res. 90, 3145–3150.

Fysh, S.A. and Clark, P.E. (1982) Aluminous hematite. A Mössbauer study. Phys. Chem. Minerals 8, 257–267.

Gaffney, J.S., Marley, N.A. and Janecky, D.R. (1989) Comment on "The dissolution of quartz as a function of pH and time at 70°C" by K.G. Knauss and T.J. Wolery. Geochim. Cosmochim. Acta 53, 1469–1470.

Garrison, R.E., Douglass, R.B., Pisciotto, K.E., Isaacs, C.M., Ingle, J.C., Eds. (1981) The Monterey Formation and Related Siliceous Rocks of California. Soc. Econ. Paleontol. Mineral., Pacific Section, Los Angeles, California, 327 p.

Garrone, E., Bolis, V., Fubini, B. and Morterra, C. (1989) Thermodynamic and spectroscopic characterization of heterogeneity among absorption sites—Co on anatase at ambient temperature. Langmuir 5, 892.

Gibbs, R.E. (1926) The polymorphism of silicon dioxide and the structure of tryidymite. Proc. Royal Soc. London A113, 351–368.

Gieré, R. (1990) Hydrothermal mobility of Ti, Zr and REE: Examples from the Bergell and Adamello contact aureoles (Italy). Terra Nova 2, 60–67.

Gislason, S.R., Veblen, D.R. and Livi, K.J.T. (1993) Experimental meteoric water–basalt interactions: Characterization and interpretation of alteration products. Geochim. Cosmochim. Acta (in press).

Glass, B.P. (1984) Tektites. J. Non-Crystalline Solids 67, 333–344.

Goldsmith, D.F., Winn, D.M. and Shy, C.M., Eds. (1986) Silica, Silicosis, and Cancer. Praeger, New York, 536 p.

Graetsch, H., Flörke, O.W. and Miehe, G. (1985) The nature of water in chalcedony and opal-C from Brazilian agate geodes. Phys. Chem. Minerals 12, 300–306.

Graetsch, H., Flörke, O.W. and Miehe, G. (1987) Structural defects in microcrystalline silica. Phys. Chem. Minerals 14, 249–257.

Griggs, D.T. and Blacic, J.D. (1965) Quartz: Anomalous weakness of synthetic crystals. Science 147, 292–295.

Gutierrez, C. and Salvador, P. (1986) Mechanisms of competitive photoelectrochemical oxidation of L- and H_2O at normal – TiO_2 electrodes: a kinetic appraoch. J. Electrochem. Soc. 133, 924–929.

Guthrie, G.D. (1992) Biological effects of inhaled minerals. Amer. Mineral. 77, 225–243.

Haggerty, S.E. 1991. Oxide mineralogy of the upper mantle. In: Oxide Minerals: Petrologic and Magnetic Significance, D.H. Lindsley, Ed. Rev. Mineral. 25, 355-416.

Hartwig, B. and Hench, L.L. (1972) The epitaxy of poly-L-alanine on l-quartz and a glass-ceramic. J. Biomed. Materials Res. 6, 413–424.

Hashimoto, K., Kawai, T. and Sakata, T. (1984) Photocatalytic reactions of hydrocarbons and fossil fuels with water: Hydrogen production and oxidation. J. Phys. Chem. 88, 4083–4088.

Hatch, D.M. and Ghose, S. (1991) The α–β phase transition in cristobalite, SiO_2. Phys. Chem. Minerals 17, 554–562.

Hazen, R. and Jeanloz, R. (1984) Wüstite ($Fe_{1-x}O$): A review of its defect structure and physical properties. Rev. Geophys. Space Phys. 22, 37–46.

Heaney, P.J. (1993) A proposed mechanism for the growth of chalcedony. Contrib. Mineral. Petrol. (in press).

Heaney, P.J. and Post, J.E. (1992) The widespread distribution of a novel silica polymorph in microcrystalline quartz varieties. Science 255, 441–443.

Heaney, P.J. and Veblen, D.R. (1991a) An examination of spherulitic dubiomicrofossils in Precambrian banded iron formations using the transmission electron microscope. Precambrian Res. 49, 355–372.

Heaney, P.J. and Veblen, D.R. (1991b) Observations of the α–β phase transition in quartz: A review of imaging and diffraction studies and some new results. Amer. Mineral. 76, 1018–1032.

Heaney, P.J. and Veblen, D.R. (1991c) Observation and kinetic analysis of a memory effect at the α–β quartz transition. Amer. Mineral. 76, 1459–1466.

Hein, J.R., Ed. (1987) Siliceous sedimentary rock-hosted ores and petroleum. Van Nostrand Reinhold, New York, 304 p.

Hemenway, D.R., Absher, M., Landesman, M., Trombley, L. and Emerson, R.J. (1986) Differential lung response following silicon dioxide polymorph aerosol exposure. In: D.F. Goldsmith, D.M. Winn and C.M. Shy, Eds. (1986) Silica, Silicosis, and Cancer. Praeger, New York, p. 105–116.

Hill, R.J., Newton, M.D. and Gibbs, G.V. (1983) A crystal chemical study of stishovite. J. Solid State Chem. 47, 185–200.

Hyde, B.G, Bagshow, A.N., O'Keefe, M. and Andersson, S. (1974) Defect structures in crystalline solids. Ann. Rev. Materials Sci. 4, 43–92.

Hyde, B. G. and Andersson, S. 1989. Inorganic crystal structures. John Wiley & Sons, New York, 430 p.

Hynes, A.J. (1980) Carbonatization and mobility of Ti, V and Zr in Ascot formation metabasalts, SE Quebec. Contrib. Mineral. Petrol. 75, 79–87.

Iler, R.K. (1979) The chemistry of silica: Solubility, polymerization, colloid and surface properties, and biochemistry. John Wiley & Sons, New York, 866 p.

Ingrin, J. and Gillet, P. (1986) TEM investigation of the crystal microstructures in a quartz–coesite assemblage of the western Alps. Phys. Chem. Minerals 13, 325–330.

International Agency for Research on Cancer (1987) Silica and some silicates. IARC Monographs on the Evaluation of the Carcinogenic Risk of Chemicals to Humans 42, 39–143.

Ishikawa, Y. and Akimoto, S. (1957) Magnetic properties of $FeTiO_3$–Fe_2O_3 solid solution series. J. Phys. Soc. Japan 12, 1083–1098.

Ishikawa, Y. and Syono, Y. (1963) Order-disorder transformation and reverse thermoremanent magnetism in the $FeTiO_3$–Fe_2O_3 system. J. Phys. Chem. Solids 24, 517–528.

Jacobs, J.D. and Martens, J.A. (1987) Synthesis of high silica aluminosilicate zeolites. Elsevier, Amsterdam, 390 p.

Jahnke, R.A. (1984) The synthesis and solubility of carbonate fluorapatite. Amer. J. Sci. 284, 58–78.

Johnsson, P.A., Hochella, M.F., Parks, G.A. and Sposito, G. (1992) Direct nanometer-scale observation of hematite dissolution in citric acid. Abstr. Amer. Chem. Soc. 203, 5.

Jones, J.B. and Segnit, E.R. (1971) The nature of opal. I. Nomenclature and constituent phases. J. Geol. Soc. Australia 18, 57–68.

Jones, J.B. and Segnit, E.R. (1972) Genesis of cristobalite and tridymite at low temperatures. J. Geol. Soc. Australia 18, 419–422.

Jørgensen, S.S. (1967) Dissolution kinetics of silicate minerals in aqueous catechol solutions. J. Soil Sci. 27, 183–195.

Kanazawa, T., Ed. (1989) Inorganic Phosphate Materials. Elsevier, Amsterdam, 288 p.

Kaufman, P.B., Dayanandan, P., Takeoka, Y., Bigelow, W.C., Jones, J.D. and Iler, R. (1981) Silica in shoots of higher plants. In: Simpson, T.L. and Volcani, B.E., Eds., Silicon and Siliceous Structures in Biological Systems. Springer-Verlag, New York, p. 409–449.

Kikkawa, H. O'Regan, B. and Anderson, M.A. (1991) The photoelectricochemical properties of Nb-doped TiO_2 semiconducting membranes. J. Electroanalyt. Chem. 309, 91–101.

Kirschvink, J.L. and Chang, S.-B.R. (1984) Ultrafine magnetite in deep-sea sediments. Geology 12, 559–562.

Klein, C. and Hurlbut, C.S. (1993) Manual of Mineralogy, 21st Ed. John Wiley & Sons, New York, 681 p.

Knauss, K.G. and Wolery, T.J. (1988) The dissolution kinetics of quartz as a function of pH and time at 70°C. Geochim. Cosmochim. Acta 52, 43–53.

Kraynov, S.R. (1971) The effect of acidity-alkalinity of ground waters on the concentration and migration of rare elements. Geochem. Int'l 8, 828–836.

Kubicki, J.D. and Lasaga, A.C. (1988) Molecular dynamics simulations of SiO_2 melt and glass: Ionic and covalent models. Amer. Mineral. 73, 941–955.

Lally, J.S., Nord, G.L., Jr., Heuer, A.H. and Christie, J.M. (1978) Transformation-induced defects in α-cristobalite. Proc. 9th Int'l Congress on Electron Microscopy, Electron Microscopy 1, 476–477.

Langer, K. and Flörke, O.W. (1974) Near infrared absorption spectra (4,000–9,000 cm^{-1}) of opals and the role of "water" in these $SiO_2 \cdot nH_2O$ minerals. Fortschr. Mineral. 52, 17–51.

Lawson, C.A. and Nord, G.L. (1984) Remanent magnetization of a "paramagnetic" composition in the ilmenite–hematite solid solution series. Geophys. Res. Lett. 11, 197–200.

Leadbetter, A.J. and Wright, A.F. (1976) The α–β transition in the cristobalite phases of SiO_2 and $AlPO_4$. I. X-ray studies. Phil. Mag. 33, 105–112.

Lee, J.B. (1981) Elevated temperature potential – pH diagrams for $Cr–H_2O$, $Ti–H_2O$, and $Pt–H_2O$ system. Corrosion 37, 467–481.

LeGeros, R.Z. and LeGeros, J.P. (1984) Phosphate minerals in human tissues. In: J.O. Nriagu and P.B. Moore, Eds., Phosphate Minerals. Springer-Verlag, Berlin, p. 351–385.

Levien, L. and Prewitt, C.T. (1981) High-pressure crystal structure and compressibility of coesite. Amer. Mineral. 66, 324–333.

Li, H.C. and De Bruyn, P.L. (1966) Electrokinetic and adsorption studies on quartz. Surface Science 5, 203–220.

Li, Y.-K. (1956) Superexchange interactions and magnetic lattices of the rhombohedral sesquioxides of the transition element and their solid solutions. Phys. Rev. 102, 1015–1020.

Liebau, F. and Böhm, H. (1982) On the co-existence of structurally different regions in the low-high quartz and other displacive phase transformations. Acta Crystallogr. A38, 252–256.

Lindsley, D.H. (1976) The crystal chemistry and structure of oxide minerals as exemplified by the Fe-Ti oxides. Rev. Mineral. 3, L1–L51.

Lindsley, D.H. (1991) Experimental studies of oxide minerals. Rev. Mineral. 25, 69–100.

Litter, M.I. and Blesa, M.A. (1992) Photodissolution of iron oxides. 4. A comparative study of the photodissolution of hematite, magnetite, and maghemite in EDTA media. Canad. J. Chem. 70, 2502–2510.

Litter, M.I., Baumgartner, E.C., Urrutia, G.A. and Blesa, M.A. (1991) Photodissolution of iron oxides. 3. Interplay of photochemical and thermal processes in maghemite carboxylic acid systems. Environ. Sci. Tech. 25, 1907–1913.

Maata, K. and Arstila, A.W. (1975) Pulmonary deposits of titnium dioxide in cytogenic and lung biopsy specimens. Lab. Investigation 33, 342–346.

Mann, S. and Perry, C.C. (1986) Structural aspects of biogenic silica. In: D. Evered and M. O'Connor, Eds., Silicon biochemistry. John Wiley & Sons, Chichester, England, p. 40–58.

Marshall, W.L. and Chen, C.A. (1982) Amorphous silica solubilities—VI. Postulated sulfate-silicic acid solution complex. Geochim. Cosmochim. Acta 46, 367–370.

Marshall, W.L. and Warakomski, J.M. (1980) Amorphous silica solubilities—II. Effect of aqueous salt solutions at 25°C. Geochim. Cosmochim. Acta 44, 915–924.

Mason, B. and Moore, C.B. (1982) Principles of Geochemistry, 4th edition. John Wiley & Sons, New York, 344 p.

Massry, S.G., Ritz, E. and Jahn, H., Eds. (1980) Phosphate and minerals in health and disease. Plenum Press, New York, 675 p.

Mann, S., Frankel, R.B. and Blakemore, R.P. (1984) Structure, morphology, and crystal growth of bacterial magnetite. Nature 310, 405–407.

Marchand, R., Brohan, L. and Tournoux, M. 1980. TiO_2(B) a new form of titanium dioxide and the potassium octatitanate $K_2Ti_8O_{17}$. Mat. Res. Bull. 15, 1129–1133.

Mazer, J.J., Bates, J.K., Bradley, J.P., Bradley, C.R. and Stevenson, C.M. 1992. Alteration of tektite to form weathering products. Nature 357, 573–576.

McCammon, C.A. and Liu, L.-G. (1984) The effects of pressure and temperature on nonstoichiometric wüstite, Fe_xO: The iron-rich phase boundary. Phys. Chem. Minerals 10, 106–113.

McConnell, D. (1973) Apatite: Its crystal chemistry, mineralogy, utilization, and geologic and biologic occurrences. Springer-Verlag, New York, 111 p.

McLaren, A.C. and Phakey, P.P. (1966) Electron microscope study of Brazil twin boundaries in amethyst quartz. Phys. Status Solidi 13, 413–422.
McLaren, A.C. and Pitkethly, D.R. (1982) The twinning microstructure and growth of amethyst quartz. Phys. Chem. Minerals 8, 128–135.
Meagher, E.P. and Lager, G.A. (1979) Polyhedral thermal expansion in the TiO_2 polymorphs: Refinement of the crystal structures of rutile and brookite at high temperatures. Canad. Mineral. 17, 77–85.
Metikos-Hukovic, M. and Ceraj-Ceric, M. 1988. Investigations of chromium doped ceramic rutile electrodes. Mat. Res. Bull. , 1535–1544.
Miehe, G. and Graetsch, H. (1992) Crystal structure of moganite: A new structure type for silica. Eur. J. Mineral. 4, 693–706.
Miehe, G., Graetsch, H. and Flörke, O.W. (1984) Crystal structure and growth fabric of length-fast chalcedony. Phys. Chem. Minerals 10, 197–199.
Miser, D. E. and Buseck, P. R. 1988.Defect microstructures in oxygen-deficient rutile from Fe-bearing xenoliths from Disko, Greenland. Geol. Soc. Amer. Abstr. 20, A101.
Moore, P.B. (1984) Crystallochemical aspects of the phosphate minerals. In: J.O. Nriagu and P.B. Moore, Eds., Phosphate Minerals. Springer-Verlag, Berlin, p. 155–170.
Moriya, T. (1960) Anisotropic superexchange interaction and weak ferromagnetism. Phys. Rev. 120, 91–98.
Morse, J.W. and Casey, W.H. (1988) Ostwald processes and mineral paragenesis in sediments. Amer. J. Sci. 288, 537–560.
Mossman, B.Y. and Craighead, J.E. 91982) Comparative cocarcinogenic effects of crocidolite asbestos, hematite, kaolin and carbon in implanted tracheal organ cultures. Ann. Occup. Hygiene 26, 553–567.
Murphy, J.B. and Hynes, A.J. (1986) Contrasting secondary mobility of Ti, P, Zr, Nb, and Y in two metabasalts suites in Appalacians. Canad. J. Earth Sci. 23, 1138–1144.
Nabivanets, B.I. and Lukchina,V.V. (1964) Hydroxy complexes of titanium (IV). Inorganic Phys. Chem. 1123–1128.
Nagel, W. and Böhm, H. (1982) Ionic conductivity studies on $LiAlSiO_4$–SiO_2 solid solutions of the high quartz type. Solid State Comm. 42, 625–631.
Nakano, T., Takahara, H. and Nishida, N. (1989) Intracrystalline distribution of major elements in zoned garnet from skarn in the Chichibu mine, central Japan: Illustration by color-coded maps. Canad. Mineral. 27, 499–507.
Nancollas, G.H., Lore, M., Perez, L., Richardson, C. and Zawacki, S.J. (1989) Mineral phases of calcium phosphate. Anatomical Rec. 224, 234–241.
Naray-Szabo, S. (1930) The structure of apatite $(CaF)Ca_4(PO_4)_3$. Zeits. Kristallogr. 75, 387–388.
Nash, T., Allison, A.C. and Harington, J.S. (1966) Physio-chemical properties of silica in relation to its toxicity. Nature 210, 259–261.
Nash, W.P. (1984) Phosphate minerals in terrestrial igneous and metamorphic rocks. In: J.O. Nriagu and P.B. Moore, Eds., Phosphate Minerals. Springer-Verlag, Berlin, p. 215–241.
Navrotsky, A., Weidner, D.J., Liebermann, R.C. and Prewitt, C.T. (1992) Materials science of the Earth's deep interior. Mat. Res. Soc. Bull. 17, 30–37.
Nelson, D.G.A. and Barry, J.C. (1989) High resolution electron microscopy of nonstoichiometric apatite crystals. Anatomical Rec. 224, 265–276.
Nickel, E.H. (1958) The composition and microtexture of an ulvöspinel–magnetite intergrowth. Canad. Mineral. 6, 191–200.
Nord. G.L., Jr. (1992) Imaging transformation-induced microstructures. Rev. Mineral. 27, 455–508.
Nolan, R.P., Langer, A.M., Harington, J.S., Oster, G. and Selikoff, I.J. (1981) Quartz hemolysis as related to its surface functionalities. Environ. Res. 26, 503–520.
Nriagu, J.O. and Moore, P.B. (1984) Phosphate minerals, 442 p. Springer-Verlag, Berlin.
O'Keefe, J.A. (1984) Natural glass. J. Non-Crystalline Solids 67, 1–17.
O'Neill, C., Jordan, P., Bhatt, T. and Newman, R. (1986) Silica and oesophageal cancer. In: D. Evered and M. O'Connor, Eds., Silicon Biochemistry. John Wiley & Sons, Chichester, England, p. 214–230.
Orians, K.J., Boyle, E.A. and Bruland, K.W. (1990) Dissolved titanium in the open ocean. Nature 348, 322–325.
Petrovic, I., Heaney, P.J. and Navrotsky, A. (1993a) Calorimetric study of the silica polymorph moganite. (abstr.) EOS, Trans. Amer. Geophys. Union 74, 160.
Petrovic, I., Navrotsky, A., Davis, M.E. and Zones, S.I. (1993b) Thermochemical study of the stability of frameworks in high silica zeolites. Chemistry of Materials, submitted.
Pettijohn, F.J. (1975) Sedimentary Rocks, 3rd edition. Harper & Row, New York, 628 p.
Pezerat, H., Zalma, R. and Guignard, J. (1989) Production of oxygen radicals by the reduction of oxygen arising from the surface activity of mineral fibres. In: Non-occupational Exposure to Mineral Fibres, J. Bignon, J. Petro and R. Saracci, Eds., International Agency for Research on Cancer, Lyon, France.

Posner, A.S., Blumenthal, N.C. and Betts, F. (1984) Chemistry and structure of precipitated hydroxyapatites. In: J.O. Nriagu and P.B. Moore, Eds., Phosphate Minerals. Springer-Verlag, Berlin, p. 330–350.
Post, J. E. and Burnham, C. W. 1986. Ionic modeling of mineral structures and energies in the electron gas approximation: TiO_2 polymorphs, quartz, forsterite, diopside. Amer. Mineral. 71, 1142–150.
Pott, F. and Friedrichs, K.H. (1972) Tumoren der ratte nach i. p.-injektion faserförmiger Stäube. Naturwiss. 59, p. 318.
Pott, F., Huth, F. and Friedrichs, K.H. (1972) Tumorigenic effect of fibrous dusts in experimental animals. Environ. Health Persp. 9, 313–315.
Price, G.D. (1979) Microstructures in titanomagnetites as guides to cooling rates of a Swedish intrusion. Geol. Mag. 116, 313–318.
Price, G. D. (1981) Subsolidus phase relations in the titanomagnetite solid solutin series. Amer. Mineral. 66, 751–758.
Pye, L.D., O'Keefe, J.A. and Fréchette, V.D., Eds. (1984) Natural Glasses. North-Holland Physics Publishing, Amsterdam, 662 p.
Reardon, E.J. (1979) Complexing of silica by iron(III) in natural waters. Chemical Geol. 25, 339–345.
Rhode, L.E., Ophus, E.M. and Gylseth, B. (1981) Massive pulmonary deposition of rutile after titanium dioxide exposure. Acta Path. Microbiol. Immunol. Scand. Sec. A, 89, 455–461.
Rimstidt, J.D. and Barnes, H.L. (1980) The kinetics of silica-water reactions. Geochim. Cosmochim. Acta 44, 1683–1699.
Rosner, D. and Markowitz, G. (1991) Deadly dust, silicosis and the politics of occupational disease in twentieth century America. Princeton University Press, Princeton, New Jersey, 229 p.
Ross, N.L., Shu, J.-F., Hazen, R.M. and Gasparik, T. (1990) High-pressure crystal chemistry of stishovite. Amer. Mineral. 75, 739–747.
Sabate, J., Anderson, M.A., Kikkawa, H., Hill, C.G. (1990) A kinetic study of the photocatalytic degradation of 3-chlorsalicylic acid over TiO_2 membranes supported on glass. J. Catalysis, 127, 167–177.
Sabate, J., Anderson, M.A., Kikkawa, H., Xu, Q., Cevera-March, S. and Hill, C.G. (1992) Nature and properties of pure and Nb-doped TiO_2 ceramic membranes affecting the photocatalytic degradation of 3-chlorsalicylic acid as a model of halogenated organic compounds. J. Catalysis, 134, 36–46.
Saffiotti, U. (1986) The pathology induced by silica in relation to fibrogenesis and carcinogenesis. In: D.F. Goldsmith, D.M. Winn and C.M. Shy, Eds., Silica, Silicosis, and Cancer. Praeger, New York, p. 287–307.
Sangster, A.G. and Hodson, M.J. (1986) Silica in higher plants. In: D. Evered and M. O'Connor, Eds., Silicon Biochemistry. John Wiley & Sons, Chichester, England, p. 90–111.
Schilling, R.D. and Vink, B.W. 1967 Stability relations of some titanium-minerals (sphene, perovskite, rutile, anatase). Geochim. Cosmochim. Acta 31, 2399–2411.
Schindler, P.W. and Stumm, W. (1987) The surface chemistry of oxides, hydroxides, and oxide minerals. In: Aquatic Surface Chemistry, W. Stumm, Ed., John Wiley & Sons, New York, p. 83.
Schlinger, C.M., Smith, R.M. and Veblen, D.R. (1986) Geologic origin of magnetic volcanic glasses in the KBS tuff. Geology 14, 959–962.
Schlössin, H.H. and Lang, A.R. (1965) A study of repeated twinning, lattice imperfections and impurity distribution in amethyst. Phil. Mag. 12, 283–296.
Schmidbauer, E. (1987) ^{57}Mossbaüer spectroscopy and magnetization of cation deficient Fe_2TiO_4 and $FeCr_2O_4$. II: Magnetization data. Phys. Chem. Minerals 15, 201–207.
Schrader, R. and Büttner, G. (1963) Untersuchungen über geisen (III)-oxid. Z. Anorg. Allgem. Chemie 320, 205–219.
Schulz, H. and Tscherry, V. (1972) Structural relations between the low- and high-temperature forms of β-eucryptite ($LiAlSiO_4$) and low and high quartz. I. Low temperature form of β-eucryptite and low quartz. Acta Crystallogr. B28, 2168–2173.
Schwartzentruber, J., Fürst, W. and Renon, H. (1987) Dissolution of quartz into dilute alkaline solutions at 90°C: A kinetic study. Geochim. Cosmochim. Acta 51, 1867–1874.
Seifert, H. (1966) Epitaxy of Macromolecules on Quartz Surfaces. Proc. Int'l Conf. Crystal Growth. Pergamon Press, New York.
Selikoff, I.J. (1977) Carcinogenic potential of silica compounds. In: G. Bendz and I. Lindqvist, Eds. Biochemistry of Silicon and Related Problems. Plenum Press, New York, p. 311–336.
Selverstone, J. (1990) Fluids at high pressure: Inferences from 20 Kbar eclogites and associated veins in the Tauern Window, Austria (abstr.). Goldschmidt Conference, 81.
Shelley, D. (1993) Igneous and Metamorphic Rocks Under the Microscope. Chapman and Hall, London, 445 p.
Shropshire, J., Keat, P.P. and Vaughan, P.A. (1959) The crystal structure of keatite, a new form of silica. Zeits. Kristallogr. 112, 409–413.

Sidhu, P.S., Gilkes, R.J., Cornell, R.M., Posner, A.M. and Quirk, J.P. (1981) Dissolution of iron oxides and oxyhydroxides in hydrochloric acid and perchloric acids. Clays and Clay Minerals 29, 269–276.
Simpson, D.R. (1965) Carbonate in hydroxylapatite. Science 147, 501–502.
Sinclair, W. and Ringwood, A.E. (1978) Single crystal analysis of tthe structure of stishovite. Nature 272, 714–715.
Slansky, M. (1986) Geology of Sedimentary Phosphates. North Oxford Academic, London, 210 p.
Smelik, E.A. and Reeber, R.R. (1990) A study of the thermal behavior of terrestrial tridymite by continuous X-ray diffraction. Phys. Chem. Minerals 17, 197–206.
Smith, P.P.K. (1979) The observation of enantiomorphic domains in natural maghemite. Contrib. Mineral. Petrol. 69, 249–254.
Spackman, M.A., Hill, R.J. and Gibbs, G.V. (1987) Exploration of structure and bonding in stishovite with Fourier and pseudoatom refinement methods using single crystal and powder X-ray diffraction data. Phys. Chem. Minerals 14, 139–150.
Sosman, R.B., (1965) The Phases of Silica. Rugers University Press, New Brunswick, New Jersey, 388 p.
Stanton, M. F., Layard, M., Tegeris, A., Miller, E., May, M., Morgan, E., & Smith, A. (1981). Relation of particle dimension to carcinogenicity in amphibole asbestoses and other fibrous minerals. J. Nat'l Cancer Inst. 67, 965–975.
Stanton, R.L. (1972) Ore Petrology. McGraw Hill, New York, 713 p.
Stebbins, J.F. (1991) NMR evidence for five-coordinated silicon in a silicate glass at atmospheric pressure. Nature 351, 638–639.
Stebbins, J.F. and McMillan, P.F. (1989) Five- and six-coordinated Si in $K_2Si_4O_9$ liquid at 1.9 GPa and 1200°C. Amer. Mineral. 74, 965–968.
Stone, A.T. and Morgan, J.J. (1988) Reductive dissolution of metal oxides. In: Aquatic Surface Chemistry, W. Stumm, Ed. Wiley-Interscience, New York, p. 221–252.
Thompson, A.B. and Wennemer, M. (1979) Heat capacities and inversions in tridymite, cristobalite, and tridymite–cristobalite mixed phases. Amer. Mineral. 64, 1018–1026.
Torres, R., Blesa, M.A. and Matejevic, E. (1990) Interactions of metal hydrous oxides with chelating agents. 8. Dissolution of hematite. J. Colloidal Interface Sci. 131, 567–579.
Trestmann-Matts, A., Dorris, S.E., Kumarakrishnan, S. and Mason, T.O. (1983) Thermoelectric determinatin of cation distributions in Fe_3O_4–$FeTiO_4$. J. Amer. Ceramic Soc. 66, 829–834.
Tunesi, S. and Anderson, M.A. (1987) Photocatalysis of 3,4-DCB in TiO_2 aqueous suspensions; Effects of temperature and light intensity; CIE-FTIR interfacial analysis. Chemosphere 16, 1447–1456.
Tunesi, S. and Anderson, M.A. (1991) The influence of chemisorption of the photodecomposition of salicylic acid and related compounds using suspended TiO_2 ceramic membranes. J. Phys. Chem. 95, 3399–2405.
Tunesi, S. and Anderson, M.A. (1992) Surface effects in photochemistry: An *in situ* CIR–FTIR investigation of the effect of ring substitutes on chemisorption onto TiO_2 ceramic membranes. Langmuir 8, 487–495.
Turnock, A.C. and Eugster, H.P. (1962) Fe–Al oxides: phase relationships below 1000°C. J. Petrol. 3, 533–565.
Vali, H. and Jirschvink, J.L. (1989) Magnetofossil dissolution in a palaeomagnetically unstable deep-sea sediment. Nature 339, 203–206.
Van Tendeloo, G., Van Landuyt, J. and Amelinckx, S. (1976) The α–β phase transition in quartz and $AlPO_4$ as studied by electron microscopy and diffraction. Physica Status Solidi 33, 723–735.
Vereway, E.J.W. and deBoer, J.H. (1936) Cation rearrangement in a few oxide crystal structures of the spinel type. Recueil des Travaux Chemiques des Pays-Bas, 4th Ser., 55, 531–540.
Vlassopoulos, D., Rossman, G. and Haggerty, S.E. (1990) Hydrogen in natural and synthetic rutile (TiO_2): Distribution and possible controls on its incorporation. EOS, Trans. Amer. Geophys. Union 71, 626.
Vorwald, A.J. and Karr, J.W. (1938) Pneumonoconiosis and pulmonary carcinoma. Amer. J. Pathol. 14, 49–57.
Walther, J.V. and Helgeson, H.C. (1977) Calculation of the thermodynamic properties of aqueous silica and the solubility of quartz and its polymorphs at high pressures and temperatures. Amer. J. Sci. 277, 1315–1351.
Waychunas, G.A. (1991) Crystal chemistry of oxides and oxyhydroxides. Rev. Mineral. 25, 11–61.
Wedepohl, K.H. (1969) Handbook of Geochemistry. Springer-Verglag, Berlin.
Weiss, A. and Herzog, A. (1977) Isolation and characterization of a silicon-organic complex from plants. In: G. Bendz and I. Lindqvist, Eds., Biochemistry of Silicon and Related Problems. Plenum Press, New York, p. 109–127.
Weissner, J.H., Henderson, J.D., Jr., Sohnle, P.G., Mandel, N.S. and Mandel, G.S. (1988) The effect of crystal structure on mouse lung inflammation and fibrosis. Amer. Rev. Respir. Dis. 138, 445–460.

Wiessner, J.H., Mandel, N.S., Sohnle, P.G., Hasegawa, A. and Mandel, G.S. (1990) The effect of chemical modification of quartz surfaces on particulate-induced pulmonary inflammation and fibrosis in the mouse. Amer. Rev. Respir. Dis. 141, 111–116.

Wennemer, M. and Thompson, A.B. (1984) Ambient temperature phase transitions in synthetic tridymites. Schweizer. mineral. petrograph. Mitt. 64, 355–368.

White, W.B. and Minser, D.G. (1984) Raman spectra and structure of natural glasses. J. Non-Crystalline Solids, 67, 45–59.

Williams, L.A. and Crerar, D.A. (1985) Silica diagenesis, II. General mechanisms. J. Sed. Petrol. 55, 312–321.

Williams, L.A., Parks, G.A. and Crerar, D.A. (1985) Silica diagenesis, I. Solubility controls. J. Sed. Petrol. 55, 301–311.

Willis, M.A., Cambell, A. and Phillips, R. (1989) High salinity fluids associated with allanite mineralization, Capitan Mountains, New Mexico. Geol. Soc. Amer. Abstr. 101, 79.

Willis, M.A. and Shock, E.L. (1991) Titanium speciation in hydrothermal metamorphic and subduction fluids. EOS, Trans. Amer. Geophys. Union 72 (abstr.).

Wirth, G.S. and Gieskes, J.M. (1979) The initial kinetics of the dissolution of vitreous silica in aqueous media. J. Colloid Interface Sci. 68, 492–500.

Woodcock, L.V., Angell, C.A. and Cheeseman, P. (1976) Molecular dynamics studies of the vitreous state: Ionic systems and silica. J. Chem. Phys. 65, 1565–1567.

Wright, A.C., Desa, J.A.E., Weeks, R.A., Sinclair, R.N. and Bailey, D.K. (1984) Neutron diffraction studies of natural glasses. J. Non-Crystalline Solids 67, 35–44.

Wright, A.F. and Lehmann, M.S. (1981) The structure of quartz at 25 and 590°C determined by neutron diffraction. J. Solid State Chem. 36, 371–380.

Yamazaki-Nishida, Nagano, K.J., Phillips, L.A., Cervera-March, S. and Anderson, M.A. (1993) Photocatalytic degradation of trichloroethylene in glass phase by using titanium dioxide pellets. J. Photchem. Photobiol. (in press)

Yau, Y.C., Peacor, D.R. and Essene, E.J. (1987) Authigenic anatase and titanite in shales from the Salton Sea geothermal field, California. N. Jb. Mineral. Mh. 10, 441–452.

Zitting, A. and Skytta, E. (1979) Biological activity of titanium dioxides. Int'l Arch. Occup. Environ. Health 43, 93–97.

Zoltai, T. and Buerger, M.J. (1959) The crystal structure of coesite, the dense, high-pressure form of silica. Zeits. Kristallogr. 111, 129–141.

Zoltai, T. and Buerger, M.J. (1960) The relative energies of rings of tetrahedra. Zeits. Kristallogr. 114, 1–8.

CHAPTER 6

PREPARATION AND PURIFICATION OF MINERAL DUSTS

Steve J. Chipera, George D. Guthrie, Jr., and David L. Bish

Geology and Geochemistry Group, EES–1
Los Alamos National Laboratory
Los Alamos, New Mexico 87545 U.S.A.

INTRODUCTION

A critical aspect of any good experiment is the control of variables. In research on the biological effects of minerals, both biological and mineralogical variables must be controlled. Consequently, modern *in vivo* procedures require the use of well-characterized strains of animals (even of a particular sex) that exhibit specific responses under various conditions. Likewise, modern *in vitro* procedures use well-controlled conditions, inasmuch as they typically require the use of a particular cell type or cell line or a very specific chemical environment (e.g., a purified type of DNA or mixture of certain chemicals). The use of such restrictive procedures results in experiments that are more easily interpretable in terms of associating an observation with a specific variable. Stringent criteria should also be applied to the choice of a mineral sample used in a biological experiment, so that observations (i.e., biological response) can be linked directly to mineral species or other mineralogical properties. Unfortunately, very few mineral standards are available commercially, so the acquisition of a well characterized mineral sample is not as simple as the purchase of a reagent-grade chemical. Mineral species are the naturally occurring constituents of rocks, and as such, most mineral species occur in association with several mineral species. So, for example, when one purchases a sample identified as clinoptilolite (a zeolite-group mineral; see Chapter 4), the sample may actually contain several mineral species, including various feldspars, silica minerals, other zeolites, and clays. In some cases, such a polymineralic assemblage will be obvious, as there will be grains that have different physical appearances and that are sufficiently large to see with the unaided eye. However, polymineralic assemblages can be very difficult to recognize if the material is fine grained, if the contaminants are present in minor amounts, or if the materials have similar physical appearances. In these cases, various techniques are necessary to evaluate the mineral content of the sample (some of these will be discussed in Chapter 7). In this chapter, we discuss several approaches that can be used to obtain pure phases from a polymineralic assemblage. Because many starting materials may be lithified (i.e., they will not consist of loose grains), we discuss the procedures necessary to disaggregate a material, thereby reducing the particle size to the required dimensions. We also outline a general procedure for preparing and purifying a fine-grained mineral sample from a polymineralic starting material using water sedimentation, a technique we use in our laboratory. Finally, we discuss methods for preparing fractions within specific size ranges.

SOURCES FOR MINERAL SPECIMENS

There are numerous sources for mineral samples that can be used for research purposes (e.g., see Appendix I), including commercial suppliers and museums.[1] In addition to these sources, other potential sources for mineral samples include many university geology departments and mineral trade shows (e.g., the annual mineral show in Tucson, Arizona, U.S.A.). Finally, many materials can be collected easily, and museums and local geology departments can often provide information as to which localities are appropriate for a specific mineral need.

It cannot be overemphasized, however, that even samples from reputable sources, such as those listed in Appendix I, are often polymineralic. The preparation, purification, and characterization of the samples are the responsibility of the researcher, and it is critical that these details are documented as part of any study.

PREPARATION OF MINERAL DUST SPECIMENS

Disaggregation

The first step in the preparation of a mineral dust is normally disaggregation,[2] because most minerals are obtained from rocks. Disaggregation is typically done in stages, starting with a hammer to break off manageable pieces and finishing with a mortar and pestle to prepare a final dust. Because sample purity may be a critical component in most studies, it is often useful to separate the sample into pure and impure aliquots at each stage. For macro-crystalline minerals, this can easily be done visually by hand picking, and this effort will greatly facilitate the purification steps further along the line.

For the final preparation of a dust, a mortar and pestle or micronizing mill can be very effective for reducing particles to micrometer-sizes. Mortars and pestles are available in a wide range of materials, and it is important to select a material that will be much harder[3] than the material being prepared. Normally, corundum is preferred, because it is harder (i.e., $H = 9$) than most rock-forming minerals, thus it can be used on a wide range of rocks. Agate is also commonly used for mortars and pestles. Although agate (a type of microcrystalline quartz) is hard (i.e., $H = 7$), rock samples frequently contain minerals of equal or greater hardness (e.g., quartz). When the hardness of the mortar or pestle is close to or less than the hardness of the material being prepared (or any component of the material being

---

[1] Most mineral museums maintain a research collection from which samples may be obtained. Such samples are typically available only in quantities that would be sufficient for most *in vitro* assays and *in vivo* assays that require small quantities. Quantities sufficient for inhalation studies would likely be difficult to obtain from a museum source.

[2] Several types of equipment are available to disaggregate rock (e.g., see Appendix I). Examples of equipment that is commonly used to prepare the amounts of material which are typically needed for biological research include: jaw crusher, ball mill, impact mortar and pestle, and shatter box. Other equipment is available for preparation of large (commercial) quantities of material.

[3] Hardnesses of minerals are typically reported using the Mohs hardness scale, which ranges from $H = 1$ to $H = 10$ (softest to hardest). Talc has a hardness of 1, and diamond has a hardness of 10. Other common minerals are used to define the remaining integral hardness values. Hardnesses of minerals are tabulated in mineralogy texts, such as Klein and Hurlbut (1993).

Table 1

Conversion of ASTM mesh numbers to particle size

Mesh number	Particle size (µm)	Mesh number	Particle size (µm)
10	2000	140	106
18	1000	170	90
40	425	200	75
60	250	230	63
80	180	270	53
100	150	325	45
120	125	400	38

prepared), contamination of the sample by the mortar and pestle can occur. Most hand-held mortars and pestles are *capable* of producing micrometer-sized material, but it is generally very difficult to reduce the particle size to less than 10 to 20 µm. For most biological purposes, the particle size must be reduced further, and this generally requires the use of an automated mortar and pestle or micronizing mill. Several models are available (e.g., see Appendix I), and these are very effective at reducing the particle size to <10 µm in a matter of minutes. Separation of individual mineral species from a polymineralic assemblage generally requires that the material is reduced to particle sizes less than the sizes of the individual mineral crystallites, e.g., often <5 µm.

Purification

Sieving. Minerals in a dust sample generally exhibit a broad range in particle size, but often specific minerals will have particle-size ranges that do not overlap completely. Thus, a crude method for separating minerals is sieving. Sieving rarely results in aliquots of pure minerals, but it can be an effective first step in mineral separation when used in conjunction with other methods such as those listed below. Sieving is most often done under dry conditions, but wet sieving is normally more efficient. Sieving is a mechanical process involving the passing of a dust sample through screens of decreasing size. The screen size is often given as a mesh number, and Table 1 gives the conversion of mesh numbers to particle size for common meshes. The stack of sieves is generally tapped or shaken to facilitate the process, and mechanical shakers are available to improve the process further.

Magnetic separation. Minerals exhibit a large range in magnetic properties. The most common elemental constituent responsible for magnetism in minerals is iron, which is the third most abundant metal in the earth's continental crust (constituting ~7 wt %) (Faure, 1991). The iron content of minerals can range from extremely low (as in many quartz samples) to extremely high (as in hematite and magnetite), and in general, minerals can be separated based on this variation. Many iron-bearing minerals have densities greater than iron-poor minerals, so these may be further separated based on this property, as discussed below. However, separation based on magnetic properties can offer the advantage of dry preparation, thereby avoiding the potential alteration of the mineral surface by density-separation liquids (discussed below). Magnetic separation can be achieved using an instrument such as a Frantz Separator, which functions based on a very simple principle: the competition between gravity and magnetic attraction. A mineral-dust preparation is allowed to slide down a trough (aided by mechanical vibration); the trough is placed in a magnetic field, so that magnetic minerals move to one side of the trough; and a septum divides the magnetic portion of the sample from the non-magnetic portion. Minerals with differing magnetic properties can be separated sequentially by varying the dip and cant of the trough and the intensity of the magnetic field.

Another method for the separation of magnetic and non-magnetic material involves the use of a strong bar magnet. This technique can easily remove strongly magnetic material (e.g., magnetite) from non-magnetic material, particularly if the sample is fine-grained and suspended in water. To facilitate removal of the sample from the magnet, the magnet can be placed inside a plastic bag.

Density separation. Minerals have specific gravities[4] (**G**) within the range ~1.5 to 10.5 (Klein and Hurlbut, 1993); however, values of **G** for many common rock-forming minerals typically fall within a much more restricted range (i.e., ~2 to 4) (Klein and Hurlbut, 1993). This range in **G** allows minerals to be separated using liquids of various densities. Density separation is *most* effective for separating materials with very different densities, and careful separation can lead to relatively pure end products. Traditionally, heavy liquids used for density separations consisted of organic compounds, such as bromoform, tetrabromoethane, and methylene iodide. These liquids were diluted to lower densities by mixing with organic solvents, such as benzene, carbon tetrachloride, acetone, and alcohols (Muller, 1967). Many of these organic compounds, however, are carcinogenic and require special care and environmental control. Furthermore, it may be undesirable to separate a mineral for carcinogenic evaluation using a carcinogenic liquid, which may sorb to the mineral surface. Recently, a non-toxic sodium polytungstate liquid has been used, and it has the advantages of miscibility with water (thereby providing a mechanism for preparation of liquids over a wide range in density), a relatively low viscosity at specific gravities below ~3.0, and easy reclamation and reprocessing (by filtration to remove suspended matter and evaporation to remove excess water) (Callahan, 1987; Torresan, 1987).

Clearly, one disadvantage of this separation technique is that the material must contact a fluid that may alter the material's surface properties. As mentioned above, the use of organic liquids risks contaminating the surface with a carcinogen. However, even the use of sodium polytungstate liquids may alter some minerals. For example, minerals with a finite cation-exchange capacity (e.g., clays and zeolites) may experience some sodium exchange. Also, it is extremely difficult (if not impossible) to use density to separate many minerals with similar or

---

[4] Specific gravity is a unitless measure of density and is equivalent to the ratio between the density of a substance relative to the density of water at 4°C (i.e., the maximum density of water). Specific gravity is often determined by comparing the weight of a material in air with the weight of a material in water; hence, it is sometimes defined as the ratio of a material's weight to the weight of an equivalent volume of water (see Klein and Hurlbut, 1993, p. 256–257).

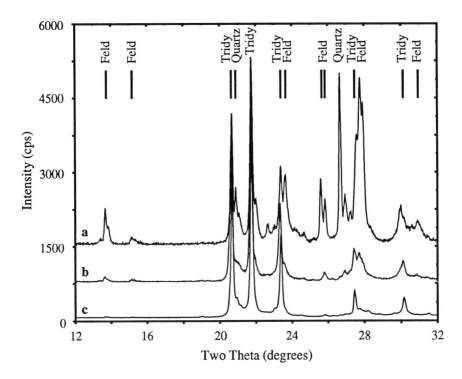

Figure 1: X-ray diffraction patterns for material from the Bandelier tuff. (a) Original tuff; (b) the G = 2.29–2.40 fraction; (c) G < 2.29 fraction. Note the significant increase in purity of the density fractions relative to the parent material. The G < 2.29 fraction is relatively pure tridymite, although some feldspar is still present. "Tridy" = tridymite; "Feld" = feldspar. (All diffraction patterns presented in this paper were collected using CuK_α radiation.)

overlapping densities, such as quartz and feldspar with values of G = 2.65 and G = 2.56 to 2.76, respectively (Klein and Hurlbut, 1993).

Figure 1 shows the results for density separation of tridymite (G = 2.26) using sodium polytungstate heavy liquids. The starting material was derived from Bandelier Tuff (New Mexico), which contains numerous other constituents, including feldspar (G = 2.56 to 2.76) and quartz (G = 2.65). The sample was initially crushed and separated into size fractions using sieving and settling-velocity separation (see below). The ~3 to 15-µm size fraction was subsequently used for additional separations with heavy liquids. The X-ray diffraction patterns of the original tuff, the G = 2.29 to 2.40 fraction, and the G < 2.29 fraction show that a relatively pure separate of tridymite was obtained. However, slight feldspar impurities still existed in the G < 2.29 fraction, suggesting that an additional separation with a slightly less dense liquid (e.g., G = 2.26) would be needed to produce pure tridymite. Additional grinding of the sample may also help by disaggregating any tridymite–feldspar agglomerates, such as those likely responsible for the feldspar peaks in Figure 1b.

Separation based on settling velocity. A very effective method for separating fine-grained minerals is through the use of settling velocities (see Appendix II for a detailed procedure). In simple terms, a dust preparation is allowed to fall through a medium (typically water), and the sediment is removed at specific time intervals. The material collected within a given time interval represents all material with a specific settling velocity.

Minerals exhibit settling velocities that are primarily a function of specific gravity and particle size/shape. This process is described by Stokes' law (Jackson, 1979),

$$v = \frac{g(s_p - s_l)}{18\eta} D^2 \qquad (1)$$

where v is the settling velocity (in cm/sec in the following treatment); g is the acceleration due to gravity (980.6 cm/sec^2); s_p and s_l are the specific gravities of the particle and liquid, respectively; D is the effective particle diameter in cm; and η is the viscosity of the liquid in poises. Viscosity and specific-gravity values for water at various temperatures are presented in Table 2.

Equation 1 can be readily expressed as either (a) the time, t, required for all particles of a specific effective diameter to fall a specific distance (h), as shown in Equation 2,

$$t = \frac{18\eta h}{g(s_p - s_l) D^2} \qquad (2)$$

or (b) the effective minimum particle diameter, D, that will fall a distance, h, in a time, t, as shown in Equation 3,

$$D = \sqrt{\frac{18\eta h}{g(s_p - s_l) t}} \qquad (3)$$

The particle diameter (D) in the above equations is an "effective" diameter, in that it strictly applies only to spherical particles with smooth surfaces. However, this effective diameter can be converted into actual particle size/shape parameters for a given experimental setup. In other words, those particles that are collected over an interval D–ΔD will correspond to a particular size/shape fraction, albeit the absolute dimensions of the fraction may not correspond to the values predicted from Equation 3.

The finite height of the column used for settling introduces some overlap in particle size for various settling times. Particles settled from the top of the column will have traveled farther than particles settled from the bottom of the column; hence, the lower parts of the column will introduce a smaller particle size than the upper parts. Figure 2a shows the results of settling a sample with a specific gravity of 2.50, in a 10-cm high suspension of water at 25°C for 5 minutes. All of the particles with an effective diameter of 19 µm have settled out of suspension. However, 1% of the particles with an effective diameter of 2 µm have also settled out of suspension. The mean diameter of the particles settled out of the

Table 2

Viscosity and specific-gravity for water

Temperature (°C)	Viscosity (poise)	Specific gravity	Temperature (°C)	Viscosity (poise)	Specific gravity
16	0.01111	0.999	36	0.00709	0.994
17	0.01082	0.999	37	0.00695	0.993
18	0.01056	0.999	38	0.00681	0.993
19	0.01030	0.998	39	0.00669	0.993
20	0.01005	0.998	40	0.00656	0.992
21	0.00981	0.998	41	0.00644	0.992
22	0.00958	0.998	42	0.00632	0.991
23	0.00936	0.998	43	0.00621	0.991
24	0.00914	0.997	44	0.00610	0.991
25	0.00894	0.997	45	0.00599	0.990
26	0.00874	0.997	46	0.00588	0.990
27	0.00855	0.997	47	0.00578	0.989
28	0.00836	0.996	48	0.00568	0.989
29	0.00818	0.996	49	0.00559	0.989
30	0.00801	0.996	50	0.00549	0.988
31	0.00784	0.995	51	0.00540	0.988
32	0.00768	0.995	52	0.00532	0.987
33	0.00752	0.995	53	0.00523	0.987
34	0.00737	0.994	54	0.00515	0.986
35	0.00723	0.994	55	0.00506	0.986

Source: Tanner and Jackson (1947)

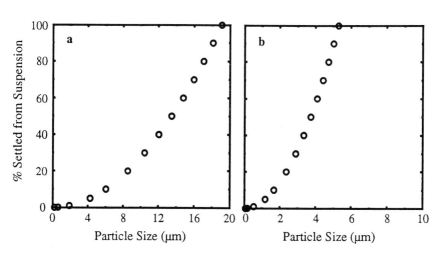

Figure 2: Stokes' Law of settling size distribution for a sample with a specific gravity of 2.50 suspended in a 10-cm column of 25°C water. (a) 5 min; (b) 65 min.

suspension during a specific time interval, however, cannot be predicted, because it is a function of particle-size distribution of the original sample. Likewise, Figure 2b shows the results for the same sample suspension and settling conditions after 65 minutes: all of the particles greater than 5.3 µm and 1% of the 0.5-µm particles have settled out of suspension. If tighter particle size distributions are required, one can use a smaller column height or repeatedly separate the same sample.

Figures 3 to 6 show X-ray diffraction patterns of clay samples separated using settling-velocity separation. The method works well for removing quartz, feldspar, and calcite from minerals such as smectite or clinoptilolite, or for separating quartz and smectite from palygorskite. The method does not work very well, however, when the physical properties of the impurity closely match those of the mineral of interest (e.g., clinoptilolite and mordenite) or where the mineral of interest and the impurity are intimately intergrown (e.g., clinoptilolite and opal-CT).

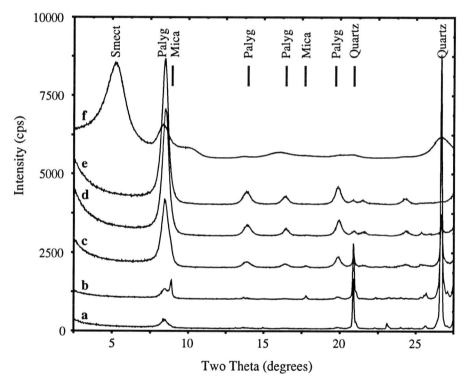

Figure 3: X-ray diffraction patterns for various size fractions of PFl-1 palygorskite (from Gadsden County, Florida, U.S.A.) separated using a 0.01-M sodium hexa-metaphosphate solution. (a) >18-µm fraction (gravity settled for 5 min); (b) 1.5–18-µm fraction (gravity settled for an additional 16 hours); (c) 0.35–1.5-µm fraction (centrifuged using a Sorvall GSA head at 3000 rpm for 5 min); (d) 0.15–0.35-µm fraction (centrifuged using a Sorvall GSA head at 8000 rpm for 5 min); (e) 0.05–0.15-µm fraction (centrifuged using a Sorvall GSA head at 8000 rpm for 40 min); (f) <0.05-µm fraction (after evaporation of water from the suspension). "Smect" = smectite; "Palyg" = palygorskite.

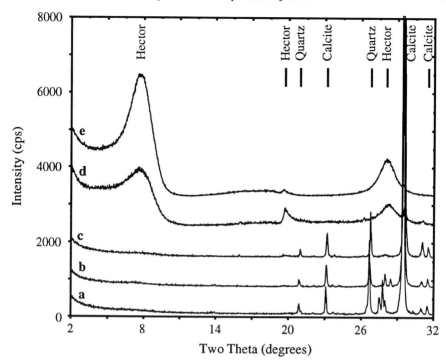

Figure 4: X-ray diffraction patterns for various size fractions of SHCa-1 hectorite (from San Bernardino County, California, U.S.A.) separated using deionized water only. (a) >18-μm fraction; (b) 5–18-μm fraction; (c) 0.25–5-μm fraction; d) 0.05–0.25-μm fraction; e) <0.05-μm fraction. "Hector" = hectorite.

One particularly important implication of Stokes' law is that even under ultra-centrifuge conditions (e.g., 10^4 g) submicrometer particles will not be removed from a suspension without very long centrifugation. This is fortuitous, in that submicrometer-size fractions can be readily prepared using centrifugation techniques. However, it also serves as a warning to those who attempt to remove particles from suspension: If the material being used contains small particles, these may remain suspended following short-duration centrifugation! In order to avoid problems with fine particles remaining in suspension, the most prudent procedure is to pre-centrifuge and decant the sample before beginning any experiment. This will effectively rid the sample of any very-fine-grained material.

Combined separation techniques. A recent innovation in separation technology is the Mag-stream Separator, which works on both magnetic and density principles by use of a magnetic suspension. Density separations are conducted by applying both an external magnetic field and rotation to provide centrifugal force. Lighter particles migrate to the center, and heavier particles migrate to the outside. Fluid density is effectively controlled by varying the magnetic field, speed of rotation, or fluid concentration. Pure magnetic separations can be conducted by applying a magnetic field with no rotation, in

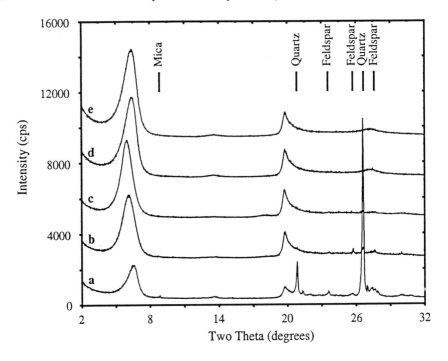

Figure 5: X-ray diffraction patterns for various size fractions of SAz-1 montmorillonite (from Cheto Mine, Arizona, U.S.A.) separated using deionized water only. (a) >18-μm fraction; (b) 5–18-μm fraction; (c) 1.5–5-μm fraction; (d) 0.25–1.5-μm fraction; (e) 0.05–0.25-μm fraction.

which case particles with high magnetic susceptibilities migrate outward (along the gradient of the magnetic field). A large range of separation conditions can be achieved by varying the relative contributions of the magnetic and centrifugal forces.

A more general class of combined separation techniques is represented by field-flow fractionation (Giddings, 1993). Field-flow fractionation uses settling velocities in reverse: particles are placed in a fluid flowing along a channel while other forces are exerted normal to the flow direction. Particles that are affected strongly by the normal forces move to the edges of the flowing fluid, which have lower velocities than the center portions (due to drag from the channel edges). This class of separation techniques has enormous potential, because the normal forces can be induced by numerous types of fields (e.g., flow field, thermal field, magnetic field, electric field, etc.), and it has proven useful at isolating particles ranging in size from ~1 nm to >100 μm.

Size fractionation

It is clear that many of the above methods (e.g., sieving and settling-velocity separation) can be used to separate a specimen into size fractions. An effective procedure for producing size fractions might combine several of these techniques

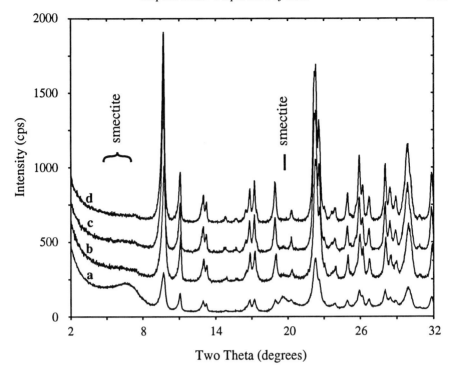

Figure 6: X-ray diffraction patterns for various size fractions of #27032 clinoptilolite (Castle Creek, Idaho, U.S.A.). Sample was suspended and gravity settled in deionized water to remove the fine-grained smectite from the more coarse-grained clinoptilolite. Note that the 1.5–15.0-μm fraction is composed of relatively pure clinoptilolite, whereas the successively finer sized fractions have increasing smectite abundances. (a) 1.5–15-μm fraction; (b) 0.7–1.5-μm fraction; (c) 0.3–0.7-μm fraction; (d) 0.1–0.3-μm fraction.

and involve alternating disaggregation and crushing with a separation technique, such that some size fractions are recrushed to produce finer fractions. Ultimately, small size fractions must be prepared by settling-velocity separation, field-flow fractionation, or a similar technique. For any of these techniques, it is imperative that the material is adequately disaggregated and dispersed (e.g., with an ultrasonic probe), so that the proper size fractions are obtained. Often small particles will aggregate to form particles with larger effective diameters, producing an apparently large-diameter size fraction.

SUMMARY

Although most geological materials are polymineralic, pure mineral separates can be obtained from naturally occurring geologic samples using the techniques described above. These separation techniques can significantly improve the purity of samples without significantly altering the physical or chemical properties of the materials. However, it should be realized that if the material is placed in any medium (including water), there is a potential to alter the surface chemistry and/or properties of the material.

Methods for purifying the mineral content of a polymineralic material generally rely on different physical properties of minerals (e.g., density, magnetic susceptibility, particle size, particle morphology, and surface properties). Specific methods for density or magnetic separation or field-flow fractionation are provided in the literature (e.g., Muller, 1967; Giddings, 1993). A generic procedure has been provided in Appendix II for the separation and size fractionation of fine-grained materials using settling-velocity separation.

REFERENCES

Callahan, J. (1987) A non-toxic heavy liquid and inexpensive filters for separation of mineral grains. J. Sedimentary Petrol. 57, 767-766.

Faure, G. (1991) Principles and Applications of Inorganic Geochemistry. Macmillan, New York.

Giddings, J.C. (1993) Field-flow fractionation: Analysis of macromolecular, colloidal, and particulate materials. Science 260, 1456–1465.

Jackson, M.L. (1979) Soil Chemical Analysis--Advanced Course. 2nd Edition, 11th printing. Published by the author, Madison, Wisconsin 53705 U.S.A.

Klein, C. and Hurlbut, C.S., Jr. (1993) Manual of Mineralogy, 21st Edition, John Wiley & Sons, New York, 681 p.

Muller, L.D. (1967) Laboratory methods of mineral separation. In: Physical Methods in Determinative Mineralogy. J. Zussman, Ed., Academic Press, New York, pp. 1–30.

Tanner, C.B. and Jackson, M.L. (1947) Monograph of sedimentation times for soil particles under gravity, Soil Science Society of America Proc. 12, 60–65.

Torresan, M. (1987) The use of sodium polytungstate in heavy mineral separations. U.S. Geol. Survey Open-File Report 87-509.

APPENDIX I

Following is an abbreviated list of potential sources for mineral specimens and materials for the preparation of geological materials. This list should not be considered an endorsement of any particular organization. Rather it is intended to provide a starting point for the researcher unfamiliar with geological-science sources. Numbers with the format xxx-xxx-xxxx are U.S. telephone numbers.

Mineral Specimens

Ward's Scientific Supply—5100 W. Henrietta Rd., Rochester, NY 14692. 800–962–2660.

Minerals Research—P.O. Box 591, Clarkson, NY 14430.

Minerals Unlimited—P.O. Box 877, Ridgecrest, CA 93556. 619–375–5279.

David New Minerals—P.O. Box 278, Anacortes, WA 98221. 206–293–2255.

Pure Minerals (W. Kleck)—P.O. Box 1914, Costa Mesa, CA 92628. 714–641–0958.

The Clay Minerals Society Source Clays Project—Dr. W.D. Johns, Dept. of Geology, University of Missouri, Columbia, MO 65211.

The Smithsonian Institute—Dept. of Mineral Science, Smithsonian Institute, Constitution Ave., Washington, D.C. 20560.

The Harvard University Mineralogical Museum—24 Oxford Street, Cambridge, MA 02138.

The American Museum of Natural History—Central Park West at 79th, New York, NY 10024.

Royal Ontario Museum—100 Queen's Park, Toronto, ON M5S 2C6 Canada.

Equipment/Supplies

Brinkmann Instruments—Cantiaque Rd., Westbury, NY 11590. 800-645-3050. Source for grinding equipment, particularly an automated mortar and pestle and spray drier.

Chemplex Industries, Inc.—160 Marbledale Road, Tuckahoe, NY 10707. 914-337-4200. Source for shatterbox.

Geoliquids—1618 Barclay Blvd., Buffalo Grove, IL 60089. 708-215-0938. Source for heavy liquids, particularly sodium polytungstate.

Heat Systems MedSonic Inc.—1938 New Highway, Farmingdale, NY 11735. 800-645-9555. Source for ultrasonic probes.

Magstream—Intermagnetics General Corp., New Karner Rd., P.O. Box 566, Guilderland, NY 12084. 518-456-5456. Source for magnetic separation equipment.

McCrone Accessories—850 Pasquinelli Dr., Westmont, IL 60559. 708-887-7100. Source for grinding equipment, including a micronizing mill.

Rocklabs—187 Morrin Rd., Knox Industrial Estate, P.O. Box 18-142, Auckland 6, New Zealand. Source for grinding equipment.

S.G. Frantz Company, Inc.—P.O. Box 1138 Trenton, NJ 08606. 609-882-7100. Source for magnetic separators.

Spex Industries—3880 Park Ave., Edison, NJ 08820. 908-549-7144. Source for grinding equipment.

Struers, Inc.—20102 Progress Dr., Cleveland, OH 44136. 216-238-2380. Source for grinding equipment.

Waring Products Division of the Dynamics Corporation of America—283 Main St., New Hartford, CT 06057. 203-379-0731. Source for laboratory blenders.

Yamato USA, Inc.—1955 Shermer Road, Suite 400, Northbrook, IL 60062. 718-498-4440. Source for spray driers.

APPENDIX II: PROCEDURE FOR SEPARATION BY SETTLING VELOCITY

Principle

This procedure involves the separation of a sample into various fractions using the principles embodied in Stokes' law. The sample is dispersed in deionized water and the sediment is collected at specific times. This procedure exploits differences in particle size/shape and density, and it is very effective at separating fine-grained materials (e.g., clays and zeolites) from the coarser materials with which they commonly co-exist (e.g., quartz, feldspar, and calcite).

Equipment and materials

Deionized water: Deionized water is typically used, because it minimizes the amount of chemical disturbance to the sample during the purification process (e.g., cation exchange).

Beakers: Large (1000-ml) beakers are normally used, because they provide a long settling distance and they can accommodate large samples (e.g., ~50 g). Smaller beakers can be used with smaller sample sizes.

Ultrasonic probe: Ideally, the sample should be disaggregated fully. An ultrasonic bath can be used to disaggregate the sample in suspension, but it is less effective than an ultrasonic probe. An ultrasonic probe is placed directly in a beaker of suspended material, and it is capable of introducing more ultrasonic energy than a bath. Hence, disaggregation with an ultrasonic probe is more rapid and efficient. One potentially deleterious outcome for some minerals is that the sample becomes heated by the ultrasonic energy.

Hot plate: A hot plate is used to dry the various size fractions and to concentrate the final supernatant suspension. The temperature should be set to facilitate drying without causing thermal degradation of the sample (e.g., between 40 and 50°C).

Centrifuge: A centrifuge is used to remove the finer particle sizes from suspension. It is important to know the geometry of the centrifuge and the resulting g-forces so that the particle size range can be estimated using Stokes' law.

Industrial blender: A blender can be used to disaggregate a sample and to disperse the material in suspension before ultra-sonic treatment.

Methodology

Below is a step-by-step procedure that can be used to produce mineral separates and size fractions using settling velocities. This method is most effective for preparing separates of clays and zeolites, because these minerals are typically very fine grained, and relatively pure aliquots of these materials can be prepared by settling velocities alone.

1. Approximately 50 g of sample and 800 ml of deionized water are placed in a 1000-ml beaker, giving a column height of 10 cm. The amount of sample may be adjusted depending on the sample's properties. For example, smectite-bearing material requires less sample due to electrostatic interactions between the particles. Hence, if 50 g of material does not produce a good separate, lesser amounts of material can be reprocessed.

2. The sample is disaggregated. This can be accomplished by some combination of vigorous stirring and/or use of a blender, and it should be followed by a 10-min treatment with an ultrasonic probe operated at ~200W. This disaggregation is merely a fine-scale disaggregation to ensure that the material is adequately suspended. It is not intended to break a large single crystal or to disaggregate a well-lithified rock, which should be done using other techniques.

3. The sample is allowed to settle undisturbed[5] for 5 min to remove the coarse particle fraction. (If a standard 1000-ml beaker is used, using Stokes' law of settling, all particles with an effective diameter of ≥ 19 μm will settle out of suspension (assuming $T = 25°C$; $G = 2.5$; and a depth of 10 cm). After 5 min, the supernatant is decanted into a separate beaker. The sediment is dried on a hot plate.

4. The supernatant is allowed to settle for an additional 60 min. All particles with an effective diameter of ≥ 5 μm should now have settled. The supernatant is decanted into a separate beaker, and the sediment is dried on a hot plate.

5. The supernatant is allowed to settle for an additional 16 hr (overnight). All particles with an effective diameter of ≥ 1.5 μm should have settled. The supernatant is decanted into a separate beaker, and the sediment is dried on a hot plate.

6. The sample remaining in suspension can be further separated using centrifugation techniques or using longer settling times. However, the final material can be removed from suspension by one of several possible processes: (a) the sample suspension can be centrifuged until all size fractions of interest are removed[6]; (b) the sample suspension can be filtered; (c) the sample suspension can be dried on a hot plate; or (d) the suspension can be spray-dried.[7] Centrifugation is preferable, because it is less likely to alter the physical characteristics of the sample and it can produce additional size fractions. Although filtration can be used to remove suspended material without altering its physical characteristics, it is not as effective in producing distinct particle size ranges, because filters remove finer and finer particles as they become clogged.

There are numerous instances in which deionized water alone does not allow the sample to disperse adequately. This results when the coarser material alone does not settle out of suspension, but, instead, the sample slowly settles en masse, forming turbid forms at the bottom and a layer of clear water at the top. In these cases, sodium hexametaphosphate $(NaPO_3)_6$ can be added to the suspension to act as a dispersant. In general, sufficient sodium hexametaphosphate is added to the suspension to produce a ~0.01 M solution. However, it should be realized that the use of any dispersant may result in a small amount of cation exchange with some samples, particularly zeolites and some clays. Also, each fraction removed from suspension should be washed with deionized water at least three times to remove all traces of the dispersant.

---

[5] During the settling process, the beakers should be covered to avoid any dust contaminants introduced by the atmosphere.
[6] For example, this requires ~1 hr at 8000 rpm in the Du Pont Sorvall SS-3 using the GSA head.
[7] Spray drying involves injecting the suspension into a hot cyclone chamber where the water is evaporated and the sample collects at the bottom. Our limited experience suggests that spray drying results in a significant loss of fine-grained material from the sample.

CHAPTER 7

MINERAL CHARACTERIZATION IN BIOLOGICAL STUDIES

George D. Guthrie, Jr.

Geology and Geochemistry
Mail Stop D 462
Los Alamos National Laboratory
Los Alamos, New Mexico 87545 U.S.A.

INTRODUCTION

The thorough characterization of mineral samples is a critical aspect of studies on mineral-induced pathogenesis. The variation in biological activity exhibited by different mineral species and different samples of specific mineral species imply that the pathogenicity of minerals is related (at least in part) to their physical and chemical properties. Unlike reagent-grade chemicals, mineral samples are typically heterogeneous in mineral content, mineral composition, and/or mineral structure, and this can affect their biological properties. This heterogeneity may be a predominant part of the sample, or it may be a relatively minor component that requires careful observation to detect. It is easy to understand why heterogeneities that dominate a sample may be important in determining a sample's properties. For example, a sample that contains 75% clinoptilolite and 25% erionite would likely have a much greater potential to induce mesotheliomas than a sample that contains 100% clinoptilolite.[1] However, even minor heterogeneity can profoundly affect a sample's properties. For example, minor amounts of asbestiform tremolite have been suggested as the cause for mesotheliomas among individuals exposed to chrysotile ore; minor amounts of metals enhance the catalytic properties of many zeolites; and minor deviations from the ideal structure (i.e., the microstructure or defect structure) of a mineral or material can determine its electromagnetic properties (e.g., conductivity). Hence, a thorough characterization of the mineral samples should be integral to every report on the biological properties of a mineral.

Mineral content must be determined for every sample to be used in a biological assay, particularly for samples derived from natural sources. **This is true even for samples labeled as a specific mineral by the supplier.** The determination of mineral content provides information on the bulk structure and ideal composition of phases in the sample. However, this information is generally insufficient for a complete description of a sample. Any *departures from the bulk structure and ideal composition* (e.g., type and distribution of defects; nature of

---

[1] Clinoptilolite is a non-fibrous zeolite, and erionite is a fibrous zeolite that has been linked to mesotheliomas in humans and animals. The structures of these minerals are discussed in Chapter 4, and the biological properties of these minerals are discussed in Chapters 12 and 15.

substitutions; or the type and abundance of trace elements) will vary between samples, so these should be determined. *Surface characteristics* will also vary between samples and may change following any sample preparation. Hence these, too, should be determined. Mineralogical aspects are as important in a study as are the cell type and culture conditions, strain/sex of the animal, etc. A thorough characterization of the mineralogical variables will eventually allow the mineralogical mechanisms of pathogenesis to be identified. Many of the techniques necessary to provide this level of characterization are not routine and would be considered "cutting-edge" mineralogy, but this makes the mineralogical side of mineral-induced pathogenesis challenging and exciting.

This chapter briefly discusses how one might characterize many of the aspects of minerals that may be important in determining biological activity. These aspects include bulk properties (such as mineral content, bulk structure, and bulk composition). However, because interactions with a mineral occur at the mineral surface, surface properties are also important in pathogenesis, and some techniques for the evaluation of mineral surfaces are discussed. The determination of particle size/shape distributions is not discussed in this chapter, because this type of mineral characterization is already well established in the biological community and, in fact, has partially overshadowed the importance of other mineralogical factors in pathogenesis. Instead, this chapter focuses on a few aspects of mineral characterization that are not well-established in the biological community.

This chapter is not intended as a comprehensive treatment of mineral analysis, nor is it intended to encompass all of the important techniques that have been applied to minerals. Detailed discussions of these techniques are presented in numerous books, many of which are dedicated to an individual technique. Instead, this chapter is meant to provide the reader who is unfamiliar with the analysis of minerals with a starting point for some of the more common techniques that can be used to characterize a mineral sample. This chapter introduces mineral characterization by addressing the questions: What needs to be characterized? And, what are some of the techniques that can be used to provide this information? Most of the techniques are discussed only very briefly, but extended references are provided. X-ray-diffraction and electron-beam methods are the primary techniques discussed, because they provide much of the basic information needed for mineral characterization and are readily available at many institutions. Any reader interested in more detailed information on characterization of minerals can begin with one of the several *Reviews in Mineralogy* volumes on analytical methods in mineralogy:

1. *Spectroscopic Methods* (Vol. 18), F.C. Hawthorne, Ed., 1988—Detailed chapters on infrared spectroscopy, Raman spectroscopy, inelastic neutron scattering, optical spectroscopy, Mössbauer spectroscopy, nuclear magnetic resonance spectroscopy (including magic-angle spinning NMR), X-ray absorption spectroscopy (including extended X-ray absorption fine structure, or EXAFS, and X-ray absorption near-edge structure, or XANES), electron paramagnetic resonance (EPR, also called electron spin resonance, or ESR), Auger electron spectroscopy (AES), X-ray photoelectron spectroscopy (XPS),

X-ray emission spectroscopy, and luminescence spectroscopies (including cathodoluminescence, photoluminescence, and chemiluminescence).

2. *Modern Powder Diffraction* (Vol. 20), D.L. Bish and J.E. Post, Eds., 1989—Detailed chapters on powder X-ray diffraction (including the Rietveld method), powder synchrotron diffraction, and powder neutron diffraction.

3. *Mineral-Water Interface Geochemistry* (Vol. 23), M.F. Hochella, Jr. and A.F. White, Eds., 1990—Applications of several spectroscopic methods. Also has brief discussions on several surface techniques.

4. *Minerals and Reactions at the Atomic Scale: Transmission Electron Microscopy* (Vol. 27), P.R. Buseck, Ed., 1992—Detailed chapters on transmission electron microscopy (TEM), including high-resolution TEM and electron diffraction.

Several other excellent sources are available that describe minerals analysis, some of which are from the Mineralogical Association of Canada short-course series:

5. *Mineralogical Techniques of Asbestos Determination*, R.L. Ledoux, Ed., 1979—Detailed chapters on the analysis of asbestos using techniques such as optical microscopy, scanning electron microscopy (SEM), TEM, and electron diffraction.

6. *Neutron Activation Analysis in the Geosciences*, G.K. Muecke, Ed., 1980.

7. *Environmental Geochemistry*, M.E. Fleet, Ed., 1984—Chapter 7 discusses several techniques, including secondary ion mass spectrometry (SIMS).

8. *Application of Electron Microscopy in the Earth Sciences*, J.C. White, Ed., 1985—Similar to #4 but also discusses SEM.

9. *Fundamentals of Surface and Thin Film Analysis*, L.C. Feldman and J.W. Mayer, 1986—Textbook on the theoretical and practical aspects of several surface analytical techniques, including SIMS, electron energy-loss spectroscopy (EELS), low energy electron diffraction (LEED), EXAFS, XPS, AES, and proton- (and helium-) induced X-ray emission (PIXE).

10. *Modern Techniques of Surface Science*, D.P. Woodruff and T.A. Delchar, 1986—Textbook similar to #9.

11. *Spectroscopic Characterization of Minerals and Their Surfaces*, L.M. Coyne et al., Eds., 1990—Chapters on techniques and applications in mineralogy.

12. *Principles of Analytical Electron Microscopy*, D.C. Joy et al., Eds., 1986—Several detailed chapters on chemical analysis using TEM (including energy-dispersive spectroscopy, or EDS, and EELS).

13. *Scanning Electron Microscopy and X-ray Microanalysis*, J.I. Goldstein et al., Eds., 1981—Textbook on chemical analysis using SEM.

14. *High-Resolution Transmission Electron Microscopy and Associated Techniques*, P.R. Buseck et al., Eds., 1988—Includes chapters on imaging, diffraction, and applications.

15. *Instrumental Methods of Chemical Analysis*, G.W. Ewing, 1985—Textbook with survey chapters on numerous analytical methods.

16. Optical microscopy of minerals—Phillips (1971), Nesse (1991), and Deer et al. (1992).

Table 1

Acronyms for several techniques

Acronym	Technique
AEM	Analytical electron microscopy (using TEM)
AFM	Atomic force microscopy
EDS	Energy-dispersive spectroscopy
EELS	Electron energy-loss spectroscopy
EMPA	Electron probe microanalysis
LEED	Low energy electron diffraction
PIXE	Proton-induced X-ray emission
RIR	Reference-intensity-ratio
SEM	Scanning electron microscopy
SIMS	Secondary ion mass spectrometry
STM	Scanning tunneling microscopy
TEM	Transmission electron microscopy
WDS	Wavelength dispersive spectroscopy
XPS	X-ray photoelectron spectroscopy
XRD	X-ray diffraction

Many of the analytical techniques are referred to by acronyms. Those discussed frequently in this chapter are listed in Table 1.

Mineral species

A fundamental aspect of this chapter is the concept of mineral species. A mineral species is a crystalline material with a well-defined structure and with a composition that either is fixed or varies within well-defined bounds. This general definition encompasses the relevant characteristics that most geoscientists attribute to mineral species: a specific structure and composition. The significance of this definition cannot be overstated, because it also embodies most of the characteristics that ultimately affect a mineral's properties, including its biological activity. Therefore, these mineralogical properties must be essential components of any comprehensive model for mineral-induced disease.

Mineral species are defined in much the same way as are animal/plant species and cell types. As indicated, structure and composition are the two characteristics used to define mineral species. Each of these characteristics is necessary, because many distinct mineral species share a similar composition or a similar structure with other mineral species. For example, quartz and stishovite share the same composition (SiO_2), but each has a unique structure (see Chapter 5). Likewise, stishovite and rutile share the same structure, but each has a unique composition (SiO_2 and TiO_2, respectively). An appreciation of the significance of the concept of mineral species and a rigorous application of mineral-species nomenclature are essential for anyone working with minerals. For example, when a particular study reports results on chrysotile, these results do not apply to the asbestos minerals in general. Likewise, results obtained on a sample of quartz do not apply to all crystalline silica polymorphs. The terms *asbestos* and *crystalline silica* are not as specific as mineral species names, rather they are more analogous to terms such as *rodents*.

The bulk structure and composition that are used to define mineral species, however, are merely foundations upon which one can build an adequate description of a sample. In fact, most samples of a particular mineral species deviate from the ideal structure and composition for that species. Consequently, a thorough description of a sample should include the mineral species and their abundances *and* any deviations each mineral species in the sample exhibits with respect to the ideal mineral species. These deviations include those discussed above, namely deviations in structure (e.g., microstructures and defects) and deviations in composition (e.g., substitutions and trace-element content). Finally, surface characteristics should be included in the sample description, because these will vary between samples of a mineral species and they may change following sample preparation. Hence, although mineral species should be a critical component of any description of a sample, it is clear that other factors are also necessary to consider when evaluating the biological activity of a mineral.

MINERAL CONTENT

Probably the most basic level of characterization of a sample before use in a biological experiment is the determination of mineral content. Most mineral samples are derived from rocks, which are polymineralic (i.e., they consist of more than one type of mineral species). As discussed in Chapter 6, it is often possible to process such samples such that the purity of the mineral of interest is increased. Nevertheless, the mineral content of the sample to be studied must be verified, and the abundances of each species must be determined, and this applies even to samples obtained from mineral suppliers. In general, the requirements for the quantification technique are that (1) each of the mineral species present can be unambiguously identified, (2) the number of particles analyzed is sufficiently large to provide a statistically significant representation of the sample, and (3) the amounts of each mineral species can be quantified. The first two of these criteria are essentially never fully satisfied by any of the common techniques used for mineral-content quantification. However, important information on mineral content can be provided by several techniques, although each has its limitations.

X-ray diffraction

X-ray diffraction is probably the most appropriate technique for determining the types and abundances of mineral species in a powder, particularly if statistically significant results are required. X-ray diffractometers are available at numerous institutions, and the technique can be applied to sample amounts ranging from $\sim 10^{-4}$ g (in microdiffractometers) to 10^0 g. One requirement for the technique is that the sample is powdered (to $<\sim 10$ μm for the best quantitative results); however, this is generally not a problem for samples used in biological assays, because this corresponds to the useful size range for these studies as well. Numerous recent advances have made powder XRD extremely effective for quantitative analysis as well as structural analysis of the samples.

Materials (e.g., minerals) that possess periodic structures will diffract X-rays at specific angles such that when the intensity of X-rays diffracted from a powder is measured as a function of diffraction angle, maxima in intensity (diffraction peaks) are observed. The integrated intensity from a specific diffraction peak is related to the structure, composition, relative abundance of the species giving rise to that peak, and other factors. Thus, the positions and relative intensities of the diffraction peaks provide a fingerprint for crystalline materials, so XRD data can be used to identify the types of crystalline phases present in a sample. The positions and relative intensities for diffraction peaks for most minerals are tabulated in the Powder Diffraction File (JCPDS, 1987). In addition to providing information on the identification of phases, XRD data provide the ability to determine mineral abundances quantitatively.

Two general approaches to quantitative analysis using X-ray diffraction (XRD) data have been commonly used in studies relating to the quantitative analysis of minerals of biological interest: the reference-intensity-ratio (RIR) method and the Rietveld method. Quantitative analysis of polyphase assemblages using powder XRD is reviewed in detail by Snyder and Bish (1989).

The RIR method, also known as the Chung method (1974a,b; 1975), uses the integrated intensities from specific diffraction peaks to quantify the amounts of material present. The effects on intensity that arise from factors other than the weight percent of the phase are accounted for by using a reference-intensity ratio, $(I/I_S)\alpha$, which is the ratio of the integrated intensities for the strongest diffraction peak (I) from phase α to the strongest diffraction peak (I_S) for a phase chosen as a standard. Typically, corundum is chosen as the standard phase, and the ratios $(I/I_C)\alpha$ are determined by measuring the intensities from known mixtures of the minerals to be studied and corundum. Reference-intensity ratios with corundum as the standard are also reported in the Powder Diffraction File (JCPDS, 1987), but these are generally not as accurate as those measured using the specific instrumental conditions and minerals with compositions closely approximating the compositions in the "unknown" mixture. Once these reference-intensity ratios are determined, the mineral content of "unknown" mixtures can be determined by measuring the diffraction-peak intensities from a sample doped with a known amount of the standard phase (i.e., called the internal standard method), provided the crystalline phases present in the "unknown" mixture have been idenitified. Certain simplifying assumptions allow the quantitative analysis of mixtures without the addition of an internal standard. Variations on the RIR method have been used successfully in numerous studies on the quantitative analysis of materials of biological interest (e.g., Blount and Vassiliou, 1983; Davis, 1990; Bish and Chipera, 1991).

The sensitivity of the RIR method can be very high under optimum conditions. Davis (1990) reported detection limits in the 0.5 to 2.0 wt % range for the serpentine and amphibole asbestos minerals in various mixtures with calcite, gypsum, montmorillonite, and quartz. Bish and Chipera (1991) reported detection limits between 100 and 500 parts per million (i.e., 0.01 to 0.05 wt %) for erionite in mixtures with analcime and illite; however, their methods involved collection of XRD data for at least 120 s every 0.02° 2θ over the range 6 to 9°

2θ.[2] Erionite frequently occurs with other zeolites (see Chapter 4), and the quantification of mineral abundances in such mixtures can be extremely difficult due to severe overlap of the diffraction peaks. However, the identification of minor amounts of erionite in zeolite-bearing samples may be critical, because erionite appears to be extremely pathogenic in humans (e.g., Baris et al., 1987a,b) and laboratory animals (e.g., Maltoni et al., 1982; Suzuki and Kohyama, 1984; Wagner et al., 1985; Johnson and Wagner, 1989).

Figure 1 shows an X-ray diffraction pattern for a synthetic mixture consisting of 0.14 wt % erionite + 0.43 wt % illite in analcime. At a normal scale, the region 6 to 9° 2θ appears devoid of any peaks. However, expansion of the scale clearly reveals diffraction peaks arising from erionite and illite. These XRD data were sufficient for quantitative analysis of this mixture containing small amounts of erionite.

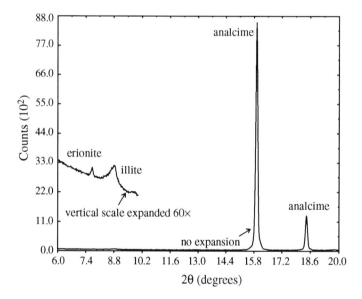

Figure 1. XRD pattern of 0.14 wt % erionite in analcime. The sample also contained small amounts of illite. After Bish (1993).

One limitation of the RIR method is that one must determine the integrated intensities for peaks from the phases of interest. This can be difficult or even impossible for complex mixtures of phases with large unit cells, due to severe overlap of diffraction peaks. An alternative approach, the Rietveld method, has recently been modified for quantitatively determining the amounts of phases from powder diffraction data (Bish and Howard, 1988; Post and Bish, 1989). Peak overlap is not a problem for the Rietveld method, because all possible reflections from a phase are considered explicitly during the analysis. The Rietveld method uses least-squares refinement to optimize the fit between the observed diffraction

---

[2] Counting times are often in the 1–5-s range for most XRD studies.

pattern and the pattern calculated from an input model. The input model includes contributions for both instrument- and sample-related effects, and any portion of the measured diffraction pattern can be used. The contributions to the intensity at each step in the diffraction pattern are calculated for every peak of every phase in the model. Thus, the Rietveld method explicitly accounts for peak overlap.

The Rietveld method works well for quantitative analysis of complex mixtures, including mixtures of mordenite–clinoptilolite. Analysis of standard mixtures of mordenite–clinoptilolite (Fig. 2) shows that the Rietveld method can be used to determine quantitatively the mineral content of these complex mixtures, despite severe problems with peak overlap and preferred orientation.[3]

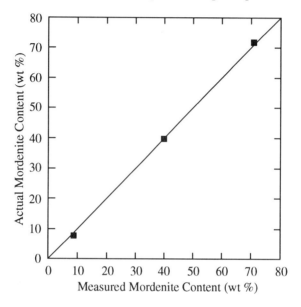

Figure 2. Results of quantitative analysis of synthetic mixtures of mordendite–clinoptilolite using the Rietveld method. Data were collected 2 to 70° 2θ at 2 s per 0.02° 2θ step. Once a refinement procedure for this assemblage was determined, the actual refinement for each mixture could be accomplished in ~15 min. After Guthrie and Bish (1991).

We have used this approach to determine the mineral content of the two mordenite-bearing samples that were used in comparative studies on the biological effects of mordenite and other fibrous minerals (mordenite A—Hansen and Mossman, 1987; Palekar et al., 1988; Mossman and Sesko, 1990; mordenite B—Suzuki, 1982; Suzuki and Kohyama, 1984). Rietveld analysis of these mordenite samples shows that the materials are, in fact, complex mixtures of several minerals (Table 2). Hence, the biological data apply specifically to these mixtures and not to pure mordenite. These results underscore the importance of

---

[3] The fibrous shape of mordenite and tabular shape of clinoptilolite lead to crystals that are not randomly oriented in the XRD mount. This affects the relative intensities of the diffracted peaks.

Table 2

Mineral content of mordenite samples used to assess toxicity

	Mordenite A	Mordenite B
mordenite	49.9%	39.2%
clinoptilolite	35.8%	28.7%
sanidine	4.4%	20.4%
albite	2.8%	4.2%
opal–CT	4.6%	4.1%
quartz	n.d.*	0.7%
gypsum	n.d.	0.5%

Source: Guthrie, Bish, and Mossman (unpublished data)
* Not detected in sample.

Table 3

Rietveld analysis of synthetic mixtures of erionite + clinoptilolite

Wt % erionite	
Known	Measured
20	16.4 ± 0.3
10	8.4 ± 0.1
5	5.2 ± 0.2
2.5	2.7 ± 0.1
2.0	2.2 ± 0.1
1.0	1.2 ± 0.2

Source: Guthrie and Bish (unpublished data)

determining the mineral content for samples to be used in biological assays. It is not known whether polymineralic assemblages behave antagonistically, additively, or synergistically in the induction of disease, so the interpretation of the biological data obtained on mordenites A and B is not straightforward.

The Rietveld method appears to be equally effective at analyzing complex mixtures containing erionite. Table 3 shows the results of a Rietveld analysis of standard mixtures containing various amounts of erionite. The comparatively poor match between the known and measured wt % for the samples containing higher amounts of erionite is probably due to preferred orientation. Samples containing significant amounts of erionite produce XRD mounts with the erionite oriented with its fiber axis in the plane of the sample mount, and this changes the relative intensities in the diffraction peaks. However, the degree of preferred orientation likely diminishes in samples containing small amounts of erionite. The effects of preferred orientation were not considered in the model used to analyze these samples, but the inclusion of a preferred orientation term would probably improve the results for the samples containing greater amounts of erionite.

Transmission electron microscopy

Transmission electron microscopy can also be used to determine the mineral content of a sample, because it can provide information on both the structure (via electron diffraction) and composition (via energy-dispersive spectroscopy or electron energy-loss spectroscopy). However, the collection and evaluation of these two types of data can be very time consuming, often requiring minutes per particle. Consequently, TEM data often do not provide the same degree of precision on the determination of mineral content as do powder diffraction data. TEM data do provide information on individual particles (e.g., structure, composition, size, and morphology), so they are complementary to XRD data. The combined use of the two techniques can result in a much greater level of detail than the use of either technique alone. The important point to remember about TEM data, however, is that they do not provide absolute identification of a mineral species unless both structural and compositional data are used. For example, Champness et al. (1976) discuss the procedures necessary for the accurate identification of asbestos using TEM. When only morphological and/or compositional data are used, the identification of mineral species is based on assumptions that can be significant, e.g., that there is only one type of fibrous mineral present. In some cases, compositional data from the TEM can provide sufficient information to identify a particle, but this is generally only when other data are available to substantiate the assumptions. For example, if XRD data are available on a sample and it is known that the only detectable crystalline silica polymorph is quartz, then the identification of a particle as quartz can be made by the presence of Si in an EDS analysis and a rapid check for crystallinity by bright-field imaging or electron diffraction. Detailed discussions of the TEM techniques that can be used to identify minerals are found in Hirsch (1977) and Spence (1988) for imaging and electron diffraction and Joy et al. (1986) for compositional analysis.

MINERAL STRUCTURES

Minerals exhibit translational periodicity (i.e., their structures repeat over large atomic distances), and they are said to be crystalline. Non-crystalline or amorphous materials, such as glass, are not minerals in a strict sense, because their detailed structures do not repeat in a regular, long-range fashion. They are often referred to as *mineraloids*. The distinction between minerals and mineraloids is very important, because a periodic structure influences many of a mineral's properties. For example, a periodic structure influences a mineral's interactions with photons, which allows minerals to be differentiated by their optical and diffraction properties; structure controls a mineral's dissolution characteristics, which is one determinant of the residence time for a mineral in the lung; and a periodic structure can impart a periodicity to the surface (i.e., a mineral surface can function as a template). These latter two examples illustrate potential mechanisms by which a mineral's structure may influence its biological activity. Any adequate model for mineral-induced pathogenesis must account for the role of mineral structure. Table 4 lists several techniques for the determination of various structural aspects of minerals.

Deviations from ideal structure

As discussed above, mineral structure is one component used to define mineral species, so the determination of mineral content for a sample will provide some information on mineral structure. However, the structures of most mineral specimens deviate from the ideal structure associated with a mineral species. These deviations range from slight distortions of the coordination polyhedra to the presence of defects. This latter type of deviation, in particular, can significantly affect the properties of a mineral. These deviations will vary from sample to sample, so even samples of the same mineral may exhibit a variation in some properties if their microstructures are significantly different.

Bulk structural properties (e.g., the average positions of atoms, site occupancies, etc.) can be determined using diffraction methods. The Rietveld method has proven very effective at structural studies using powder diffraction data, and it is probably the best current method for studying bulk properties of mineral-dust samples, such as those used in biological studies. Pack mounts provide the best presentation of the sample for the X-ray beam, and such mounts require ~0.5 to 1.0 g of material (see Bish and Reynolds, 1989, for a discussion on sample preparation for XRD). However, microdiffraction can be used to obtain diffraction data from much smaller amounts of material (e.g., <1 mg). The Rietveld method is described in detail by Post and Bish (1989).

The determination of microstructure can be a time-consuming endeavor. In general, TEM is probably the most useful technique for determining the types of microstructures that might be important in pathogenesis. Applications of TEM to the study of microstructures in amphibole asbestos, for example, can be found in a number of studies (e.g., Hutchison et al., 1975; Alario Franco et al., 1977; Veblen, 1980; Cressey et al., 1982; Ahn and Buseck, 1991). The use of TEM to identify and characterize microstructures in minerals is discussed in detail by Buseck and Veblen (1988), Baronnet (1992), and Veblen (1992).

MINERAL COMPOSITIONS

The determination of mineral compositions can be made using several techniques, particularly when sufficient homogeneous material is available. The techniques for determining the composition of a bulk material, however, will not be discussed here. Instead a few of the techniques that can be applied to individual particles are discussed below. Table 5 lists several techniques for the determination of mineral compositions.

Electron probe microanalysis

Scanning electron microscopy is one of the principle techniques used by geologists to determine the composition of individual mineral grains in a rock

Table 4

Techniques for characterizing mineral structure

Technique	Information provided	Comments on technique	Comments on sample	Reference
Diffraction Methods				
powder X-ray diffraction (XRD)	average structural information: mineral type, lattice parameters, mineral abundance, atomic position	relatively large amount of material sampled, so provides good statistical information; no information on individual particles; no information on defects or non-periodic features	smear mounts: ~10-mg pack mounts: ~0.5–1.0 g	Bish and Post (1989)
single-crystal XRD	average structural information on one crystal	ability to solve crystal structure; higher-quality structural information than powder XRD; only one crystal sampled; crystal must be "perfect," so no information on defects or non-periodic features	one crystal approximately 10s of µm on a side	Woolfson (1970); Glusker and Trueblood (1985)
micro XRD	analogous to powder XRD but on small amounts of material	provides powder XRD data on small amount (e.g., tenths of mg) of material or even on single crystal; less material sampled than traditional powder XRD, so poorer statistical information	~0.1–5 mg	
neutron diffraction	analogous to XRD methods	sensitive to hydrogen and can discriminate between elements with similar atomic numbers; requires access to reactor or pulsed neutron source	1–10 g	Von Dreele (1989)

Table 4 (cont'd)

Technique	Information provided	Comments on technique	Comments on sample	Reference
Spectroscopic Methods				
infrared spectroscopy	local environment around an atomic site; some configurations allow sorption processes to be studied	provides information on local environment of atoms, including hydrogen; can be used to study mineral–fluid interactions at mineral surface	~1–5 mg of sample sufficient to study IR characteristics of bulk; 10^1–10^2 mg may be required for some studies	McMillan and Hofmeister (1988)
Raman spectroscopy	local environment around an atomic site; some configurations allow sorption processes to be studied	provides information on local environment of atoms		McMillan and Hofmeister (1988)
Mössbauer spectroscopy	oxidation state and local environment for some elements	sensitive to local environment and oxidation state of iron and various other elements	~10^1–10^3 mg, depending on the composition of the material	Hawthorne (1988)
Microscopic Methods				
transmission electron microscopy	electron diffraction and direct imaging of structure	time consuming and small % of crystals examined	individual particles	White (1985); Buseck (1992)
scanning electron microscopy	electron diffraction	not a standard SEM technique	thick specimen (>~10 μm) with polished, flat surface	Goldstein et al. (1981); Lloyd (1985)
optical microscopy	effect of structure on optical properties	difficult to use on very small particles	individual particles	Phillips (1971); Nesse (1991); Deer et al. (1992)

Table 5
Techniques for characterizing mineral composition

Analysis technique	Information provided	Comments on technique	Comments on sample	Reference
Electron Microprobe (EPMA) (uses SEM)		capable of quantitative analysis; requires good standard materials for high-quality quantitative analysis; difficult to use on particles smaller than ~5 μm	excitation volume is about 1–2 μm in diameter on standard EPMA of bulk specimens; analysis of small particles not routine	Armstrong and Buseck (1975); Armstrong (1978, 1980); Goldstein et al. (1981)
wavelength dispersive spectroscopy (WDS)	type and abundance of elements heavier than ~Li	quantitative results on major and minor elements; time-consuming relative to EDS		Goldstein et al. (1981)
energy dispersive spectroscopy (EDS)	type and abundance of elements heavier than Na for normal detectors and heavier than Li for windowless or thin-windowed detectors	rapid; qualitative to semi-quantitative		Goldstein et al. (1981); Joy et al. (1986)
analytical TEM (AEM)		analyzes much smaller volumes than SEM; allows collection of structural information; poor statistical information due to small amount of material studied; less quantitative than SEM analysis using WDS	analysis of thin areas from individual particles	
EDS	same as above under EPMA			Joy et al. (1986); Peacor (1992)
electron energy loss spectroscopy (EELS)	investigation of light elements and information on oxidation state and bonding environment for some atoms			Joy (1979); Somlyo and Shuman (1982); Buseck and Self (1992)

Table 5 (cont'd)

Analysis technique	Information provided	Comments on technique	Comments on sample	Reference
X-ray fluorescence (XRF)	major to select trace element abundances	solid sample irradiated by X-ray beam, characteristic fluorescence X-rays measured by EDS or WDS	milligrams of sample required; synchrotron XRF allows microanalysis	Johansson and Campbell (1988)
proton induced X-ray emission (PIXE)	trace element abundance for high atomic number (>20) elements	capable of trace element sensitivity (5 ppm) for many elements in mineral matrices; solid sample bombarded with high-energy protons	micro-PIXE excites a <10-μm × 30-μm cylinder in standard thin sections	Reed (1989); Shimizu and Hart (1982)
secondary ion mass spectrometry (SIMS)	low detection limits; capable of determining ratios between isotopes of a given element	requires similar-matrix standards; <10-μm spatial resolution	destructive; <10-μm × 5-μm pit is eroded into sample	Muecke (1980)
neutron activation (NA)	element concentrations with low detection limits for a limited number of trace elements	neutron flux excites γ-ray emission characteristic of element	powdered samples; neutron fluxed in reactor then counted directly	Reed (1989)
ICP mass spectroscopy	major to trace element concentrations of solid and liquid samples	ions extracted from plasma analyzed in quadropole mass spectrometer; parts per billion detection limits	geologic samples generally must be fluxed and dissolved; laser ablation also possible	Thompson and Walsh (1988)
atomic emission (AE)	major to trace element concentrations of solid and liquid samples	optical emission of excited ions detected using grating spectrometer; 1 ppm detection limits are typical for solids	samples generally must be fluxed and dissolved	

(Goldstein et al., 1981; Joy et al., 1986). Specialized SEMs equipped to perform wavelength dispersive spectroscopy (WDS) can routinely provide analytical information with precisions down to 0.5% relative on elements with atomic numbers greater than Na. In electron probe microanalysis (EPMA), an electron beam penetrates the specimen and generates X-rays within a volume with a diameter of ~1 to 2 µm. The intensities of X-rays from each element can be related to concentration of that element, but the conversion requires corrections for absorption and fluorescence. Most correction routines are applicable to data collected from samples with a flat, polished surface and with a thickness greater than several micrometers. Hence, these routines cannot be used on individual particles with an irregular surface and/or with thickness <1 µm. Quantitative analysis of individual particles can be achieved using specialized procedures that account for particle size/shape (Armstrong and Buseck, 1975; Armstrong, 1978, 1980). These procedures appear to work extremely well on particles of respirable size. For example, Armstrong (1978) analyzed over 1500 particles with known compositions and compared the elemental concentrations determined by standard reduction procedures with those determined by the Armstrong-Buseck procedure (which accounts for particle size/shape). Elemental concentrations determined by the Armstrong-Buseck method exhibited mean values identical to the known values. In contrast, elemental concentrations determined by the standard reduction procedures exhibited mean values with errors as high as 40% relative to the known values. Hence, compositional analysis of individual particles is possible using EPMA, but the techniques necessary to produce accurate analyses of small particles differ from those typically done for geological materials.

Analytical electron microscopy

Transmission electron microscopes can be equipped to perform compositional analysis by two basic techniques: energy-dispersive spectroscopy, or EDS, and electron energy-loss spectroscopy, or EELS. Each of these analytical techniques assesses elemental composition based on the interactions between the electron beam of the TEM and the sample. In EDS, the X-rays generated by this process are used to determine elemental composition, as in EPMA. In EELS, the energy distribution of the primary beam is used to evaluate the energy-loss that occurs during the interaction between the electron beam and the sample, and this energy-loss is converted to elemental composition.

Compositional analysis by EDS can provide compositional data with errors that are typically > ±5% relative, primarily due to poor X-ray counting statistics (Goldstein et al., 1986). However, the accuracy in EDS analyses can still be quite good. Peacor (1992) presents data (from Jiang et al., 1990) comparing the accuracy of EDS analyses of a muscovite and a pyrophyllite with the compositions as determined by EPMA, and the concentrations for all elements reported (Si, Al, Ti, Fe, Mg, Ca, Na, and K) were well within 2σ of the concentrations determined by EPMA. Peacor (1992) also presents data clearly demonstrating that EDS can be applied to elements with concentrations at least as low as ~0.1 wt %. One drawback for EDS is that light elements are difficult and, in some cases, impossible to detect or to analyze quantitatively. The detector lighter than Na by many detectors. Thin-windowed and windowless detectors, however, are capable of detecting elements at least as low as Li. Joy et al. (1986) present a detailed discussion of EDS.

Electron energy-loss spectroscopy does not suffer from the light-element limitations of EDS. In fact, EELS can even detect the presence of hydrogen. Buseck and Self (1992) offer a nice comparison of EDS and EELS. In summary, EDS is normally better for quantification of heavy elements, whereas EELS is better for light elements. However, EELS offers the ability to evaluate the chemical and physical properties of the material by providing information on oxidation state and bonding environment for atoms. Joy (1979) and Somlyo and Shuman (1982) also discuss EELS.

PIXE and SIMS

Many analytical techniques for determining minor and trace element abundances in bulk mineral samples are used routinely in geochemical investigations. Prominent among these are X-ray fluorescence, instrumental neutron activation analysis (Muecke, 1980), and inductively coupled plasma spectrometry (Thompson and Walsh, 1988). However, it is often difficult to obtain high-purity mineral separates for bulk analyses, particularly when considering fine-grained biologically active materials. Thus, this section concentrates on two microanalytical methods—proton induced X-ray emission (PIXE) and secondary ion mass spectrometry (SIMS)—frequently used for characterization of element concentrations on a few micrometer scale. Other exciting new techniques for trace-element microanalysis include synchrotron radiation X-ray fluorescence (Lu et al., 1989).

Proton-induced X-ray emission (PIXE). PIXE microanalysis of geologic specimens is reviewed in articles by Benjamin et al. (1988), Annegarn and Baumann (1990), Sie et al. (1991), Ryan and Griffin (1993), and Campbell et al. (1993). PIXE microanalysis is capable of quantifying trace-element abundances in flat, polished mineral mounts with 5-μm lateral resolution. In PIXE, either a thin- or thick-target geological sample is bombarded with a focused beam of high energy (generally 2 to 3 MeV) protons, and the characteristic X-ray spectra of the bombarded material is measured. PIXE analyses are non-destructive and routinely yield ppm-level detection limits for many elements (atomic numbers 26 to 47) in Fe-bearing minerals. A few other elements can be analyzed with slightly higher detection limits, particularly in low-Fe matrices. These detection limits are significantly improved over those achievable with EPMA, because of the smaller background contributions. Two advantages of PIXE microanalysis are that (1) the trace-element concentrations can be quantified with 5 to 10% accuracies without resorting to trace-element working standards similar in composition to unknown minerals (Campbell et al., 1993) and (2) concentrations of a wide-range of elements are determined in a single analysis (which typically requires 5 to 15 minutes). Two disadvantages of PIXE microanalysis are (1) the limited number of elements that can easily be analyzed using this method, the REE being particularly difficult to measure in many minerals and (2) the excitation by the proton beam of a sample volume that encompasses depths greater than typical 30-μm thin-sections used for most geological analysis. The second drawback complicates PIXE analysis of the fine-grained minerals typically encountered in many toxicological studies.

Secondary ion mass spectrometry (SIMS). Two reviews of applications of secondary ion mass spectrometry (SIMS) to geological problems are provided by Shimizu and Hart (1982) and Reed (1989). SIMS, also known as ion microprobe analysis, is a highly sensitive method for determining trace-element abundances with spatial resolution of <10 µm. A polished sample is bombarded with an energetic primary ion beam, generally either O^- or Cs^+ in geologic applications. Resultant sputtered secondary ions are extracted into a doubly focusing mass spectrometer, and counted. The principal strengths of SIMS microanalysis are the low detection limits (sub-ppm for many elements), the wide range of elements that can be analyzed, including light elements such as hydrogen, and the ability to examine surface and near-surface compositions using depth-profiling. The principal drawbacks of quantitative analysis using the technique are molecular ion interferences, which complicate interpretation of the secondary-ion spectra of geologic materials, and the requirement that homogeneous trace-element standards of similar matrix to unknowns be readily available. In recent years, the former problem has been minimized by examining high-energy secondary ions (energy filtering) (Shimizu et al., 1978) and by using high-mass resolution ion microprobes, such as the SHRIMP. The latter problem is particularly severe for fine-grained, potentially heterogeneous minerals, such as many of those studied by biologists.

MINERAL SURFACES

Ultimately, mineral surfaces interact with components in the biological medium (cell, fluid, proteins, amino acids, etc.), so the nature of these surfaces is likely to play an important role in determining the pathogenic properties of a mineral. A review of surface analytical techniques is well beyond the scope of this chapter, but numerous texts are available on the subject, e.g., Feldman and Mayer (1986) and Woodruff and Delchar (1986). Hochella (1990) discusses several other surface techniques as applied to minerals. A few of the common techniques are discussed below. The measurement of surface charge and surface area is not discussed, but the interested reader should consult Sposito (1984).

Mineral surfaces may possess a structure and composition that is significantly different from that of the bulk. In some cases, the surface may consist of a material that is *very* different from the bulk, i.e., a surface coating. In other cases, the surface may simply differ slightly from the bulk, due to relaxation of the chemical bonds. (Chapter 8 discusses the nature of mineral surfaces in detail.) The characterization of the mineral surface necessarily involves an evaluation of the surface structure and surface composition both before and after interaction with the biological system. The analysis of the surface before interaction provides information on the properties of that material and can be used to explain the biological response. The analysis of the surface following the interaction can be used to evaluate the mineral's response to the interaction with the biological medium. For example, if iron is released from the mineral surface to participate in a Fenton reaction, then this process may be recorded in the mineral by a depletion of surface Fe. Likewise, if the biochemical process involves oxidation/reduction induced by the mineral surface, then the mineral surface may record information about the oxidation/reduction process.

Surface structure

Several techniques that can be applied to the study of surface structure are described by Feldman and Mayer (1986, Ch. 7) and Yagi (1988), including X-ray and electron diffraction methods and TEM. In addition, there are several techniques for directly imaging surfaces in plan view (e.g., AFM, STM, and SEM). As discussed in Chapter 8, both surface structure and microtopography can affect the surface properties. Hence, a combination of techniques providing both plan view and cross-section view as well as high resolution (i.e., atomic-level) and lower resolution (e.g., nanometer- to micrometer-range) detail is often necessary for a complete picture of the surfaces of minerals.

Scanning electron microscopy. Scanning electron microscopy has a tremendous potential for providing important information on mineral surface topography, particularly when the instrument used is equipped with a field-emission gun. A field-emission gun provides an extremely bright beam that can be focused to a small spot, thereby enabling a better resolution for secondary electron images. Resolutions for modern SEMs can be as low as ~1 nm (Joy and Pawley, 1992). The field-emission gun can also be operated at much lower currents, thereby eliminating the need to coat non-conductive materials (e.g., silicates). This is very important for studies on mineral surfaces, because it is the uncoated mineral that ultimately must be characterized. A detailed discussion of SEM imaging capabilities is given by Goldstein et al. (1981).

Scanning probe microscopies. The application of atomic force microscopy (AFM) and scanning tunneling microscopy (STM) to mineral surfaces is a relatively recent phenomenon. Much of the initial work focused on minerals such as galena or pyrite, because STM can only be used on conductive materials. This precludes the use of STM on silicates. However, AFM can be used on mineral surfaces regardless of their electrical properties, and AFM can even be used on surfaces in contact with a fluid. The resolution of AFM and STM makes them excellent complementary techniques to SEM (particularly SEM using a field-emission gun), because the resolution of AFM and STM ranges from atomic level up to the range covered by SEM. The power of AFM for applications relating to biologically important minerals is well illustrated by the recent images of the hydroxyl surface of lizardite (Wicks et al., 1992). These images provide atomic-level information on the probable nature of the cylinder surface of chrysotile (see Chapters 3 and 8). Numerous applications of AFM and STM to mineral surfaces can be found in Hochella (1990).

Low energy electron diffraction. Low energy electron diffraction (LEED) is another technique applicable to mineral-surface structure. As a diffraction technique, LEED provides information on the overall periodicity of the mineral surface. Hence, LEED provides information that is complementary to atomic-level information as obtain by a microscopy, such as AFM or TEM. A general discussion of LEED can be found in Feldman and Mayer (1986) and Woodruff and Delchar (1986).

Transmission electron microscopy. Transmission electron microscopy can be applied to the study of mineral surfaces, by observing the surface in cross section. Some of the aspects of amphibole surfaces, for example, can be extracted from micrographs presented in Ahn and Buseck (1991). The use of TEM for investigating mineral surfaces is discussed in detail by Yagi (1988).

Determination of surface composition

Several techniques offer the ability to study the surface compositions of minerals. X-ray photoelectron spectroscopy (XPS) and secondary ion mass spectroscopy (SIMS) can provide compositional information on mineral surfaces, and XPS has the added ability to provide information on oxidation state of some elements. Hochella (1988) discusses the use of XPS on minerals. Analytical TEM also has the ability to investigate the compositions of minerals surfaces in cross-section. The minimum probe diameter for most TEMs is generally on the order of ~10 nm, but TEMs equipped with a field-emission gun or dedicated scanning TEMs are capable of analytical probes with diameters down to ~1 to 2 nm which can provide analytical resolutions of ~5 to 10 nm for thin samples using EDS (Williams et al., 1992) and analytical resolutions of <0.5 nm for EELS (Batson, 1992). References on compositional analysis using TEM are given above under "Analytical electron microscopy."

ACKNOWLEDGMENTS

I am grateful to B. Carey and D. Hickmott for providing critical reviews of the manuscript and to D. Hickmott for providing input to the presentations of several of the techniques.

REFERENCES

Annegarn, H.J. and Bauman, S. (1990) Geological and Mineralogical Applications of PIXE: A Review. Nucl. Inst. Methods Phys. Res. B49, 264–270.
Ahn, J.H. and Buseck, P.R. (1991) Microstructures and fiber formation mechanisms of crocidolite asbestos. Amer. Mineral. 76, 1467–1478.
Alario Franco, M., Hutchison, J.L., Jefferson, D.A. and Thomas, J.M. (1977) Structural imperfection and morphology of crocidolite (blue asbestos). Nature 266, 520–521.
Altree-Williams, S., Byrnes, J.G. and Jordan, B. (1981) Amorphous surface and quantitative X-ray powder diffractometry. Analyst 106, 69–75.
Armstrong, J.T. (1978) Methods of quantitative analysis of individual microparticles with electron beam instruments. SEM/1978/I, 455.
Armstrong, J.T. (1980) Rapid quantitative analysis of individual microparticles using the α-factor approach. In: Microbeam Analysis—1980, D.B. Wittry, Ed., San Francisco Press, San Francisco, California, p. 193–198.
Armstrong, J.T. and Buseck, P.R. (1975) Quantitative chemical analysis of individual microparticles using the electron microprobe. Analyt. Chem. 47, 2178.
Baris, I., Simonato, L., Artvinli, M., Pooley, F., Saracci, R., Skidmore, J. and Wagner, C. (1987) Epidemiological and environmental evidence of the health effects of exposure to erionite fibres: a four-year study in the Cappadocian region of Turkey. Int'l J. Cancer 39, 10–17.

Baris, Y.I., Artvinli, M., Sahin, A.A., Sebastien, P. and Gaudichet, A. (1987) Diffuse lung fibrosis due to fibrous zeolite (erionite) exposure. Eur. J. Resp. Dis. 70, 122–125.

Baronnet, A. (1992) Polytypism and stacking disorder. Rev. Mineral. 28, 231–288.

Batson, P.E. (1992) Spatial resolution in electron energy loss spectroscopy. Ultramicroscopy 47, 133–144.

Benjamin, T.M., Duffy, C.J. and Rogers, P.S.Z. (1988) Geochemical Utilization of Nuclear Microprobes. Nucl. Inst. Methods Phys. Res. B30, 454-458.

Bish, D.L. (1993) Studies of clays and clay minerals using the Rietveld method. In: Computer Applications in Clay Mineralogy, R.C. Reynolds, Jr. and J. Walker, Eds., CMS Workshop Lectures, Clay Minerals Society, Boulder, Colorado, p. 80–121.

Bish, D.L. and Chipera, S.J. (1991) Detection of trace amounts of erionite using X-ray powder diffraction: Erionite in tuffs of Yucca Mountain, Nevada, and central Turkey. Clays & Clay Minerals 39, 437–445.

Bish, D.L. and Howard, S.A. (1988) Quantitative phase analysis using the Rietveld method. J. Appl. Crystallogr. 21, 86–91.

Bish, D.L. and Post, J.E., Eds. (1989) Modern Powder Diffraction. Reviews in Mineralogy 20, Mineralogical Society of America, Washington, D.C., 369 p.

Bish, D.L. and Reynolds, R.C., Jr. (1989) Sample preparation for X-ray diffraction. Rev. Mineral. 20, 73–100.

Blount, A.M. and Vassiliou, A.H. (1983) Identification of chlorite and serpentine in cosmetic or pharmaceutical talc. Envir. Health Pers. 51, 379–385.

Buseck, P.R., Ed. (1992). Minerals and Reactions at the Atomic Scale: Transmission Electron Microscopy. Reviews in Mineralogy 27, Mineralogical Society of America, Washington, D.C. 508 p.

Buseck, P.R. and Self, P. (1992) Electron energy-loss spectroscopy (EELS) and electron channelling (ALCHEMI). Rev. Mineral. 28, 141–180.

Buseck, P.R. and Veblen, D.R. (1988) Mineralogy. In: High-Resolution Transmission Electron Microscopy and Associated Techniques, P.R. Buseck, J. Cowley and L. Eyring, Eds., Oxford University Press, New York, 308–377.

Buseck, P.R., Cowley, J. and Eyring, L., Eds. (1988) High-Resolution Transmission Electron Microscopy and Associated Techniques. Oxford University Press, New York, 645 p.

Campbell, J.L, Higuchi, D., Maxwell, J.A. and Teesdale, W.J. (1993) Quantitative PIXE microanalysis of thick specimens. Nucl. Instru. Methods Phys. Res. B77, 95-109.

Champness, P.E., Cliff, G. and Lorimer, G. W. (1976) The identification of asbestos. J. Micros. 108, 231–249.

Chung, F.H. (1974a) Quantitative interpretation of X-ray diffraction patterns. I. Matrix-flushing method of quantitative multicomponent analysis. J. Appl. Crystallogr. 7, 519–525.

Chung, F.H. (1974b) Quantitative interpretation of X-ray diffraction patterns. II. Adiabatic principle of X-ray diffraction analysis of mixtures. J. Appl. Crystallogr. 7, 526–531.

Chung, F.H. (1975) Quantitative interpretation of X-ray diffraction patterns. III. Simultaneous determination of a set of reference intensities. J. Appl. Crystallogr. 8, 17–19.

Cressey, B.A., Whittaker, E.J.W., and Hutchison, J.L. (1982) Morphology and alteration of asbestiform grunerite and anthophyllite. Mineral. Mag. 46, 77–87.

Davis, B.L. (1990) Quantitative analysis of asbestos minerals by the reference intensity X-ray diffraction procedure. Amer. Indus. Hygiene Assoc. 51, 297–303.

Deer, W.A., Howie, R.A. and Zussman, J. (1992) An introduction to the rock forming minerals. 2nd edition, Longman, London, 696 p.

Ewing, G.W. (1985) Instrumental Methods of Chemical Analysis, 5th edition. McGraw-Hill, New York, 538 p.

Feldman, L.C. and Mayer, J.W. (1986) Fundamentals of Surface and Thin Film Analysis. North-Holland, New York, 352 p.

Fleet, M.E., Ed. (1984) Environmental Geochemistry. Mineralogical Association of Canada, Toronto, Ontario, 306 p.

Goldstein, J.I., Newbury, D.E., Echlin, P., Joy, D.C., Fiori, C. and Lifshin, E., Eds. (1981) Scanning Electron Microscopy and X-ray Microanalysis. Plenum Press, New York, 448 p.

Goldstein, J.I., Williams, D.B. and Cliff, G. (1986) Quantitative X-ray analysis. In: Principles of Analytical Electron Microscopy, D.C. Joy, A. Romig D., Jr. and J.I. Goldstein, Eds., Plenum Press, New York, 155–218.

Guthrie, G.D., Jr. and Bish, D.L. (1991) Quantitative phase analysis of mordenite samples using the Rietveld method, Amer. College Chest Phys., 4th Int'l Conf. on Environmental Lung Disease (abstract).

Hansen, K. and Mossman, B.T. (1987) Generation of superoxide (O_2^-) from alveolar macrophages exposed to asbestiform and nonfibrous particles. Cancer Res. 47, 1681–1686.

Hawthorne, F.C., Ed. (1988) Spectroscopic Methods in Mineralogy and Geology. Reviews in Mineralogy 18, Mineralogical Society of America, Washington, D.C. 698 p.

Hirsch, P., Howie, A., Nicholson, R.B., Pashley, D.W. and Whelan, M.J. (1977) Electron Microscopy of Thin Crystals, 2nd edition. Robert E. Krieger Publishing Co., Malabar, Florida.

Hochella, M.F., Jr. (1988) Auger electron and X-ray photoelectron spectroscopies. Rev. Mineral. 18, 573–638.

Hochella, M.F., Jr. (1990) Atomic structure, microtopography, composition, and reactivity of mineral surfaces. Rev. Mineral. 23, 87–132.

Hochella, M.F., Jr. and White, A.F., Eds. (1990) Mineral–Water Interface Geochemistry. Reviews in Mineralogy 23, Mineralogical Society of America, Washington, D.C. 603 p.

Hutchison, J.L., Irusteta, M.C. and Whittaker, E.J.W. (1975) High-resolution electron microscopy and diffraction studies of fibrous amphiboles. Acta Crystallogr. A31, 794–801.

Jiang, W.-T., Essene, E.J. and Peacor, D.R. (1990) Transmission electron microscopic study of coexisting pyrophyllite and muscovite: Direct evidence for the metastability of illite. Clays & Clay Minerals 38, 225–240.

Johnson, N.F. and Wagner, J.C. (1989) Effect of erionite inhalation on the lungs of rats. Biological Interaction of Inhaled Mineral Fibers and Cigarette Smoke, 355–372.

JCPDS, Joint Committee for Powder Diffraction Standards (1987) Powder Diffraction File, Alphabetical Index and Search Manual, Inorganic Phases. Int'l Centre for Diffraction Data, Swarthmore, Pennsylvania.

Joy, D.C. (1979) The basic principles of electron energy loss spectroscopy. In: Introduction to Analytical Electron Microscopy, J.J. Hren, J.I. Goldstein and D.C. Joy, Eds., Plenum Press, New York,

Joy, D.C. and Pawley, J.B. (1992) High-resolution scanning electron microscopy. Ultramicroscopy 47, 80–100.

Joy, D.C., Romig, A.D., Jr. and Goldstein, J.I., Eds. (1986) Principles of Analytical Electron Microscopy. Plenum Press, New York, 448 p.

Kirkpatrick, R.J. (1988) MAS NMR spectroscopy of minerals and glasses. Rev. Mineral. 18, 341–404.

Ledoux, R.L., Ed. (1979) Mineralogical Techniques of Asbestos Determination. Mineralogical Association of Canada, Toronto, Ontario, 279 p.

Lu, F.-O., Smith, J.V., Sutton, S.R., Rivers, M.L., and Davis, A.M. (1989) Synchrotron X-ray fluorescence analysis of rock-forming minerals. Chem. Geol. 75, 123-143.

Maltoni, C., Minardi, F. and Morisi, L. (1982) Pleural mesothelioma in Sprague–Dawley rats by erionite: first experimental evidence. Envir. Res. 29, 238–244.

McMillan, P.F. and Hofmeister, A.M. (1988) Infrared and Raman spectroscopy. Rev. Mineral. 18, 99–160.

Mossman, B.T. and Sesko, A.M. (1990) *In vitro* assays to predict the pathogenicity of mineral fibers. Toxicol. 60, 53–61.

Muecke, G.K., Ed. (1980) Neutron Activation Analysis in the Geosciences. Mineralogical Association of Canada, Toronto, Ontario, 279 p.

Nesse, W.D. (1991) Introduction to Optical Mineralogy. Oxford University Press, New York, 335 p.

O'Connor, B.H. and Chang, W.-J. (1986) The amorphous character and particle size distributions of powders produced with the micronizing mill for quantitative X-ray powder diffractometry. X-ray Spectros. 15, 267–270.

Palekar, L.D., Most, B.M. and Coffin, D.L. (1988) Significance of mass and number of fibers in the correlation of V79 cytotoxicity with tumorigenic potential of mineral fibers. Envir. Res. 46, 142-152.

Peacor, D.R. (1992) Analytical electron microscopy: X-ray analysis. Rev. Mineral. 28, 113–140.

Phillips, W.R. (1971) Mineral Optics—Principles and Techniques. W.H. Freeman, San Francisco, 249 p.

Post, J.E. and Bish, D.L. (1989) Rietveld refinement of crystal structures using powder X-ray diffraction data. Rev. Mineral. 20, 277–308.

Puledda, S. and Marconi, A. (1990) Quantitative X-ray diffraction analysis of asbestos by the silver membrane filter method: application to chrysotile. Amer. Indus. Hygiene J. 51, 107–114.

Reed, S.J.B. (1989) Ion microprobe analysis—a review of geological applications. Mineral. Mag. 53, 3–24.

Ryan, C.G. and Griffin, W.L. (1993) The nuclear microprobe as a tool in geology and mineral exploration. Nucl. Inst. Methods Phys. Res. B77, 381–398.

Sie, S.H., Griffin, W.L., Ryan, C.G., Suter, G.F. and Cousens, D.R. (1991) The proton microprobe: a revolution in mineral analysis. Nucl. Inst. Methods Phys. Res. B54, 284–291.

Shimizu, N., Semet, M.P. and Allegre, C.J. (1978) Geochemical applications of quantitative ion-microprobe analysis. Geochim. Cosmochim. Acta, 42, 1321–1334.

Shimizu, N. and Hart, S.R. (1982) Applications of the ion microprobe to geochemistry and cosmochemistry. Ann. Rev. Earth Planet. Sci. 10, 483–526.

Somlyo, A.P. and Shuman, H. (1982) Electron probe and electron energy loss analysis in biology. Ultramicroscopy 8, 219–234.

Snyder, R.L. and Bish, D.L. (1989) Quantitative analysis. Rev. Mineral. 20, 101–144.
Spence, J.C.H. (1988) Experimental High-Resolution Electron Microscopy. 2nd edition. Oxford University Press, Oxford.
Sposito, G. (1984) The Surface Chemistry of Soils. Oxford University Press, New York, 234 p.
Stebbins, J.F. (1988) NMR spectroscopy and dynamic processes in mineralogy and geochemistry. Rev. Mineral. 18, 405–430.
Suzuki, Y. (1982) Carcinogenic and fibrogenic effects of zeolites: preliminary observations. Envir. Res. 27, 433–445.
Suzuki, Y. and Kohyama, N. (1984) Malignant mesothelioma induced by asbestos and zeolite in the mouse peritoneal cavity. Envir. Res. 35, 277–292.
Thompson, M. and Walsh, J.N. (1988), Handbook of ICP Spectrometry, 2nd edition. Blackie, Glasgow, 320 p.
Veblen, D.R. (1980) Anthophyllite asbestos: Microstructures, intergrown sheet silicates, and mechanisms of fiber formation. Amer. Mineral. 65, 1075–1086.
Veblen, D.R. (1992) Electron microscopy applied to nonstoichiometry, polysomatism, and replacement reactions in minerals. Rev. Mineral. 28, 181–230.
Wagner, J.C., Skidmore, J.W., Hill, R.J. and Griffiths, D.M. (1985) Erionite exposure and mesotheliomas in rats. Brit. J. Cancer 51, 727-730.
White, J.C., Ed. (1985) Applications of Electron Microscopy in the Earth Sciences. Mineralogical Association of Canada, Toronto, Ontario, 213 p.
Wicks, F.J., Kjoller, K. and Henderson, G.S. (1992) Imaging the hydroxyl surface of lizardite at atomic resolution with the atomic force microscope. Can. Mineral. 30, 83–91.
Wiles, D.B. and Young, R.A. (1981) A new computer program for Rietveld analysis of X-ray powder diffraction patterns. J. Appl. Crystallogr. 14, 149–151.
Williams, D.B., Michael, J.R., Goldstein, J.I. and Romig, A.D., Jr. (1992) Definition of the spatial resolution of X-ray microanalysis in thin foils. Ultramicroscopy 47, 121–132.
Woodruff, D.P. and Delchar, T.A. (1986) Modern Techniques of Surface Science. Cambridge University Press, Cambridge, 453 p.
Yagi, K. (1988) Surfaces. In: High-Resolution Transmission Electron Microscopy and Associated Techniques, P.R. Buseck, J. Cowley and L. Eyring, Eds., Oxford University Press, New York, p. 568–606.

CHAPTER 8

SURFACE CHEMISTRY, STRUCTURE, AND REACTIVITY OF HAZARDOUS MINERAL DUST

Michael F. Hochella, Jr.

Department of Geological Sciences
Virginia Polytechnic Institute and State University
Blacksburg, Virginia 24061 U.S.A.

INTRODUCTION

It would be naive to think that the harmful effects of mineral dust on the human respiratory system is solely dependent on the physical shape of the dust particles. One could even argue in various cases that the particle shape is only of minor importance. Most researchers working in this field today realize that there is a complex disease-generating interaction between mineral dust particles and the biological system with which they are in contact. Certainly one of the key factors in this interaction is the mineral–cell interface. The pathogenicity of this situation is probably dependent upon many possible reactivity pathways, but ultimately the mineral dust particle must impart its influence on the biological system via its surface properties, chemical and electrical as well as mechanical/dimensional.

This chapter deals primarily with the mineral side of the mineral–cell interface and the interface region itself. What is the nature of the mineral surfaces to which cells react? Are there clues on the mineral surface which may lead to a better understanding of mineral dust-generated disease? We will attempt to look into these and related questions in this chapter after a brief introduction to the general field of mineral surface science.

Scientists began studying the nature of mineral surfaces, interfaces, and their chemical reactivity over a century ago. This author has traced studies of the dissolution (corrosion) of minerals back to, among others, Ebelmen (1847) and Daubrée (1867), who give particularly interesting accounts of some of the earliest experiments in this field. In addition, some of the first work on oxidation–reduction reactions at mineral–solution interfaces was performed in the 1870's by Newbury, Skey, and Wilkinson (see Liversidge, 1893, for an intriguing account of this work). In these early studies, the only way to study surfaces was indirectly, that is by observing/measuring changes in the fluid (gas or liquid) composition in contact with the surface as a function of time. These methods were, and still are, very powerful. However, today, we have the additional resources, in the form a surface and interface sensitive spectroscopic and microscopic tools, to study these processes directly. These tools include X-ray photoelectron spectroscopy (XPS), Auger electron spectroscopy (AES) and scanning Auger microscopy (SAM), secondary electron microscopy (SEM), atomic force microscopy (AFM), and scanning tunneling microscopy (STM), just to name a few. In addition, some techniques generally used for bulk analysis have been adapted for surface research, including diffuse reflectance Fourier transform infra-red spectroscopy (FTIR), X-ray absorption spectroscopy (XAS), and transmission electron microscopy (TEM). Most importantly, all of the methods listed above have provided direct observations of either mineral–fluid interfaces or the mineral surface before and after reaction. However, some have never been used in the study of mineral–cell interaction.

There have been revolutionary advances in the field of surface science in just the last 10 to 20 years using the tools mentioned above. Perhaps the most fundamentally important discovery that has been made is that, in contrast to bulk properties which are derived from a combination of the composition and atomic structure of the material, the properties of a surface depend on its composition, atomic structure, surface charge, and microtopography. Here, we define microtopography as the morphology of the surface on the general dimensional scale of the chemical interactions that will take place there, usually from a few tenths to perhaps a few nanometers. Below, we will take a look at the general nature of oxide and silicate surfaces via these three critical factors, as well as surface charge which is a function of these factors, and review the role that each plays in the reactivity of a mineral surface.

The next section of this chapter presents a review of the literature which covers the biological activity dependence on mineral surface composition, structure, microtopography, and charge. (Biological activity in this chapter refers to those processes, no matter how fundamental, that may lead to mineral dust-generated disease.) Finally, we will look at the specific surface structure and properties of the two minerals most studied in mineral dust research, chrysotile and crocidolite. In many cases, connections are made to the biological/medical literature in order to begin to relate findings in pathogenic studies to what is known about the characteristics and reactive nature of the surface of these minerals.

THE GENERAL NATURE OF MINERAL SURFACES

Mineral surfaces are complex in terms of atomic structure, composition, shape and, as a result, reactivity. This results from mineral surface heterogeneity, which is driven ultimately by the natural "impurity" of minerals, as well as by the presence of complex microtopography. Therefore, reactions that occur at the mineral–fluid interface can proceed along a number of chemical fronts, and as a result the most important of these (that which determines the overall reaction rate) may be the fastest reaction, not the slowest. Whether one wishes to study mineral surfaces for the purposes of industrial heterogeneous catalysis, environmental geochemistry, or as in this case, mineral–cell interaction, one should be cognizant of this complexity and of the tools, both theoretical, experimental, and instrumental, needed for advanced characterization and understanding. In addition, the upshot of this complexity is that mineral surfaces are not static or inert, even in the context of biological systems. Instead, they are dynamic and highly interactive with their surroundings.

The following four subsections lay the foundation for understanding mineral surface chemistry and chemical reactivity in terms of composition, atomic structure, microtopography, and surface charge.

Surface composition

Fresh surfaces, exposed by fracture, are highly reactive due to the presence of under-coordinated surface atoms and the "dangling" bonds which accompany them. Only in ultra-high vacuum (10^{-9} torr or better) can one expect to maintain a clean, uncontaminated surface, and then only for a relatively short time (minutes to hours). Under atmospheric conditions, molecules in the overlying medium will immediately react with the unsatisfied surface atoms, and the surface will be covered with a layer of adventitious (i.e., foreign) material in as little as 10^{-9} sec. Therefore, surface compositions are essentially never representative of the bulk. Specifically, oxide and silicate mineral surfaces exposed to air or an aqueous solution immediately take up hydrogen, oxygen, usually carbon, and sometimes nitrogen (from O_2, H_2O, CO_2, hydrocarbons, and N_2 and/or NO_x in the air or solution) (Fig. 1). Often, adventitious molecules in the first monolayer dissociate (break apart) during adsorption, and this process is critical in the characteristics and reactivity of the surface as we will see below (see section entitled "Surface charge" below).

The build-up of adventitious materials "passivates" the clean highly reactive surface as the dangling bonds are satisfied. Therefore, self-contamination of surfaces slows and essentially halts after just of few layers (call monolayers) are added. Further contamination at this point may or may not occur; this depends more on the surrounding environment than on the actual surface.

The adventitious contamination of surfaces as described above, for all practical purposes, should be considered instantaneous and ubiquitous. Perhaps

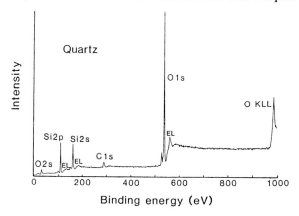

Figure 1. X-ray photoelectron spectroscopy (XPS) spectrum of a quartz surface which has been exposed by fracture and let stand in room air for 1 hour. Besides expected peaks for oxygen and silicon, a carbon line is also present. In addition, the oxygen photoelectron lines are more intense than expected for stoichiometric SiO_2; this is due to the addition of adventitious oxygen. Adventitious hydrogen is also present on this surface (as determined with FTIR), but hydrogen is not detectable with XPS. Peaks labeled "EL" are due to energy loss phenomena, and "O KLL" is an oxygen Auger line. From Hochella (1988).

this is one of the reasons why it is often ignored in geo- and biochemical surface studies. Another reason that adventitious contamination has not been studied more is the fact that the contaminants in question are general only a few monolayers thick, and consist of relatively light elements, both making for difficult characterization. More details on the natural composition of mineral surfaces can be found in Hochella (1990).

There are two other common categories of surface compositional modification. These are the sorption of other species besides the ubiquitous adventitious contaminants, and the desorption of components in the near-surface region of the mineral. Both processes may even act in concert, as they do in ion exchange reactions. All of these processes, separately or together, may affect only the top layer of atoms of the mineral surface, or they may modify the mineral to considerable depths due to near-surface solid state diffusion. For those interested in further reading on these subjects, the following references, and papers cited therein, are recommended: Davis and Kent (1990), Schindler (1990), Brown (1990) provide reviews of sorption processes on mineral surfaces; recent reviews of leaching and dissolution can be found in Schott (1990), Casey and Bunker (1990), and Hering and Stumm (1990).

One common aspect of mineral surface chemistry which plays an important role in surface reactivity is lateral compositional heterogeneity. This is to say that, generally speaking, mineral surfaces do not react homogeneously with their environment, and they develop compositional variations across their surfaces as surface reactions proceed. Evidence, both direct and circumstantial, for surface compositional heterogeneity due to sorption can be found throughout the literature (see Hochella, 1990, and references therein). It is clear from these types of studies that various sites on a surface have different reaction potentials with, for example, species in aqueous solution. Similarly, different sites on mineral surfaces will respond to dissolution reactions in different ways, and high resolution surface analysis of dissolving mineral surfaces show that dramatic heterogeneities can be present. For example, we have mapped the surface Ca concentration (depth of analysis approximately 10 Å) on a relatively flat plagioclase (a very common Na,Ca aluminosilicate) surface after dissolution under hydrothermal conditions (deionized water at 350°C for 57 days) using AES/SAM with a lateral resolution of 0.25 μm (Hochella, 1990). While a quantitative assessment is difficult, some areas show only slight leaching of Ca, while others have very little Ca left in the near-surface region. At least in this case, the style of dissolution varies depending most likely on a number of factors, including the variations in surface composition of the pre-dissolved surface, the starting microtopography of the surface, and the variation of the flow of fluids over different parts of the surface.

Surface atomic structure

The atomic structure on a surface is generally modified from the equivalent structure in the bulk by virtue of the fact that the entire surface, by definition and in reality, is a defect. On one side of the surface, the nearest neighbor atoms are the same as in the bulk; on the other side, everything is different. But there are even changes on the bulk side of the surface plane. The atoms on the surface, and those up to several monolayers deep, assume slightly different atomic positions (called surface relaxation), and if heated to a few hundred degrees or more, may assume dramatically different positions (called surface reconstruction). Relaxation involves subtle changes in surface and near-surface bond angles and lengths, but no bonds are broken. Reconstruction involves bond breakage and reformation, as the surface atoms choose a new configuration to minimize configurational entropy. For a growth surface or a surface exposed simply by fracture, the atomic structure across the surface (i.e., laterally) is likely to be only relaxed (not reconstructed) with respect to equivalent atoms in the bulk, especially if it is a low symmetry structure (e.g., a feldspar surface; see Hochella et al., 1990). Mineral structures with intermediate surface symmetry, like the corundum structure adopted by hematite, show subtle lateral atomic relaxation (Eggleston and Hochella, 1992). Higher symmetry structures like the rocksalt structure of galena may show no lateral relaxation at all (Hochella et al., 1989). However, in all cases from the highest to the lowest surface and near-surface symmetry, some relaxation perpendicular to the surface is expected. Specifically, most clean surfaces (i.e., in ultra-high vacuum) relax inward toward the bulk, with the spacing between the first and second atomic layers being reduced by up to 15% (see Somorjai, 1981; 1990, and references therein). This results from unsatisfied dangling bonds, resulting in the shortening of existing bonds to increase overall overlap populations around each atom. By eliminating the dangling bonds via surface sorption of additional atoms, the underlying bonds will again relax and expand. The expansion may even exceed the original bond lengths depending on the sorbate, resulting in a further structural weakening of the surface.

An important aspect of surface structure to emphasize, as inferred above, is the positional response that surface atoms undergo when they react with adsorbing or otherwise interacting species. In the vacuum-based surface science literature, there are many examples which conclusively show substrate surface atoms nearest to the sorbate shifting into new positions to better accommodate the sorbed species (e.g., Somorjai, 1990, and references therein). This generally only occurs when the sorbate-surface interaction is strong, i.e., the sorbate is chemically reactive with the surface and strong (specific) bonding occurs. Therefore, it should be emphasized that surfaces should not be viewed as static. Surfaces are dynamic structurally as well as chemically as they respond to interacting species.

Surface microtopography

Although under very special conditions (e.g., annealing in ultra-high vacuum) surfaces can be made to be atomically flat, most natural surfaces, including those of nearly all minerals, are rough on the molecular scale. We have found recently that even crystal planes that are reported to have "perfect" cleavage or growth faces, although they may appear perfectly smooth even under close SEM inspection, can have considerable roughness on the atomic scale (e.g., single atom high steps), if not relief of several tens to hundreds of Angstroms (Sunagawa, 1987; Hochella, 1990; Eggleston and Hochella, 1992).

The surface roughness described above provides reaction flexibility that would not be available with a perfectly flat surface (see Hochella, 1990, for a review). This is to say that the electronic structure of a rough surface is far more varied than that of a perfectly smooth surface. As a result, inasmuch as the reactivity of the surface with its particular surroundings is ultimately dependent on both the electronic structure of the surface and the reactant above the surface, rough surfaces are generally found to be more reactive than atomically flat surfaces.

Currently, the most useful tools for characterizing microtopography are the scanning probe microscopes, including scanning tunneling microscopy (STM) and atomic or scanning force microscopy (AFM or SFM, respectively). Their development and use on minerals has been reviewed by Hochella (1990). The STM is a relatively new and powerful tool, able to "probe" surfaces with atomic resolution. Complications include the fact that surface atom electronic and positional parameters cannot be easily separated. Also, STM, as the technique currently stands, will not work on insulating surfaces, making it useless for most mineral dust/health effects studies. AFM is an ultra-sensitive mechanical technique which resulted from the development of the STM. It can "image" any solid surface (conducting or insulating) and has atomic resolution in height and, under special circumstances, nearly atomic resolution laterally. However, it does not collect electronic information like STM, and so is constrained in that aspect. STM and AFM images will be shown below.

Before the advent of STM and AFM, models describing the microtopography of surfaces had been developed from studies dealing with sorption, crystal growth, dissolution (corrosion), and heterogeneous catalysis over many decades (see the review by Somorjai, 1981). The general model of surface shape that had been developed from these studies includes flat areas, called terraces, which are separated by steps (Fig. 2). Steps can be from one to many atomic layers high. Kink sites appear where a step changes direction twice in a short distance along itself. An atom added to a step or a terrace is called a step or terrace adatom, respectively. Likewise, an added molecule is call an admolecule. An atomic or molecular-sized hole in a terrace is called a vacancy. Importantly, observations with phase contrast optical microscopy, TEM, STM, and AFM have all generally supported this model. This is not to say that especially STM and AFM have not recently (last 5 to 10 years) made a huge contribution in this area. They have revealed the frequency and spatial distribution of these important surface topographic features, and STM is now being used to determine the electronic structure on and around topographic features. This will eventually unlock the mysteries of the specific reactivity of each of these sites.

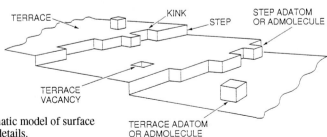

Figure 2. General schematic model of surface topography. See text for details.

Qualitatively, the importance of surface microtopography can be best explained as follows. Atoms that make up terraces are relatively common and generally have higher coordination (with neighboring surface and near-surface atoms) than other surface sites. An example of a surface atom with higher coordination than on a terrace would be one underlying a single atom surface vacancy. At the top edge of a step, the number of nearest-neighbors is reduced, and atoms at the outer corners of kink sites have even fewer neighbors. Terrace adatoms generally have the fewest nearest-neighbors on a clean surface of all the different surface sites.

The reason that the features described above are important is because a number of highly specific and controlled vacuum and non-vacuum based studies have generally shown that surface sites with lower coordination are more chemically reactive (e.g., Iwasawa et al., 1976; Christmann and Ertl, 1976; Somorjai et al., 1988). Unfortunately, none of these detailed studies has yet been performed on a mineral surface. However, we are starting to observe variations in electronic and atomic structure at sites of surface roughness with atomic resolution on minerals. Figure 3 shows an STM image of the (001) surface of hematite, α-Fe_2O_3. In this image, each bump represents a surface oxygen atom (the surface

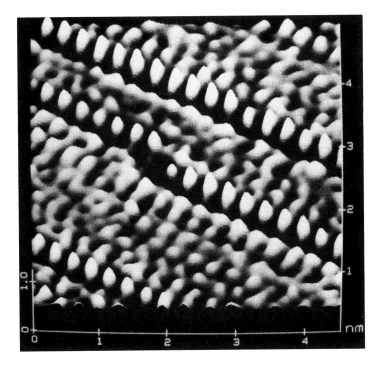

Figure 3. Scanning tunneling microscopy (STM) image of the (001) surface of hematite (α-Fe_2O_3). Scale is in nanometers (1 nm = 10 Å). The image is displayed in a three-dimensional block tilted back from the viewer by 30° for added perspective. Each bump in the image represents a single surface oxygen atom. Terraces step down from upper right to lower left. See text for more details. From Eggleston and Hochella (1992).

iron atoms are not observed under the microscope conditions used to collect this image). There are several terraces in the image frame separated by monoatomic steps. (Also note the kink site along a step near the center of the image.) The height of each step is approximately 2.3 Å. The height of each bump represents the electronic population in a valence orbital (at a particular energy) belonging to the oxygen atoms. The height of the bump is directly proportional to the population density of the electronic states that are being imaged. It should also be noted that these particular states represent a portion of a "frontier" orbital, important in the reactivity of these oxygens as a reactant approaches. Note that there is a higher state population on atoms at the step edge than on the terraces. It is important not to overinterpret these results, as many more state populations from both the oxygen and iron atoms need to be collected, and the difference in bump heights may also in part be due to oxygen displacement at the step edge. Nevertheless, our observations strongly suggest with direct evidence that step and terrace oxygens have different reactivity potentials.

Surface charge

It has been known for some time that all surfaces are electrically charged (see, e.g., Parks, 1990, for a review), except in the special case where positive and negative species on the surface balance, resulting in a net neutrality. Surface charge may be an important factor to consider in the reactivity of mineral dust in biological systems. Here, we define and characterize the notion of surface charge.

When a fresh surface is exposed to ambient conditions, as described above, dangling bonds are quickly satisfied as species in the surrounding medium attach (bond) to these highly reactive sites. For purposes of simplicity and to demonstrate the origin and characteristics of surface charge, let us now specifically consider a silicate surface in contact with water. Infrared and ultraviolet photoelectron spectroscopies show that the first monolayer of water molecules on an oxide surface dissociate, that is, they break down into H^+ and OH^- (Henrich, 1985; Parks, 1990; Hochella, 1990, and references therein). This means that for each siloxane bridge (Si–O–Si linkage) on the surface of the silica, one reacting water molecule results in the formation of two silanol groups (Si–OH) according to the following reaction:

$$Si-O-Si + H_2O \rightarrow 2(Si-OH) \tag{1}$$

When such a surface is submersed in water, each silanol group can react in one of two ways:

$$SiOH \rightarrow SiO^- + H^+ \tag{2}$$

or

$$SiOH + H^+ \rightarrow SiOH_2^+ \tag{3}$$

Reaction 2, in which a hydrogen is given off to solution, creates a negative surface site and decreases the pH of the surrounding solution. The latter reaction, in which a hydrogen in solution is sorbed, creates a positive surface site and increases the pH of the surrounding solution. These reactions are dependent on the pH of the solution, and therefore there exists a pH (called pH_{crit}) where the surface will be neutral. At pH's lower than pH_{crit}, as the activity of H^+ increases, the latter reaction is favored and the surface becomes positive; at pH's higher than pH_{crit}, as the activity of H^+ decreases, the former reaction is favored and the surface takes on a negative charge. Therefore, there must be a pH for each solid such that, when it is introduced to water, no pH change occurs as both reactions are equally favored and the surface remains neutral. This special condition is called the point of zero charge (PZC) (Parks, 1975), or to be more exact, the point of zero net proton charge (PZNPC) (Sposito, 1984). As one moves further away on either side of the PZNPC, the surface becomes more highly charged.

The above discussion has been focused on quartz. In fact, ionizable groups form on the surfaces of all silicates, although exact speciation and population depends on the specific material in question. The PZNPC's of silicates have been found to be near the average of the PZNPC's of the constituent oxides when each is weighed relative to the composition of the silicate (Parks, 1967). The interested reader is directed to Parks (1967, 1975) and Yoon et al. (1979) for a review of the compositional and structural factors which affect the PZNPC.

Now let us consider other species in the surrounding aqueous solution. As the ionic strength of the solution is increased, one can measure a reduction in the surface charge but not its elimination. Apparently, labile counterions are attracted to the charged surface, remaining in solution in a diffuse layer whose countercharge falls off exponentially the farther one gets from the surface. Therefore, this so-called electrical double layer (EDL) consists of the static (bonded) charge on the mineral surface and the diffuse layer of counter ions adjacent to this in solution (not directly bonded). This model, which has withstood the test of time and much experimentation relatively intact, was first proposed by Guoy and Chapman in the early part of this century (see Adamson, 1982).

An indirect way to measure net charge in an electrolytic solution which requires EDL theory for understanding is via the electrophoretic mobility measurement method, or electrophoresis for short (see Adamson, 1982, for a review and a description of the technique). In such a device, the direction and velocity of immersed particles placed between two oppositely charged plates is measured with the aid of a microscope. The fact that particles (which have some charge associated with their surface) in an electrolytic solution (which contains the oppositely charged diffuse layer) have a net charge at all is due to the movement of the particle in the solution. This movement causes the outer portions of the diffuse layer to slip away (called the slipping plane), resulting in a net non-zero charge. This charge can be related to the charge at the slipping plane which is called the zeta potential. Just like surface charge, the zeta potential is a function of pH, and it is also a function of ionic strength and adsorption density of solutes.

Finally, this leads to well substantiated theories on the interaction of charged species in solution to charged surfaces. There are basically two extremes in behavior which essentially encompass all sorption phenomena. In non-specific adsorption, species in solution sorb to any surface with the opposite charge, and sorption densities decrease with increasing ionic strength due to competition effects. The coulombic interactions which control non-specific adsorption allow negatively charged species to adsorb at low pH (i.e., below PZNPC for that surface) and positively charged species to adsorb at high pH (i.e., above PZNPC for that surface). In specific or chemical adsorption, species in solution do not adsorb according to the surface charge. Species may adsorb to surfaces of like charge or zero charge. Ionic strength of an indifferent (non-sorbing) electrolyte does not affect sorption density. In this case, specific chemical bonding is achieved between the sorbate and sorbent.

RELATIONSHIP BETWEEN MINERAL SURFACES AND THEIR BIOLOGICAL ACTIVITY

Assuming that the surfaces of mineral dust particles play at least some active role in their deleterious effects on human health, we have seen above that one should be able to relate this reactivity specifically to mineral surface composition, structure, microtopography, and surface charge. Perhaps this is most easily shown for surface composition, as several studies have shown the dependence of mineral-biological system interaction on this (see below). In fact, if one considers mineral surface composition in terms of adventitious and other sorption phenomena, as well as desorption reactions and the heterogeneity that can result from all of these processes, the dependence of cell interaction with surface composition could be quite complex. One could easily postulate from this that, given what otherwise you would assume to be a homogeneous mineral, a portion of the surface may be highly active and potentially toxic to cells, while other portions or sites on the surface may be relatively passive. Alternatively, a mineral surface may be highly reactive (and potentially pathogenic) in contact with a cell, while that same mineral with a slightly modified surface composition may be only partially active, or even inactive, in the same environment. Basically, this has been the approach taken by some researchers working in the field of the health effects of mineral dust. A brief review of some of these studies follows, along with the roles that surface structure, microtopography, and surface charge play in mineral dust reactivity in biological systems.

Evidence for activity dependence on surface composition

Modification of the surface chemistry of chrysotile fibers via interaction with cells *in vivo* and *in vitro*, as well as with important biologic molecules, has been the subject of a number of studies, including Langer et al. (1972a,b), Johan et al. (1976), Thomassin et al. (1976), and Jaurand et al. (1977, 1983). These studies determined the chemical modification of the fibers directly using electron microprobe analysis as well as XPS. All of these studies, as well as our own work described below, describe surface chemical modification of chrysotile fibers in contact with cells or other biological agents. Generally, the most apparent modification is the leaching of Mg from the fiber. Depending on the extent of reaction, this Mg depletion can even be observed with the electron microprobe which has an analysis depth of 1 to 2 μm. (For comparison, XPS has an analysis depth of a few tens of Ångströms.)

The study by Jaurand et al. (1977) is particularly interesting in that they investigated the leaching of chrysotile asbestos in both human lungs and in the presence of rabbit alveolar macrophages *in vitro*. In both cases, the Mg:Si ratio (measured by an energy dispersive electron microprobe) indicated that Mg had been selectively leached, and that the amount of leaching was variable from fiber to fiber as well as along the length of a single fiber. The variation in Mg leaching among different fibers probably depends on a number of factors, including the original chemical and structural state of the fiber surfaces, the exact "microenvironment" surrounding each fiber, and their residence time. Analogous arguments can be made for leaching variations along a single fiber. In turn, the surface reactivity of these fibers will be heterogeneous as suggested above.

There have also been a number of groups that have modified the surface chemistry of chrysotile fibers and silica dust to reduce their cytotoxicity. The nonionic polymer polyvinyl-2-pyridine N-oxide (PVPNO), when sorbed to silica particles even in relatively small amounts, passivates them and dramatically reduces their cytotoxic effects on macrophages *in vitro* (Marchisio and Pernis, 1963; Schlipköter and Beck, 1965) and inhibits the development of fibrosis in lungs of experimental animals and humans (Schlipkoter et al., 1963; Prügger et al., 1984). Although PVPNO does not passivate chrysotile, some acidic, water-soluble polymers, such as carboxymethylcellulose (CMC) are effective (see, e.g., Schnitzer, 1974, and references therein). For example, Klockars et al. (1990) studied the role of CMC in blocking the production of reactive oxygen metabolites by human polymorphonuclear leukocytes and macrophages *in vitro*. These metabolites include superoxide and hydroxyl radicals as well as hydrogen peroxide which have been implicated in important (i.e., potentially disease initiating) cell inflammatory processes (Rossi et al., 1985; see also section on chrysotile and crocidolite below).

Brown et al. (1990) published a report in which they describe the modification of the surfaces of amosite fibers (mostly cummingtonite–grunerite, occasionally tremolite–actinolite) with octyldimethylchlorosilane (C_8) or octadecyldimethylchlorosilane (C_{18}). In *in vitro* tests, fibers modified with both hydrocarbon chains were less cytotoxic. In *in vivo* studies using laboratory animals, fibers coated with C_8 chains were not significantly different than untreated fibers in producing mesotheliomas, whereas fibers coated with C_{18} chains were dramatically less active at producing tumors. Brown et al. (1990) speculate that the longer carbon chain may be a more effective pacifying agent because its linear dimension is similar to that of cell membranes.

One surface modifying agent that has received considerable attention recently is deferoxamine, a well-known iron chelator. The reasoning behind this direction of research comes from the current popular theories about the catalytic generation (on fiber surfaces) of hydroxyl and superoxide free radicals. Because of the presence of reduced iron on crocidolite surfaces with Lewis base character, it is possible that surface iron plays an important role in this catalysis. In fact, deferoxamine has inhibited the catalytic activity of asbestos fibers in its ability to generate hydroxyl radicals from hydrogen peroxide (Weitzman and Graceffa, 1984) and to catalyze lipid peroxidation (Weitzman and Weitberg, 1985). This inhibition is reported to occur either when deferoxamine is present in the aqueous medium of the experiment or when the asbestos is prewashed with deferoxamine, rinsed, and re-introduced to the experiment. Further, deferoxamine has been shown to reduce the cytotoxicity of asbestos fibers *in vitro* (Goodglick and Kane, 1986). Finally, Weitzman et al. (1988) have shown that deferoxamine binds quickly and strongly to crocidolite both *in vitro* and *in vivo*. Taken together, the above results suggest that deferoxamine may leach iron from the near-surfaces of fibers, or bind strongly to iron surface sites, either way inhibiting its cell-damaging catalytic ability. Our own work with deferoxamine is reported below in the section concerning crocidolite.

Finally, Guthrie et al. (1992) determined the cytotoxicity of erionite, a fibrous zeolite which has been shown to have an extremely high pathogenic activity in humans, as a function of the exchangeable cation (Na, K, Ca, Fe). Although, strictly speaking for zeolites, this is not just a surface chemical modification, cation exchange does hold all physiochemical factors constant except composition on one site in the crystal structure. In this study, the *in vitro* work was performed on rat-lung epithelial cells, and varying the exchangeable cation has no significant affect on the cytotoxicity of erionite. The authors concluded that this result supports the idea that the cell killing ability of erionite *in vitro* stems from the acid/base sites on the mineral surface, presumably having to do with the aluminosilicate framework of the zeolite and not the exchangeable site composition.

Evidence for activity dependence on surface atomic structure

There have been several studies on the role of mineral surface atomic structure in biological interactions, especially with regards to silica dust and the pathogenesis of silicosis (*in vitro* studies: Stalder and Stober, 1965; Ottery and Gormley, 1978; Kozin, 1982; *in vivo* studies: King et al., 1953; Attygalle et al., 1956; Brieger and Gross, 1966, 1967; Pratt, 1983). In these studies, cytotoxic and pathogenic observations have been made as a function of the silica polymorph used in each study. Although there are various apparent contradictions among these studies, in general quartz, tridymite, and cristobalite are more biologically active than coesite, stishovite, and amorphous silica. Some of the problems in this area of research have been recently addressed by Wiessner et al. (1988). In this study, both *in vitro* (red blood cell lysis) and *in vivo* (lung fibrosis in laboratory mice) experiments were performed. All particles of the various silica polymorphs used in their study were carefully chosen to be in the same size range. BET surface area measurements were made to assure that each experiment utilized the same crystal surface area for direct comparison of results. Trace metal oxide contaminants were removed by an acid pretreatment. Wiessner et al. found that quartz, tridymite, and cristobalite were actively lytic to red blood cells *in vitro* and were also active in mice lungs leading to inflammatory responses and eventually fibrosis. On the other hand, coesite was less active in its lytic capacity *in vitro* and gave negative fibrosis results *in vivo*. An inflammatory response was not even observed. These results lead Wiessner et al. to conclude that surface crystal structure does significantly contribute to biological activity, at least in this case. They postulate that the atomic packing density is essentially the only variable which is left to explain the difference reactivity of these minerals in their experiments. Quartz, cristobalite, and tridymite are stable under relatively low pressure, high temperature conditions, and their packing density is also relatively low. Coesite is a high pressure silica polymorph, and it's packing density is relatively high. Therefore, their work suggests that the biological activity of silica increases as the atomic packing density of the structure (and therefore the surface) decreases.

The studies of Reeves et al. (1974) and Langer et al. (1978) on chrysotile also support the notion that the biological activity of mineral dust depends, at least in part, on surface atomic structure. These workers determined the hemolytic activity and free radical generation of chrysotile as a function of grinding time of the starting chrysotile dust. With increased grinding time, the fibers obviously became shorter, and the *in vitro* activity was significantly reduced. The only explanation of this behavior came from extensive characterization of the fibers before and after various grinding times using TEM and selected area electron diffraction, X-ray diffraction, infrared spectroscopy, and electron spin resonance spectroscopy. These techniques clearly show that grinding progressively reduces fiber crystallinity as Si–O and Mg–O bonds are disrupted. This, of course, reduces the ordering of the chrysotile fiber surface, which in turn is probably at least partially responsible for the reduction of its biological activity. Had not these extensive efforts been made in this study to structurally characterize the chrysotile fibers as a function of grinding time, fiber dimensionality would have been the only variable available to explain their results.

Evidence for activity dependence on surface microtopography

This author is not aware of any study dealing with health effects of mineral dust which address the issue of surface microtopography directly. However, in the pure surface science literature, there are many such studies. Several of the general conclusions from these studies have already been discussed in the section on the general nature of mineral surfaces above, and it has been shown that microtopography often plays an important role, and sometimes a critical role, in reactions that occur on surfaces. Certainly, if the surface chemistry of mineral dusts truly plays any role in lung disease, then surface microtopography is highly likely to play a role in this surface chemistry. However, it would be very difficult to design an experiment which would test this particular surface attribute. Ideally, one would set up *in vitro* and *in vivo* experiments in which different cultures/groups were exposed to two sets of particles. The particles from both sets would essentially be identical except for surface roughness. Preferably, one set of particles would have predominantly smooth surfaces, while the other would have rough surfaces on the molecular scale. Unfortunately, generating two such groups, and insuring that surface chemistry and overall particle dimensions were the same, would be very difficult.

Although the work by Langer et al. (1978) was performed for a different purpose, their study indirectly illustrates the potential problems of understanding the role of surface roughness in reactions in biological systems. As described earlier, they used mechanical milling to produce chrysotile fibers of variable lengths. This technique is commonly used to study the role of fiber dimension in *in vitro* and *in vivo* studies of fiber-induced lung disease. Milling may also affect fiber surface roughness. Unfortunately, as Reeves et al. (1974) and Langer et al. (1978) discovered, grinding and milling also has many other effects on fibers.

With increasing grinding or milling time, as mentioned briefly in the previous section, fiber crystallinity is progressively reduced as Si–O and Mg–O bonds are disrupted. In concert with this, the biological activity of the fibers, in terms of hemolytic potency and free radical generation, is also reduced. Obviously, a surface roughness effect can not be extracted from these experiments.

The protocol for an experiment (perhaps a bioassay) which effectively tests the role of surface roughness of mineral particles in biological systems remains to be developed.

Evidence for activity dependence on surface charge

At the outset, it is important to make a clear distinction between surface charge and electrostatic charge associated with mineral dust. Surface charge, originating from chemical interactions between a surface and its surrounding environment, is described in some detail above. Electrostatic charge, which will not otherwise be discussed in this chapter because it is not a material specific surface property, is due to free electron build-up on a particle; the charge can be drained through any appropriate ground. Davis et al. (1988) provide a relatively recent review and study of electrostatic charge on the pathogenicity of chrysotile asbestos *in vivo*. In their work, laboratory animals were exposed to normally dispersed airborne fibers which carried an electrostatic charge, as well as airborne discharged fibers. Results show that animals exposed to electrostatically charged dust retained significantly more fibers in their lungs compared with the group breathing uncharged dust. Subsequently, the former group developed more pulmonary fibrosis and tumors.

Turning to surface charge, there have been a few studies, especially in relation to silica dust, supporting the notion that this may be an important property in disease generation. The ideas proposed center on the Brønsted acidity of the surface silanol groups, that is to say their hydrogen donating capability, as first proposed by Nash et al. (1966) and reviewed and expanded by Nolan et al. (1981) and references therein. Such a process may be capable of cracking (breaking) biological macromolecules. For quartz, the PZNPC is at high pH, so that under nearly all biological conditions, the surface is negatively charged with Si–O$^-$ groups dominating over Si–OH$_2^+$ groups. The remaining sites are silanol groups (Si–OH) with a hydrogen donor functionality. When a proton-accepting polymer such as PVPNO (also mentioned above) interacts with the quartz surface, the silanol site is deactivated and the quartz looses its hemolytic ability (Nolan et al., 1981). Also, when quartz surfaces are reacted with strong Lewis acids like AlCl$_3$ or FeCl$_3$ which react with the Si–O$^-$ sites, the biological activity, in terms of lysing red blood cells, is also reduced. The reduction in negative surface charge has been followed by zeta potential measurements, and when the charge has been neutralized by adsorption of metallic cations, the hemolytic activity is nearly completely quenched. Nolan et al. (1981) have therefore concluded that the quartz surface is clearly bifunctional, with both surface sites (Si–OH, Si–O$^-$) active in hemolysis and blockable in different ways.

THE SURFACES OF CHRYSOTILE AND CROCIDOLITE

The preceding portions of this chapter have dealt primarily with the nature and reactivity of mineral surfaces, both outside and within biological systems. Using this important background material, we now explore the mineral surface science dealing specifically with the two most studied minerals in terms of hazardous dusts, chrysotile and crocidolite. What is the nature of the surfaces of these minerals? What is it about these surfaces that give them cytotoxic, carcinogenic, and/or other biologically damaging properties? Can we identify the surface sites that are biologically active, and if so, what is their nature? Can these sites be modified to a more passive or biologically inert state? Does the simple dimensionality of the particle play a critical role in their activity in biologic systems? We will attempt to shed as much light on these questions as possible mostly from the standpoint of mineral surface science.

Chrysotile

General surface description. Chrysotile, $(Mg_3Si_2O_5(OH)_4)$, a common member of the serpentine mineral family, is a sheet silicate consisting of layers of polymerized silica tetrahedra (SiO_4 groups) alternating with layers of polymerized magnesium octahedra ($MgO_2(OH)_4$ groups) (see Fig. 8a in Chapter 2 by Klein, and Chapter 3 by Veblen and Wylie). These layers are perpendicular to the c crystallographic axis. The repeat unit along c consists of 1 octahedral sheet (o) and 1 tetrahedral sheet (t). However, there is a dimensional misfit between these two layers. The connecting points on the o sheet are spread too far apart for the matching connecting point on the t sheet. As a result, the connecting side of the o sheet becomes concave to reduce its dimensions, while the connecting side of the t sheet becomes convex to expand. The sheets then fit together, but the net result is a structure that literally rolls up resulting in the fibrous nature of chrysotile (Fig. 4). Due to this arrangement, the fiber surface is made up of the "back-side" of the Mg octahedral sheet consisting of hydroxyl groups on the surface, each bonded to

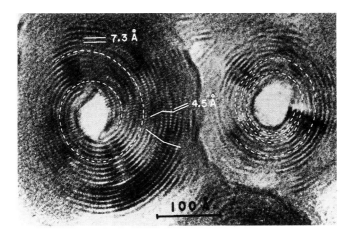

Figure 4. Transmission electron microscopy (TEM) image of chrysotile fibers looking down the length of the fiber (the fiber axis). Dashed lines follow single sheets. The 7.3 Å and 4.5 Å spacings shown are for the (001) and (020) planes, respectively. From Yada (1967).

Figure 5. AFM image of the octahedral layer surface of lizardite, essentially showing what the equivalent surface of chrysotile looks like. The image is just over 1.2 nm (= 12 Å) on a side, and has been tilted back from the viewer by 30° to give perspective. The highest (lightest) bumps in the image, forming an hexagonal array with 3.1 Å spacing, represent hydroxyls. The slightly lower (grayer) maxima in the middle of an hydroxyl triangle represent Mg positions. From Wicks et al. (1992).

three Mg atoms. A more precise way to look at this is that each surface oxygen is bonded to one hydrogen (the $\vec{OH}$ vector pointing away from the surface) and three Mg atoms in the next layer down. Note that for the surface oxygen, Pauling's second rule is satisfied (the sum of the bond strengths going to the oxygen are equal to the charge on the oxygen), and the overall configuration (without additional surface ligands) is charge neutral.

Recently, Wicks et al. (1992) have been able to image the octahedral sheet surface of lizardite, a mineral closely related to chrysotile, with AFM at nearly atomic resolution. Lizardite is essentially isostructural with chrysotile, except that the octahedral–tetrahedral sheets remain flat and are more amenable to imaging. A high resolution AFM image of the outer octahedral sheet is shown in Figure 5. Each high node represents a hydroxyl in this image, each separated from one another by approximately 3.1Å. The bands coming off each high node intersect at Mg positions in the next layer down. Ideally, this is the surface that the surrounding biological system "sees" on the outside of the chrysotile fiber.

Surface site character and surface charge: As just described, by far the dominant chrysotile fiber surface site is a hydroxyl group on the back-side of the octahedral sheet and bonded to three Mg atoms in the next layer down. Minority surface sites include the ends of the rolled sheets, that is, in the a and b crystallographic plane (see Fig. 8a in Chapter 2). In this plane, we are dealing with the "edges" of both Si tetrahedral and Mg octahedral sites. In the case of the Si tetrahedra on the edge of the sheet, we have an Si–O$^-$ group which would most likely immediately complete its coordination with a H$^+$ from solution. Therefore, this surface site would carry a weak Brønsted acid character. On the other hand, octahedral sites on the surface would most likely complete their Mg coordination spheres with H$_2$O from air or the solution to which the surface is exposed. These sites probably also possess a weak Brønsted acidity.

The surface charge of chrysotile (measured by zeta potential) over the pH range of interest in biological systems is positive (Fig. 6). This can be rationalized

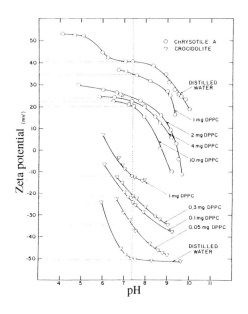

Figure 6. Zeta potentials for chrysotile and crocidolite in distilled water and in the presence of dipalmitoyl phosphatidylcholine (DPPC), a surfactant, as a function of pH. From Light and Wei (1977).

by the preceding discussion of surface sites. In an acid to neutral aqueous solution, the hydroxylated surface sites on the back-side of the octahedral sheet mean that there is little likelihood of surface Si–O$^-$ groups. On the other hand, the charge associated with the edge sites in the *a-b* crystallographic plane are probably dominated by the octahedral sites which consist of near-surface Mg^{2+} ions coordinated with oxygens and hydroxyl below and water molecules above (instead of more negatively charged oxygens and hydroxyls). Therefore, these sites have a residual positive charge over a considerable pH range. At very high pH's (>10, see Fig. 6), surface water molecule ligands would be progressively replaced by hydroxyls, eventually producing an overall negative surface charge.

Surface site reactivity. It is difficult to get at the nature of the reactivity of each surface site on a chrysotile fiber (or on any mineral surface). The thermodynamic measurement of surface charge only provides a hint as to how the surface will react with species in solution (see text concerning surface charge above). One way to characterize the reactive nature of surface sites is by observing the uptake characteristics of "probe" molecules, that is, specific molecules that are known to react with surface sites possessing Lewis acidic or basic character. For chrysotile (and crocidolite, see below), such work has been performed by Bonneau et al. (1986) by studying the uptake of pyridine (reactive with acidic sites) and CO_2, benzoic acid, and fluoride and phosphate compounds (reactive with basic sites). On chrysotile fibers, pyridine was primarily physisorbed (as evidenced by IR spectra and heating experiments); there was no evidence of chemisorption, implying that the existence of acceptor (acidic) sites is minor. On the other hand, the donor (basic) site probe molecules were chemisorbed with different apparent coverages depending on their electron accepting affinity. Due to uptake amounts, Bonneau et al. (1986) concluded that the active basic site on the chrysotile fiber is the hydroxyl site on the "outside" of the octahedral sheet, by far the most abundant sites on the surface of this fibrous mineral.

Dissolution behavior. The dissolution of the relatively insoluble silicate minerals is a surface chemistry controlled phenomenon. Whether dissolving in pure water or fluids in the lung, the dissolution of chrysotile will occur as surface and near-surface species are hydrolyzed and carried out into solution.

There have been a number of studies on the dissolution of chrysotile (Gronow, 1987; Churg et al., 1984, Thomassin et al., 1977; Jaurand et al., 1977; and references therein), but total dissolution rate (fiber lifetime) cannot be calculated from these studies because silicon release rates were not determined. The ultimate dissolution depends on hydrolyzing silicon–oxygen bonds because the linked silica tetrahedra are the most chemically resistant component in the mineral. Recalling the structural makeup of a chrysotile fiber, the host solution must first attack the outermost magnesium octahedral layer. Due to the relatively high solubility of the magnesium oxyhydroxide layer, the underlying silica tetrahedral layer will be reached relatively quickly. The fiber lifetime should depend on the silica release rate from this layer, since this will be the slowest reaction in the dissolution process. This simple model ignores dissolution at the end of the fiber and also at microcracks and other imperfections that intersect the surface. These factors may be important in the overall dissolution rate of the mineral.

Fortunately, a chrysotile dissolution study with biodurability in mind has recently been performed by Hume and Rimstidt (1992). The complete dissolution reaction in the pH range of lung fluids is

$$Mg_3Si_2O_5(OH)_4 + 6\,H^+ \rightarrow 3\,Mg^{2+} + 2\,H_4SiO_4 + H_2O \qquad (4)$$

As expected from the nominally low concentration of Mg and Si in lung fluids, chrysotile is undersaturated and will dissolve. Hume and Rimstidt measured the silica release rate from dissolving chrysotile at pH's of 2 to 6 and 37°C (approximately body temperature). The result, 5.9 (± 3.0) x 10^{-10} mol/m^{2}s, was independent of pH over the range measured. From this, they could predict that a chrysotile fiber 1 µm in diameter would completely dissolve in the human lung in 9 (± 4.5) months.

Implications for biological activity. The surface chemistry discussed above, in terms of specific surface site character and reactivity, as well as dissolution rate, all have implications for the pathogenicity of chrysotile in the human lung. Presumably, for chrysotile to initiate a lung disease, fibers must promote chemical reactions that either directly or indirectly lead to cellular or biological structure malfunction and disease. In addition, it seems logical that the efficiency of disease generation would increase as the lifetime of the fiber in the lung increases. However, it is becoming increasingly clear that chrysotile in lung tissue in far less dangerous than crocidolite (Mossman et al., 1990, and references therein) even though they both have similar surface site character and can promote or catalyze the same reactions that may eventually lead to disease (see Bonneau et al., 1986, and below). Therefore, we will discuss these suspected damaging reactions in the text on crocidolite below. But first, we briefly discuss the *in vivo* evidence for the biodurability of chrysotile and the implications that this has on disease generation.

Jones et al. (1989) showed that continuous exposure of rats to airborne chrysotile at first produces an increase of lung fiber content, but that this trend levels out with time. Wagner et al. (1982) also reported that chrysotile amounts found in the lung tissue of asbestos workers is less than expected. (This rapid clearance of chrysotile relative to amphibole asbestos is discussed in detail in Chapters 13 and 14.) These studies indirectly imply what the biodurability study described above tells us directly. It should also be noted that although chrysotile dissolves relatively quickly, the onset of disease is typically many (20 to 60) years after the first exposure (see Chapters 11 and 13). Therefore, because of the likelihood of a very short lifetime of an individual chrysotile fiber in lung tissue, and if chrysotile (and not minor amounts of amphibole contamination) is truly the cause of severe lung disorders, at least the first step in disease generation must take place relatively quickly.

Crocidolite

General surface description. Crocidolite is the fibrous variety of the amphibole riebeckite [$Na_2Fe^{3+}_2Fe^{2+}_3Si_8O_{22}(OH)_2$], although Mg can substitute completely for the ferrous iron; see Chapter 3). Unlike chrysotile, there is no misfit between structural units, and therefore amphibole fibers are simply straight single crystals with extremely high aspect ratios. Individual fiber diameters are generally between a few hundreds of a micrometer up to nearly one micrometer. The long axis of the fiber, in this case the c crystallographic direction, is the direction of the primary structural unit of amphiboles, a flat double chain of silica tetrahedra (see Fig. 3a,b in Chapter 2). Sandwiched between these double chain groups are the Fe,Mg octahedral sites and the 8-coordinated Na site, arranged such that neighboring sets of double chains are not bonded directly to one another.

During the growth of a crocidolite single crystals, small defects can develop (often due to the formation of triple or higher order silica chains in the midst of the usual double chains) which may propagate or help develop major planar defects parallel to the c axis. There is often a general crystalline misorientation across one of these planar defects, resulting in distinct subgrain boundaries (Fig. 7). These semicoherent to incoherent boundaries can be further disturbed by geologic (tectonic) stress which may serve to break up these subgrains into individual, very narrow, fibers.

From the above, it is readily apparent that the surface of these fibers consists of one of several possible crystallographic planes which are parallel to the fiber (c) axis, as well as the terminal planes which cut the fiber axis. We will first discuss the surfaces parallel to the length of the fiber. TEM observations (Alario Franco et al., 1977; Ahn and Buseck, 1991) show that the principle planes which generate surfaces parallel to c are (100), (110), and (010), although there are others. SEM and AFM images (Figs. 8 and 9a,b, respectively) show that these faces are exposed as subgrains, or fibrils, which together make up a complete fiber. The surface sites on these planes will provide the vast majority of all surface sites on a single fiber.

One possibility for the (100) surface can be achieved by breaking the structure continually between opposing sets of silica double chains creating an up and down stepped surface (Fig. 3a in Chapter 2). The silica chains exposed at the surface are fully coordinated, that is one does not have to break through individual silica tetrahedra to create this surface. Alternatively, a flat (100) surface would consist of strips of half vacant octahedral sites (the missing half is on the matching surface now removed) with the fully coordinated silica double chain surfaces in between.

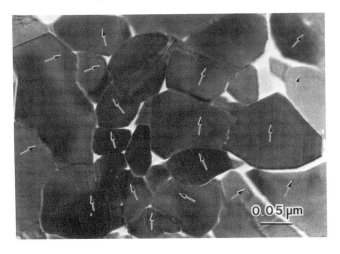

Figure 7. Low magnification transmission electron microscopy (TEM) image of crocidolite fibrils looking down the fiber axes. The arrows are all pointing in the a^* crystallographic direction for structure orientation reference. The outer surface of this fibril packet looks like what is shown in Figures 8 (SEM) and 9 (AFM). From Ahn and Buseck (1991).

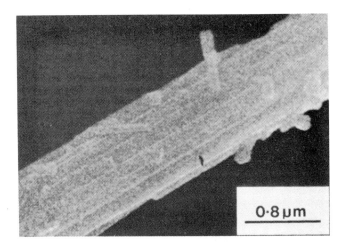

Figure 8. Scanning electron microscopy (SEM) image of a crocidolite fiber showing the characteristic fluted surface (due to fibrils, cf. Fig. 9) and attached debris. The debris is probably smaller crocidolite fiber sections electrostatically attached to the larger fiber. From Crawford (1980).

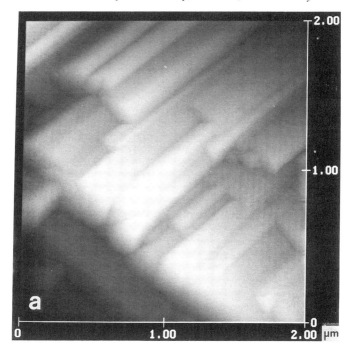

Figure 9. (a) Atomic force microscopy (AFM) image of the surface of a crocidolite fiber showing the characteristic fibrils which make up a whole fiber. Fibril terminations, with respect to one another, are irregular. However, individual terminations seem to be distinct, and fibril surfaces along their length appear smooth and should correspond to one of three crystallographic planes (see text).

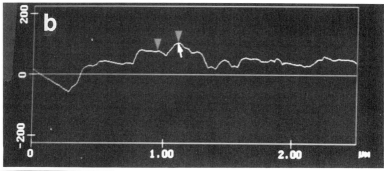

Figure 9. (b) Cross-section along line in image. Left end of cross-section starts at upper left of image. The height difference between the two markers on the cross-section line is 260 Å.

The (010) surface consists exclusively of M4 sites (relatively large cation sites with oxygen coordinations of 6 to 8), again half vacant, and the edges of silica chains. The density of non-tetrahedral sites on this surface is considerably less than on the second alternative of the (100) surface discussed above. The (110) surface is, in actuality, not planar, but a stepped (100)/(010) surface which runs on average along (110) (Fig. 3b in Chapter 2). The crystallographic plane or planes that cut the end of a crocidolite fiber are not specifically known.

Surface site character and surface charge: Each of the three surfaces parallel to the long direction of the fiber discussed above provides surface octahedral sites, with three oxygens (or two oxygens and one hydroxyl) in the next layer down (in the case of the (100) and (110) surfaces), or with two oxygens in the next layer down and a third on the surface with the metal site (in the case of the (010) surface). In air or solution, these surface sites will complete their coordination with hydroxyls and should present basic character (see below). As for the fiber ends, because silica chains must be cut, the surface terminating sites are likely to be similar to those described above for chrysotile. In the case of the Si tetrahedra on the end of a silica chain, we have a Si–O$^-$ group which may accept, for example, a H$^+$ from solution if available giving the site weak Brønsted acid character.

Although other amphiboles will not be specifically addressed in this chapter, it should be noted that anthophyllite and amosite (a trade/commercial name for fibrous varieties of the amphiboles cummingtonite, grunerite, tremolite, and/or actinolite) have similar textual features to crocidolite which makes them fibrous (see, e.g., Veblen, 1980, and Cressey et al., 1982). Therefore, their surface sites will be structurally similar (and often chemically dissimilar) to those described above for crocidolite.

Light and Wei (1977) show that crocidolite has a net negative surface charge in solution over much of the pH range of biological interest. Although electrophoresis measurements were not made below a pH of 6, the PZNPC is probably close to pH 5. In addition, electrophoretic mobility measurements of blocky vs. fibrous amphiboles by Schiller et al. (1980) support the idea of both positively and negatively charged sites on these surfaces. Blocky specimens have a smaller net negative charge compared to the fibrous variety, as would be expected given the presumed distribution of basic and acidic sites on the surface of this mineral.

Surface site reactivity. As with chrysotile (see above), Bonneau et al. (1986) used probe molecules to characterize the surface sites on crocidolite fibers. The results are very similar to those obtained for chrysotile. Pyridine (a base used to probe acidic sites) was only physisorbed, indicating that acidic sites on the surface are few or weak, or both. However, fluoride, phosphate and benzoic acid, in particular, chemisorbed to crocidolite, presumably to surface hydroxyls primarily on (100), (110), and (010) surfaces as described above. Therefore, the vast majority of reactive sites on crocidolite fibers, like chrysotile, have definite Lewis base character.

Dissolution behavior. It is known that in geological environments, amphiboles generally dissolve more rapidly than sheet silicates (e.g., Krauskopf, 1979). This suggests that crocidolite should dissolve faster than chrysotile. However, depending on the exact nature of the amphibole and sheet silicate and the dissolving environment, relative dissolution rates may be reversed. Unfortunately, there are as yet no laboratory dissolution rate data available for crocidolite equivalent to what was done for chrysotile by Hume and Rimstidt (1992) (reviewed above). Such a study would allow for the prediction of the lifetime of crocidolite fibers in lung tissue for a direct comparison with chrysotile.

There are a few geochemical studies concerning the dissolution of various amphiboles and amphibole-related minerals. Work has been done on the Fe-free amphibole tremolite (e.g., Schott et al., 1981), as well as the Fe-containing amphibole hornblende (e.g., Mogk and Locke, 1988; Velbel, 1989). In addition, there is a study of the dissolution of the Fe-containing pyroxene bronzite (Schott and Berner, 1983). We are interested in this latter study because pyroxenes and amphiboles have related atomic structures. Collectively, these studies show that amphiboles dissolve by first becoming relatively depleted in their non-tetrahedrally coordinated cations (Na, Mg, Ca, Fe) due to leaching from near-surface layers. In laboratory studies using surface analytic and depth profiling techniques, these leached layers generally remain quite thin (a few monolayers up to a few tens of Ångstroms, although under certain conditions they can be up to thousands of Angstroms deep). XPS and solution composition data also show that the leached layer will eventually achieve a steady state thickness when the silica rich residual surface dissolves as fast as the leaching front moves into the crystal (the leaching front movement slows as the leached layer becomes thicker until steady state is reached). If the Fe is oxidized during the leaching process, it quickly reprecipitates as a ferric oxyhydroxide in the vicinity of the surface of the crystal.

We do know some specifics about the mechanism of crocidolite dissolution in biological systems from the work of Crawford (1980). In this study, crocidolite fibers were exposed to human blood serum *in vitro* and rodent lung tissue *in vivo*. TEM observation of the fibers after exposure, in both cases, showed two intriguing features. First, approximately a quarter of the fibers examined from both the *in vitro* and *in vivo* exposure showed enhanced dissolution features at the intersection of (100) planar defects and the fiber surface (Fig. 10a,b). Second, approximately half of the fibers from both exposures showed semi-amorphous to amorphous surfaces layers (presumably leached) up to 10 nm in thickness (Fig. 11a,b). Leached layers of comparable or even greater thickness have been observed before on feldspar minerals with TEM (Casey et al., 1989; Casey and Bunker, 1990), XPS (Hellmann et al., 1990), and SIMS (Nesbitt and Muir, 1988), as well as on pyroxenes and other amphiboles as noted above. These studies help support the idea that the outer amorphous-looking layers seen on the exposed crocidolite surfaces are indeed due to leaching. By analogy with previous applicable silicate dissolution studies, it is likely that these leached layers are nearly stripped of Na and Mg, with reduced quantities of Fe in an amorphous, hydrous, and semi-coherent silica network. This could be confirmed by an XPS study of these grains. The leached surface will, of course, have completely different site character and

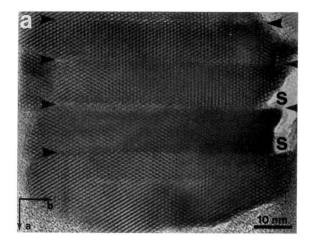

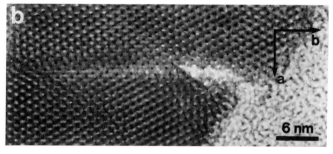

Figure 10. (a) High resolution TEM image of a crocidolite fibril looking down the fiber axis. The arrows point out (100) planar defects. The "S" labels point out where planar defects intersect the fiber surface, generating pits and steps on the surface. (b) High resolution TEM image of a fiber, in the same orientation as (a), that has been subjected to human blood serum. It appears as if there has been preferential etching on the fiber at the intersection of the (100) planar defect, horizontal in the image as in (a), and the surface. From Crawford (1980).

charge relative to the unreacted crocidolite. The leached layer will not necessarily continue to thicken, as the diffusion of species being removed through the leached layer slows this process until the movement of this front matches the removal rate of the outer silica rich surface. In addition, the narrow pits along the intersection between (100) planar defects and the fiber surface will generate a variety of surface sites which may be important in pathogenic reactions.

From epidemiological and *in vivo* studies, it is already apparent that crocidolite fibers have a significantly longer lifetime in lung tissue than chrysotile. As mentioned above, Jones et al. (1989) showed that lung tissue accumulation of crocidolite in exposed rats continues to rise with time, whereas accumulation of chrysotile rises initially and then levels off. Wagner et al. (1982), Gardner et al. (1986), and Churg (1988), among others, have shown that workers in the chrysotile industries can display unexpectedly high concentrations of amphibole fibers in their lungs (relative to chrysotile) even when exposure to amphiboles had

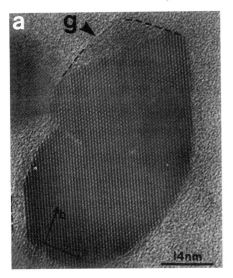
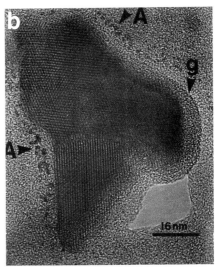

Figure 11. (a) High resolution TEM image of a small crocidolite fiber after exposure to human blood serum looking down the fiber axis. The amorphous-looking layer on the outside of the fiber, labeled g, is probably due to leaching/dissolution. See text for further explanation. (b) Same as in (a), except also showing ultra-small particles (labeled A) which are probably ferroprotein aggregates. From Crawford (1980).

been relatively minor in the workplace. Finally, Bellman et al. (1986) have demonstrated the high biodurability of amphibole fibers through *in vivo* experimentation.

Implications for biological activity. It is readily apparent that crocidolite is highly pathogenic. There is strong evidence that it persists for long periods of time in lung tissue, presumably due to very sluggish dissolution kinetics (see above). In addition, the surface sites described above, like those on chrysotile, have significant basic character and could potentially promote or catalyze reactions that eventually lead to fibrogenesis and bronchopulmonary carcinogenesis. There are several groups that now suggest that these reactions involve the generation of active oxygen species, such as the hydroxyl and superoxide radicals ($OH^{\bullet}$ and O_2^-, respectively) on the surfaces of asbestos fibers and iron oxides. Weitzman and Graceffa (1984) and Eberhardt et al. (1985) were among the first to show that asbestos catalyzes the formation of both hydroxyl and superoxide radicals in the presence of hydrogen peroxide (H_2O_2). This line of research became more intriguing when it was shown by Hansen and Mossman (1987) that rodent alveolar macrophages produce increased amounts of superoxide in the presence of asbestos fibers *in vitro*. Shatos et al. (1987) showed that hydroxyl and superoxide scavengers prevented asbestos induced cytotoxicity in cultures of lung fibroblasts and tracheal epithelial cells. Zalma et al. (1987a,b, 1989) took these studies a step further by showing that one can generate the toxic hydroxyl radical in the presence of free oxygen which is readily available in biological systems. The mechanism proposed involves the reduction of oxygen via the oxidation of iron on the mineral surface (although, presumably, one must also consider the oxidation of iron in solution). This general class of reactions was first proposed by Fenton (1894; Fenton and Jackson, 1899). The entire reaction sequence can be written (Zalma et al., 1987a,b) as follows:

$$O_2 + e^- \rightarrow O_2^- \tag{4}$$

$$O_2^- + e^- + 2H^+ \rightarrow H_2O_2 \tag{5}$$

$$H_2O_2 + e^- \rightarrow OH^\bullet + OH^- \tag{6}$$

$$OH^\bullet + e^- + H^+ \rightarrow H_2O \tag{7}$$

and, in the presence of organics (RH), one can postulate

$$OH^\bullet + RH \rightarrow R^\bullet + H_2O \tag{8}$$

The products of Equations 4 and 5 can also be combined to provide an alternative to Equation 6 as first suggested by Haber and Weiss (1934):

$$O_2^- + H_2O_2 \rightarrow O_2 + OH^\bullet + OH^- \tag{9}$$

To include iron explicitly in the reaction as the electron source, one can write reactions of the following form in place of, for example, Equation 6:

$$H_2O_2 + Fe^{2+} \rightarrow OH^\bullet + OH^- + Fe^{3+} \tag{10}$$

The viability of these reactions depends on a number of chemical factors if they are to occur on surfaces. Ferrous iron must be present on or near the surface of the mineral, and there must be a redox potential driving force to promote the electron transfer, in this case, for the reduction of free oxygen. (It is interesting to note here that White and Yee (1985) have shown that Fe^{2+}, on the surface of the amphibole hornblende, is a better reducing agent than Fe^{2+} in aqueous solution.) Lateral surfaces and near-surfaces of crocidolite have both ferrous and ferric iron in octahedral coordination (described above in the surface structure section). Once the near-surface ferrous iron has been oxidized to some depth and electron transfer is no longer probable, new near-surface sites must be generated if the reaction is to continue (note that this is not a true catalytic reaction). This can be achieved as the mineral slowly dissolves, but this is hindered by the fact that ferric oxides has far lower solubilities than ferrous oxides. Another hindrance could occur if the redox site is effectively blocked by the attachment of relatively unreactive organic molecules from the system. In the case of chrysotile, because only small amounts of iron are likely to substitute for Mg in the octahedral layer, such reactions as discussed above would probably depend on the presence of associated ferrous iron-containing phases which have been reported in the past such as nemalite $[(Mg,Fe^{2+})(OH)_2]$ and magnetite (Fe_3O_4, or $Fe_2^{3+}Fe^{2+}O_4$). However, magnetite appears to be relatively ineffective at driving such redox reactions relative to chrysotile alone. This suggests that another process may also affect the generation of oxygen radicals. Finally, for the radicals to be biologically harmful, they must be in very close proximity to their target molecules as radical lifetimes are exceptionally short.

Treatment with deferoxamine. One of the promising approaches both to the understanding of the role of crocidolite fibers in pathogenicity and in methods to reduce or inhibit their toxicity has been to chemically treat fiber surfaces. Brown et al. (1990) and Klockars et al. (1990) are just two recent examples of studies in which surface chemically modified amphibole fiber surfaces have been shown to have reduced pathogenic properties both *in vivo* and *in vitro*. As discussed above, deferoxamine has been shown to inhibit radical formation and cytotoxicity *in vitro*.

Recently, we have looked at the deferoxamine-induced surface chemical modification of crocidolite and amosite fibers directly with XPS. XPS gives a semi-quantitative to quantitative compositional analysis (for all elements except H and He) of the top several tens of Angstroms of the surface. Figure 12 shows XPS survey spectra for crocidolite and amosite fibers before deferoxamine treatment. The surface chemistry of these fibers is as expected from their bulk composition, with the exception of the detection of adventitious carbon and oxygen. Survey scans of these fibers after treatment (soaked in deferoxamine solution for 24 hours at room temperature, then thoroughly rinsed in distilled water) are not shown because changes are much more easily seen in high resolution narrow scans as shown below. Also, although photoelectron peak intensities are directly proportional to elemental quantity, raw intensity measurements vary from sample to sample (depending on instrument conditions and the exact amount of sample surface in the analytic area). However, this variation can be eliminated by using the ratio of the peak intensities for the element of interest to a conservative element, that is an element that is probably not affected by the treatment. We chose silicon in this case due to its well-known sluggish reactivity on silicate surfaces. Therefore, Table 1 shows the Fe/Si and Mg/Si intensity ratios for crocidolite and amosite before and after treatment.

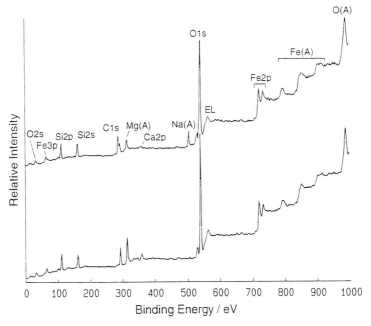

Figure 12. X-ray photoelectron spectroscopy (XPS) survey spectra of crocidolite (top) and amosite (bottom) fibers before treatment with deferoxamine. Small carbon lines are due to a carbonaceous mounting material (see Fig. 13) and adventitious carbon. The peak labeled EL is due to an energy loss phenomenon.

Table 1

XPS intensity ratios for crocidolite and amosite fibers before and after treatment with deferoxamine

	Crocidolite		Amosite	
	Untreated	Treated	Untreated	Treated
$\frac{Fe2p}{Si2p}$	9.27	7.64	7.60	6.82
$\frac{Mg1s}{Si2p}$	0.99	0.49	1.57	1.06

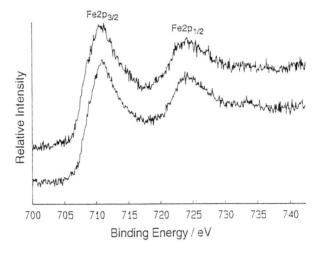

Figure 13. High energy resolution XPS spectra of the Fe 2p region of crocidolite before (bottom) and after (top) deferoxamine treatment. The spectra have been shifted to correct for charging. The Fe near-surface concentration, relative to Si, is down about 18%. The spectral line shapes, essentially identical for both, show primarily Fe^{2+} character (Hochella, 1988). At least on the surfaces of these crocidolite fibers, there is little Fe^{3+} present.

Figure 13 shows narrow scans for the Fe2p photoelectron lines for crocidolite before and after treatment. These lines, which generally have the shape characteristics of ferrous iron (Hochella, 1988) before treatment, do not change in shape after treatment, indicating that the surface iron has not undergone any major oxidation state change. The Fe2p/Si2p intensity ratios, with expected errors of approximately ±5%, show a moderate reduction in near-surface iron. Interestingly, however, the magnesium-to-silicon ratios change dramatically (Table 1, Fig. 14). In fact, the near-surface concentration of Mg in crocidolite fibers after this deferoxamine treatment was reduced by a factor of two. Not surprisingly, the results for amosite are similar although not as dramatic (Table 1; spectra not shown). The surface Fe concentration dropped slightly and the Fe oxidation state did not change after deferoxamine treatment, but the concentration of near-surface Mg was reduced by approximately one-third.

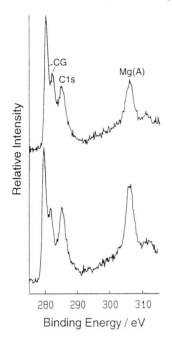

Figure 14. High energy resolution XPS spectra of the C1s and MgKLL Auger lines of crocidolite before (bottom) and after (top) deferoxamine treatment. The spectra have been shifted to correct for charging. However, the lines labeled CG are due to C1s lines from the colloidal graphite used to mount the sample. Being a conductor, the lines emanating from the graphite were not charge shifted in the first place, so with the shift to correct the line positions of the C1s and Mg(A) peaks from crocidolite, the graphite lines are now in the wrong position. Note that, after deferoxamine treatment, the Mg(A) line is reduced by about a factor of two. See text and Table 1 for details.

These examples show the dangers of speculating about surface mechanisms and processes without direct analysis or verification. While the inhibition of biological activity of asbestos fibers by deferoxamine is interesting and potentially useful, the mechanism by which this occurs is still very much unknown. If ferrous iron is important in the generation of potentially damaging radicals and cell-killing observed *in vitro*, the above results suggest that the inhibition by deferoxamine is more likely due to Fe-site blockage rather than surface Fe removal. Surprisingly, deferoxamine leaches Mg from the surface of these fibers more efficiently than Fe.

SUMMARY AND CONCLUSIONS

This chapter has reviewed a number (certainly not all) of the pertinent studies dealing with the health effects of mineral dust from a surface chemistry perspective. There are many aspects of this area of research which can be suitably explained or rationalized via a surface chemistry approach. Other aspects, which may have surface chemistry implications, remain a mystery.

Listed below are a few of the more important points of this chapter, reworded, repeated, and in some cases expanded, for clarification and emphasis.

(1) Mineral surfaces are not static, inert systems upon which chemistry happens. They are dynamic and tailor themselves to their surroundings. They are often much more complex and heterogeneous (in terms of structure, composition, and microtopography) than most of our models allow. It is also important to remember that this complexity includes minute amounts of other phases that are concentrated (and often go undetected) on the surfaces of the primary phase.

(2) Samples of the same mineral from different localities may have quite different biological activities. The trace and minor element chemistry and physical properties of the same mineral can vary significantly, which may lead to differences in their cytotoxicities and possibly carcinogenicities. Evidence for this has already been found by Pott et al. (1974), Palekar et al. (1979), and Nolan et al. (1991), among others. For example, Nolan et al. (1991) show that palygorskite specimens collected from various geologic localities show a significant range of carcinogenic activity in laboratory animals. Therefore, they argue persuasively that mineral identity and morphology alone are poor prognosticators of the *in vitro* and *in vivo* activity of fibrous mineral dust. The biological activity of mineral dust depends on a number of factors (morphology and durability among them), some of which are subtle and/or overlooked (see below).

(3) The toxicity and pathogenicity of mineral dust is almost certainly dependent on (a) a combination of mechanical/dimensional properties (shape as well as stiffness, or bending strength, and fracture characteristics) and (b) chemical properties (dissolution mechanisms and kinetics, heterogeneous catalytic activity, etc., these being related to surface properties). Concerning the former, the Stanton hypothesis (Stanton et al., 1977, 1981) states that the carcinogenicity of fibers depends on dimension and durability rather than physicochemical properties. Perhaps as important in this realm of thinking is work by, for example, Wang et al. (1987), who show conclusively that fibers have direct, intricate, and presumably damaging physical interactions directly with chromosomes. The latter point has been the subject of this chapter. The following is a list of the factors which are thought to be most important in this. They are given in no particular order because each factor's relative importance changes depending on the mineral in question and its biological environment. Also, most of these factors are not entirely independent from one another.

—surface and near-surface composition
—surface atomic structure
—surface microtopography
—surface charge and its dependence on pH and surrounding solutions
—durability (dissolution rate)
—associated minor or trace elements
—associated minor or trace phases

Note that the first four items listed are not static, but variable as the mineral dissolves in lung fluids. Further, Lund and Aust (1990, 1991) have argued that it may be more important to concentrate on what leaches out of a fiber, particularly iron and how it interacts with the cell, rather than what is left behind. This underscores the need for additional work on the dissolution properties of these minerals under physiological conditions.

(4) The probable primary reasons that glass fibers are not dangerous, relative to a number of minerals, to human health are: their mechanical integrity (stiffness, strength) is not that of the more toxic mineral dusts, and their rate of dissolution is very high, dramatically faster than crystalline grains or fibers.

ACKNOWLEDGMENTS

The author is grateful to George Guthrie for developing his interest in this field, for sharing many ideas, and for constructively reviewing and improving the original manuscript, to Edward Gabrielson for supplying the samples in the deferoxamine study and for several enlightening discussions, to Andrew Werner and Jodi Junta for assistance with the AFM instrumentation, image processing and references, and to Sharon Chiang and Ron Wirgart for technical assistance in manuscript preparation. Financial support from the National Science Foundation (EAR-9105000), the Petroleum Research Fund, administered by the American Chemical Society (PRF 22892-AC5.2), Virginia Polytechnic Institute and State University, and the Center for Materials Research at Stanford University is also greatly appreciated.

This chapter is dedicated to the memory of Dr. Andrew Gratz (1962–1993) and his many contributions to mineral surface science.

REFERENCES

Adamson, A.W. (1982) Physical Chemistry of Surfaces, 4th ed., John Wiley & Sons, New York, 664 p.
Ahn, J.H. and Buseck, P.R. (1991) Microstructures and fiber-formation mechanisms of crocidolite asbestos. Amer. Mineral. 76, 1467–1478.
Alario Franco, M., Hutchison, J.L., Jefferson, D.A. and Thomas, J.M. (1977) Structural imperfection and morphology of crocidolite (blue asbestos). Nature 266, 520–521.
Attygalle, D., King, E.J., Harrison, C.V. and Nagelschmidt, G. (1956) The action of variable amounts of tridymite, and of tridymite combined with coal, on the lungs of rats. Brit. J. Indus. Med. 13, 41–50.
Bellman, B., Konig, H., Muhle, H. and Pott, F. (1986) Chemical durability of asbestos and of man-made mineral fibers in vivo. J. Aerosol Sci. 17, 341–345.
Bonneau, L., Suquet, H., Malard, C. and Pezerat, H. (1986) Studies on surface properties of asbestos. I. Active sites on surface of chrysotile and amphiboles. Environ. Res. 41, 251–267.
Brieger, H. and Gross, P. (1966) On the theory of silicosis. I. Coesite. Arch. Environ. Health 13, 38–43.
Brieger, H. and Gross, P. (1967) On the theory of silicosis. III. Stishovite. Arch. Environ. Health 15, 751–757.
Brown, G.E., Jr. (1990) Spectroscopic studies of chemisorption reaction mechanisms at oxide-water interfaces. In: Mineral-Water Interface Geochemistry, Hochella, M.F., Jr. and White, A.W., eds., Rev. Mineral. 23, 309–364.
Brown, R.C., Carthew, P., Hoskins, J.A., Sara, E. and Simpson, C.F. (1990) Surface modification can affect the carcinogenicity of asbestos. Carcinogenesis 11, 1883–1885.
Casey, W.H., Westrich, H.R., Arnold, G.W. and Banfield, J.F. (1989) The surface chemistry of dissolving labradorite feldspar. Geochim. Cosmochim. Acta 53, 821–832.
Casey, W.H. and Bunker, B. (1990) Leaching of mineral and glass surfaces during dissolution. In: Mineral-Water Interface Geochemistry, Hochella, M.F., Jr. and White, A.W., eds., Rev. Mineral. 23, 397–426.
Christmann, K. and Ertl, G. (1976) Interaction of hydrogen with Pt(111): The role of atomic steps. Surf. Sci. 60, 365–384.
Churg, A. (1988) Chrysotile, tremolite, and melignant mesothelioma in man. Chest 93, 621–628.
Churg, A., Wiggs, B., Depaoli, L., Kampe, B. and Stevens, B. (1984) Lung asbestos content in chrysotile workers with mesothelioma. Amer. Rev. Resp. Diseases 130, 1042–1045.
Crawford, D. (1980) Electron microscopy applied to studies of the biological significance of defects in crocidolite asbestos. J. Microscopy 120, 181–192.
Cressey, B.A., Whittaker, E.J.W. and Hutchison, J.L. (1982) Morphology and alteration of asbestiform grunerite and anthophyllite. Mineral. Mag. 46, 77–87.
Daubrée, M. (1867) Expériences sur les décompositions chimiques provoquées par les actions mécaniques dans divers minéraux tels que le feldspath. Comptes Rendus 339–345.
Davis, J.A. and Kent, D.B. (1990) Surface complexation modeling in aqueous geochemistry. In: Mineral-Water Interface Geochemistry, Hochella, M.F., Jr. and White, A.W., eds., Rev. Mineral. 23, 177–260.
Davis, J.M.G., Bolton, R.E., Douglas, A.N., Jones, A.D. and Smith, T. (1988) Effects of electrostatic charge on the pathogenicity of chrysotile asbestos. Brit. J. Indus. Med. 45, 292–299.

Ebelmen, J.J. (1847) Recherches sur la décomposition des roches. Annal. des Mines 12, 627–654.
Eberhardt, M.K., Roman-Franco, A.A. and Quiles, M.R. (1985) Asbestos-induced decomposition of hydrogen peroxide. Environ. Res. 37, 287–292.
Eggleston, C.M. and Hochella, M.F., Jr. (1992) The structure of hematite {001} surfaces by scanning tunneling microscopy: Image interpretation, surface relaxation, and step structure. Amer. Mineral. 77, 911–922.
Fenton, H. J. H. (1894) Oxidation of tartaric acid in presence of iron. J. Chem. Soc. 65, 899–910.
Fenton, H. J. H. and Jackson, H. (1899) The oxidation of polyhydric alcohols in presence of iron. J. Chem. Soc. 75, 1–11.
Gardner, M.J., Winter, P.E., Pannett, B. and Powell, C.A. (1986) Follow up study of workers manufacturing chrysotile asbestos cement products. Brit. J. Indus. Med. 43, 726–732.
Goodglick, L.A. and Kane, A.B. (1986) Role of reactive oxygen metabolites in crocidolite asbestos toxicity to mouse macrophages. Cancer Res. 46, 5558–5566.
Grim, R.E. (1968) Clay Mineralogy. McGraw-Hill, New York, 596 p.
Gronow, J.G. (1987) The dissolution of asbestos fibers in water. Clay Minerals 22, 21–35.
Guthrie, G., McLeod, K., Johnson, N. and Bish, D. (1992) Effect of exchangeable cation on zeolite cytotoxicty. V. M. Goldschmidt Conference Program and Abstracts, A-46. (abstract)
Haber, F. and Weiss, J. (1934) The catalytic decomposition of hydrogen peroxide by iron salts. Proc. Royal Soc. London, Series A, 147, 332–351.
Hansen, K. and Mossman, B.T. (1987) Generation of superoxide (O_2^-) from alveolar macrophages exposed to asbestiform and nonfibrous particles. Cancer Res. 47, 1681–1686.
Hellmann, R., Eggleston, C.M., Hochella, M.F., Jr. and Crerar, D.A. (1990) The formation of leached layers on albite surfaces during dissolution under hydrothermal conditions. Geochim. Cosmochim. Acta 54, 1267–1281.
Henrich, V.E. (1985) The surfaces of metal oxides. Reports Progress Phys. 48, 1481–1541.
Hering, J.G. and Stumm, W. (1990) Oxidation and reductive dissolution of minerals. In: Mineral-Water Interface Geochemistry, Hochella, M.F., Jr. and White, A.W., eds., Rev. Mineral. 23, 427–466.
Hochella, M.F., Jr. (1988) Auger electron and x-ray photoelectron spectroscopies. In: Spectroscopic Methods in Mineralogy and Geology, Hawthorne, F.C., ed., Rev. Mineral. 18, 573–637.
Hochella, M.F., Jr. (1990) Atomic structure, microtopography, composition, and reactivity of mineral surfaces. In: Mineral-Water Interface Geochemistry, Hochella, M.F., Jr. and White, A.W., eds., Rev. Mineral. 23, 87–132.
Hochella, M.F., Jr., Eggleston, C.M., Elings, V.B., Parks, G.A., Brown, G.E., Jr., Wu, C.M. and Kjoller, K. (1989) Mineralogy in two dimensions: Scanning tunneling microscopy of semiconducting minerals with implications for geochemical reactivty. Amer. Mineral. 74, 1235–1248.
Hochella, M.F., Jr., Eggleston, C.M., Elings, V.B. and Thompson, M.S. (1990) Atomic structure and morphology of the albite {010} surface: An atomic-force microscope and electron diffraction study. Amer. Mineral. 75, 723–730.
Hume, L.A. and Rimstidt, J.D. (1992) The biodurability of chrysotile asbestos. Amer. Mineral. 77, 1125–1128.
Iwasawa, Y., Mason, R., Textor, M. and Somorjai, G.A. (1976) The reactions of carbon monoxide at coordinatively unsaturated sites on a platinum surface. Chem. Phys. Let. 44, 468–470.
Jaurand, M.-C., Baillif, P., Thomassin, J.-H., Magne, L. and Touray, J.-C. (1983) X-ray photoelectron spectroscopy and chemical study of the adsorption of biological molecules on chrysotile asbestos surface. J. Colloid Interface Sci. 95, 1–9.
Jaurand, M.C., Bignon, J., Sebastien, P. and Goni, J. (1977) Leaching of chrysolite asbestos in human lungs. Environ. Res. 14, 245–254.
Johan, Z.A., Goni, J., Sarcia, C., Bonnaud, G. and Bignon, J. (1976) Influence de certains acides organiques sur la stabilité du reseau du chrysotile 6th congrès int'l geoch. organique. Technip Ed. Paris, 883–903.
Jones, A.D., Vincent, J.H., McIntosh, C., McMillan, C.H. and Addison, J. (1989) The effect of fiber durability on the hazard potential of inhaled chrysotile asbestos fibers. Exper. Path. 37, 98–102.
King, E.J., Mohanty, G.P., Harrison, C.V. and Nagelschmidt, G. (1953) The action of different forms of pure silica on the lungs of rats. Brit. J. Indus. Med. 10, 9–17.
Klein, C. and Hurlbut, C. S., Jr. (1985) Manual of Mineralogy (20th ed.), John Wiley & Sons, New York, 596 p.
Klockars, M., Hedenborg, M. and Vanhala, E. (1990) Effect of two particle surface-modifying agents, polyvinylpyridine-N-oxide and carbomethylcellulose, on the quartz and asbestos mineral fiber-induced production of reactive oxygen metabolites by human polymorphonuclear leukocytes. Archives Environ. Health 45, 8–14.
Kozin, F., Millstein, B., Mandel, G. and Mandel, N. (1982) Silica-induced membranolysis: A study of different structural forms of crystalline and amorphous silica and the effects of protein adsorption. J. Colloid Interface Sci. 88, 326–337.

Krauskopf, K.B. (1979) Introduction to Geochemistry, 2nd ed., McGraw-Hill, New York, 617 p.
Langer, A.M., Rubin, I.B. and Selikoff, I.J. (1972a) Chemical characterization of asbestos body cores by electron microprobe analysis. J. Histochem. Cytochem. 20, 723–734.
Langer, A.M., Rubin, I.B., Selikoff, I.J. and Pooley, F.D. (1972b) Chemical characterization of uncoated asbestos fibers from the lungs of asbestos workers by electron microprobe analysis. J. Histochem. Cytochem. 20, 735–740.
Langer, A.M., Wolff, M.S., Rohl, A.N. and Selikoff, I.J. (1978) Variation of properties of chrysolite asbestos subjected to milling. J. Toxic. Environ. Health 4, 173–188.
Light, W.G. and Wei, E.T. (1977) Surface charge and hemolytic activity of asbestos. Environ. Res. 13, 135–145.
Liversidge, A. (1893) On the origin of gold nuggets. J. Royal Soc. New South Wales 27, 303–343.
Lund, L.G. and Aust, A.E. (1990) Iron mobilization from asbestos by chelators and ascorbic acid. Arch. Biochem. Biophys. 278, 60–64.
Lund, L.G. and Aust, A.E. (1991) Mobilization of iron from crocidolite asbestos by certain chelators results in enhanced crocidolite-dependent oxygen consumption. Arch. Biochem. Biophys. 287, 91–96.
Marchisio, M.A. and Pernis, B. (1963) The action of vinylpyridine-polymers on macrophages cultivated in vitro in presence of tridymite dust. Grundfragen Silikoseforsch 6, 245–247.
Mogk, D.W. and Locke, W.W., III (1988) Application of Auger electron spectroscopy (AES) to naturally weathered hornblende. Geochim. Cosmochim. Acta 52, 2537–2542.
Mossman, B.T., Bignon, J., Corn, M., Seaton, A. and Gee, J.B.L. (1990) Asbestos: Scientific developments and implications for public policy. Science 247, 294–301.
Mossman, B.T. and Sesko, A.M. (1990) In vitro assays to predict the pathogenicity of mineral fibers. Toxicology 60, 53–61.
Nash, T., Allison, A.C. and Harington, J.S. (1966) Physio-chemical properties of silica in relation to its toxicity. Nature 210, 259–261.
Nesbitt, H.W. and Muir, I.J. (1988) SIMS depth profiles of weathered plagioclase, and processes affecting dissolved Al and Si in some acidic soil solutions. Nature 334, 336–338.
Nolan, R.P., Langer, A.M., Harington, J.S., Oster, G. and Selikoff, I.J. (1981) Quartz hemolysis as related to its surface functionalities. Environ. Res. 26, 503–520.
Nolan, R.P., Langer, A.M. and Herson, G.B. (1991) Characterisation of palygorskite specimens from different geological locales for health hazard evaluation. Brit. J. Indus. Med. 48, 463–475.
Ottery, J. and Gormley, I.P. (1978) Some factors affecting the haemolytic activity of silicate minerals. Ann. Occup. Hyg. 21, 131–139.
Palekar, L.D., Spooner, C.M. and Coffin, D.L. (1979) Influence of crystallization habit of minerals on in vitro cytotoxicity. Annals New York Acad. Sci. 330, 673–686.
Parks, G.A. (1967) Surface chemistry of oxides in aqueous systems. In: Equilibrium Concepts in Aqueous Systems, Stumm, W., eds., Advances in Chemistry Series, 67, Amer. Chem. Society, Washington, D.C., 121–160.
Parks, G.A. (1975) Adsorption in the marine environment. In: Marine Geochemistry, 2nd ed., Riley, J.P. and Skirrow, G., eds., 1, Academic Press, New York, 241–308.
Parks, G.A. (1990) Surface energy and adsorption at mineral/water interfaces: An introduction. In: Mineral-Water Interface Geochemistry, Hochella, M.F., Jr. and White, A.F., eds., Rev. Mineral. 23, 133–176.
Pott, F., Huth, F. and Friedrichs, K.H. (1974) Tumorigenic effect of fibrous dust in experimental animals. Envir. Health Persp. 9, 313–315.
Pratt, P.C. (1983) Lung dust content and response in guinea pigs inhaling three forms of silica. Arch. Environ. Health 38, 197–204.
Prügger, F., Mallner, B. and Schlipköter, H.W. (1984) Polyvinylpyridin-N-oxid (Bay 3504, P-204, PVPNO) in de Behandlung der Silkose des Menschen. Wein Klin. Wochenschr. 96, 848–853.
Reeves, A.L., Puro. H.E. and Smith, R. (1974) Inhalation carcinogenesis from various forms of asbestos. Environ. Res. 8, 178–202.
Rossi, F., Bellavite, G., Berton, G., Grzeskowiak, M., Papini, E. (1985) Mechanism of production of toxic oxygen radicals by granulocytes and macrophages and their function in the inflammatory process. Path. Res. Pract. 180, 136–142.
Schiller, J.E., Payne, S.L. and Khalafalla, S.E. (1980) Surface charge heterogeneity in amphibole cleavage fragments and asbestos fibers. Science 209, 1530–1532.
Schindler, P.W. (1990) Co-adsorption of metal ions and organic ligands: formation of ternary surface complexes. In: Mineral-Water Interface Geochemistry, Hochella, M.F., Jr. and White, A.W., eds., Rev. Mineral. 23, 281–308.
Schlipköter, H.W. and Beck, E.G. (1965) Observations on the relationship between quartz cytotoxicity and fibrogenicity while testing the biological activity of synthetic polymers. Med. Lav. 56, 485–493.
Schlipköter, H.W., Dolgner, R. and Brockhaus, A. (1963) The treatment of experimental silicosis. Ger. Med. Month 8, 509–514.

Schnitzer, R.J. (1974) Modification of biological surface activity of particles. Environ. Health Perspectives 9, 261–266.
Schott, J. (1990) Modeling of the dissolution of strained and unstrained multiple oxides: The surface speciation approach. In: Aquatic Chemical Kinetics: Reaction Rates of Processes in Natural Waters, Stumm, W., ed., John Wiley & Sons, New York, 337–366.
Schott, J. and Berner, R.A. (1983) X-ray photoelectron studies of the mechanism of iron silicate dissolution during weathering. Geochim. Cosmochim. Acta 47, 2233–2240.
Schott, J., Berner, R.A. and Sjöberg, E.L. (1981) Mechanism of pyroxene and amphibole weathering. I. Experimental studies of iron-free minerals. Geochim. Cosmochim. Acta 45, 2123–2135.
Shatos, M.A., Doherty, J.M., Marsh, J.P. and Mossman, B.T. (1987) Prevention of asbestos-induced cell death in rat lung fibroblasts and alveolar macrophages by scavengers of active oxygen speceis. Environ. Res. 44, 103-116.
Somorjai, G.A. (1981) Chemistry in Two Dimensions, Cornell University Press, Ithaca, New York, 575 p.
Somorjai, G.A. (1990) Modern concepts in surface science and heterogeneous catalysis. J. Phys. Chem. 94, 1013–1023.
Somorjai, G.A., Van Hove, M.A. and Bent, B.J. (1988) Organic monolayers on transition-metal surfaces: The catalytically important sites. J. Phys. Chem. 92, 973–978.
Sposito, G. (1984) The Surface Chemistry of Soils, Oxford University Press, New York, 234 p.
Stalder, K. and Stober, W. (1965) Haemolytic activity of suspensions of different silica modifications and inert dusts. Nature 207, 874–875.
Stanton, M.F., Layard, M., Tegeris, A., Miller, E., May, M. and Kent, E. (1977) Carcinogenicity of fibrous glass: Pleural response in the rat in relation to fiber dimension. J. Nat'l Cancer Inst. 58, 587–603.
Stanton, M.F., Layard, M., Tegeris, A., Miller, E., May, M., Morgan, E., and Smith, A. (1981) Relation of particle dimension to carcinogenicity in amphibole asbestoses and other fibrous minerals. J. Nat'l Cancer Inst., 67, 965–975.
Sunagawa, I. (1987) Surface microtopography of crystal faces. In: Morphology of Crystals, Sunagawa, I., ed., Terra Scientific Publishing, Tokyo, 323–365.
Thomassin, H., Goni, J., Baillif, P. and Touray, J.C. (1976) Etude par spectrométrie ESCA des premiers stades de la lixiviation du chrysotile en milieu acide organique. C. R. Acad. Sci. Paris 283, 131–134.
Thomassin, J.H., Goni, J., Baillif, P., Touray, J.C. and Jaurand, M.C. (1977) An XPS study of the dissolution kinetics of chrysotile in 0.1 N oxalic acid at different temperatures. Phys. Chem. Minerals 1, 385–398.
Veblen, D.R. (1980) Anthophyllite asbestos: Microstructures, intergrown sheet silicates, and mechanisms of fiber formation. Amer. Mineral. 65, 1075–1086.
Velbel, M.A. (1989) Weathering of hornblende to ferruginous products by a dissolution-reprecipitation mechanism: Petrography and stoichiometry. Clays Clay Minerals 37, 515–524.
Wagner, J.C., Berry, G. and Pooley, F.D. (1982) Mesotheliomas and asbestos types in asbestos textile workers: A study of lung contents. Brit. Med. J. 285, 603–606.
Wang, N.S., Jaurand, M.C., Magne, L., Kheuang, L., Pinchon, M.C. and Bignon, J. (1987) The interactions between asbestos fibers and metaphase chromosomes of rat pleural mesothelial cells in culture: A scanning and transmission electron microscopic study. Amer. J. Path. 126, 343–349.
Weitzman, S.A. and Graceffa, P. (1984) Asbestos catalyzes hydroxyl and superoxide radical generation from hydrogen peroxide. Arch. Biochem. Biophys. 228, 373–376.
Weitzman, S.A. and Weitberg, A.B. (1985) Asbestos-catalyzed lipid peroxidation and its inhibition by deferoxamine. Biochem. J. 225, 259–262.
Weitzman, S.A., Chester, J.F. and Graceffa, P. (1988) Binding of deferoxamine to asbestos fibers *in vitro* and *in vivo*. Carcinogenesis 9, 1643–1645.
White, A.F. and Yee, A. (1985) Aqueous oxidation-reduction kinetics associated with coupled electron-cation transfer from iron-containing silicates at 25°C. Geochim. Cosmochim. Acta 49, 1263–1275.
Wicks, F.J., Kjoller, K. and Henderson, G.S. (1992) Imaging the hydroxyl surface of lizardite at atomic resolution with the atomic force microscope. Can. Mineral. 30, 83–91.
Wiessner, J.H., Henderson, J.D., Jr., Sohnle, P.G., Mandel, N.S. and Mandel, G.S. (1988) The effect of crystal structure on mouse lung inflammation and fibrosis. Amer. Rev. Respir. Dis. 138, 445–450.
Yada, K. (1967) Study of chrysotile asbestos by a high resolution electron microscope. Acta Crystallogr. 23, 704–707.
Yoon, R.H., Salman, T. and Donnay, G. (1979) Predicting points of zero charge of oxides and hydroxides. J. Colloid Interface Sci. 70, 483–493.
Zalma, R., Bonneau, L., Guignard, J. and Pezerat, H. (1987a) Formation of oxy radicals by oxygen reduction arising from the surface activity of asbestos. Can. J. Chem. 65, 2338–2341.
Zalma, R., Bonneau, L., Guignard, J. and Pezerat, H. (1987b) Production of hydroxyl radicals by iron solid compounds. Toxic. Environ. Chem. 13, 171–187.

CHAPTER 9

LIMITATIONS OF THE STANTON HYPOTHESIS

Robert P. Nolan and Arthur M. Langer

*Environmental Sciences Laboratory
Brooklyn College
Avenue H and Bedford Avenue
Brooklyn, New York 11210 U.S.A.*

Quantum mechanics is very impressive. But an inner voice tells me that it is not yet the real thing. The theory has produced a good deal but hardly brings us closer to the secret of the old one.
— A. Einstein (1926)

INTRODUCTION

In the mid-1970s, experimental animal studies began to suggest that a relationship existed between the ability of different types of fibers to induce mesothelioma and their morphology (Pott et al., 1974; Stanton, 1974; Pott, 1978). The fiber types evaluated in the experimental animal models varied widely in elemental composition and had both crystalline and amorphous structures. These studies raised the possibility that the association between mesothelioma and asbestos might be more general and the disease could be caused by exposure to other types of fine fiber.

The injection and implantation of fibers into experimental animals found that the activity among the different fiber types increases with decreasing diameter and increasing fiber length. The results of these studies indicate the optimum morphology for the induction of intrapleural tumors by these routes of administration is a diameter ≤ 0.25 µm and a length >8 µm (Stanton et al., 1981). The relationship between a fiber's morphology and its activity for the induction of tumors is commonly referred to as the Stanton Hypothesis. The physico-chemical properties—surface properties and elemental composition—are thought to be important only in determining if the fiber will dissolve or be translocated from the lung. It seems unlikely that morphology and durability alone could define a fiber's ability to induce a mesothelioma.

Stanton and his colleagues published the final paper of their studies relating morphology to carcinogenic potency in 1981. Since then, additional experimental animal studies and epidemiological surveys have produced results which are inconsistent with the hypothesis. This paper reviews the experimental evidence produced by Stanton and his colleagues that forms the basis for the hypothesis and describes the limitations of its predictive value.

EARLY EXPERIMENTAL ANIMAL STUDIES RELATING FIBER MORPHOLOGY TO CARCINOGENICITY

Stanton and his colleagues, in 1969, reported the results of a study to develop a consistent and rapid method for inducing tumors in experimental animals. Several varieties of asbestos had already been shown to induce pleural tumors in rats, hamsters, and mice (Wagner, 1962; Smith, et al., 1965; Roe et al., 1967; Gross, et al., 1967). The primary objective of these early studies was to demonstrate that the mesothelial tumors found in asbestos-exposed individuals could be produced by asbestos in experimental animals. At this time, questions still existed about whether the asbestos itself was causing the disease or whether the problem was caused by contaminants associated with the mineral, such as cobalt, nickel, chromium or polyaromatic hydrocarbons (Cralley, et al., 1967; Harington, 1962).

Often in the early animal studies it was difficult for the experimental pathologist to induce a significant tumor incidence in a group of animals. There was considerable uncertainty about the dose to be used, and a latency of 1 to 2 years was required for the tumors to develop. Stanton, et al. (1969) experimented with three methods:

- A concentration of chrysotile as high as possible was suspended in an inert fluorocarbon. The suspension was then injected into rats intravenously at the maximum tolerated dose. The rationale was that the fluorocarbon would block the capillaries leading to the peripheral lung causing ischemic infarctions. The regenerating tissue at the site of the infarction had been shown in previous studies to be more susceptible to some carcinogens than normal tissue. Chrysotile asbestos produced no pulmonary tumors by this method.

- A mixture of asbestos, beeswax, and tricaprylin was deposited into the lung. The mixture was first liquefied by heating to 70°C and then injected into the lung through a 20-gauge needle after thoracotomy. The mixture is a firm solid at body temperature, and the slow release of asbestos from the wax pellet over time might enhance the carcinogenicity. No tumors were found in the lungs of any of these animals.

The two methods described above were designed to evaluate the pulmonary response and produced negative results. The third method was for the evaluation of pleural response:

- A pledget weighing ~45 mg and measuring ~30-mm × ~20-mm × ~3-mm was cut from a pad of coarse fibrous glass. Forty milligrams of crocidolite asbestos were deposited on the pledget from a 10% solution of gelatin. The gelatin hardened on drying, and the pledget was implanted over the surface of the left lung and pericardium by open thoracotomy. This method of administration allowed a large section of the mesothelial surface to be covered by asbestos, and the fiber glass pad acted as a nonspecific irritant. After 23 months, >70% of the rats surviving more than one month had mesothelioma at the site of implantation, whereas the fiber-glass pledget alone produced no tumors. The high tumor incidence caused by the asbestos on the glass pads could be used for quantitative studies. Although the percentage of tumor probability can be very high, the latency period had not been shortened, and two years from implantation was required for high tumor incidence to occur.

After the implantation model was developed using the coarse fiber-glass pad as a support, Stanton and Wrench, (1972) used the model to evaluate the carcinogenicity of 7 asbestos and 6 fibrous-glass samples. The asbestos specimens were the UICC standard reference samples of crocidolite, amosite, and Zimbabwe chrysotile. Two additional crocidolite specimens were also included, and these were obtained from the northwest Cape Province, South Africa, and from J.C. Wagner. The crocidolite specimens were milled using various conditions to determine if metal fragments from the mills were contributing to the sample's carcinogenic activity.

Among the asbestos reference samples, the earliest mesothelioma occurred in the amosite treated group after 54 weeks. The first mesothelioma occurred in animals exposed to amosite rather than in the animals exposed to chrysotile or crocidolite, which was the same order as reported by Wagner (1962) and Berry and Wagner (1969). The three UICC reference samples tested were found to be potent carcinogens, the activities of which were indistinguishable under the conditions of evaluation (Table 1). The UICC, northwest Cape Province, and Wagner crocidolite samples produced almost identical results (Table 2). The UICC and northwest Cape Province samples were ball milled to reduce the fiber length as were the glass samples, which were very long. Mild milling of the northwest Cape Province sample had no detectable effect on its carcinogenicity, whereas the UICC crocidolite sample was markedly less carcinogenic after extensive milling. Analysis of the extensively milled sample by X-ray diffraction indicated the area beneath the peaks characteristic of crocidolite were "reduced." Stanton attributed the change in area to fiber size although a loss in crystallinity may also have occurred. This sample contained no particles >2.5 µm in diameter or longer than 20 µm. Approximately 85% of the particles visible by light microscope appeared as non-fibrous clumps <2.5 µm that contained short fibers visible by transmission electron microscopy.

The coarse coated fibrous glass pads alone were implanted in 90 animals as controls, 58 of which survived long enough to be at risk for mesothelioma. None of the long-term survivors developed the tumor. Two glass samples (each of narrower diameter than the coarse fibrous glass), Pyrex and old glass wool, each induced 1 mesothelioma in 25 animals (Table 3). The fine diameter AAA fiber glass was evaluated uncoated and coated with a urea-formaldehyde resin. The starting AAA glass fibers were much longer than the asbestos samples and were milled to shorten the length to one comparable with a medium-size range asbestos fiber sample. These fine diameter fiber glass samples induced 8 mesotheliomas in 54 animals. These data taken together indicate a higher mesothelioma incidence ($P < 0.05$) than found with the intact coarse glass vehicle or the milled glass fiber

Table 1

Comparison of three UICC asbestos reference samples and crocidolite from the northwest Cape Province, South Africa

Sample	Dose (mg)	Incidence of Mesothelioma	Comment
chrysotile (UICC, Zimbabwe)	40	15/26 (57.7%)	high fibrosis at implant site
amosite (UICC)	40	15/25 (60.0%)	high fibrosis at implant site
crocidolite (UICC)	40	14/23 (60.9%)	high fibrosis at implant site
crocidolite (J.C. Wagner preparation)	40	15/20 (75.0%)	high fibrosis at implant site

Source: Stanton and Wrench (1972)

Table 2

Comparison of crocidolite samples from UICC, Northwest Cape Province, and Wagner

Dose (mg)	Incidence of mesothelioma			Pleural fibrosis			Comment
	UICC	NW Cape	Wagner	UICC	NW Cape	Wagner	
1	2/25 (8.0%)	4/30 (13.3%)	—	++	+	—	
2	5/23 (21.7%)	—	—	++	—	—	
10	11/27 (40.7%)	—	—	+++	—	—	
10	9/21 (42.9%)	—	—	++++	—	—	dose administered in saline, no glass pad
20	12/25 (48.0%)	10/24 (41.7%)	—	++++	+++	—	
40	14/23 (60.9%)	18/27 (66.7%)	15/20 (75.0%)	++++	++++	++++	hand milled
40	8/25 (32.0%)	15/23 (65.2%)	—	+++	++++	—	ball milled partially pulverized

Source: Stanton and Wrench (1972)

samples of larger diameter. The higher tumor incidence was associated with more severe pleural fibrosis than was found with the large diameter glass.

The interim results of further experiments on chrysotile, crocidolite, fibrous glass, and aluminum oxide (Al_2O_3) were published by Stanton in 1973. The dose remained constant at 40 mg and was delivered on a coarse glass pad as in his earlier studies. The incidence of mesothelioma was divided into four groups: high (>40%), moderate (20 to 30%), low (10 to 20%), and negligible (<5%). The UICC crocidolite and chrysotile (Zimbabwe), size fractionated fine fibrous glass (>10 μm × 1μm) and aluminum oxide whiskers all produced a high response

(Table 4). When crocidolite and chrysotile samples occurred as fiber bundles or as long fibers, the mesothelioma incidence was reduced to moderate. A sample produced by shearing a fibrous glass sample that had caused a high tumor incidence and a different fiber glass sample with a 3-μm diameter and length >10 μm both produced a moderate response. Pulverizing the crocidolite, chrysotile, and fibrous glass samples further reduced the mesothelioma incidence to a low or negligible incidence. A non-fibrous aluminum oxide sample and whole glass fibers produced a negligible tumor incidence.

Stanton and his colleagues interpreted their results as indicating non-asbestos fibers such as glass and aluminum oxide are carcinogenic if their size

Table 3

Comparison of six fibrous glass samples

Sample	Dose (mg)	Incidence of mesothelioma	Pleural fibrosis	Comment
coarse coated	40	0/58 (<1.7%)	+	1–25 μm in diameter
coarse coated	40	1/24 (4.2%)	++	1–25 μm in diameter, partially pulverized
Pyrex, coated	40	1/25 (4.0%)	+	<10 μm in diameter, partially pulverized
old wool, uncoated	40	1/25 (4.0%)	+	~1–15 μm in diameter
fine AAA, uncoated	40	3/26 (11.5%)	+++	0.06–3 μm in diameter, ~5 μm in length, partially pulverized
fine AAA, coated	40	5/28 (17.9%)	+++	0.06–3 μm in diameter, ~5 μm in length, partially pulverized

Source: Stanton and Wrench (1972)

Table 4

Interim results of a study of various asbestos and non-asbestos fibers

Specimen	Incidence of mesothelioma			
	High (>40%)	Moderate (20%-30%)	Low (10-20%)	Negligible (<5%)
crocidolite	UICC, untreated	crude fiber	partly pulverized	fully pulverized
chrysotile, Zimbabwe	UICC, untreated	long fiber	—	fully pulverized
commercial	—	—	—	whole fibers
fine fibrous glass	crude, separated ~>10 μm x 1 μm	sheared, separated >10 μm x 3 μm	partly pulverized	Separated 3 μm x <5 μm
aluminum oxide	whiskers	—	—	non-fibrous

Source: Stanton (1973)

range approaches that of asbestos. The carcinogenicity of the aluminum oxide whiskers was taken as further evidence that fibers of diverse elemental compositions can produce mesotheliomas in experimental animals and therefore chemistry plays no role. The reduction in the carcinogenicity of crocidolite, chrysotile, and glass fiber after pulverization further emphasized the importance of the fibrous structure in tumor induction.

The experimental studies showing fibrous glass to be carcinogenic were further extended by studying the relationship between carcinogenicity and fiber dimension for 17 fibrous glass samples (Stanton et al., 1977). Three samples of borosilicate glass fibers having nominal diameters of <1 µm, <2 µm, and <5 µm were made in either a flame attenuated or rotary process. Each of the three samples was fractionated into long and short size ranges by suspending the fibers in water and then breaking them in a blender. After blending, the long and short fibers were separated by sedimentation and centrifugation. The sample with a nominal diameter <2 µm was unfractionated.

The flame-attenuated AAA borosilicate glass (with a nominal diameter range of 0.06 to 3.0 µm and which had been used in two earlier studies) was further evaluated. The unfractionated fiber, a modestly sheared sample, and four samples size reduced in a Spex ball-mill (two samples were run for the same length of time) were prepared and tested.

Three separate batches of the type commonly used commercially (with diameters in the range of 5 to 25 µm) were evaluated by implantation. All the samples required length reduction for application and ease of size measurement. These three samples included a mineral wool and the coarse glass fibers that were used for the pledget in the implantation studies.

The probability of mesothelioma induction fell into three groups: (1) high risk was associated with samples composed of intact fibers or of long fiber length fraction of fine diameter; (2) intermediate risk was associated with the finest diameter fiber type that had been shortened by milling and the long fiber preparation with diameters >5 µm; and (3) low risk was associated with large diameter fibers or very short fibers (Table 5). The results within any one of the three risk groups were the same within statistical error, although the risk groups differed significantly from each other and the low risk group was not statistically different from the treated control group.

In 1978, Stanton and Layard reported the results of studies on other non-asbestos fibers that cause pleural mesothelioma when surgically implanted on a glass vehicle. There were 7 dawsonites, 2 potassium octatitanates, 1 silicon carbide, 1 borosilicate glass, 7 aluminum oxides, 1 aluminum nitride/oxide and 1 nickel titanate. The dose was the same as used in earlier studies, 40 mg, and the probability of mesothelioma induction ranged from 0 to 100% (Table 6). At least one specimen of dawsonite, silicon carbide, aluminum oxide produced a >60% tumor probability. The size distributions for these samples are reported in Stanton et al. (1981). This paper also contains many of the results from the earlier studies in a large summary of data on 72 samples in support of the Stanton Hypothesis

Table 5

Seventeen experiments with four types of fibrous glass, three of which synthesized at different diameters, some were size fractionated

Description of sample/Stanton Code	Tumor incidence	Probability of mesothelioma	Log f/μg[†]
High Risk			
Borosilicate[1]/nominal diameter <1 μm, long length/MOL	9/17	85.3±13.2	5.16
Borosilicate[2]/AAA, nominal diameter range 0.06–3.0 μm, modest shearing long length/KL	20/29	73.9±8.5	3.59
Borosilicate[1]/nominal diameter <2 μm, long length/M6L	18/29	71.2±9.1	4.02
Borosilicate[2]/AAA, nominal diameter, range 0.06–3.0 μm, unfractionated/KW	16/25	69.3±9.6	3.00
Borosilicate[2]/nominal diameter <2 μm, unfractionated/ M6W	7/22	64.4±17.7	4.01
Intermediate Risk			
Borosilicate[2]/AAA, nominal diameter range 0.06–3.0 μm/Spex ball-milled, 30 sec/coated with heat cured urea-formaldehyde resin/KCP	5/28	21.5±8.7	2.50
Borosilicate[2]/AAA, nominal diameter range 0.06–3.0 μm/Spex ball milled, 30 sec/KUP	3/26	19.4±10.3	3.01
Borosilicate[1]/nominal diameter <5 μm, long length/M8L	2/28	14.3±9.4	1.85
Low Risk			
Borosilicate[1]/nominal diameter <1 μm, short length/MOS	2/27	8.3±5.6	—
Borosilicate[2]/AAA, nominal diameter range 0.06–3.0 μm/Spex mill, 2 min/K2P	2/28	8.1±5.5	—
Mineral Wool derived from silica slag/O2P	1/25	6.7±5.4	—
Borosilicate/AAA, nominal diameter range 0.06–3.0 μm/Spex mill, 10 min/KFP	1/27	5.9±5.5	—
Fibrous Glass/used in filtration, higher NaO and CaO than in conventional borosilicate glass/P2P	1/25	5.7±5.5	—
Modern insulation-type fiber glass/used for glass pads/Spex milled, 2 min/phenol formaldehyde coated Y2P	1/24	5.5±5.9	1.30
Borosilicate[1]/nominal diameter <5 μm, short length/M8S	1/29	4.5±4.4	—
Borosilicate[1]/nominal diameter <2 μm, short length/M6S	0/28	0	—
Modern insulation-type fiber glass/used for glass pads/phenol-formaldehyde coated/YW	0/115	0	—

Sources: Stanton et al. (1977); Stanton et al. (1981)
[†] Fibers were those particles with diameters ≤0.25 μm and lengths >8.0 μm. Log was calculated as common log.
[1] Flame attenuated or rotary processed
[2] Flame attenuated

(Table 7). Stanton et al. ((1981) tested numerous samples of several of the amphibole asbestos minerals, including crocidolite (13 samples), tremolite (2 samples) and amosite (1 sample). Data for chrysotile and anthophyllite asbestos were not reported.

Table 6

Probability of mesothelioma induction for selected non-asbestos fibers[1]

	Tumor incidence	Tumor probability (%)	Log $f/\mu g$[†]
Dawsonite V, dihydroxy sodium aluminum carbonate	26/29	100	4.94
Potassium Octatitanate I	21/29	100	4.94
Potassium Octatitanate II	20/29	100	4.70
Silicon carbide GTC #1	17/26	100	5.15
Dawsonite, dihydroxy sodium aluminum carbonate I	20/25	95	4.66
Borosilicate glass (M6D)	12/31	77	4.29
Aluminum Oxide - HC	15/24	70	3.63
Dawsonite, dihydroxy sodium aluminum carbonate VII	16/30	68	4.71
Dawsonite, dihydroxy sodium aluminum carbonate IV	11/26	66	4.01
Dawsonite, dihydroxy sodium aluminum carbonate III	9/24	66	5.73
Aluminum oxide #3	8/27	44	2.95
Aluminum oxide #4 alpha	9/27	41	2.47
Aluminum nitride + aluminum oxide #6 alpha	4/25	28	2.60
Aluminum oxide #2	4/22	22	3.73
Aluminum oxide #4	2/28	13	0.82
Dawsonite, dihydroxy sodium aluminum carbonate VI	3/30	13	—
Dawsonite, dihydroxy sodium aluminum carbonate II	2/27	12	—
Aluminum oxide #5	1/25	5	—
Aluminum oxide-LC (non-fibrous)	1/28	3	—
Nickel titanate	1/28	8±8.0	—

Sources: Stanton and Layard (1978); Stanton et al. (1981)
[†] Fibers were those particles with diameters ≤0.25 μm and lengths >8.0 μm. Log was calculated as common log.
[1] Fibrous glass samples previously reported in Stanton et al. (1977) were omitted.

Dosimetry of fibers

Generally Stanton and his colleagues used a single dose of 40 mg to determine the carcinogenic potency of each sample. The morphology was determined using a combination of light and transmission electron microscopies. A transmission electron microscopy (TEM) grid of each sample was prepared and a series of grid openings were examined by light microscopy at various magnifications. Representative grid openings were photographed at a magnification appropriate for the size distribution of the fibers present. To determine the presence of fibers below the resolution of the light microscopy (at the magnification used), the same grid opening was examined using TEM at higher magnifications. If fibers were observed by TEM but had not been seen by light microscopy, representative areas of the grid would be photographed. The size distributions of all the fibers present were determined using this combination of light and electron microscopies.

The thirty-four size categories were selected to describe the morphologies of the fibers found. Of the 34 size categories, on average 7 had an aspect ratio <3

Table 7

Selected samples from Stanton et al. (1981) grouped by mineral name[1]

Mineral	Tumor incidence	Tumor probability (%±SD)	Log f/μg[†]	Number fibers/μg ≤0.25 μm x >8 μm	Total number fibers/μg
Crocidolite					
1	18/27	94±6.0	5.21	162,181	5.88×10^6
2	17/24	93±6.5	4.30	19,952	1.68×10^6
3	15/23	93±6.9	5.01	102,329	6.57×10^6
4	15/24	86±9.0	5.13	134,896	3.06×10^5
5	14/29	78±10.8	3.29	1,950	3.54×10^5
6	9/27	63±13.9	4.60	39,811	9.69×10^5
7	11/26	56±11.7	2.65	447	8.13×10^9
8	8/25	53±12.9	—	0	1.09×10^6
9	8/27	33±9.8	4.25	17,783	8.17×10^5
10	6/29	37±13.5	3.09	1,230	6.21×10^5
11	4/29	19±8.5	—	0	6.84×10^3
12	2/27	10±.0	3.73	5,370	9.28×10^4
13	0/29	0	—	0	2.74×10^5
Tremolite					
1	22/28	100	3.14	1,380	1.41×10^5
2	21/28	100	2.84	692	6.86×10^4
Amosite					
1	14/25	93±.7.1	3.53	3,388	4.04×10^4
Wollastonite					
1	5/20	31±.12.5	—	0	2.57×10^4
3	3/21	19±.10.5	—	0	3.75×10^4
2	2/25	12±.8.0	—	0	6.38×10^4
4	0/24	0	—	0	7.1×10^2
Halloysite					
1	4/25	20±.9.0	—	0	1.79×10^7
2	5/28	23±.9.3	—	0	6.46×10^6
Talc					
1	1/26	7±.6.9	—	0	1.26×10^5
3	1/29	4±.4.3	—	0	1.26×10^5
2	1/30	4±.3.8	—	0	7.43×10^3
4	1/29	5±.4.9	—	0	1.90×10^5
5	0/30	0	—	0	2.67×10^5
6	0/30	0	3.3	1,995	3.76×10^5
7	0/29	0	—	0	9.75×10^5

Source: Stanton et al. (1981)
[†] Fibers were those particles with diameters ≤0.25 μm and lengths >8.0 μm. Log was calculated as common log.
[1] List is not complete and data reported in earlier tables was not repeated.

Table 8

Size intervals used by Stanton et al. (1981) to characterize experimental samples

Diameter	Length				
	<0.1–1.0 μm	>1.0–4 μm	>4–8 μm	>8–64 μm	>64 μm
>8.0 μm	—	—	—	P/F	P/F
>4.0–8.0 μm	—	—	P	F	F
>2.5–4.0 μm	—	P	P	F	F
>1.5–2.5 μm	—	P	F	F	F
>0.50–1.5 μm	P[1]	P	F*	F*	F*
>0.25–0.5 μm	P	F	F*	F*	
>0.10–0.25 μm	F[2]	F	F*	F*	
>0.05–0.10 μm	F	F	F*	F*	
>0.01-0.05 μm	F	F	F*	F*	

Source: Stanton et al. (1981)
* Relatively good correlation coefficients of logit of tumor probability with common logarithm or number of particles per microgram in different dimensional ranges.
P   Particle
F   Fiber

and were therefore considered particles not fibers; two categories included fibers and/or particles; and the rest could only be fibers (Table 8). The diameter and length of each fiber was measured and placed into one of the 34 size categories. Stanton et al. (1981) indicated that in most cases at least 1,000 fibers were measured for each sample. However, the average number for any single sample or the average of the series was not given.

Once the number of fibers in each category was determined it was assumed that the size distribution was normal for the diameter and length range of the category and that the particles within a given category had a diameter and length equivalent to the mean. For example, for the category with a diameter range of >1.5 to 2.5 μm and a length range of >4.0 to 8.0 μm, all of the fibers were assumed to be 2 μm × 6 μm. The mass of the average fiber was then determined by assuming it to have a circular cross-section and a given density. The number of particles or fibers in each category was not reported nor the densities used to convert the number of fibers in each interval to the number of fibers per microgram (see Table 9 for typical densities for some of the samples). Re-analysis of the data collected for any given sample was quite reproducible. However, if the samples were re-prepared, the results could be very different. Stanton et al. (1981) indicated confidence in the size data of about any order of magnitude in the number of fibers per microgram. Therefore, the results are given as the common logarithm of the number of fibers per microgram.

Table 9

Densities of commonly used fibers

Mineral fibers		Man-made vitreous fibers		Synthetic organic fibers	
Specimen	Density (g/cm³)[a]	Specimen	Density (g/cm³)[b]	Specimen	Density (g/cm³)[b]
		alumina	3.15	aramid (Nomex)	1.38
amosite	3.54	alumina-silicate-		Dolan 10	1.16
anthophyllite	3.09	zirconia	2.63	Kevlar 29	1.44
chrysotile	2.53	boron nitride	1.90	Kevlar 49	1.45
crocidolite	3.40	glass	2.50	polyacylonitrite	1.7
erionite	2.02	mineral wool	2.55		
halloysite	2.11	silica	2.19	polyvinylalcohol	1.3
palygorskite	2.4	silicon carbide	3.21	teflon	2.1
tremolite-actinolite	2.9–3.2	silicon nitride	3.18		
wollastonite	2.8–3.1	xonotlite	2.7		

[a] Joint Committee on Powder Diffraction
[b] Hodgson (1985) and references therein

Three types of control groups were used to determine the background incidence of mesothelioma in the animals so that the incidence in the exposed animal could be evaluated. The three types were untreated, open thoracotomy and the application of a non-carcinogenic material either applied to the pleura or implanted into the lung. Pleural tumors in the untreated female rats were essentially non-existent. Combining the three control groups and calculating the incidence of mesothelioma by life-table method, a background of 7.7±4.3% was determined. A tumor probability of 30% or greater in any group of experimental animals was required for the result to be of significance.

Conclusions of the implantation studies

Stanton and his colleagues, from their surgical implantation studies, generated the hypothesis that a fiber's morphology is the principle determinant of its carcinogenic activity. Of the 72 fibrous samples included in the group's final report (Stanton, 1981), 30 produced a significant increase in the tumor probability. When the 34 size categories used to describe the fiber size distribution were compressed into 11 categories, the highest correlation coefficient was for the fibers with diameters ≤0.25 µm and lengths >8 µm. Additionally a high tumor probability was correlated with the number of fibers in other categories having diameters ≤1.5 µm and lengths >4 µm (Table 10). Examination of the data indicates that as the diameter increases and the length decreases, i.e. the fiber shape is lost for a given mass, the probability of tumor induction falls (see Pott, 1978, for a discussion).

If a particular fiber sample has a size distribution associated with a high tumor probability, the durability of the fiber should be considered. Is the fiber stable enough to be able to remain in the tissue for a length of time sufficient to allow transformation to occur? Initially Stanton et al. (1981) considered lint,

Table 10

Correlation coefficients of logit of tumor probability
with common log of number of particles/μg

	Length		
Diameter	≤4 μm	> 4–8 μm	> 8 μm
> 4.0 μm	—	–0.28	–0.30
> 1.5–4.0 μm	–0.45	–0.24	0.13
> 0.25–1.5 μm	–0.01	0.45	0.68
≤ 0.25 μm	–0.20	0.63	0.80

Source: Stanton et al. (1981)

Table 11

Surface-area/fiber, mass/fiber, and aspect ratio as a function of fiber diameter
for 8-μm-long fiber with square or circular cross section

	Diameter					
	0.05	0.10	0.25	0.50	1.00	2.00
Surface area/fiber (μm^2)						
circular cross-section	1.26	2.53	6.38	12.96	26.70	56.55
square cross-section	1.61	3.22	8.13	16.50	34.00	72.00
% of surface area on fiber end	0.31	0.62	4.00	3.03	5.88	11.11
Mass/fiber (10^{-6} μg)[1]			Width			
erionite	0.040	0.160	1.00	4.00	16.00	64.00
chrysotile	0.051	0.202	1.265	5.06	20.24	80.96
crocidolite	0.068	0.272	1.700	6.80	27.20	108.80
Aspect ratio	160	80	32	16	8	4

[1] Mass given in 10^{-6} μg assuming 2.00, 2.53, and 3.40 g/cm^3 for the approximate densities of erionite, chrysotile, and crocidolite, respectively. Calculation assumed a square cross-section and a length of 8 μm.

gypsum, and carrageenan to be non-durable. Latter investigators questioned whether glass fibers and chrysotile were sufficiently non-durable to reduce their carcinogenicity *in vivo*. Even today no consensus exists concerning how long a fiber must remain in the tissue to induce a tumor (see Pott, 1987, for a discussion).

Stanton and coworkers recognized that the variations in size distribution and density could lead to a large range in surface area for the fiber samples, even though equivalent mass doses were used. Two glass samples, Glass 1 (MOL) and Glass 10 (MOS), were reported to be of the same origin, the only difference being Glass 10 (MOS) was milled to a smaller fiber length. The two samples therefore differed only in surface area which was formed by breaking across the fiber axis. For fine diameter fibers the contribution of the fiber ends to the surface area is minimal (see Table 11). Glass fiber is rarely polyfilamentous, and longitudinal splitting common in asbestos minerals seldom occurs. Glass 1 (MOL) with the long fibers produced a tumor probability of 85±13.2% while for the milled samples the probability fell to 8±5.6% (Table 5). Five crocidolite samples (6,7,11,12,13) were prepared by various milling, sedimentation, and flotation methods from a single sample of the UICC reference crocidolite (5) (Table 7). Stanton et al. (1981) concluded that all five of the samples would have similar surface areas and the variation in their percent tumor probability could be attributed to differences in morphology.

Limitations of the Stanton hypothesis

In early experimental animal studies, the inoculation of the asbestos minerals amosite, chrysotile, and crocidolite produced mesotheliomas in rats, hamsters, and mice (Wagner, 1962; Smith et al., 1965; Roe et al., 1967; Wagner and Berry, 1969). Intrapleural and intraperitoneal injection of free fibers had already been shown to produce tumors experimentally when Stanton and his colleagues began to develop their surgical implantation method (Stanton et al., 1969). The method developed by Stanton and coworkers—i.e., the implantation of a pledget with a 40-mg dose of test fibers in a matrix of hardened gelatin— has only been used by one other investigator in order to evaluate a fiber being considered for commercial use as an asbestos substitute (Nair, 1987).

The primary purpose of the pledget was to increase the effective dose by localizing it to a relatively small area of the lung surface, thereby increasing the tumor incidence. However, intraperitoneal injection of free fiber can produce a high incidence, making the pledget an unnecessary confounder (Wagner et al., 1973; Wagner et al., 1985; Pott et al., 1987; Davis et al., 1991). The administration of the dose on a pledget by surgical implantation has the following limitations:

- Encasing the fibers in a gelatin matrix may reduce any contribution of the fiber surface properties make to the carcinogenic activity. For example, actinolite asbestos with 2-polyvinylpyridine-N-oxide hydrogen bonded to its surface produced fewer mesotheliomas with a longer latency period than actinolite without the polymer (Pott et al., 1989).

- In addition to the probability of tumor induction, the time required to produce a tumor (the latency period) is also important in evaluating a fiber's carcinogenic activity. The latency period required for the production of a high tumor yield, two years after the implantation of the pledget, did not indicate that localizing the dose on the glass pad significantly enhanced the carcinogenic activity of the various test fibers. Release from the gelatin matrix may be required for the transformation process to begin. Results from the artificial route of administration (surgical implantation) are made more difficult to interpret due to the gelatin and the glass pad. It is also possible that the rates of release of the different fiber types from the pad may not be uniform.

- The number of tumors and level of fibrosis caused by surgical implantation of the pledget alone was low. The mass implanted when a fiber was tested (40-mg dose and 45 mg of pledget) was not always the mass of the negative control. The possibility of non-specific irritation caused by the pledget, interacting synergistically with the test fibers to increase the fibrogenic and possibly carcinogenic response is difficult to evaluate in the implantation model. Intraperitoneal injection of low doses of mineral fibers can produce tumors with little or no fibrosis (Pott et al., 1987). Surgical implantation of the test fiber on a pledget commonly induces high levels of fibrosis with tumor induction.

Beyond any questions concerning the route of administration, the 72 samples in the Stanton et al. (1981) report were evaluated at a single dose of 40 mg. These samples are primarily fibrous in morphology, and the size distribution and density would determine their surface areas (Table 11). It is important to consider that just as a sphere has the least surface area per unit volume, a fiber forms the largest surface area per unit volume of any shape (particularly as the diameter becomes infinitely small) (see Lowell, 1979, for discussion). The surface area per mass (or volume) of a fiber sample increases linearly as the fiber diameter decreases. For example, a sample of 1-μm diametered fibers would have a ~10-fold greater surface area per mass than a sample of 0.1-μm diametered fibers (assuming each has fibers with identical lengths). The surface area of a fiber with a circular cross-section has ~21.5% less surface area per mass than a similarly sized fiber with a square cross-section.

It is experimentally difficult to deliver a dose at a constant surface area, particularly for fiber samples with a range of size distributions and densities. Stanton et al. (1981) decided instead to evaluate the importance of surface area by taking a sample and reducing the fiber length by milling. The comparison of milled and unmilled samples would allow the evaluation of surface area, because the surface area of the two samples should be very similar (assuming fibers did not break longitudinally). It was assumed that the milling did not alter the fiber surfaces, which is probably a poor assumption (see Langer et al., 1978, for a discussion). Glass 1 (MOL) and Glass 10 (MOS) were supposed to represent unmilled and milled versions of the same material.[1] Glass 1 (MOL) produced a tumor probability of 85±13.2%, with ~145,000 fibers (≤0.25 μm × >8 μm) per μg (Table 12), whereas Glass 10 (MOS) produced 8±5.6% with no fibers (≤0.25 μm × >8 μm) being reported in the dose (Table 13).

The size distribution of Glass 1 (MOL) and Glass 10 (MOS) are shown in Tables 12 and 13. The data indicate that it is unlikely that these two sample differed only in surface area. Glass 10 (MOS) contained ~27-fold more particles than Glass 1 (MOL) and a number of particles with diameters >1.5 μm were present, whereas none were detected in Glass 1 (MOL). Greater than 85% of the particles in Glass 10 (MOS) were in the smallest of the 34 categories (a diameter of >0.01 to 0.05 μm and a length 0.1 to 4 μm), whereas in the Glass 1 (MOL) no particles were present in that category.

---

[1] Review of the description of the glass samples MOL and MOS in Stanton et al. (1977) indicates Glass 10 (MOS) was not a milled sample of Glass 1 (MOL). Although the two samples were from a single starting batch with a nominal diameter of <1 μm, it appears that a single sample was separated into a long and short fraction, and, therefore, their surface areas per mass need not be the same.

Table 12

Fiber size distribution of Glass 1 by number of particles/µg in 34 dimensional categories

Diameter	Length				
	>0.1–1.0 µm	>1–4 µm	>4–8 µm	>8–64 µm	>64 µm
>4.00–8.00 µm	—	—	—	—	—
>2.50–4.00 µm	—	—	—	—	—
>1.50–2.50 µm	—	—	—	—	—
>0.50–1.50 µm	—	1.70×10^2	3.39×10^2	1.70×10^3	1.20×10^3
>0.25–0.50 µm	—	1.20×10^3	—	8.91×10^4	
>0.10–0.25 µm	—	8.51×10^2	2.24×10^3	3.39×10^4	
>0.05–0.10 µm	—	—	8.51×10^3	6.17×10^4	
>0.01–0.05 µm	—	—	2.88×10^3	4.47×10^4	

Total concentration of particles was 248,490/µg; total concentration of fibers was 248,320/µg; total range for all categories was 170–89,100.

Table 13

Fiber size distribution of Glass 10 by number of particles/µg in 34 dimensional categories

Diameter	Length				
	>0.1–0.5 µm	>1–4 µm	>4–8 µm	>8–64 µm	>64 µm
>4.00–8.00 µm	—	—	2.88×10^2	524.8	—
>2.50–4.00 µm	—	9.33×10^2	9.77×10^2	57.5	—
>1.50–2.50 µm	—	2.69×10^3	2.34×10^2	14.8	—
>0.50–1.50 µm	7.59×10^3	8.13×10^3	5.75×10^2	2.88×10^2	—
>0.25–0.50 µm	2.19×10^4	1.05×10^4	4.90×10^3	2.34×10^3	
>0.10–0.25 µm	2.69×10^4	7.59×10^3	—	—	
>0.05–0.10 µm	7.90×10^5	1.55×10^4	—	—	
>0.01–0.05 µm	5.89×10^6	4.27×10^4	—	—	

Total concentration of particles was 6.79×10^6/µg; total concentration of fibers was 6.79×10^6/µg; total range for all categories was 14.8–5.89×10^6.

---

starting batch with a nominal diameter of <1 µm, it appears that a single sample was separated into a long and short fraction, and, therefore, their surface areas per mass need not be the same.

The experiments with UICC crocidolite (Exp. 6, 7 vs. Exp. 11, 12, 13) are referred to by Stanton et al. (1981) as providing additional experimental data that morphology rather than surface area is the primary determinant of carcinogenic activity. These five samples were described by Stanton et al. (1981) as being prepare by milling, sedimentation and flotation from a single sample of UICC reference crocidolite.

Stanton et al. (1981) indicated that Crocidolite 6 and 7 were being compared to Crocidolite 11, 12, and 13. Although not clearly specified, it is implied by their lower tumor probabilities that Crocidolite 11, 12, and 13 are the milled samples. For completeness the UICC reference starting sample Crocidolite 5 and Crocidolite 8 (also prepared from the UICC reference standard) are included in Table 7 and Figure 1. The pattern of the data is not what would be qualitatively expected, the total number of particles per µg is the lowest in the three milled samples, and the reduction in fiber length by breaking across the fiber axis and separation of the fiber bundles should increase the fiber number.

The total number of fibers in Crocidolite 7 is almost 8400-fold larger than Crocidolite 6, and no explanation is offered. An examination of data indicates the 8.13×10^9 fibers/µg are in the size category with a diameter of >0.01 to 0.50 µm and a length of >4 to 8 µm. The remainder of the size categories contain 1811 fibers/µg, and two categories contain only 18 fibers/µg. It seems unlikely that crocidolite 6 and 7 would have a surface area similar to Crocidolite 11, 12 and 13. The number of fibers ≤0.25 µm × >8 µm compared to the total number of particles present is quite small and their relationship to the percent tumor probability is not striking (Table 7 and Figure 1).

A review of the percent tumor probability of the thirteen crocidolite samples and the number of fibers with diameters ≤0.25 µm and lengths >8 µm indicates the probability remains between ~93 and 94%, whereas the number of fibers (measuring ≤0.25 µm × >8 µm) per µg increases from ~20,000 to ~166,000. Crocidolite 8 produced a greater than 50% tumor probability and was reported to contain no fibers ≤0.25 × >8 µm. For the 13 samples, on average, less than 5% (range 0 to 43.8%) of the particle population was ≤0.25 µm × >8 µm. The two tremolite asbestos samples which had a percent tumor probability of 100% had 1380 and 692 fibers ≤0.25 µm × >8 µm per µg, whereas the Talc 6 sample (with 1995 particles measuring ≤0.25 µm × >8 µm per µg) had a 0% tumor probability fibers (Fig. 1). A high correlation coefficient may exist between the number of fibers with a diameters ≤0.25 µm and lengths >8 µm and the percent tumor probability for the 72 samples tested. Although for any given sample the presence of these fibers have little predicative value for either the qualitative or quantitative determination of tumor incidence.

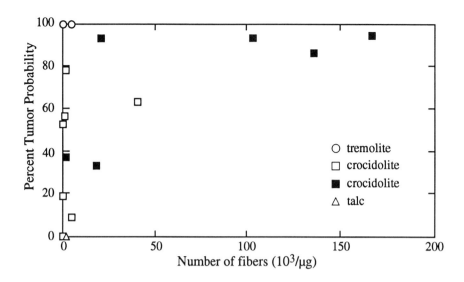

Figure 1. Number of fibers ≤0.25μm × >8μm/μg plotted against the percent tumor probability for the 13 crocidolite samples (open squares are UICC crocidolite preparations), the 2 tremolite asbestos samples, and the talc 6 sample.

REFERENCES

Cralley, L.J., Keenan, R.G., and Lynch, J.R. (1967) Exposure to metals in the manufacture of asbestos textiles products. Amer. Indus. Hygiene Assoc. J. 28, 452–461.

Davis, J.M.G., Bolton, R.E., Miller, B.G., and Niven, K. (1991) Mesothelioma dose-response following intraperitoneal injection of mineral fibres. Int'l J. Exp. Pathol. 72, 263–274.

Gross, P., deTreville, R.T.P., Tolker, E.B., Kaschak, M., and Babyak, M.A. (1967) Experimental asbestosis: The development of lung cancer in rats with pulmonary deposits of chrysotile asbestos dust. Arch. Environ. Health 18, 343–355.

Harington, J.S. (1962) Natural occurence of oils containing 3,4-benzpyrene and related substances in asbestos. Nature 193, 43–45.

Langer, A.M., Wolff, M.S., Rohl, A.N., and Selikoff, I.J. (1978) Variation of properties of chrysotile asbestos subjected to milling. J. Toxicol. Environ. Health 4, 173–188.

Lowell, S. Introduction to Powder Surface Area. Wiley-Interscience, John Wiley & Sons, New York, 1979.

Nair, R. (1987) Design and execution of a toxicity testing program for a unique fiber: Phosphate fibre. In: Fibres and Friction Materials Symposium. The Asbestos Institute, Montreal, Quebec, 124–128.

Pott, F., Huth F., and Friedrichs, K.H. (1974) Tumorigenic effect of fibrous dusts in experimental animals. Environ. Health Perspect. 9, 313–315.

Pott, F. (1978) Some aspects on the dosimetry of carcinogenic potency of asbestos and other fibrous dusts. Stub-Reinhalt Luft 38, 486–490.

Pott, F. (1987) Problems in defining carcinogenic fibres. Ann. Occup Hygiene 31, 799–802.

Pott, F., Ziem, U., Reiffer, F.J., Huth, F., Ernst, H., and Mohr, U. (1987) Carcinogenicity studies of fibres, metal compounds, and some other dusts in rats. Exp. Pathol. (JENA) 32, 129–152.

Pott, F., Roller, M. Ziem, U., Reiffer, F.J., Bellmann, B.,Rosenbruch, M., and Huth, F. (1989) Carcinogenicity studies of natural and man-made fibers with the intraperitoneal test in rats. In: Non-occupational Exposure to Mineral Fibers. Bignon, J., Peto, J., Saracci, R., Eds. IARC Sci. Publ. No. 90, Lyon, France, 173–179.

Roe, F.J.C., Carter, R.L., Walters, M.A., and Harington, J.S. (1967) The pathological effects of subcutaneous injection of asbestos fibers in mice: migration of fibers to submesothelial tissues and induction of mesotheliomata. Int'l J. Cancer, 2, 628–638.

Smith, W.E., Miller, L., Elasasser, R.E., and Hubert, D.D. (1965) Tests for carcinogenicity of asbestos. Ann. New York Acad. Sci. 132, 456–488.

Stanton, M.F., Backwell, R., and Miller, E. (1969) Experimental pulmonary carcinogenesis with asbestos. Amer. J. Indus. Hygiene Assoc. J. 30, 236–244.

Stanton, M.F. (1973) Some etiological considerations of fibre carcinogenesis. In: Biological Effects of Asbestos. Bogovski, P., Gilson, J.C., Timbrell, V., Wagner, J.C., Eds. IARC Sci. Publ. No. 8, Lyon, France, 289–294.

Stanton, M.F. (1974) Fiber Carcinogenesis: Is asbestos the only hazard? J. Nat'l Cancer Inst. 52, 633–634.

Stanton, M.F., Layard, M., Tegeris, A., Miller, E., May, M. and Kent, E. (1977) Carcinogenicity of fibrous glass. J. Nat"l Cancer Inst. 58, 587–603.

Stanton, M.F. and Layard, M. (1978) The carcinogenicity of fibrous minerals. In: Proc. Workshop on Asbestos: Definitions and Measurement Methods. Gaithersburg, Maryland, July 18–20, 1977. National Bureau of Standards Spec. Publ. No. 506, 143–151.

Stanton, M.F., Layard, M., Tegeris, A., Miller, E., May, M., Morgan, E. and Smith, A. (1981) Relation of particle dimension to carcinogenicity of amphibole asbestoses and other fibrous minerals. J. Nat'l Cancer Inst. 67, 965–975.

Wagner, J.C. (1962) Experimental production of mesothelial tumours of the pleura by implantation of dust in laboratory animals. Nature 196, 180.

Wagner, J.C. and Berry, G. (1969) Mesothelioma in rats following inoculation with asbestos. Brit. J. Cancer 23, 567–581.

Wagner, J.C., Berry, G. and Timbrell, V. (1973) Mesothelioma in rats after inoculation with asbestos and other minerals. Brit. J. Cancer 28, 173–185.

Wagner, J.C., Skidmore, J.W., Hill, R.J. and Griffiths, D.M. (1985). Erionite exposure and mesothelioma in rats. Brit. J. Cancer 51, 727–730.

CHAPTER 10

THE SURFACE THERMODYNAMIC PROPERTIES OF SILICATES AND THEIR INTERACTIONS WITH BIOLOGICAL MATERIALS

Rossman F. Giese, Jr.

*Department of Geology
State University of New York, Buffalo
Buffalo New York 14260 U.S.A.*

Carel J. van Oss

*Department of Microbiology and Chemical Engineering
State University of New York, Buffalo
Buffalo New York 14214 U.S.A.*

INTRODUCTION

Although the precise mechanisms by which a mineral induces disease in an organism are not well understood, what is clear is that the initial contact between the organism and the mineral occurs via the interface between the mineral surface as mediated by an aqueous environment (Heppleston, 1984). If this initial interaction is benign, then, presumably, nothing further happens and the mineral would be viewed as non-toxic. On the other hand, if there is an interaction which changes either the mineral surface or something in the organism (or both) then a chain of events may be initiated which may eventually lead to a toxic end result. Thus, it would seem to be of paramount importance to understand the first step in the chain of events, i.e., the interaction at the interface.

This is at once a simplifying factor (because the bulk chemical composition and crystal structure are of little importance) and a complication (because surfaces are much more difficult to study than are bulk crystalline materials). The examination of the structure of mineral surfaces is advancing rapidly due to the advent of new tools such as the atomic force microscope. However, the physical structure of a mineral surface is only part of the information needed to understand the activity of that surface. Along with the physical interactions the chemical interactions must also be accounted for. One very powerful methodology for the study of such interactions at surfaces and interfaces is the determination of the thermodynamic properties of the surface.

The purpose of this contribution is to give a brief résumé of the theory underlying the thermodynamics of surfaces and interfaces, and to summarize the present state of our knowledge of the surface thermodynamic properties of minerals. Detailed presentations on the energetics of surfaces can be found in numerous sources (e.g., Adamson, 1982), and Parks (1990) presents the subject with particular emphasis on the mineral–fluid interface.

SURFACE THERMODYNAMIC THEORY

J. D. van der Waals first observed deviations of the ideal gas law, apparently caused by interatomic or intermolecular attractions between neutral gas atoms or molecules. These non-covalent, non-electrostatic intermolecular interactions were subsequently named "van der Waals forces." Their physical nature became clear during the early part of this century. Keesom (1915; 1920; 1921a,b) described the interaction between permanent dipoles (orientation forces). Debye (1920, 1921) described the interaction between a permanent dipole and a dipole induced by it (induction forces). Finally, London (1930) showed that also in nonpolar atoms and molecules rapidly fluctuating dipoles arise, which can induce dipole moments in other atoms, and thus attract them (dispersion forces). Similar interactions exist between macroscopic bodies. Lifshitz (1955) explained how these interactions operated between macroscopic bodies, and these are denoted as "Lifshitz–van-der-Waals" or LW forces.

In the mid-1960s, it became clear that, whereas LW forces alone were adequate to explain the interactions between apolar liquids and solids (e.g., Teflon and hydrocarbons), some other kind of interaction was needed to explain the properties of polar materials. Fowkes (1966) identified these interactions as arising from electron donor and electron acceptor sites (i.e., acid/base interactions, as described by Lewis) in these materials. Subsequently, Chaudhury (1984), in collaboration with C. J. van Oss and R. J. Good (see van Oss et al., 1988, for a review), determined how these polar interactions could be accounted for at interfaces. The following is a brief summary of those aspects of surface thermodynamic theory that are relevant to this chapter.

Surface tension, interfacial tension and surface free energies

The free energy of cohesion of a solid or liquid, i, is related to the surface tension of the material in a simple manner (Good, 1966):

$$\Delta G_{ii}^{Coh} = -2\gamma_i \tag{1}$$

where γ_i is the surface tension of the material.[†] Thus, there is a proportionality between surface free energy of cohesion and surface tension. The units for surface free energy of cohesion (i.e., an energy per unit area) are mJ/m^2 (or ergs/cm^2) and those for surface tension (i.e., a force per unit length) are N/m (or dynes/cm). To simplify matters, the units used in this paper are always mJ/m^2 in view of the equality in Equation (1) (Adamson, 1982).

The free energy of cohesion of material i is the sum of two contributors.

[†] In chemical and surface thermodynamics (and in this paper), it is the convention that the free energy of interaction, ΔG, has a negative value when the interaction is attractive and a positive value when the interaction is repulsive.

The free energy of cohesion of material i is the sum of two contributors.

$$\Delta G_{ii}^{Coh} = \Delta G_{ii}^{LW} + \Delta G_{ii}^{AB} \tag{2}$$

where ΔG_{ii}^{LW} is the Lifshitz–van-der-Waals component (related to physisorption), and ΔG_{ii}^{AB} is the Lewis acid/base component (related to chemisorption).

The apolar component. In apolar systems, which experience only LW interactions, the surface free energy is related to the surface tension of the material by (see Eqn. 1):

$$\Delta G_{ii}^{LW} = -2\gamma_i^{LW} \tag{3}$$

For two different apolar materials, in a vacuum, the free energy of adhesive interaction is given by the Dupré equation (Dupré, 1869)

$$\Delta G_{12}^{LW} = \gamma_{12}^{LW} - \gamma_1^{LW} - \gamma_2^{LW} \tag{4}$$

where the apolar (LW) interfacial tension component between materials 1 and 2, γ_{12}^{LW}, is given by the Good-Girifalco-Fowkes relation (Girifalco and Good, 1957; 1960; Fowkes, 1963):

$$\gamma_{12}^{LW} = \left(\sqrt{\gamma_1^{LW}} - \sqrt{\gamma_2^{LW}}\right)^2 \tag{5}$$

The values of γ^{LW} are readily determined by straightforward experiments as will be described in a later section.

The polar component. Interactions of an electron acceptor-electron donor type (Lewis acid/base), including hydrogen bonding, are termed AB interactions. These interactions are generally asymmetrical and are described by two numerical parameters: one for the electron donor (or Lewis base) sites, γ^-, and a second for the electron acceptor (or Lewis acid) sites, γ^+. The polar AB component of the surface tension for a single material is given by

$$\gamma_1^{AB} = 2\sqrt{\gamma_1^+ \gamma_1^-} \tag{6}$$

and the AB interfacial surface tension component for the interaction between two different condensed-phase materials is

$$\gamma_{12}^{AB} = 2\left(\sqrt{\gamma_1^+\gamma_1^-} + \sqrt{\gamma_2^+\gamma_2^-} - \sqrt{\gamma_1^+\gamma_2^-} - \sqrt{\gamma_2^+\gamma_1^-}\right) \tag{7}$$

The AB component of the free energy of cohesive interaction between two objects of the same type is:

$$\Delta G_{11}^{AB} = -2\gamma_{11}^{AB} \tag{8}$$

and the Dupré equation, also applicable to adhesive AB interactions *in vacuo*, yields:

$$\Delta G_{12}^{AB} = \gamma_{12}^{AB} - \gamma_1^{AB} - \gamma_2^{AB} \tag{9}$$

As $\gamma = \gamma^{LW} + \gamma^{AB}$, one can combine γ_{12}^{LW} and γ_{12}^{AB} to yield the total interfacial tension between two different materials, expressed as:

$$\gamma_{12} = \left(\sqrt{\gamma_1^{LW}} - \sqrt{\gamma_2^{LW}}\right)^2 + 2\left(\sqrt{\gamma_1^+\gamma_1^-} + \sqrt{\gamma_2^+\gamma_2^-} - \sqrt{\gamma_1^+\gamma_2^-} - \sqrt{\gamma_2^+\gamma_1^-}\right) \quad (10)$$

The AB contribution to the surface tension or the interfacial tension between an oxide (i.e., a silicate) surface and a polar liquid such as water is very important and to ignore the polar interactions, such as is done in the classical DLVO calculation, can lead to important errors (van Oss et al., 1990).

The LW and AB components of the surface free energy are additive (Eqn. 2). Expanding ΔG_{ij}^{LW} and ΔG_{ij}^{AB} with Equations. (5), (6), (8), (9) and (10) gives a general expression for the interfacial (IF) free energy in terms of the individual surface tension components and parameters of materials i and j.

$$\Delta G_{ij}^{IF} = -2\left(\sqrt{\gamma_i^{LW}\gamma_j^{LW}} + \sqrt{\gamma_i^+\gamma_j^-} + \sqrt{\gamma_i^-\gamma_j^+}\right) \quad (11)$$

Free energy of adsorption

We can now examine the interaction of particles by surface thermodynamic forces (interfacial forces, denoted by IF) when both are immersed in a liquid (neglecting for the moment any electrostatic interaction which will be treated in a later section). The convention is that if ΔG^{IF} is negative, there is a net attraction between the particles in the presence of the liquid, and a positive sign indicates a repulsion. If both particles are the same material, 1, and we denote the liquid by 2, then the relevant interaction energy is (van Oss et al., 1988):

$$\Delta G_{121}^{IF} = -2\gamma_{12} = -2\gamma_{12}^{LW} - 2\gamma_{12}^{AB} \quad (12)$$

where γ_{12} is specified in Equation (10).

Of more interest to the interaction between mineral particles and biological material (e.g., cells, biopolymers, proteins) is the free energy of interaction between the mineral particle, 1, and another particle, 2, immersed in a liquid, 3. The Dupré equation then is:

$$\Delta G_{132}^{IF} = \gamma_{12} - \gamma_{13} - \gamma_{23} \quad (13)$$

The individual interfacial tensions in this expression can be derived from Equations (5) and (7) to give the total interfacial interaction energy:

$$\Delta G_{132}^{IF} =$$

$$\left(\sqrt{\gamma_1^{LW}} - \sqrt{\gamma_2^{LW}}\right)^2 - \left(\sqrt{\gamma_1^{LW}} - \sqrt{\gamma_3^{LW}}\right)^2 - \left(\sqrt{\gamma_2^{LW}} - \sqrt{\gamma_3^{LW}}\right)^2$$

$$+ 2\sqrt{\gamma_3^+}\left(\sqrt{\gamma_1^-} + \sqrt{\gamma_2^-} - \sqrt{\gamma_3^-}\right)$$

$$+ 2\sqrt{\gamma_3^-}\left(\sqrt{\gamma_1^+} + \sqrt{\gamma_2^+} - \sqrt{\gamma_3^+}\right)$$

$$- 2\left(\sqrt{\gamma_1^+ \gamma_2^-} + \sqrt{\gamma_1^- \gamma_2^+}\right) \tag{14}$$

Contact angles and the Young equation

The values of γ^{LW}, γ^+, and γ^- of a solid can be derived from measurements of the contact angle, θ, formed by a drop of liquid on a smooth surface of the solid, provided that the surface tension properties of the liquid are known. The minimum solid surface needed for an accurate contact angle measurement is approximately 5 mm × 5 mm, although the measurement process proceeds much more rapidly for larger surfaces. Three or more different liquids must be used of which at least two must be polar. Contact angles obtained with apolar liquids (for which γ_L^- and γ_L^+ are zero so that $\gamma_L = \gamma_L^{LW}$) give directly the value of the LW surface tension component, γ_S^{LW}, for the solid. Contact angles obtained with polar liquids yield values for γ_L^+ and γ_L^- by the solution of a set of simultaneous (Young) equations (Eqn. 15), with the number of equations equal to the number of polar liquids. The relevant thermodynamic data for the liquids commonly used for direct contact angle measurements and for wicking are given in Table 1. The relation between the contact angles and the surface tension components and parameters is given by the extended form of the Young equation (S and L refer to solid and liquid, respectively) (van Oss et al., 1988):

$$(1 + \cos\theta)\,\gamma_L = 2\left(\sqrt{\gamma_S^{LW} \gamma_L^{LW}} + \sqrt{\gamma_S^+ \gamma_L^-} + \sqrt{\gamma_S^- \gamma_L^+}\right) \tag{15}$$

With these techniques, it is possible to measure contact angles with a standard deviation of 1°, which translates into errors for the γ values of about 2% (unpublished data).

As an example of the determination of the surface tension components of a mineral, the contact angles on a single crystal of muscovite were measured and are shown in Table 2.

The set of Young's equations was solved using the contact angle values in Table 2 to yield the values of the surface tension components and parameters. These values show that muscovite has a strong Lewis base activity ($\gamma^- = 55.5$ mJ/m²) a modest Lewis acid activity ($\gamma^+ = 1.1$ mJ/m²) and $\gamma^{LW} = 40.6$ mJ/m².

Table 1

Surface tension components, parameters, and viscosities of test liquids normally used for contact angle measurements and thin layer wicking

Liquid	γ_L mJ/m²	γ^{LW} mJ/m²	γ^+ mJ/m²	γ^- mJ/m²	viscosity, η (Poise)
hexane	18.40	18.40	0.0	0.0	0.00326
heptane	20.14	20.14	0.0	0.0	0.00409
octane	23.83	23.83	0.0	0.0	0.00542
decane	23.83	23.83	0.0	0.0	0.00907
dodecane	25.35	25.35	0.0	0.0	0.01493
hexadecane	27.47	27.47	0.0	0.0	0.03451
cis-decaline	32.18	32.18	0.0	0.0	0.03381
α-bromonaphthalene	44.4	44.4	0.0	0.0	0.0489
diiodomethane	50.8	50.8	0.0	0.0	0.028
water	72.8	21.8	25.5	25.5	0.010
ethylene glycol	48.0	29.0	1.92	47.0	0.199
formamide	58.0	39.0	2.28	39.6	0.0455
glycerol	64.0	34.0	3.92	57.4	14.90

Values reported in mJ/m² and determined at 20°C

Table 2

Observed contact angles for test liquids on cleavage surface of muscovite, at 20° C

Liquid	Observed contact angle
α-bromonaphthalene	24.4°
diiodomethane	38.0°
water	3.6°
formamide	7.8°
glycerol	29.3°

Table 3

Wicking data for a talc sample from Luzenac, France

Liquid	$2h^2\eta/t$	γ_L mJ/m²	$\cos\theta$	θ
α-bromonaphthalene	4.73	44.4	0.691	46.3
diiodomethane	3.83	50.8	0.489	60.7
water	1.61	72.8	0.144	81.7
ethylene glycol	2.40	48.0	0.324	71.1
formamide	3.25	58.0	0.363	68.7

Sources: Li et al. (1993); Li (1993)

Powdered materials and the Washburn equation

Although measuring the contact angles of liquid drops in contact with a smooth surface of the solid is the most reliable procedure, it is not possible to use this approach with minerals that do not have sufficiently large and smooth crystal faces. In the case of the smectite clay minerals (e.g., montmorillonite), a pseudo-crystal surface can be created by fabricating the clay mineral into a self-supporting, smooth film suitable for contact angle measurements. Many other (non-swelling) fine-grained minerals however, do not form suitable films and do not occur as large, perfect single crystals with well-developed faces. For these materials, the contact angles can be determined indirectly by measuring the capillary flow rate of a liquid through a thin, uniform layer of powdered material deposited on a smooth glass plate. This procedure is referred to as thin layer wicking (Costanzo et al., 1991; Giese et al., 1991; van Oss et al., 1992). The capillary flow rate is related to the contact angle by the Washburn equation (Washburn, 1921):

$$h^2 = \frac{t R \gamma_L \cos\theta}{2\eta} \qquad (16)$$

where h is the height of the capillary rise of the liquid in time t, γ_L the surface tension of the liquid, R the average pore radius, and η the viscosity of the liquid. The value of R is obtained by measuring the rate of capillary rise with a series of low-energy liquids (hydrocarbons ranging from decane to hexadecane) which wet the solid, so that $\cos\theta = 1$ (see van Oss et al., 1992, for a discussion). Note that glycerol can not be used for wicking because of its very high viscosity.

Thin layer wicking is a very useful technique for the study of any kind of porous and permeable solid or powder. The one limitation is that the contact angle of the test liquid with the solid must be less than 90°; for θ values greater than or equal to 90°, the liquid will not rise in the solid. This is normally not a problem but for very hydrophobic materials, the number of test liquids which can be used is reduced (Li et al., 1993; Li, 1993). Equation (16) can be written as:

$$\frac{2 h^2 \eta}{t} = R \cos\theta \, \gamma_L \qquad (17)$$

which is seen to be linear when $2h^2\eta/t$ is plotted against γ_L for the spreading liquids for which $\cos\theta = 1$. Further, the straight line must go through the origin, leaving only one parameter to be determined, the slope of the line. This slope yields the average pore diameter, R, which is needed for the determination of the contact angles, θ, for the non-spreading liquids. Data from the study of talc by Li (1993) and Li et al. (1993) are shown in Figure 1 along with the regression line determined by the spreading liquids along with the experimental values for the non-spreading liquids. The values of $2h^2\eta/t$ and the derived values of θ for this experiment are listed in Table 3 (above).

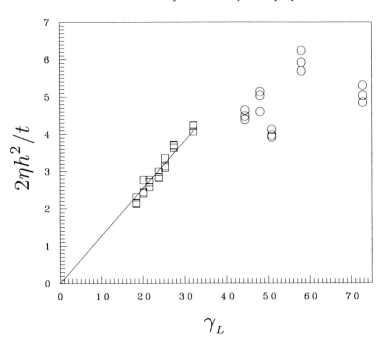

Figure 1. A wicking plot for the talc described in Table 3. The open squares are the low-energy apolar hydrocarbon liquids for which $\cos\theta = 1$; the open circles are the non-spreading apolar and polar liquids.

Hydrophobic and hydrophilic surfaces

The adjectives hydrophobic and hydrophilic are in common use, although in a very qualitative manner. It is not clear under what conditions a hydrophobic surface becomes hydrophilic and vice versa. The terms can be put on a quantitative basis by examining the interactions between two particles of the same material immersed in water. Excluding electrostatic interactions, the free energy of interaction for the system is ΔG^{IF}_{1w1} (Eqn. 12). When ΔG^{IF}_{1w1} is exactly zero, a balance is achieved in which the attraction of the particles for each other is balanced by attraction between the particles and water and between the water molecules themselves. Thus, the material has no net preference for water or another particle of the same material. This is a logical boundary separating the hydrophilic and hydrophobic states. If this condition does not exist, then ΔG^{IF}_{1w1} has either a negative or a positive sign, corresponding to a net attraction between the particles (negative sign) or a net repulsion between the particles (positive sign). Therefore, a negative value for ΔG^{IF}_{1w1} corresponds to a hydrophobic surface and a positive value indicates a hydrophilic surface. Moreover, the magnitude of ΔG^{IF}_{1w1} is a useful indicator of the degree of hydrophobicity or hydrophilicity. For example, the talc described in Table 3 is strongly hydrophobic because $\Delta G^{IF}_{1w1} = -35.2$ mJ/m², whereas muscovite (Table 2) is strongly hydrophilic with $\Delta G^{IF}_{1w1} = 32.6$ mJ/m².

ELECTROSTATIC INTERACTIONS

When a solid material comes into contact with a polar liquid such as water, the surface of the materials typically manifests an electrostatic charge, expressed as the surface potential, ψ_o. The origin of this charge is diverse and includes a differential dissolution of the surface, ionization of the surface, a permanent structural imbalance in charge (e.g., the layer charge of clay minerals) a permanent electrostatic charge on the surface resulting from broken bonds, and specific adsorption of charged species from the liquid (Everett, 1988). The particles' surface potential is not measurable directly. The structure of the water molecules in the close vicinity of the solid surface is perturbed by the presence of the electrostatic charge and are attracted to the surface and become part of the surface. The influence of the surface decreases with distance from the surface so that eventually the water molecules are no longer bound to the solid; this transition occurs at the slipping plane. In addition to perturbing the water molecules, the charged surface alters the physical distribution of dissolved ions. This redistribution, compared to the bulk liquid, is expressed as the diffuse ionic double layer, with $1/\kappa$, the Debye length.

Experimentally, the potential (the ζ-potential or zeta-potential) on the particles' surface at the slipping plane can be measured by electrokinetic experiments. For relatively small values of the ζ-potential and a spherical geometry, the potential, ψ_o, can be derived from the ζ-potential by (Overbeek, 1952):

$$\psi_o = \zeta \left(1 + \frac{z}{a}\right) e^{\kappa z} \tag{18}$$

where z is the distance from the particle surface to the slipping plane and a is the radius of the spherical particle. Values of $1/\kappa$ are tabulated for a number of common 1:1, 1:2, and 2:2 electrolytes by van Oss (1975). The value for z is normally taken to lie between 3 and 5 Å.

ζ-potential

Measurement of the ζ-potential can be accomplished in a number of different ways. A commonly used technique (electrophoresis) is to measure the velocity, U, of a colloidal-sized particle immersed in an aqueous electrolyte solution of known ionic strength. There are two limiting cases for the relation between U and the ζ-potential. The first refers to the situation where the double layer is thick and insubstantial, and the second is for a thin and dense double layer. Under the conditions normally encountered in electrophoresis of mineral particles, the thin double layer applies (Overbeek and Wiersema, 1967; van Oss, 1975):

$$U = \varepsilon \zeta V / 4\pi \eta \tag{19}$$

where η is the viscosity of the liquid, ε is the dielectric constant of the liquid, and V is the applied electric field.

Total interaction energy between two particles

The total interaction energy between two particles of the same material immersed in water, ΔG_{1w1}^{TOTAL}, is the sum of the interfacial and the electrostatic interaction energies:

$$\Delta G_{1w1}^{TOTAL} = \Delta G_{1w1}^{LW} + \Delta G_{1w1}^{AB} + \Delta G_{1w1}^{EL} \tag{20}$$

The total interaction energy can be calculated as a function of the distance between the particle surfaces by using decay relations appropriate to the geometry of the particles. When only the first and third terms of Equation (20) are used, the resultant energy versus distance curve is the classical DLVO plot. However, neglect of the AB contribution to the total interaction energy can lead to erroneous conclusions (van Oss et al., 1990). Inclusion of the LW, AB and EL contributions as a function of interparticle distance yields an extended DLVO (or XDLVO) plot.

For two identical spherical particles, the decay relation for the LW contribution is:

$$\Delta G_1^{LW} = -A R / 12l \tag{21}$$

where R is the radius of the particles, A is the Hamaker constant for the material in question (A is expressed in Joules and is equal to $\gamma^{LW} \times 1.86 \times 10^{-21}$ [van Oss et al., 1988], and l is the interparticle distance).

The decay relation for the AB interaction energy is:

$$\Delta G_1^{AB} = \pi R \lambda \Delta G_{l_0}^{AB"} \exp\left[\frac{l_0 - l}{\lambda}\right] \tag{22}$$

where λ is the correlation length of water (~10 Å, although, for very hydrophobic surfaces this may be larger by a factor of 10 or more), and $\Delta G_{l_0}^{AB"}$ is the interaction energy at contact (i.e., $l = l_0 \approx 1.57$ Å).

For the EL interaction energy, the relevant relation is:

$$\Delta G_1^{EL} = \frac{1}{2} \varepsilon R \psi_0^2 \ln\left[1 + \exp(-\kappa l)\right] \tag{23}$$

For the materials of interest to the present subject, i.e., silicate minerals, immersed in an *aqueous medium*, it should be emphasized that the AB contribution to ΔG^{TOTAL} is of extreme importance and usually dominates the LW and EL contributions.

SURFACE TENSION OF MINERALS AND RELATED MATERIALS

A number of factors have been suggested as having importance in determining the toxicity of minerals. These include both chemical (e.g., acid/base interactions, the production of hydroxyl and oxygen radicals, exchangeable cations) and structural (e.g., periodicities in surface structure) (Mossman and

Marsh, 1989; Hansen and Mossman, 1987; Guthrie, 1992; Kennedy et al., 1989; Appel et al., 1988).

Unfortunately, studies of the cytotoxicities of minerals, as an example, do not allow an easy evaluation of the surface thermodynamic properties of the mineral samples utilized.

Measurement of the surface thermodynamic properties of minerals can provide information about the chemical activity of mineral surfaces and also about the forces acting between mineral particles and biological materials. The application of this approach to surface thermodynamic theory of minerals is relatively recent, the first mineral examined being the clay mineral hectorite (van Oss et al., 1990). Subsequently, most of the studies of minerals have concentrated on the smectites, and, although these are not generally considered to be health hazards (e.g., see references in Guthrie, 1992), the smectites do allow the examination of a number of chemical aspects that may be of importance to more toxic materials such as the asbestos minerals. These include exchangeable cations, a variable structural charge, a structure which is closely related to the amphibole minerals, and the ability to apply carefully controlled organic coatings to the silicate structure.

Table 4 is a summary of the surface tension component values for the minerals discussed in this paper.

Asbestos

Although mineral and inorganic dust in general may have deleterious effects on living organisms, fibrous varieties are especially suspect (e.g., Stanton et al., 1981; Woodworth et al., 1982). Only two modest studies of surface thermodynamic properties have been carried out on fibrous minerals, and neither can be considered to be more than a first attempt to understand what surface properties are conferred on a material by its fibrous habit (i.e., those surface properties that are not characteristic of macroscopic crystals of the same mineral). Li (1993) has examined several fibrous 1:1 layer silicates and amphiboles as part of a larger ongoing study aimed at understanding the relationships between particle morphology and surface thermodynamic properties. In this study, the surface thermodynamic properties of fibrous and non-fibrous (e.g., single crystals or massive material) varieties of similar chemical composition are being determined by wicking or contact-angle measurements, respectively. Preliminary results show that the chrysotile (Royal Ontario Museum #M85782) is moderately hydrophobic whereas massive tremolite (Royal Ontario Museum #M19605) is hydrophilic. There seems to be little if any change in the surface properties of the tremolite upon grinding. Further measurements should allow the determination of whether the differences in the surface properties of the chrysotile and tremolite are related to their different morphologies, to their different structures and compositions, or to both.

Sepiolite and palygorskite

In the same study, Li (1993) examined samples of sepiolite and palygorskite. Data in Table 4 for sepiolite (SepNev-1) and palygorskite (PFL-1) were obtained from standards in the Clay Mineral Repository of the Clay Minerals Society. The sepiolite is hydrophobic and the palygorskite is moderately hydrophilic.

Table 4
Surface tension component values for a number of silicates and related materials

Material	Technique	γ^{LW}	γ^{AB}	γ^+	γ^-	Reference
smectite nat.	CA	41.2	14.3	1.54	33.3	1
smectite (low)	CA	40.0	12.6	1.95	20.2	1
smectite (high)	CA	42.9	12.5	0.9	43.3	1
quartz, crystal	CA	38.7	10.8	1.1	26.3	2
quartz, ground	W	39.8	0.97	0.05	46.7	3
quartz, ground	W	46.5	0.0	0.0	53.5	3
quartz, ground	W	33.4	0.0	0.0	32.8	3
silica gel	W	43.3	0.0	0.0	69.4	4
muscovite	CA	36.5	6.8	0.2	57.7	5
illite	W	40.2	0.0	0.0	19.1	6
palygorskite	W	29.5	0.0	0.0	28.7	7
sepiolite	W	30.5	0.2	4.2	17.3	8
chrysotile	W	38.3	0.0	0.0	23.9	9
tremolite (massive)	CA	35.8	0.0	0.0	37.5	9
tremolite (ground)	W	35.3	10.4	0.8	35.0	10
zeolite 4A	W	25.1	3.5	0.1	29.9	11
glass	CA	34.0	15.8	1.0	64.2	12
glass powder	W	31.1	7.9	0.4	37.1	12
volcanic ash 1	W	38.7	4.3	0.1	45.6	13
volcanic ash 2	W	35.4	10.1	1.6	15.8	13
talc	W	31.5	5.1	2.4	2.7	14
pyrophyllite	W	34.4	4.7	1.7	3.2	14
HSA	CA	26.8	35.2	6.0	51.5	15
erythrocytes	CA	35.2	1.36	0.01	46.2	16
Streptococcus thermophilus	CA	35.5	11.6	0.6	56.1	17

CA = contact-angle measurement: W = thin layer wicking: all values reported in mJ/m^2 and determined at 20°C
1   Wyoming montmorillonite (SWy-1) in the untreated form (nat.) and saturated with Ba or Mg (Norris, 1993)
2   Single crystal specimen from Herkimer Co., New York; measurements made on a rhombohedral face
3   Samples from Neversummer Range, Jackson Co., Colorado; samples were ground for 15 m, 120 min and 480 min, respectively (Wentzek, 1993)
4   Synthetic 90:10 silica:K-silicate monolithic gel (Kerch et al., 1990; unpublished data)
5   Single crystal of ruby muscovite from India
6   Silver Hill illite (SMt-1) (Li, 1993)
7   Palygorskite (PFl-1) (Li, 1993)
8   Sepiolite from Nevada (SepNev-1) (Li, 1993)
9   Specimen from the Royal Ontario Museum (M8572) (Li, 1993)
10   Tremolite from the Royal Ontario Museum (M19605) (Li, 1993)
11   Commercial zeolite from the Ethyl Corporation (Costanzo et al., 1991)
12   Glass petrographic microscope slides
13   Ash 1 is alkaline-phonolite ash cloud deposit from the 1631 eruption at Mt Vesuvius (VES 8412); Ash 2 is from the 1902 *nuée ardente* at Mt. Pelée (PEL 8811) (Li, 1993)
14   Giese et al. (1991)
15   Human serum albumin hydrated with 2 water layers (Costanzo et al., 1990)
16   van Oss et al. (1992)
17   Skvarla (1993)

2:1 layer silicates

Clay minerals are typically very fine-grained materials so that it is not possible to measure the contact angle of a liquid in contact with a crystal surface. Two approaches have been taken. The smectite clay minerals will form uniform, relatively non-porous films by slow evaporation of an aqueous suspension on a suitable substrate. These films will support a liquid drop for a sufficiently long period of time (tens of seconds to several minutes) so that the contact angle can be measured. Initially, self-supporting films were examined (Giese et al., 1990; Norris et al., 1992) but subsequent work used films supported on glass plates. The latter technique yields more robust surfaces.

The measurement of the surface thermodynamic properties of fine-grained materials such as the clay minerals yields values which represent an average of all the exposed surface. For platey materials whose dominant surfaces are identical, as is the case for the 2:1 minerals, these values can be readily related to the actual surface structure. The extreme case is that of muscovite where the contact angle liquids interact with a surface which is as well defined as any. On the other hand, the 1:1 structures, such as kaolinite, present platey surfaces which are, on the same crystal, very different, i.e., one surface is the basal oxygen surface of the tetrahedral sheet and the other surface is hydroxyl groups bonded to aluminum (or some other cation). For these asymmetrical structures, the surface thermodynamic properties which one can measure are only an average of these two potentially very different surfaces, and as such they should be treated with care. For these reasons no values have been entered for kaolinite in Table 4.

Smectite. Preliminary data on a number of smectites have appeared in the literature (Giese et al., 1990; Norris et al., 1993) and the definitive study will shortly be published (Norris, 1993). It is not possible here to summarize these results adequately. There are a number of generalizations which can be made; smectites, in common with most materials, are primarily Lewis base monopoles (γ^+ is small or zero), the value of the Lewis base parameter, γ^-, varies from approximately 20 mJ/m^2 to more than 45 mJ/m^2, depending on the type of clay (and the magnitude of the layer charge) and the nature of the exchangeable cation. The two values in Table 4 are the extremes observed for the same smectite (SWy-1) saturated with barium (small γ-) or magnesium (large γ-). Not all smectite clays exhibit such a wide range of Lewis basicities. Thus, these materials can be either hydrophobic or hydrophilic depending on the value of the Lewis base parameter. Most natural smectites are hydrophilic. This variability in surface properties is mirrored in the variable response observed by Gormley and Addison in their study of the cytotoxicity of standard smectite minerals (Gormley and Addison, 1983).

Muscovite and illite. Some of the complexity in the surface properties of the 2:1 minerals is illustrated by muscovite and illite. Both have the same structure (ignoring the layer stacking) and differ principally in illite having a lesser aluminum substitution for silicon in the tetrahedral sites, thereby reducing the layer charge. Thus, illite lies (chemically) between mica and the dioctahedral smectite minerals, yet the surface properties of illite are not colinear with those of the two end members in this series, muscovite (layer charge = 1) and pyrophyllite (layer charge = 0).

Talc and pyrophyllite. Talc and pyrophyllite are 2:1 layer silicates having essentially no cationic substitution and a zero layer charge (Newman, 1987). Talc and, by implication, pyrophyllite have widely recognized hydrophobic properties. These are the basis for the inclusion of this material in various cosmetic and personal hygiene consumer products. Neither mineral forms a self-supporting film, so the determination of their surface thermodynamic properties was accomplished via thin layer wicking (Giese et al., 1991). As seen in Table 4, talc and pyrophyllite are unlike the other materials in the table in having very small values for the Lewis basicity, γ^-, and rather larger values for the Lewis acidity, γ^+, than the other minerals in the table, even though their values for γ^+ are generally small.

Quartz

Exposure to fine particles of quartz or silica are known to induce a biological response (Adamis and Timar, 1978; Campbell, 1940; Le Bouffant et al., 1982; Heppleston, 1984). Quartz has poor cleavage characteristics, so the external surfaces of fine particles may not be well defined crystallographically or structurally. As such, different samples of fine particulate silica may have very different properties depending on how they have been prepared and treated. The measurements reported in Table 4 (Wentzek, 1993) are for a single crystal of quartz and quartz samples which have been ground for different times (15, 120 and 480 min) under the same set of experimental conditions (ground dry in an automatic agate mortar-and-pestle). The grinding was sufficiently energetic to produce micrometer-sized particles in the shortest time period (15 min). The single-crystal sample and the three powder samples are all essentially Lewis base monopoles. The single-crystal sample is hydrophobic, but this same material becomes strongly hydrophilic after 15 min of grinding and even more hydrophilic after 120 min of grinding. Curiously, the value of γ^- decreases when grinding continues beyond 120 min to 480 min. The LW surface tension components of the single-crystal sample and the 15-min-ground sample are essentially the same, but longer grinding (120 min) *increases* γ^{LW} and eventually (480 min) γ^- decreases and becomes less than the original crystal or the briefly ground material. In contrast to quartz, the massive and ground tremolite samples have very similar properties (see Table 4). Clearly, grinding mineral material that has poorly developed cleavage is a complex process, and the study of Wentzek was not designed to be a definitive study of quartz grinding.

Glass, natural and synthetic

Recent volcanic explosions have placed very large quantities of fine-grained volcanic ash in the atmosphere, and there have been some questions as to the health problems that may result from breathing such material (Merchant et al., 1982; Green et al., 1982). A recent study of a suite of 9 volcanic ashes from Italy, Arizona and Martinique with silica contents in the range 53 to 75% (Li et al., 1993) found that ash samples display a wide range of surface properties, from very hydrophilic to moderately hydrophobic. Furthermore, Li et al. concluded that the chemical composition of the ash did not correlate with the surface properties. What did appear to control the surface properties was the access (if any) of water to the magma during and prior to eruption. This wide range in the surface properties suggests that volcanic ash may exhibit a similarly wide range in biological response. Thus, although some volcanic ash has been shown to be only mildly toxic (e.g., the Mount St. Helen's—Merchant et al., 1982), other volcanic ash may not pose such a low risk.

Two samples of synthetic glass are included in Table 4. These are common microscope glass slides whose surface properties were measured by direct contact angle measurement on the original slide and on the same material after grinding to less than 325 mesh (i.e., <45 µm). Both materials are hydrophilic, but there are differences between the slide and the ground glass. Grinding affected the glass differently than it did the quartz: The intact glass slide is a stronger Lewis base than is the ground material, and the LW surface tension component of the slide is greater than the ground material.

EFFECTS OF SURFACE-SORBED ORGANIC MATERIAL

Oxide surfaces—including silicates and non-oxide materials with oxidized surfaces—are normally strong Lewis bases and, therefore, hydrophilic. This can be seen in Table 4 by the hdrophilic nature of the majority of the inorganic materials. Strongly oriented surface water sorbed onto a surface which is a Lewis base also is a strong Lewis base, because the hydrogen atoms of the water molecules are directed toward the surface. However, when more complex molecules sorb to a mineral surface, they can present a surface to the fluid that differs significantly from surface of the pristine mineral. Frequently, mineral surfaces are contaminated by various organic materials. In soils, for example, these organic coatings are likely to be humic or fulvic acids; *in vivo*, the coatings are likely to be proteinaceous. Depending on the nature of the organic coating, the surface properties may be very different from the uncoated mineral.

An extreme situation is presented by the organo-clays which are of considerable interest as sorbents for toxic organic material (Mortland et al., 1986; Costanzo et al., 1990). The aim is to make the external surfaces of the organo-clays hydrophobic (organophilic), and this is accomplished by exchanging the inorganic cations already on the clay surfaces with alkylammonium. A second, although less common, approach is to sorb a neutral amine on the surface of a material that has little or no cation exchange capacity.

The surface thermodynamic properties of these two types of organo-clay have been examined (Norris et al., 1992; Li, 1993; Li et al., 1993), and some of these data are shown in Table 5. The origin of the hydrophilicity of natural smectite clay minerals (see, e.g., SWy-1 in Table 4) is the high value of γ^-. Therefore, it is not surprising that treatment of the surface with either an amine or an alkylammonium cation reduces the Lewis basicity. This reduction is understandable because the part of the amine and alkylammonium ion exposed to the fluid is the hydrocarbon moiety of the molecule, an essentially apolar material. In contrast, there is every reason to expect that humic acids, which are rich in

Table 5

Comparison of surface properties of montmorillonite and talc untreated and treated with several different alkylammonium cations and a C_{18}-amine, respectively

Material	γ^{LW}	γ^{AB}	γ^{+}	γ^{-}	Reference
montmorillonite	41.7	10.1	0.7	36.2	1
-C_{10}	40.5	4.7	0.3	15.8	1
-C_{15}	37.7	0.4	0.3	0.1	1
-HDTMA	40.0	4.0	0.5	8.7	1
talc	28.5	2.2	0.1	11.6	2
-C_{18}	19.3	0.0	0.0	0.0	2

Values reported in mJ/m^2 and determined at 20°C
Sources:
1   Ammonium Wyoming montmorillonite (SWy-1); Norris et al. (1992); Norris (1993)
2   Impure talc; C_{18} coverage at 1 wt % (Li et al., 1993; Li, 1993)

carboxyl, hydroxyl and C=O groups, will have an appreciable Lewis basicity (Chen and Schnitzer, 1978).

SURFACE PROPERTIES OF BIOLOGICAL MATERIALS

Most native biological materials are hydrophilic, e.g., blood serum albumin, erythrocytes, and most bacteria. They typically have a high γ^{-} value ($\gamma^{-} > 40$ mJ/m^2) and, in the non-hydrated form, a low γ^{+} value and a γ^{LW} value of 35–43 mJ/m^2. In the hydrated form, the γ^{LW} approaches more closely the value for water ($\gamma^{LW} = 21.8$ mJ/m^2), usually remaining at around 27 mJ/m^2 (see values for HSA in Table 4) (van Oss, 1992).

INTERACTIONS BETWEEN MINERAL SURFACES AND BIOLOGICAL MATERIALS

Relatively few mineral–biological interactions have been studied with respect to surface thermodynamic properties. As an example of these interactions, the case of adsorption of human serum albumin (HSA) onto clean glass surfaces will be described.

The three-dimensional configuration of HSA was elucidated by He and Carter (1992): for the purposes of the following calculations, the six sub-domains are taken to be 40-Å long cylinders, with R = 13 Å. The "elbows" are taken to be half-spheres, R = 13 Å. When HSA is dissolved in 0.1M NaCl at pH = 7.0, its ψ_o-potential is –31.8 mV (Abramson et al., 1964). Hydrated HSA has the surface properties listed in Table 4 (van Oss, 1992). For these calculations, a glass surface different from that listed in Table 4 was used: a ψ_o-potential of –54.4 mV was assumed, $\gamma^{LW} = 37.4$, $\gamma^{+} = 1.82$ and $\gamma^{-} = 43.5$ mJ/m^2 were calculated from contact angle measurements. The aggregate ψ_o-potential for the HSA-glass interaction then is –41.6 mV.

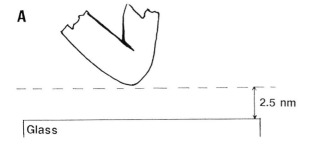

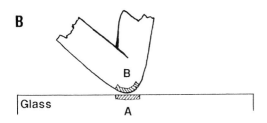

Figure 2. Schematic diagrams of the average non-specific repulsion (A) and the discrete-site attraction (B) between a section of a protein molecule (e.g., human serum albumin) and a high-energy surface (e.g., clean glass). After van Oss (1994).

Average non-specific repulsion

The average non-specific interaction energies (1 is glass, 2 is HSA and w is water, at "contact," pH = 7.0 and an ionic strength of 0.1) are:

$$\Delta G^{LW}_{1w2} = -0.47 \text{ kT}$$
$$\Delta G^{AB}_{1w2} = +25.4 \text{ kT}$$
$$\Delta G^{EL}_{1w2} = +5.0 \text{ kT}$$

At a separation of 25 Å, the total repulsion is +1 kT; at 10 Å it exceeds +10 kT. Owing to the strong hydrophilicity (high γ^-) of both high energy mineral surfaces (e.g., clean glass, muscovite, most smectite clay minerals) and of most native biological materials (e.g., blood serum proteins), there is an average non-specific repulsion, which, if it were operating alone, would prevent proteins, dissolved in aqueous media, from making any contact at all with clean glass surfaces; see Figure 2B.

Discrete site attraction

However, there are discrete sites on glass surfaces, e.g., sites with encrusted Ca^{2+} ions. These can locally not only attract a (generally negatively charged) protein molecule at any part of its surface, but they will also attract the predominantly electron-donor protein surface via the Ca^{2+} ions' electron-acceptor capacity. Both types of attraction can most readily interact with protein at their sites of greatest curvature (i.e., at the smallest value of the radius of curvature, R), because it is the sites of the smallest R that can most readily pierce the average repulsion field; see Figure 2B. The HSA-glass interaction at a discrete site attraction (a positively charged site on glass and negatively charged site on HSA) corresponds to about −12.5 kT (Gabler, 1978), so that when one assumes two charge attractions per site, $\Delta G^{EL}_{1w2} = -25$ kT per site. Concomitantly with the double electrostatic attraction, one hydrogen-bonding attraction ($\Delta G^{AB}_{1w2} = -7.5$ kT per site) is also assumed to occur; the decay with distance for such site-site interactions follows the same rules as for macroscopic interactions.

The adsorption of proteins onto high energy surfaces usually is more or less strongly pH-dependent (MacRitchie, 1972; Norde, 1986), but gives rise to no or little protein denaturation (Morissey and Stromberg, 1974).

Low energy surfaces

When mineral surfaces are "hydrophobized," with amines or tertiary or quaternary ammonium bases (see Table 5), their outer surfaces become more or less completely apolar (i.e., $\gamma^+ = \gamma^- \approx 0$). Immersed in water, they will then not only attract hydrophobic compounds, but they will even bind to fairly hydrophilic organic materials which are dissolved in the water (Costanzo et al., 1990). Such hydrophobized surfaces will readily sorb, e.g., proteins over their entire surfaces (no privileged attractor sites play a role here; see van Oss, 1991). The adsorption of proteins by low-energy surfaces usually is an order of magnitude greater than the adsorption of proteins by high energy surfaces (MacRitchie, 1972). The sorption of proteins onto low-energy surfaces tends to have a denaturing effect on the protein (Kim and Lee, 1975), but it is not very pH-dependent (van Oss and Singer, 1966; MacRitchie, 1972).

REFERENCES

L.S.Abramson, H.A., Moyer, L.S. and Gorin, M.H. (1964) Electrophoresis of Proteins. Hafner, New York.
Adamis, Z. and Timar, M. (1978) Studies on the effect of quartz, bentonite and coal dust mixtures on macrophages in vitro. Brit . J. Exp. Path. 59, 411–415.
Adamson, A.W. (1982) Physical Chemistry of Surfaces. Wiley-Interscience, New York.
Appel, J.D., Fasy, T.M., Kohtz, D.S., Kohtz, J.D. and Johnson, E.M. (1988) Asbestos fibers mediate transformation of monkey cells by exogeneous plasmid DNA. Proc. Nat'l Acad. Sci. 85, 7670–7674.
Campbell, J.A. (1940) Effects of precipitated silica and of iron oxide on the incidence of primary lung tumours in mice. Brit. Med. J. 2, 275–280.
Chaudhury, M.K. (1984) Short Range and Long Range Forces in Colloidal and Microscopic Systems. Ph.D. dissertation, State University of New York at Buffalo.
Chen, Y. and Schnitzer, M. (1978) The surface tension of aqueous solutions of soil humic substances. Soil Science 125, 7–15.
Costanzo, P.M., Giese, R.F. and van Oss, C.J. (1990) Determination of the acid-base characteristics of clay mineral surfaces by contact angle measurements - implications for the adsorption of organic solutes from aqueous media. J. Adhesion Sci. Technol. 4, 267–275.
Costanzo, P.M., Giese, R.F. and van Oss, C.J. (1991) The determination of surface tension parameters of powders by thin layer wicking. In: Advances in Measurement and Control of Colloidal Processes, Int'l Symposium on Colloid and Surface Engineering, R.A. Williams and N.C. de Jaeger, Eds., Butterworth Heinemann, London, 223–232.
Debye, P. (1920) van der Waals cohesion forces. Phys. Zeits. 21, 178–187.
Debye, P. (1921) Molecular forces and their electrical interpretation. Phys. Zeits. 22, 302–308.
Dupré, A. (1869) Theorie Mecanique de la Chaleur. Gauthier-Villars, Paris.
Everett, D.H. (1988) Basic Principles of Colloid Science. Royal Society of Chemistry, London.
Fowkes, F.M. (1963) Additivity of intermolecular forces at interfaces. I. Determination of the contribution to surface and interfacial tensions of dispersion forces in various liquids. J. Phys. Chem. 67, 2538–2541.
Fowkes, F.M. (1966) Role of acid-base interfacial bonding in adhesion. J. Adhesion Sci. Technol. 1, 7–27.
Gabler, R. (1978) Electrical Interactions in Molecular Biophysics. Academic Press, New York.

Giese, R.F., van Oss, C.J., Norris, J. and Costanzo, P.M. (1990) Surface energies of some smectite minerals. In: Proc.9th Int'l Clay Conference, Strasbourg, France, Y. Tardy and V.C. Farmer, Eds., p. 33–41.

Giese, R.F., Costanzo, P.M. and van Oss, C.J. (1991) The surface free energies of talc and pyrophyllite. Phys. Chem. Minerals. 17, 611–616.

Girifalco, L.A. and Good, R.J. (1957) A theory for the estimation of surface and interfacial energy. I. Derivation and applicat ion to interfacial tension. J. Phys. Chem. 64, 904–909.

Good, R.J. (1966) Physical significance of parameters γ, γ_s and Φ, that govern spreading on adsorbed films. SCI Monographs 25, 328–356.

Good, R.J. and Girifalco, L.A. (1960) A theory for estimation of surface and interfacial energies. III. Estimation of surface e nergies of solids from contact angle data. J. Phys. Chem. 64, 561–565.

Gormley, I.P. and Addison, J. (1983) The in vitro cytotoxicity of some standard clay mineral dusts of respirable size. Clay Minerals 18, 153–163.

Green, F.H.Y., Bowman, L., Castranova, V., Dollberg, D.D., Elliot, J.A. et al. (1982) Health implications of the Mount St. Helens eruption: Laboratory investigations. Ann. Occup. Hygiene 26, 921–933.

Guthrie, G.D., Jr. (1992) Biological effects of inhaled minerals. Amer. Mineral. 77, 225–243.

Hansen, K. and Mossman, B.T. (1987) Generation of superoxide (O^{2-}) from alveolar macrophages exposed to asbestiform and nonfibr ous particles. Cancer Res. 47, 1681–1686.

He, X.M. and Carter, D.C. (1992) Atomic structure and chemistry of human serum albumin. Nature 358, 209–214.

Heppleston, A.G. (1984) Pulmonary toxicology of silica, coal and asbestos. Environ. Health Persp. 55, 111–127.

Keesom, W.M. (1915) The second virial coefficient for rigid spherical molecules whose mutual attraction is equivalent to that of a quadruplet placed at its center. Proc. Royal Acad. Sci. Amsterdam, 18, 636–646.

Keesom, W.M. (1920) Quadrupole moments of the oxygen and nitrogen molecules. Proc. Royal Acad. Sci. Amsterdam, 23, 939–942.

Keesom, W.M. (1921a) van der Waals attractive force. Phys. Zeits. 22, 129–141.

Keesom, W.M. (1921b) van der Waals attractive force. Phys. Zeits. 22, 643–644.

Kennedy, T.P., Dodson, R., Rao, N.V., Ky, H., Hopkins, C., Maser, M., Tolley, E. and Hoidal, J.R. (1989) Dusts causing pneumoconiosis generate OH and produce hemolysis by acting as Fenton catalysts. Arch. Biochem. Biophys. 269, 359–364.

Kerch, H.M., Gerhardt, R.A. and Grazul, J.L. (1990) Quantitative electron microscope investigation of the pore structure in 10:90 colloidal silica/potassium silicate sol-gels. J. Amer. Ceram. Soc. 73, 2228–2237.

Kim, S.W. and Lee, R.G. (1975) Adsorption of blood proteins onto polymer interfaces. Advan. Chem. Ser. 145, 218–229.

Le Bouffant, L., Daniel, H., Martin, J.C. and Bruyere, S. (1982) Effect of impurities and associated minerals on quartz toxicity. Ann. Occup. Hygiene 26, 625–634.

Li, Z. (1993) Surface Free Energy of Some Phyllosilicate Minerals and Fibrous Minerals: The Application of Thin Layer Wicking to the Measurement of the Surface Free Energy of Solids. Ph.D. dissertation, State University of New York at Buffalo.

Li, Z., Giese, R.F., van Oss, C.J., Yvon, J. and Cases, J. (1993a) The surface thermodynamic properties of talc treated with octa decylamine. J. Colloid Interface Sci. 156, 279–284.

Li, Z., Giese, R.F., Wu, W., Sheridan, M.F. and van Oss, C.J. (1993b) The surface thermodynamic properties of some volcanic ash colloids. Colloids and Surfaces (submitted).

Lifshitz, E.M. (1955) Zh. Eksp. Teor. Fiz. 29, 94.

London, F. (1930) Theory and systematics of molecular forces, Zeits. Phys. 63, 245–279.

MacRitchie, F. (1972) The adsorption of proteins at the solid/liquid interface. J. Colloid Interface Sci. 38, 484–488.

Merchant, J.A., Baxter, P., Bernstein, R., McCawley, M., Falk, H., Stein, G., Ing, R. and Attfield, M. (1982) Health implications of the Mount St. Helens eruption: Epidemiological considerations. Ann. Occup. Hygiene 26, 911–919.

Morissey, B.W. and Stromberg, R.R. (1974) The conformation of adsorbed blood proteins by infrared bound fraction measurements. Colloid Inter. Sci. 46, 152–164.

Mortland, M.M., Shaobi, S. and Boyd, S. (1986) Clay–organic complexes as adsorbents for phenol and chlorophenols. Clays & Clay Minerals 34, 581–585.

Mossman, B.T. and Marsh, J.P. (1989) Evidence supporting a role for active oxygen species in asbestos-induced toxicity and lung disease. Environ. Health Persp. 81, 91–94.

Newman, A.C.D. (1987) Chemistry of Clays and Clay Minerals. Mineralogical Society (U.K.), Wiley, New York.

Norde, W. (1986) Adsorption of proteins from solutions at the solid–liquid interface. Ad. Colloid Interface Sci. 25, 267–340.
Norris, J. (1993) The Surface Free Energy of Smectite Clay Minerals. Ph.D. dissertation, State University of New York at Buffalo.
Norris, J., Giese, R.F., van Oss, C.J., and Costanzo, P.M. (1992) Hydrophobic nature of organo-clays as a Lewis acid-base phenomenon. Clays & Clay Minerals 40, 327–334.
Norris, J., Giese, R.F., Costanzo, P.M. and van Oss, C.J. (1993) The surface energies of cation substituted Laponite. Clay Minerals 28, 1–11.
Overbeek, J.ThG. (1952) Electrokinetics. In: Colloid Science. H.R. Kruyt, Ed., Elsevier, Amsterdam, p. 194–244.
Overbeek, J.Th.G. and Wiersema, P.H. (1967) Interpretation of electrophoretic mobilities. In: Electrophoresis. M. Bier, Eds., Academic Press, New York, p. 1–52.
Parks, G.A. (1990) Surface energy and adsorption at mineral/water interfaces: An introduction. Rev. Mineral. 23, 133–175.
Skvarla, J. (1993) A physico-chemical model of microbial adhesion. J. Chem. Soc. Faraday Trans. 1, 89, 88–90.
Stanton, M.F., Layard, M., Tegeris, A., Miller, E., May, M., Morgan, E. and Smith, A. (1981) Relation of particle dimension to carcinogenicity in amphibole asbestoses and other fibrous minerals. J. Nat'l Cancer Inst. 67, 965–975.
van Oss, C.J. (1975) The influence of the size and shape of molecules and particles on their electrophoretic mobility. Separ. Purif. Meth. 4, 167–188.
van Oss, C.J. (1991) The forces involved in bioadhesion to flat surfaces and particles—Their determination and relative roles. Biofoul. 4, 25–35.
van Oss, C.J. (1992) Hydrophilic and hydrophobic interactions. In: Protein Interactions. H. Visser, Ed., VCH, Weinheim, New York, p. 25–55.
van Oss, C.J. (1994) Interfacial Forces in Aqueous Media (in press). Marcel Dekker, New York.
van Oss, C.J. and Singer, J.M (1966) The binding of immune globulins and other proteins by polystyrene latex particles. J. Ret . Soc. 3, 29–40.
van Oss, C.J., Chaudhury, M.K. and Good, R.J. (1988) Interfacial Lifshitz–van der Waals and polar interactions in macroscopic systems. Chem. Rev. 88, 927–941.
van Oss, C.J., Giese, R.F. and P.M. Costanzo (1990) DLVO and non-DLVO interactions in hectorite. Clays & Clay Minerals. 38, 151–159.
van Oss, C.J., Giese, R.F., Li, Z., Murphy, K., Norris, J., Chaudhury, M.K. and Good, R.J. (1992a) Determination of contact angles and pore sizes of porous media by column and thin layer wicking. J. Adhesion Sci. Technol. 6, 413–428.
van Oss, C.J., Giese, R.F. and Norris, J. (1992b) Interaction between advancing ice fronts and erythrocytes. Cell Biophys. 20, 253–261.
Washburn, E.W. (1921) The dynamics of capillary flow. Phys. Rev. 17, 273–283.
Wentzek, R. (1993) Surface Characterization of Cominuted Silicate Particles, M.A. thesis, State Univeristy of New York at Buffalo, 45 p.
Woodworth, C.D., Mossman, B.T. and Craighead, J.E. (1982) Comparative effects of fibrous and non-fibrous minerals on cells and liposomes. Environ. Res. 27, 190–205.

CHAPTER 11

EPIDEMIOLOGY AND PATHOLOGY OF ASBESTOS-RELATED DISEASES

Agnes B. Kane

Department of Pathology and Laboratory Medicine
Brown University
Providence, Rhode Island 02912 U.S.A.

HISTORICAL INTRODUCTION

Asbestos is a term derived from the Greek meaning "not quenchable" or "inextinguishable," and it is used to describe naturally occurring, fibrous silicates with flexibility, strength, and a high resistance to heat and chemicals. These properties have been exploited for almost 4500 years, resulting in a wide range of human exposures to asbestos. Currently, chrysotile (serpentine asbestos) is the most common form of asbestos used commercially (95%), followed by amosite and crocidolite (two types of amphibole asbestos). Fibrous serpentine and amphiboles are widely distributed at the earth's surface (see Chapter 2), so, in addition to occupational exposures, a background exposure to these minerals can be significant in some locations. Furthermore, numerous other naturally-occurring minerals can exhibit a fibrous morphology, including some zeolites (e.g., erionite), wollastonite, palygorskite ("attapulgite"), and sepiolite. Many of these minerals are also used commercially.

Despite the widespread distribution of these natural mineral fibers, the health hazards associated with exposure to fibrous minerals were not recognized until the twentieth century when extensive commercial use of asbestos fibers began. In fact, recognition of the link between occupational exposure to asbestos and specific lung diseases was a relatively late development in occupational epidemiology. Asbestos fibers have been used since antiquity. For example, anthophyllite was used in pottery in Finland in 2500 B.C., and asbestos was used for lamp wicks and burial cloths in the first century. Asbestos tablecloths and clothing were described by Charlemagne in 800 A.D. and Marco Polo in 1250 A.D. (Liddell, 1991). During these centuries, asbestos fibers were not considered harmful, although diseases resulting from occupational exposure to other mineral dusts were recognized. In 79 A.D., Pliny recorded disease associated with exposure to mercuric sulfide dust. Agricola in 1556 reported increased death rates for miners in Bohemia, and Paracelsus described miner's sickness in 1567. Ramazzini, the father of occupational medicine, cataloged numerous occupational diseases in 1700 (Checkoway et al., 1989). The pathologic definition of the term pneumoconiosis—"the non-neoplastic reaction of the lung to the accumulation of inhaled dust"—was formulated by Zenker in 1867 (Thurlbeck, 1988). Major industrial use of asbestos began in 1878 with mining of chrysotile in Quebec, followed by crocidolite mining in 1910 and amosite mining in 1916 in South

Africa. It was not until 1907 when asbestos-related diseases were first recognized: Murray described pulmonary fibrosis in a textile worker in Britain (Craighead et al., 1982). In 1955, Doll reported an increased risk of lung cancer in asbestos workers (Doll, 1955). In 1960, the association between asbestos exposure and malignant mesothelioma was described by König (1960) in Germany and by Wagner in the Union of South Africa (Wagner et al., 1960). Concern about the effects of environmental exposure to mineral fibers is more recent. Disease associated with environmental exposure to erionite was not reported until the 1980s (Baris et al., 1987). Currently, there is ongoing concern about the health risks associated with exposure to asbestos in buildings and urban environments (Health Effects Institute, 1991).

Between 1920 and 1950, isolated case reports described lung fibrosis, cancer, and malignant mesotheliomas occurring in asbestos workers (Lemen et al., 1980). Widespread acceptance of the association between exposure to asbestos and these diseases did not occur until longer case series and epidemiologic studies of cohorts of asbestos workers were published in the 1950s and 1960s (Checkoway et al., 1989). These key epidemiologic studies will be discussed in the next section. The evidence from occupational epidemiology was supported by production of similar lung diseases by asbestos fibers in laboratory animals, which have primarily been rodents (discussed in detail in Chapter 15; see also: Pott and Friedrichs, 1972; Wagner et al., 1973; Davis, 1974). These laboratory models investigated in the 1960s and 1970s raised initial concern that other mineral fibers besides asbestos could cause malignant mesothelioma (Stanton et al., 1981). This potential health hazard will be discussed in the final section of this chapter.

KEY EPIDEMIOLOGIC STUDIES OF ASBESTOS-RELATED DISEASES

Major diseases associated with asbestos exposure

Exposure to asbestos fibers is associated with the following diseases: diffuse fibrous scarring of the lungs (asbestosis), localized fibrous scars lining the space surrounding the lungs (pleural plaques), rare malignant tumors arising from the pleural, pericardial, or peritoneal linings (diffuse malignant mesotheliomas), and an increased risk of lung cancer (bronchogenic carcinoma) especially in cigarette smokers (Craighead et al., 1982). The discovery of the relationship between these diseases and exposure to asbestos is an important lesson in occupational epidemiology. In fact, these epidemiologic studies are frequently used as prototypes in teaching occupational health.

Several different methodologies have been used to establish the relationships between diseases and asbestos exposure. The choice of methodology can vary depending on the nature of the disease being studied. For a disease like lung cancer (with an annual incidence in the United States of 500 to 2000 cases per million people; Checkoway et al., 1989), the disease can be studied using *cohort studies*, in which the incidence of disease is followed over time in a large group of individuals with similar occupations or other characteristics (a cohort). In contrast, diffuse malignant mesothelioma is a very rare tumor, approximately 1 to 10 cases per million people are diagnosed in the United States each year (Checkoway et al., 1989). Such rare diseases are difficult to study using a cohort approach. Instead, epidemiological studies of mesothelioma typically use a "case" approach, in which individuals are studied to elucidate the relationships between disease and pathogenic factors (e.g., asbestos exposure). *Case studies* report the histories (e.g., asbestos exposure history, smoking history) of individuals with a disease. *Case control studies* compare the histories of a disease-bearing group with the histories of a non-disease-bearing control group. Finally, a cluster of a rare disease associated with a common exposure is published as a *case series*.

Studies in occupational epidemiology of asbestos-related diseases

Case series. The classic case series in the history of asbestos-related diseases is a paper by J.C. Wagner, C.A. Sleggs, and P. Marchand in the *British Journal of Industrial Medicine* in 1960. This paper described 33 cases of diffuse malignant mesothelioma clustered in the North West Cape Province in the Union of South Africa, revealing several important aspects of the relationships between this rare tumor and exposure to amphibole asbestos. After diligent investigation of the occupational and environmental exposures of these patients to the blue asbestos (or crocidolite) that was mined in this region, they discovered that most of these patients had worked in these mines or transported crude asbestos (Wagner, 1991). Thus, this landmark paper established a connection between occupational exposures to crocidolite asbestos and diffuse malignant mesothelioma. In addition, Wagner et al. (1960) noted a connection between pleural plaques and asbestos exposure in these workers. However, a few patients had no occupational exposure to asbestos, but they did live close to asbestos mills or tailings dumps. Several patients were also exposed as children while they played on these dumps. Hence, this paper presented the first examples of mesotheliomas resulting from environmental exposure to crocidolite asbestos. Finally, on the basis of thorough occupational histories, the long latent period between the first exposure to asbestos and the diagnosis of malignant mesothelioma was discovered—an average of 44 years in this case series (Wagner et al., 1960).

Case control studies. The results of the Wagner et al. (1960) study strongly suggest that occupational and high environmental exposures to crocidolite asbestos are associated with the development of diffuse malignant mesothelioma. In order to test this hypothesis more critically, a different epidemiologic study design is required. For such a rare disease as malignant mesothelioma, a very large number of asbestos workers or exposed individuals would need to be followed for several decades in a conventional prospective longitudinal study of a cohort. An alternative approach is a case-control study that estimates the relative risks of an exposed worker developing a specific disease in comparison to an unexposed worker. By choosing appropriate unexposed workers as controls, a case-control study can account for the effects of complicating factors (e.g., smoking history), thereby isolating the effects of exposure to mineral dusts. This approach may also provide information about the levels of exposure associated with an increased risk of developing that disease (Checkoway et al., 1989). An example of this risk of developing that disease (Checkoway et al., 1989). An example of this approach is the nested case-control study conducted by Berry and Newhouse (1983) of workers involved in the manufacture of asbestos friction materials. This manufacturing plant predominantly used chrysotile asbestos, except for a brief period when a small number of workers used crocidolite asbestos. Berry and Newhouse identified ten cases of malignant mesothelioma in workers at this plant. In order to determine whether these tumors were associated with crocidolite asbestos exposure, they matched each of these cases with four workers of the same sex, age, and length of employment, assuming that these criteria would properly select for all possible complicating factors. This nested case-control analysis revealed that 8/10 of the workers who developed malignant mesothelioma had a history of occupational exposure to crocidolite asbestos compared with only 3/40 of the matched controls (Checkoway et al., 1989).

Cohort studies. Numerous case-control studies and longitudinal cohort studies have clearly established that workers exposed to amphibole asbestos (e.g., crocidolite and amosite) have a greater risk of developing malignant mesothelioma than those exposed to chrysotile asbestos. These epidemiologic studies have also confirmed the long latent period (20 to 60 years) characteristic of malignant mesothelioma; a lack of association between the development of malignant mesothelioma and cigarette smoking or asbestosis; and a higher incidence of this rare tumor in certain industries, such as shipbuilding, insulation, and assembly of gas masks (Craighead et al., 1982; Churg and Green, 1988). However, despite these intensive epidemiologic studies, other important questions are difficult to answer by these approaches. For example, some cohorts exposed to chrysotile have a higher incidence of lung cancer than other cohorts. Is this discrepancy due to different levels or duration of exposure to chrysotile, to past or concurrent exposure to amphiboles, or to exposure to other co-carcinogens? Exposure conditions, especially in historical cohort studies, are difficult to ascertain because different techniques were used to monitor airborne levels of fibers in the past. In addition, some chrysotile mines are contaminated with tremolite as well as other iron-containing minerals such as nemalite. Complete characterization of the fibers in use at any given time is frequently not available (Checkoway et al., 1989). Consequently, a direct determination of fiber exposure for an individual or cohort can be problematic. An alternative approach is to count and to characterize the fibers present in lung tissue obtained by biopsy or at autopsy (see the detailed discussion in Chapter 13). This technique measures lung fiber burden and reflects the dose of fibers delivered to the target tissue in each specific individual. However, there are several problems associated with the use of lung fiber burden as an indirect assessment of exposure. This technique is very demanding and time-consuming for proper identification of fibers. The best current approach requires tedious fiber counts using transmission electron microscopy and verification of fiber types using sophisticated techniques, such as electron diffraction and X-ray microanalysis. This approach is not routine, and results can vary between laboratories.

Analyses of lung fiber burdens have provided useful and interesting information, despite the limitations discussed above. Quite surprisingly, workers exposed to chrysotile asbestos for many years often have low levels of fibers in their lungs when examined at autopsy (Roggli, 1990). Hence, it appears that chrysotile fibers dot not remain in the lung over time, possibly because they split and fragment within lung tissue (Roggli and Brody, 1984) or they dissolve (Hume and Rimstidt, 1992). In contrast, amphibole fibers apparently persist in the lung (Bellman et al., 1987). This observation raises a question about the length of time required for a critical number of fibers to interact with the target tissue before disease develops (Pott, 1987). Unfortunately, this important question cannot be answered by epidemiologic approaches or measurement of lung fiber burdens alone. New information based on animal models and *in vitro* mechanistic studies is required in order to answer this important question. Lung fiber burden studies also cannot address the issue of confounding exposures, including cigarette smoking or exposure to polycyclic aromatic hydrocarbons and other chemicals (Checkoway et al., 1989).

The association between diffuse fibrosis of the lungs and exposure to asbestos was recognized in the 1930s. The Chief Medical Inspector of Factories in the United Kingdom, E.R.A. Merewether, examined 363 workers in an asbestos textile factory and found lung fibrosis in 25%. In an historical analysis of these cases, he established a correlation between fibrosis (asbestosis) and the intensity and duration of exposure to asbestos (Checkoway et al., 1989). He also described a latent period between exposure and the development of asbestosis; the latent period for asbestosis (7 to 25 years) is shorter than that for diffuse malignant mesothelioma. On the basis of these observations, Asbestos Industry Regulations were established in the United Kingdom in 1933 in order to suppress dust exposure in this industry and prevent the development of asbestosis. As a result of this key epidemiologic study, exposure limits for asbestos fibers have been decreased progressively and asbestosis is becoming less common.

Using a similar approach, Doll (1955) observed increased mortality from lung cancer in British asbestos workers. He described a latent period of 20 years or more between onset of exposure and the development of lung cancer. All of these initial associations between various types of lung disease and occupational exposure to cancer have been confirmed by formal cohort studies. A cohort study follows a population of workers over several decades in order to compare their development of disease relative to the background occurrence of this disease in the general population. There are two types of cohort studies: historical and prospective. Both types require a large number of subjects and a long period for follow-up. Two large historical cohort studies have followed the development of disease in asbestos workers. J.C. McDonald's study of chrysotile asbestos miners and millers (a cohort of 11,000 workers) in Quebec began in 1977 and I.J. Selikoff's study of insulation workers (a cohort of 17,800) in the United States and Canada began in 1967. An increased incidence of lung cancer was observed in the chrysotile miners and millers, whereas the insulation workers had an increased incidence of asbestosis, lung cancer, malignant mesothelioma, and other cancers (McDonald and McDonald, 1980; Checkoway et al., 1989; Selikoff et al., 1979; Lidell, 1991). Data obtained from Selikoff's cohort of insulation workers revealed a synergistic association between exposure to asbestos and cigarette smoking. In a key paper, Selikoff et al. (1964) reported the risk of asbestos workers developing lung cancer is 50 times greater if they also smoked cigarettes. Lung cancer is responsible for the majority of deaths caused by asbestos-related diseases (Checkoway et al., 1989).

Limitations of epidemiologic studies. Although these large cohort studies provided important information about a synergistic risk associated with cigarette smoking and lung cancer, they raised numerous other questions that could not be answered by an epidemiologic study: Why did the cohort of insulation workers have higher mortality and incidence of disease than chrysotile miners and millers? What is the dose-response relationship between asbestos exposure and the development of asbestos-related diseases? What is the magnitude of these workers' past exposures? Can the risks of developing asbestos-related diseases be extrapolated from these high, historical exposure levels (~10 fibers/ml) to the current low levels from occupational exposure (~0.1 fibers/ml), or even background environmental exposure levels (<0.0002 to 0.008 fibers/ml) (Health Effects Institute, 1991, Table 4-8)? How much risk is attributable to asbestos and how much to cigarettes in the development of lung cancer? Do workers with asbestosis have a higher risk of developing lung cancer? Epidemiologic studies alone are unlikely to provide the answers for these complex issues (Checkoway et al., 1989). Additional information is required from pathologic observations and mechanistic studies.

CLINICAL AND PATHOLOGIC FEATURES OF ASBESTOS-RELATED DISEASES

Pulmonary diseases

Asbestosis. Diffuse bilateral scarring or fibrosis of the lungs can be produced by occupational exposure to all types of asbestos, usually at high levels for long periods of time. The scarring process is slowly progressive, even after exposure has ceased. Gradually, patients develop shortness of breath during physical exertion and a dry, nonproductive cough. The scarring is visible on a chest x-ray as linear and nodular densities predominantly involving the lower lobes of both lungs. The diagnosis is based on the clinical and pathologic or radiologic features combined with an occupational history of asbestos exposure. Many other factors (e.g., drug or allergic reactions, collagen diseases, or sarcoidosis[1]) can produce this pattern of diffuse interstitial fibrosis, so the presence of fibrosis does not necessarily reflect exposure to asbestos (Craighead et al., 1982).

The histopathologic features of asbestosis are focal areas of scarring around the terminal bronchioles spreading into the walls of the alveoli. A chronic inflammatory reaction and increased numbers of alveolar macrophages may be present. Although it is difficult to resolve respired asbestos fibers using the light microscope, some amphibole fibers become coated with iron and mucopolysaccharides and are easily recognized as asbestos (or ferruginous) bodies (Churg and Green, 1988).

---

[1] Sarcoidosis is a disease of unknown etiology that affects the lymph nodes, lungs, and several other organs. It is characterized by granulomatous lesions showing little or no necrosis (dead cells).

There is no effective therapy for asbestosis. Unfortunately, this disease progresses, leading to early death from lung or heart failure. These patients may also have an increased risk of developing lung cancer (Churg and Green, 1988).

Lung cancer. Workers exposed to asbestos fibers have a greatly increased risk of developing the same types of lung cancer as cigarette smokers. As discussed earlier, the interaction between cigarette smoke and asbestos fibers is synergistic. Patients with lung cancer may have no symptoms, and the disease is diagnosed incidentally as a mass visible on chest x-ray. Frequently, the tumor obstructs the lumen (or interior space) of the bronchus leading to persistent cough or repeated episodes of pneumonia. Some types of lung cancer spread widely throughout the lungs and other organs. Patients with disseminated cancer may develop weight loss, anemia, and other systemic symptoms. Unless diagnosed and removed at an early stage when the tumor is still localized, survival beyond 1 to 2 years is unusual, although some patients may initially respond to radiation or chemotherapy (Churg and Green, 1988).

Lung tumors arise from epithelial cells lining the bronchi and bronchioles, the conducting airways of the lungs. These tumors usually grow in large masses, but they frequently spread to lymph nodes and throughout the lungs. Eventually, the tumor cells disseminate to the brain, bone marrow, and other organs. The tumors have different histologic appearances: they consist of squamous, glandular, and small or large undifferentiated cells. No histologic pattern or markers specific for lung cancer produced by asbestos exist (Churg and Green, 1988). In other words, it is not possible to determine if a lung cancer were caused by asbestos exposure or exposure to another agent.

Pleural reactions

The lungs are surrounded by a potential space lined by a single layer of mesothelial cells. This potential space is called the pleural cavity or pleural space, and the tissues lining the pleural space are called the visceral pleura (adjacent to the lungs) and the parietal pleura (adjacent to the ribs). By an unknown route, asbestos fibers migrate towards this pleural space to produce a range of pleural diseases. The most common condition associated with asbestos exposure is a localized fibrotic scar called a pleural plaque. Up to 40 to 60% of some cohorts of asbestos workers show evidence of pleural plaques. These develop several years or decades after exposure to asbestos as well as to other types of mineral fibers. Environmental exposure to tremolite and erionite is also associated with pleural plaques. After many years, the fibrous scars may become calcified and, therefore, visible on chest x-ray. If they are extensive, these plaques may interfere with lung function; however, they usually do not cause severe disability or premature death. When examined histologically, pleural plaques contain abundant collagen fibers, frequently covered by a single layer of mesothelial cells. There is no association between the presence of pleural plaques and the development of malignant mesothelioma (Craighead et al., 1982; Craighead, 1987).

Other less common reactions of the pleura to asbestos fibers are the accumulation of fluid in the pleural space (effusions or pleurisy) and diffuse fibrosis of the pleura surrounding the lungs. Patients with pleural effusions may have no symptoms or they may complain of chest pain or difficulty in breathing. Pleural effusions usually appear within 10 years after initial exposure to asbestos fibers, followed later by the appearance of pleural plaques. Pleural effusions usually disappear spontaneously after a few months. Some of these patients may eventually develop diffuse scarring of the pleura resulting in impaired breathing (Churg and Green, 1988).

The least common, yet most rapidly fatal, complication of asbestos exposure is diffuse malignant mesothelioma. This tumor usually arises from the pleural lining, but may also develop in the pericardial sac or peritoneal lining of the abdominal cavity. This tumor has the longest latent period (20 to 60 years) of all of the asbestos-related diseases. Cases have appeared following exposure in childhood (Craighead, 1987). Consequently, there is concern that children exposed to asbestos fibers in schools are at increased risk of developing this tumor as adults (Health Effects Institute, 1991). As described earlier, exposure to amphibole fibers is associated with a higher risk for development of diffuse malignant mesothelioma. Unlike lung cancer, there is no increased risk in workers with asbestosis or with smoking histories. The most common cause of malignant mesothelioma is asbestos exposure, although this tumor has been found in association with chronic inflammation or radiation exposure. A very low background incidence of unknown etiology also exists (Peterson et al., 1984).

Patients with diffuse malignant mesothelioma usually have extensive spread of this tumor when they develop symptoms of weight loss, chest pain, or difficulty breathing. Complete surgical removal of this tumor is usually impossible and malignant mesothelioma does not respond well to radiation or chemotherapy. Death usually occurs 6 to 18 months after diagnosis (Alberts et al., 1988).

Other diseases associated with asbestos exposure

Various epidemiologic studies of asbestos workers have revealed a slightly elevated risk for development of cancers of the larynx, stomach, colon, and lymphoid tissues (lymphoma) (Churg and Green, 1988). These correlations have not been confirmed by all investigators.

Pathogenesis of asbestos-related diseases

A wide range of pathologic reactions to asbestos fibers (from fibrosis to cancer) is found in the lungs and pleura. However, a common feature of all of these reactions is a latent period of at least 10 years. This long latent period suggests that fibers may need to persist in the lungs and pleura for several years in order to cause injury to the target cells (Pott, 1987). This is an interesting, important hypothesis that has implications for the design of asbestos fiber substitutes. Theoretically, if fibers could dissolve or be effectively cleared from the lungs and pleura before initiating a biochemical response, these chronic diseases would not develop.

In addition to durability and persistence in lung tissue, other physical and chemical properties of fibers determine their biological reactivity. Some of these parameters have been identified in animal models and *in vitro* systems (discussed in detail in Chapter 17). The classic experiments conducted by Stanton and his colleagues in the 1960s led to formulation of the Stanton hypothesis: Long, thin mineral fibers, regardless of their composition, are more potent in inducing mesotheliomas than short fibers (Stanton et al., 1981). It is speculated that long, thin amphibole fibers are more easily translocated to the pleura where they interact with macrophages, phagocytic cells in the pleural space, and mesothelial cells. However, Stanton et al. applied fibers directly to the pleural surface, so translocation of particles was not a factor in their experiments. The physical interaction between fibers and these target cells perturbs the cellular membrane and organization of subcellular organelles. Macrophages attempting to phagocytize long asbestos fibers continuously release reactive oxygen species such as hydrogen peroxide and superoxide anion extracellularly (Goodglick and Kane, 1986; Shatos et al., 1987). These oxidants may damage adjacent normal cells: mesothelial cells lining the pleural space, fibroblasts in the alveolar walls, or epithelial cells lining the bronchi and bronchioles. According to this mechanism, the particle's shape determines its bioactivity, because a particle must be sufficiently large to prevent complete phagocytosis by a macrophage.[2] The chemical composition of asbestos fibers also determines their cytotoxicity. Surface Mg^{2+} ions of chrysotile asbestos contribute to their toxicity and carcinogenicity (Monchaux et al., 1981), and trace amounts of Fe^{2+} and Fe^{3+} ions in natural or manmade mineral fibers can catalyze the formation of the highly reactive hydroxyl radical (OH^-) (Weitzman and Graceffa, 1984; Aust and Lund, 1991). The pathogenicity of amphibole fibers may result from their high iron content and/or their prolonged persistence in tissue (Bellman et al., 1987).

The acute toxic effects of fibers that persist within these target cell populations in the lung are healed by the host. Injury inflicted by reactive oxygen species in the mesothelial lining or alveolar walls is repaired by proliferation of fibroblasts and deposition of collagen, just as wounds to the skin are healed by fibrous scarring. These fibrous scars are characterized histologically by the accumulation of extracellular collagen, the formation of pleural plaques beneath the mesothelial lining, and the development of diffuse interstitial fibrosis or asbestosis in the lungs (Craighead et al., 1982; McClellan et al., 1992). Both of these fibrotic reactions develop slowly but progressively, perhaps in response to repeated episodes of injury inflicted by fibers that persist at these sites. The biochemical mechanisms responsible for these chronic, progressive fibrotic reactions are not completely known. Recently, toxicologists have discovered that macrophages exposed to asbestos fibers release a variety of chemical mediators that perpetuate inflammation and stimulate fibroblasts to proliferate and deposit collagen (McClellan et al., 1992). A better understanding of these biochemical mechanisms is required in order to develop effective therapies for these fibrotic disorders.

The mechanisms responsible for the development of lung cancer and malignant mesothelioma after exposure to asbestos fibers are more complex.

---

[2] Macrophages are typically ~10–14 μm in diameter (Fawcett, 1986) and routinely phagocytize foreign bodies ~1–2 μm in diameter (e.g., bacteria). However, macrophages apparently have difficulty completely engulfing fibrous particles >~5–10 μm in length.

Asbestos and cigarette smoke interact synergistically to cause lung cancer. Several experimental studies suggest that cigarette smoke interferes with clearance of fibers from the tracheobronchial epithelial lining. Fibers facilitate transport of carcinogenic chemicals in cigarette smoke into target cells at this site (McClellan et al., 1992). *In vitro* studies (described in Chapter 17) have revealed a variety of interactions between asbestos fibers and tracheobronchial epithelial target cells that may promote the development of lung cancer at this site.

The mechanism leading to development of malignant mesothelioma is independent of cigarette smoke. It is hypothesized that persistent release of reactive oxygen species causes repeated injury to the mesothelial lining of the pleura. The mesothelium responds to injury by proliferation of adjacent, uninjured cells (Moalli et al., 1987). Reactive oxygen species or direct physical interference of fibers with the mitotic apparatus could cause genetic damage in this proliferating target cell population (Barrett et al., 1989). Proliferating mesothelial cells may also produce their own growth stimulatory factors. Repeated episodes of injury followed by proliferation and genetic damage may give rise to a tumor that proliferates autonomously (Antoniades, 1992).

A better understanding of these complex mechanisms and the properties of fibers responsible for toxicity and carcinogenicity is urgently required. As discussed in detail below, recognition of the health hazards associated with exposure to asbestos fibers has resulted in the development of asbestos fiber substitutes. Are these man-made mineral fibers also hazardous (McClellan et al., 1992)? Although commercial use of asbestos has been banned in the United States and Europe, these fibers persist in buildings and in the urban environment. Is exposure to in-place asbestos hazardous (Health Effects Institute, 1991)?

Implications of epidemiologic studies for prevention of asbestos-related diseases

Recognition of asbestos-related diseases was based on publication of isolated case reports, followed by case series and cohort studies as described in the Introduction to this chapter. These studies are hallmarks in the history of occupational epidemiology. As described in a new textbook on this discipline, "the data derived from occupational epidemiology research are used for decision making. Regulatory agencies rely heavily on epidemiologic data when proposing occupational and nonoccupational exposure limits" (Checkoway et al., 1989). Have epidemiologic studies of asbestos-related diseases fulfilled these important objectives? These studies eventually triggered strict occupational exposure limits and led to bans on commercial use of asbestos fibers in the United States and Europe (see Brodeur, 1985, for a discussion of the belated response of the asbestos industry to this information). These epidemiologic studies have been successful in sharply decreasing the incidence of at least one asbestos-related disease (asbestosis) among workers. Unfortunately, this success is limited to the United States and Europe. Asbestos and asbestos-containing products are shipped to developing countries with the emergence of asbestos-related diseases in these countries. Better occupational hygiene should be applied in developing countries in order to prevent these diseases in workers (Levy and Seplow, 1992).

While epidemiologic studies and effective occupational hygiene have hopefully reduced future asbestos-related diseases among workers, more complex questions about the health effects of in-place asbestos, background environmental exposure, and exposure to man-made mineral fibers remain. The shortcomings inherent in epidemiologic studies, especially in assessing risks at levels of exposure several orders of magnitude lower than those studied in historical cohorts (Peto, 1985; Health Effects Institute, 1991) were discussed above. Past epidemiologic studies alone cannot adequately address the complex issues associated with current exposure to in-place asbestos. Additional information from animal models and *in vitro* mechanistic studies is required to answer many of these questions (Kane, 1992).

Questions about the safety of asbestos fiber substitutes is a more complicated issue. The International Agency for Research on Cancer (IARC) evaluated the toxicologic and epidemiologic evidence for potential carcinogenicity of man-made mineral fibers currently in use in 1987. Four classes of man-made fibers were considered to be possible human carcinogens: glass, rock, and slag wool and ceramic fibers (IARC, 1987). The epidemiologic studies, especially for exposure to slag wool, are in question because these workers may have been exposed to other carcinogens, especially metals. These epidemiologic studies of past exposures to man-made mineral fibers suffer from the same shortcomings as studies of asbestos workers. Current exposures to man-made mineral fibers are lower than past exposures to asbestos fibers. A new group of workers at potential risk has now been identified: end-users who apply or remove products containing asbestos and man-made fibers. There is current concern that their workplace exposures are not carefully monitored or controlled (Huncharek, 1992; McClellan et al., 1992).

New fibers are currently under development, including synthetic whiskers and polymer matrix composites. The physical properties of these new fibers resemble asbestos, yet few of these fibers have been evaluated for potential toxicity and carcinogenicity. Ideally, if the physical and chemical determinants of fiber toxicity and carcinogenicity were identified, then development of a safe asbestos fiber substitute should be feasible. Clearly, meeting this challenge requires an understanding of both the biological and mineralogical aspects of pathogenesis. Until this information is available, a strategy for testing man-made fibers should be developed that incorporates the knowledge based on epidemiologic and toxicologic studies of asbestos-related diseases. A workshop was held in the United States in 1991 to develop such a strategy and a consensus report was published in *Regulatory Toxicology and Pharmacology* in 1992 (McClellan et al. 1992). This interdisciplinary workshop included representatives from industry, unions, government regulatory agencies, occupational health physicians and epidemiologists, toxicologists, and basic scientists. Ongoing dialogue among these interdisciplinary groups (including mineralogists) is required in order to address current concerns about the safety of asbestos-fiber substitutes and to prevent future outbreaks of occupational lung disease in this industry.

ACKNOWLEDGMENTS

The skillful secretarial assistance of Jeanne Ramos and Ann Baxter is gratefully acknowledged. Research reported from the author's laboratory was supported by grants (NIH R01 ES 03721 and NIH R01 ES 05712) from the National Institute of Environmental Health Sciences.

REFERENCES

Alberts, A.S., Falkson, G., Goedhals, L., Vorobiof, D.A., and van der Merwe, C.A. (1988) Malignant pleural mesothelioma: a disease unaffected by current therapeutic maneuvers. J. Clin. Oncol. 6, 527–535.

Antoniades, H.N. (1992) Linking cellular injury to gene expression and human proliferative disorders: examples with the PDGF genes. Mol. Carcinogen. 6, 175–180.

Aust, A.E. and Lund, L.G. (1991) Iron mobilization from crocidolite results in enhanced iron-catalyzed oxygen consumption and hydroxyl radical generation in the presence of cysteine. In Mechanisms in Fibre Carcinogenisis, R.C. Brown, J.A. Hoskins, and N.F. Johnson, eds., Plenum Press, New York, 397–405.

Baris, I., Simonato, L., Artvinli, M., Pooley, F., Saracci, R., Skidmore, J., and Wagner, C. (1987) Epidemiological and environmental evidence of the health effects of exposure to erionite fibers: A four-year study in the Cappadocian region of Turkey. Int'l J. Cancer. 39, 10–17.

Barrett, J.C., Lamb, P.W., and Wiseman, R.W. (1989) Multiple mechanisms for the carcinogenic effects of asbestos and other mineral fibers. Environ. Health Persp. 81, 81–89.

Bellman, B., Konig, H., Pott, F., Kloppel, H., and Spurny, K. (1987) Persistence of man-made mineral fibers (MMMF) and asbestos in rat lungs. Ann. Occup. Hygiene 31, 693–709.

Berry, G. and Newhouse, M.L. (1983) Mortality of workers manufacturing friction materials using asbestos. Brit. J. Indus. Med. 40, 1–7.

Brodeur, P. (1985) Outrageous Misconduct: The Asbestos Industry on Trial. Pantheon Books, New York.

Checkoway, H., Pearce, N., and Crawford-Brown, D.J. (1989) Research Methods in Occupational Epidemiology. Oxford University Press, New York, 3–17, 46–71.

Churg, A. and Green, F.H.Y., Eds. (1988) Pathology of Occupational Lung Disease. Igaku-Shoin, New York, Chapters 7, 8, 9.

Craighead, J.E. (1987) Current pathogenetic concepts of diffuse malignant mesothelioma. Human Pathol. 18, 544–557.

Craighead, J.E., Abraham, J.L., Churg, A. Green, F.H.Y., Kleinerman, J., Pratt, P.C., Seemayer, T.A. Vallayathan, V., and Weill, H. (1982) The pathology of asbestos-associated diseases of the lungs and pleural cavities: Diagnostic criteria and proposed grading schema. Arch. Pathol. Lab. Med. 106, 544–596.

Davis, J.M.G. (1974) Histogenesis and find structure of peritoneal tumors produced in animals by injection of asbestos. J. Natl. Cancer Inst. 52, 1823–1837.

Doll, R. (1955) Mortality from lung cancer in asbestos workers. Brit. J. Indus. Med. 12, 81–86

Fawcett, D.W. (1986) A Textbook of Histology. 11th Ed., W.B. Saunders Co., Philadelphia, 1017 p.

Goodglick, L.A. and Kane, A.B. (1986) The role of reactive oxygen metabolites in crocidolite asbestos toxicity to macrophages. Cancer Res. 46, 5558–5566.

Health Effects Institute (1991) Asbestos in Public and Commercial Buildings: A Literature Review and Synthesis of Current Knowledge. Health Effects Institute–Asbestos Research, Cambridge, 366 p.

Hume, L.A. and Rimstidt, J.D. (1992) The biodurability of chrysotile asbestos. Amer. Mineral. 77, 1125–1128.

Huncharek, M. (1992) Changing risk groups for malignant mesothelioma. Cancer 69, 2704–2711.

IARC (1987) International Agency for Research on Cancer. IARC Monographs on the Evaluation of Carcinogenic Risks to Humans. In Overall Evaluation of Carcinogenicity: An Updating of IARC Monographs, IARC, Lyon.

Kane, A.B. (1992) Animal models of mesothelioma induced by mineral fibers: Implicatios for human risk assessment. In Relevance of Animal Studies to the Evaluation of Human Cancer Risk. Wiley-Liss, Inc., New York, 37–50.

König, A. (1960) Über die asbestose. Arch. Gewerbepath. 18, 15–19.

Landrigan, P.J. and Kazemi, H., Eds. (1991) The Third Wave of Asbestos Disease: Exposure to Asbestos in Place. New York Acad. Sci., New York., 459 p.

Lemen, R.A., Dement, J.M., and Wagoner, J.K. (1980) Epidemiology of asbestos-related diseases. Environ. Health Persp. 34, 1–11.

Levy, B.S. and Seplow, A. (1992) Asbestos-related hazards in developing countries. Environ. Res. 59, 167–174.

Liddell, D. (1991) Exposure to mineral fibers and human health: Historical background. In Mineral Fibers and Health, D. Liddell and K. Miller, Eds. CRC Press, Boca Raton, Florida, Chapter 1.

McClellan, R.O., Miller, F.J., Hesterberg, T.W., Warheit, D.B., Bunn, W.B., Kane, A.B., Lippmann, M., Mast, R.W., McConnell, E.E., and Reinhardt, C.F. (1992) Approaches to evaluating the toxicity and carcinogenicity of man-made fibers: Summary of a workshop, Nov. 11–13, 1991, Durham, North Carolina. Reg. Toxicol. Pharmacol, 16, 321–364.

McDonald, A.D. and McDonald, J.C. (1980) Malignant mesothelioma in North America. Cancer 46, 147–154.

Moalli, P.A., Macdonald, J.L., Goodglick, L.A., and Kane, A.B. (1987) Acute injury and regeneration of the mesothelium in response to asbestos fibers. Amer. J. Pathol. 128, 425–445.

Monchaux, G., Bignon, J., Jaurand, M.C., Lafuma, J., Sebastien, P., Masse, R., Hirsh, A., and Goni, T. (1981) Mesotheliomas in rats following inoculation with acid-leached chrysotile asbestos and other mineral fibers. Carcinogenesis 2, 229–236.

Peterson, J.T., Greenberg, S.D., and Buffler, P.A. (1984) Non-asbestos-related malignant mesothelioma. Cancer 54, 951–960.

Peto, J. (1985) Problems in dose-response and risk assessment: The example of asbestos. In Risk Quantitation and Rgulatory Policy, Banbury Rept. 19, Cold Spring Harbor Lab., New York, 89–101.

Pott, F. (1987) Problems in defining carcinogenic fibres. Ann. Occup. Hygiene 31, 799–802.

Pott, F. and Friedrichs, K.H. (1972) Tumours in rats after intraperitoneal injection of asbestos dusts. Naturwissenschaften 59, 318–332.

Roggli, V. (1990) Human disease consequences of fiber exposures—A review of human lung pathology and fiber burden data. Environ. Health Persp. 88, 295–304.

Roggli, V.L. and Brody, A.R. (1984) Changes in numbers and dimensions of chrysotile asbestos fibers in lungs of rats following short-term exposure. Exp. Lung Res. 7, 133–147.

Selikoff, I.J., Churg, J., and Hammond, E.C. (1964) Asbestos exposure and neoplasia. J. Amer. Med. Assoc. 188, 22–26.

Selikoff, I.J., Hammond, E.C., and Seidman, H. (1979) Mortality and experience of insulation workers in the United States and Canada, 1943–1976. Ann. New York Acad. Sci. 330, 91–116.

Shatos, M.A., Doherty, J.M., Marsh, J.P., and Mossman, B.T. (1987) Prevention of asbestos-induced cell death in rat lung fibroblasts and alveolar macrophages by scavengers of active oxygen species. Environ. Res. 44, 103–116.

Stanton, M.F., Layard, M., Tegeris, A., Miller, E., May, M., Morgan, E., and Smith, A. (1981) Relation of particle dimension to carcinogenicity in amphibole asbestoses and other fibrous minerals. J. Natl. Cancer Inst. 67, 965–975.

Thurlbeck, W.M., Ed. (1988) Pathology of the Lung. Thieme Medical Publishers, New York, Chapter 23.

Wagner, J.C. (1991) The discovery of the association between blue asbestos and mesotheliomas and the aftermath. Brit. J. Indus. Med. 48, 399–403.

Wagner, J.C., Berry, G., and Timbrell, V. (1973) Mesotheliomata in rats after inoculation with asbestos and other materials. Brit. J. Cancer 28, 173–185.

Wagner, J.C., Sleggs, C.A., and Marchand, P. (1960) Diffuse pleural mesothelioma and asbestos exposure in North Western Cape Province. Brit. J. Indus. Med. 17, 260–271.

Weitzman, S.A. and Graceffa, P. (1984) Asbestos catalyzes hydroxyl and superoxide radical release from hydrogen peroxide. Arch. Biochem. Biophys. 288, 373–376.

CHAPTER 12

HEALTH EFFECTS OF MINERAL DUSTS OTHER THAN ASBESTOS

Malcolm Ross

U.S. Geological Survey, M.S. 959
Reston, Virginia 22092 U.S.A.

Robert P. Nolan and **Arthur M. Langer**

Environmental Sciences Laboratory
Brooklyn College
Avenue H and Bedford Avenue
Brooklyn, New York 11210 U.S.A.

W. Clark Cooper

8315 Terrace Drive
El Cerrito, California 94530 U.S.A.

INTRODUCTION

Many workers, including miners, millers, quarry workers, sandblasters, stone masons, tunnel drivers, and agricultural workers, are occupationally exposed to mineral dusts. Exposure is derived from a variety of sources, particularly from powders arising from the fragmentation of rocks, and their constituent minerals, and occasionally from soils. These workers may develop pneumoconiosis and, in some instances, malignant neoplasms, particularly lung cancer, as a result of such exposures, provided the exposures are of sufficient intensity and duration. Diseases may also be found in working populations that process or handle rock and mineral products in secondary capacities, for example, in the smelter, foundry, and construction environments. For a detailed description of minerals to which workers may commonly be exposed the reader is referred to chapters in this volume, to Langer (1986), and Skinner et al. (1988). As background to understanding the relationship between inhalation of mineral particles and lung disease, Skinner et al. (1988) and Netter (1979) review the structure and function of the human respiratory system.

Although mankind has been exposed to many different types of mineral dusts, there are only a few mineral exposures which have been linked with definitive medical evidence indicating an etiological association between injury and exposure. The nature of the disease or lesion in the exposed worker is defined at first through clinical observation. Statistical proof of the association is sought through epidemiological study of groups of workers exposed to the same type or types of mineral dust. The question asked is whether or not these exposed workers develop some specific disease at a statistically significant rate greater than is found in a control or in the general population (unexposed workers). Epidemiological study establishes association between agent and disease and a dose-response relationship between the amount of exposure and degree of injury. The basic principles of epidemiology are discussed by Morris (1975), and Shy (1986)

Miners are generally exposed to rock dusts containing mixtures of minerals—for example, granite workers are exposed to large amounts of quartz, feldspars, amphiboles, pyroxenes, and micas; coal miners are exposed to quartz, carbonaceous material, and lesser amounts of many other minerals including various clays, carbonates, sulfides, and silicates. These multiple exposures often make it difficult to ascribe the exact agent in the etiology of the worker's disease. For example, there is still some discussion as to the relative importance of quartz versus carbonaceous compounds to the development of coal workers pneumoconiosis (black lung) and quartz versus layer silicates (especially fibrous sericite) to the origin of slate workers pneumoconiosis. Since workers are often exposed to more than one type of dust, comparative epidemiology and pathology are necessary to distinguish among the effects of the dusts responsible for injury. For example, if workers exposed to only quartz dust (such as the Gauley Bridge tunnel drivers working in nearly pure quartzite rock, Cherniack, 1986) show the same health effects as granite workers (normalizing exposures on the basis of percent concentration), then quartz can be considered to be the important contributor to disease, the other minerals in the granite (feldspars, amphiboles, micas, etc.) being much less important.

The following text is a general review of the health effects from occupational exposure to mineral dusts. Citations include those providing both clinical evidence of disease and support by epidemiological data. In many instances these associations have been studied experimentally to establish the etiological agents. The health effects of the following mineral and rock dusts will be reviewed: the three polymorphs of silica (quartz, cristobalite, and tridymite), coal (predominantly carbonaceous material plus minor amounts of other minerals), sepiolite, palygorskite (attapulgite), kaolinite, talc, vermiculite, non-fibrous amphiboles, mica, chlorite, erionite (a fibrous zeolite), wollastonite, and fuller's earth (generally a mixture of minerals including the clay minerals, montmorillonite and palygorskite).

THE SILICA MINERALS AND AMORPHOUS SILICA

Mineralogy of silica

The seven naturally occurring crystalline silica minerals are composed predominantly of silicon and oxygen (SiO_2), with only trace amounts of Al, Fe, Mn, Mg, Ca, and Na in the crystal structures. These minerals, generally referred to in the mineralogical literature as the *silica minerals*, are: quartz (including chalcedony), cristobalite, moganite, tridymite, melanophlogite, coesite, and stishovite. Microcrystalline varieties of quartz include chalcedony, chert, jasper, flint, onyx, prase, and agate. The silica minerals are all composed of SiO_4 tetrahedra, each linked to four like tetrahedra to form the three-dimensional crystal structure. However, the orientation of the tetrahedra are different in each of these seven minerals. Quartz is the second most common mineral (after feldspar) in the earth's crust. As an example, common rocks such as granite contain 25 to 40% quartz, shales average 22% quartz, and sandstones 67% quartz (Clark, 1924). Cristobalite and tridymite are much less common than quartz and occur in cavities in volcanic rocks and as the devitrification products of volcanic glasses. Occasionally, they may occur as devitrification products of diatomaceous earth when the latter is calcined during industrial manipulation. Moganite, melanophlogite, coesite, and stishovite are rare minerals and need not be considered here. General reviews on the nature of the crystalline silica minerals are given in other chapters in this volume and by Frondel (1962), Ampian and Virta (1992), and USBM (1992a).

In the health reports of SiO_2-exposed workers the mineral names quartz, cristobalite, etc. are seldom used, instead the term *silica, crystalline silica*, or less often (and improperly) *free silica* is substituted for the correct mineral name. In most of the health studies of silica-exposed workers the exposure has been to quartz, but since the mineral name is often not explicitly stated in the reports, we must in this review continue to use the term *silica*. There are very few studies of workers exposed to cristobalite (predominantly those exposed to calcined diatomaceous earth or fire brick dusts); there are no studies of workers exposed only to tridymite.

The non-crystalline forms of silica (*amorphous silica*) include natural glasses found in various volcanic rocks and synthetic glasses such as fume silica, fiber glass, and mineral wool. Opal is a natural occurring hydrated form of silica; it is usually amorphous or nearly amorphous and appears in a variety of geologic localities, especially in hot spring deposits. Opaline materials characterize diatomaceous earth deposits and are also found in bentonite clays. Many opals have no crystallinity (amorphous) and are sometimes referred to as amorphous silica or as opal A. Devitrification in some opals may produce very small inclusions of poorly crystallized cristobalite; such material is sometimes referred to as opal C or opal CT. Amorphous silica may, when heated to high temperature, be converted to cristobalite. Dusts composed of amorphous silica (with the exception of fiber glass) have not been implicated with human disease and thus will not be considered further in this review.

Diseases related to exposure to silica dust

Silicosis. Three types of silicosis are defined:

1) *Chronic silicosis* (also referred to as *classic* or *nodular silicosis*) is a progressive lung disease characterized by the development of fibrotic (scar) tissue. In response to inhalation of quartz particles (and perhaps other forms of crystalline silica) in the median size range of 0.5–0.7 µm, macrophages phagocytose the particles and produce cytokines, proteins and growth factors. These in turn stimulate fibroblasts which produce collagen, one of the essential components of scar tissue (Craighead, 1988). The hilar lymph nodes tend to become enlarged, with macrophages laden with particles, and silica-containing fibrotic nodules develop in the region of the small airways (which form confluent masses

2–3 mm in diameter). The nodules may appear in large numbers in the apical (top) and posterior (back) regions of the upper and lower lobes of the lung. As silicosis progresses, the fibrotic nodules may coalesce into large confluent masses and these may contain the remnants of blood vessels and bronchi. These lesions vary in size from about 3 mm to a size involving one third of the lung (up to 15cm across) (Lapp, 1981). Heart or respiratory failure (cor pulmonale) is the ultimate consequence of silicosis.

2) *Acute silicosis* (also referred to as *silicotic alveolar proteinosis*) develops in workers exposed to exceptionally high concentrations of fine particles of silica, usually quartz dust. Here the lining of the airways are damaged and a lipid-rich protein accumulates, obliterating the air spaces. Progressive massive fibrosis appears, usually in the upper, apical, regions of the lung, and superinfection by mycobacterial organisms (tuberculosis) generally occurs. In acute silicosis the lungs are often very "heavy" and "rigid" and microscopic examination of lung parenchyma on autopsy shows that the air spaces are filled with a finely granular substance (Craighead, 1988).

3) *Accelerated silicosis* is a condition whereby the progression of disease is intermediate between chronic and acute silicosis. This form of silicosis develops after five to ten years of heavy exposure to crystalline silica dust and is especially seen in sandblasters who were exposed to predominantly small fragments (<1.0 µm) of almost pure quartz. The victim of accelerated silicosis often shows no clinical abnormalities other than breathlessness, but X-ray shows upper chest irregular fibrosis associated with numerous nodules. This form of silicosis is progressive with a continuing decrease of lung function even in the absence of further dust exposure. Death by cardiopulmonary failure within ten years of onset of symptoms is often the outcome of this form of silicosis (Craighead, 1988).

Silicotuberculosis. There has long been noted an association between silicosis and *tuberculosis* and in the past tuberculosis was a major cause of death among silicotics; death certificates often listed the cause of death as due to *silicotuberculosis*. For example, in the 1976 *Vital Statistics of the United States*, it is reported that there were 215 silicosis deaths (ICD.8, 515.0) and 92 silico-tuberculosis deaths (ICD.8, 010); in contrast, only 54 *asbestosis* (ICD.8, 515.2) deaths were listed for 1976. More recently, due to better dust control and chemotherapy, tuberculosis associated with silicosis has much decreased except in the less developed areas of the world. Experimental evidence shows that the presence of silica promotes the growth of *M. tuberculosis* in macrophage cultures. Silicosis appears to modify the progress of tuberculosis but may also change the character of the tuberculosis lesions. Epitheloid cell proliferation, Langhan's giant cell formation, and the lymphocytic reaction seen in the usual tuberculosis patient may be suppressed by the silicosis (Lapp, 1981).

Cancer. Recently, the crystalline silica minerals have been implicated in the pathogenesis of bronchogenic carcinoma. In 1986 a working group of the International Agency for Research on Cancer (IARC) reviewed the scientific data that suggested a relationship between exposure to crystalline silica dust and cancer induction. Their review was published in IARC Monograph 42 (IARC, 1987a), in which it was concluded that there was sufficient evidence for carcinogenicity in experimental animals and limited evidence for carcinogenicity in humans. Monograph 42 was followed by IARC's 1987 publication of Supplement 7 (IARC, 1987b) which, upon review of 628 substances, placed the crystalline silica minerals into Group 2A as *probably carcinogenic to humans* (IARC, 1987b, p. 31–32, 341–343; Simonato and Saracci, 1990) As a result of the IARC classification, the United States Occupational Safety and Health Administration (OSHA) invoked the OSHA Hazard Communication Standard of 1983 to require that *any product* containing any of the crystalline silica minerals in amounts greater than 0.1 wt % be labeled as a possible human carcinogen. Glenn (1992) and Goldsmith (1993) have presented extensive reviews of the relationship between silica exposure and disease.

The fibrogenic effects of crystalline silica in animal models are well known (Reiser and Last, 1979; Saffiotti, 1986), but there are also a significant number of experimental studies that show that tumors can be produced in rats through the inhalation and intrapleural or intratracheal injection of silica dusts (usually quartz is used in animal experiments). Wagner (1966) and Wagner and Wagner (1972) produced tumors in Alderly Park rats through single intrapleural inoculations of silica. These tumors consisted of malignant histiocytic lymphomas, reticulum cell sarcomas, and lymphoblastic and lymphocytic lymphomas. Holland et al. (1983) noted squamous cell carcinomas and adenocarcinomas in Fischer-344 rats after silica inhalation and in Sprague-Dawley rats after intratracheal instillation of silica. Johnson et al. (1987) produced tumors within alveolar type II cells of female Fischer-344 rats through inhalation of quartz dusts (Min-U-Sil-5). Such bronchioloalveolar tumors are rare in humans, causing less than 5% of all lung neoplasms with type II alveolar cell tumors comprising a small fraction of these (Johnson et al., 1987).

Saffiotti (1986), in a review of the fibrogenic and carcinogenic activity of silica-bearing dusts on experimental animals, noted that there were significant differences in the pathogenic effects on different animal species. Saffiotti states that the response of the Syrian golden hamster to the introduction of crystalline silica in the respiratory tract (either by inhalation or by intratracheal instillation) is altogether different from that in the rat. A consistent association was observed in these two species between the induction of fibrosis and the induction of tumors of the lung—the same types of silica that were effectively fibrogenic and carcinogenic in rats failed to produce either type of response in the hamsters. In regard to such animal studies, Craighead (1992) suggests that induction of cancer may be a reaction to chronic irritation and scarring of tissue rather than a neoplastic response to a carcinogen; on this subject, also see Ames and Gold (1990) and Cohen and Ellwein (1990).

Epidemiological studies of occupational cohorts exposed to crystalline silica dust

Compared to asbestos workers and coal miners, relatively few epidemiological studies of silica exposed workers have been completed. However, in the 1980s, perhaps due to the suggestion that silicosis may increase the risk of lung cancer, many new studies were initiated. Early prevalence studies initiated by the U.S. Public Health Service and the U.S. Bureau of Mines (Lanza and Higgins, 1915; Higgins et al., 1917) showed that many miners in the Joplin District of Missouri, who were exposed to high levels of quartz-bearing dusts, suffered from very high rates of nonmalignant lung disease. The first very large and comprehensive study of U.S. hard rock miners exposed to quartz dusts was that of Flinn et al. (1963). This study included over 14,000 employees working at 50 metal mines. The mine dust concentrations (based on 14,480 impinger sample measurements) varied from 0 to 50 million particles per cubic foot of air (mppcf) and quartz content of the dust varied from 2 to 95%. The medical history, lung function, chest X-rays, occupational histories, and mine dust control methods, were recorded. A strong relationship was found between duration of exposure to silica and prevalence of silicosis. The health of those having less than 5 years exposure was unaffected whereas 60% of those exposed for 30 or more years showed evidence of silicosis. This and earlier studies clearly demonstrated the need to limit exposure to silica-bearing dust and were the basis to institute more comprehensive state and federal regulations to protect the worker.

One of the problems of measuring the effect of silica dust is that miners, millers, foundry workers, etc. may also be exposed to a variety of natural and man-made dusts, organic fumes, radon daughter products, arsenicals, as well as other biologically active agents. Thus, it has been very difficult to mount an epidemiological study of such workers where the degree and types of exposure to dusts, chemicals, radon, etc. are accurately evaluated. More recently, as funds were made available and exposure and epidemiological techniques improved, scientists have made progress in defining the effects of these more complex workplace exposures. In the following we will review some of the newer studies of silica exposure in specific occupational groups.

Minnesota iron ore miners (magnetite-bearing rock). Taconite is a term used particularly in the Lake Superior region of Minnesota for certain rock from the Biwabik Iron-formation. A high-grade iron ore concentrate is obtained from commercial grade taconite that contains enough magnetite (Fe_3O_4) to be economically processed by fine grinding and wet-magnetic separation. During the period from 1989 to 1991 about 43 million metric tons of taconite iron ore concentrate was processed each year in Minnesota (USBM, 1992b). Taconite is a hard, dense, fine-grained metamorphic rock that contains major amounts of quartz (20 to 50%), and magnetite (10 to 20%), in addition to various other mineral constituents including hematite, carbonates, amphiboles (principally of the cummingtonite-grunerite series although actinolite and hornblende also occur), greenalite, chamosite, minnesotaite, and stilpnomelane (French, 1968). The average mineral composition the Biwabik Iron-formation taconite is: quartz (31.9%), minnesotaite (19.3%), magnetite (18.4%), siderite (9.3%), stilpnomelane

Table 1

Cohort categories by job and extent of exposure to mineral dust at mine and mill sites, Reserve Mining Company, Babbitt and Silver Bay, Minnesota

Group	Processing activity	Dust composition[*]		
		Quartz (wt %)	Silicates (wt %)	Magnetite (wt %)
1	Mining, crushing	25–40	25–35	25–40
2	Pelletizing, shipping	2–5	6–25	65–90
3	Railroad, power plant	0	0	0
4	None, school system employees	0	0	0

Source: Clark et al. (1980)
[*] Percentages based on total airborne dust data obtained from industrial hygiene surveys conducted by the Trudeau Institute, Saranac Lake, New York.

(8.7%), plus minor amounts of other silicates and carbonates (Lepp, 1972, p. 275).

The mineral content of the dusts emitted from the mining and milling of taconite rock collected at the mine and mill sites of the Reserve Mining Company, near Babbitt and Silver Bay, Minnesota, respectively, are given in Table 1. Clark et al. (1980) report that the quartz content of the rock dust for two groups of Reserve miners varied from 0.21 million to 7.74 million particles per cubic foot with a mean of 2.7 mppcf. Based on the quartz content of the rock, the threshold limit value (TLV) for total dust in the work area is given as 5 mppcf. The Minnesota taconite miners work in open pits thus there should be no health effects due to inhalation of radon gas.

Analysis of the mortality among men who were employed by the Reserve Mining Company from 1952 to 1976 has been reported by Higgins et al. (1983). This study was initiated in the 1970s in response to suggestions that "asbestos" was released into the air and water during processing of the taconite rock and posed a risk to the miners as well as to the general public. The town and city of Silver Bay and Duluth obtained their drinking water from Lake Superior, into which Reserve had deposited pulverized waste rock "tailings" from the mill. The Justice Department thought this was a particular hazard to the residents. It was alleged that the amphibole in the waste rock (cummingtonite-grunerite series) was "asbestos" and this asbestos would cause gastrointestinal cancer through ingestion and lung cancer from inhalation of the air-born particles (a complete review of this controversy and the ensuing court case is given by Schaumburg, 1976; see also Langer et al., 1979, and Ross, 1984, p. 72–78).

The Reserve cohort consisted of 5751 men, 298 were deceased, and 907 had worked for the company for over 20 years. The men were exposed to respirable dust concentrations from a low of 0.02 mg/m^3 to a high of 2.75 mg/m^3, the modal range being 0.2 to 0.6 mg/m^3. The mineral "fiber" content of the dust was occasionally higher than 0.5 "fibers"/ml in the crushing department, but usually concentrations were much lower (these "fibers" were actually elongate cleavage fragments of cummingtonite and grunerite amphibole, see Langer, et al., 1991). The observed and expected deaths and standardized mortality ratios (SMRs) for

Table 2

Selected causes of mortality for men who worked one year or longer for the Reserve Mining Company

Cause of death	ICD*	Deaths		SMR**
		Observed	Expected	
All causes	000–E999	298	343.65	87
Cardiovascular disease	402, 404, 410–429	112	123.79	90
All cancer	140–209	58	63.38	92
Respiratory cancer	160–163	15	17.94	84
Digestive cancer	150–159	20	17.57	114
Urinary cancer	188–189	3	2.97	101
Genital cancer	180–187	3	3.31	91
Selected non-malignant respiratory diseases	470–474, 480–486, 490, 491, 493, 510–519	4	6.80	59
Most trauma	E800–E978	76	72.76	104
Motor vehicle accidents	E810–E823	38	31.12	122

Source: Higgins et al. (1983)
* International Classification of Diseases, 8th Revision.
** Standardized Mortality Ratio, based on white male mortality in Minnesota, 1952–1976.

all men who had worked more than one year or longer from 1952 to 1976 are given in Table 2.

Overall mortality was less than expected (SMR=87) when compared to the male mortality in Minnesota, as was mortality from cardiovascular disease, all cancer, respiratory cancer, and selected non-malignant respiratory diseases including pneumoconiosis. There was no relationship between mortality and lifetime exposure to silica dust which was as high as 1000 mg/m³ x years, nor was there any suggestion that deaths from cancer increased after 15 to 20 years of latency. No deaths from mesothelioma or asbestosis occurred.

A second epidemiological study of Minnesota taconite miners and millers, who were employed by the Erie Mining Company (Erie mine) and the U.S. Steel Corporation (Minntac mine), is reported by Cooper et al. (1992). This study cohort, followed from 1947 through 1988 with a minimum observation period of 30 years for all participants, was composed of 3,431 men of which 1,058 were deceased. Dust levels in the two mines are reported as containing 28 to 40% crystalline silica in one and 20% in the other (Sheehy and McJilton, 1987). Mineral "fiber" counts at the two mines were nearly all below 2 fibers/ml and nearly all were shorter than 5 μm in length (Wylie, 1990). The total number of deaths (Table 3) of these taconite workers was significantly fewer than expected, SMR=83 (based on U.S. male rates) and 91 (based on Minnesota male rates). SMRs for all cancer, respiratory cancer, diseases of the circulatory system, and nonmalignant respiratory disease was also fewer than expected when compared to both control groups. Slightly elevated, but not statistically significant, SMRs were found for colon cancer, cancer of the kidney, and lymphopoietic cancer. There was one reported case of mesothelioma in a 62-year old worker whose

Table 3

Deaths by major causes (1948–1988) in taconite miners and millers*
exposed for 3 months or more prior to 1959

Cause of death (ICD, 7th Revision, 1955)	Deaths Observed	Deaths Expected	SMR
All causes (001–998)	1058	1272.5	83
All malignant neoplasms (140–205)	232	267.7	87
Digestive organs and peritoneum (150–159)	66	70.5	94
Stomach (151)	11	12.0	92
Large intestine (153)	26	23.9	109
Respiratory system (160–164)	65	97.0	67
Bronchus, trachea, lung (162–163)	62	92.2	67
Kidney (180)	12	6.8	177
Lymphopoietic (200–205)	29	25.8	112
All diseases of circulatory system (400–468)	477	575.1	83
Arteriosclerotic heart disease (420)	368	481.8	76
Cirrhosis of liver (581)	24	35.5	68
Nonmalignant respiratory disease (470–527)	55	77.2	71
All external causes of death (800–998)	114	112.3	102
All accidents (800–962)	79	74.4	106
Motor vehicle accidents (810–835)	32	33.4	96
Suicides (963, 970–979)	32	27.3	117
Cause unknown	19		
Number of workers	3,431		
Number of person–years	101,055		
Deaths per 1,000 person–years	10.5		
Adjustment of cause-specific SMRs for missing certificates	+1.8%		

Source: Cooper et al. (1992)
* Miners and millers employed by the Erie Mining Co. and the U.S. Steel Corp.

exposure to taconite had begun only 11 years before his death. He had previously been employed in the railroad industry as a locomotive fireman and engineer, an occupational environment where both amosite and crocidolite asbestos were known to have been used as insulation. Analyses of the mortality of workers with varying lengths of service, exposure to varying amounts of dust, and with a minimum potential latency period of 30 years, provided no evidence to support any association with exposure to quartz or amphibole "fibers" with lung cancer, non-malignant respiratory disease, or any other specific cause.

Iron ore miners (hematite-bearing rock). The Biwabik iron formation of Minnesota is composed of both hard rock taconite (the present source of most Minnesota iron ore as discussed above) and soft rock, which formed by late-stage alteration of the primary taconite. Oxidation changed the iron-bearing minerals of the taconite to hematite, Fe_2O_3, and goethite, $FeO(OH)$. This soft ore was a primary source of U.S. iron ore in the past and was mined in the area of the Mesabi Range located between the towns of Hibbing and Mesaba. At present only a small amount of the soft ore is mined. The Mesabi ore contains major amounts of iron oxides, and several percent fine-grained quartz as well as minor amounts of other minerals, including kaolinite, sulfides, residual carbonates, and other silicates (Gruner, 1946). Lawler et al. (1985) studied a population of 10,403 workers employed by a large steel company engaged in mining hematite iron ore in St. Louis County, Minnesota. Of these, 4,708 worked underground and 5,695 worked aboveground. The interval of follow-up was from 1937 through 1978 and at the end of this period 2,642 underground and 2,057 aboveground workers were deceased. Quartz content of the ore was 7–10% and radon daughter levels were low (< 60 pCi/l or 0.3 working levels). This mortality study (Table 4) found no excess risk of lung cancer in the total cohort (SMR=94), in the underground miners (SMR=100), or in the aboveground miners (SMR=88), contrary to the reports of excess of this disease in hematite miners from Sweden, France, and England. In the Minnesota hematite miners there was also a deficiency of nonmalignant respiratory disease (SMR=79, total cohort; 72, underground; 89, aboveground). Significant excess mortality due to stomach cancer was observed in both groups of miners when compared to U.S. males, but this excess disappeared, except for Finnish-born miners, when comparison was made to local county rates. Lawler et al. (1985) suggest that the apparent lack of significant radon exposure, underground smoking prohibition, and absence of underground diesel fuel (and therefore diesel combustion fumes) may explain why the underground miners do not show the cancer risk seen in hematite miners of Europe. However, since both groups of miners have very similar health histories (Table 4), any additional health effects of underground mining are not apparent.

European iron ore miners. Epidemiological studies have reported excess malignant and non-malignant lung disease in European iron ore miners — for example, the studies of the iron ore miners of Kiruna, Sweden (Jörgensen, 1984); of Kiruna and Gällivare, Sweden (Damber and Larsson, 1985); of Cumberland, England (Boyd et al., 1970); and of Lorraine, France (Pham et al., 1983). The excess disease in the English and French studies was variously attributed to radon, silica, and iron oxides, but the quality of the studies was not sufficient to define the etiology of the miners' diseases. The recent Swedish reports, however, implicate radon exposure and tobacco use as the important factors in causing excess lung cancer in the Swedish underground iron ore miners. There were 15 cases of lung cancer (4.6 expected) among the Kiruna miners who were in the past exposed to radon concentration greater than 60 pCi/l or 0.3 working levels (Jörgensen, 1984). The average radon level reported in the study of Damber and Larsson (1985) was 50 pCi/l. There were 42 lung cancer cases (38 were smokers); the lung cancer risk was calculated by Damber and Larsson to be about 45% from radon exposure during underground mining and about 80% from smoking. No attempt was made at calculating a synergy effect of radon and cigarettes. Possible effect of mineral dusts to both cohorts of Swedish miners was not mentioned. Stokinger (1984), in a review of the world literature on exposure to iron oxides in the underground mining environment, finds that ionizing radiation "might be an etiological factor." He exonerates iron oxides as carcinogenic, both in the mine and factory workplace. Stokinger also brings to question the nature of the disease "siderosis," a pneumoconiosis thought to be caused by inhalation of iron oxide-containing fumes and dusts. Nonmalignant disease attributed to these dusts may be due to exposure to other dusts, especially

Table 4

Deaths of total cohort of St. Louis County, Minnesota, iron ore (hematite) miners, 1937–1978

Cause of death	ICD*	Deaths Obs.	Exp.	SMR
All causes	001–999	4,699	5,058.6	93
All cancers	140–209	854	879.1	97
Respiratory system	160–163	230	242.3	95
Larynx	161	12	13.7	88
Lung	162–163	212	225.8	94
Digestive organs	150–159	329	295.5	111
Tuberculosis	010–019	33	74.0	45
Arteriosclerotic heart disease	410–413	1,783	1,743.3	102
Vascular lesions of CNS	430–438	405	444.7	91
Respiratory disease	460–519	234	295.5	79
Pneumonia	480–486	95	133.1	71
Emphysema	492	59	63.4	93
Asthma	493	8	13.5	59
Digestive system	520–577	178	228.8	78
Cirrhosis	571	69	92.2	75
Genitourinary system	580–629	46	118.9	39
Chronic nephritis	582	19	39.9	48
Symptoms, senility, and ill-defined	780–799	17	43.0	40
Accidents	800–949	297	274.9	108
Motor vehicle	810–827	97	112.2	86
Suicide	950–959	102	87.4	117

Source: Lawler et al. (1985)
* ICD, 8th revision

those which contain quartz. Iron oxide particles, being opaque to x-rays produce a chest x-ray showing a profusion of opacities (reticulation) but without the presence of diseased tissues (fibrosis). These workers are also essentially symptom free (Stokinger, 1984, p. 129). Among underground iron ore miners of Cumberland, England there were, over a 20 year period, 42 deaths attributed to lung cancer; 21 deaths from this disease were expected. There was a substantial radon hazard associated with mining of the hematite-bearing ore in Cumberland for radon in the mine air averaged 100 pCi/l. In addition, some of the miners developed silicosis indicating that quartz dust was an etiological factor, as was tobacco use, in promoting disease. It would appear that a combination of high radon levels, cigarette smoking, and quartz-bearing mine dust were cofactors in the increased health risks to these miners. This situation was not unlike that found for uranium miners in the four-corner region of the Colorado Plateau.

Gold miners (South Dakota). The Homestake Gold Mine, located in Lead, South Dakota, has been operating nearly continuously since 1878. The total production of gold, by 1965, was nearly 28 million troy ounces (Slaughter, 1968). The gold-bearing rock lies within the Homestake Formation which is composed, in part, of metamorphosed siderite-quartz and cummingtonite-quartz schists. The gold-bearing ore contains large masses of quartz, numerous quartz veins, abundant chlorite, amphibole (of the cummingtonite-grunerite series), and siderite, as well as lesser amounts of sulfides (including pyrrhotite, pyrite, arsenopyrite, galena sphalerite, and chalcopyrite), calcite, ankerite, biotite, garnet, fluorite, iron oxides, and gypsum (Noble, 1950).

Four major health studies (Flinn et al., 1963; Gillam et al., 1976; McDonald et al., 1978; and Brown et al., 1986) have been made of the Homestake miners. The 1963 health study covered workers employed at 50 metal mines and included the Homestake gold miners (Flinn et al., 1963, discussed previously). A significant silicosis mortality was found in the total mining population, but the 1963 report does not give specific details about the individual mines studied. In the second study, Gillam et al. (1976) reexamined a "1960 U.S. Public Health Service cohort of Homestake miners that was assembled during a silicosis survey of a hard rock gold mine in Lead South Dakota." We assume that the Gillam cohort was part of the 1963 Flinn cohort, but since Gillam et al. (1976) do not cite this study in their report we cannot be sure. The 1976 cohort was composed of 440 individuals who had been employed at least 60 months in underground mining at Homestake and had never mined underground elsewhere. It was found that the total mortality, total cancer mortality, lung cancer mortality, and mortality from pneumoconiosis was greater than expected (Obs./Exp.=71/52.9, 15/9.7, 10/2.7, and 5/1.9, respectively). Dust levels in the underground mine were reported to be 1.7 mppcf with a "free silica" (meaning quartz) content of 39%, which was said to be below the OSHA TLV of 5 mppcf. Amphibole "fiber" levels were reported to be 0.25 and 4.82 fibers/ml by light optical and electron optical methods, respectively. Concluding, Gillam et al. (1976) state that the observed excess of malignant respiratory disease can be ascribed to asbestos, with a possible additive role from low exposure to free silica dust. The authors of this study state that their findings indicate a need to reevaluate the adequacy of the OSHA standard for asbestos exposure.

The third study of miners who had worked 21 or more years at the Homestake mine was reported by McDonald et al. (1978). The cohort was composed of 1,358 persons, of which 1,312 were traced (96.6%). Of the 1,312 workers traced, 660 were deceased. Historical dust levels were carefully evaluated. During the period 1937 to 1952 the average dust counts varied from 11.0 to 24.6 mppcf, from 1952 through 1960 dust averaged 4.0 to 9.7 mppcf, and since 1960 dust varied from 2.0 to 5.0 mppcf. The study cohort was divided into five job categories depending on average dust levels: very low, low, moderate, high, and very high, giving a lifetime exposure for each miner based his job positions.

The mortality data for this Homestake cohort (Table 5) indicates a slight excess in overall mortality (SMR=115), an extreme excess of respiratory tuberculosis and silico-tuberculosis (SMR=1083), and 35 silicosis deaths (none were expected). Also, silicosis was mentioned on the death certificates in 28 of

Table 5

Selected deaths of the total cohort of
Homestake Gold Miners for the period 1937–1973

Cause of death	ICD*	Obs.	Exp.	SMR**
Malignant neoplasms (total)	140–205	93	90.5	103
Respiratory	160–164	17	16.5	103
Gastrointestinal	150–159	39	35.1	111
Other	140–149 165–205	37	38.9	95
Vascular lesions of the CNS	330–334	64	63.0	102
Diseases of the heart	400–443	264	232.5	114
Pneumoconiosis (total)	523–524	37		
Silicosis	515	35	0	
Respiratory tuberculosis (including silico-tuberculosis)	001–008	39	3.6	1083
Accidents	800–999	19	28.3	67
All other causes		115	131.8	87
Total deaths		631	549.7	115

Source: McDonald et al. (1978)
* ICD, 7th revision
** SMR based on male mortality rates in South Dakota

the 264 deaths coded to heart disease and in almost one half of those coded to tuberculosis and silico-tuberculosis. One of the 17 respiratory cancers was recorded as a possible mediastinal mesothelioma (unfortunately no autopsy was performed) in a man who did not work underground and thus was not exposed to much dust. McDonald et al. (1978) conclude by stating "past exposure in this mine produced substantial excess mortality, primarily from silicosis, tuberculosis, and silico-tuberculosis, but not from any significant increase in malignant disease." There was a clear relationship between degree of dust exposure and the incidence of pneumoconiosis and tuberculosis; those men in the low and very low dust categories showed no disease risk, whereas those in the high to very high categories had a relative risk of 19.9 and 16.0 for pneumoconiosis and tuberculosis, respectively. The conclusions given by Gillam et al. (1976) are not supported by the results of the McDonald study.

The fourth study of the Homestake miners was reported by Brown et al. (1986). They considered the possible environmental factors that might have effected the miners health, including dust levels, silica, arsenic and "fiber" content of the dusts, and radon levels. They noted that before the early 1950s dust levels were sufficiently high and contained enough respirable silica, averaging 13.1%, to cause silicosis. However, beginning in 1952 most of the dust levels were below the now current OSHA standard. Arsenic levels (the form of the arsenic was not given) averaged 1.17 mg/m^3 and radon daughter levels varied from 0 to 0.130 working levels; both exposures were said to be within the current occupational standard. Analysis of airborne samples showed that 84% of the airborne "fibers" were identified as amphibole; 69% belonging the cummingtonite-grunerite series, 15% belonging to the tremolite-actinolite series, and the remainder unidentified. The average exposure was 0.44 fibers/ml. The study cohort was composed of 3,328 men of whom 3,143 were traced, and 861 had died; mortality data are given in Table 6. Overall mortality was slightly elevated (SMR=112), lung cancer was exactly what was expected (SMR=100), but nonmalignant respiratory disease and respiratory tuberculosis were highly elevated (SMR=279 and 364, respectively). There was no evidence of an increased risk of lung cancer with either increased dust exposure or increased latency. Of the 36 tuberculosis deaths, 32 occurred in men first employed before 1930; the significant decline in this disease is stated to be a reflection of a higher standard of living and better health care in the post-1930 period. Brown et al. (1986) failed to cite the three pervious studies (Flinn et al., 1963; Gillam et al., 1976; and McDonald et al., 1978) of the Homestake miners in their paper, nor do they refer to the role of amphibole fibers in the lung cancer incidence.

Table 6

Selected deaths of the total cohort of Homestake Gold Miners for the period prior to 1930 through 1964

Cause of death	ICD*	Obs.	SMR**
Malignant neoplasms (total)	140–205	137	97
Buccal cavity and pharynx	140–148	6	125
Stomach	151	5	57
Intestine	152–153	10	79
Rectum	154	6	120
Pancreas	157	8	103
Trachea, bronchus, and lung	162–163	43	100
Other parts of respiratory sys.	160,164	3	500
Prostate	177	11	143
Lympho- and reticulosarcoma	200	6	150
Hodgkin's disease	201	2	95
Leukemia and aleukemia	204	10	169
Diseases of the circulatory sys.	400–468	285	84
Pneumonia	490–493	16	96
Bronchitis	500–502	2	80
Other non malignant respiratory disease, including silicosis	510–527	53	279
Respiratory tuberculosis	001–008	36	364
Cirrhosis of the liver	581	15	71
Transportation accidents	800–866	35	104
Accidental falls	900–904	11	164
Other accidents	910-936 960–962	58	331
All deaths		861	112

Source: Brown et al. (1986)
* ICD revision in effect at time of death
** SMR based on U.S. National male rates

However, in a longer unpublished version of this same report (Brown et al., 1985), the three previous studies of Homestake miners were briefly referred to and the subject of the possible health effects of amphibole particles treated at length. Brown et al. (1985) stated that the cummingtonite-grunerite amphibole in the Homestake rocks is *nonfibrous* and that the study "provides no evidence that either exposure to CG (cummingtonite-grunerite) fibers or to silica from Homestake Mine is associated with an excess risk of lung cancer."

Summarizing, the studies of McDonald et al. (1978) and Brown et al. (1986) are in general agreement with regard to health risk of underground mining at the Homestake Gold Mine. Both studies show a significant increase in risk of silica induced pneumoconiosis and respiratory tuberculosis. Silicosis incidence is directly related to degree and duration of exposure to quartz dust and to latency. In neither study was there any evidence that lung cancer incidence was elevated because of quartz, arsenic, amphibole particles, or radon. Tuberculosis incidence appeared mostly in those entering employment prior to 1930 at a time of inferior working and living conditions and health support. The study of Gillam et al. (1976) not been corroborated by the subsequent research.

Gold miners (Kalgoorlie, Western Australia). Armstrong et al. (1979) followed a cohort of 1,974 underground gold miners from Kalgoorlie, Western Australia, in which 500 deaths had occurred. Only 34 workers were untraced. No information was given concerning the type of rock being mined nor the silica content of the mine dust. Smoking history of the cohort was carefully evaluated, arsenic levels averaged 49 ppm, and radon concentrations averaged far below 0.045 working levels. Mortality data, given in Table 7, shows an elevated risk of lung cancer (SMR=140), but which is weakly and inconclusively related to the extent of underground mining experience. The authors state that the excess of lung cancer may be explained by the prevalence of cigarette smoking (66.3% in the miners versus 53.2% in male Western Australian controls). There was radiographic evidence of pneumoconiosis in 21.7% of the Kalgoorlie miners and this is reflected in large excess (SMR=640) of this disease, stated to be entirely silicosis. Surprisingly, the total nonmalignant respiratory disease reported for these miners is less than expected (57 deaths observed compared to 63.8 expected).

Granite workers (Vermont). The granite industry located in the Barre area of central Vermont includes 60 quarrying and manufacturing companies employing more than 1,700 workers (USBM, 1989, p. 490). The average mineral composition (mineral mode) of the Barre granite is as follows (Chayes, 1952): quartz (27.2 vol %), potash feldspar (19.1%), plagioclase (35.2%), biotite (8.1%), muscovite (8.3%), opaque accessories (metal oxides and sulfides, 0.2%), and non opaque accessories (0.8%). This area has a long history of granite quarrying, milling, and carving–occupations that have employed a large portion of the residents of Barre township. For many years, especially in the 1920s and 1930s, there was a great prevalence of silicosis and tuberculosis among the Vermont granite workers. Dust control measures were instituted between 1937 and 1940, so that granite dust levels were reduced from average values of 40 to 60 mppcf to

Table 7

Deaths for 1961–1975 in cohort of 1,974 gold miners working in Kalgoorlie, Western Australia

Cause of death	ICD*	Deaths Obs.	Exp.	SMR**
Tuberculosis	010–019	4	1.6	250
Stomach cancer	151	4	9.8	40
Colorectal cancer	153–154	9	11.0	80
Pancreatic cancer	157	7	5.8	120
Respiratory cancer	161–163	59	40.8	140
Melanoma of skin	172	1	2.0	50
Bladder cancer	188	2	3.2	60
Lymphoma, leukemia, etc.	200–209	7	8.1	90
Other cancers		22	21.0	100
Ischaemic heart disease	410–414	178	172.7	100
Other heart disease		17	17.1	100
Cerebrovascular disease	430–438	40	39.6	100
Total respiratory disease		57	63.8	89
Bronchitis	490–491	23	23.5	100
Pneumoconiosis	415–416	11	1.7	640
Other respiratory disease		23	38.6	60
Industrial accidents and falls	916–921 923–928	13	4.5	290
Other accidents		21	18.9	110
Suicide and homicide	950–969	12	8.9	130
All other causes of death		47	42.3	110
Total number of deaths		500	471	124

Source: Armstrong et al. (1979)
* ICD, 8th revision
** SMR based on mortality experience of Western Australian men

levels below 10 mppcf (Russell, 1941).

Costello and Graham (1986) followed a cohort of 5,414 Vermont granite workers from 1950 through 1982 to determine the causes of mortality. Many of these workers were employed well prior to 1940 when dust levels were very high. Of the total cohort, 1,532 were deceased by 1982. Preliminary results show that silicosis, tuberculosis, and lung cancer accounted for 24.4% of the deaths, with silicosis alone accounting for 8% of the deaths. Lung cancer was the cause of death for 102 workers, with an additional 25 lung cancers cited as an "additional cause of death." The Standard Mortality Ratio (SMR) for the total cohort is: silicosis (586.6), tuberculosis (473.8), and lung cancer (104.9). High quartz dust levels did not appear to be a significant factor in the induction of lung cancer in these workers. Costello and Graham (1986) also examined the mortality data for those employed after 1940 when dust levels were greatly reduced. As can be seen in Table 8, the health experience for post-1940 workers is superior to

Table 8

Selected causes of death and SMR for Vermont granite workers by date of first employment

Cause of death *	Year of first employment									
	Before 1930		1930–1939		1940–1949		1950–1959		1960–1969	
	Obs.	SMR	Obs.	SMR	Obs.	SMR	Obs.	SMR	Obs.	SMR
All causes of death	891	96	209	78	250	70	135	54	38	42
All tuberculosis	116	764	6	137	2	40	0	—	0	—
All malignant neoplasms	152	96	54	105	49	73	28	58	9	54
Respiratory cancer	53	128	21	124	20	89	9	54	5	88
Lung cancer	49	128	20	125	20	95	8	51	5	91
All circulatory-system diseases	399	75	83	58	120	66	37	31	9	25
Arterial and coronary heart diseases	266	80	63	64	82	64	25	29	6	22
All respiratory diseases	86	147	17	102	10	49	7	53	1	24
All pneumonia	10	38	2	32	3	38	1	21	0	—
Emphysema	19	166	4	97	5	104	2	60	0	—
Silicosis	34	919	4	421	1	90	0	—	0	—
Suicide	16	132	7	133	7	83	10	129	2	45

Source: Costello and Graham (1986)   * ICD, 8th revision

that of the control group (SMR for all causes < 100). The SMRs for silicosis, tuberculosis, and lung cancer are all less than 100 indicating that the post-1940 dust control was effective in minimizing these diseases as important causes of death.

Slateworkers (Vermont, North Wales, and Germany). Slate, the compact, fine-grained metamorphic rock, is particularly characterized by a well-defined cleavage—thus its usefulness for such purposes as roofing shingles. The two most important minerals contained in slate are mica (usually muscovite or sericite) and quartz; lesser amounts of chlorite, hematite, carbonates, rutile, pyrite, magnetite, and carbonaceous matter are often found in slate. Dale (1914) gives the following mineral percentages for slate from Ardennes, France: muscovite (38-40%), quartz (31-45%), Chlorite (6-18%), hematite (3-6%), and rutile (1-1.5%). In the United States slate is quarried particularly in Virginia, Maryland, Pennsylvania, New York, and Vermont. About 35% of the total work force of the U.S. slate industry is employed in a 250 square mile area of western Vermont and adjacent New York. Black slate from Benson, Vermont has the following mineral constituents in order of abundance: muscovite, quartz, chlorite, carbonate, rutile, pyrite, and magnetite (Dale, 1914, p. 144). The quartz in this slate is very fine-grained with particles varying in greatest dimension from 13 to 30μm. The workers who quarry and process this slate are thus exposed to very fine dust particles.

Craighead et al. (1992) made a mineralogic and pathologic examination of lung tissue of 12 western Vermont slateworkers who had developed pneumoconiosis while employed in their trade. The clinical diagnoses of the 12 workers revealed the following diseases: silicosis, silicotuberculosis, various carcinomas, pneumoconiosis, pulmonary embolism, heart disease, emphysema, and hemorrhage. Histological study revealed the presence of perivascular and peribronchial lesions, interstitial fibrosis, and macules, the latter scattered diffusely in the lung tissue. These lesions were associated with a variable number of silicotic nodules. X-ray diffraction analysis of four lung tissue ash residues showed them to be composed mostly of quartz and muscovite. Energy dispersive X-ray spectrographic analysis of inorganic residues obtained on other tissue samples showed the presence of major amounts of various aluminum silicates and silica, consistent with the X-ray diffraction results.

Pneumoconiosis in slateworkers has been reported sporadically in many parts of the world. Many of these cases were diagnosed as silicosis clinically and histologic examination of lung tissue revealed a large number of silicotic lesions. In a comparative study, however, Craighead et al. (1992) found that silicotic lesions were more prominent in the lungs of the Welsh slateworkers (Glover et al., 1980) than in those of the Vermont slateworkers. In addition, chest X-rays of the Vermont slateworkers reveal a diffuse interstitial pulmonary disease, not the nodular lesions characteristic of silicosis. Concluding, Craighead et al. (1992) state "slateworkers are exposed to respirable airborne dust that has the capacity to produce a pneumoconiosis that differs from the classic silicosis." It appears that excessive exposure to mica dusts can cause a pneumoconiosis that is somewhat different from silicosis.

A study of slate workers in the German Democratic Republic (Mehnert et al., 1990) showed an elevated mortality from nonmalignant diseases of the respiratory system (SMR=226) and 40 deaths due to silicosis. No clear overall excess of lung cancer for the total cohort was observed (obs. 27/exp. 24.71, SMR=109). However, for those workers with over 20 years employment there was an excess of lung cancer (obs. 17/exp. 10.83, SMR=157)

Diatomaceous earth workers. Diatomaceous earth (also referred to as diatomite) is an earthy (sometimes chert-like) material derived in the most part from the skeletal remains of diatoms. The earth also may contain other impurities such as the skeletal remains sponges and radiolarians, clay minerals, quartz sand, and an assortment of other minerals. Typically, diatomaceous earth (DE) contains 0.1 to 4% quartz and up to 90% amorphous, or nearly amorphous silica (opal A, C, or CT), a substance that makes up most of the fossil fragments. In the processing for commercial use, DE is crushed, dried, sorted to remove contaminants, and then heated, with or without a chemical flux ("calcined" or "flux-calcined"), to temperatures of 800 to 1000°C. The original material is thus converted to a porous, lightweight, thermally and chemically resistant, substance. However, during this heating process the original amorphous opal is partly converted to cristobalite producing a final product that contains 10 to 25% of this polymorph of silica. In the United States approximately 600,000 metric tons of DE is produced each year and is used in many applications such as for filters, filler in paints, plastics, cements, roofing plasters, and rubber compounds, as an anti-caking agent, and as an abrasive (Bates and Jackson, 1982).

Over forty years ago silicosis was identified as a hazard in the DE industry (Vigliani and Motlura, 1948; Smart and Anderson, 1952; Abrams, 1954; and Caldwell, 1958), however, studies conducted by the U.S. Public Health Service in the 1950s, found that the prevalence of silicosis among DE workers has lessened over time as a result of improved control measures. The first complete epidemiological study of diatomaceous earth workers was only recently accomplished by Checkoway and coworkers (Checkoway et al., 1992). They evaluated the health histories of the DE workers at the Johns-Manville Corporation and Great Lakes Carbon Company plants located in Lompoc, California, DE workers at the Grefco Company plant in Basalt, Nevada, and DE workers at the Witco Corporation plant in Quincy, Washington. The Nevada and Washington State cohorts are very small compared to the Lompoc cohorts, thus only the Lompoc groups are reviewed here. Analysis was made of a combined cohort of 2,570 Lompoc workers from the two plants and included: smoking history of 1,113 members of the combined cohort; classification of the death certificates according to the ICD codes in effect at the times of death (5th through 9th revisions); exposure assessment to natural, calcined, or flux-calcined material; and calculation of standard mortality ratios (SMRs) with regard to both national and regional reference populations. Dust exposures were classified into four levels according to job title and product type (natural, calcined, or flux-calcined). Representative respirable dust levels in these four categories was: heavy (0.274 to 0.613 mg/m^3), intermediate (0.131 to 0.204 mg/m^3), light (0.113 to 0.167 mg/m^3), and "none" (less than limit of detection).

The observed and expected deaths for major diseases for the combined Lompoc cohorts is given in Table 9. Mortality from all causes was slightly elevated (SMR=112) and this overall excess was primarily explained by increased rates of nonmalignant respiratory disease excluding pneumonia and infectious diseases (SMR=259) and lung cancer (SMR=143). Relative risks for both nonmalignant respiratory disease and lung cancer generally increased with increasing exposure level. Lung cancer risk, however, may be partly confounded by lack of complete smoking histories (smoking data was available for only 1,113 of the 2,570 cohort members). Also, there may be a synergy between DE exposure and tobacco use. An additional problem arose concerning the accuracy of the reported silicosis incidence. The death certificate information was incomplete with regard to identifying the specific form of nonmalignant respiratory disease causing death. Not withstanding some of the weaknesses of the DE health study, there appears to be a definite excess of lung cancer and nonmalignant respiratory disease (NMRD) among the long-time diatomaceous earth workers when compared to both national and regional populations. The disease trends among these workers are consistent with both length of time in the trade and degree of exposure to DE dust (noted as cumulative exposure). It is not certain that quartz or cristobalite is the cause of malignant disease in the Lompoc cohort. Quartz is known to cause silicosis in man, and both quartz and cristobalite are known to cause silicosis in experimental animals (see Langer, 1978 for review of early animal studies).

In one of the summary statements in the report of Checkoway et al. (1992, p. vi) an optimistic forecast was made concerning the future health of the Lompoc

Table 9

Deaths by major causes for combined cohort of 2,570 white males[†] at two diatomaceous earth plants in Lompoc, California

Cause of death	Deaths		SMR**
	Obs.	Exp.	
All causes	628	563	112
All cancers	132	121	109
Cancer of the buccal cavity and pharynx	3	3.62	83
Cancer of the colon	13	10.9	119
Cancer of the larynx	2	1.74	115
Cancer of the stomach	5	5.39	93
Cancer of the lung	59	41.4	143
Diabetes mellitus	8	7.89	101
Ischemic heart disease	159	187	85
Cerebrovascular disease	30	30.9	97
Non malignant digestive disease	21	28.5	74
All non-malignant respiratory disease	77	34.0	227
Acute upper respiratory infections	1	0.23	435
Influenza	1	0.60	166
Pneumonia	19	11.5	165
Chronic bronchitis	1	1.45	69
Emphysema	12	6.69	180
Asthma	2	1.02	196
Pneumoconiosis + other resp. diseases*	41	12.5	329
Non-malignant respiratory disease except pneumonia and infectious diseases	56	21.6	259
Nervous system diseases	5	5.81	86
Accidents	51	42.4	120

Source: Checkoway et al. (1992)
[†] Employed for at least 12 months cumulative service in the industry and for at least one day between 1 January 1942 and 31 December 1987.
* Includes 5 deaths defined as "silicosis," 5 deaths defined as "diatomaceous earth pneumoconiosis," and 7 deaths defined as "pneumoconiosis."
** SMRs based on rates for U.S. white males, 1942–1987.

workers hired since 1960: "The results for lung cancer and NMRD (nonmalignant respiratory disease) indicate that the excesses were most likely attributable to relatively intense exposures encountered during the 1930s and 1940s, before dust control measures were implemented on a wide-scale basis in the industry. At present it cannot be said with certainty that lung cancer and NMRD risks have been reduced to baseline levels experienced by the population at large. However, it is noteworthy that there has been no excess of lung cancer among Lompoc cohort workers hired since 1960, and there have been no deaths attributed to silicosis among cohort members hired since 1950. These trends are strongly suggestive of reduced hazards, probably related to improved environmental dust control and the increased use of respiratory protective devices by the workforce."

Sandblasting The occupation of sandblasting exposes the worker, unless carefully protected, to large quantities of respirable quartz dust. The health danger of this trade was particularly brought to the attention of medical scientists in the Gulf Coast area of Louisiana, when the emerging petrochemical and ship building industries required a great increase in sandblasting to protect metal surfaces. Respirable quartz dust levels were measured at 318 times the threshold limit value (TLV) in samples taken outside protective hoods during sandblasting (Samimi et al., 1974). Hughes et al. (1982) described the health history of a cohort of silicotic sandblasters. There were 83 patients with a mean age of 44 years and an average silica exposure of 11.3 years. Complicated disease, as defined by the presence of large opacities in the lung, distortion of the intrathoracic organs, or tuberculosis, was present in 64% of the cohort. By 1982, 11 members of the cohort had died. The authors further stated that accelerated silicosis, and the now rare acute silicosis, kills many Gulf Coast sandblasters in the 30 to 40 year age group. In a review of several health studies of those in the sandblasting trades, Jones et al. (1986) state that this occupation can be extremely hazardous, in theory it could be safe, but in practice it is unsafe. The extreme hazard of sandblasting, it might be noted, was recognized in Great Britain in the 1930s. Hunter (1955) reported that the following sign appeared in a Coventry workplace where the practice was common: "Join the Navy and see the world. Become a sandblaster and see the next."

Cohort studies of certified silicotics. The classification of the crystalline silica minerals as "probably carcinogenic to humans" presented as a conclusion in a report of the International Agency for Research on Cancer (IARC, 1987b) has generated much additional concern about the health effects of these minerals. Conventional epidemiological studies of occupational cohorts, such as those presented above, do not show that workers exposed to silica dust alone have a significantly increased risk of death from lung cancer. However, in the last six years there have been 15 or more studies of *certified silicotics*, groups of workers within a specific regional area who were occupationally exposed to silica and are (or were) drawing workers' compensation for diagnosed silicosis. For example, Kurppa et al. (1986) reported on 961 diagnosed cases of silicosis in Finnish men for the period 1935 and 1977. In this Finnish cohort, as expected, there was a very significant excess of nonmalignant respiratory disease (silicosis, tuberculosis, bronchitis, etc.), but also there was a three-fold excess of lung cancer. Mortality data for this cohort is given in Table 10. In a similar study of 2,399 certified silicotics in Switzerland, it was estimated that the risk of lung cancer was 2.2 times that expected from national mortality statistics (Schüler and Rüttner, 1986). In addition, Westerholm et al. (1986) found a 2- to 5-fold risk of lung cancer in Swedish silicotics, Zambon et al. (1986) reported a 2-fold risk of lung cancer in silicotics from the Veneto region of Italy, and a 2-fold risk of this disease was reported by Finkelstein et al. (1986) for miners from Ontario, Canada. In contrast to these studies, Hessel and Sluis-Cremer (1986) reported on a case-control study of 127 pairs of South African gold miners; the paired cases consisted of miners who died of lung cancer and miners who died of other causes. A carefully documented history of smoking habits for both cases and controls was made. The results of this study indicated that there was no association between lung cancer and silicosis.

Table 10

Mortality data for 961 Finnish men diagnosed as having silicosis

Cause of death	Deaths		SMR
	Obs.	Exp.	
Total deaths, all causes	667	335.2	199
Cancer, all sites (ICD 140–209)	122	68.9	177
Lung (trachea, bronchus, pleura) (ICD 162)	80	25.6	312
Esophagus, stomach, colon-rectum (ICD 150–154)	21	19.9	106
Larynx (ICD 161)	—	1.4	—
Oropharynx (ICD 140–149)	1	0.8	128
Leukemia, lymphosarcoma and other neoplasms of lymphatic and haemapoietic system (ICD 200–207)	6	3.7	161
All other sites	20	19.5	103
All cardiovascular disease (ICD 390–458)	203	183.1	111
All pulmonary disease (ICD 460–519)	165	23.4	704
Pneumoconiosis (silicosis) (ICD 515)	120	0.0	∞
Pneumonia, influenza (ICD 470–474, 480–486)	13	10.2	127
Chronic bronchitis, emphysema, and asthma (ICD 490–493)	27	5.6	478
Other respiratory disease	5	7.5	66
All renal disease (ICD 580–593)	7	5.9	117
Accidents, violent deaths (ICD 800–999)	20	23.9	84
Tuberculosis (ICD 010–012)	130	17.6	738
All other causes	20	24.2	83

Source: Kurppa et al. (1986)

Because many conventional cohort studies do not show a significant risk of lung cancer in silica-exposed workers, it is difficult to interpret the results of studies which suggest that such a relationship between silicosis and lung cancer exists. There is a difficulty in properly evaluating the contribution of tobacco use to disease within an occupational cohort if smoking habits are unknown or if the regional or national populations used as controls do not have the same smoking habits as the occupational cohort. For example, cigarette smoking was prevalent among the U.S. male blue collar trades workers and commonly 75 to 85% of the workers smoked. On the other hand, only about 50% of the total male population of the United States were smokers. If 80% of the men in a particular occupational cohort smoke, but who have no other health risks, the proportional mortality for lung cancer for the total cohort is about 8.5%. If 50% of the control cohort are smokers, the lung cancer proportional mortality is about 5.5% (Ross, 1984, Fig. 2). Thus, if this occupational cohort is compared to this control group, an incorrect proportional mortality ratio (PMR) would be calculated (8.5/5.5 = 1.55); the 55% excess lung cancer would be attributed to some type of occupational hazard rather than to differences in smoking habits.

Other problems with interpreting the studies of silicotics is that they are not true cohorts in that they are not well defined occupational groups for the men usually come from several industries. Also, the types and lengths of exposure to dusts, chemicals, radon, etc. were generally unknown. Craighead (1992) gives cogent arguments for being cautious in the interpretation of these studies of silicotics as does the Working Group on health effects of silica (IARC, 1987a, p. 108). However, it should be noted that Huges and Weill (1991) give data to indicate that asbestosis is a precursor to lung cancer. Can silicosis also be a precursor to lung cancer?

COAL

Mineralogy

Several dozen minerals are reported to occur in coals, although most occur only sporadically or in trace amounts. Most of the minerals in coal fall into four groups: (1) aluminum silicates, (2) carbonates, (3) sulfides, and (4) silica, mainly quartz. Aluminum silicates commonly found in coal are clay minerals, including montmorillonite, illite-sericite, kaolinite, halloysite, chlorite, and mixed-layer clays. Principle sulfide minerals are: pyrite, marcasite, galena, chalcopyrite, pyrrhotite, arsenopyrite, and millerite. Carbonate minerals include calcite, dolomite, siderite, ankerite, and witherite. In a study of 65 Illinois coals, Rao and Gluskoter (1973) found the mineral matter content to vary from 9.4 to 22.3 % with 15% being an average value. Harvey and Ruch (1986) give the following variation in mineral content of Illinois Basin coals: quartz (1.2 to 3.1 wt. %), calcite (0.9 to 2.3%), pyrite (2.8 to 5.9%), and clay minerals (6.6 to 11.2%). Extensive reviews of the mineralogy and petrology of coal is given by Gluskoter et al. (1981) and Stach et al. (1982).

Diseases related to exposure to coal dust

Coal workers' pneumoconiosis. *Coal workers' pneumoconiosis* is caused, most importantly, by fine-grained coal dust composed of carbonaceous material. One form of this disease, *simple coal workers' pneumoconiosis*, is characterized by the coal macule, a lesion 1 to 4 mm in diameter and composed of dust laden macrophages that form a mantle around the first and second order respiratory bronchioli. These bronchioles dilate as the macule enlarges causing focal emphysema. In lung sections the macules appear as black areas, the smaller macules are usually circular, the larger ones are more irregular and often stellate. The cause of simple pneumoconiosis is considered to be an overwhelming of normal lung particle clearance mechanisms by the large amounts of coal dust entering the respiratory passages. The current term used to describe this physiological state is "pulmonary overload." The fibrosis and emphysema may be due to the damaging effects of enzymes released by the macrophages (Merchant et al., 1986; Lapp, 1981).

A second type of lesion, a nodular lesion similar to that seen in the chronic silicotic, also occurs in coal workers' pneumoconiosis. These nodules are gray or black in color, are often rounded, but may have irregular prolongations penetrating the surrounding tissue and may be associated with scar emphysema. The major difference between the macular and nodular lesions is that hyalinized collagen is present in the latter. The nodular lesions, which are thought to have developed from the macules, contain bundles of collagen that are usually arranged in an irregular pattern. This arrangement of collagen is useful for distinguishing these lesions, formed by an accumulation of carbonaceous particles, from the fibrotic nodules of the silicotic where the collagen is concentrically arranged. Merchant et al. (1986, p. 354–361) present optical photomicrographs of lung sections that contain macules and nodules caused by the inhalation of carbonaceous dust and sections that contain silicotic nodules caused by inhalation of quartz dust. *Complicated coal workers' pneumoconiosis* or *progressive massive fibrosis* is characterized by extensive fibrosis of the lung tissue. Often, small, separate nodular lesions coalesce to form a single one which replaces a major portion of the lobe of the lung in which it occurs. There is a strong association between tuberculosis and progressive massive fibrosis, particularly in the coal miners of South Wales (Merchant et al., 1986).

The onset of simple coal workers' pneumoconiosis, which takes many years to develop, is related to the amount and duration of coal dust exposure, but according to Lapp (1981), unlike silicosis there is no convincing evidence of progression of this disease in the absence of further exposure. Complicated coal workers' pneumoconiosis is a more serious condition in that it is generally progressive even after exposure to coal dust ceases. Severe cases of "black lung" disease lead to airway obstruction and the coughing of large volumes of inky black sputum; death generally occurs in about 10 to 15 years as a result of cardiopulmonary failure.

Coal workers' silicosis. Coal workers' lung disease may be complicated by exposure to other minerals associated with coal, but particularly from exposure to quartz dust in coal and (more importantly) from quartz in the country rock enclosing the coal beds. Evidence for silicosis has been found particularly in those who are primarily engaged in surface drilling of quartz-bearing country rock (Lapp, 1981). Merchant et al. (1986) observed silicotic nodules in the lungs of some coal workers (their Fig. II-29), but they state that good evidence is lacking that quartz dust plays a significant role in coal miner's disease.

Epidemiological studies of occupational cohorts exposed to coal dust

Historical accounts of miners' "black lung" were reported as early as 1831 and since then many studies have been published which described coal workers' pneumoconiosis and its associated morbidity and mortality. Merchant et al.

(1986) review twenty five of the mortality studies and thirteen morbidity studies of coal miner cohorts which were published between 1936 and 1981.

Pneumoconiosis mortality and morbidity. Coal workers' pneumoconiosis (CWP), often referred to as "black lung" or "anthracosilicosis," is prevalent in coal miners and is described in a vast literature. This is a chronic and incapacitating disease due to progressive loss of lung function. There is a clear dose-response relationship between the severity of lung disease and the number of years working underground—the surrogate for dose. Jacobsen et al. (1971) give the probability of developing simple CWP over a 35-year working life with mine dust levels varying from an average of 1 to 7 mg/m^3. For a dust level of 6 mg/m^3 the probability of developing simple to more severe pneumoconiosis is ~10%. Bronchitis and associated respiratory infections are associated with CWP and seriously impaired lung function is often found in coal workers who have advanced pneumoconiosis, especially progressive massive fibrosis. One of the largest and most extensive mortality study of coal miners is that of Rockette (1977). He examined a cohort of 22,998 miners, a ten percent sample of the qualified members of the United Mine Workers of America Health and Retirement Fund (UMWA cohort). The major findings of the study are given in Table 11. The overall Standard Mortality Ratio (SMR) is 101.6 and is not significantly different from the control value of 100. The SMRs for influenza (189.6), emphysema (143.7), asthma (174.9), and tuberculosis (145.5) were significantly elevated. However, major cardiovascular diseases were not elevated in this cohort (SMR=95.2).

Cancer mortality. Overall cancer mortality in the UMWA cohort of Rockette (1977) was slightly less than that of the control (the 1965 total male U.S. population) with an SMR of 97.7 (Table 11). There was a small but statistically insignificant increase in stomach cancer and lung cancer. A study by Enterline (1964) of U.S. coal miners gives an elevated SMR for lung cancer (1.92), however many other health studies of coal miners do not demonstrate that lung cancer is a significant factor in coal miner mortality (Merchant et al., 1986, Table II-20; Bridbord et al., 1979, Table 1). The report by Bridbord et al. (1979) reviews nine epidemiological studies of coal miners; four cohorts show an increased risk of stomach cancer, one shows an increased risk of prostate cancer, one shows an increased risk of lung cancer (Enterline, 1964, cited above) and six show a decreased risk of lung cancer. Rosmanith and Schimanski (1986) report that bronchial cancer seldom develops in cases of severe coal workers' pneumoconiosis in coal miners from the Czechoslovakian and West German coalfields (Ostrava and Ruhr). Breining (1986) reports that during the last ten years they examined the lung tissue of more than 1,000 diseased coal workers from the Ruhr area of West Germany who had anthracosilicosis, of these only five cases were found with silicotic scar carcinomas. As yet, there does not appear to be compelling evidence that coal workers have a significant risk for increased lung cancer incidence.

Table 11

Deaths for cohort of 22,998 U.S. coal miners for selected causes

Cause of death	Deaths Observed	Expected	SMR
All causes	7,628	7,506.1	101.6
All malignant neoplasms	1,223	1,252.2	97.7
Benign and unspecified Neoplasms	14	14.4	97.5
Major cardiovascular diseases	4,285	4,501.2	95.2
Bronchitis	27	31.5	84.8
Influenza	28	14.8	189.6
Pneumonia	217	232.3	93.4
Emphysema	170	118.3	143.7
Asthma	32	18.3	174.9
Tuberculosis	63	43.3	145.5
Syphilis	16	13.1	122.3
Other infective and parasitic disease	13	17.6	74.1
Diabetes mellitus	64	110.2	58.1
Peptic ulcer	42	58.7	71.6
Cirrhosis of the liver	64	104.9	61.0
Cholelithiasis, cholecystitis, and cholangitis	22	16.7	132.0
Nephritis and nephrosis	42	46.2	91.0
Accidents	408	283.0	144.2
Suicides	81	81.3	99.6
Homicides	30	26.1	115.1
Ill-defined causes	162	86.2	187.9
All other causes	625	459.5	136.0

Source: Rockette (1977)

THE SILICATE MINERALS (OTHER THAN ASBESTOS)

Mineralogy of the silicates

The silicates form the largest chemical class in the mineral system and they comprise a large portion of the earth's crust. Silicates are structurally and chemically complex and are characterized by the presence of essential amounts of tetrahedrally coordinated silicon. These minerals also have essential amounts of one or more other elements which are found in various coordination schemes within the crystal structures. In addition to Si, the most common elements found in the silicates are Na, K, Ca, Al, Mg, and Fe. Important mineral groups within the silicate class include feldspars, pyroxenes, amphiboles (including amphibole asbestos), micas, chlorites, serpentines (including chrysotile asbestos), the aluminum silicates (sillimanite, kyanite, and andalusite), talcs, clay minerals, zeolites, nephelines, garnets, humites, olivines, and epidotes. Workers, particularly miners and millers, are exposed to the dusts of many of these minerals as they process various types of rocks to extract particular mineral commodities. Some of these silicate minerals are mined for a particular use; examples are talc, wollastonite, kaolinite, attapulgite, amphibole and serpentine asbestos, bentonite, mica, vermiculite, and zeolites. Exposure to a few of the silicates (especially to asbestos) by occupational groups has been great enough to produce clinical evidence of disease and to permit epidemiological studies to be made. The silicates most implicated with diseases (chrysotile and amphibole asbestos are discussed in another chapter) are reviewed below.

Health effects of selected silicate minerals

Talc and pyrophyllite. Talc [$Mg_3Si_4O_{10}(OH)_2$], which commonly forms by hydrothermal alteration of ultrabasic rocks rich in magnesium, is mined in many countries and processed in numerous manufacturing industries for use in paints, ceramics, rubber products, roofing materials, paper, insecticides, cosmetics, and pharmaceuticals (Langer et al., 1990). In the United States 1,172,000 metric tons of talc was mined in 1989 (Virta, 1991). There have been numerous health studies of talc workers (Gamble, 1986; IARC, 1987a,b), but the results of these studies have often been ambiguous. Talc workers exposed to talc dust may exhibit symptoms of *talc pneumoconiosis*, sometimes referred to as *talcosis*, as well as bronchitis, emphysema, abnormal chest X-rays, and increased risk of tuberculosis. Clinically, talc workers' pneumoconiosis resembles silicosis or asbestosis and since the talc exposed worker may have been exposed to quartz and other silicates, including chrysotile, anthophyllite, and/or tremolite asbestos, the true etiology of this disease is difficult to describe. One study (Hogue and Mallette, 1949) reported that rubber workers exposed to high levels of talc dust show no disease and thus they concluded that talc dust was benign. However, examination of lung tissues of some of the persons exposed to talc show diffuse pleural thickening, fibrous adhesions, pleural plaques, and large fibrotic masses (Gamble, 1986). The several epidemiological studies of talc workers are in disagreement over whether talc causes lung cancer. Excess cancer may be related to underestimation of smoking habits or to previous exposure to commercial asbestos. Rubino et al. (1976) studied 1,514 miners and millers from the Piedmont in Italy who were exposed to asbestos-free quartz-bearing talc dust and found elevated pneumoconiosis (described as silicosis) and associated tuberculosis. Lung cancer mortality was much less than expected. A study of New York talc miners and millers (Brown et al., 1979) indicated that there was excess lung cancer (SMR=270); the SMR for nonmalignant lung disease was 277. Stille and Tabershaw (1982) studied nearly the same cohort of New York miners and millers as examined by Brown et al. (1979) and found a lesser risk of lung cancer (SMR=157). The epidemiological studies of talc workers thus far completed have not proven that talc is a human carcinogen for the cohorts studied were not large enough to produce statistically significant data, smoking habits are not well defined, and the workers were often exposed to other mineral dusts that could produce disease. There is no doubt, however, that heavy exposure to talc dust can cause nonmalignant respiratory disease.

Pyrophyllite [$Al_2Si_4O_{10}(OH)_2$] has similar uses as talc but is produced in much smaller quantities (U.S. production was 81,000 metric tons in 1989; Virta, 1991). Little health data have been obtained on pyrophyllite workers. Hogue and Mallette (1949) report no apparent disease in pyrophyllite-exposed rubber workers.

Kaolinite. Kaolinite [$Al_2Si_2O_5(OH)_4$], a common clay mineral (also known as "china clay"), is mined for many uses and particularly for ceramics and for filler in papers, paints, and plastics. The kaolinite-bearing rock (referred to as "kaolin") generally contains variable amounts of other minerals including quartz. In 1989 the United States produced 8,973,669 metric tons of this mineral (Ampian, 1991). Kaolinite workers who have been heavily exposed to kaolinite dust may develop a pneumoconiosis, often referred to as *kaolinosis*. Simple kaolinosis is similar to other mineral dust pneumoconiosis that are characterized by the presence of rounded opacities in the lung. Complicated kaolinosis is similar to progressive massive pneumoconiosis of the coal worker (Gamble, 1986). If the kaolinite worker is also exposed to silica dust, as often may be the case, the lung disease may appear to be typical silicosis. An increased lung cancer risk has not been reported in kaolinite worker cohorts.

The clinical aspects of kaolinosis are discussed in a recent prevalence study of kaolin workers in east central Georgia. The study (Morgan, et al., 1990) reported that in a randomly selected sample of 2,000 workers, there were 90 subjects with simple and 18 with complicated pneumoconiosis, yielding an overall prevalence of 3.2% and 0.63%, respectively. The nature of the lesion caused by exposure to pure kaolinite has been studied by exposing rats to intensely high (300 mg/m^3) concentrations of kaolinite for 3 months. The rats in this study were serially sacrificed to describe the development of the lesion (Wastiaux and Daniel, 1990). The experimental animal study helps us to understand the role of kaolin in the human pathology, but which can be complicated by the individual's exposure to quartz and other minerals in kaolin-exposed individuals (see Gibbs, 1990 for a discussion).

Bentonite. Bentonite is a soft rock, plastic when wet, and is composed primarily of silicates belonging to the montmorillonite subgroup of clay minerals, which has the general composition $(Na,Ca)_{0.33}(Al,Mg)_2Si_4O_{10}(OH)_2 \cdot nH_2O$. Bentonite deposits are generally associated with other minerals including very fine-grained quartz and amorphous silica. Bentonite is used as drilling mud, as a bleaching clay, as a foundry sand bond, as an iron ore pelletizer, and for many other uses. In 1989 the United States mined 3,112,365 metric tons of bentonite (Ampian, 1991). A study of a random sample of Wyoming bentonite workers revealed that 44% had silicosis, including 2 cases of progressive massive fibrosis (Phibbs et al., 1971). Silica content (which included both quartz and cristobalite) of the Wyoming bentonite clays ranged from 0 to 24%. Surveys of bentonite processing plants showed that silica composed between 5 and 10% of the mineral matter in the airborne dust and the TLV for silica was exceeded 3 to 10 times (Gamble, 1986).

Palygorskite/attapulgite and sepiolite. Palygorskite (synonymous with attapulgite) has the approximate composition $(Al,Mg)_4(Si,Al)_8O_{20}(OH)_4 \cdot 2H_2O$ and is structurally related to sepiolite, $(Mg,Fe)_8Si_{12}O_{30}(OH)_4 \cdot 4H_2O$. Palygorskite, sepiolite and loughlinite (sodium sepiolite) form the a mineral subgroup of the clay mineral group. Palygorskite can also occur with montmorillonite and quartz and is referred to as "fuller's earth." Although the commercial deposits generally contain more than 50% palygorskite, the ores are beneficiated to 60 to 90% purity and marketed in various grades. Palygorskite occurs naturally with sepiolite, phosphate minerals, carbonate minerals, opal, quartz, cristobalite, montmorillonite and other clay minerals (Galan and Castillo, 1984; Clarke, 1985; Nolan, et al., 1991b).

The three major producing countries are the U.S. (with four U.S. companies producing approximately 85% of the world market), Senegal, and Spain (Haas, 1972; Clarke, 1985; Clarke, 1990). In 1989 the U.S. production of palygorskite was 1,881,511 metric tons (Ampian, 1991). The naturally high porosity and sorptivity of the minerals are the industrially important properties. These properties are often enhanced by heating the mineral. Palygorskite is used in drilling muds (particularly salt-water drilling because the mineral does not lose swelling capacity in salt water), oil/grease absorbents, pesticide carriers, and pet litter products (IARC, 1987a; Clarke, 1990). The markets are similar for palygorskite and sepiolite although the latter has a greater surface area (300 m^2/g), internal channel/pore surface (400 m^2/g), and higher temperature stability.

Concern has been raised about these fibrous clay minerals which are being used as asbestos substitutes (Bignon, et al., 1980; Bignon, 1990). A health study incorporating chest x-rays of 701 U.S. workers, with an average exposure of 11 years mining and milling palygoskite clay (generally referred to as attapulgite clay in the southern U.S. deposits) in Florida and Georgia, were found to have a 6.4% pneumoconiosis prevalence ($\geq$ 1/0 in the 1980 ILO classification). The unilateral and bilateral pleural thickening was found in 3.2% and 4.2% of the workers, respectively. The total dust concentrations range from 0.6 to 3.1 mg/m^3 in mining occupations and from 0.1 to 23 mg/m^3 in milling and shipping occupations. The average the respirable dust concentration was < 5mg/m^3 (Gamble et al., 1986).

A mortality study was conducted of 2,302 men employed at least 1 month between 1940–1975 and who worked in a production plant milling the ore from the same deposit in Florida and Georgia discussed above (Waxweiler et al., 1988). The respirable dust levels were between 0.02 and 0.32mg/m^3 and respirable silica was generally below the 50$\mu g/m^3$ NIOSH standard. The airborne palygorskite (attapulgite) fibers had a mean length of 0.4μm and a mean diameter of 0.07μm; no fibers greater than 2.5μm in length were found (Zumwalde, 1976). The palygorskite from the Georgia-Florida deposit is generally short-fibered (Fig. 1), although the mineral can occur with considerably greater fiber lengths depending on the geological locale. Transmission electron microscopy shows the mean fiber length of the bulk sample from Attapulgus, Georgia is < 0.5μm with a range from 0.1 to 2.5μm (Nolan, et al., 1991a). This cohort had a deficit of both overall mortality and mortality from nonmalignant respiratory disease. However, a statistically significant excess of lung cancer (16obs./8.3exp. SMR=193) was found among the long term millers who had high level exposures. No data on

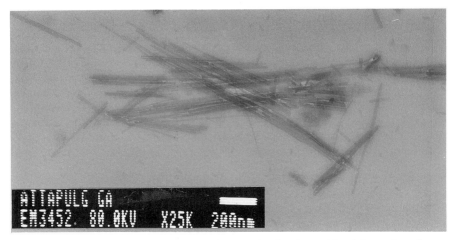

Figure 1. Transmission electron photomicrograph of palygorskite (also referred to as attapulgite) from Attapulgus, Georgia, U.S.A. Approximately 83% of the fibers are <1 μm in length (Nolan et al., 1991a).

past cigarette smoking habits were available. The IARC evaluation of the data concluded there is inadequate evidence for the carcinogenicity of palygorskite in humans (IARC, 1987a).

Micas. The common minerals of the mica group are muscovite [$KAl_2AlSi_3O_{10}(OH)_2$], phlogopite [$K(Mg,Fe)_3AlSi_3O_{10}(OH)_2$], and biotite [$K(Fe,Mg)_3AlSi_3O_{10}(OH)_2$]. The micas have numerous uses, particularly in electrical devises and as fillers in plastics, tiles, etc. Mica can cause pneumoconiosis in workers exposed to high concentrations of mica dust from various occupational categories. A health study of a few individuals involved in packing and grinding of mica have been made. Pathological examination of lung tissues showed that the lesions present in these workers' lungs were similar to the lesions found in those with kaolin pneumoconiosis, except that foreign body giant cells and ferruginous bodies were not present (Gibbs, 1990). Gibbs concluded that, "Although rare, I believe that pneumoconiosis can result from the inhalation of large quantities of mica in the absence of other silicates or quartz." It should be noted, however, that, in the majority of mica-exposed individuals, mixed dust exposures, particularly involving quartz, are expected (Bignon, 1990). Gibbs' observation concerning the apparent health effect of very intense exposures to pure, or nearly pure, mica dust suggests the correctness of the observation of Craighead et al. (1992)—that the Vermont slate workers pneumo-coniosis appears to be enhanced by exposure in mica dust.

Vermiculite. The mineral vermiculite is essentially a hydrated biotite or phlogopite mica with water molecules located between the silicate layers, and its generalized formula is $(Mg,Ca)_{0.35}(Mg,Fe,Al)_3(Al,Si)_4O_{10}(OH)_2 \cdot nH_2O$. Upon heating, this mineral expands to form a light weight product used for insulation, as a soil conditioner, and filler. Vermiculite is often associated with various amphibole minerals and these rather than vermiculite itself has been the focus of the health concerns. Four groups of vermiculite exposed workers have been studied. One group working in a vermiculite exfoliation plant and three involved in mining and milling. None of these studies produced much of a discussion on the health effects of vermiculite but rather focused on the presence of amphibole minerals associated with the ore.

The pulmonary hazards of fibrous tremolite occurring at low levels in vermiculite were evaluated among a group working in a vermiculite exfoliating plant. Twelve cases of benign pleural effusion occurred in the group over a 12- year period. The cause for this unusually high number of cases could not be established and a study was designed (Lockey et al., 1983) to determine the prevalence of pulmonary abnormalities among the vermiculite exposed workers. The study population consisted of 513 current employees with a history of exposure to vermiculite containing fibrous tremolite. The control group was made up of employees who were not exposed to dust. About 44% of the cohort were current smokers and 20% ex-smokers with no other significant differences among the exposure groups. The medical component of the cross-sectional epidemiological study was correlated with exposure by job category, cumulative fiber exposure, and time from first exposure to the vermiculite dust. A statistically significant association was found between cumulative dust exposure and symptoms, i.e., shortness of breath with wheezing, dyspnea on exertion, and pleuritic chest pain. The radiographic survey involved 501 of the 513 individuals in the exposed group. Pleural and/or parenchymal changes were found in 22 workers, eleven with costophrenic angle blunting and eleven with pleural abnormalities. The mean cumulative exposure for those with an abnormal x-ray was 8.74±11.71 (fibers/ml) x yr. The vermiculite ore processed in this plant, mostly obtained from Libby, Montana, was reported to contain from 0.006% to 0.41% fibrous tremolite (Banks, 1980; Lockey et al., 1984; Moatamed, et al., 1986). In examining of a specimen of the Libby, Montana deposit by analytical transmission electron and polarized light microscopy both tremolite asbestos and richterite asbestos were found (see Langer et al., 1991; Nolan et al., 1991b for the criteria for mineral identification and further discussion).

The health effects of exposure to vermiculite containing fibrous tremolite on miners and millers in Libby, Montana has been studied by two groups (McDonald et al., 1986a,b; Amandus et al., 1987a,b; Amandus and Wheeler, 1987). Each study involved a cohort mortality study and a cross-sectional radiographic survey. Slightly different criteria were used to define each cohort and, therefore, the McDonald cohort contained 406 men with 165 deaths and the Amandus cohort contained 575 men with 161 deaths. Both research groups used historical air samples to estimate an exposure index for each cohort member. The older measurements were all made with the midget impinger and conversion from million particles per cubic foot (mppcf) was made to give the approximate the number of fibers per milliliter (f/ml). The exposures in the dry mill, before the installation of dust control equipment in 1964, were estimated by McDonald et al., (1986a) and Amandus et al. (1987a) to be ~100 fibers/ml and ~168 fibers/ml, respectively. Dust levels between 1965 and the closure of the dry mill in 1974 was estimated by these authors to be ~20 f/ml and ~33 f/ml respectively. These were the highest exposures measured except for dust levels developed during floor sweeping in the dry mill which were ~20% higher. The McDonald cohort has an SMR for total mortality of 117 with 23 respiratory cancer cases (SMR=245) and 4 mesotheliomas (3 pleural and 1 peritoneal). The SMR for total mortality of the Amandus cohort was 110 with 20 respiratory cancer cases (SMR = 223) and 2 mesotheliomas. The lung cancer SMR for more than 20 years since first employment at *all* exposure levels was, 242 and 279, for the McDonald and Amandus cohorts, respectively. Both cohorts had an SMR of ~2.5 for nonmalignant respiratory disease.

The mortality experienced in Montana cohorts was compared to vermiculite miners and millers in the Enoree region of South Carolina where the ore contains trace amounts of fibrous tremolite (McDonald et. al., 1988). This cohort was made up of 194 men. Fifty-one deaths had occurred among the men with >15 years of exposure. Of these, 4 were from respiratory cancer (SMR=121), and 3 of these 4 lung cancers appeared in lowest exposure group (<1f/ml) × yr.). There were no mesotheliomas or pneumoconiosis deaths. The SMR for total mortality was 117, similar to the values given for the Montana cohorts. All the air samples, except one, were analyzed by phase contrast microscopy (using the NIOSH reference method for asbestos) which showed mean fiber concentrations of <0.01f/ml (N = 58). The analysis of the same samples by analytical transmission electron microscopy were higher and the mean concentrations varied from 0.01 to 0.32 fibres/ml for fibers >5µm in length. The average fiber diameter was 1.1µm and the average length was 12.7µm. It is noted that less than half the fibers examined had elemental compositions consistent with the tremolite-actinolite series.

Addison and Davies (1990) measured the amphibole content of 57 vermiculite samples and found that 26 contained detectable amphibole with a range of 0.02 to 6.4%. A sample of tremolite separated from vermiculite from the Enoree, South Carolina locality and examined by light and transmission microscopy showed the tremolite present to be in the form of non-asbestiform cleavage fragments—unlike the Libby, Montana ore where fibrous amphibole was present. A photomicrograph of Enoree tremolite cleavage fragments is compared in Figure 2 to one showing the asbestiform fibers found in the Libby, Montana samples (see Langer et al., 1991; Nolan et al., 1991b for the criteria for mineral identification and further discussion)

The Palabora vermiculite deposit, located about 300 km northeast of Johannesburg, is formed from a hydrated phlogopite rather than a more iron-rich hydrobiotite found in the Libby, Montana deposit. Serpentine, pyroxene, diopside, biotite, and phlogopite are reported to be associated with this ore. Although Moatemed et al.(1986) found 0.4% amphibole in the unexpanded ore, none was detected in a similar study of the mill dust (Sluis-Cremer and Hessel, 1990). A prevalence study was made of 172 Palabora miners having an average length of exposure of 15.3 years. The control group was 354 miners from a nearby copper mine operated by the same company. A respiratory symptom questionnaire, lung function test, and chest X-ray revealed no evidence that the dust from the Palabora vermiculite ore had any deleterious effects on respiratory health of the miners, either due to vermiculite, the associated minerals, or to the small content of tremolite (Sluis-Cremer and Hessel, 1990; Hessel and Sluis-Cremer, 1989). Intrapleural injection of vermiculite powder from Palabora, South

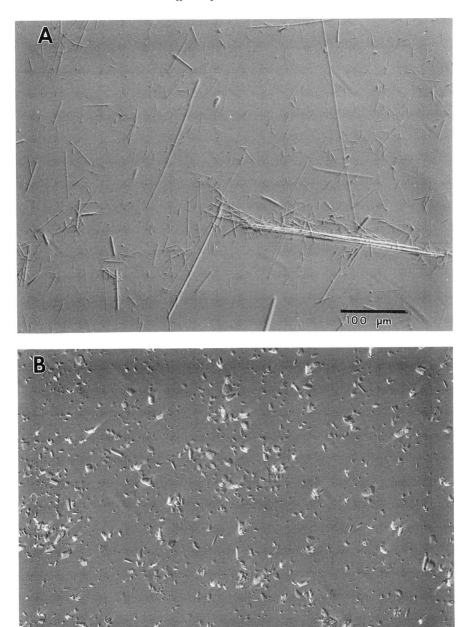

Figure 2. Light photomicrographs taken with Hoffman modulation optics of (A) the asbestiform fiber present in the vermiculite ore from Libby, Montana, U.S.A. and (B) the tremolite cleavage fragments present in the vermiculite ore from Enoree, South Carolina, U.S.A.

Africa into rats produced no tumors (Hunter and Thomson, 1973; see Johnson, 1993 for a discussion of animal models).

Wollastonite. Wollastonite ($CaSiO_3$), as with most naturally occurring silicates, rarely occurs as a pure mineral. Major ore deposits contain from 17 to 97% wollastonite, which is often associated with calcite, quartz, garnet and diopside. The Willsboro, New York wollastonite deposit was the first to enter large scale commercial production in the early 1950s. Two mines, one in the U.S. and one in Finland, are the major world producers, although one smaller mine is in operation in each of the following countries: India, Japan, Kenya, Mexico and New Zealand (Coope, 1982). Wollastonite is marketed in a variety of grades, including a long-fibered grade (aspect ratio varies from 15:1 to 20:1) which is sold as a filler and as an asbestos substitute. Wollastonite fiber with lower aspect ratios is more common and is primarily used in the ceramics and plastics industry. The uses of this mineral include production of mineral wool and in paints, coatings, heat-containment panels, ceiling and floor tiles, brake linings, and welding fluxes for metal coating.

Chest radiographs of 46 Finnish quarry workers, having an average of 21.5 years of exposure, revealed seven cases of small irregular parenchymal opacities and seven cases of pleural and parenchymal changes (Huuskonenen, et al., 1983a). A mortality study was made of 238 workers from a Finnish limestone-wollastonite quarry, the workers having at least one year of exposure between 1920 and 1980. The total number of deaths in the cohort were less than expected (79 obs./96 exp., SMR=82) as was the mortality from lung cancer. The exposure during mining and milling was 1 to 45 fiber/ml and the mean level of respirable quartz <10 $\mu g/m^3$ (Huuskonen et al., 1983b). These exposure levels were similar to those reported for Willsboro wollastonite workers (IARC, 1987a).

Zeolites. Among this large group of hydrated aluminum silicates are several fibrous varieties. One of them, erionite, $NaK_2MgCa_{1.5}(Al_8Si_{28}O_{72}) \cdot 28H_2O$, is implicated in respiratory disease in residents of certain villages located in the Cappadocia region of Turkey. This subject will be discussed in the next section.

ENVIRONMENTAL EXPOSURES TO FIROUS MINERALS

Environmental exposure to mineral fiber in Turkey, Cyprus, and Greece

An unusual series of tumors of the lung and pleura occurring in certain villages in southeast Turkey have been studied retrospectively in the Diyarbakir Chest Hospital (Yazicioglu et al., 1980). Between 1968 and 1976, 177 malignant lung cancers and 44 pleural tumors patients were admitted to the Chest Hospital. The geographic distribution of the tumors suggested an etiology related to an environmental asbestos exposure. Of the 44 mesotheliomas found in the group of 221 malignant tumors, ten (6 females, 4 males) occurred among the 20 to 40-year-old age group. The early age at which these tumors appeared, and the near equal distribution between the sexes, indicates the exposure was environmental and began in early childhood. Interestingly, many cases of benign pleural effusion were found as well. A similar observation would later be reported by Lockey et al. (1983) among the workers in a vermiculite exfoliation plant in the United States.

A radiographic cross-sectional survey (Yazicioglu et al., 1980) of 7,000 individuals living southeast Turkey revealed that 6.6% (461) had pleural thickening and calcification and 1.5% (103) had interstitial pulmonary fibrosis. By age 65, 50% of this group would have radiographic evidence of pleural abnormalities. The exposure starts at birth and continues as long as the individual remains in this environment, which may explain why the changes occur so early and are very extensive. The clinical effects of these childhood exposures are found years later in individuals who moved away from the exposure areas.

The health effects described in Turkey were also noted by Constantopoulus et al. (1987) in their observations pertaining to geographic distribution of plaques and tumors in Greece. An earlier report attributed the malignant tumors, particularly those of the pleura, to chrysotile exposure. The source of the asbestos around the town of Cermik, Turkey, was later found to be from the numerous outcroppings of tremolite asbestos, which was used locally to make a whitewash or stucco for the walls, floors, and roofs of the houses (Yazicioglu et al., 1980). The whitewash contained fibrous tremolite and the non-fibrous minerals talc, chlorite, antigorite, and lizardite. Although occasional chrysotile fibers were found in the environment, the investigators attributed the pleural changes, pulmonary fibrosis, and the malignant tumors of the lung and pleural occurring in the Cermik region to tremolite asbestos.

Apparent asbestos-like diseases of the chest, including mesothelioma, have also been reported in the small Anatolian Village of Caparkayi in Turkey (Baris et al., 1988a,b). Four cases of pleural mesothelioma were reported in a population of 425 over a three year period. All of the tumors occurred in women between 26 and 40 years of age. Again, the non-occupational nature of the exposure is indicated by the tumors occurring at a young age and in women. In a radiographic cross-sectional survey of 167 individuals over 20 years old from this village, 63 abnormalities were found. Due to a migration of the younger people, the village population has a higher percentage of older individuals than would generally be expected (51% are over 20 years of age). Approximately 15% of the 167 individuals surveyed had calcified pleural plaques, interlobar fissure thickening, and/or diffuse interstitial fibrosis. Although there is no asbestos mine near this village, the villagers commonly used white stucco described by Baris et al. (1988b) as "rich in tremolite asbestos including some very fine fibre." The Baris report indicates that the high incidence of mesothelioma and some of the pleural and parenchymal abnormalities in the village are associated with exposure to tremolite fibers.

Located in the central mountains of the island of Cyprus is a large chrysotile mine which has been in commercial operation since 1904. Initially it was thought that the site would provide an opportunity to study human mesothelioma from exposure only to chrysotile. The first mesothelioma case was found in a woman who had never worked in the mine, although she lived in a nearby village. The lung tissue in this woman contained asbestos bodies and amphibole asbestos. In all, 14 apparent mesothelioma cases were reported from the onset of the study in 1969 to March, 1986. No tremolite was found in the chrysotile sample taken from the mine, although tremolite and chrysotile were found to be present in an environmental dust sample taken from the roof eaves of the local houses (McConnochie et al., 1987). Of the 14 apparent cases of mesothelioma identified in Cyprus asbestos mining area, 7 were confirmed by a panel of pathologists. Analysis of the inorganic fiber in the lung tissue of both humans and sheep (residing within 5 miles of the mine) showed chrysotile and tremolite to be present. The tremolite found was in a form that included long, thin fibres having a similar size range to that of crocidolite asbestos. Further study (McConnochie et al., 1989) has positively identified a total of 13 mesothelioma cases, 5 of which occurred in persons unconnected with the local asbestos mine. A stucco used in the region contained fine fibrils of chrysotile and long, thin tremolite fibers. The various reports describing the distribution of tumors relative to the local occurrence of tremolite asbestos and its use as stucco, suggests the chrysotile asbestos mine is not the major cause of pleural disease in this region but rather environmental exposure to tremolite asbestos (McConnochie et al., 1989).

In northwest Greece, six deaths from malignant pleural mesothelioma out of the 7 reported (3 males and 4 females) have occurred among residents in the villages of Milea, Metsovo, Anilio, and Votonosi (Constantopoulos et al., 1987). Bilateral pleural plaques, pleural thickening, restrictive lung function, and mesotheliomas constitute the cluster of disease referred to as Metsovo lung (Constantopoulos et al., 1985). Before 1940 virtually all of the inhabitants of the area painted their homes with a whitewash containing tremolite asbestos (Langer et al., 1987). Exposure to the tremolite asbestos contained in the whitewash has been associated with Metsovo lung. A hypothesis for expecting such findings to be regional and to occur in other parts of the world has been proposed (Constantopoulos et al., 1987).

Fibrous zeolites and the Karain experience

In the early 1970s the government of Turkey conducted a nation-wide survey to locate and treat residual pockets of tuberculosis which existed primarily in rural areas. The Department of Thoracic Medicine of the Hacetteppe University, Ankara, participated in this survey. One of the areas evaluated was the central mountainous Anatolian plateau, called Cappadocia. Chest x-rays obtained on inhabitants of several small villages on this plateau showed the presence of lesions which were interpreted as tuberculous in origin. The x-rays revealed both pleural fibrosis and calcification (scarring of the outer lining of the lung and the chest wall) which resembled healed tuberculosis scars. However, bilateral pleural thickening was present, which is not normally associated with tuberculosis. Interviews with the persons examined indicated they had also experienced recurrent benign pleural effusions—an experience also noted in the vermiculite workers and Greek residents environmentally exposed to tremolite asbestos, as discussed previously. The majority of the Turkish villagers showed no evidence of active tuberculosis.

One of the cases of pleural effusion did not clinically resolve upon treatment with drugs and when the person was taken to hospital it was discovered he was suffering from malignant pleural mesothelioma (Baris et al., 1978; Artvinli and Baris, 1979; Baris et al., 1979; Baris, 1991). This tumor, a highly malignant cancer which arises on the mesothelial surface of the chest cavity, was in the same tissue in which plaques had been found in other inhabitants. Mesothelioma had previously been considered as a signal tumor for asbestos exposure (Wagner et al., 1971; Wagner and Pooley, 1986).

In the years which immediately followed, 1970 to 1974, 24 new cases of mesothelioma occurred in village of Karain (population of 575). The 24 new cases had occurred during a time period when only 56 deaths had occurred in the entire village. The proportional mortality from mesothelioma had accounted for an extraordinary 42.9% of the deaths. No asbestos cohort had ever experienced such a mesothelioma mortality. The highest mesothelioma mortality for a cohort of asbestos-exposed workers in the United States is 9.75%—recorded for the insulation workers exposed to high concentrations of mixed fiber types, over their working lifetime (Selikoff and Seidman, 1991). In the three years which followed, 1974-1977, Karain experienced an additional 18 cases of mesothelioma, bringing the total to 42 (Saracci et al., 1982). And, according to recent figures (Baris, 1987) there has occurred a total of 62 mesothelioma deaths out of about 175 deaths in this village.

Baris and colleagues extended their studies to other towns on the Cappadocian plateau. During the same time period that Karain experienced 62 mesotheliomas, the town of Tuzköy, with a population of approximately 3,000, experienced 27 mesotheliomas, Sarihidir, a town of approximately 900 inhabitants, experienced 3 mesotheliomas, and the towns of Karlik and Kizilköy experienced 2 mesotheliomas among its inhabitants. In all, 94 mesotheliomas were discovered over a period of approximately 17 years among a total population of about 5,000 persons. Fifty six of these mesotheliomas occurred in males while 38 occurred in females. The average age at which the disease occurred was approximately 50 years with the age at diagnosis ranging from 27 to 71 years. The time from the onset of symptoms until death was only one and a half to two years.

The Cappadocian plateau is made up of geologically recent volcanic ash deposits (tuffs). The tuffaceous sediments are largely unconsolidated and are easily worked with hand tools. The inhabitants of these villages sometimes have carved their homes and other structures out of soft tuff outcroppings, or have cut "dimension stone" from the tuffs and transported them to sites where buildings are erected. Occasionally, tuffs are ground and used as stucco pastes and whitewashes for use in the interior or exterior of buildings (Rohl et al., 1982).

Mineral assay of the Cappadocian soils and rocks showed that the tuffs which had previously consisted of volcanic glasses, feldspars, and other silicate minerals, had been locally altered by ground water, and within local alkali lakes, into montmorillonite and zeolite minerals. The assemblage of zeolites included clinoptilolite, chabazite, and erionite. The erionite occurs, morphologically, as fine needles and these same fibers were found in lung tissues obtained at biopsy from the patients with mesothelioma (Pooley, 1979; Sébastien, et al., 1981). The environmental exposure begins at birth. The analysis of outdoor air samples by transmission electron microscopy have found lower fiber levels of erionite in the effected village (Table 12). However, the levels found do not necessarily represent past exposures, nor do they indicate that individuals within the village did not experience higher fiber exposure during certain activities (see Baris, 1991 for a discussion).

Experimental animal studies corroborate the human studies

A number of animal studies were initiated to verify the hypothesis that erionite fiber was the agent involved in the induction of the malignant mesotheliomas observed on Cappadocian plateau. The most important of the animal studies was an inhalation study in rats (Wagner, et al., 1985; see Guthrie, 1992, and Johnson, 1993 for recent reviews). A previous inhalation study of commercial varieties of asbestos produced 11 mesotheliomas in 648 rats. Using erionite, 27 mesotheliomas were produced in 28 rats. Crocidolite, the positive control, produce no mesotheliomas in 28 rats (Wagner, et al., 1985) An evaluation of such animal studies in relation to human environmental exposure is given by Simonato et al. (1989). The size distribution of an erionite sample from Rome, Oregon and crocidolite were indistinguishable by transmission electron microscopy (Table 13). The fine-fibered nature of erionite from Rome, Oregon is shown in Figures 3 and 4.

Table 12

Mineral content and fiber concentration of outdoor air samples[†] collected from three Turkish villages with endemic mesothelioma and from Karlik, a nearby control village

Village	Concentration[*]	Description of fibers and particles present[**]
Karain	0.002–0.010	80% zeolite + calcium oxide and sulfate
Sarihidir	0.001–0.029	60% zeolite+calcite, quartz, volcanic glass, and tremolite and chrysotile fibers
Tuzköy	0.005–0.025	85% zeolite+quartz, volcanic glass, and aluminum silicate
Karlik	0.002–0.006	20% zeolite+calcium oxide and sulfate

Sources: Baris (1987); Baris, (1991); Simonato, et al. (1989)
[†] Determined by transmission electron microscopy of N > 150 samples.
[*] Air concentration in fibers/ml for fibers > 5 μm in length.
[**] From elemental composition, the zeolite fibers are probably mostly erionite. "Calcium oxide and sulfate" may refer to the elemental composition of gypsum ($CaO \cdot SO_3 \cdot 2H_2O$).

Figure 3. Transmission electron photomicrograph of erionite from Rome, Oregon, U.S.A., showing many fibers with high (length/width) aspect ratios.

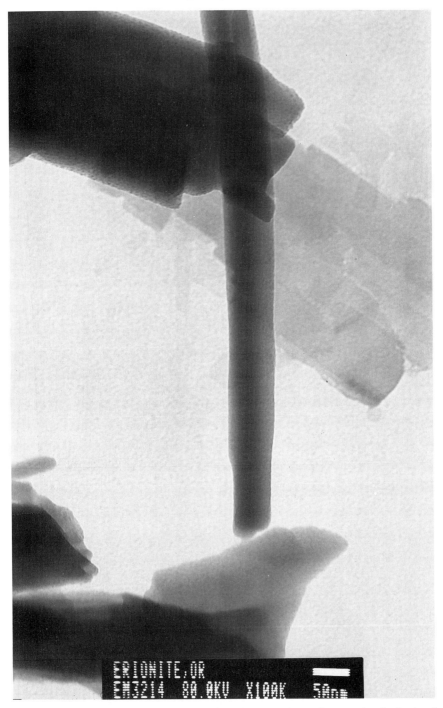

Figure 4. High magnification transmission electron photomicrograph of erionite from Rome, Oregon, U.S.A. An electron-transparent fiber (diameter of 60 nm) appears in the center of the photograph.

Table 13

Percentage of mineral fibers falling within four diameter size ranges as determined by tranamission electron microscopy (average magnification 4000×)

Specimen	N	< 0.1μm	<0.1–0.5μm	<0.5–2.0μm	> 2.0μm
crocidolite Wittenoom Gorge, Australia	521	46.1	51.8	2.1	0.0
crocidolite Cape Province, S. Africa*	277	46.2	53.1	0.7	0.0
erionite Rome, Oregon	162	17.3	37.7	40.7	4.3
erionite Needle Peak, Nevada	508	40.2	52.4	7.1	0.4
anthophyllite Paakila, Finland**	172	3.5	29.1	62.2	5.2

* National Institute of Environmental Health Sciences sample.
** Union International Control Cancer (UICC) sample.

REFERENCES

Abrams, H.K. (1954) Diatomaceous earth pneumoconiosis. Amer. J. Public Health 44, 592-599.

Addison, J. and Davis, L.S.T. (1990) Analysis of amphibole asbestos in chrysotile and other minerals. Ann. Occup. Hygiene 14, 159-175.

Amandus, H.E. and Wheeler, R. (1987) The morbidity and mortality of vermiculite miners and millers exposed to tremolite-actinolite: Part II Mortality. Amer. J. Indus. Hygiene 11, 15-26.

Amandus, H.E., Wheeler, R., Jankovic, J., and Tucker, J. (1987a) The morbidity and mortality of vermiculite miners and miller exposed to tremolite-actinolite: Part I Exposure estimates. Amer. J. Indus. Hygiene 11, 1-14.

Amandus, H.E., Althouse, R., Morgan, W.K.C., Sargent, N., and Jones, R. (1987b) The morbidity and mortality of vermiculite miners and millers exposed to tremolite-actinolite: Part III Radiographic findings. Amer. J. Indus. Hygiene 11, 27-37.

Ames, B.N. and Gold, L.S., 1990, Chemical carcinogenesis: too many rodent carcinogens. Proc. Nat'l Acad. Sciences 87, 7772-7776.

Ampian, S.G. (1991) Clays. In Minerals Yearbook, v. 1, Metals and Minerals, U.S. Dept. of the Interior, Bureau of Mines, Washington, D.C., 271-304.

Ampian, S.G. and Virta, R.L. (1992) Crystalline Silica Overview: Occurrence and Analysis. Information Circular IC 9317, U.S. Dept. of the Interior, Bureau of Mines, Washington, D.C., 27 p.

Armstrong, B.K., McNulty, J.C., Levitt, L.J., Williams, K.A., and Hobbs, M.S.T. (1979) Mortality in gold and coal miners in Western Australia with special reference to lung cancer. Brit. J. Indus. Med. 36, 199-205.

Artvinli, M. and Baris, Y.I. (1979) Malignant mesotheliomas in a small village in the Anatolian region of Turkey: an epidemiologic study. J. Natl. Cancer Inst. 63, 17-22.

Banks, W. (1980) Asbestiform and/or Fibrous Minerals in Mines, Mills, and Quarries. U.S. Dept. of Labor, Mine Safety and Health Administration. Informational Report IR 1111.

Baris, Y.I. (1987) Asbestos and Erionite Related Chest Diseases. Semih Ofset Matbaacilik Ltd, Ankara, Turkey, 169 p.

Baris, Y.I. (1991) Fibrous zeolite (erionite) related diseases in Turkey. Amer. J. Indus. Med. 19, 373-378.

Baris, Y.I., Sahin, A.A., Ozesmi, M., Kerse, I., Ozen, E., Kolacan, B., Altinors, M., and Goktepeli, A. (1978) An outbreak of pleural mesothelioma and chronic fibrosing pleurisy in the village of Karain/Urgup in Anatolia. Thorax 33, 181.

Baris, Y.I., Artvinli, M., and Sahin, A.A. (1979) Environmental mesothelioma in Turkey. Ann. New York Acad. Sci. 330, 423-432.

Baris, Y.I., Artvinli, M., Sahin, A.A., Bilir, N., Kalyoncu, F., Sébastien, P. (1988a) Non-occupational asbestos related chest diseases in a small Anatolian village. Brit. J. Indus. Med. 45, 841-842.

Baris, Y.I., Bilir, N., Artvinli, M., Sahin, A.A., Kalyoncu, F., Sébastien, P. (1988b) An epidemiological study in an Anatolian village environmentally exposed to tremolite asbestos. Brit. J. Indus. Med. 45, 838-840.

Bates, R.L. and Jackson, J.A. (1982) Our Modern Stone Age. William Kaufmann, Inc., Los Altos, California, 132 p.
Bignon, J. (1990) Health Related Effects of Phyllosilicates. Springer-Verlag, New York, 447 p.
Bignon, J., Sébastien, P., Gaudichet, A., and Jaurand, M.C. (1980) Biological Effects of Attapulgite. In: Biological Effects of Asbestos, Vol. 1. J.C Wagner, Ed. Int'l Agency for Research on Cancer, Lyon 163-181.
Boyd, J.T., Doll, R., Faulds, J.S., and Leiper, J. (1970) Cancer of the lung in iron ore (haematite) miners. Brit. J. Indus. Med. 27, 97-105.
Breining, H. (1986) Pneumoconiosis and lung cancer: a review of 5 cases (abstr.). In: Silica, Silicosis, and Cancer, D.F. Goldsmith, D.M. Winn, and C.M. Shy, Eds. Praeger Publishers, New York, 533.
Bridbord, K., Costello, J., Gamble, J., Groce, D., Hutchison, M., Jones, W., Merchant, J., Ortmeyer, C., Reger, R., and Wagner, W.L. (1979) Occupational safety and health implications of increased coal utilization. Environ. Health Persp., 33, 285-302.
Brown, D.P., Dement, J.M., and Wagoner, J.K. (1979) Mortality patterns among miners and millers occupationally exposed to asbestiform talc. In: Dusts and Disease, J.M. Dement and R. Lemen, Eds, Pathtox, Park and Forest South, Illinois, 317-324.
Brown, D.P., Kaplan, S.D., Zumwalde, R.D., Kaplowitz, M., and Archer, V.E. (1985) Retrospective cohort mortality study of underground gold mine workers. Unpublished paper, Centers for Disease Control, NIOSH, Cincinnati, OH, available from the National Technical Information Service, 72 p.
Brown, D.P., Kaplan, S.D., Zumwalde, R.D., Kaplowitz, M., and Archer, V.E. (1986) Retrospective cohort mortality study of underground gold mine workers. In: Silica, Silicosis, and Cancer, D.F. Goldsmith, D.M. Winn, and C.M. Shy, Eds. Praeger Publishers, New York, 335-350.
Caldwell, D.M. (1958) The coalescent lesion of diatomaceous earth pneumoconiosis. Amer. Rev. Tuberculosis 77, 644-661.
Chayes, F. (1952) The finer grained Calc-alkaline granites of New England. J. Geol. 60, 207-254.
Checkoway, H., Heyer, N.J., Demers, P.A., and Breslow, N.E. (1992) A cohort mortality study of workers in the diatomaceous earth industry. Unpublished final report from the University of Washington School of Public Health and Community Medicine, Seattle, WA: Submitted to the International Diatomite Producers Assoc., 133 p.
Cherniack, M. (1986) The Hawk's Nest Incident: America's Worst Industrial Disaster. Yale Univ. Press, New Haven, Connecticut, 194 p.
Clark, F.W. (1924) The Data of Geochemistry. U.S. Geol. Survey Bull. 770, Washington, D.C., 841 p.
Clark, T.C., Harrington, V.A., Asta, J., Morgan, W.K.C., and Sargent, E.N. (1980) Respiratory effects of exposure to dust in taconite mining and processing. Amer. Rev. Respir. Dis. 121, 959-966.
Clarke, G.M. (1985) Special Clays. Indus. Mineral. 216, 25-51.
Clarke, G.M. (1990) Phyllosilicates as industrial minerals. In: Health Related Effects of Phyllosilicates, J. Bignon, Ed. Springer-Verlag, New York, 31-45.
Cohen, S.M. and Ellwein, L.B., 1990, Cell proliferation in carcinogenesis. Science 249, 1007-1011.
Constantopoulos, S.H., Goudevenos, J.A., Saratzis, N., Langer, A.M., Selikoff, I.J., and Moutsopoulos, H.M. (1985) Pleural calcification and restrictive lung function in northwestern Greece-Environmental exposure to mineral fiber as etiology. Environ. Res. 38, 319-331.
Constantopoulos, S.H., Langer, A.M., Saratzis, N., and Nolan, R.P. (1987) Regional findings in Metsovo lung. The Lancet ii, 452-453.
Coope, B. (1982) Industrial Minerals Directory, A World Guide to Producers of Nonmetallic Minerals, Non-Fuel Minerals. 2nd Ed., Surrey Metal Bulletin Books, Worcester Park, London, p. 634-635.
Cooper, W.C., Wong, O., Trent, L.S., and Harris, F. (1992) An updated study of taconite miners and millers exposed to silica and non-asbestiform amphiboles. J. Occup. Med. 34, 1173-1180.
Costello, J. and Graham, W.G.B. (1986) Vermont granite workers' mortality. In: Silica, Silicosis, and Cancer, D.F. Goldsmith, D.M. Winn, and C.M. Shy, Eds. Praeger Publishers, New York, 437-438.
Craighead, J.E. (1988) Diseases associated with exposure to silica and nonfibrous silicate minerals. Report of the Silicosis and Silicate Disease Committee, J.E. Craighead, Chairman. Arch. Pathol. Lab. Med. 112, 673-720.
Craighead, J.E. (1992) Do silica and asbestos cause lung cancer? Arch. Pathol. Lab. Med. 116, 16-20.
Craighead, J.E., Emerson, R.J., and Stanley, D.E. (1992) Slateworker's pneumoconiosis. Human Pathology 23, 1098-1105.
Dale, T.N. (1914) Slate in the United States. U.S. Geol. Surv. Bull. 586, 220 p.
Damber, L. and Larsson, L-G. (1985) Underground mining, smoking, and lung cancer: a case-control study in the iron ore municipalities in Northern Sweden. J. Natl. Cancer Inst., 74, 1207-1213.
Enterline, P.E. (1964) Mortality rates among coal miners. Amer. J. Public Health 54, 758-768.
Felius, R.O. (1990) Mineralogy of phyllosilicates. In: Health Related Effects of Phyllosilicates, J. Bignon, Ed. Springer-Verlag, New York, 3-14.
Finkelstein, M.M., Muller, J., Kusiak, R. and Suranyi, G. (1986) Follow-up of miners and silicotics in

Ontario; In: Silica, Silicosis, and Cancer, D.F. Goldsmith, D.M. Winn, and C.M. Shy, Eds. Praeger Publishers, New York, 321-325.
Flinn, R.H., Brinton, H.P., Doyle, H.N., Cralley, L.J., Harris, R.L., Westfield, J., Bird, J.H., and Berger, L.B. (1963) Silicosis in the Metal Mining Industry; a Revaluation 1958-1961. U.S. Dept Health Education Welfare, Public Health Service and Dept. of Interior, Bureau of Mines, Washington, D.C., 238 p.
French, B.M. (1968) Progressive Contact Metamorphism of the Biwabik Iron-formation, Mesabi Range, Minnesota. Minn. Geol. Surv. Bull. 45, Univ. Minnesota Press, Minneapolis, MN, 103 p.
Frondel, C. (1962) Silica Minerals, v. 3, The System of Mineralogy. John Wiley, New York, 334 p.
Galan, E. and Castillo, A. (1984) Sepiolite - palygorskite in Spanish tertiary basins: generic patterns in continental environments. In: Palygorskite - sepiolite: Occurrences, Genesis and Uses. Elsevier, Amsterdam, 87-124.
Gamble, J.F.(1986) Silicate pneumoconiosis. In: Occupational Respiratory Diseases, J.A. Merchant, Ed., DHHS (NIOSH) Publication 86-102, U.S. Dept Health Human Services, Washington, D.C., 243-285.
Gamble, J.F., Sieber, W.K., Wheeler, R.W., Reger, R., and Hall, B. (1986). A cross-sectional study of U.S. attapulgite workers. Ann. Occup. Hygiene 32, 475-481.
Gibbs, A.R. (1990) Human pathology of kaolin and mica pneumoconiosis. In: Health Related Effects of Phyllosilicates, J. Bignon, Ed. Springer-Verlag, New York, 217-226.
Gillam, J.D., Dement, J.M., Lemen, R.A., Wagoner, J.K., Archer, V.E., and Blejer, H.P. (1976) Mortality patterns among hard rock gold miners exposed to an asbestiform mineral. Ann. New York Acad. Sci. 271, 336-344.
Glenn, R.E. (1992) Health effects of crystalline silica. In: 10th "Industrial Minerals" Int'l Congress, J.B. Griffiths, Ed. Industrial Minerals Division of Metal Bulletin plc, May 1992, Surrey, U.K., 112–119.
Glover, J.R., Bevan, C., and Cotes, J.E. (1980) Effects of exposure to slate dust in North Wales. Brit. J. Indus. Med. 37, 152-162.
Gluskoter, H.J., Shimp, N.F., and Ruch, R.R. (1981) Coal analysis, trace elements and mineral matter. In: Chemistry of Coal Utilization, 2nd Supplementary Vol., M.A. Elliot, Ed. John Wiley, New York, 369-422.
Goldsmith, D.F. (1993) Silica exposure and pulmonary cancer. In: Epidemiology of Lung Cancer, J.M. Samet, Ed. Marcell Dekker, New York, in press.
Gruner, J.W. (1946) Mineralogy and Geology of the Mesabi Range. Iron Range Resources and Rehabilitation, State Office Bldg., St. Paul, Minnesota, 127 p.
Guthrie, G.D. (1992) Biological effects of inhaled minerals. Amer. Mineral. 77, 225-243.
Haas, C.Y. (1972) Attapulgite clays for industrial mineral markets. Indus. Miner. 63, 45, 47.
Harvey, R.D. and Ruch, R.R. (1986) Mineral matter in Illinois and other U.S. coals. In: Mineral Matter and Ash in Coal, K.S. Vorres, Ed. ACS Symposium Series 301, Amer. Chem. Soc., Washington, D.C., 10-40.
Hessel, P.A. and Sluis-Cremer, G.K. (1986) Case-control study of lung cancer and silicosis. In: Silica, Silicosis, and Cancer, D.F. Goldsmith, D.M. Winn, and C.M. Shy, Eds. Praeger Publishers, New York, 351-355.
Hessel, P.A., Sluis-Cremer, G.K. (1989) X-ray findings, lung function, and respiratory symptoms in Black South African vermiculite workers. Amer. J. Indus. Med. 15, 21-29.
Higgins, E., Lanza, A.J., Laney, F.B., and Rice, G.S. (1917) Siliceous Dust in Relation to Pulmonary Disease Among Miners in the Joplin District, Missouri. Bull. 132, U.S. Dept. of Interior, Bureau of Mines, Washington, D.C., 116 p.
Higgins, I.T.T., Glassman, J.H., Oh, M.S., and Cornell, R.G. (1983) Mortality of Reserve Mining Company employees in relation to taconite dust exposure. Amer. J. Epidem. 118, 710-719.
Hogue, W.L. and Mallette, F.S. (1949) A study of workers exposed to talc and other dusting compounds in the rubber industry. J. Indus. Hyg. Toxic. 31, 359-364.
Holland, L.M., Gonzales, M., Wilson, J.S., and Tillery, M.I. (1983) Pulmonary effects of shale dusts in experimental animals. In: Health Issues Related to Metal and Nonmetallic Mining, W.L. Wagner, W.N. Rom, and J.A. Merchant, Eds. Butterworth Publishers, Boston, 485-496.
Hughes, J.M. and Weill, H. (1991) Asbestosis as a precursor of lung cancer: a prospective mortality study. Brit. J. Indus. Med. 48, 229-233.
Hughes, J.M., Jones, R.N., Gilson, J.C., Hammad, Y.Y., Samimi, B., Hendrick, D.J., Turner-Warwick, M., Doll, N.J., and Weill, H. (1982) Determinants of progression in sandblasters silicosis. Ann. Occup. Hygiene 26, 701-712.
Hunter, B. and Thomson, C. (1973) Evaluation of the tumorgenic potential of vermiculite by intrapleural injection in rats. Brit. J. Indus. Med. 30, 167-173.
Hunter, D. (1955) The Diseases of Occupation. English University Press, Ltd., London, 1100 p.
Huuskonen, M.S., Tossavainen, A., Koskinen, H., Zitting, A., Korhonen, O., Nickel, J, Korhonen, K., and Vaaranen, V. (1983a) Wollastonite exposure and lung fibrosis. Environ Res. 30, 291-304.

Huuskonen, M.S., Järvisalo, J., Koskinen, H., Nickels, J, Räsänen, J., and Asp, S. (1983b) Preliminary results from a cohort of workers exposed to wollastonite in a Finnish limestone quarry. Scand. J. Work Environ. Health 9, 169-175.

IARC (1987a) IARC Monographs on the Evaluation of the Carcinogenic Risk of Chemicals to Humans: Silica and Some Silicates, v. 42, World Health Organization, Int'l Agency for Research on Cancer, Lyon, 288 p.

IARC (1987b) IARC Monographs on the Evaluation of the Carcinogenic Risk of Chemicals to Humans, Overall Evaluations of Carcinogenicity: An Updating of IARC Monographs 1 to 42, Supplement 7, World Health Organization, Int'l Agency for Research on Cancer, Lyon, 440 p.

Jacobsen, M., Rae, S., Walton, W.H., and Rogan, J.H. (1971) The relation between pneumoconiosis and dust-exposure in British coal mines. In: Inhaled Particles III, v. 2, W.H. Walton, Ed. Unwin Brothers, London, 903-917.

Johnson, N.F. (1993) The limitation of inhalation, intratracheal and intracoelomic routes of administration for identifying hazardous fibrous materials. In: Fiber Toxicology. David B. Warheit, Ed. Academic Press, New York, 43-72.

Johnson, N.F., Smith, D.M., Sebring, R., and Holland, L.M. (1987) Silica-Induced Alveolar Cells Tumors in Rats. Amer. J. Indus. Med. 11, 93-107.

Jones, R.N., Hughes, J.M., Hammad, Y.Y., and Weill, H. (1986) Sandblasting and silicosis. In: Silica, Silicosis, and Cancer, D.F. Goldsmith, D.M. Winn, and C.M. Shy, Eds. Praeger Publishers, New York, 71-75.

Jörgensen, H.S. (1984) Lung cancer among underground workers in the iron ore in Kiruna on the basis of 30 years of observation (abstr.). Scand. J. Work Environ. Health 21, 128.

Kurppa, K., Gudbergsson, H., Hannunkari, I., Koskinen, H., Hernberg, S., Koskela, R.-S., and Ahlman, K. (1986) Lung cancer among silicotics in Finland. In: Silica, Silicosis, and Cancer, D.F. Goldsmith, D.M. Winn, and C.M. Shy, Eds. Praeger Publishers, New York, 311-319.

Langer, A.M. (1978) Crystal faces and cleavage planes in quartz as templates in biological processes. Quart. Rev. Biophysics 11, 534-575.

Langer, A.M. (1986) Characterization and measurement of the industrial environment: Mineralogy. In: Occupational Respiratory Diseases, J.A. Merchant, Ed. Dept Health Human Services (NIOSH) Publ. 86-102, Washington, D.C., 3-40.

Langer, A.M., Maggiore, C.M., Nicholson, W.J., Rohl, A.N., Rubin, I.B., and Selikoff, I.J. (1979) The contamination of Lake Superior with amphibole gangue minerals. In: Health Hazards of Asbestos Exposure, I.J. Selikoff and E.C. Hammond, Eds. Ann. New York Acad. Sci. 330, 349-372.

Langer, A.M., Nolan, R.P., Constantopoulos, S.H., and Moutsopoulos, H.M. (1987) Association of Metsovo lung and pleural mesotheliomas with exposure to tremolite-containing whitewash. The Lancet i, 965-967.

Langer, A.M., Nolan, R.P., and Pooley, F.D. (1990) Phyllosilicates: Associated Fibrous Minerals. In: Health Related Effects of Phyllosilicates. Bignon, J., Ed. Springer-Verlag, New York, 59-74.

Langer, A.M., Nolan, R.P., and Addison, J. (1991) Distinguishing between amphibole asbestos fibers and elongate cleavage fragments of their non-asbestos analogues. In: Mechanisms in Fibre Carcinogenesis, R.C. Brown et al., Eds. Plenum Press, New York, 253-267.

Lanza, A.J. and Higgins, E. (1915) Pulmonary Disease Among Miners in the Joplin District, Missouri and Its Relation to Rock Dust in the Mines: A Preliminary Report. Tech. Paper 105, U.S. Dept. of Interior, Bureau of Mines, Washington, D.C., 48 p.

Lapp, N.L. (1981) Lung disease secondary to inhalation of nonfibrous minerals. Clinics in Chest Medicine 2, 219-233.

Lawler, A.B., Mandel, J.S., Schuman, L.M., and Lubin, J.H. (1985) A retrospective cohort mortality study of iron ore (hematite) miners in Minnesota. J. Occup. Med. 27, 507-517.

Lepp, H. (1972) Normative mineral composition of the Biwabik formation: a first .approach. In: Studies in Mineralogy and Precambrian Geology, B.R. Doe and D.K. Smith, Eds. Geol. Soc. Amer. Memoir 135, 265-278.

Lockey, J., Jarabek, A.M., Carson, A., McKay, R., Harber, P., Khoury, P., Morrison, J., Wiot, J., Spitz, H., and Brooks, S.M. (1983) Pulmonary hazards of vermiculite exposure. In: Health Studies Related to Metal and Non-metal Mining, W.L. Wagner, W.N. Rom, and J.A. Merchant, Eds. Butterworth Publishers, Boston, 303-315.

Lockey, J., Brooks, S.M., Jarabek, A.M., Khoury, P.R., McKay, R.T., Carson, A., Morrison, J.A., Wiot, J.F., Spitz, H.B. (1984) Pulmonary changes after exposure to vermiculite contaminated with fibrous tremolite. Amer. Rev. Respir. Dis. 129, 952-958.

McConnochie, K., Simonato, L., Mavrides, P., Christofides, P., Pooley, F.D., and Wagner, J.C. (1987) Mesothelioma in Cyprus: the role of tremolite. Thorax 42, 342-347.

McConnochie, K., Simonato, L., Mavrides, P., Christofides, P., Mitha, R., Griffiths, D.M., and Wagner, J.C. (1989) Mesothelioma in Cyprus. In: Non-Occupational Exposure to Mineral Fibres, J. Bignon, J.

Peto, and R. Saracci, Eds. IARC Scientific Publ. 90, 411-419.
McDonald, J.C., Gibbs, G.W., Liddell, F.D.K., and McDonald, A.D. (1978) Mortality to cummingtonite-grunerite. Amer. Rev. Respir. Dis. 118, 271-277.
McDonald, J.C., McDonald, A.D., Armstrong, B., and Sébastien, P. (1986a) Cohort study of mortality of vermiculite miners exposed to tremolite. Brit. J. Indus. Med. 43, 436-444.
McDonald, J.C., Sébastien, P., Armstrong, B. (1986b) Radiological survey of past and present vermiculite miners exposed to tremolite. Brit. J. Indus. Med. 43, 445-449.
McDonald, J.C., McDonald, A.D., Sébastien, P., and Moy, K. (1988) Health of vermiculite miners exposed to trace amounts of fibrous tremolite. Brit. J. Indus. Med. 45, 630-634.
Mehnert, W.H., Staneczek, W., Möhner, M., Konetzke, G., Müller, W., Ahlendorf, W., Beck, B., Winkelmann, R., and Simonato, L. (1990) A mortality study of a cohort of slate quarry workers in the German Democratic Republic. In: Occupational Exposure to Silica and Cancer Risk, L. Simonato, A.C. Fletcher, R. Saracci, and T.L. Thomas, Eds. Int'l Agency for Research on Cancer, Lyon, 55-64.
Merchant, J.A., Taylor, G., and Hodous, T.K. (1986) Coal workers' pneumoconiosis and exposure to other carbonaceous dusts. In: Occupational Respiratory Diseases, J.A. Merchant, Ed., Dept. Health Human Services, DHHS (NIOSH) Publ. 86-102, Washington, D.C., 329-384.
Moatamed, F., Lockey, J.E., and Parry, W.T. (1986) Fibre contamination of vermiculite: A potential occupational and environmental health hazard. Environ. Res. 41, 207-218.
Morgan, W.K.C., Donner, A., Higgins, I.T.H., Pearsons, M.G., and Rawlings, W. (1990) Clinical aspects of kaolin pneumoconiosis. In: Health Related Effects of Phyllosilicates, J. Bignon, Ed. Springer-Verlag, Berlin Heidelberg New York, 191-201.
Morris, J.N. (1975) Uses of Epidemiology, 3rd edition. Churchill Livingstone, New York, 318 p.
Netter, F.H. (1979) Respiratory System. The Ciba Collection of Medical Illustrations, v. 7, Ciba Medical Education Division, Summit, New Jersey.
Noble, J.A. (1950) Ore mineralization in the Homestake Gold Mine, Lead, South Dakota. Geol. Soc. Amer. Bull., 61, 221-252.
Nolan, R.P., Langer, A.M., and Herson, G.B. (1991a) Characterization of palygorskite specimens from different geological locales for health hazard evaluation. Brit. J. Indus. Med. 48, 463-475.
Nolan, R.P., Langer, A.M., Oechsle, G.W., Addison, J., and Colflesh, D.E. (1991b) Association of tremolite habit with biological potential. In: NATO Advanced Research Workshop on Mechanisms in Fibre Carcinogenesis, R.C. Brown, J. Hoskins, N. Johnson, Eds. Albuquerque, New Mexico, October 22-25, 1990, 231-251.
Pham, Q.T., Gaertner, M., Mur, J.M., Braun, P., Gabiano, M., and Sadoul, P. (1983) Incidence of lung cancer among iron miners. Europ. J. Respir. Dis., 64, 534-540.
Phibbs, B.P., Sundin, R.E., and Mitchell, R.S. (1971) Silicosis in Wyoming bentonite workers. Amer. Rev. Respir. Disease 103, 1-17.
Pooley, F.D. (1979) Evaluation of fibers samples taken from the vicinity of two villages in Turkey. In: Dusts and Disease, R. Lemen and J. Dement, Eds. Pathotox, Park Forest South, Illinois, p. 41-44.
Rao, C.P. and Gluskoter, H.J. (1973) Occurrence and distribution of minerals in Illinois coals. Illinois State Geol. Survey Circ. 476, 56 p.
Reiser, K.M. and Last, J.A. (1979) Silicosis and fibrogenesis: fact and artifact. Toxicology 13, 51-72.
Rockette, H. (1977) Mortality among coal miners of the UMAW Health and Retirement Funds. DHEW NIOSH Publication No. 77-155, U.S. Dept. Health Education and Welfare, Washington, D.C.
Rohl, A.N., Langer, A.M., Moncure, G., Selikoff, I.J., and Fischbein, A. (1982) Endemic pleural disease associated with exposure to mixed fibrous dust in Turkey. Science 216, 518-520.
Rosmanith, J. and Schimanski, P. (1986) Epidemiological studies on the relationship between coal workers pneumoconiosis and lung cancer (abstr.) In: Silica, Silicosis, and Cancer, D.F. Goldsmith, D.M. Winn, and C.M. Shy, Eds. Praeger Publishers, New York, 535 p.
Ross, M. (1984) A survey of asbestos-related disease in trades and mining occupations and factory and mining communities as a means of predicting health risks of nonoccupational exposure to fibrous minerals. In: Definitions for Asbestos and Other Health-related Silicates, B. Levadie, Ed. ASTM STP 834, Amer. Soc. Testing Materials, Philadelphia, PA, 51-104.
Rubino, G.F., Scansetti, G, Piolatto, G., and Romano, C.A. (1976) Mortality of talc miners and millers. J. Occup. Med. 18, 186-193.
Russell, A.E. (1941) The health of workers in dusty trades: restudy of a group of granite workers. U.S. Public Health Service Bull. 269, Washington, D.C.
Saffiotti, U. (1986) The pathology induced by silica in relation to fibrogenesis and carcinogenesis. In: Silica, Silicosis, and Cancer, D.F. Goldsmith, D.M. Winn, and C.M. Shy, Eds. Praeger Publishers, New York, 287-307.
Samimi, B., Weill, H., and Ziskind, M. (1974) Respirable silica dust exposure of sandblasters and associated workers in steel fabrication yards. Arch. Environ. Health 29, 61-66.
Saracci, R., Simonato, L., Baris, Y., Artvinli, M., and Skidmore, J. (1982) The age-mortality curve of

associated workers in steel fabrication yards. Arch. Environ. Health 29, 61-66.
Saracci, R., Simonato, L., Baris, Y., Artvinli, M., and Skidmore, J. (1982) The age-mortality curve of endemic pleural mesothelioma in Karain, Central Turkey. Brit. J. Cancer 45, 147-149.
Schaumburg, F.D. (1976) Judgment Reserved. Reston Publishing Co., Reston, Virginia, 265 p.
Schüler, G. and Rüttner, J.R. (1986) Silicosis and lung cancer in Switzerland. In: Silica, Silicosis, and Cancer, D.F. Goldsmith, D.M. Winn, and C.M. Shy, Eds. Praeger Publishers, New York, 357-366.
Sébastien, P., Gaudichet, A., Bignon, J., and Baris, Y.I. (1981) Zeolite bodies in human lungs from Turkey. J. Lab. Investig., 44, 420-425.
Selikoff, I.J. and Seidman, H. (1991) Asbestos-associated deaths in asbestos among insulation workers in the United States and Canada, 1967-1987. Ann. New York Acad. Sci., 643, 1-14.
Sheehy, J.W. and McJilton (1987) Development of a model to aid in reconstruction of historical silica dust exposures in the taconite industry. Amer. Indus. Hygiene J. 48, 914-918.
Shy, C.M. (1986) Epidemiologic principles and methods for occupational health studies. In: Occupational Respiratory Diseases, J.A. Merchant, Ed. Dept Health Human Services, DHHS (NIOSH) Publ. 86-102, Washington, D.C., 103-136.
Simonato, L. and Saracci, R. (1990) Epidemiological aspects of the relationship between exposure to silica dust and lung cancer. In: Occupational Exposure to Silica and Cancer Risk, L. Simonato, A.C. Fletcher, R. Saracci, and T.L. Thomas, Eds. Int'l Agency for Research on Cancer, Lyon, 1–5.
Simonato, L., Baris, I., Saracci, R., Skidmore, J., and Winkelmann, R. (1989) Relation of environmental exposure to erionite fibres to risk of respiratory cancer. In: Non-Occupational Exposure to Mineral Fibres. Bignon, J., Peto, J., and Saracci, R., Ed. Lyon, Int'l Agency for Research on Cancer 90, 398-404.
Skinner, H.C.W., Ross, M., and Frondel, C. (1988) Asbestos and Other Fibrous Minerals. Oxford Univ. Press, New York, 204 p.
Slaughter, A.L. (1968) The Homestake Mine. In: Ore Deposits of the United States, 1933-1976, v.2, J.D. Ridge, Ed. Amer. Inst. Mining, Metallurgical, Petroleum Engineers, Inc., New York, 1436-1459.
Sluis-Cremer, G.K., Hessel, P.A. (1990) Palabora Vermiculite In: Health Related Effects of Phyllosilicates, J. Bignon, Ed. Springer Verlag, New York, 227-234.
Smart, R.H. and Anderson, W.M. (1952) Pneumoconiosis due to diatomaceous earth: clinical and x-ray aspects. Indus. Med. Surgery 21, 509-518.
Stach, E., Mackowsky, M.-TH., Teichmüller, M., Taylor, G.H., Chandra, D, and Teichmüller, R. (1982) Coal Petrology. Gebrüder Borntraeger, Berlin, 535 p.
Stille, W.T. and Tabershaw, I.R. (1982) The mortality experience of upstate New York talc workers. J. Occup. Med. 24, 480-484.
Stokinger, H.E. (1984) A review of world literature finds iron oxides noncarcinogenic. Amer. Indus. Hygiene Assoc. J. 45, 127-133.
USBM, 1989, Minerals Yearbook, v. II, Area Reports - Domestic: U.S. Dept. Interior, Bureau of Mines, Washington, D.C., 533 p.
USBM (1992a) Crystalline Silica Primer. U.S. Dept. Interior, Bureau of Mines, Washington, D.C., 49 p.
USBM (1992b) State Mineral Summaries 1992. U.S. Dept. Interior, Bureau of Mines, Washington, D.C., 69-71.
Vigliani, E.C. and Motlura, G. (1948) Diatomaceous earth silicosis. Brit. J. Indus. Med. 5, 148-350.
Virta, R.L. (1991) Talc and pyrophyllite. In: Minerals Yearbook, v. 1, Metals and Minerals, U.S. Dept. Interior, Bureau Mines, 1053-1059.
Wagner, J.C. (1966) The induction of tumors by the intrapleural inoculations of various types of asbestos dust. In: Proc. 3rd Quadrennial Int'l Conference on Cancer, L. Severi, Ed. Perugia, Italy, 589-606.
Wagner, J.C. and Pooley, F.D. (1986) Mineral fibres and mesothelioma. Thorax 41, 161-166.
Wagner, J.C., Gilson, J.C., Berry, G., and Timbrell, V. (1971) Epidemiology of asbestos cancer. Brit. Med. Bull. 27, 71-76.
Wagner, J.C., Skidmore, J.W., Hill, R.J., and Griffiths, D.M. (1985) Erionite exposure and mesothelioma in rats. Brit. J. Cancer,51, 727-730.
Wagner, M.M.F. and Wagner, J.C. (1972) Lymphomas in the Wistar rat after intrapleural inoculation of silica. J. Natl. Cancer Inst. 49, 81-91.
Wastiaux, A., and Daniel, H. (1990) Pulmonary toxicity of kaolin in rats exposed by inhalation. In: Health Related Effects of Phyllosilicates. J. Bignon, Ed. Springer-Verlag, New York, 405-414.
Waxweiler, R.J., Zumwalde, R.D., Ness, G.O., and Brown, D.P. (1988) A retrospective cohort mortality study of males mining and milling attapulgite clay. Amer. J. Indus. Med. 13, 305-315.
Westerholm, P., Ahlmark, A., Maasing, R., and Segelberg, I. (1986) Silicosis and lung cancer: a cohort study. In: Silica, Silicosis, and Cancer, D.F. Goldsmith, D.M. Winn, and C.M. Shy, Eds. Praeger Publishers, New York, 327-333.
Wylie, A.G. (1990) Discriminating amphibole cleavage fragments from asbestos: rationale and methodology. In: Proc. VIIth Int'l Pneumoconiosis Conf. 2, 1065-1069.

Yazicioglu, S., Ilcayto, R., Balci, K., Sayli, B.S., and Yorulmaz, B. (1980) Pleural mesotheliomas and bronchial cancers caused by tremolite dust. Thorax 35, 564-569.

Zambon, P., Simonato, L., Mastrangelo, G., Winkelmann, R., Saia, B., and Crept, M. (1986) A mortality study of workers compensated for silicosis during 1959 to 1963 in the Veneto region of Italy. In: Silica, Silicosis, and Cancer, D.F. Goldsmith, D.M. Winn, and C.M. Shy, Eds. Praeger Publishers, New York, 367-374.

Zumwalde, R. (1976) Industrial Hygiene Study. Engelhard Minerals and Chemicals Corporation, Attapulgus, Georgia (NIOSH 00106935), Cincinnati, OH, National Institute for Occupational Safety and Health.

CHAPTER 13

ASBESTOS LUNG BURDEN AND DISEASE PATTERNS IN MAN

Andrew Churg

Department of Pathology and University Hospital
University of British Columbia
Vancouver, British Columbia V6T 2B5 Canada

The development of methods for microanalytical evaluation of asbestos fibers in human lung has permitted detailed examination of the relationship between fiber burden taken in its broadest sense (e.g., fiber concentration, fiber type, fiber size, and fiber distribution) and the pattern of asbestos-related disease in man. These studies have shown that everyone carries a fairly large number of asbestos fibers, the bulk of which are short fibers of chrysotile, yet this burden is not associated with any evident pathologic abnormality. Higher burdens are found in some very specialized general populations, such as residents of the chrysotile mining townships of Quebec, but again this burden of chrysotile and its contaminant tremolite appears to have no adverse effect on health. Workers with occupational exposure to asbestos carry several orders of magnitude greater lung burdens, and reproducible patterns of disease and fiber concentration can be discerned within such groups. These patterns differ between chrysotile and commercial amphibole-asbestos (e.g., amosite and crocidolite) exposure. For chrysotile, both asbestosis and mesothelioma require the same, extremely high, fiber burden in the lung, whereas plaques are seen at much lower fiber burdens. However, for amosite and crocidolite, mesothelioma and plaques occur at about the same fiber burden, which is several orders of magnitude less than the fiber burden seen in cases of asbestosis. These findings are of considerable importance for public health, because they imply that even a moderate level of exposure to chrysotile will not produce mesothelioma. The relationships between fiber size and disease are unclear. Animal studies show a strong association between long thin fibers and the development of both mesothelioma and asbestosis. However, the few human studies which have looked at fiber length and asbestosis have not confirmed these predictions. Moreover, there is no clear evidence that long thin fibers are more important than any other type of fiber in producing mesothelioma in humans. The only consistent association of fiber size and disease has been the finding that for both chrysotile- and amphibole-exposed populations, high aspect ratio fibers seem to be associated with pleural plaques. The relationship of fiber distribution and disease pattern within the lung is even less well understood. The distributions actually found in autopsied lungs do not correspond at all with theoretical models and completely fail to explain the geographic distribution of asbestos related disease within the lung.

INTRODUCTION

Although it has been clear for many years that there is a relationship between the amount of mineral dust found in the lungs of workers with occupational dust exposure and the development of specific diseases, it is only with the dissemination of practical methods of analytical electron microscopy (AEM) that it has been possible to understand the relationship between mineral particle burden taken in its broadest sense (e.g., particle concentration, type, size, and distribution) and detailed patterns of pathologic response. This relationship can be investigated by documenting the lung burdens of asbestiform and non-asbestiform minerals in the general population and in various occupationally-exposed groups. As discussed in Chapter 2, the general population receives a background exposure to both asbestiform and non-asbestiform minerals arising from natural and anthropogenic sources. Special occupationally-exposed groups can receive an additional (sometimes large) exposure to specific minerals. Hence, one can compare lung burdens found in the general population with those found in ocupationally-exposed groups and then relate these lung burdens to disease patterns.

In this chapter, I discuss the relationship between disease and asbestos burden in the lung. I also address the differences in this relationship for the various types of asbestos. Finally, I emphasize some of the outstanding issues, in particular discrepancies between data derived from animal experiments, predictions based on mathematical models, and data derived from actual analysis of autopsied human lungs. The various types of asbestos are discussed in Chapter 3, and the diseases accepted as caused by asbestos are listed in Table 1.

Table 1

Diseases caused by asbestos

Carcinoma of the lung (in association with asbestosis)

Malignant mesothelioma of the pleura and peritoneum

Non-neoplastic lung disease
   asbestosis
   small airways disease

Non-neoplastic pleural disease
   pleural plaques
   pleural effusions
   pleural fibrosis
   rounded atelectasis

ANALYTICAL METHODS

As used here, the term analytical electron microscope refers to a scanning or transmission electron microscope (SEM or TEM) capable of X-ray microanalysis using an energy-dispersive spectrometer (EDS). These instruments allow morphological and chemical characterization of mineral particles extracted from lung. The TEM provides higher resolution (which is important for accurate detection of small fibers) and electron diffraction capabilities (which are necessary to identify the mineral species of a fiber uniquely) (Champness et al., 1976; Churg, 1989). In addition to identifying fibers, most laboratories routinely measure the fiber size, and most have special criteria for choosing the fiber sizes which will be included in any counts (Gylseth et al., 1985).

In theory, it is possible to use electron-microscope tissue sections (thickness on the order of 600 Å) to search for asbestos fibers in lung. However, because of the very small volume of tissue included in such sections, even lungs with high fiber concentrations may show no fibers in the sections. As well, fiber distribution within the lung is known to be markedly heterogeneous (see below), and sections exaggerate this problem.

Most investigators instead use a digestion procedure that extracts mineral particles from large volumes of lung.[†] A variety of techniques for tissue digestion are available including bleach, strong base, and activated oxygen plasmas. Each has advantages (reviewed in Churg, 1989). We have routinely employed bleach digestion because it does not affect fiber morphology nor lead to fiber breakage, an artifact which spuriously increases fiber counts.

To obtain the dust sample, a suitable portion of lung is selected; a digest is prepared; and the digest is centrifuged to sediment the mineral fibers. The mineral fibers are then either directly deposited on an SEM stub, or they are collected on a membrane type filter and transferred to a coated electron microscope grid. One can then calculate the concentration of fibers found in the lung from (1) the weight of the lung originally digested, (2) the amount of the digest used to prepare the electron microscope sample, and (3) the areal concentration of fibers observed on an electron microscope grid or stub (Churg, 1989).

In terms of asbestos fiber identification, these methods on the whole are quite reliable, but a few problems should be noted. The differentiation between tremolite and actinolite using EDS data can be difficult. However, this is not a serious problem, since there are no clear distinctions in biologic effect between the two minerals. In contrast, the differentiation of talc and anthophyllite is an important aspect in such studies, because these two minerals have very different biological activities. Talc and anthophyllite can be very difficult to differentiate using EDS data alone. Generally, these two minerals can be distinguished based on morphological criteria; however, even this can lead to ambiguous results. Although talc generally exhibits a tabular morphology, some occurences of talc exhibit an acicular morphology. Furthermore, in some orientations, even tabular talc can appear acicular, so each particle must be observed in several orientations. Of course, talc and anthophyllite can be differentiated using electron diffraction, but this is a time consuming process. Nevertheless, if other means fail, electron diffraction data can be exploited.

[†] Typical lung sections used in digestion studies have a wet-weight of several grams or a volume of ~5–10 ml. For comparison, the dry-weight for lung sections is ~0.1 of the original wet-weight, and two healthy lungs have a dry-weight of ~40 g.

The main problem associated with lung burden determinations made by electron microscopy is that results can vary between laboratories. In a study performed by Gylseth et al. (1985) in which the same sample was analyzed by seven different laboratories, it was found that all of the laboratories reported the "high" samples as "high" and the "low" samples as "low." But the absolute fiber concentrations from each laboratory were markedly different. It is unclear to what extent this variation reflects differences in the type of microscope used (i.e., SEM vs. TEM), in the choice of fiber sizes to be counted, or in the dissolution and collection procedure used. The important point to remember is that absolute values of fiber concentration cannot be compared between laboratories. However, one can evaluate the data in broad terms by comparing the lung burdens found in the general population (arising from exposures to background sources of asbestos) with the lung burdens found in individuals with various diseases (such as asbestosis, lung cancer, or mesothelioma). This type of analysis allows one to unravel the relationships between pathologic response and exposure level.

THE RELATIONSHIP BETWEEN PHYSICAL/CHEMICAL PROPERTIES OF THE ASBESTOS MINERALS AND BIOLOGICAL BEHAVIOR

One of the most important discoveries to come out of the use of analytical electron microscopy on human lungs has been the realization that serpentine asbestos (i.e., chrysotile) behaves in a fashion quite different from amphibole asbestos. The basis of this difference appears to be related to physical and chemical differences in these two mineral groups. Chrysotile is extremely soluble in acid solutions, resulting in the leaching of magnesium from the fiber surface (Morgan et al., 1977b; Jaurand et al., 1984). In fact, recent data on the dissolution properties of chrysotile suggest that a 0.5-µm diameter chrysotile fiber would completely dissolve in ~2 months under lunglike conditions (Hume and Rimstidt, 1992). Whether these processes actually occurs *in vivo* is unclear. Lastly, it is clear from animal studies that chrysotile fibers are extremely fragile in the environment of the lung and tend to fragment laterally, breaking into shorter segments that are easily phagocytized by macrophages and, hence, removed from the lung (Churg et al., 1989a).

In contrast, asbestiform amphiboles appear to be relatively stable (chemically) in lunglike environments. Hence they are present in the lung long after the exposure, and analyses of amphibole fibers recovered after decades of residence in the lung show no evidence of elemental loss. Furthermore, asbestiform amphiboles tend to fragment longitudinally, breaking into thinner fibers of the same length. Consequently, they often remain too long to be phagocytized by macrophages, so this clearance mechanism is limited.

These physical and chemical differences translate, by mechanisms that are not entirely clear, into major differences in the accumulation of chrysotile compared to amphiboles in the lung. Wagner et al. (1974) noted nearly twenty years ago that the lungs of rats continuously exposed to amphibole asbestos accumulate fibers in a more or less linear fashion. In contrast, exposure to

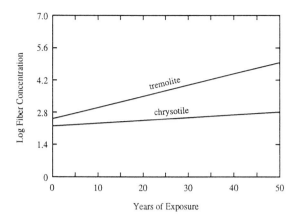

Figure 1. Lung concentration of tremolite and chrysotile as a function of years of exposure in a series of 91 chrysotile miners and millers. Tremolite, which occurs as a contaminant of the chrysotile ore, steadily accumulates with increasing exposure, whereas chrysotile does not. The chrysotile regression line is actually not statistically different from zero.

chrysotile produces only a small initial increase in chrysotile recoverable from the lung, and thereafter the amount of recoverable chrysotile does not increase despite continuing exposure. This observation from animal studies has been confirmed in numerous human studies (Fig. 1; reviewed in Churg, 1993), all of which show that for any human exposure to both chrysotile and amphibole asbestos, the residual fiber burden always contains amphibole at a greater relative abundance than found in the original dust to which the individual was exposed. The magnitude of this effect can be appreciated from studies of lungs from chrysotile miners in the Thetford mines region of Quebec. The chrysotile ore in this region typically contains only a few percent tremolite (Sébastien et al., 1986b), but the lungs of long-term chrysotile miners and millers contain about 80% tremolite and only about 20% chrysotile fibers (Churg et al., 1984).

The exact reason for this difference in fiber accumulation is still disputed. It has been suggested that because of their curled shape, chrysotile fibers impact high in the bronchial tree and hence never reach the parenchyma (Timbrell, 1965). However, animal studies suggest that, in fact, the deposition fraction for chrysotile and amphiboles is similar (Roggli and Brody 1984; Roggli et al., 1987; Coin et al., 1992), and studies of the peripheral and central portions of human lung have shown unequivocally that chrysotile fibers are found all the way out to the pleura (Churg et al., 1984). It has also been proposed that chrysotile fibers dissolve in the acidic environment of alveolar macrophage phagosomes (Morgan et al., 1977b; Jaurand et al., 1984). Although this is theoretically possible, careful studies of the composition of chrysotile fibers recovered from human lung have generally failed to provide conclusive evidence for acid induced fiber dissolution (Sébastien et al., 1986b; Langer et al., 1972; Jaurand et al., 1977; Churg and DePaoli 1988). Hume and Rimstidt (1992) argue that chrysotile is undersaturated in lung fluid at pH < 8, so it should dissolve. Based on their experimental dissolution data, Hume and Rimstidt (1992) concluded that a 0.5-µm diameter fiber of chrysotile should dissolve in ~2 months. Lastly, there is experimental evidence that chrysotile fibers readily fragment into very short segments, which are easily phagocytized and removed by macrophages (Churg et al., 1989a).

Whatever the exact mechanism, studies of the time course of chrysotile compared with amphibole clearance show that it is possible to derive clearance half times measured in years or decades for amosite- and crocidolite-exposed populations (Churg, 1993). When the same process is applied to chrysotile-exposed populations, most studies fail to show any clear correlation between time since last exposure and residual fiber burden (Churg, 1993). This observation, combined with the fact that at any time point chrysotile is under-represented compared to amphibole in lung tissue, suggests that in fact chrysotile clearance must be extremely fast, probably a matter of weeks to months after exposure.

This difference in behavior of chrysotile and amphibole asbestoses in terms of accumulation in lung can be linked to differences in the general pathogenicity of chrysotile compared to amphibole asbestoses for a variety of diseases. Most notably, in the case of mesothelioma, it is now clear that amphibole asbestoses (e.g., amosite and crocidolite) are much more pathogenic than serpentine asbestos (i.e., chrysotile) in man (McDonald and McDonald, 1986a). The latter also appears to be true for carcinoma of the lung and asbestosis, although here the differences in fiber type are not as dramatic (McDonald and McDonald, 1986b; McDonald, 1990). These observations have lead to the concept that, particularly for mesothelioma, fiber durability is extremely important in the genesis of disease. Additional support for this idea comes from the observation that fiberglass, which also shows very poor durability in lung tissue, has never been epidemiologically linked to the development of mesothelioma in man (McDonald, 1990).

ASBESTOS FIBERS IN THE LUNGS OF THE GENERAL POPULATION

One of the most surprising discoveries made through the use of electron microscopy to study mineral burdens in human lungs has been the magnitude of the asbestos fiber burden carried by everyone in the general population. In many senses the credit for this observation goes to Thomson et al. (1963), who reported that 25% of a series of routine autopsy lungs from Cape Town, South Africa, contained asbestos bodies. Asbestos bodies are fibers of asbestos coated by an iron-rich material derived from a protein (probably ferritin or hemosiderin). These structures are readily visible in the light microscope, and if one digests lung and collects the sediment in a counting chamber or on a membrane filter, it is possible to quantitate asbestos bodies much in the way one quantitates fibers by electron microscopy.

The report of Thomson et al. (1963) set off a flurry of investigation, with the result that asbestos bodies were reported in the lungs of virtually every population in all westernized countries. These observations then lead to the idea that asbestos must be present in urban air, which was subsequently confirmed by a large number of studies (Health Effect Institute-Asbestos Research, 1991). The source of asbestos contamination in the air is not entirely clear in all cases, since there is considerable serpentinite (one host rock for chrysotile and other serpentine minerals) outcropping throughout North America. However, amosite- and crocidolite-rich rocks (particularly of ore-grade) are not common in North America, suggesting either that some of the atmospheric contaminant of amphibole asbestos arises from release of commercial sources or that dust from major natural sources (e.g., in South Africa or Australia) is transported extremely long distances. Chapter 2 presents a thorough discussion of natural sources of asbestos dust.

As is true of any set of fiber counts generated by analytical electron microscopy, the actual magnitude of the asbestos fiber burden in the general population varies from laboratory to laboratory. Table 2 shows data from my laboratory (Churg and Wiggs, 1986) for asbestos fiber levels in the population of Vancouver, British Columbia (see Roggli, 1990, for a summary of data from other laboratories). Several points are of interest. Firstly, the bulk of the fibers are chrysotile and tremolite, the latter, a common contaminant of chrysotile ore. Secondly, the amount of amosite and crocidolite is relatively small. Lastly, a simple calculation shows that appreciable numbers of asbestos fibers occur in lungs with no evidence of a pathologic response: Two normal lungs weigh approximately 40 grams when dried; using the 95th percentile data in Table 2, the lung of an urban dweller in Vancouver, British Columbia, might contain as many as 40,000,000 fibers of chrysotile, 40,000,000 fibers of tremolite, and 400,000 fibers of amosite and crocidolite asbestos. Nonetheless, there is no evidence to suggest that this fiber burden is producing any type of disease.

Table 2

Asbestos fiber content of the lungs of the general population of Vancouver, British Columbia

Fiber type	Mean (10^6 fibers/g of dry lung)	Median (10^6 fibers/g of dry lung)	95th percentile
chrysotile	0.3	0.2	1.1
tremolite	0.4	0.2	1.2
amosite + crocidolite	0.001	0.0	0.01

Source: Churg and Wiggs (1986).

Because of both local mining activities and extensive contamination of the soil with chrysotile and tremolite, the ambient fiber content of air in the Quebec mining townships is several hundred fold that of urban areas in North America (Sébastien et al., 1986). Analysis of the asbestos content of the lungs of those who have lived many years or whole lifetimes in the townships, but have never been employed in the mining or milling industry, indicates that such persons carry five to tenfold greater amounts of chrysotile and tremolite than ordinary urban dwellers (Churg, 1986; Case and Sébastien, 1987). However, a number of epidemiologic studies have looked at these types of individuals and failed to find any evidence of an excess of mesothelioma or lung cancer (Churg, 1986; McDonald, 1985).

These data thus indicate that the lung can tolerate considerably higher burdens of chrysotile and tremolite (at least the type of short-fiber tremolite found in these chrysotile localities) than typically found in the lungs of urban dwellers, without the appearance of disease. These findings should provide reassurance not only about the innocuous effects of the general ambient background level of asbestos, but also about the effects of low level chrysotile release in public buildings, which is at a level very close to that of ambient air. The latter issue has been a source of considerable regulatory worry and litigation, without any actual evidence that this type of exposure causes disease (Health Effect Institute-Asbestos Research, 1991).

Fiber Concentration in the Lungs of Occupationally-Exposed Groups

A fairly substantial number of studies have now been published on the fiber burdens found in groups of workers occupationally exposed to various types of asbestos (for a listing of individual reports see references Churg, 1993; Roggli, 1990; Churg, 1991). A number of comments about the limitations of such studies and the precautions needed in interpreting them are in order. Firstly, most working populations have been exposed to mixture of chrysotile and amosite or crocidolite. Populations with pure exposure to only one of these fibers are relatively few. However, as indicated previously, there is every reason to believe that chrysotile and the commercial amphiboles have markedly different biologic effects. Therefore studies which fail to separate fiber type and merely provide a total fiber burden are of relatively little value in understanding the relationship between fiber concentration and disease.

A second problem is that, as noted, chrysotile fails to accumulate in the lung. Therefore a fiber analysis may incorrectly suggest that a working population has only been exposed to amphiboles, despite quite substantial chrysotile exposure. For example, in a series of 168 shipyard and insulation workers analyzed in our laboratory, all of whom were known to have had substantial mixed amosite and chrysotile exposure in the past, the median amosite concentration at autopsy was 830,000 fibers/gm dry lung; the median chrysotile concentration 200,000 fibers/gm dry lung, and the median tremolite concentration 180,000 fibers/gm dry lung (A. Churg and S. Vedal, unpublished data).

To a certain extent the tremolite contaminant of chrysotile ore may serve as a substitute marker for chrysotile, since it, like other amphiboles, tends to be preferentially retained in lung (Churg, 1988). On the other hand, it appears that some variable but unknown proportion of tremolite is removed from the ore during milling, and analysis of different worker populations suggests that different secondary uses or different chrysotile products contain quite variable proportions of tremolite (Churg, 1988). For example, the proportion of tremolite in the lungs of chrysotile miners and millers is extremely high, and that in the lungs of chrysotile textile workers is also substantial, although relatively lower than in chrysotile miners, but if one looks at shipyard and insulation workers known to have used extensive amounts of chrysotile in the past, both their tremolite and chrysotile burdens tend to be extremely low (Churg, 1988). Thus one must be cautious in interpreting analyses which fail to detect chrysotile or only detect background levels.

Lastly, fiber analysis of autopsy lungs is performed at only one point in time, but lung burden by definition is the result of a dynamic process reflecting both deposition and clearance. To the extent that disease reflects what may have been present in the lung in the remote past rather than what is present in the lung at the time of autopsy, then fiber burden studies in theory could be extremely misleading. On the other hand, what fiber-burden data do provide is a substitute for actual exposure data. These data are an important substitute in the sense that they are the only measure of what is actually in the lung. The fact that distinct and reproducible correlations with disease patterns are found by fiber burden studies (Churg, 1993; Roggli, 1990; Churg, 1991) suggests that the use of fiber-burden data as a substitute for actual exposure data is a reasonable approach.

Several groups have performed systematic studies of workers with exposure both to the commercial amphiboles amosite and crocidolite and to chrysotile (Roggli, 1990; Churg, 1991; Wagner et al., 1988; Roggli et al., 1986). One of the consistent findings in these studies has been the very low levels of chrysotile found in the lung, as indicated above. Thus I have chosen to treat these populations as if they really had been exposed only to the amphibole. This is a necessary approximation in order to discern relationships between amphibole exposure and disease, but by definition this approach may miss chrysotile's contribution to disease in these workers.

A number of general conclusions can be drawn regarding the relationship between amosite and crocidolite concentration and disease patterns. For example, disease appears at several orders of magnitude greater fiber burden than is found in the lungs of the general population. In regard to both benign and malignant pleural disease, this difference is approximately some 2 to 3 orders of magnitude. There is yet another similar increase in fiber burden comparing cases of benign and malignant pleural disease to cases of asbestosis. These studies leave no doubt that asbestosis requires relatively a very high burden of fibers before it appears. If one further evaluates asbestosis by either general or local grade of fibrosis, then higher grades of asbestosis are found to be associated quite consistently with higher fiber burdens (Wagner et al., 1988; Roggli et al., 1986; Churg et al., 1989b; Churg et al., 1990)

These observations emphasize the ideas that there is a distinct fiber concentration–disease relationship (as shown in Fig. 2), that the pleura is relatively sensitive to amosite or crocidolite exposure, and that the parenchyma is relatively insensitive. They do not, however, explain why some individuals develop benign pleural disease and some malignant pleural disease (see below under **EFFECTS OF FIBER SIZE** for additional comments on this topic).

There are relatively few working populations with even claimed exposure to chrysotile asbestos alone, and one of the most important observations that has come out of studies of lung burden has been the extent to which many such populations have had widespread but apparently occult exposure to amosite or

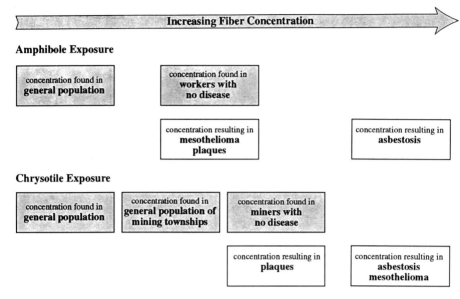

Figure 2. The relationship of fiber concentration to disease for amphibole (amosite or crocidolite) or chrysotile exposure. The distances along the horizontal axis are not equivalent to actual fiber concentrations but merely show the relative concentrations.

crocidolite as well (Churg, 1988; Wagner et al., 1982). Retrospective review of factory records suggest that this situation arose because many factories which used predominantly chrysotile either incorporated amphibole in a very limited range of products, or tried out relatively small amounts of amphibole from time to time for test purposes (for more detail see Churg, 1988). Because amosite and crocidolite are so much more potent mesothelial carcinogens than chrysotile, this type of occult exposure can provide extremely misleading data about the danger of chrysotile exposure. Thus estimates of risk, and subsequent regulatory rules promulgated on such estimates, may be incorrect (e.g., Peto, 1978).

Mineralogical analyses of lungs from Quebec chrysotile miners and millers have shed light on the important issue of whether chrysotile induces mesothelioma in man and if so, what dose is required. We (Churg et al., 1984; Churg and Wright, 1989) analyzed a series of such workers with mesothelioma and found that, although a few cases had had amosite exposure, apparently from an amosite processing plant which operated in the region for several years, most of the mesothelioma cases had only chrysotile and tremolite in their lungs, and the fibers were always present in very high concentration. Thus these data unequivocally indicate that chrysotile ore components, in sufficient concentration, can induce mesothelioma in man.

Whether the chrysotile or the tremolite component is the actual agent of mesothelioma in these workers remains an unresolved question, but a number of lines of evidence favor the idea that tremolite actually may be the culprit. Firstly, if one compares the tremolite-to-chrysotile ratio in the lungs of various occupational groups, the groups with the highest ratio, namely chrysotile miners and millers, have by the far the highest incidence of mesothelioma, whereas those with the lower ratios have the lower incidences of such tumors (Churg, 1988). Secondly, a formal multiple regression analysis performed on the data from a series of nearly 100 chrysotile miners and millers analyzed by our laboratory showed that disease in general could be correlated quite well with tremolite fiber concentration, but that correlations with chrysotile concentration vanished after adjusting for the presence of tremolite (A. Churg and S. Vedal, unpublished data). This type of analysis must be viewed with caution, because, by definition, most of the chrysotile which was once present has long disappeared, so a role for extremely high but temporary burdens of chrysotile is difficult to rule out. Nonetheless, at this point, a reasonable case can certainly be made that, at least in regard to mesothelioma, tremolite seems more likely to be the important agent than does chrysotile.

As is true for amosite and crocidolite exposure, chrysotile exposure results in asbestosis only at very high chrysotile/tremolite fiber concentrations, based on analysis of occupationally-exposed individuals (Churg, 1991; Churg and Wright, 1989; Green et al., 1986; Churg et al., 1989b). Likewise, there is a good correlation between grade of asbestosis and total fiber concentration (Green et al., 1986; Churg et al., 1989b). What is most interesting, however, is that mesothelioma and asbestosis appear at about the same extremely high fiber burden in chrysotile-exposed populations (Churg, 1991; Churg and Wright, 1989), in marked contradistinction to the situation for amosite- or crocidolite-exposed populations. Pleural plaques in these workers, however, appear at a considerably lower burden. Thus, whereas the relationships between fiber concentration, pleural plaques, and asbestosis are the same for both chrysotile exposure and amosite or crocidolite exposure, it is quite clear that the burdens associated with chrysotile-induced mesothelioma are distinctly different (Fig 2). An important corollary of this observation is that the types of chrysotile exposures sufficient to produce asbestosis are now matters of historic interest and would never occur at any current or recent regulated exposure level. These findings further suggest that there is no risk of chrysotile induced mesothelioma from low level or environmental level chrysotile exposure.

An additional finding which emerges from these studies, and which is not evident from the scheme shown in Figure 2, is that for any given disease, there are marked differences in the absolute fiber concentration seen in workers with amosite or crocidolite exposure compared to chrysotile/tremolite exposure. Thus, in our laboratory, chrysotile-induced asbestosis appears at about 4 times the fiber burden seen in amosite-induced asbestosis (Table 3). For mesothelioma the difference is even more dramatic: chrysotile-induced mesothelioma appears at about 400 times the median fiber burden of amosite-induced mesothelioma (Table 3; Churg and Wright, 1989).

This can be looked at in another way. If one examines the fiber burden seen in long-term chrysotile miners and millers with no asbestos-induced disease, we have found that the mean chrysotile burden is in the order of 20,000,000 fibers/gm dry lung and the mean tremolite burden 50,000,000 fibers/gm dry lung

Table 3

Median fiber concentration found in age- and exposure-period-matched workers with chrysotile or amphibole (amosite) exposure

	Mesothelioma (10^6 fibers/g of dry lung)	Asbestosis (10^6 fibers/g of dry lung)
chrysotile workers	290[a]	110[a]
amosite workers	0.7[b]	26[b]

Source: Churg and Wiggs (1986)
[a] chrysotile/tremolite fibers × 10^6/gm dry lung
[b] amosite fibers × 10^6/gm dry lung

(Churg, 1983). In contrast, mean values in the general population of Vancouver are approximately 300,000 fibers/gm dry lung for chrysotile and approximately 400,000 fibers/gm dry lung for tremolite (Churg and Wiggs, 1986). Were these fiber concentrations reflecting amosite exposure instead of chrysotile/tremolite exposure, the levels cited above would be well into the range that is associated with mesothelioma! (Churg and Wright, 1989). These findings again reinforce the idea that, relatively, an extraordinary burden of chrysotile and its contaminant tremolite can be tolerated without the appearance of disease. These observations suggest that carefully regulated exposure to chrysotile may produce no health hazard. This idea is reinforced by epidemiologic studies of various working populations exposed in the range of 1-5 fibers/ml, amongst whom no excess of any type of asbestos related disease has been observed (McDonald et al., 1984; Gardner et al., 1986; Neuberger and Kundi, 1990).

The preceding comments have concerned only the lung parenchyma. Two different groups have examined fiber concentration in the pleura (Sébastien et al., 1980; Dodson et al., 1990). Both have come to the surprising conclusion that the predominant fiber in the pleura is chrysotile, even in cases where the parenchyma shows considerable amosite or crocidolite burden. It is difficult to reconcile these findings with the accepted observation that amosite and crocidolite are much more potent mesothelial carcinogens than is chrysotile.

We have recently begun to examine the asbestos fiber concentration in the airway mucosa (all tissues between the airway lumen and the bronchial cartilage). Although the airway mucosa contains fairly high concentrations of non-asbestos particles (on the order of several hundred million per gram dry tissue), we were essentially unable to demonstrate amosite fibers in the mucosa in a series of never smoking long term shipyard workers and insulators (Churg and Stevens, 1993). These observations raise questions about the possible role of asbestos in producing bronchogenic carcinomas by direct action on the airway epithelium, rather than, as we and others have suggested, via its effects in causing asbestosis (Browne, 1986; Churg and Green, 1988).

EFFECTS OF FIBER DISTRIBUTION

Most studies analyzing the fiber content of human lung have been concerned only with fiber concentration, and relatively little information exists about fiber distribution. Since asbestos related diseases have a very distinct distribution, most notably the lower zonal predominance of asbestosis and benign pleural disease, it appears logical to assume that these effects can be related to the distribution of fibers within the lung. Similarly it appears equally reasonable to assume that, since mesothelioma originates in the pleural mesothelium, patients with mesothelioma should show different distributions of central to peripheral fibers compared to patients with other asbestos diseases.

In practice, it has thus far proved impossible to provide any clear correlations of fiber distribution and disease in man, or even to arrive at agreed upon descriptions of what the distribution in man might be. One major problem is that there appears to be marked heterogeneity of fiber concentration within the lung. Several different groups have shown that as much as ten-fold differences in fiber concentration can be seen over distances as small as 1 cm immediately under the pleura (Churg and Wood, 1983; Morgan and Holmes, 1983; Morgan and Holmes, 1984). A further problem is that, as noted previously in regard to fiber concentration, autopsy lungs sample fiber distribution at only one time point, but there is extremely suggestive data from studies in rats that fiber distribution may change over time, and that there may be subpleural accumulation of fibers in selected hotspots (Morgan et al., 1977a). Lastly, it is crucial that one examines fiber distribution by specific type of fiber, because different fibers have different size characteristics, and these will clearly influence where they end up in the lung.

We have attempted to approach this problem by using exactly comparable mid-sagittal slices of lungs from workers with similar exposures, and by examining only one type of fiber, namely amosite, because of its relative persistence within lung tissue. As shown in Figure 3 (Churg, 1990), this approach does in fact reveal that there are runs of high and low fiber concentration immediately under the pleura in the periphery of the lung. However, these distributions do not match the predictions made from theories of fiber deposition (Asgharian and Yu, 1988). For example, such theories predict that the shorter the airway path length to any point in the periphery of the lung, the higher the resulting fiber concentration. Thus one might explain the relatively high fiber concentration shown in Figure 3 in the apex of the lung on the basis that this area of the lung is served by airways of short path length, but the superior segment of the lower lobe is served by airways of equally short path length and fails to accumulate fibers.

Theory also predicts that in general fiber concentration should be higher in the central portions of the lung and decreased toward the periphery of the lung because of progressive impaction and deposition of fibers as the airways narrow (Timbrell, 1965; Asgharian and Yu, 1988). Again, however, the human data fail to match prediction in any consistent fashion. For example, we have found that there is no consistent difference between the central and peripheral portions of the lung in regard to fiber concentration (Churg et al., 1984).

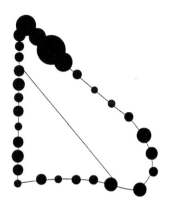

Figure 3. Distribution of amosite concentration in the periphery of a series of anatomically normal lungs from heavily exposed shipyard workers and insulators. The drawing represents a mid-sagittal slice of lung with the apex at the top and the straight line as the major fissure separating upper and lower lobe. The diameters of the circles are scaled to fiber concentration. Note that there are distinct regions of high and low concentration (from Churg, 1990, by permission).

Overall, the question of fiber distribution needs considerable more investigation and in particular detailed mapping of human lungs to determine exactly where fibers are in each portion of the lung.

EFFECTS OF FIBER SIZE

A large body of experimental animal data as well as theoretic models have lead to predictions about the sizes of fibers that should cause disease in man. Unfortunately, as is true of theories of fiber distribution, there are major discrepancies between the models and animal data, and what has been found in human lungs.

One of the basic tenants of asbestos related diseases is that long high aspect ratio fibers are considerably more dangerous than short low aspect ratio fibers. Certainly in experimental animals exposure to long fibers is associated with much greater degrees of fibrosis (asbestosis) than exposure to short fibers (Adamson and Bowden, 1987a; Adamson and Bowden 1987b; Davis et al., 1986; Davis and Jones, 1986). Exposure to long fibers also results in much greater incidences of mesothelioma than does exposure to short fibers (Davis et al., 1986; Davis and Jones, 1986; Stanton et al., 1981).

The limited data available from man on fiber size are at best contradictory. We have evaluated the detailed correlations of fiber size and local degree of fibrosis in a series of workers with asbestosis and either amosite or chrysotile/tremolite exposure and concluded that much better correlations were found between the concentration of short fibers and fibrosis than between the concentration of long fibers and fibrosis (Churg et al., 1990; Churg et al., 1989b). There are several possible explanations for this observation. One, and the most obvious, is that the animal data are incorrect and that short fibers are more pathogenic then is commonly believed. There is some experimental support for this idea since Adamson and Bowden (Adamson and Bowden, 1990) have recently shown that short fibers have just as great a potential to evoke fibroblast growth factors from macrophages as do long fibers, but that short fibers ordinarily cannot gain access to the interstitial tissues. This lead us to note that all of our patients in the studies were smokers, and smoking is known to increase the retention and penetration of the fibers into the pulmonary tissues (McFadden et al., 1986). Thus it is possible that in smoking humans, short fibers gain access to interstitial macrophages and are markedly fibrogenic. Coin et al. (1992) have suggested an alternate explanation, namely that fiber clearance is in general inversely proportional to fiber length and that sufficiently long fibers are not cleared at all. Thus they postulate that long fibers may produce fibrosis and interfere with local clearance mechanisms, with secondary accumulation of short fibers.

Equally confusing is the issue of fiber size and mesothelioma in man. The results one obtains on this question depends to a critical degree on how one structures one's studies. McDonald et al. (1989) recently performed a case control study in which patients with mesothelioma (largely occupationally exposed) were compared to controls without mesothelioma. They found that mesothelioma was associated with fibers greater than 8 µm length and that shorter fibers played no role. They also found that only tremolite, amosite and crocidolite fibers appeared to be of importance. Rogers et al. (1991) performed a similar study and concluded that while long fibers were very important, short fibers of chrysotile also played a role in the genesis of mesothelioma. However, both these studies may be biased because they compare occupationally exposed persons to the general population, and studies from our own labs suggest that persons in the general population tend to have relatively short asbestos fibers in their lungs (Churg and Wiggs, 1986).

We have examined this question in a different fashion by performing multiple regression analyses accounting for the presence of multiple diseases in workers with heavy exposure to chrysotile/tremolite or amosite (A. Churg and S. Vedal, unpublished data). In both groups it was clear that patients with any asbestos induced disease had longer fibers than exposed patients with no asbestos induced disease; however, there was no evidence at all that patients with mesothelioma had longer fibers than patients with asbestosis, pleural plaques, lung cancers, or asbestos induced airway fibrosis.

The one finding which did emerge from these latter two studies was the intriguing observation that pleural plaques were associated with high aspect ratio fibers whereas no other disease was associated with high aspect fibers. As well, measures of fiber length, fiber width, fiber surface, fiber mass, and total fiber size parameters (in other words the aggregate length, width, aspect ratio, etc., of all the fibers in the sample) failed to show any associations with disease, despite predictions in the literature that total fiber size parameters can be used to explain the presence of disease (Timbrell et al., 1988). In fact, there was a suggestion that mesotheliomas were if anything associated with lower aspect ratio fibers than the other diseases.

Lastly, to further confound prediction, we have examined the length of fiber found in various portions of the central and peripheral areas of the lungs of heavily exposed amosite workers and found that while the lower lobe showed longer fibers in the center compared to the periphery of the lung, as predicted by deposition theory, in the upper lobe the longest fibers were found in the posterior portion of the periphery (Churg and Wiggs, 1987).

In short, therefore, studies of fiber size have failed to provide any consistent correlations with disease and at this point contradict both animal studies and experiments. This again is an area in need of further examination.

ACKNOWLEDGMENTS

Supported by Grants from the Medical Research Council and National Cancer Institute of Canada.

REFERENCES

Adamson, I.Y.R. and Bowden, D.H. (1987a) Response of mouse lung to crocidolite asbestos. 1. Minimal fibrotic reaction to short fibers. J. Pathol. 152, 99–107.

Adamson, I.Y.R. and Bowden, D.H. (1987b) Response of mouse lung to crocidolite asbestos. 2. Pulmonary fibrosis after long fibers. J. Pathol. 152, 109–117.

Adamson, I.Y.R. and Bowden, D.H. (1990) Pulmonary reaction to long and short asbestos fibers is independent of fibroblast growth factor production by alveolar macrophages. Amer. J. Pathol. 137, 523–529.

Asgharian, B. and Yu, C.P. (1988) Deposition of inhaled fibrous particles in the human lung. J. Aerosol Med. 1, 37–50.

Browne, K. (1986) Is asbestos or asbestosis the cause of the increased risk of lung cancer in asbestos workers. Brit. J. Indus. Med. 43, 145–149.

Case, B.W. and Sébastien, P. (1987) Environmental and occupational exposures to chrysotile asbestos: A comparative microanalytic study. Arch. Environ. Health 1987; 42, 85–191.

Champness, P.E., Cliff, G., and Lorimer, G.W. (1976) The identification of asbestos. J. Micros. 108, 231–249.

Churg, A. (1983) Asbestos fibre content of the lungs of chrysotile miners with and without asbestos airways disease. Amer. Rev. Respir. Dis. 127, 470–473.

Churg, A. (1986) Lung asbestos content in long-term residents of a chrysotile mining town. Amer. Rev. Respir. Dis. 134, 125–127.

Churg, A. (1988) Chrysotile, tremolite, and mesothelioma in man. Chest 93, 621–628.

Churg, A. (1989) Quantitative methods for analysis of disease induced by asbestos and other mineral particles using the transmission electron microscope. In P. Ingram, J.D. Shelburne, and V V.L. Roggli, eds., Microprobe Analysis in Medicine. Hemisphere Publishing Company, New York, 79–96.

Churg, A. (1990) The distribution of amosite asbestos in the periphery of the normal human lung. Brit. J. Indus. Med. 47, 677–681.

Churg, A. (1991) Analysis of lung asbestos content. Brit. J. Indus. Med. 48, 649–652

Churg, A. (1993) Persistence of natural mineral fibers in human lung. Environ. Hlth. Perspect, in press.

Churg, A. and DePaoli, L. (1988) Clearance of chrysotile from human lung. Exptl. Lung Res. 14, 567–574.

Churg, A. and Green, F.H.Y. (1988) Pathology of Occupational Lung Disease. New York, Igaku-Shoin Medical Publishers.

Churg, A. and Stevens, B. (1993) Absence of amosite asbestos in airway mucosa of nonsmoking longterm workers with occupational asbestos exposure. Brit. J. Indus. Med., in press.

Churg, A. and Wiggs, B. (1986) Fiber size and number in users of processed chrysotile ore, chrysotile miners, and members of the general population. Amer. J. Indus. Med. 9, 143–152.

Churg, A., and Wiggs, B. (1987) Accumulation of long asbestos fibers in the peripheral upper lobe in cases of mesothelioma in man. Amer. J. Indus. Med. 11, 563–570.

Churg, A., and Wood, P.L. (1983) Observations on the distribution of asbestos fibers in human lung. Environ. Res. 31, 374–80.

Churg, A., and Wright, J. (1989) Fibre content of lung in amphibole vs chrysotile induced mesothelioma: Implications for environmental exposure. In J. Bignon et al., eds., Non-occupational Exposure to Mineral Fibres. IARC, Lyon, 314–318.

Churg, A., DePaoli, L., Kempe, B., and Stevens, B. (1984) Lung asbestos content in chrysotile workers with mesothelioma. Amer. Rev. Respir. Dis. 130, 1042–1045.

Churg, A., Wright, J.L., DePaoli, L., and Wiggs, B. (1989b) Mineralogic correlates of fibrosis in chrysotile miners and millers. Amer. Rev. Respir. Dis. 139, 891–896.

Churg, A., Wright, J.L., Gilks, B., and DePaoli, L. (1989a) Rapid short term clearance of chrysotile compared to amosite asbestos in the guinea pig. Amer. Rev. Respir. Dis. 139, 885–890.

Churg, A., Wright, J., Wiggs, B., and DePaoli, L. (1990) Mineralogic parameters related to amosite asbestos induced fibrosis in man. Amer. Rev. Respir. Dis. 142, 1331–1336.

Coin, P.G., Roggli, V.L. and Brody, A.R. (1992) Deposition, clearance, and translocation of chrysotile asbestos from peripheral and central regions of the lung. Environ. Res. 58, 97–116.

Davis, J.M.G., Addison, J., Bolton, R.E. and Donaldson, K., et al. (1986) The pathogenicity of long vs short fibre samples of amosite asbestos administered to rats by inhalation and intraperitoneal injection. Brit. J. Expt. Path. 67, 415–430.

Davis, J.M.G. and Jones, A.D. (1986) Comparisons of the pathogenicity of long and short fibre chrysotile asbestos in rats. Brit. J. Expt. Path. 69, 717–737.

Dodson, R.F., Williams, M.G., Corn, C.J., Brollo, A. and Bianchi, C. (1990) Asbestos content of lung tissue, lymph nodes, and pleural plaques from former shipyard workers. Amer. Rev. Respir. Dis. 142, 843–847.

Gardner, M.J., Winter, P.D., Pannett, B. and Powell, C.A. (1986) Follow up study of workers manufacturing chrysotile asbestos cement products. Brit. J. Indus. Med. 43, 726–732.

Green, F.H.Y., Harley, R., Vallyathan, V., Dement, J., Pooley, F. Althouse, R. (1986) Pulmonary fibrosis and asbestos exposure in chrysotile asbestos textile workers: Preliminary results. Accomplishments in Oncology; 1, 59–68.

Gylseth, B., Churg, A. and Davis, J.M.G. et al. (1985) Analysis of asbestos fibers and asbestos bodies in human lung tissue samples. An international laboratory trial. Scand. J. Work Environ. Hlth. 11, 107–110.

Health Effect Institute-Asbestos Research (1991) Asbestos in Public and Commercial Buildings: A Literature Review and Synthesis of Current Knowledge, Health Effects Institute, Cambridge, MA.

Hume, L.A. and Rimstidt, J.D. (1992) The biodurability of chrysotile asbestos. Am. Mineral. 77, 1125–1128.

Jaurand, M.C., Bignon, J., Sebastien, P. and Goni, J. (1977) Leaching of chrysotile asbestos in human lungs: Correlation with in vitro studies using rabbit alveolar macrophages. Environ. Res. 14, 245–254.

Jaurand, M.C., Gaudichet, A., Halpern, S., and Bignon, J. (1984) In vitro biodegradation of chrysotile fibres by alveolar macrophages and mesothelial cells in culture: Comparison with a pH effect. Brit. J. Indus. Med. 41, 389–395.

Langer, A.M., Rubin, I.B. and Selikoff, I.J. (1972) Chemical characterization of asbestos body cores by electron microprobe analysis. J. Histochem. Cytochem. 20, 723–734.

McDonald, J.C. (1990) Cancer risks due to asbestos and man-made fibres. Recent Results Cancer Res 120, 122–133.

McDonald, J.C. (1985) Health implications of environmental exposure to asbestos. Environ. Hlth. Perspect 62, 319–328.

McDonald, J.C. and McDonald, A.D. (1986a) Epidemiology of malignant mesothelioma. In Asbestos-Related Malignancy, edited by K. Antman and J. Aisner. New York Grune and Stratton. Pages 57–79.

McDonald, J.C. and McDonald, A.D. (1986b) Epidemiology of asbestos-related lung cancer. In K. Antman and J. Aisner, eds., Asbestos-Related Malignancy, Grune and Stratton, New York, 31–56.

McDonald, A.D., Fry, J.S., Woolley, A.J., and McDonald, J.C. (1984) Dust exposure and mortality in an American chrysotile asbestos friction products plant. Brit. J. Indus. Med. 41, 151–157.

McDonald, J.C., Armstrong, B., and Case, B. et al. (1989) Mesothelioma and asbestos fiber type. Evidence from lung tissue analysis. Cancer 63, 1544–1547.

McFadden, D., Wright, J., Wiggs, B., and Churg, A. (1986) Smoking increases the penetration of asbestos fibers into airway walls. Amer. J. Pathol. 123, 95–99.

Morgan, A. and Holmes, A. (1983) Distribution and characteristics of amphibole asbestos fibers, measured with the light microscope, in the left lung of an insulation worker. Brit. J. Indus. Med. 40, 45–50.

Morgan, A. and Holmes, A. (1984) The distribution and characteristics of asbestos fibers in the lungs of Finnish anthophyllite mine-workers. Environ. Res. 33, 62–75.

Morgan, A., Evans, J.C. and Holmes, A. (1977a) Deposition and clearance of inhaled fibrous minerals in the rat. Studies using radioactive tracer technique. In W.H. Walton and B. McGovern, eds., Inhaled Particles IV, Pergamon Press, New York, 259–274.

Morgan, A., Davies, P, Wagner, J.C., Berry, G., and Holmes, A. (1977b) The biological effects of magnesium-leached chrysotile asbestos. Brit. J. Exptl. Pathol. 58, 465–473.

Neuberger, M. and Kundi, M. (1990) Individual asbestos exposure, smoking and mortality--A cohort study in the asbestos cement industry. Brit. J. Indus. Med. 47, 615–620.

Peto, J. (1978) The hygiene standard for chrysotile asbestos. Lancet 1, 484–488.

Rogers, A.J., Leigh, J., Berry, G., Ferguson, D.A., Mulder, H.B. and Ackad, M. (1991) Relationship between lung asbestos fiber type and concentration and relative risk of mesothelioma. Cancer 67, 1912–1920.

Roggli, V.L. (1990) Human disease consequences of fiber exposures--A review of human lung pathology and fiber burden data. Environ. Health Perspect. 88, 295–303.

Roggli, V.L., Pratt, P.C. and Brody, A.R. (1986) Asbestos content of lung tissue in asbestos associated disease: A study of 110 cases. Brit. J. Indus. Med. 43, 18–29.

Roggli, V.L. and Brody, A.R. (1984) Changes in numbers and dimensions of chrysotile asbestos fibers in lungs of rats following short-term exposure. Expt. Lung Res. 7, 133–147.

Roggli, V.L., George, M.H. and Brody, A.R. (1987) Clearance and dimensional changes of crocidolite asbestos isolated from lungs of rats following short-term exposure. Environ. Res. 42, 94–105.

Sebastien, P., Plourde, M., and Robb, R., et al. (1986a) Ambient air asbestos survey In Quebec Mining Towns: Part 2--Main study, Environment Canada Report 5/AP/RQ-2E.

Sebastien, P., Begin, R., Case, B.W. and McDonald, J.C. (1986b) Inhalation of chrysotile dust. In J.C. Wagner, ed., The Biological Effects of Chrysotile, J.B. Lippincott, Philadelphia, PA, 19–29.

Sebastien, P., Janson, Z. and Gaudichet, A. et al. (1980) Asbestos retention in human respiratory tissues. Comparative measurements in lung parenchyma and in parietal pleura. In J.C. Wagner, ed., Biological Effects of Mineral Fibers, IARC, Lyon, 237–246.

Stanton, M.F., Layard, M., and Tegeris, A., et al. (1981) Relation of particle dimension to carcinogenicity in amphibole asbestoses and other fibrous minerals. J.N.C.I. 67, 965–975.

Thomson, J.G., Kaschula, R.O.C. and MacDonald, R.R. (1963) Asbestos as a modern urban hazard. So. Afr. Med. J. 37, 77–81.

Timbrell, V. (1965) The inhalation of fibrous dusts. Ann. New York Acad. Sci. 132, 255–273.

Timbrell, V., Ashcroft, T., Goldstein, B. et al. (1988) Relationships between retained amphibole fibres and fibrosis in human lung tissue specimens. Ann. Occup. Hyg. 32 (Supp 1) 323–340.

Wagner, J.C., Berry, G., Skidmore, J.W. and Timbrell, V. (1974) The effects of the inhalation of asbestos in rats. Brit. J. Cancer 29, 252–269

Wagner, J.C., Berry, G. and Pooley, F.D. (1982) Mesotheliomas and asbestos type in asbestos textile workers: A study of lung contents. Brit. Med. J. 285, 603–606.

Wagner, J.C., Newhouse, M.L., Corrin, B., Rossiter, C.E.R. and Griffiths, D.M. (1988) Correlation between fibre content of the lung and disease in East London asbestos factory workers. Brit. J. Indus. Med. 45, 305–308.

CHAPTER 14

DEFENSE MECHANISMS AGAINST INHALED PARTICLES AND ASSOCIATED PARTICLE-CELL INTERACTIONS

Bruce E. Lehnert

Pulmonary Biology/Toxicology Program
Cell Growth, Damage and Repair Group
Life Sciences Division
Los Alamos National Laboratory
Los Alamos, New Mexico 87545 U.S.A.

INTRODUCTION

From the moment of birth, we continuously breath air that is contaminated with particles, the mass concentrations and hazardous characteristics of which vary from one environmental setting to another. It is now becoming increasingly recognized that the deposition in the respiratory tract of virtually all inhaled particles can result in the development of numerous pathologic disorders, e.g., alveolitis, emphysema, fibrosis, and cancer. The likelihood of development and the extent of expression of such disorders, however, is a function of the site(s) of particle deposition along the respiratory tract, the amounts of particles that are deposited, the physicochemical characteristics of the particles, particle-cell interactions and their attendant outcomes, and the post-depositional disposition of the particles. While some attention will be given in this chapter to the physical factors that govern the deposition of particles within the total respiratory tract, the focus of this report is primarily on the lower respiratory tract and particle-associated phenomena that occur there, with the main objective being to provide the reader with an introductory understanding of: (1) the fates of particles that deposit in the conducting airways and some typical responses that are associated with particle deposition in this anatomical compartment of the lung, and (2) the fates of particles and responses they can elicit following particle deposition in the lung's alveolar region. Specific emphasis is given to lung defense mechanisms against inhaled particles as well as some particle-cell interactions that may be involved in the development of particle-induced lung diseases. The interested reader is referred to the following recent publications from which various topic items in the present chapter have been adopted (Lehnert et al., 1990b; Lehnert and Oberdörster, 1993; Oberdörster and Lehnert, 1991; Lehnert, 1992).

DEPOSITION OF PARTICLES IN THE RESPIRATORY TRACT

The mammalian respiratory tract can be viewed as being divided into two general anatomical regions: the upper respiratory tract, or nasopharyngeal region that is composed by the nose, nasal cavity, nasopharynx, and oropharynx, and the lower respiratory tract, which includes the conducting airways, or the

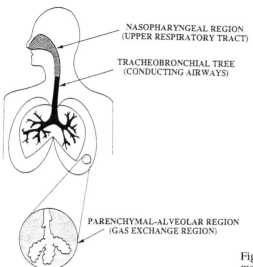

Figure 1. Generalized anatomical arrangement of the respiratory tract.

tracheobronchial region, and the pulmonary-alveolar region where gas exchange occurs (Figs. 1 and 2). The first line of defense against the deposition of inhaled particles in the respiratory tract is an anatomical filtration system that basically serves to limit the penetration of airborne particles into the lower respiratory tract. Such filtration is accomplished by the structural design of the respiratory tract and various physiologic aspects of airflow, which collectively favor particle deposition in the nasopharyngeal region and larger airways, while tending to spare the more peripheral pulmonary-alveolar region. The efficiency of the aerodynamic filtering mechanism and factors that determine the sites of particle deposition along the respiratory tract are complex and dependent on numerous factors, including the geometric dimensions, densities, and physicochemical characteristics of the inhaled particles, the geometric structural configuration of the respiratory tract (e.g., complex nasopharyngeal structures, lower respiratory tract airway branching and daughter branch angulations), and an individual's breathing pattern. Aerodynamic filtration along inhaled air pathways, however, may be incomplete, thus allowing particles that escape more proximal deposition to remain airborne for deposition at more distal sites. Thus, the luminal surfaces of the structural components of our respiratory tracts potentially can more or less receive a virtually continuous particulate shower when atmospheres found in most environmental settings are inhaled.

The main physical processes that govern the deposition of particles are: (1) inertial impaction, (2) gravitational settling or sedimentation, (3) Brownian diffusion, and (4) interception (Brain and Valberg, 1979). Inertial impaction occurs when a particle leaves the airstream and irreversibly collides with a stationary anatomical component of the airways, as may happen when the airstream follows a bend in its path. The probability that a particle will deposit by inertial impaction is proportional to the velocity of the airstream in which it is contained, the airway branching angle through which air is flowing, and the

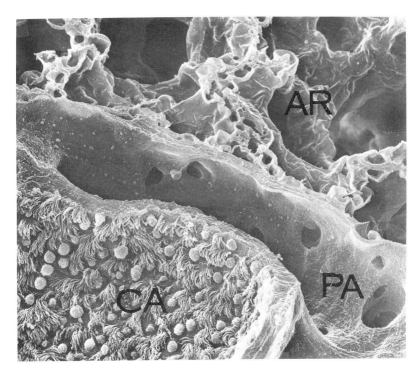

Figure 2. Scanning electron micrograph of a rat's lung. CA: a conducting airway; note the presence of ciliated cells on the surface of the airway. PA: a nearby pulmonary arteriole that carries venous blood that originates from the right ventricle. A pulmonary arterial tree accompanies the tracheobronchial tree, and it ultimately delivers blood to the network of capillaries that course the walls of the lung's terminal air sacs. AR: alveolar region consisting of individual air sacs or alveoli. The average diameter of a rat alveolus is ~86 µm.

aerodynamic diameter of the particle.[†] Relatively large particles, particles with high densities, high airstream velocities like those that occur during exercise and labored breathing, large branching angles in the conducting airways, and small airway radii all favor deposition by impaction. Generally, impaction is most pronounced in the nasopharyngeal region where the inhaled airstream passes through a complex array of nasal turbinates against which particles can collide. Impaction also plays a dominating role in particle deposition in the upper portions of the tracheobronchial tree where airflow velocities are relatively high. However, some studies with animals have suggested that impaction may also be involved in mediating particle deposition in more peripheral regions of the lungs, including the bifurcations of alveolar ducts (Warheit and Hartsky, 1990) from which the alveoli or terminal air sacs emanate. Deposition by the impaction mechanism usually involves particles that exceed 0.2 µm in diameter.

Deposition by settling or gravitational sedimentation governs the deposition pattern of inhaled particles that are in the 0.05 to 20 µm aerodynamic size range.

[†] The aerodynamic diameter is a descriptor of the particle's aerodynamic behavior. It takes into account the particle's size, mass, and shape, and it is inversely proportional to the radius of the airway structure through which the airstream is passes.

A particle's terminal settling velocity, which is that velocity caused by the force of gravity opposed by the viscous resistance of the air through which the particle is falling, is a major determinant of deposition by this mechanism. Overall, the rate of particle settling is proportional to the square of the particle's diameter. The distance a particle may transverse along a horizontal axis until deposition from the airstream occurs onto the surface of an airway structure is proportional to the terminal settling velocity and the velocity of airflow through the conducting airway. Hence, the residency time for a particle in a given airway structure can play a decisive role in terms of whether or not the particle will deposit in the structure. Predominant anatomical sites where gravitational settling of particles occurs include the mid-sized and smaller bronchioles, as well as the lung's alveolar region.

Deposition by Brownian diffusion becomes the most prevalent deposition mechanism for particles that are <0.5 µm in diameter. When these small particles enter the respiratory tract, they are displaced by the random bombardment of gas molecules, which can cause particles to collide with the surfaces along the pathway of the inhaled airstream. Particle displacement by Brownian diffusion is inversely proportional to the viscosity of air and the diameter of the particle, and it is directly proportional to the residency time of the particle in a given air space. In the nasal region, Brownian diffusion plays a major role in limiting the penetration of ultrafine particles (<0.2 µm) into the lower respiratory tract. Deposition by diffusion is also an important mechanism for small particles in conducting airways with low cross-sectional areas, and in the peripheral alveolar regions of the lung.

Finally, interception is another main deposition mechanism that pertains mainly to particles with elongated shapes, i.e., fibers. In this case, an airborne particle may travel until it reaches a small airway structure where the cross-sectional distance across the airway is less than the particle's size. Some evidence suggests that interception is the predominant mechanism for the deposition of long fibers that exceed an equivalent mass diameter of 0.1 µm (Asgharian and Yu, 1988).

The above overview of the numerous, complex factors that determine particle deposition in the respiratory tract has been greatly simplied, and it should be noted that a wide variety of other factors can also affect particle deposition in the lung, e.g., particle shape factors, charge, and hydroscopicity, gravitational force, mouth versus nose breathing, and breathing patterns. Further information about these factors and the physics of particle deposition generally is available to the interested reader elsewhere (Task Group on Lung Dynamics, 1966; Brain and Valberg, 1979; Stuart, 1984; Lehnert et al., 1985a). What should be emphasized here is that *the site(s) of deposition of particles in the respiratory tract and the amounts at which particles are deposited at particular regional sites prominantly figure into the subsequent post-depositional fates of and untoward or pathological responses to inhaled particles.*

PARTICLES IN THE CONDUCTING AIRWAYS

Local defensive mechanisms against deposited particles

The human tracheobronchial tree is composed of several orders of irregularly, dichotomously branching airways that range from as few as 8 to as many as 23 generations before reaching the alveoli, or the lung's terminal air sacs (Horsefield and Cumming, 1968). The alveoli are the anatomical structures where most gas exchange occurs. Particles that deposit in these conducting airways first encounter what has been variously called the "mucociliary apparatus," the "mucociliary ladder," or the "mucociliary escalator." The mucociliary apparatus is composed of a specialized type of epithelium that is more or less covered by mucus (Figs. 3 and 4). Its basic function is to entrap deposited particles and subsequently transport them out of the respiratory tract along with the airway mucus, thereby protecting the cells that line the airways from particles. The specialized epithelium in the larger tracheobronchial airways is populated mainly with ciliated columnar cells and goblet cells at a ratio of 5:1 (Rhodin, 1966). The main function of the ciliated cells is to propel overlying mucus in the cephalad direction toward the pharynx. Each of the ciliated cells projects ~200 cilia from its apical surface into the airway lumen. The cilia shafts in the larger airways are ~6 µm long, but their lengths progressively decrease as airways become increasingly smaller (Rhodin, 1966; Breeze and Turk, 1984). The beat frequency

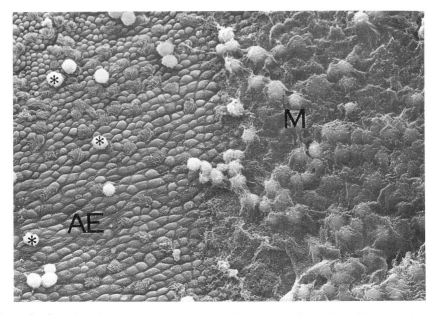

Figure 3. Scanning electron micrograph of a portion of the rat's trachea. M: mucus layer. AE: airway epithelial cells that are visible in a region where the mucus has been removed, probably due to the tissue processing procedure. *: free cells, at least some of which are airway macrophages. The airway macrophages have an average volume of ~1000 µm^3 or an average diameter of ~12 µm.

Figure 4. Scanning electron micrograph of epithelial cells that populate the rat's trachea. CC: ciliated cells; GC: goblet cells. As indicated in the text, the ciliary shafts are ~6-μm long.

of the cilia, which ranges from ~15 to 26 Hz, increases from the peripheral airway generations to the trachea (Iravani and van As, 1972). In the laboratory rat, which is frequently used to assess the toxicity of inhaled dusts, the ciliary beat frequency is ~17 Hz in the main lobar bronchi, but only ~7 Hz in the more distal airways (Iravani, 1967). The mucous lining component of the mucociliary apparatus consists of two layers. The lower sol phase or hypophase phase, which is less viscous than the upper layer, bathes the airway epithelial cells, whereas the higher viscosity of the upper layer, or gel phase, serves as an efficient trap for depositing particles and probably helps prevent particles from gaining direct access to the underlying epithelial cells. The actual propulsion of mucus is related primarily to the biphasic pattern of ciliary motion. The cilia beat in one plane with a fast effective stroke that moves in the cephalad direction followed by a slower, retrograde recovery stroke (Sleigh, 1977). The cilia sweep through the periciliary fluid, or hypophase, of the mucous lining during the effective stroke as rigid, slightly curved rods that bend mainly at their bases. The apices of the ciliary shafts project into the underside of the epiphase layer of mucus when perpendicularly extended during the effective stroke. Coupling of the ciliary action to the epiphase is augmented by the tips of the cilia, which ultrastructurally appear to be "claw-like" (Jeffrey and Reid, 1975). The slower recovery stroke proceeds in the less viscous hypophase in the opposite direction of the effective stroke. Cranially directed movement of the periciliary fluid is also likely under the influence of ciliary motion, but its velocity is thought to be less than that of the epiphase (Sleigh, 1977). Thus, particles that penetrate into the sol phase of mucus may not be removed from the airway structure as readily as particles that are on or in the gel phase.

The mucous lining, or what has also been called the "mucous blanket," is composed of ~95% water, and it contains an array of constituents derived mainly from tracheobronchial and submucosal glands, goblet cells, and alveolar lining materials, including surfactants. Approximately 10 to 100 ml of the mucous lining material is transported up the respiratory tract daily for disposal either by swallowing or expectoration. Based on observations made with animals and humans, it has been concluded that the rate at which mucus is transported in the conducting airways toward the pharynx depends on its location in the tracheobronchial tree. In the dog, mucus moves at a rate of ~0.4 mm/min in small bronchioles and ~13 mm/min in the trachea (Asmundsson and Kilburn, 1970). Similarly, particle clearance data have indicated that the mucus velocity in the terminal bronchioles of the human is ~0.1 to 0.6 mm/min (Morrow et al., 1967), whereas the flow rate of mucus in the human trachea ranges from ~5 to 20 mm/min (Yeates et al., 1975; Yeates et al., 1981; Wood et al., 1975).

The quality and viscoelastic properties of the airway lining material varies according to its location in the tracheobronchial tree, with the viscosity of the mucus being less in the more peripheral airways (Sleigh, 1977). Also, instead of simply being a protective "mucous blanket" *per se*, the mucus lining may actually be focally discontinuous (van As and Webster, 1974; van As, 1980. and occur as "islands" or "sheets." This arrangement would be expected to have important consequences in airway defense against deposited particles in that epithelial cells at sites where overlaying mucus is absent would not have a physical barrier against particles (i.e., mucus) that is otherwise present elsewhere. Along the same line, discontinuous mucus would be expected to decrease regionally the mucociliary rate of particle transport up the respiratory tract, and, hence, further increase the probability of particle–epithelial cell interactions. On the other hand, some investigators have found no evidence for discontinuity in the mucous layer, although some observations indicate that the gel phase, at least, is not a prominent component of the airway lining fluid in the more peripheral conducting airways (Gil and Weibel, 1971; Luchtel, 1982). Instead, the lining fluid of the peripheral terminal and respiratory bronchioles that lead to the alveolar ducts consists of a serous fluid of relatively low viscosity that is continuous with alveolar lining fluid.

In addition to the mucociliary apparatus, epithelial cell defense against particles is also afforded by the presence of macrophages that are present on the surfaces of the conducting airways, Figures 3 and 5. These cells—which in healthy lung evidently are alveolar macrophages (to be discussed) that have translocated from the lung's alveoli via the mucociliary escalator and which are capable of ingesting or phagocytizing freshly deposited particles (Geizer et al., 1988; Lehnert et al., 1990b)—are found atop the mucous lining of the airways, submerged in the mucous lining, and beneath the mucous lining in close apposition to underlying airway epithelial cells (Sorokin and Brain, 1975; Brain et al., 1984). The cephalad transport rate of at least some fraction of the airway

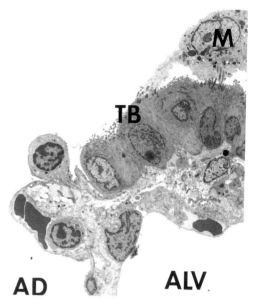

Figure 5. Transmission electron micrograph of macrophage (M) that is on the surface of a terminal bronchiole (TB). The macrophage is in intimate contact with underlying cilia. AD: alveolar duct from which alveolar sacs anatomically project. ALV: an alveolar space.

macrophages probably occurs passively with mucous flow, although histologic evidence has suggested that the residency time(s) of some of these phagocytes may be relatively prolonged (Sorokin and Brain, 1975).

The numbers of intra-luminal phagocytes can be substantially increased following the deposition of aerosols of some types of particles in the conducting airways. This response initially involves mainly the recruitment of another phagocytic cell type (called polymorphonuclear leukocytes or PMN) from blood vessels present in the airways. The types of materials that have been shown to elicit this type of response are diverse, e.g., simple colloidal carbon (Bowden and Adamson, 1984), vegetable dusts (Kilburn et al., 1973), aldehyde-treated carbon particles (Kilburn and McKenzie, 1978), endotoxin (Venaille et al., 1989), cigarette smoke (Hulbert et al., 1981b), and vapor-free cigarette smoke particles (Kilburn et al., 1975). This rapid response, which in all likelihood can occur following a sufficient deposition of all types of cytotoxic particles, has presumably evolved to provide the surfaces of the conducting airways with additional cells as needed to phagocytize high local burdens of deposited particles. The intra-luminal influx of PMN occurs prior to the free cell response in the alveolar space compartment by several hours (Bowden and Adamson, 1984) (to be discussed). As in the alveolar region, the influx of additional phagocytes into the airways is thought to be due to PMN responding to a gradient of chemotactic factor(s), which provide directional signals to these mobile phagocytes so that they migrate from the the peribronchiolar vasculature onto the airway epithelial surface. The source(s) of the chemotactic factor(s) has not been well-investigated. Regardless, in response to chemotactic signals, the PMN initially migrate into the peribronchiolar spaces through junctions between the endothelial cells present on the surfaces of small blood vessels, and then they continue to migrate between the airway epithelial cells in seemingly single file through "solitary" pathways onto the epithelial surface (Hulbert et al., 1981a). Within a day or two following particle deposition on the airways, the initial influx of PMN can be further supplemented by the migration from the vasculature of another type of mononuclear phagocyte, blood monocytes, that are also capable of ingesting particles (Bowden and Adamson, 1984); this mononuclear response, however, is generally not as pronounced as is the earlier PMN response.

Particle-airway epithelial cell interactions

Some types of particles may escape the protective actions of airway secretions and mucociliary transport in the conducting airways and gain direct access to the epithelial cells that line the conducting airways. Outcomes of such interactions may include particle-induced damage to the epithelial cells, cell killing, and/or the penetration of the particles into subepithelial sites. Particle–airway-epithelial-cell encounters may be especially pronounced in the more peripheral conducting airways. Asbestos fibers, for example, have been found in the walls of airways in chrysotile miners, with the more peripheral respiratory bronchioles having higher fiber concentrations than the more proximal, larger airways (Churg and Wright, 1988). Moreover, following the inhalation of chrysotile asbestos by rats, McGavran and Brody (1989) found no evidence of an enhancement of cell proliferation in the trachea after fiber deposition, whereas cell proliferation (as an index of the biological response to the fibers) was observed in airways 8 to 12 mm below the carina (where the trachea bifurcates into the mainstem bronchi), with the most substantial labeling occurring in more distal regions that contained terminal bronchioles. Although fiber deposition apparently can occur in all airways (Morgan et al., 1975), the possibility that these findings may have been due to a preferential deposition of fibers in the more peripheral conducting airways (Brody and Roe, 1983), however, has not been ruled out.

Nevertheless, the anatomical pattern of fiber penetration into smaller versus larger airway walls may reflect differences in the quality and viscoelastic properties of proximal and peripheral airway lining fluid. The lining fluid atop the epithelial cells in the terminal and respiratory bronchiolar region lacks a gel phase and consists of a serous fluid of relatively low viscosity, as previously described. Also as discussed earlier, another factor favoring the penetration of particles across the epithelial barrier of the more peripheral airways more so than the epithelial cells lining the larger conducting airways is that the rate of particle clearance in the former is substantially slower than in the latter (Morrow et al., 1967; Lippmann et al., 1980). Thus, the residence time of particles in the peripheral airways may be relatively prolonged, which, in turn, increases the likelihood of direct particle encounters with epithelial cells. Although the more peripheral airway epithelial cells are phenotypically different from the epithelial cells in the larger airways (more cuboidal vs. pseudostratified and columnar), and some evidence indicates different cell types can be more or less sensitive to some types of particles (Haugen et al., 1982), the types of cells in the larger and the smaller airways may not account for a preferential uptake of particles in the latter. Numerous investigators have commonly reported that asbestos fibers can penetrate the tracheal epithelium *in vitro* (e.g., Topping et al., 1980; Haugen et al., 1982; Mossman and Craighead, 1979). Epithelial cells in larger airways, accordingly, are clearly capable of internalizing (or endocytizing) fibers. Unlike the *in vivo* setting, however, a protective, intact mucous barrier and airway phagocytes would no longer be in place under the experimental conditions used in tracheal explant studies.

How particles initially gain contact with the epithelial cells for subsequent penetration into the airway walls is unclear. One possibility is that airway lining fluid–particle interactions may actually favor particle–airway-epithelial-cell contacts due to the surface tension lowering activities of a surfactant layer on airway mucus (Gehr et al., 1990). With this mechanism, deposited particles become coated by surfactant that is a constituent of airway lining fluid, and physical forces cause the particles to become submerged in the fluid in a manner that positions the particles in close apposition with airway epithelial cells. The actual process of transepithelial uptake and transport of particles into the airway's submucosa once they have come in contact with the airway epithelial cells also has not yet been fully characterized. An understanding of such mechanisms in the context of the qualities of particles is of obvious importance inasmuch as the penetration of particles into airway walls may be fundamentally related to the development of bronchogenic carcinoma (Churg and Stevens, 1988).

Several lines of evidence have suggested that the penetration of particles into airway walls involves several cell-mediated and acellular mechanisms. These include the endocytosis or engulfment of particles by airway epithelial cells (Watson and Brain, 1979; Mossman et al., 1978; Mossman et al., 1977), the passage of deposited agents between epithelial cells (Mossman et al., 1978; Mossman et al., 1977; Richardson et al., 1976), and the migration of particle-containing airway macrophages across the epithelium into subepithelial sites (Stirling and Patrick, 1980). The most important pathways for insoluble particles is probably epithelial cell engulfment (Watson and Brain, 1979) and transcellular transport. The endocytosis of particles by airway epithelial cells is related to particle size. The process has been demonstrated for submicron size particles (Watson and Brain, 1976), whereas no evidence of epithelial penetration by endocytosis or other processes for that matter was obtained in a study that examined the fate of 7.9 μm size spheroidal particles in the airways of guinea pigs (Valasquez and Morrow, 1984). We have observed polystyrene microspheres as large as 2 μm in diameter, however, in rat airway epithelial cells following their intra-pulmonary deposition (Lehnert, 1990b), and some types of fibers can be engulfed as well. Virtually no information is currently available as to how the surface characteristics of different materials may affect their endocytosis by the airway epithelial cells.

It is also possible that particle–epithelial cell encounters may result from the release (Riley and Dean, 1978) or perhaps the transfer of particles (Aronson, 1963) from particle-containing macrophages that are in close apposition with the airway epithelial cells. Reports of prolonged retentions of particles in the airways beyond what would otherwise be expected by totally efficient mucociliary transport (e.g., Stirling and Patrick, 1980; Gore and Patrick, 1978) may perhaps be explained by the translocation of a fraction of the deposited particles into slowly clearing or perhaps non-clearing subluminal sites. The extent to which particles may penetrate into the airway walls apparently can be affected by other factors aside from the physicochemical characteristics of the particles that are transported. For example, cigarette smoke has been demonstrated to increase the penetration of asbestos fibers into the airway walls of guinea pigs (McFadden et al., 1986). One important outcome of the penetration of mineral fibers, including asbestos, across the epithelial barrier of the conducting airways is fibrosis (excessive fibroblasts and collagen) in the interstitium of the airway walls (Churg et al., 1985).

Solublization of particles in the conducting airways

The above processes are primarily applicable to particles that are relatively insoluble *in vivo*. Soluble particles, on the other hand, can undergo dissolution upon deposition in the tracheobronchial tree. Particle-derived solutes that do not chemically react with constituents in the mucous lining or airway epithelial cells, or otherwise adsorb to mucous gel-phase components, can be removed from the luminal surface by transepithelial pathways that result in the translocation of solutes into the airway interstitium. Once in the interstitium, the solutes can pass into the vasculature and be further transported in the blood for distribution to other body sites. The epithelial pathways consist of the transcellular transport of molecules in epithelial cell endocytic vesicles and the passage of solutes through intercellular junctions (e.g., Bhalla and Crocker, 1986; Ranga et al., 1980). Some evidence suggests that pinocytotic vesicular transport may be of greater importance for the transepithelial passage of more lipid-soluble molecules across the epithelial barrier (Ranga et al., 1980), whereas the transepithelial passage of more hydrophilic materials is mainly through the paracellular route, i.e., between cells (Boucher, 1980). Lower molecular weight solutes generally clear the respiratory tract faster than high molecular weight solutes, suggesting that the rate of transepithelial clearance of a material is related to the tight junctional dimensions of the epithelial cells. The inhalation of a variety of toxic agents such as nitrogen dioxide, ozone, and cigarette smoke have been shown to have profound effects on the absorptive properties or permeability characteristics of the epithelial barrier in the conducting airways. Alterations in the airway epithelium in response to these substances result in increases in the translocation of soluble materials across the airway epithelium (e.g., Bhalla and Crocker, 1986; Ranga et al., 1980; Bhalla et al., 1986; Hulbert et al., 1981a; Boucher et al., 1980). Solutes that are not transported across the epithelial barrier probably are removed together with the routine removal of airway lining fluid. It should be noted that clearance of soluble particles by dissolution and diffusive transport is not restricted to the conducting airways. These same processes can also occur throughout the highly vascularized lower respiratory tract.

Kinetics of particle clearance from the conducting airways

The removal of relatively insoluble particles that land in the tracheobronchial tree has generally been assumed to be complete as of about 24 hrs after they have been deposited. Some recent experimental evidence, however, has indicated, that the process may be of longer duration (Stahlofen et al., 1986; Smaldone et al.,

1988). The tracheobronchial component of lung clearance, which is frequently referred to as "rapid" or "early" phase clearance, has typically been mathematically modeled simply as a first order process or as the sum of two or more negative exponential components (Morrow and Yu, 1985). Other modeling approaches have considered time dependency in the clearance of particles from the conducting airways by applying a power function to experimentally-derived particle retention data (Morrow and Yu, 1985).

Expressing the kinetics of tracheobronchial clearance as the sum of several components with varying rates may have corresponding physiologic bases (Oberdörster, 1988). For example, the rates of particle removal from larger, more proximally positioned airways are higher than the rates of particle clearance from more distally located airway sites (Morrow et al., 1967; Yeates et al., 1981; Wilkey et al., 1980). These observations are consistent with the previously cited findings that the rates of mucociliary transport in larger airways are faster than those in the smaller bronchioles. These differing rates, in turn, appear to be related to the more rapid ciliary beat frequencies in the larger airways relative to the slower ciliary beat frequencies in the smaller, more peripheral airways. Another factor that may account for differences in particle clearance rates from the smaller and larger airways is the lack of a high viscosity epiphase in the fluid lining smaller bronchioles, which would be expected to decrease the efficiency of the action of cilia on mucus. A further explanation for slower mucociliary transport in the more peripheral airway generations may be due anatomical airway branching. Transporting mucus may encounter numerous tributary bronchial openings as it is propelled cranially up the tracheobronchial tree. These openings offer obstructions to the pathway of mucous flow, and mucus must be diverted around them for continued passage up the respiratory tract. As revealed by Hilding (1959), retardations in mucous flow and even stasis of mucous flow can occur where smaller airways enter larger airways. Some of these factors, as well as perhaps a relatively slow removal of particle laden airway phagocytes and/or the uptake of particles by epithelial cells and the penetration of particles into subepithelial sites, may contribute to the previously mentioned observations that tracheobronchial clearance is incomplete as of 24 hrs after particle deposition in the airways.

Aside from the mucociliary transport of particles, coughing is also an effective mechanism for eliminating mucus and particles from the larger conducting airways, especially in individuals with airway hypersecretions (Kohler et al., 1986; Camner, 1981). The cough mechanism is initiated with a deep inspiration which is followed by a forced exhalation against a closed glottis. Immediate decompression of the trachea and larger bronchi following the sudden opening of the glottis causes a high transbronchial pressure with an immediate reduction in diameter of the larger airways. Under these conditions, the diameters of the trachea and bronchi may be reduced to a fraction of their original calibers. As a result, the airflow through these structures can approach sonic velocity and thereby apply considerable shearing force to mobilize mucus. Coughing is considered to be primarily effective in mobilizing mucus from the first few generations of the larger airways (Mossberg, 1980), although some experimental modeling of coughing suggests that its effects may be significant down to the 12th generation of the conducting airways (Scherer, 1981). While coughing can enhance airway clearance, some evidence suggests that repetitive coughing may impair clearance mediated by the mucociliary apparatus (Smaldone, 1986; Smaldone et al., 1979).

PARTICLES IN THE PULMONARY-ALVEOLAR REGION

Defense against particles by alveolar phagocytes

The adult human's pulmonary region contains ~300 million alveoli or air sacs that collectively provide a large surface area (about the size of a tennis court). Contained in these alveoli is a cell type called the alveolar macrophage (AM) (Figs. 6 and 7), which reside in contact with the surfactant lining of the alveolar epithelium (Weibel and Gil, 1968). These phagocytic cells are the first line of cell-mediated defense against particles that deposit in the alveolar region. On average, each alveolus probably contains at least one AM (Lehnert et al., 1985c), but the number of these cells in individual alveoli can certainly be higher, depending on the exposure history of the individual. Using lung tissue from a subject who smoked, for example, Parra and co-workers (1986) observed an average of 24 AM per alveolus in each of two alveoli that were studied extensively by serial sectioning and three-dimension reconstruction, which is consistent with other observations that the lungs of smokers have more AM than nonsmokers (Plowman, 1982).

A host of diverse materials are readily phagocytized by the AM shortly after deposition and, in this manner, the AM serve to remove particles from the extracellular fluid lining the alveoli and thereby limit particle encounters with sensitive alveolar epithelial cells. Under normal conditions, the AM represent ~3 to 5% of all the cells in the gas exchange region of mammalian species, including humans (Lehnert et al., 1985c; Pinkerton et al., 1982), and the total number of AM in the lung is generally closely matched to the number of alveoli present, as previously indicated (Ferin, 1982b; Lehnert et al., 1985c). The actual number of AM in an individual's lung under steady state conditions, however, is related to the quality of air that has been inhaled. In the healthy human lung, the AM constitute ~85% of the lung free cells washed from the lungs by a technique called bronchoalveolar lavage, with lymphocytes and PMN composing the remaining retrieved cell types (Reynolds, 1989). Usually, a fraction of the AM undergo transport from the lung as airway macrophages on a continual basis via the mucociliary apparatus (Lehnert et al., 1989, 1990b). Macrophages that are removed by this process are believed to be homeostatically replenished by the migration of blood monocytes[†] into the alveoli (Bowden and Adamson, 1982; Bowden and Adamson, 1980); by the local proliferation of resident AM (e.g., by cell division) (Tarling and Coggle, 1982; Shellito et al., 1987); and perhaps by the migration of pulmonary interstitial macrophages into the alveoli (Bowden and Adamson, 1980; Bowden and Adamson, 1978). The number of AM that are removed from the adult human lung and replenished daily is not known. Estimates made with the laboratory rat broadly range from a virtually complete turnover of the resident AM population on a daily basis (Spritzer et al., 1968) to

---

[†] A blood monocyte presumably can mature further in the alveoli to become an alveolar macrophage.

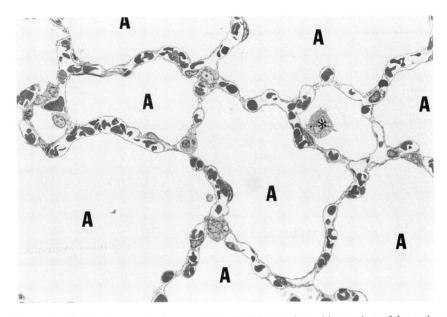

Figure 6. Light micrograph of several alveoli (A) seen in a thin section of lung tissue. *: an alveolar macrophage that is present in one of the alveoli.

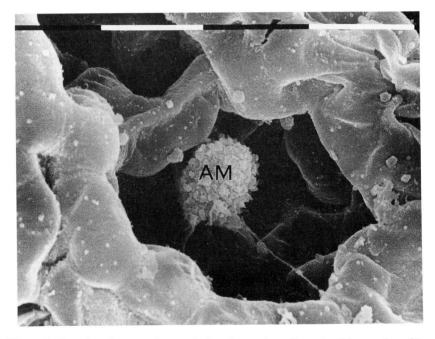

Figure 7. Scanning electron micrograph that gives a three dimensional impression of how an alveolar macrophage (AM) appears in an alveolus. Several filapodia are observed to be projecting from the cell. These extensions are thought to be involved in the migration of the AM on the alveolar surface and in the surveillance of particles for ingestion.

the removal of as little as ~1.5% of the lung's population of AM daily (Shellito et al., 1987; Lehnert and Morrow, 1985a).

The primary function of the AM is to phagocytize or engulf the particles upon deposition. This process requires particle-AM encounters as the first step (Fig. 8). These encounters may occur by just the random mobility of AM on the alveolar surface, but theoretical considerations of the efficiency or rate at which AM may encounter particles by this process alone strongly suggest that a directed migration of AM to particles by locally generated chemotactic factors is necessary to account for the rapidity by which deposited particles are phagocytized (Fisher et al., 1988). Newly deposited burdens of particles commensurate with typical environmental exposures are virtually all phagocytized by AM shortly after deposition, i.e., within hours (Lehnert and Morrow, i985a).

Numerous *in vitro* studies have shown that the mobility of the AM can be provided directionality by chemoattractant factors. Some of these chemotactic factors are formyl peptides analogous to substances produced by bacteria (Marasco et al., 1984), complement-derived components such as C5a (Richards et al., 1984), and some less than well characterized constituents in alveolar extracellular fluid lining (Schwartz and Christman, 1979). Other chemoattractants can be produced by various immune cell types (Bitterman et al., 1981; Kagan et al., 1983; Miller et al., 1980). Several types of particles (e.g., asbestos fibers, fly ash from coal combustion, some aluminum and zirconium compounds, carbonyl iron) are capable of activating serum complement, and thereby generate chemotactic factors, including C5a, upon interaction with complement components present in alveolar lining fluid (Warheit et al., 1985; Saint-Remy and Cole, 1980; Hill et al., 1982; Wilson et al., 1977, Ramanathan et al., 1979; Warheit et al., 1988; Robertson et al., 1976; Kolb et al., 1981). Whether the abilities of particles to activate complement upon deposition is due to direct interactions with the various complement components in the alveolar fluid or if activation of complement is secondary to the initial adsorption of complement-activating proteins, e.g., immunoglobulins M and G (IgM and IgG), onto the particles (Valerio et al., 1986, 1987) remains unclear. In addition to the above chemoattractants, a wide variety of phagocytic stimuli can cause AM to elaborate chemotactic factors such as leukotriene B4 and macrophage inflammatory protein 2 (Migliorisi et al., 1987; Driscoll et al., 1993) to recruit more phagocytes from extra-alveolar space sources to the site where the particles are deposited.

Particle deposition in the alveoli can cause a relatively prompt expansion in the size of the AM population, and, in a manner similar to what has already been described for the conducting airways, an influx of PMN onto the alveolar surface via the local generation of chemotactic factors (Fig. 8). Such a particle-associated increase in lung phagocyte numbers is frequently referred to as the "free cell response" primarily because the response to particle deposition in animal studies is frequently measured quantitatively by counts and differential analyses of cells that are susceptible to being washed from the lungs by bronchoalveolar lavage. The magnitude of the free cell response is dependent in part on the particulate load of a material deposited, with more phagocytes being recruited as the burden

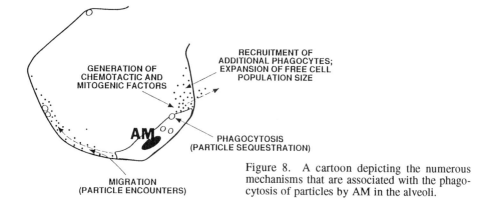

Figure 8. A cartoon depicting the numerous mechanisms that are associated with the phagocytosis of particles by AM in the alveoli.

of particles deposited increases (Brain, 1971). Particle size also is an important determinant of the magnitude of the lung's free cell response. Smaller and larger particles delivered to the lung in equivalent masses result in greater and lesser changes in the free cell population size, respectively (Brain, 1971; Adamson and Bowden, 1981).

Typically, the lung's response to the acute deposition of particles is initially characterized by an influx of PMN from the pulmonary vasculature onto the alveolar surface (e.g., Bowden and Adamson, 1978; Gross et al., 1969; Adamson and Bowden, 1978; Lipscomb et al., 1983; Lehnert et al., 1985b) (see Fig. 9). These PMN migrations occur in response to chemotactic stimuli generated upon particle deposition and/or by chemotactic factors released by resident AM (e.g., leukotriene B_4, macrophage inflammatory protein 2, tumor necrosis factor, interleukin-8) that have already encountered and phagocytized particles (Kazmierowski et al., 1977; Dauber and Daniele, 1978; Hunninghake et al., 1978; Fels et al., 1982; Figari and Palladino, 1987; Ming et al., 1987; Driscoll et al., 1993) (see Fig. 8). Shortly after this early response, which begins within hours after the particles are deposited, AM numbers increase primarily because of a migration of blood monocytes from the pulmonary vasculature into the alveoli. Unlike the PMN response, which is usually immediate and transient (lasting for only a few days or so), the expansion of the AM population size is longer lived. Using laboratory mice, this more persistent macrophagic response has been shown to be biphasic (Bowden and Adamson, 1978; Adamson and Bowden, 1978). The early phase of increases in AM numbers, which is mainly attributable to the recruitment of blood monocytes, is subsequently sustained by the later migration of interstitial macrophages into the alveoli. To what extent influxes of interstitial macrophages or an intra-alveolar replication of AM that are present at the time of particle deposition may play in elevating AM numbers during the free cell response requires further study. Regardless, while the overall magnitude of the free cell response to particles is clearly influenced by the intensity of particle deposition, numeric increases in alveolar phagocytes does not appear to be finely tuned by the deposited particulate burden (Lehnert et al., 1985b). Except in instances where deposited burdens of particles exceed the lung's ability to mount a free cell response with enough phagocytes to phagocytize the particles (Adamson and Bowden, 1981), which can occur under some extraordinary experimental

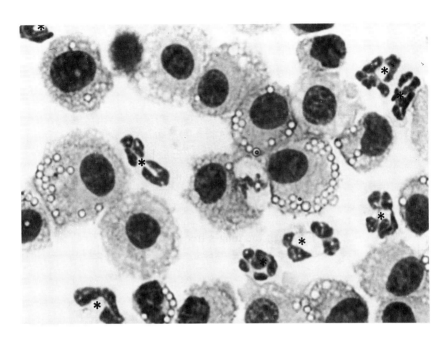

Figure 9. Light micrograph of a cytocentrifuged slide preparation of lung free cells lavaged from a rat's lung one day after the intratracheal instillation of polystyrene microspheres. The larger, mononucleated cells are lavaged macrophages, many of which have phagocytized the 2-µm diameter microspheres. The smaller cells (*) with lobulated-segmented nuclei are polymorphonuclear leukocytes (PMN) that were recruited into the lung after particle deposition. Normally, virtually all of the cells that are lavaged from the lungs of healthy rats (specific-pathogen-free) are macrophages. In the case shown in the figure, the PMN represented ~30% of all of the lavaged cells.

conditions, the recruitment of macrophages and PMN appears to be exaggerated so that more phagocytic cells are made available than are required to sequester the particles (Lehnert et al., 1985b). This generalization is illustrated by Figure 9 in which most of the PMN that were recruited within 24 hrs after the deposition of particles into a rat's lung are particle-free.

It should be noted that the alveolar deposition of particulate agents does not necessarily result in a readily detectable expression of the free cell response as described above, at least in terms of elevations in AM numbers. Instead, deposition of some types of materials can lead to a decrease in the numbers of AM present in the lung. Particle-associated decreases in AM numbers have been observed with several types of materials, including asbestos (Tucker and Frank, 1976; Harrington, 1976; Jaurand et al., 1978), silica (Allison et al., 1966; Civil and Heppleston, 1979), Mn_3O_4 and MnO_2 (Adkins et al., 1980; Bergstrom and Rylander, 1977), $CdCl_2$, CdO, and Cd fumes (Koshi et al., 1978; Gardner et al., 1977; Bouley et al., 1977), Sb fumes (Koshi et al., 1978), and lead sequioxide (Bingham et al., 1968). Decreases in AM numbers following the deposition of these materials is likely due to their direct cytotoxic effects. The cytotoxic actions of these agents may also lead to the destruction of newly recruited cells, and/or these materials may inhibit the migration of new mononuclear phagocytes into the alveoli under some conditions (Gardner, 1984).

The phagocytosis of particles by AM (Fig. 11) and PMN is viewed as having both specific and nonspecific bases. In the former case, or opsonin-mediated[†] phagocytosis, the surface of a particle is coated with an opsonin, which in turn serves as a ligand that can attach to sites or receptors on the phagocyte's outer membrane. In essence, opsonization of a particle can make particles that otherwise may not be ingested more palatable to a phagocyte. Attachment of the particle to a phagocyte via bridging by the opsonin is the recognition step for phagocytosis to proceed (Fig. 10). Several immune and nonimmune opsonins exist in the alveolar lining fluid. The most important of these is probably immunoglobulin G (IgG), given the relative abundance of this immunoglobulin present in lavage fluid (Reynolds, 1988; Reynolds, 1989). (IgM is also present in lung fluid, but only in small amounts; Reynolds, 1989.) The source of the IgG in the respiratory tract appears to be due to its steady transudation onto the alveolar surface from the lung's vasculature and to its local production by lung B lymphocytes, which are involved in mediating the so called humoral immunity to antigenic materials that deposit in the lung. The binding of IgG to antigenic sites on a particle's surface involves the $F(ab')_2$ portion of the immunoglobulin molecule, whereas the Fc portion of IgG is the site on the immunoglobulin that makes it useful as an opsonin for promoting the binding of a particle to phagocytes that have Fcγ receptors present on their surfaces. Binding of ligand to the cell's surface receptors causes the plasmalemmal membrane to progressively creep over a particle's surface until the particle is internalized; this process has been modeled as a "zipper" mechanism (Griffin et al., 1975).

The reaction of particle or antigen-bound IgG or IgM with components of complement provides another mechanism for particle opsonization. Receptors for major fragments of the third component of complement, e.g., C3b, generated from the antigen-complement reaction is present on human AM (Daughaday and Douglas, 1976). These receptors mainly play a role in enhancing the stability of particle binding to phagocytes but not particle ingestion unless the cells are stimulated or activated (Bianco et al., 1975; Ehlenberger and Nussenzweig, 1977). In addition to Fcγ and complement receptors, studies have also demonstrated that AM have receptors for IgA (FcαR) (Sibille et al., 1989) and IgE (FcϵR) (Boltz-Nitulescu and Spiegelberg, 1982; Boltz-Nitulescu et al., 1982; Joseph et al., 1980). Although the role(s) these receptors may play in human defense against deposited particles requires further study, it is noteworthy that human monocytes and AM can phagocytize particles opsonized with IgA (Maliszewski et al., 1985; Reynolds et al., 1975). Still other constituents in the complex fluid that lines the alveoli that may promote particle binding and phagocytosis include albumin (Hof et al., 1980; Verbrugh et al., 1982), fibronectin or fibronectin fragments (Saba et al., 1973; Czop et al., 1982a,b), and probably additional as yet to be determined substances.

[†] An opsonin is a material that binds to a foreign body, thereby facilitating phagocytosis. Immune opsonins are also called immunoglobulins (abbreviated Ig) or antibodies, and they possess specific binding sites that match to specific binding sites on the foreign body's surface. This relationship is known as an antigen–antibody relationship.

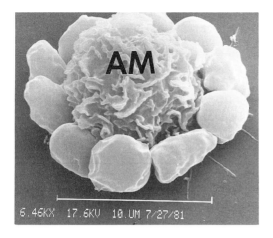

Figure 10. After particles are encountered, the next step in the process of phagocytosis is particle binding. This scanning electron micrograph shows an AM that bounds several erythrocytes *in vitro* that were opsonized with IgG that specifically recognized an antigen on the surfaces of the erythrocytes. The attachment of the erythrocytes around the AM was mediated by the AM's Fcγ receptors. Phagocytosis of the bound particles was inhibited by performing the binding reaction at 4°C. Otherwise, many, if not all, of the bound erythrocytes would have been phagocytized within minutes.

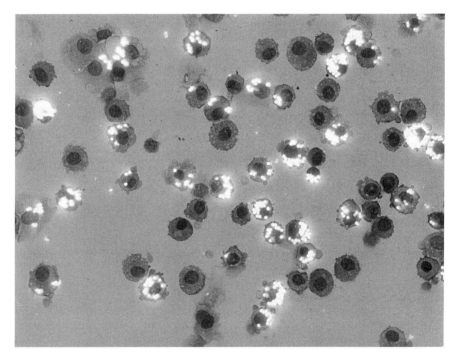

Figure 11. The bulk of retained lung burdens of particles are usually contained in alveolar macrophages that have phagocytized the particles. This light-fluorescence micrograph shows the presence of 2-μm diameter, fluorescent microspheres in alveolar macrophages that were lavaged from a rat's lung 7 days after the particles were originally deposited. By this postdepositional time, the polymorphonuclear leukocyte response had subsided.

In the case of opsonin-independent phagocytosis, nonspecific sites or generally poorly characterized sites on the surface of the phagocyte mediate the attachment and ingestion of some types of particles in the absence of specific opsonins. The lack of an absolute requirement of opsonins for the phagocytosis of widely diverse particulate materials (e.g., polystyrene microspheres, tanned erythrocytes, colloidal gold, some microorganisms) by phagocytic cells has been well demonstrated *in vitro* (van Oss and Stinson, 1970; van Oss and Gillman, 1972; Valberg et al., 1982; Lehnert and Ferin, 1983; Lehnert and Tech, 1985). The ability of lung phagocytes to endocytize a wide range of foreign materials may be especially important for the phagocytosis of some types of particles that land in the alveoli, although it can be argued that many if not all relatively insoluble particles that deposit on the alveolar surface may become nonspecifically opsonized by one extracellular fluid lining constituent or another. Numerous *in vitro* studies have demonstrated the non-antigenic binding of IgG onto both abiotic particles and bacteria markedly enhances their ingestion by phagocytes are numerous (e.g., van Oss and Stinson, 1970; van Oss and Gillman, 1972; Absolom et al., 1982).

Several factors affect the rate or extent to which particles are engulfed once they are bound to the AM's surface. Opsonin-mediated phagocytosis is limited by the amount of an opsonin that is bound to a particle, by the availability of functional receptors for the opsonin on the phagocytes' surfaces, and by the functional well being of a cell as a professional phagocyte. In the case of opsonin-independent phagocytosis, it appears that the engulfment process is related to the relative hydrophobicities of the particle and the phagocyte; more hydrophobic particles are more readily ingested (e.g., van Oss and Stinson, 1970; van Oss and Gillman, 1972; Absolom et al., 1982). Particle size is also an important factor influencing particle ingestion rate. Spherical particles 1.5 to 3 µm in diameter are more rapidly phagocytized than are smaller or larger particles (Holma, 1967; Hahn et al., 1977). As shown by Allison (1973), relatively short fibers with a length of <5 µm can be completely phagocytized by macrophages, whereas much longer fibers (>30 µm) can not be totally engulfed, i.e., part of the fibers' shafts continued to remain extracellular. The inability of AM to phagocytize long fibers may be the most important underlying factor for the observations that longer fibers are selectively retained in the lung (e.g., Morgan et al., 1978; Roggli and Brody, 1984) and that they do not appear to be as susceptible to removal by AM-mediated clearance (to be discussed).

In vitro studies have indicated that the phagocytosis of particles by a macrophage can be influenced by the macrophage's prior phagocytic history (Werb and Cohn, 1972; Schmidt and Douglas, 1972; Daughaday and Douglas, 1976). One explanation for a decrease in the phagocytosis of particles after prior particle ingestion is that the receptors required for mediating phagocytosis are internalized with the cell membrane that is required to encapsulate engulfed particles. Another factor that may impact on the phagocytic activities of the AM is related to their phagocytic capacity, or the volume of a material they can maximally engulf. Information about the phagocytic capacity of AM *in situ*, however, is limited, and it may not be reliably predicted from *in vitro* analyses. Lavaged AM containing hundreds of microspheres 2 µm in diameter have been observed after rats were administered high lung burdens of these particles (Lehnert, 1990a), whereas far lesser numbers of the microspheres are phagocytized *in vitro* by AM in monolayer even when particles are present in high abundance (Lehnert and Tech, 1985). Some investigators have also investigated how particle deposition in the lung can alter subsequent AM phagocytic function (e.g., Brain and Cockery, 1977; Levens et al., 1977; Kavet and Brain, 1977; Lehnert and Morrow, 1985b). Results from these studies collectively indicate that particle-AM interactions that occur *in vivo* can modify the particle-AM interactions that occur following a second particle challenge *in vitro*. When a second particle challenge is given to AM shortly after an initial episode of particle deposition in the lung, phagocytosis appears to be depressed. At later times after initial particle deposition, phagocytic function of AM is either restored or even enhanced. Similar results have been obtained when the phagocytic function of AM was assessed *in vivo* following particle deposition in the lung. Factors other than a particle-associated compromise in AM phagocytic activities, however, could have contributed at least in part to the results obtained in these latter *in vivo* studies. Particles initially deposited on the alveolar surface, for instance, may become nonspecifically opsonized by alveolar fluid constituents such as IgG and thereby deplete opsonin availability for coating particles subsequently deposited. Along the same line, alveolar-fluid-lining components required for the formation of chemotactic factors could be consumed by a prior bout of particle deposition. These possibilities require further experimental examination.

Detrimental aspects of the phagocytic protective mechanism

When AM and PMN phagocytize IgG-opsonized microorganisms, they undergo what is called a "respiratory burst." During the respiratory burst, oxygen is enzymatically converted by the phagocytes to superoxide anion, which in turn leads to the formation of hydrogen peroxide, hydroxyl radical, and probably other reactive products (Babior, 1984; Johnson, 1981). These reactive species, which normally provide phagocytes with the necessary means to kill rapidly microorganisms (Nathan, 1983), are released into phagocytic vacuoles and to the local environment external to the cells. Although this defensive mechanism normally contributes to maintaining sterile conditions in the lung, it is increasingly being viewed as a double-edged sword. Reactive oxygen species released during phagocytic metabolism can peroxidize cell lipids (which are major constituents in cell membranes) and produce cytotoxic metabolites such as malonyldialdehyde. Moreover, some lines of evidence indicate that oxygen species released from phagocytic cells can react with DNA and produce genomic disturbances (Weitzman and Stossel, 1981; Phillips et al., 1984). Other evidence points to an important role of phagocyte-derived oxidants and/or oxidant products in the metabolic activation of procarcinogens to their ultimate carcinogenic form (Kensler et al., 1987). The respiratory burst also occurs in AM upon the phagocytosis of a wide variety of other particulate materials, e.g., silica, metal oxide-coated fly ash, polymethylmethacrylate beads, chrysotile asbestos, fugative dusts, carboxylated microspheres, glass and latex beads, uncoated fly ash, and fiberglass (Hatch et al., 1980; Beall et al., 1977; Hoidal et al., 1978; Drath, 1985; Lowrie and Aber, 1977). PMN have also been demonstrated to increase their production of superoxide radical, hydrogen peroxide, and hydroxyl radical in response to membrane-reactive agents and particles (e.g., Sibille and Reynolds, 1990).

Regurgative exocytosis of hydrolytic enzymes is another outcome of the phagocytosis of particles by AM (Sandusky et al., 1977). In this process, lysosomal proteases are released to the exterior of the phagocytes as they ingest particles. These enzymes may be detrimental to the well being of alveolar epithelial cells, especially Type I pneumocytes that cover over 90% of the alveolar surface. The release of proteolytic enzymes by phagocytes in the lungs of smokers may be a major contributing factor to the development of emphysema. Other substances associated with various lung disease processes that can be elaborated by phagocytic AM include, but certainly are not limited to, cytokines that affect the proliferative and functional activities of lung cells and numerous proinflammatory eicosanoids (Nathan, 1987; Kelley, 1990; Holtzman, 1991).

Alveolar macrophage-mediated particle clearance

In addition to sequestering particles from other lung constituents via phagocytosis, AM are also fundamentally involved in mediating the removal of particles from the alveoli by way of the conducting airways. Evidence for such AM-mediated particle clearance comes from numerous sources. Following the alveolar deposition of relatively insoluble iron oxide particles, Lehnert and Morrow (1985a) found that >90% of the retained lung burden could be accounted for as being contained in AM over the course of alveolar clearance. Also, airway macrophages containing phagocytized particles at times well after lung deposition are observed on the lumina of the conducting airways (Lehnert et al., 1990b), and the post-depositional frequency distributions of particles contained in AM and airway macrophages have been observed to be virtually identical during alveolar phase clearance (Lehnert et al., 1990b). Still other, less direct evidence for the AM-mediated particle clearance process comes from observations that the removal of relatively insoluble particulate agents from the lower respiratory tract can be quantitatively accounted for in the gastrointestinal tract and feces (Gibb and Morrow, 1962; Kenoyer et al., 1981), as would be expected if AM with their particulate burdens are removed from the lung via mucociliary transport and subsequently swallowed with airway mucus.

Specific details of the mechanism(s) involved in AM-mediated particle clearance have not been well delineated. It is presently unknown as to how particle-containing AM actually gain access to the mucociliary escalator at the level of the respiratory and terminal bronchioles for subsequent transport up the respiratory tract (Fig. 12) although several explanations have been proposed over the last several decades. Many of these explanations have been discounted for a variety of theoretical and experimental reasons too detailed to discuss here. Two currently postulated mechanisms by which AM encounter the ciliated airways are related to the ameboid movement of AM on alveolar surfaces (Morrow, 1988). According to one of these postulates, AM gain access to the mucociliary escalator by directed ameboid migration or by a directed physical transport of the AM to the terminal airways. A directed migration implies the necessity of a chemo-

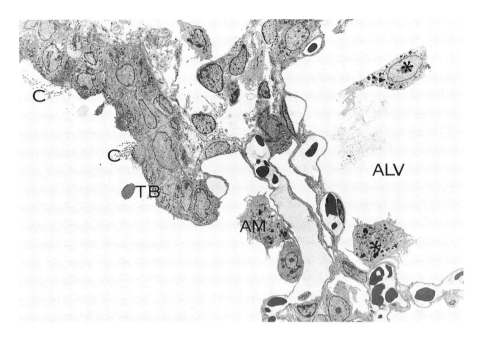

Figure 12. Transmission electron micrograph showing a particle (TiO$_2$)-containing AM in close proximity to a terminal bronchiole (TB). How such AM gain acess to these terminal airways for subsequent translocation up the respiratory tract by mucociliary transport is as yet poorly understood. C: cilia. Other particle-laden macrophages (∗) are also present in an adjacent alveolus (ALV).

attractant gradient in the alveolar-terminal bronchiolar region for AM to follow. Evidence for such a gradient or for an identifiable source of chemoattractant(s) in the peripheral airways during alveolar phase clearance is presently nonexistent. A potential physical process that may favor the directed movement of AM along alveolar surfaces toward the ciliated epithelium, on the other hand, may be related to alveolar lining fluid currents generated during the respiratory cycle. This long-held possibility, however, has been recently seriously questioned (Morrow, 1988). According to the second postulate, a normal, random migration of AM in the alveoli provides a stochastic basis for these cells to encounter the mucociliary transport apparatus. Such a process, as with chemotactically-directed migration, would be expected to behave as a first order process resembling first order mathematical depictions of the kinetics of alveolar clearance (e.g., Lehnert and Morrow, 1985a; Lehnert et al., 1989; Snipes et al., 1988).

Another process that likely impacts on the retention of insoluble particles in the alveolar space compartment is the lack of fidelity in the containment of particles in AM until the AM are translocated out of the lung (Lehnert et al., 1992). This lack of fidelity, which has been observed with even noncytotoxic particles, is manifested by a gradual redistribution of retained particles among the lung's AM population concurrent with the removal of AM via the conducting airways. With more cytotoxic particles, e.g., silica, the particle redistribution phenomenon is likely more pronounced. Potential mechanisms for such a redistribution of particles among the AM (Lehnert et al., 1992) may include: (1) the *in situ* division of particle-containing AM and the allocation of the parent AM's particles to daughter cells, (2) the *in situ* autolysis of particle-containing AM and the uptake of particles by other AM with pre-existing or no particle burdens, (3) the exocytosis of particles by AM and the subsequent phagocytosis of the particles by other AM, (4) the phagocytosis of effete, particle-containing AM by other AM, and/or (5) the direct transfer of particles from one AM to another. The particle redistribution phenomenon may have important significance in the context of dust-induced diseases. If the overall process substantially involves the release of free particles from phagocytes that have previously engulfed them, such particles would again be available to interact on the alveolar surface with epithelial cells and alveolar phagocytes at times well after the particles were initially deposited. Additionally, free particles, however temporary prior to being rephagocytized, would again be afforded the opportunity to gain access to the lung's interstitium via a Type I cell transepithelial transport mechanism (to be discussed). The rephagocytosis of freed particles at times well after their initial deposition in the lung would also be expected to have some other potentially detrimental consequences, including the intra-alveolar elaboration of mediators of inflammation by the phagocytes, e.g., reactive oxygen species, proteases, proinflammatory cytokines, etc. Conceivably, the particle redistribution phenomenon may explain the development of chronic inflammatory processes that occur at times well after inhalation exposure to injurious dusts has ceased (Lehnert et al., 1992).

Although AM-mediated particle clearance via the conducting airways is the main mechanism by which insoluble particles are removed from the lung, the ability of AM to dissolve some mineral particles after they are phagocytized is also an important AM-mediated lung clearance mechanism. The dissolution of various types of particles, especially some metallic particles (e.g., Lundborg et al., 1984; Marafante et al., 1987) that may not otherwise be readily soluble in simple aqueous solutions, has been attributed mainly to the acidic environment in the phagolysosomes of AM. Similar to the pH in the phagocytic vacuoles of PMN and macrophages from other anatomical sites, the pH in the phagolysosomes of AM is on the acidic side, i.e., ~pH 5 (Nilsen et al., 1988). Particle dissolution in AM can result in faster rates of removal of susceptible particles from the lung than the rate afforded by the translocation of particle-laden AM up the mucociliary escalator alone. Non-reactive solutes derived from the dissolution process presumably can leave the lung during the routine removal of lung lining fluid and/or by uptake into the pulmonary blood and draining lymphatic fluid following transepithelial passage. On the other hand, some evidence indicates that the dissolution of some types of particles in phagolysosomes can result in the binding of solubilized constituents to other AM structural components (Godleski et al., 1988). The durability of fibers, or their *in vivo* solubility, is a recognized determinant of their retention characteristics in the lung, as well as evidently being of importance in the development of pathological responses to fibers. Bellman et al. (1990) observed that glass fibers of low durability (high solubility in tissue) cleared from rat lungs with a biological half-time of ~40 days, whereas more highly durable (low solubility) glass fibers with initially equivalent diameters and lengths were removed from the lung with a substantially longer half-time, i.e., ~240 days. When these same fiber samples were examined in an intraperitoneal carcinogenicity test, tumors were only induced by the more durable fiber type (Pott et al., 1990). Recent dissolution experiments by Hume and Rimstidt (1992) show that dissolution of chrysotile should be rapid in the lung and that the dissolution is independent of pH within the pH range 2 to 6.

Other mechanisms of particle clearance and retention

While the phagocytosis of particles by lung free cells and the mucociliary clearance of the cells with their particulate burdens represent the most prominent mechanisms that govern the fate of insoluble particles deposited in the alveoli, other mechanisms exist that can impact both on the retention characteristics of such particles and on the lung clearance pathways for the particles. These mechanisms generally become most significant under conditions in which relatively high burdens of particles are deposited in the alveoli, although the fates of some types of cytotoxic particles may involve these processes even when deposition is less intense.

One extra-AM mechanism is the endocytosis of particles by Type I pneumocytes (Adamson and Bowden, 1981; Adamson and Bowden, 1978) that normally cover most of the surface area of the alveoli (Crapo et al., 1983) (Fig. 13). The Type II pneumocyte apparently lacks the ability to endocytize most types of particles, even though it is the progenitor cell of the Type I cell

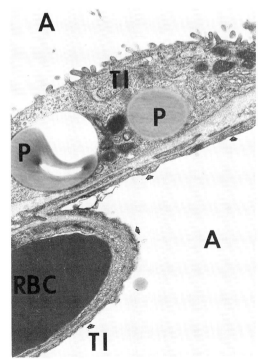

Figure 13. Occurrence of polystyrene microspheres (P) in a Type I pneumocyte, a cell type that normally covers ~95% of the alveolar surface. These cells (TI, small arrows) are generally well-spread on the alveolar surface so that the distance between the alveolar surface and red blood cells (RBC) in capillaries is minimal. The size of the erythrocyte as sampled in the micrograph is ~4 µm. A: alveoli.

(Adamson and Bowden, 1974). Particle internalization by Type I pneumocytes may be related to both particle size and the number of particles that deposit on the alveolar surface. Adamson and Bowden (Adamson and Bowden, 1981), for example, found that with increasing loads of carbon particles (0.03 µm in diameter) instilled into the lungs of mice, more free particles were observed in the alveoli within a few days thereafter, and the relative abundance of particles endocytized by Type I epithelial cells also increased with increasing lung burdens of the particles. These same investigators demonstrated that instilled latex particles with a diameter of 0.1 µm were endocytized by Type I pneumocytes when delivered at high lung burdens, but latex microspheres with a diameter of 1.0 µm were rarely observed to be engulfed by epithelial cells. Although the endocytic ability of Type I cells must have an as-yet undetermined upper particle size limitation, the infrequent appearance of the 1.0 µm particles in the Type I pneumocytes may have been related to the fact that the larger particles were administered in the above study at the same mass concentration as were the smaller particles, and hence, at lower numbers than the smaller particles. Even so, Adamson and Bowden (1981) have postulated that when high lung burdens of particles exceed the capacity of resident and recruited phagocytes to engulf the particles, the presence of free particles in the alveoli increases the likelihood of particle encounters with Type I cells. A hypothetical extension of this postulate is that some fraction of particles deposited in the alveoli gain entry into Type I epithelial cells even when deposited lung burdens are low, or otherwise when the mechanism of particle phagocytosis is not overwhelmed. Alternatively, the process of particle phagocytosis, i.e., particle encounters and engulfment, by resident AM and newly recruited phagocytes seemingly would have to be kinetically accomplished with 100% efficiency within a post-depositional time span that would preclude particle uptake by the Type I cells. Unfortunately, detailed information about the kinetics of particle endocytosis by Type I pneumocytes is nonexistent. However, the above extension of the Adamson and Bowden postulate is consistent with some existing observations. For example, even when relatively low burdens of particulate agents are deposited in the lungs, some fraction of the particles usually appear over time in the regional lymph nodes (e.g., Lehnert et al., 1989). As will be discussed, the endocytosis of particles by Type I pneumocytes is an initial, early step involved in the passage of particles to the lymph nodes. Assuming particle phagocytosis by AM is not sufficiently rapid or perfectly efficient, increasing numbers of particles would be expected to gain entry into the Type I epithelial cell compartment essentially continuously during chronic aerosol exposures. Additionally, when particles are released on a continual basis by alveolar phagocytes that initially sequestered them after lung deposition, as discussed earlier, a fraction of these free particles may also undergo passage from the alveolar surface into the Type I pneumocytes.

As previously indicated, the endocytosis of particles by Type I cells represents only the first step in a process that can ultimately result in an accumulation of particles in the lung's interstitial compartment and the subsequent translocation of some fraction of the particles to the regional lymph nodes that receive drainage of interstitial fluids from the lung. A vesicular transport mechanism in the Type I pneumocyte can transfer particles from the air surface of the alveolar epithelium into the interstitial compartment, or tissue region

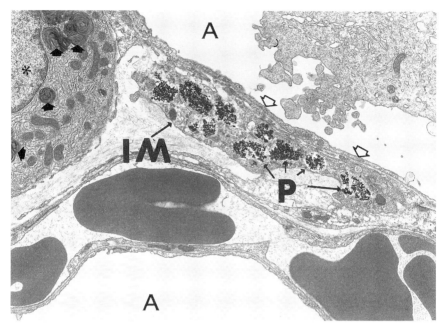

Figure 14. Transmission electron micrograph that shows a particle (P)-containing interstitial macrophage (IM) beneath a Type I pneumocyte (small arrows). The equivalent spheroidal diameter of the IM is ~7 to 8 µm. A: alveoli. ∗: a nearby Type II pneumocyte. These cells, which usually reside in the corners of alveoli, produce the lung surfactant that lines the alveolar surface and serves to maintain the patency of the alveolar sacs. They are identified by the presence of lamellar bodies that contain surfactants (dark arrows).

that is present between epithelial and endothelial cell barriers. Here the particles may be phagocytized by interstitial macrophages (Fig. 14), which normally represent ~40% of the lung's total macrophage population (Lehnert et al., 1985c), or the transferred particles may remain in a "free" state for some ill-defined period of time. The duration of this "free" state may be dependent upon the physico-chemical characteristics of the particle. Regardless, the lung's interstitial regions represent an anatomical site for the retention of particles in the lung. Whether or not alveolar phagocytes that have engulfed particles also contribute to the particle translocation process into the interstitium remains undecided, but some investigators believe that once phagocytes assume residency on the alveolar surface, they do not migrate across the alveolar epithelial cell lining back into interstitial sites. Evidence for this view mainly comes from repeated failures to find ultrastructural evidence in support of such a cell-translocating mechanism (Adamson and Bowden, 1981; Adamson and Bowden, 1978; Lauweryns and Baert, 1977). Other studies have indicated that if AM do translocate across the alveolar epithelial barrier, the process may be restricted by the magnitude of the particulate burdens they contain (Lehnert et al., 1986).

Factors that govern the fate of free particles that enter the lung's interstitial compartment are not well understood. Some particles, as previously indicated, may be phagocytized by interstitial macrophages, whereas others can apparently remain in a free state in the interstitium for some time without being engulfed by phagocytes. What fraction of the interstitial macrophages may subsequently migrate over time into the alveoli with their engulfed burdens and thereby contribute to the size of the resident AM population during alveolar clearance remains problematic. Moreover, no investigations have been conducted to date to assess the influence that an interstitial macrophage's burden of particles may have on its ability to enter the alveolar space compartment from the interstitium. Of relevance, accumulating evidence is indicating that macrophage-derived factors released following the phagocytosis of fibrogenic particles by interstitial macrophages may play a central role in the proliferative and collagen-synthesizing activities of lung fibroblasts, which are underlying features of dust-induced pulmonary fibrosis (Adamson et al., 1989; Adamson et al., 1991).

Some particles that gain entry into the interstitial compartment can further translocate to the extrapulmonary regional lymph nodes after entering lymphatic capillaries. This process can involve the passage of "free" particles as well as particle-containing cells (Fig. 15) to the lymph nodes via afferent lymphatic channels in the lungs (Lee et al., 1985; Lehnert et al., 1986; Leak, 1980; Harmsen et al., 1985). Free particles can enter the lymphatic channels through clefts between the endothelial cells that form the lymphatic pathways and also enter the lymphatic channels after they are endocytized by the lymphatic endothelial cells (Lauweryns and Baert, 1977; Leak, 1980). The overall process of the passage of particles to the lymph nodes is particle-size limited. Snipes and co-workers (Snipes et al., 1988), for example, found that particles up to 7 μm in diameter translocated from the lungs of animals to the regional lymph nodes whereas spheroidal particles larger than 9 μm did not. Further investigations are required to determine if transfer of the larger, spheroidal particles is due to their failure to undergo transepithelial transport, or whether there is a size-associated limitation in the transport of large particles and phagocytes that contain high volumetric burdens particles to and through lymphatic channels. It remains possible that the mobility of the interstitial macrophages may be particle burden-dependent, and, under conditions of high cellular burden-associated decreases in macrophage mobility, a greater fraction of particles that accumulate in the lymph nodes under high lung burden conditions may reach these sites as free particles. Unlike spherical particles, it is interesting to note that fibers with lengths as long as 16 μm have been found in the regional lymph nodes after lung deposition (Oberdörster et al., 1988), even though the lymphatic transport of fibers preferentially involves shorter fibers. However particles are ultimately transferred to the regional lymph nodes that receive drainage from the lung, the fraction of an initial alveolar burden of particles that may be transferred to the lymph nodes nodes over time can be substantial. Generally, the mass of an initially deposited lung burden that is translocated to the regional lymph nodes over time increases as the mass of the deposited lung burden is increased (Ferin and Feldstein, 1978). In addition to the translocation of particles to the lymph nodes, particle-containing macrophages also appear to accumulate over time in perivascular and peribronchiolar regions by an as yet to be understood mechanism.

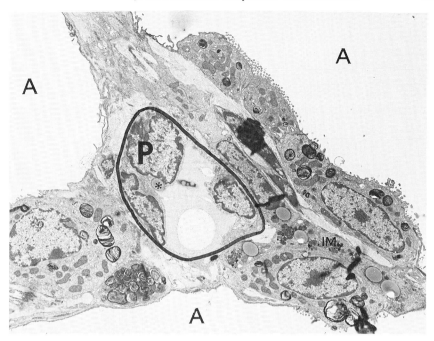

Figure 15. Transmission electron micrograph showing a particle (∗)-containing phagocyte (P) in a peripheral lymphatic channel, which is outlined in the micrograph. A particle-laden interstitial macrophage (IM) is also present. The polystyrene particles shown in the cells are 2 μm in diameter. A: alveoli.

It should be further emphasized that particles in the alveoli and the interstitium can undergo dissolution concurrently with the above processes. This can be the case for particles that otherwise are considered to be relatively insoluble *in vivo*. Such particle solubilization can become the most prominant lung clearance mechanism at times well after particle deposition (Kreyling et al., 1988).

Kinetics of particle clearance from the alveoli

The removal of relatively insoluble particles from the alveolar region requires a much more prolonged period of time than does clearance of the conducting airways via mucociliary transport in that half-times of alveolar phase clearance can range from weeks to thousands of days for the major fraction of deposited particles (Bailey et al., 1982; Bohning et al., 1982; Philipson et al., 1985). The pattern of particle removal from the alveoli is usually complex with faster rates of particle clearance occurring shortly after deposition and longer rates being better resolved as the post-depositional period increases. The removal characteristics for insoluble particles deposited in the alveolar region have been depicted descriptively by numerous investigators as a multicompartment or multi-component process with each component following simple first order kinetics (e.g., Lehnert et al., 1989; Snipes et al., 1988). While often suitable as a mathematical approach for describing lung retention data, actual physiologic mechanisms that may form the bases for differing empirically resolvable compartments or components with correspondingly different rates of particle clearance have not been well characterized. Indeed, it remains questionable whether functionally separate "compartments" with attributable linear kinetic properties *per se* are even involved in the clearance of particles from the lung. Nevertheless, the multiphasic pattern of alveolar clearance has recently been observed to be manifested in the rates of disappearance of particle-containing macrophages from the lung's AM population (Lehnert et al., 1989). This observation suggests that the central role of AM in mediating particle clearance from the lung may be influenced by factors yet to be revealed.

Particle "overloading"

Several animal studies have revealed that impairment of alveolar clearance can occur following chronic aerosol exposures to broadly diverse, relatively insoluble materials, e.g., titanium dioxide, carbon black, particulate diesel exhaust, and chromium dioxide[†] (Griffis et al., 1983; Wolf et al., 1987; Vostal et al., 1982; Lee et al., 1986; Lee et al., 1988). Because high lung burdens of different types of particles result in diminutions of normal clearance kinetics, or result in what is now called particle "overloading," this outcome appears to be more related to the mass and/or volume, or even, perhaps, to the surface areas of particles in the lung than to the physicochemical attributes of the particles *per se*. An outcome of a particle overloading condition is that it modifies the dosimetry for particles in the lung by prolonging their residency times, which, in turn, can alter the usual response(s) to the particles (Morrow, 1988). For example, excessive lung burdens of particulate titanium dioxide, a substance that has been considered to be a relatively innocuous nuisance dust, result in the development of pulmonary fibrosis and lung cancer (Lee et al., 1986). Similar findings have been reported for numerous other types of particles under particle overloading conditions, e.g., particulate diesel exhaust, carbon black, silica, CrO_2, and particulate oil shale (Lee et al., 1988; Mauderly et al., 1987; Kawabata, 1986; Holland et al., 1985).

The mechanisms underlying particle-load-dependent retardations in lung clearance, a condition that is functionally expressed as the emergence of an abnormally clearing compartment or "sequestration" compartment, are likely numerous. As previously discussed, one potential site for particle sequestration is the extra-macrophagic containment of particles in the lung's Type I epithelial cells. Unfortunately, information about the retention kinetics for particles endocytized by the Type I cells is non-existent, and no information as to how the vesicular transport of particles across the Type I cell may be exhausted or otherwise modified during particle overload is currently available. Another anatomical region in the lung that may clear slowly during particle overload is the lung's interstitial compartment. Again, little is known about the kinetics of removal of "free" particles or particle-containing macrophages from the lung's interstitial spaces or about what fraction of a retained burden of particles may be contained therein during the particle overload condition. Yet, it is evident from histologic observations of lungs with excessive burdens of particles that particles indeed are

---

[†] Chromium dioxide (CrO_2) is a very toxicologically stable compound (Lee et al., 1988) and is a common constituent of some magnetic tapes.

present in Type I cells and in the lung's interstitium and lymphatic channels. Consistent with particles contained in an extra-macrophagic compartment(s), Strom (1984) has shown that lesser percentages of retained burdens of diesel particles were lavageable from the lungs of animals chronically exposed to the particles compared to those percentages of the retained lung burdens that were lavageable after acute exposures.

Although the above mentioned anatomical sites must be considered as potential contributors to the diminutions of lung clearance during particle overload, disturbances in particle associated AM-mediated particle clearance is undoubtedly a predominant cause in that the AM are the primary reservoirs of deposited particles. The mechanism underlying a failure of removal of particle-laden AM from lungs with high particulate burdens remains obscure, although a hypothesis concerning the process has been offered. Morrow (1988) has proposed that dust overloading is caused by a loss in the mobility of particle engorged AM and that such an impediment is related to the cumulative volumetric load of particles in the AM. Morrow further estimated that the clearance of an AM is impaired when the particulate burden it contains is of a volumetric size equivalent to 60% of the AM's normal volume. Experimental evidence in support of this hypothesis is mounting. Oberdörster and co-workers (1992) assessed the alveolar clearance of smaller (3.3 µm in diameter) and larger (10.3 µm in diameter) polystyrene particles, the larger of which were volumetrically equivalent to ~60% of the average normal volume of an AM, after they were intratracheally instilled into the lungs of rats. Whereas both sizes of particles were found to be phagocytized by AM as of one day after deposition, only minimal lung clearance of the larger particles was observed over an ~200-day post-depositional period. In another study, Lehnert (1990a) also found that a condition of particle overloading brought about by intratracheally instilling a high burden of polystyrene particles into the lungs of rats could be accounted for by a failure of removal of AM that contained a volumetric load of particles equivalent to $\sim \geq 60\%$ of their original volumes. Also directionally consistent with the above hypothesis, *in vitro* analyses of the migratory behaviors of particle-laden AM have indicated that both unstimulated migration and migration in response to a chemotactic stimulus are impaired, with such impairments increasing with increases in the particle loads contained in the AM (Lehnert et al., 1990a).

Other processes may also be involved in preventing particle-laden AM from leaving the alveolar compartment via the tracheobronchial route under conditions of particle overload. Clusters or aggregates of particle-laden AM (Fig. 16) are typically found in the alveoli of animals that have been exposed to high concentrations of insoluble particles (e.g., Lee et al., 1985; Lehnert, 1990a, b; White, and Garg, 1981). The aggregation of AM may represent a major, if not the primary reason particle-laden AM are unable to leave the alveoli. The mechanism(s) responsible for the agglutination of the AM has not as yet been determined. Possibly, the clustering of AM may be related to the Type II cell hyperplasia, Figure 17, seen in lungs that have received high burdens of particles (e.g., Lehnert et al., 1993) and to an abnormal abundance of Type II cell-derived products, e.g., phospholipids, in the alveoli. Ferin (1982a) observed that the clearance of titanium dioxide from the rat's lung ceased when a condition of

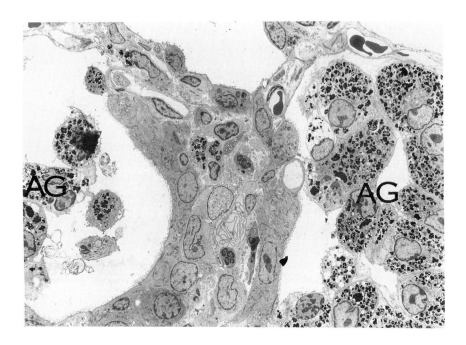

Figure 16. Transmission electron micrograph of two nearby alveoli that contain aggregates (AG) of AM, which in turn contain high cellular burdens of particles (TiO_2).

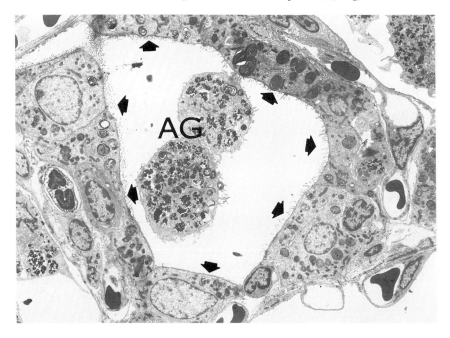

Figure 17. Extensive Type II cell hyperplasia (arrows) is often times observed in alveoli that contain aggregates (AG) of particle-containing AM.

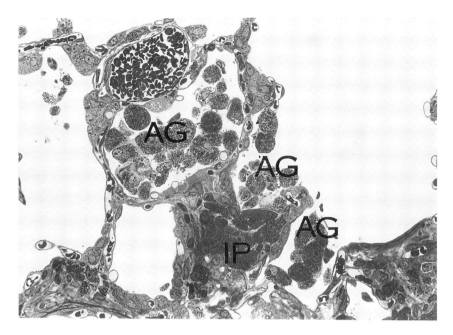

Figure 18. Interstitialized particles (IP) that are contained in phagocytes present in a region adjacent to alveoli that contain aggregated AM (AG).

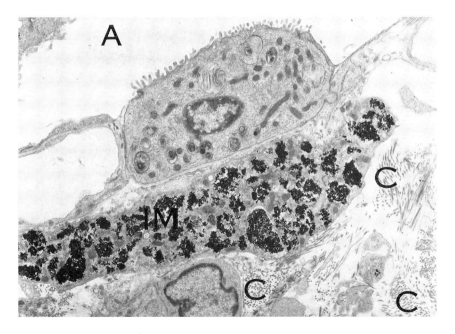

Figure 19. Collagen (C), an extracellular matrix material produced by lung fibroblasts, is observed in close proximity to an interstitial macrophage (IM) that is heavily loaded with particles (TiO_2). The adjacent alveolus (A) contained an aggregate of particle-laden AM.

phospholipidosis was experimentally induced, but lung clearance again resumed when the phospholipidotic agent administered to the animals was withdrawn. Other potential factors that may be result in the aggregation of AM include the clustering effect is a continual alteration in the permeability of the alveolar epithelial barrier during particle overload (Creutzenberg et al., 1989) in a manner that allows the passage of constituents in the lung's capillaries to leak out into the alveoli. In this case, aggregating factors such as fibronectin (Godfry et al., 1984) originating from the blood compartment or locally elaborated by AM could contribute to AM adhering to one another. Still another potential mechanism for AM agglutination may be the release of agglutinating mediators by lymphocytes or even by the AM themselves (Garcia et al., 1987; Garcia-Moreno and Myrivik, 1977). Some evidence, however, suggests that cytokines, such as macrophage migration inhibition factor, may not be the dominant mechanism responsible for AM aggregation in the air spaces in that increases in lymphocytes appear not to be a prerequisite for AM aggregation to occur (Strom, 1984; Lee et al., 1986). It is noteworthy that AM lavaged from the lungs of animals exposed to diesel exhaust particulates continue to demonstrate the propensity to aggregate *in vitro*, which further suggests the additional possibility that that particle-overloaded macrophages may have an over expression of adhesion molecules (Hynes, 1992) on their surfaces.

Regardless of the mechanism underlying the aggregation of AM, several general responses that have been associated with such aggregation are the proliferation of Type II cells on the surface of the alveoli that contain the aggregates (Fig. 17) marked interstitializations of particles at sites where AM aggregation occurs (Fig. 18) and the appearance of excessive lung collagen, Figure 19, which is a feature of fibrogenic lung disease.

ACKNOWLEDGMENTS

The author extends his gratitude to Mr. R.J. Sebring for preparing the micrographs shown herein. This report was produced under the auspices of the U.S. Department of Energy, and its preparation was funded by a DOE project entitled "Mechanisms of Pulmonary Damage."

REFERENCES

Absolom, D.R., van Oss, C.J., Zingg, W. and Neumann, A.W. (1982) Phagocytosis as a surface phenomenon: Opsonization by aspecific adsorption of IgG as a function of bacterial hydrophobicity. J. Reticuloendothel. Soc. 31, 59-70.

Adamson, I.Y.R. and Bowden, D.H. (1978) Adaptive responses of the pulmonary macrophagic system to carbon. II. Morphologic studies. Lab. Invest. 38, 430-438.

Adamson, I.Y.R., and Bowden, D.H. (1981) Dose response of the pulmonary macrophagic system to various particulates and its relationship to transepithelial passage of free particles. Exp. Lung Res. 2, 165-175.

Adamson, I.Y.R. and Bowden, D.H. (1974) The type 2 cell as progenitor of alveolar epithelial regeneration. Lab. Invest. 30, 35-42.

Adamson, I.Y.R., Letourneau, H.L. and Bowden, D.H. (1991) Comparison of alveolar and interstitial macrophages in fibroblast stimulation after silica and long and short asbestos. Lab. Invest. 64, 339-344.

Adamson, I.Y.R., Letourneau, H.L. and Bowden, D.H. (1989) Enhanced macrophage-fibroblast interactions in the pulmonary interstitium increases fibrosis after silica injection to monocyte-depleted mice. Amer. J. Pathol. 134, 411-418.

Adkins, B., Luginbuhl, G.H., Miller, F.J. and Gardner, D.E. (1980) Increased pulmonary susceptibility to streptococcal infection following inhalation of manganese oxide. Environ. Res. 23, 110-120.

Allen, R.C. and Loose, L.D. (1976) Phagocytic activation of a luminol-dependent chemiluminescence in rabbit alveolar and peritoneal macrophages. Biochem. Biophys. Commun. 69, 245-252.

Allison, A.C. (1973) Experimental methods-cell and tissue culture: Effects of asbestos particles on macrophages, mesothelial cells and fibroblasts. In: Biological Effects of Asbestos, Proc. Working Conf. Int'l Agency for Research on Cancer, Lyon, France, October 1972. P. Bogovski, J.C. Gilson, V. Timbrell, J.C. Wagner, eds., Int'l Agency for Research on Cancer Scientific Publ. No. 8, Lyon, France, p. 89-93.

Allison, A.C., Harrington, J.C. and Birbeck, M. (1966) An examination of the cytotoxic effects of silica on macrophages. J. Exp. Med. 124, 141-154.

Aronson, M. (1963) Bridge formation and cytoplasmic flow between phagocytic cells. J. Exp. Med. 118, 1084-1102.

Asgharian, B. and Yu, C.P. (1988) Deposition of inhaled fibrous particles in the human lung. J. Aerosol. Med. 1, 37-50.

Asmundsson, T. and Kilburn, K.H. (1970) Mucociliary clearance rates at various levels in dog lungs. Amer. Rev. Respir. Dis. 102, 388-397.

Babior, B.M. (1984) The respiratory burst of phagocytes. J. Clin. Invest. 73, 599-601.

Bailey, M.R., Fry, F.A. and James, A.C. (1982) The long-term clearance kinetics of insoluble particles from the human lung. Ann. Occup. Hyg. 26, 273-290.

Beall, G.D., Repine, J.E., Hoidal, J.R. and Rasp, F.L. (1977) Chemiluminescence by human alveolar macrophages: Stimulation with heat killed bacteria or phorbal myristate acetate. Infect. Immunol. 17, 117-120.

Bellmann, B., Muhle, H. and Pott, F. (1990) Study on the durability of chemically different glass fibers in lungs of rats. Zentralbl. Hyg. 190, 310-314.

Bergstrom, R. and Rylander, R. (1977) Pulmonary injury and clearance of MnO_2 particles. In: Pulmonary Macrophages and Epithelial Cells. C.L. Sanders, P. Schneider, G.E. Dagle and H.A. Ragan, eds. Technical Info. Center, ERDA, Springfield, VA 523-532.

Bhalla, D.K. and Crocker, T.T. (1986) Tracheal permeability in rats exposed to ozone. An electron microscopic and autoradiographic analysis of the transport pathway. Amer. Rev. Respir. Dis. 134, 572-579.

Bhalla, D.K., Mannix, R.C., Kleinerman, M.T. and Crocker, T.T. (1986) Relative permeability of nasal, tracheal, and bronchoalveolar mucosa to macromolecules in rats exposed to ozone. J. Toxicol. Environ. Health 17, 269-283.

Bianco, C., Griffin, F.M. and Silverstein, S.C. (1975) Studies of the macrophage complement receptor: Alteration of receptor function upon macrophage activation. J. Exp. Med. 141, 1278-1290.

Bingham, E., Pfitzer, E.A., Barkley, W. and Radford, E.P. (1968) Alveolar macrophages: reduced numbers in rats after prolonged inhalation of lead sesquioxide. Science 62, 1297-1299.

Bittermann, P.B., Rennard, S.I. and Crystal, R.G. (1981) Environmental lung disease and the interstitium. Clin. Chest. Med. 2, 393-412.

Bohning, D.E., Atkins, H.L. and Cohn, S.H. (1982) Long-term particle clearance in man: Normal and impaired. Ann. Occup. Hyg. 26, 259-271.

Boltz-Nitulescu, G., Plummer, J.M. and Spiegelberg, H.L. (1982) Fc receptors for IgE on mouse macrophages and macrophage-like cell lines. J. Immunol. 28, 2265-2268.

Boltz-Nitulescu, G. and Spiegelberg, H.L. (1982) Receptors specific for IgE on rat alveolar and peritoneal macrophages. Cell Immunol. 59, 106-114.

Boucher, R.C. (1980) Chemical modulation of airway epithelial permeability. Environ. Health Perspect. 35, 3-12.

Boucher, R.C., Johnson, J., Inoue, S., Hulbert, W. and Hogg, J.C. (1980) The effect of cigarette smoke on the permeability of guinea pig airways. Lab. Invest. 43, 94-100.

Bouley, G., Dubreuil, A., Despaux, N. and Bouden, C. (1977) Effects of cadmium microparticles on the respiratory system. Scand. J. Work Environ. Health 3, 116-121.

Bowden, D.H. and Adamson, I.Y.R. (1978) Adaptive response of the pulmonary macrophagic system to carbon: I. Kinetic studies. Lab. Invest. 38, 422-429.

Bowden, D.H. and Adamson, I.Y.R. (1982) Alveolar macrophage response to carbon in monocyte-depleted mice. Amer. Rev. Respir. Dis. 126, 708-711.

Bowden, D.H. and Adamson, I.Y.R. (1984) Pathways of cellular efflux and particulate clearance after carbon instillation to the lung. J. Pathol. 143, 117-125.

Bowden, D.H. and Adamson, I.Y.R. (1980) Role of monocytes and interstitial cells in the generation of alveolar macrophages. I. Kinetic studies of normal mice. Lab. Invest. 42, 511-517.
Brain, J.D. (1971) The effects of increased particles on the number of alveolar macrophages. In: Inhaled Particles, W.H. Walton, ed., Vol. III, Unwin Brothers, Old Woking, U.K., 209-223.
Brain, J.D. and Cockery, G.C. (1977) The effect of increased particles on the endocytosis of radiocolloids by pulmonary macrophages in vivo: competitive and toxic effects. In: Inhaled Particles and Vapours W.H. Walton, ed., Vol. IV, Pergamon Press, London 551-564.
Brain, J.D., Gehr, P. and Kavet, R.I. (1984) Airway macrophages: The importance of the fixation method. Amer. Rev. Respir. Dis. 129, 823-826.
Brain, J.D. and Valberg, P.A. (1979) Deposition of aerosol in the respiratory tract. Amer. Rev. Respir. Dis. 120, 1325-1373.
Breeze, R. and Turk, N. (1984) Cellular structure, function and organization in the lower respiratory tract. Environ. Health Perspect. 55, 3-24.
Brody, A.R. and Roe, M.W. (1983) Deposition pattern of inorganic particles at the alveolar level in the lungs of rats and mice. Amer. Rev. Respir. Dis. 128, 724-729,.
Camner, P. (1981) Studies on the removal of inhaled particles from the lungs by voluntary coughing. Chest 80, 824-827.
Churg, A. and Stevens, B. (1988) Association of lung cancer and airway particle concentration. Environ. Res. 45, 58-63.
Churg, A. and Wright, J. (1988) Mineral particle concentration in the walls of small airways in long term chrysotile miners. In: Inhaled Particles VI, J. Dodgson, R.I. McCallum, M.R. Bailey, and D.R. Fisher, eds., Pergamon Press, New York 173-180.
Churg, A., Wright, J.L., Wiggs, B., Pare, P.D. and Lazar, N. (1985) Small airways disease and mineral dust exposure: Prevalence, structure, and function. Amer. Rev. Respir. Dis. 131, 139-143.
Civil, G.W. and Heppleston, A.G. (1979) Replenishment of alveolar macrophages in silicosis: Implication of recruitment by lipid feed-back. Brit. J. Exper. Pathol. 60, 537-547.
Crapo, J.D., Young, S.L., Fram, E.K., Pinkerton, K.E., Barry, B.E. and Crapo, R.O. (1983) Morphometric characteristics of cells in the alveolar region of mammalian lungs. Amer. Rev. Respir. Dis. 128, S42-S46.
Creutzenberg, O., Muhle, H., Bellmann, H., Kilpper, R., Mermelstein, R. and Morrow, P. (1989) Reversability of biochemical and cytological alterations in bronchoalveolar lavagate upon cessation of dust exposure. The Toxicologist 9, A305.
Czop, J.K. and Austen, K.F. (1985) A β-glucan inhibitable receptor on human monocytes: Its identity with the phagocytic receptor for particulate activators of the alternative complement pathway. J. Immunol. 134, 2588-2593.
Czop, J.K. and Austen, K.F. (1980) Functional discrimination by human monocytes between their C3b receptors and their recognition units for particulate activators of the alternative complement pathway. J. Immunol. 125, 124-128.
Czop, J.K., Fearon, D.T. and Austen, K.F. (1978) Opsonin-independent phagocytosis of activators of the alternative complement pathway by human monocytes. J. Immunol. 120, 1132-1138.
Czop, J.K., Kadish, J.L. and Austen, K.F. (1982a) Purification and characterization of a protein with fibronectin determinants and phagocytosing-enhancing activity. J. Immunol. 129, 163-167.
Czop, J.K., McGowan, S.E. and Center, D.M. (1982b) Opsonin-independent phagocytosis by human alveolar macrophages: Augmentation by human plasma fibronectin. Amer. Rev. Respir. Dis. 25, 607-609.
Dauber, J.H., Daniele, R.P. (1978) Chemotactic activity of guinea pig alveolar macrophages. Amer. Rev. Respir. Dis. 117, 673-684.
Daughaday, C.C. and Douglas, S.D. (1976) Membrane receptors on rabbit and human pulmonary alveolar macrophages. J. Reticuloendothel. Soc. 19, 37-45.
Drath, D.B. (1985) Enhanced superoxide release and tumoricidal activity by a post lavage, in situ pulmonary macrophage population in response to activation by *Mycobacterium bovis* BCG exposure. Infect. Immun. 49, 72-75.
Driscoll, K.E., Simpson, L., Carter, J., Hassenbein, D. and Leikauf, G.D. (1993) Osone inhalation stimulates expression of a neutrophil chemotactic protein, macrophage inflammatory protein 2. Toxicol. Appl. Pharmacol. 119, 306-309.
Ehlenberger, A.G. and Nussenzweig, V. (1977) The role of membrane receptors C3b and C3d in phagocytosis. J. Exp. Med. 145, 357-371.
Evans, M.J., Shami, S.G. and Martinez, L.A. (1986) Enhanced proliferation of pulmonary alveolar macrophages after carbon instillation in mice depleted of blood monocytes by strontium-89. Lab. Invest. 54, 154-159.
Fels, A.O.S., Pawlowski, N.A., Cramer, E.B., King, T.K.C., Cohn, Z.A. and Scott, W.A.(1982) Human alveolar macrophages produce leukotriene B4. Immunol. 79, 7866-7870.

Ferin, J. (1982a) Alveolar macrophage mediated pulmonary clearance suppressed by drug-induced phospholipidosis. Exp. Lung Res. 4, 1-10.
Ferin, J. (1982b) Pulmonary alveolar pores and alveolar macrophage-mediated particle clearance. Anat. Rec. 203, 265-272.
Ferin, J. and Feldstein, M.L. (1978) Pulmonary clearance and hilar lymph node content in rats after particle exposure. Environ. Res. 6, 342-352.
Figari, I.S. and Palladino, M.A. (1987) Stimulation of neutrophil chemotaxis by recombinant tumor necrosis factors alpha and beta. Fed. Proc. 44, 300-304.
Fisher, E.S., Lauffenburger, D.A. and Daniele, R.P. (1988) The effect of alveolar macrophage chemotaxis on bacterial clearance from the lung surface. Amer. Rev. Respir. Dis. 137, 1129-1134.
Garcia, J.G.N., Noonan, T.C., Jubiz, W. and Malik, A.B. (1987) Leukotrienes and the pulmonary microcirculation. Amer. Rev. Respir. Dis. 136, 161-169.
Garcia-Moreno, L.F. and Myrvik, Q.N. (1977) Macrophage-agglutinating factor produced *in vitro* by BCG-sensitized lymphocytes. Infect. Immunol. 17, 613-620.
Gardner, D.E. Alterations in macrophage functions by environmental chemicals. (1984) Environ. Health Perspect. 55, 343-358.
Gardner, D.E., Miller, F.J., Illing, J.W. and Kirtz, J.M. (1977) Alterations in bacterial defense mechanisms of the lung induced by inhalation of cadmium. Bull. Europ. Physiopath. Respir. 13, 157-174.
Geiser, M., Im Hof, V., Gehr, P. and Cruz-Orive, L.M. (1988) Histological and stereological analysis of particle deposition in the conductive airways of hamster lungs. J. Aerosol. Med. 1, 197A.
Gehr, P., Schurch, S., Berthiaume, Y., Im Hof, V. and Geiser, M. (1990) Particle retention in airways surfactant. J. Aerosol Med. 3:27-44.
Gibb, F.R. and Morrow, P.E.. (1962) Alveolar clearance in dogs following inhalation of iron-59 oxide aerosol. J. Appl. Physiol. 17, 429-432.
Gil, J. and Weibel, E.R. (1971) Extracellular lining of bronchioles after perfusion fixation of rat lungs form electron microscopy. Anat. Rec. 169, 185-200.
Godfrey, H.P., Channabasappa, V.A., Wolstencroft, R.A. and Bianco, C. (1984) Localization of macrophage agglutination factor activity to the gelatin-binding domain of fibronectin. J. Immunol. 133, 1417-1423.
Godleski, J.J., Stearns, M.R., Katler, M.R. and Brain, J.D. (1988) Particle dissolution in alveolar macrophages assessed by electron energy loss analysis using the Zeiss CEM902 electron microscope. J. Aerosol. Med. 1, 198-199.
Gore, D. J. and Patrick, G. (1978) The distribution and clearance of inhaled UO_2 particles on the first bifurcation and trachea of rats. Phys. Med. Biol. 23, 730-737.
Griffin, F.M., Griffin, J.A., Leider, J.E. and Silverstein, S.C. (1975) Studies on the mechanism of phagocytosis. I. Requirements for circumferential attachment of particle-bound ligands to specific receptors on the macrophage plasma membrane. J. Exp. Med. 142, 1263-1282.
Griffis, L.C., Wolff, R.K., Henderson, R.F., Griffith, W.C., Mokler, B.V. and McClellen, R.O. (1983) Clearance of diesel soot particles from rat lung after subchronic diesel exhaust exposure. Fund. Appl. Toxicol. 3, 99-103.
Gross, P., de Treville, R.T.P., Tolker, E.B., Kaschak, M. and Babyak, M.A. (1969) The pulmonary macrophage response to irritants: An attempt at quantitation. Arch. Environ. Health 18, 174-185.
Hahn, F.F., Newton, G.J. and Bryant, P.L. (1977) In vitro phagocytosis of respirable-sized monodispersed particles by alveolar macrophages. In: Pulmonary Macrophages and Epithelial Cells. C.L. Sanders, R.P. Sneider, G.E. Dagle, H.A. Ragan, eds., Nat'l Technical Info. Service, U.S. Dept. of Commerce 424-435.
Harmsen, A.G., Muggenburg, B.A., Snipes, M.B. and Bice, D.B. (1985) The role of macrophages in particle translocation from lungs to lymph nodes. Science 230, 1277-1280.
Harrington, J.S. (1976) The biological effects of mineral fibers, especially asbestos, as seen from *in vitro* and *in vivo* studies. Ann. Anat. Pathol. 21, 155-198.
Hatch, G.E., Gardner, D.E., Menzel, D.B. (1980) Stimulation of oxidant production in alveolar macrophages by pollutant and latex particles. Environ. Res. 23, 121-136.
Haugen, A., Schafer, P.W., Lechner, J.F., Stoner, G.D., Trump, B.F. and Harris, C.C. (1982) Cellular ingestion, toxic effects, and lesions observed in human bronchial epithelial tissue and cells cultured with asbestos and glass fibers. Int'l J. Cancer 30:265-272.
Hilding, A.C. (1959) Ciliary streaming in the lower respiratory tract. J. Thoracic Surg. 37, 108-117.
Hill, J.O., Rothenberg, S.J., Kanapilly, G.M., Hanson, R.L. and Scott, B.R. (1982) Activation of immune complement by fly ash particles from coal combustion. Environ. Res. 28, 113-122.
Hof, D.G., Repine, J.E., Peterson, P.K. and Hoidal, J.R. (1980) Phagocytosis by human alveolar macrophages and neutrophils: Qualitative differences in the opsonic requirements for uptake of *Staphylococcus aureus* and *Streptococcus pneumoniae in vitro*. Amer. Rev. Respir. Dis . 121, 65-71.

Hoidal, J.R., Repine, J.E., Beall, G.D., Rasp, F.L. and White, J.G. (1978) The effect of phorbol myristate on metabolism and ultrastructure of human alveolar macrophages. Amer. J. Pathol. 91, 469-482.

Holland, L.M., Wilson, J.S., Tillery, M.I. and Smith, D.M. (1985) Lung cancer in rats exposed to fibrogenic dusts. In: Silica, Silicosis and Cancer. D.Goldsmith, D.Winn, and C. Shy, eds., Praeger, New York 267-279.

Holma, B. (1967) Lung clearance of mono- and di-disperse aerosols determined by profile scanning and whole body counting: A study on normal and SO_2 exposed rabbits. Acta Med. Scand. Suppl. 473, 1-102.

Holtzman, M.J. (1991) Arachidonic acid metabolism: Implications of biological chemistry for lung function and disease. Amer. Rev. Respir. Dis. 143, 1988-203.

Horsefield, K. and Cumming, G. (1968) Morphology of the bronchial tree in man. J. Appl. Physiol. 24, 373-383.

Hulbert, W.C., Walker, D.C. and Hogg, J.C. (1981a) The site of leukocyte migration through the tracheal mucosa in the guinea pig. Amer. Rev. Respir. Dis. 124, 310-316.

Hulbert, W.C., Walker, D.C., Jackson, A. and Hogg, J.C. (1981b) Airway permeability to horseradish peroxidase in guinea pigs: the repair phase after injury by cigarette smoke. Amer. Rev. Respir. Dis. 123, 320-326.

Hume, L.A. and Rimstidt, J.D. (1992) The biodurability of chrysotile asbestos. Amer. Mineral. 77, 1125-1128.

Hunninghake, G.W., Gallin, J.I. and Fauci, A.S. (1978) Immunologic reactivity of the lung: The *in vivo* and *in vitro* generation of a neutrophil chemotactic factor by alveolar macrophages. Amer. Rev. Respir. Dis. 117, 15-23.

Iravani, J. (1967) Flimmerbewegung in den intrapulmonalen Luftwegene der Ratte. Plugers. Arch. 297, 221-237.

Iravani, J. and van As, A. (1972) Mucus transport in the tracheobronchial tree of normal and bronchitic rats. J. Pathol. 106, 81-93.

Jaurand, M.C., Bignon, J., Gaudichet, A., Magne, L. and Oblin, A. (1978) Biological effects of chrysotile after SO_2 sorption. Environ. Res. 17, 216-227.

Jeffrey, P.K. and Reid, L.M. (1975) New features of rat airway epithelium: a quantitative and electron microscopic study. J. Anat. 120, 295-320.

Joseph, M., Tonnel, A.B., Capron, A. and Voisin, C. (1980) Enzyme release and superoxide anion production by human alveolar macrophages stimulated with immunoglobulin E. Clin. Exp. Immunol. 40, 416-422.

Johnston, R.B. (1981) Enhancement of phagocytosis-associated oxidative metabolism as a manifestation of macrophage activation. Lymphokines 3, 33-56.

Kagan, E., Oghiso, Y. and Hartmann, D-P. (1983) Enhanced release of a chemoattractant for alveolar macrophages after asbestos inhalation. Amer. Rev. Respir. Dis. 128, 680-687.

Kavet, R.I. and Brain, J.D. (1977) Phagocytosis: quantitation of rates and intracellular heterogeneity. J. Appl. Physiol. Respirat. Environ. Exercise Physiol. 42(3), 432-437.

Kawabata, Y. (1986) Effects of diesel soot on unscheduled DNA synthesis of tracheal epithelium and lung tumor formation. In: Carcinogenic and mutagenic effects of diesel engine exhaust. N Ishinishi, ed., Elsevier Sci. Pub., B.V. 213-222.

Kazmierowski, J.A., Gallin, J.I., and Reynolds, H.Y. (1977) Mechanism for the inflammatory response in primate lungs. Demonstration and partial characterization of an alveolar macrophage-derived chemotactic factor with preferential activity for polymorphonuclear leukocytes. J. Clin. Invest. 59, 273-281.

Kelley, J. (1990) Cytokines of the lung. Amer. Rev. Respir. Dis. 141, 765-788.

Kensler, T.W., Egner, P.A., Moore, K.G., Taffe, B.G., Twerdok, L.E. and Thrush, M.A. (1987) Role of inflammatory cells in the metabolic act of polycyclic aromatic hydrocarbons in mouse skin. Toxicol. Appl. Pharm. 90, 337-346.

Kenoyer, J.L., Phalen, R. and Davis, J.R. (1981) Particle clearance from the respiratory tract as a test of toxicity. Exp. Lung Res. 2, 111-120.

Kilburn, K.H., Lynn, W.S., Tres, L.L. and McKenzie, W.N. (1973) Leukocyte recruitment through airway walls by condensed vegetable tannins and quercetin. Lab. Invest. 28, 55-59.

Kilburn, K.H. and McKenzie, W. (1975) Leukocyte recruitment to airways by cigarette smoke and particle phase in contrast to cytotoxicity of vapor. Science 189, 634-637.

Kilburn, K.H. and McKenzie, W.N. (1978) Leukocyte recruitment to airways by aldehyde-carbon combinations that mimic cigarette smoke. Lab. Invest. 38, 134-142.

Kohler, D., App, E., Schmitz-Schumann, M., Wurtemberger, G. and Matthys, H. (1986) Inhalation of amiloride improves the mucociliary and cough clearance in patients with cystic fibrosis. Eur. J. Respir. Dis. 69, 319-326.

Kolb, W.P., Kolb, L.M., Wetsel, R.A., Rogers, W.R. and Shaw, J.O. (1981) Quantitation and stability of the fifth component of complement (C5) in bronchoalveolar lavage fluids obtained from non-human primates. Amer. Rev. Respir. Dis. 123, 226-231.

Koshi, K., Homma, K. and Sakabe, H. (1978) Damaging effect of cadmium oxide dust to the lung and its relation to solubility of the dust. Indus. Health 16, 81-89.

Kreyling, W.G., Schumann, G., Ortmaier, A., Ferron, G.A. and Karg, ed. (1988) Particle transport from the lower respiratory tract. J. Aerosol. Med. 1, 351-370.

Lauweryns, J.M. and Baert, J.H. (1977) Alveolar clearance and the role of the pulmonary lymphatics. Amer. Rev. Respir. Dis. 115, 625-683.

Leak, L.V. (1980) Lymphatic removal of fluids and particles in the mammalian lung. Environ. Health Perspect. 35, 55-76.

Lee, K.P., Henry III, N.W., Trochimowicz, H.J. and Reinhardt, C.F. (1986) Pulmonary response to impaired lung clearance in rats following excessive TiO_2 dust deposition. Environ. Res. 41, 144-167.

Lee, K.P., Trochimoicz, H.J. and Reinhardt, C.F. (1985) Transmigration of titanium dioxide (TiO_2) particles in rats after inhalation exposure. Exp. Molecular Pathol. 42, 331-343.

Lee, K.P., Ulrich, C.E., Geil, R.G. and Trochimowicz, H.J. (1988) Effects of inhaled chromium dioxide dust on rats exposed for two years. Fund. Appl. Toxicol. 10, 125-145.

Lee, P.S. Chan, T.L., and Hering, W.E. (1983) Long-term clearance of inhaled diesel exhaust particles in rodents. J. Toxicol. Environ. Health 12, 801-813.

Lehnert, B.E. (1990a) Alveolar macrophages in a particle "overload" condition. J. Aerosol. Med. 3, S9-S30.

Lehnert, B.E. (1990b) Lung defense mechanisms against deposited dusts. In: Problems in Respiratory Care, Lippincott Series on Respiratory Public Health, Vol. 3, No. 2, p.130-162.

Lehnert, B.E. (1992) Pulmonary and thoracic macrophage subpopulations and the clearance of relatively insoluble particles from the lung. *Environ. Health Perspect.* 97, 17-46.

Lehnert, B.E. and Ferin, J. (1983) Particle binding and phagocytosis and plastic substrate adherence characteristics of alveolar macrophages from rats acutely treated with chlorphentermine. J. Reticuloendothel. Soc. 33, 293-303.

Lehnert, B.E. and Morrow, P.E. (1985a) Association of ^{59}iron oxide with alveolar macrophages during alveolar clearance. Exp. Lung Res. 9, 1-16.

Lehnert, B.E. and Morrow, P.E. (1985b) Characteristics of alveolar macrophages following the deposition of a low burden of iron oxide in the lung. J Toxicol Environ Health 16, 855-868.

Lehnert, B.E. and Oberdörster, G. (1993) Fate of fibers in the lower respiratory tract. In: Fiber Toxicology. D.B. Warheit, ed., Academic Press, Orlando, FL (in press).

Lehnert, B.E., Ortiz, J.B., London, J.E., Valdez, Y.E., Cline, A.F., Sebring, R.J. and Tietjen, G.L. (1990a) Migratory behaviors of alveolar macrophages during the alveolar clearance of light to heavy burdens of particles. Exp. Lung Res. 16, 451-479.

Lehnert, B.E., Ortiz, J.B., Steinkamp, J.A., Tietjen, G.L., Sebring, R.J. and Oberdörster, G. (1992) Mechanisms underlying the "particle redistribution phenomenon". J. Aerosol Med. 4, 261-277.

Lehnert, B.E., Sebring, R.J. and Oberdörster, G. (1993) Pulmonary macrophages: Phenomena associated with the particle "overload' condition. In: Proceedings of the Toxic and Carcinogenic Effects of Solid Particles in the Respiratory tract, 4th Inhalation Symposium. U. Mohr, ed. Int'l Life Sciences Institute, Washingon, D.C., in press.

Lehnert, B.E., Smith, D.M., Holland, L.M., Tillery, M.I. and Thomas, R.G. (1985a) Aerosol deposition along the respiratory tract at zero gravity: A theoretical study. In: Lunar Bases and Space Activities of the 21st Century. W.W. Mendell, ed., Lunar and Planetary Institute, Houston, TX, p. 671-677.

Lehnert, B.E. and Tech, C. (1985b) Quantitative evaluation of opsonin-independent phagocytosis by alveolar macrophages in monolayer using polystyrene microspheres. J. Immunol. Meth. 78, 337-344.

Lehnert, B.E., Valdez, Y.E. and Bomalaski, S.H. (1985c) Lung and pleural "free-cell responses" to the intrapulmonary deposition of particles in the rat. J. Toxicol. Environ. Health 16, 823-839.

Lehnert, B.E., Valdez, Y.E. and Holland, L.M. (1985) Pulmonary macrophages: Alveolar and interstitial populations. Exp. Lung Res. 9, 177-190.

Lehnert, B.E., Valdez, Y.E. and Tietjen, G.L. (1989) Alveolar macrophage-particle relationships during lung clearance. Amer. J. Respir. Cell Mol. Biol. 1, 145-154.

Lehnert, B.E., Valdez, Y.E., Sebring, R.J., Lehnert, N.M., Saunders, G.C. and Steinkamp, J.A. (1990b) Airway intra-luminal macrophages: Evidence of origin and comparisons to alveolar macrophages. Amer. J. Respir. Cell Mol. Biol. 3, 377-391.

Lehnert, B.E., Valdez, Y.E. and Stewart, C.C. (1986) Translocation of particles to the tracheobronchial lymph nodes after lung deposition: Kinetics and particle-cell relationships. Exp. Lung Res. 10, 245-266.

Levens, D., Nicholas, F., Brain, J.D. and Huber, G.L. (1977) The effect of iron oxide inhalation on the antibacterial defense system of the lung. Amer. Fed. Clin. Res. 24, 633A.

Lippmann, M., Yeates, D.B. and Albert, R.E. (1980) Deposition, retention, and clearance of inhaled particles. Brit. J. Indus. Med. 37, 337-362.

Lipscomb, M.F., Onofrio, J.M., Nash, E.J., Pierce, A.K. and Toews, G.B. (1983) A morphological study of the role of phagocytes in the clearance of *Staphylococcus aureus* from the lung. J. Reticuloendothel. Soc. 33, 429-442.

Lowrie, D.B. and Aber, V.R. (1977) Superoxide production by rabbit alveolar macrophages. Life Sci. 21, 1575-1584.

Luchte, D.L. (1982) Mucociliary interactions in rabbit intrapulmonary airways. Cell Motility Suppl. 1, 77-81.

Lundborg, M., Lind, B. and Camner, P. (1984) Ability of rabbit alveolar macrophages to dissolve metals. Exp. Lung Res. 7, 11-22.

Maliszewski, C.R., Shen, L. and Fanger, M.W. (1985) The expression of receptors for IgA on human monocytes and calcitriol-treated HL-60 cells. J. Immunol. 35, 3878-3881.

Marafante, E., Lundborg, M., Vahter, M. and Camner, P. (1987) Dissolution of two arsenic compounds by rabbit alveolar macrophages *in vitro*. Fund. Appl. Toxicol. 8, 382-388.

Marasco, W.A., Phan, S.H. and Krutsch, H. (1984) Purification and identification of formyl-methionyl-leucyl-phenylalanine as the major neutrophil chemotactic factor produced by *Escherichia coli*. J. Biol. Chem. 259, 5430-5439.

Mauderly, J.L., Jones, R.K., Griffith, W.C., Henderson, R.F. and McClellan, R.O. (1987) Diesel exhaust is a pulmonary carcinogen in rays exposed chronically by inhalation. Fund. Appl. Toxicol. 9, 208-221.

McFadden, D., Wright, J., Wiggs, B. and Churg, A. (1986) Cigarette smoke increases the penetration of asbestos fibers into airway walls. Amer. J. Pathol. 123, 95-99.

McGavran, P.D. and Brody, A.R. (1989) Chrysotile asbestos inhalation induced tritiated thymidine incorporation by epithelial cells of distal bronchioles. Amer. J. Respir. Cell. Mol. Biol. 1:231-235.

Migliorisi, G., Folkes, E., Pawlowski, N. and Cramer, E.B. (1987) *In vitro* studies of human monocyte migration across endothelium in response to leukotriene B4 and f-met-leu-phe. Amer. J. Pathol. 127, 157-167.

Miller, K., Calverley, A. and Kagan, E. (1980) Evidence of a quartz-induced chemotactic factor for guinea pig alveolar macrophages. Environ. Res. 22, 31-39.

Ming, W.J., Bersani, L. and Mantovani, A. (1987) Tumor necrosis factor is chemotactic for monocytes and polymorphonuclear leukocytes. J. Immunol. 138, 1469-1474.

Morgan, A., Talbot, R.J., Holmes, A. (1978) Significance of fibre length in the clearance of asbestos fibres from the lung. Brit. J. Indus. Med. 35, 146-153.

Morgan, A., Evans, J.C., Evans, R.J., Hounam, R.F., Holmes, A. and Doyle, S.G. (1975) Studies on the deposition of inhaled fibrous material in the respiratory tract of the rat and its subsequent clearance usinf radioactive tracer techniques. Environ. Res. 10, 196-207.

Morrow, P.E. (1988) Possible mechanisms to explain dust overloading of the lungs. Fund. Appl. Toxicol. 10, 197-207.

Morrow, P.E., Gibb, F.R. and Gazioglu, K.M. (1967) A study of particulate clearance from the human lungs. Amer. Rev. Respir. Dis. 96, 1209-1221.

Morrow PE and Yu CP. (1985) Models of aerosol behavior in airways. In: Aerosols in Medicine, Principles, Diagnosis and Therapy. F. Moren, M.T. Newhouse, M.B. Dolovich, eds., Elsevier Sci. Pub. 149-191.

Mossberg, B. (1980) Human tracheobronchial clearance by mucociliary transport and cough. Eur. J. Respir. Dis. 61, Suppl. 107, 51-58.

Mossman, B.T., Adler, K.B. and Craighead, J.E. (1978) Interaction of carbon particles with tracheal epithelium in organ culture. Environ. Res. 16, 110-122.

Mossman, B.T. and Craighead, J.E. (1979) Use of hamster tracheal organ cultures for assessing the cocarcinogenic effects of inorganic particulates on the respiratory epithelium. Prog. Exp. Tumor Res. 24, 37-74.

Mossman, B.T., Kessler, J.B., Ley, B.W. and Craighead, J.E. (1977) Interaction of crocidolite asbestos with hamster respiratory mucosa in organ culture. Lab Invest 36, 131-139.

Nathan, C.F. (1983) Mechanisms of macrophage antimicrobial activity. Trans. Royal Soc. Trop. Med. Hyg. 77, 620-630.

Nathan, C.F. (1987) Secretory products of macrophages. J. Clin. Invest. 79, 319-326.

Nilsen, A., Nyberg, K. and Camner, P. (1988) Intraphagosomal pH in alveolar macrophages after phagocytosis in vivo and in vitro of fluorescein-labeled yeast particles. Exp. Lung Res. 14, 197-207.

Nitulescu et al. (1982) cited p. 13

Oberdörster, G. (1988) Lung clearance of inhaled insoluble and soluble particles. J. Aerosol. Med. 1, 289-330.

Oberdörster, G., Ferin, J. and Morrow, P.E. (1992) Volumetric loading of alveolar macrophages (AM): A possible basis for diminished AM-mediated particle clearance. Exp. Lung Res. 18, 87-104.

Oberdörster, G., Ferin, J., Morse, P., Corson, N.M., and Morrow, P. (1988) Volumetric alveolar macrophage (AM) burden as a mechanism of impaired AM mediated particle clearance during chronic dust overloading of the lung. J. Aerosol. Med. 1, A207.

Oberdörster, G. and Lehnert, B.E. (1991) Toxicological aspects of the pathogenesis of fiber-induced pulmonary effects. In: Mechanisms in Fibre Carcinogenesis. R.C. Brown, J.A. Hoskins, N.F. Johnson, eds., NATO ASI Series, Plenum Press, NY, p. 157-179.

Oberdörster, G., Morrow, P.E. and Spurney, K. (1988) Size dependent lymphatic short term clearance of amosite fibres in the lung. Ann. Occup. Hyg. 32, 149-156.

Parra, S.C., Burnette, R., Price, H.P. and Takaro, T. (1986) Zonal distribution of alveolar macrophages, type II pneumocytes, and alveolar septal connective tissue gaps in adult human lungs. Amer. Rev. Respir. Dis. 133, 908-912.

Phillips, B.J., James, T.E.B. and Anderson, D. (1984) Genetic damage in CHO cells exposed to enzymatically generated active oxygen species. Mut. Res. 126, 265-271.

Philipson, K., Falk, R. and Camner, P. (1985) Long-term lung clearance in humans studied with Teflon particles labelled with ^{51}Cr. Exp. Lung Res. 9, 31-42.

Pinkerton (1982) p. 10

Plowman, P.N. (1982) The pulmonary macrophage population of human smokers. Amer. Occup. Hygiene 25, 393-405.

Pott, F., Roller, N., Rippe, R.M., Germann, P.G. and Bellman, B. (1991) Tumours by the intraperitoneal and intrapleural routes and their significance for the classification of mineral fibers. In: Mechanisms in Fibre Carcinogenesis. R.C. Brown, J.A. Hoskins, N.F. Johnson, eds., NATO ASI Series, Plenum Press, London, p. 547-565.

Ramanathan, V.D., Badenoch-Jones, P. and Turk, J.L. (1979) Complement activation by aluminium and zirconium compounds. Immunol. 37, 881-888.

Ranga, V., Kleinerman, J., Ip, P.C. and Collins, A.M. (1980) The effect of nitrogen dioxide on tracheal uptake and the transport of horseradish peroxidase in the guinea pig. Amer. Rev. Respir. Dis. 122, 483-490.

Reynolds, H.Y. (1989) Pulmonary host defenses: State of the art. Chest 95, 223S-230S.

Reynolds, H.Y. (1988) Immunoglobulin G and its function in the human respiratory tract. Mayo Clinic Proc. 63, 161-174.

Reynolds, H.Y., Kazmierowski, J.A. and Newball, H.H. (1975) Specificity of opsonic antibodies to enhance phagocytosis of *Pseudomonas aeruginosa* by human alveolar macrophages. J. Clin. Invest. 56, 376-385.

Reynolds, H.Y. and Thompson, R.E. (1973a) Pulmonary host defenses. I. Analysis of protein and lipids in bronchial secretions and antibody responses following vaccination with *Pseudomonas aeruginosa* and alveolar macrophages. J. Immunol. 111, 358-368.

Reynolds, H.Y. and Thompson, R.E. (1973b) Pulmonary host defenses. II. Interaction of respiratory antibodies with Pseudomonas aeruginosa and alveolar macrophages. J. Immunol. 111, 369-380.

Rhodin, J.A.G. (1966) Ultrastructure and function of the human tracheal mucosa. Am Rev. Respir. Dis. 93, 1-15.

Richards, S.W., Peterson, P.K., Verburgh, H.A., Nelson, R.D., Hammerschmidt, D.E. and Hoidal, J.R. (1984) Chemotactic and phagocytic responses of human alveolar macrophages to activated complement components. Infect. Immunol. 43, 775-778.

Richardson, J., Bouchard, T. and Ferguson, C.C. (1976) Uptake and transport of exogenous proteins by respiratory epithelium. Lab. Invest. 35, 307-314.

Riley, P.A. and Dean, R.T. (1978) Phagocytosis of latex particles in relation to the cell cycle in 3T3 cells. Exp. Cell. Biol. 46, 367-373.

Robertson, J., Caldwell, J.R., Castle, J.R. and Walden, R.H. (1976) Evidence for the presence of components of the alternative (properdin) pathway of complement activation in respiratory secretions. J. Immunol. 117, 900-903.

Roggli, V.L., and Brody, A.R. (1984) Changes in numbers and dimensions of chrysotile asbestos fibers in lungs of rats following short-term exposure. Exp. Lung Res. 7, 133-147.

Saba, T.M., Blumenstock, F.A., Weber, P. and Kaplan, J.E. (1973) Physiological role of cold insoluble globulin in systemic host defense: Implication of its characterization as the opsonic γ-2 surface binding glycoprotein. Ann. NY Acad. Sci. 32, 43-55.

Saint-Remy, J.M.R., and Cole, P. (1980) Interactions of chrysotile asbestos fibres with the complement system. Immunol. 41, 431-437.

Sandusky, C.B., Cowden, M.W. and Schwartz, S.L. (1977) Effect of particle size on regurgative exocytosis by rabbit alveolar macrophages. In: Pulmonary Macrophage and Epithelial Cells. CONF-760927, Nat'l Technical Info. Service, U.S. Dept. Commerce, Washington, DC 85-105.

Scherer, P.W. (1981) Mucus transport by cough. Chest 80, 830-833.
Schmidt, M.E. and Douglas, S.D. (1972) Disappearance and recovery of human monocyte IgG receptor activity. J. Immunol. 109, 914-917.
Schwartz, L.W. and Christman, C.A. (1979) Alveolar macrophage migration: influence of lung lining material and acute lung insult. Amer. Rev. Respir. Dis. 120, 429-439.
Shellito, J., Esparza, C. and Armstrong, C. (1987) Maintenance of the normal alveolar cell population. Amer. Rev. Respir. Dis. 135, 78-82.
Sibille, Y., Chatelain, B., Staquet, P., Merril, W.W., Delacroix, D.L., and Vaerman, J-P. (1989) Surface IgA and Fc-alpha receptors on human alveolar macrophages from normal subjects and from patients with sarcoidosis. Amer. Rev. Respir. Dis. 139, 740-747.
Sibille, Y. and Reynolds, H.Y. (1990) Macrophages and polymorphonuclear neutrophils in lung defense and injury. Amer. Rev. Respir. Dis. 141, 471-501.
Sleigh, M.A. (1977) The nature and action of repiratory tract cilia. In: Respiratory Defense Mechanisms, Part I, J.D. Brain, D.F. Proctor, and L.M. Reid, eds., Marcel Dekker, New York 247-288.
Smaldone, G.E. (1986) Lung mechanics and mucociliary clearance. In: Aerosols: Formation and Reactivity. Pergamon Jour. Ltd., UK 175-181.
Smaldone, G.E., Itoh, H., Swift, D.L. and Wagner, H.N. (1979) Effect of flow-limiting segments and cough on particle deposition and mucociliary clearance in the lung. Amer. Rev. Respir. Dis . 120, 747-758.
Smaldone, G.E., Perry, R.J., Bennett, W.D., Messina, M.S., Zwang, J. and Ilowite, J. (1988) Interpretation of "24 hour lung retention" in studies of mucociliary clearance. J. Aerosol. Med. 1, 11-20.
Snipes, M.B., Olson, T.R., and Yeh, H.C. (1988) Deposition and retention patterns for 3-, 9-, and 14-µm latex microspheres inhaled by rats and guinea pigs. Exp. Lung Res. 14, 37-50.
Sorokin, S.P. and Brain, J.D. (1975) Pathways of clearance in mouse lungs exposed to iron oxide aerosols. Anat. Rec. 181, 151-163.
Spritzer, A.A., Watson, J.A., Auld, J.A. and Guetthoff, M.A. (1968) Pulmonary macrophage clearance: The hourly rates of transfer of pulmonary macrophages to the oropharynx of the rat. Arch. Environ. Health 17, 726-730.
Stahlofen, W., Gebhart, J., Rudolf, G., Scheuch, G. and Philipson, K. (1986) Clearance from the human airways of particles of different sizes deposited from inhaled aerosol boli. In: Aerosols: Formation and Reactivity. Pergamon Jour. Ltd. UK 192-208.
Stirling, C., and Patrick, G. (1980) The localisation of particles retained in the trachea of the rat. J. Pathol. 131, 309-320.
Strom, K.A. (1984) Response of pulmonary cellular defenses to the inhalation of high concentrations of diesel exhaust. J. Toxicol. Environ. Health 13, 919-944.
Stuart, B.O. (1984) Deposition and clearance of inhaled particles. Environ. Health Perspect. 55, 369-390.
Tarling, J.D. and Coggle, J.E. (1982) Evidence for the pulmonary origin of alveolar macrophages. Cell Tissue Kinet. 15, 577-584.
Task Group on Lung Dynamics. (1966) Deposition and retention models for internal dosimetry of the human respiratory tract. Health Phys. 12, 173-208.
Topping, D.C., Nettesheim, P. and Martin, D.H. (1980) Toxic and tumorigenic effects of asbestos on tracheal mucosa. J. Environ. Pathol. Toxicol. Oncol. 3, 261-275.
Tucker, R.W. and Frank, A.L. (1976) Asbestos cytotoxicity in a long-term macrophage-like cell culture. Nature 264, 44-46.
Valasquez, D.J. and Morrow, P.E. (1984) Estimation of guinea pig tracheobronchial transport rates using a compartmental model. Exp. Lung Res. 7, 163-176.
Valberg, P.A., Bing-Heng, C. and Brain, J.D. (1982) Endocytosis of colloidal gold by pulmonary macrophages. Exp. Cell. Res. 141, 1-14.
Valerio, F., Balduccio, D., and Lazzarotto, A. (1987) Adsorption of proteins by chrysotile and crocidolite: Role of molecular weight and charge density. Environ. Res. 44, 312-320.
Valerio, F., Balducci, D. and Scarabelli, L. (1986) Selective adsorption of serum proteins by chrysotile and crocidolite. Environ. Res. 41, 432-439.
van As, A. (1980) Pulmonary airway defense mechanisms: An appreciation of integrated mucociliary activity. Eur. J. Respir. Dis. 61, Suppl. 111, 21-24.
van As, A. and Webster, I. (1974) The morphology of mucus in mammalian pulmonary airways. Environ. Res. 7, 1-12.
van Oss, C.J. and Gillman, C.F. (1972) Phagocytosis as a surface phenomenon. II. Contact angles and phagocytosis of encapsulated bacteria before and after opsonization by specific antiserum and complement. J. Reticuloendothel. Soc. 12, 497-502.
van Oss, C.J. and Gillman, C.F. (1973) Phagocytosis as a surface phenomenon. III. Influence of C1423 on the contact angle and on the phagocytosis of sensitized encapsulated bacteria. Immunol. Commun. 2, 415-419.

van Oss, C.J. and Stinson, M.W. (1970) Immunoglobulins as aspecific opsonins. I. The influence of polyclonal and monoclonal immunoglobulins on the in vitro phagocytosis of latex particles and staphylococci by human neutrophils. J. Reticuloendothel. Soc. 8, 397-405.

Venaille, T., Snella, M-C., Holt, P.G. and Rylander, R. (1989) Cell recruitment into lung wall and airways of conventional and pathogen-free guinea pigs after inhalation of endotoxin. Amer. Rev. Respir. Dis. 139, 1356-1360.

Verbrugh, H.A., Hoidal, J.R., Nguyen, B.T., Verhoef, J., Quie, P.G. and Peterson, P.K. (1982) Human alveolar macrophage cytophilic immunoglobulin G-mediated phagocytosis of protein A-positive staphylococci. J. Clin. Invest. 69, 63-74.

Vostal, J.J., Schreck, R.M., Lee, P.S., Chan, T.L. and Soderholm, S.C. (1982) Deposition and clearance of diesel particles from the lung. In: Toxicological Effects of Emissions from Diesel Engines, J. Lewtas, ed. Elsevier Biomedical, New York 143-159.

Warheit, D.B., George, G., Hill, L.H., Snyderman, R. and Brody, A.R. (1985) Inhaled asbestos activates a complement-dependent chemoattractant for macrophages. Lab. Invest. 52, 505-514.

Warheit, D.B. and Hartsky, M.A. (1990) Species comparisons of proximal alveolar deposition patterns of inhaled particulates. Exp. Lung Res. 16, 83-99.

Warheit, D.B., Overly, L.H., George, G. and Brody, A.R. (1988) Pulmonary macrophages are attracted to inhaled particles through complement activation. Exp Lung Res 14, 51-66.

Watson, A.Y. and Brain, J.D. (1979) Uptake of iron oxide aerosols by mouse airway epithelium. Lab. Invest. 40, 450-459.

Weibel, E.R. and Gill, J. (1968) Electron microscopic demonstration of an extracellular duplex lining layer of alveoli. Respir. Physiol. 4, 52-57.

Weitzman, S.A. and Stossel, T.P. (1981) Mutation caused by human phagocytes. Science 212, 546-547.

Werb, Z. and Cohn, Z.A. (1972) Plasma membrane synthesis in the macrophage following phagocytosis of polystyrene latex particles. J. Biol. Chem. 247, 2439-2446.

White, H.J. and Garg, B.D (1981) Early pulmonary response of the rat lung to inhalation of high concentration diesel particles. J. Appl. Toxicol. 1, 104-110.

Wilkey, D.D., Lee, P.S., Hass, F.J., Gerrity, T.R., Yeates, D.B. and Lourenco, R.V. (1980) Mucociliary clearance of deposited particles from the human lung: Intra- and inter-subject reproducibility, total and regional clearance, and model comparisons. Arch. Environ. Health 35, 294-303.

Wilson, M.R., Gaumer, H.R. and Salvaggio, J.E. (1977) Activation of the alternative pathway and generation of chemotactic factors by asbestos. J. Allergy Clin. Immunol. 60, 218-222.

Wolff, R.K., Henderson, R.F., Snipes, M.B., Griffith, W.C., Mauderly, J.L., Cuddihy, R.G., and McClellan, R.O. (1987) Alterations in particle accumulation and clearance in lungs of rats chronically exposed to diesel exhaust. Fund. Appl. Toxicol. 9, 154-166.

Wood, R.E., Wanner, A., Hirsch, J. and Farrel, P.M. (1975) Tracheal mucociliary transport in patients with cystic fibrosis and its stimulation by terbutaline. Am Rev Respir Dis 111, 733-738.

Yeates, D.B., Aspin, N., Levison, H., Jones, M.T. and Bryan, A. (1975) Mucociliary tracheal transport rates in man. J. Appl. Physiol. 39, 487-495.

Yeates, D.B., Pitt, B.R., Spektor, D.M., Karron, G.A. and Albert, R.E. (1981) Coordination of mucociliary transport in human trachea and intrapulmonary airways. J. Appl. Physiol.: Respirat. Environ. Exer. Physiol. 51, 1057-1064.

CHAPTER 15

IN VIVO ASSAYS TO EVALUATE THE PATHOGENIC EFFECTS OF MINERALS IN RODENTS

John M. G. Davis

*Institute of Occupational Medicine Ltd.
8 Roxburgh Place
Edinburgh EH8 9SU Scotland*

METHODS OF EXPOSURE

Concern for the pathogenic effects of minerals centers on two pathological conditions: fibrosis and neoplasia (or the growth of abnormal tissue). For many years it has been known that some dusts (e.g., quartz and coal-mine dusts) can produce nodular pulmonary fibrosis (pneumoconiosis) when inhaled by humans. Following the Second World War, it was realized that some fibrous dusts (e.g., asbestos) could also cause cancer in addition to a diffuse form of pulmonary fibrosis (asbestosis). In order to assess the pathogenicity of various dusts and to learn more about the mechanisms of this pathogenicity, a number of *in vivo* assays have been developed. This chapter focuses on the principles underlying *in vivo* methods and reviews some of the data and significant discoveries derived by these techniques.

A general principle of experimental toxicology using *in vivo* methods is that, where possible, substances being tested for possible hazard should be administered by the physiological route(s) relevant to human exposure. For mineral dusts, the predominant exposure route is inhalation, and experimental inhalation studies have been very productive in producing information on pathogenesis in the respiratory tract and associated organs (e.g., the pleura). A secondary exposure route in humans is ingestion, because much inhaled dust is removed from the lung by the mucociliary escalator and swallowed (see Chapter 14). Hence, experimental ingestion studies have been used to determine if there is a potential hazard to the gut linings.

Long term inhalation studies are extremely expensive to undertake, and only a few laboratories have the necessary equipment. Furthermore, this type of study requires a relatively large amount of material (e.g., up to kg quantities), and it can sometimes be useful to screen materials before allocating resources that can be tied up for years. For these reasons, many techniques using other exposure routes have been developed, and these methods are generally cheaper, require less material, and produce results more rapidly. However, these techniques employ artificial exposure mechanisms that are not directly relevant to human exposure. Most of these exposure techniques involve injection of the material into the target organ. For example, intratracheal injection can be used to deposit dusts directly into the lung, and this exposure technique is the most relevant injection method for administering finely divided minerals. However, as with studies of the harmful nature of many chemicals, injection into quite abnormal sites has also been used. Intramuscular injection, intravenous injection, and intrarenal injection have all been reported. Injection directly into the pleural and peritoneal cavities has been the most common injection technique used to evaluate fibrous dusts, due to the realization that some asbestos varieties can cause tumors in the cells lining these body cavities (mesotheliomas). In life, dust must reach the body cavities by transport from the pulmonary parenchyma (i.e., lung), but dust injection bypasses this route. By doing so, intrapleural and intraperitoneal injection techniques speed delivery of the material to the pleura, but they also eliminate the translocation process. This translocation process may be important in pathogenesis in humans (e.g., by determining which types of particles are delivered to the pleura and peritoneum or by allowing the particles to interact with cells and fluid before reaching the target organ).

For experimental studies with all mineral dusts, the following points are important to note. First, small laboratory rodents (particularly rats) constitute the vast majority of test animals, but the smaller dimensions of their bronchial tubes make the penetration of large dust particles more difficult than in man. Whereas particles with a diameter of 10 μm can penetrate to the pulmonary parenchyma of man with some ease, the corresponding figure for rats is only 5 μm. For fibers, the critical diameter for man is 3 μm, but, in the rat, few fibers >1 μm in diameter can penetrate as far as the ends of the terminal bronchioles. This means that if the harmful potential of a mineral is to be tested, the dust cloud used must contain high levels of "rat respirable" material, and this material will not necessarily be identical to "human respirable" material for the same sample. Second, the principle pathological conditions known to be caused by dusts (pulmonary fibrosis and tumors) develop slowly and only after there has been considerable build up of dust in the lungs. These two conditions require that high dose levels must be administered for a long period and that the experimental animal must be followed for its full potential lifespan. With rats, few pulmonary tumors have been reported following dust inhalation in animals less than two years old. Finally, where a non-physiological method of dust application is to be used, the same restrictions on particle and fiber size should apply if one is interested in testing materials directly *relevant to human exposure*. In such cases, there is little point in depositing in tissues material that could never reach any tissue by physiological routes. However, in cases where one is interested in evaluating specific mineralogical properties, it may be useful to test materials not directly relevant to human exposure.

PATHOGENIC POTENTIAL OF AMPHIBOLES AND SERPENTINES

The harmful potential of the amphiboles and serpentines should be considered together, since the asbestiform varieties of these minerals are the most widely studied of all minerals. Commercially, the amphibole minerals crocidolite and amosite have been of major importance, but tremolite and anthophyllite have been mined as well (see Chapter 3). Chrysotile asbestos, a serpentine mineral, has always represented over 90% of commercially used asbestos. Both clinico-pathological assessment and epidemiology has demonstrated that all asbestos varieties can produce asbestosis and pulmonary carcinomas in humans if exposure levels are high. Mesotheliomas on the other hand appear much more associated with the amphibole minerals, particularly crocidolite (Wagner et al., 1960). Whether chrysotile fibers on their own can ever cause mesotheliomas is still debated, since many chrysotile occurrences contain tremolite (Pooley, 1976; Churg, 1988; see Chapter 3 for the mineralogy of chrysotile occurences and Chapter 13 for data on lung burdens in asbestos-exposed individuals). Experimental animal studies with asbestos commenced in the 1930s at a time when asbestosis was believed to be the only asbestos-related industrial hazard. Preliminary work was undertaken in Gardner's laboratory at Saranac Lake in the United States using chrysotile dust samples, and these experiments continued for several years. The final results were published after Gardner's death (Vorwald et al., 1951). Gardner's methods of dust generation were primitive by today's standards, and no real sizing or quantification of the dust cloud was undertaken. Nonetheless recognizable pulmonary asbestosis developed in rats, mice, rabbits, and cats. The major finding of Gardner's studies was that very finely ground chrysotile asbestos was much less fibrogenic than dust containing longer fibers. Gardner failed to demonstrate the carcinogenic potential of asbestos fibers mainly because his rats, now regarded as the most susceptible species of experimental animal, did not live long enough.

A major impetus to experimental studies with asbestos was Wagner's discovery that crocidolite asbestos was responsible for the development of pleural mesotheliomas[1] in asbestos miners in South Africa (Wagner et al., 1960). Wagner quickly demonstrated that finely divided asbestos could produce mesotheliomas in experimental animals if injected directly into the pleural cavity (Wagner et al., 1962), but most major work in the subsequent few years explored the pathogenicity of asbestos-related disease by administering asbestos dust by inhalation. Gross et al. (1967) confirmed that rats could develop pulmonary asbestosis and were the first to record the presence of pulmonary carcinomas in experimental studies. These findings were confirmed by Reeves et al. (1974) who reported the development of asbestosis in rats, mice, and gerbils treated with asbestos; however, only rats developed pulmonary tumors. A very large and important study was reported by Wagner et al. (1974) in which the five major types of asbestos were each administered to rats by inhalation at a high-dose level for periods between 1 day and two years. Chrysotile (both Canadian and Rhodesian), amosite, crocidolite, and anthophyllite all produced pulmonary fibrosis and pulmonary tumors, and the incidence and severity of these conditions increased with the length of exposure. Only Rhodesian chrysotile produced no mesotheliomas, but the number of animals developing these tumors from the other mineral samples was always low, a response now recognized as characteristic of the rat. Davis et al. (1978) also demonstrated that pulmonary fibrosis and tumor production was dose related in rats treated by inhalation with different levels of chrysotile, crocidolite, and amosite. They also attempted to assess the relative pathogenicities of the minerals by administering doses with equal fiber numbers. Fibers were, however, estimated as in the factory environment (by light microscopy), and this technique fails to see many of the thinner fibers. Results

---

[1] Mesotheliomas are normally extremely rare in humans (see Chapter 11).

showed that dust clouds of chrysotile asbestos examined in this way were more pathogenic than the amphiboles, although it is now realized that the numbers of very thin fibers of chrysotile used were in fact much higher. Both Wagner et al. (1980) and Davis et al. (1986a) have amplified the importance of the numbers of very fine fibers in determining chrysotile pathogenicity. An examination of the harmful potential of different asbestos types was continued by Davis et al. (1985), who demonstrated that asbestiform tremolite was extremely potent in producing both pulmonary fibrosis and tumors in rats.

Non-pulmonary pathogenesis

Following a suggestion by Selikoff (1964) that the incidence of gastrointestinal cancers was increased in asbestos workers, numerous experimental ingestion studies have been undertaken in which laboratory animals have been fed large doses of asbestos with their diets. These studies have been extensively reviewed (Condie, 1983; Davis, 1993), and none demonstrated a causal link between gastrointestinal cancers and fiber injection. Hence, unless the intestines of small animals differ markedly from those of humans in their susceptibility to damage by fibers, asbestos is unlikely to be a hazard when ingested.

Importance of fiber geometry

Perhaps one of the major findings from experimental studies on the pathogenesis of asbestos-related disease concerns the importance of fiber geometry. Gardner's original observation that short fibers cause less fibrosis than long fibers was confirmed by King et al. (1946) using intratracheal injection in rabbits. Furthermore, a large series of studies using intrapleural or intraperitoneal injection have demonstrated that long thin fibers are the most effective at producing mesotheliomas once they are within the body cavities (Pott et al., 1972; Pott, 1976; Stanton et al., 1972, 1977). The combined studies by Pott and Stanton suggested that the most dangerous fibers were >8 µm in length and <0.25 µm in diameter. Davis et al. (1986b, 1988) confirmed that long fibers of amosite and chrysotile are more carcinogenic and fibrogenic than short fibers of these minerals following inhalation. A specially prepared short fiber sample of amosite, with all fibers below 5 µm in length, produced neither fibrosis nor pulmonary tumors, whereas a long fiber sample was extremely fibrogenic and carcinogenic. Similar studies were undertaken with long and short fiber samples of chrysotile but attempts to produce chrysotile, with all fibers shorter than 5 µm, had been less successful than the studies of amosite. Nonetheless the "short" chrysotile dust produced many fewer tumors and much less fibrosis than the long. Comparing data from three amosite studies, Davis (1989) suggested that with respect to the induction of pulmonary tumors and fibrosis by inhalation, fibers longer than 20 µm are more potent than fibers shorter than 10 µm.

The importance of fiber dimensions is particularly emphasized when the same mineral exhibits both fibrous and nonfibrous morphologies, such as the case for tremolite. Asbestiform tremolite is an extremely pathogenic material. However, non-asbestiform tremolite is very widely found in the earth's crust, so there is a background component of tremolite to which many are exposed (e.g., see

Chapters 2 and 3). Furthermore, tremolite can appear as a contaminant in many mineral products. Davis et al. (1991a) demonstrated that although asbestiform tremolite is extremely carcinogenic when injected into the peritoneal cavities of rats, non-asbestiform tremolite samples have little or no carcinogenic potential. These observations suggest that tremolite contamination of any material poses a concern only if thin asbestiform fibers are present. This may be particularly confusing for minerals such as tremolite, however, because tremolite can occur in massive, fibrous (but non-asbestiform), and asbestiform habits (see Chapter 3). Each of these morphologies has different physical properties, and it is not systematically understood how pathogenic potential relates to these variations (e.g., Is the pathogenic potential of fibrous tremolite more similar to asbestiform tremolite or massive tremolite?).

One contentious area of asbestos-related disease where data from experimental studies may be of importance concerns the relationship between asbestosis and pulmonary cancers. There have been many suggestions that, in humans, lung cancer resulting from asbestos exposure is only found in workers whose level of pulmonary damage has also resulted in asbestosis. Certainly all experimental inhalation studies where levels of fibrosis as well as tumor production have been recorded show clearly that those dusts that produce large numbers of tumors also cause widespread fibrosis, whereas those that failed to produce tumors also failed to stimulate significant fibrosis. Davis and Cowie (1990) summarized data from a number of inhalation studies where records were available for individual rats. They showed that animals that had developed pulmonary tumors had on average twice the amount of fibrosis as those that did not. Certainly in rats a gradation of effects can be seen with advanced fibrosis being mixed with hyperplastic alveolar epithelium to produce a pattern of adenomatosis. Some early tumors can certainly be found developing from these lesions.

Relative pathogenic potential of serpentine and amphibole

One area where the experimental studies themselves have produced data that is contentious concerns the relative pathogenicity of the different asbestos varieties. Human epidemiology suggests that the amphiboles, particularly crocidolite, are more harmful than chrysotile, particularly in the causation of mesotheliomas. In animal experiments, however, both chrysotile and the amphibole varieties appear equally pathogenic, indeed in some studies chrysotile appears more dangerous than the amphibole dust. These apparent contradictions may result from differences in the durability of fibers within the lung tissue, which have been suggested by many experimental studies. Wagner et al. (1974) showed that the accumulation of amphibole in rat lungs is much more rapid than the accumulation of chrysotile (when the two are administered at the same dose). This was confirmed by Davis (1989) who reported that, in the six months following a one-year period of inhalation, only 14% of a long fiber amosite preparation was cleared from rat lungs compared to 50% for a long fiber chrysotile sample. With short fibers the difference was even more marked: Only 21% of a short fiber amosite was cleared within six months compared to 90% of short fiber chrysotile. Those experimental studies that show comparable pathogenic potentials for chrysotile and amphibole may reflect the following: At very high doses (as occurs in many *in vivo* experiments), chrysotile is able to exert its maximum possible pathogenic effect, becuase high levels of chrysotile are artificially maintained in the target organ. However, in humans exposed at doses several orders-of-magnitude lower, the rapid clearance of chrysotile makes it much more difficult for dust accumulation to reach harmful levels.

PATHOGENIC POTENTIAL MAN-MADE MINERAL FIBERS

While experimental studies with asbestos commenced with the fact that the material was harmful to humans already established, the situation with man-made mineral fibers (MMMF) has been quite different. There is still no substantive evidence that any type of MMMF has caused disease in man, yet these materials elicit concern because of the results from experimental studies on asbestos. Pott et al. (1972, 1976) and Stanton et al. (1972, 1977) included samples of glass and other types of MMMF in injection studies in rats and reported that these materials could produce mesotheliomas with similar frequency to asbestos. Fiber geometry appeared to be the only important factor. If samples contained large numbers of long thin fibers, they were highly carcinogenic. Intratracheal injection of fine glass fibers has also been shown to produce disease in rat lungs. Wright and Kuschner (1977) showed that long glass fibers injected into the lung could cause widespread pulmonary fibrosis, whereas this did not occur with short fiber samples. Pott et al. (1987) produced both pulmonary tumors and mesotheliomas by the intratracheal injection of glass fibers in rats. This set of findings stimulated much further experimental work to examine the possibility of human hazard, and long term inhalation studies have since been reported using a wide range of MMMF. Early studies (Schepers et al., 1955; Gross et al., 1970; Lee et al., 1979) suffered from difficulties in generating dust clouds of fibers from materials such as glass, because glass fibers can fracture easily to produce what is essentially a non-fibrous dust. Consequently, none of these studies demonstrated significant pathological change in animal lungs. More recent studies have overcome these difficulties, but the results have still been negative. Both Wagner et al. (1984) and McConnell et al. (1984) treated rats with very high doses of fine glass microfiber of similar dimensions to asbestos but produced neither pulmonary fibrosis nor tumors in excess of the numbers found in control animals. Similarly, Le Bouffant et al. (1987), Mühle et al. (1987), and Smith et al. (1987) all reported results with of MMMF.

Differences between the results of injection studies and inhalation studies appear once more to result from the effects of low fiber durability in lung tissue. Many MMMF types are quite soluble in lung tissue (even more soluble than chrysotile asbestos), and fiber levels high enough to cause disease are difficult to maintain. The direct injection of large numbers of fibers into the body cavities bypasses clearance mechanisms, and since fibers are packed together in masses, this technique also reduces the potential for fiber dissolution. There is a very wide range of fiber durability among MMMF,depending on dimension and chemistry, and it is now realized that experimental studies must combine data on solubility with the reports of pathological findings. That fibers other than asbestos can produce pulmonary damage if their dimensions are suitable and they are sufficiently durable has been shown with ceramic aluminum silicate fibers. Davis et al. (1984) reported both pulmonary fibrosis and tumor production with this material and this has now been confirmed in much larger studies (Hesterberg et al., 1992). Furthermore, erionite is extremely pathogenic, as discussed below under **Pathogenic Potential of Zeolites**.

Many years of experimentation with mineral fibers has resulted in the acceptance of the following requirements for inhalation studies where the potential for human hazard is being examined:

- Dust levels must be high, and fibers of dangerous length (but sufficiently small to be respirable by rats) must be present in large numbers. A minimum of 100 fibers/ml >20 µm in length and <1 µm in diameter is now considered necessary.
- Data on clearance rates and dissolution behavior of fibers in the lung tissue must be obtained so that likely survival of fibers in human lungs can be predicted.

Since some inhalation studies, even those published within the last ten years, have been defective in one or both of these requirements, many types of MMMF are being re-tested at present and the results of these studies are awaited. When this data is available it may be possible to confirm the suggestion that a fiber will be dangerous if it is long enough and thin enough and can survive in lung tissue long enough for dangerous fiber levels to accumulate.

PATHOGENIC POTENTIAL OF CLAY MINERALS

The clay minerals are a complex family that includes a number of platy and fibrous silicates (see Chapter 5). The main minerals that have been studied in bioassays are the kaolin minerals, talc, the micas, vermiculite, sepiolite, and palygorskite. Some of these are known to be pathogenic to humans, and the experimental knowledge relating to their pathogenicity is best described for each type separately.

Kaolin minerals

Kaolin or china clay is extensively mined in some parts of the world, and workers in this industry have developed pneumoconiosis (Oldham, 1983). Where workers have inhaled dust from the associated rocks in addition to kaolin, the pneumoconiotic nodules are often similar to silicosis. Where exposure has been to relatively pure kaolin, however, the pathologic response is slightly different: Much smaller stellate macules are the main lesions, but a few larger palpable nodules may also be found. There are records of massive lesions several centimeters in diameter.

The results from two recent experimental inhalation studies with kaolin dusts showed differences, but these appear to be due to differences in the dust doses used. Wagner (1990b) treated rats with two kaolin samples from the china clay industry at a dose level of 10 mg/m^3 of air for 12 months. Even after 24 months, only minimal fibrosis of a non-nodular type had occurred, but no pulmonary tumors had developed. Wastiaux and Daniel (1989) treated female rats with a china clay kaolin dust sample at a dose level of 300 mg/m^3 of air, although the exposure lasted for 3 months only. In this study alveolar lipoproteinosis was present at the end of dusting, and fibrosing macules developed as the animals aged. Three rats developed pulmonary carcinomas. This result is similar to a number of studies in the literature where massive doses of relatively non-toxic dust have been administered to rats. In this species, a massive chronic foreign-body reaction appears to lead to tumor production, especially in females, and care must be taken with dose levels if results are to be used to predict human risk.

Early inhalation studies with kaolin (Gross et al., 1960) and a number of intrapleural or intraperitoneal injection studies have demonstrated that kaolin dust will produce granulomatous lesions that undergo reticulinization, but fibrosis seldom progresses to the production of mature collagen (Attygalle et al., 1955; Martin et al., 1977; Sahu et al., 1978; Rosmanith et al., 1989).

Talc

Exposure to talc dust in industry can cause pneumoconiosis that may vary from an asymptomatic simple type to a disabling conglomerate pneumoconiosis (Miller et al., 1971; Vallyathan and Craighead, 1981). Experimental studies have, however, demonstrated that talc dust possesses only a low level of pathogenicity.

Numerous studies have been reported in which talc has been administered to rats, hamsters, and mice by intratracheal, intraperitoneal or intrapleural injection. In no case have significant numbers of tumors developed (Pott et al., 1976; Bischoff and Bryson, 1976; Stenbäck and Rowland, 1978 and Ozesmi et al., 1985).

Wehner et al. (1977, 1979) treated hamsters by inhalation of a dust cloud from talc baby powder at a dose level of 9.8 mg/m^3 but for only 30 days and for a maximum of 2 hours per day. No histological changes of any type were found. Wagner et al., 1977 treated rats by inhalation to a sample of Italian talc at a dose level of 10.8 mg/m^3 for 1 year. Only minimal fibrosis developed, and this did not progress with time after the cessation of dusting. No significant increase in pulmonary tumors above control levels was found.

A recent study from the U.S. National Toxicology program treated both rats and mice by inhalation to a cosmetic grade of talc. Dose levels were up to 18 mg/m^3 and exposure continued for 2 years. Rats developed a "granulomatous inflammation" the severity of which increased with exposure duration and concentration. This eventually resulted in interstitial fibrosis, and, in female rats, pulmonary tumors occurred. In mice the pulmonary response was less, and neither fibrosis nor tumor production occurred. This study is one where very high doses of a material not particularly expected to be carcinogenic has produced significant numbers of tumors on female rats. The problem this creates in using the data for hazard assessment in humans is described elsewhere in this chapter.

One particular problem with some deposits of talc is that they can be contaminated with tremolite. As discussed above, asbestiform tremolite is extremely carcinogenic, and significant contamination of talc by such tremolite would result in a hazardous material. Non-fibrous tremolite does not possess the same carcinogenic potential, and talc contamination with this material would be of little concern.

Mica

The mica family of minerals has shown little evidence of causing human pulmonary disease (Gibbs, 1990). Perhaps because of this, the micas have been the subject of very little experimentation.

King et al. (1947) found that untreated samples of mica (sericite) showed little fibrogenic potential following intratracheal injection into rats. However, one sample treated with acid which modified the surface properties of the dust was found to produce a fibrogenic reaction similar to quartz. Recent studies also using intratracheal injection into rats have reported that each of 4 muscovite samples produced pulmonary granulomas with eventual production of collagen (Rosmanith et al., 1990). Similarly, Sahu (1990) injected muscovite intratracheally into rats, mice, and guinea pigs. Early lesions in all of these species were only of the foreign-body type, but as the animals aged these showed some evidence of fibrosis.

Sepiolite and palygorskite ("attapulgite")

These minerals are of industrial importance being used as sorbents and lubricants, particularly in drilling muds in the oil industry. Nonetheless there only limited evidence that industrial exposure has thus far produced at most minor disease in humans (e.g., Waxweiler et al., 1988). However, these minerals are of interest from the health point of view, because they can occur with fibrous morphologies, some of which have dimensions similar to asbestos. Because of this they have been used in a number of experimental studies. In general the results have been what would have been expected for a fibrous mineral with samples having fibers >5 µm in length proving harmful, whereas material consisting entirely of short fibers is not. Stanton et al. (1981) implanted two samples of attapulgite into the pleural cavity of rats. One sample contained no fibers >4 µm in length, whereas the other contained <1% of this size. In both experiments mesothelioma production was not raised above control levels. Pott et al. (1976) gave rats intraperitoneal injections of an attapulgite sample containing 30% fibers >5 µm in length, and 77% of animals eventually developed mesotheliomas. In more recent studies Pott et al. (1990) gave rats intraperitoneal injections of 4 attapulgite samples and 2 sepiolite samples. One attapulgite and one sepiolite specimen that contained significant numbers of fibers >5 µm in length produced high levels of mesotheliomas. The other dusts produced no effect above control levels.

Wagner et al. (1987) produced essentially similar results in combined inhalation and injection studies in rats. Following intrapleural injection of 3 samples of sepiolite, 2 samples of attapulgite, and 1 of palygorskite,[2] the palygorskite and one of the attapulgite samples (both of which contained significant numbers of fibers >6 µm in length) produced large numbers of mesotheliomas. The remaining samples, which consisted of shorter fibers, produced no tumors. The results from inhalation studies were less dramatic with only the palygorskite sample (which contained the most long fibers) producing levels of fibrosis and pulmonary tumors that appeared to be just above control levels.

Vermiculite

Vermiculite is another clay mineral used by industry, and there are at present concerns regarding possible health hazards due to vermiculite exposure. Much of this concern however results from the fact that some samples of vermiculite are contaminated with tremolite, which, if asbestiform, would present a hazard in its own right. Pure vermiculite exposure appears to present little problem (Sluis-Cremer and Hessel, 1990). A few experimental studies have been undertaken with pure vermiculite materials and these have produced largely negative results. Vallyathan (1990) included a sample of vermiculite in rat inhalation studies which encompassed a number of mineral types. There were signs of early toxicity following the inhalation of vermiculite, but these were transient and did not progress to either fibrosis or the production of pulmonary tumors. Liu et al. (1988) did produce pulmonary fibrosis in rats following intratracheal injection of 50 mg of dust. This is a very high dose, however, and the authors described the results as "slight fibrogenic changes."

When administered to rats by intraperitoneal injection, vermiculite does not produce mesotheliomas (Hunter and Thompsin, 1973).

PATHOGENIC POTENTIAL OF ZEOLITES

Natural zeolites are dominantly either platy (tabular) or fibrous. One fibrous variety, erionite, is perhaps the most potentially harmful mineral known. Baris et al. (1979) reported that in some isolated villages in Turkey over 40% of the population died from mesothelioma, a hazard level never approached even by the most uncontrolled use of asbestos. The basal rocks near the affected villages were found to contain erionite, and experimental studies have confirmed the extremely hazardous nature of this material.

Wagner et al. (1985) reported that rats treated by inhalation of erionite almost all developed pleural mesotheliomas. However, he later demonstrated that fiber geometry is as important in the carcinogenicity of erionite as it is with asbestos. Rats inhaling finely ground erionite developed no tumors (Wagner, 1990a).

Several studies in which erionite has been injected into the pleural or peritoneal cavities of rats or mice have confirmed its carcinogenic potential (Maltoni et al., 1982; Suzuki, 1982; Wagner et al., 1985; and Pylev et al., 1986).

---

[2] No distinction made between the attapulgite samples and the palygorskite sample.

Davis et al. (1991) demonstrated that some factor other than fiber geometry enables erionite to exert its extremely carcinogenic effect. In this study dose response had been examined with chrysotile, crocidolite, amosite, and erionite by intraperitoneal injection into rats. All these minerals showed clear dose response effect but that for the three commercial asbestos varieties was very similar when dose was calculated by numbers of long thin fibers. At any fiber number, however, erionite was more carcinogenic than the asbestos varieties.

Only one *in vivo* study has examined the carcinogenic potential of a natural fibrous zeolite other than erionite. Suzuki (1982) administered mordenite to mice by intraperitoneal injection. Erionite was included in the same study. While mesotheliomas developed in 80% of the mice receiving erionite, none occurred in mice injected with mordenite. Other studies have used unnamed fibrous zeolites. Suzuki and Kohyama, (1984) injected erionite and synthetic zeolite 4A intraperitoneally into mice. The erionite was highly carcinogenic, but zeolite 4A produced no tumors. Wagner et al. (1985) included a non-fibrous Japanese zeolite as well as erionite in intrapleural injection studies in rats. Once again only the erionite produced mesotheliomas.

The combined experimental evidence available suggests that non-fibrous zeolites are not hazardous; however, the data are *extremely* limited. More information is needed on why fibrous erionite is such a particularly dangerous material, and this information may result from a more thorough investigation of the pathogenic affects of zeolites in general.

THE PATHOGENIC POTENTIAL OF SILICA

As with asbestos, experimental studies with silica commenced with the prior knowledge that, in humans, inhalation of this material could produce disease. Silica dust inhalation usually results in a classical picture of dense round fibrotic nodules (silicosis), although when extremely high doses are inhaled over a relatively short period, widespread pulmonary inflammation resulting in alveolar lipoproteinosis can occur. This condition is known as acute silicosis and is often fatal. Where silica is inhaled mixed with larger quantities of other dusts (as occurs in coal mines), more diffuse fibrotic nodules develop.

Experimental studies have reproduced the lesions of silicosis in animals and helped to explain the important features of disease development. Several studies have reported the development of classical silicotic nodules. These were found in rats treated with finely divided quartz within 300 days by King et al. (1950). Similar findings have been obtained in rats (Marenghi and Rota, 1953; Watanabe, 1956 and Heppleston, 1962), guinea pigs (Gardner, 1935), rabbits (Denny et al., 1939), and monkeys (Cauer and Neymann, 1953) after periods of exposure by inhalation varying from one week to two years. Other experimental studies have not resulted in the production of silicotic nodules. Heppleston (1967, 1970) reported alveolar lipoproteinosis in rats treated with pure quartz dust. Other studies have confirmed these findings in rats and indicated that this type of lesion is very similar to the alveolar lipoproteinosis found in acute silicosis in humans (Gross and de Treville, 1968; Corrin and King, 1970). The lipoproteinaceous material appears to be produced by hyperactive Type II pneumonocytes (also called Type II alveolar cells) and may be related to pulmonary surfactant.

In general, it appears that in both experimental animals and humans, very high doses of pure quartz result in toxic change to the lung, whereas lower doses result in the development of silicotic nodules. By no means are all of the factors involved in these different responses fully understood. Silica exists in a number of forms (see Chapter 5), and each of these forms appears to possess a different pathogenic potential. Amorphous silica is apparently less pathogenic than crystalline varieties, although most of the experimental work demonstrating this has been undertaken by injection exposure rather than inhalation. Silverman and Moritz (1950) injected both amorphous and crystalline silica into rabbits. Both silica varieties produced silicotic nodules, but those produced by the crystalline dust were much larger. Similar results were described by Policard and Collet (1957). Klosterkötter and Jötten (1953) exposed rats by intraperitoneal injection, intratracheal injection and inhalation to a variety of silica preparations and reported that solutions of silicic acid and silica gels were non toxic and non fibrogenic, but colloidal amorphous silica was toxic but not fibrogenic. They also found that quartz produced maximum fibrotic response. King et al. (1953a) found that following intratracheal injection into rats, tridymite was the most fibrogenic followed by cristobalite and then quartz. Similar findings were observed by Schmidt and Luchtrath (1955) and Saffiotti (1962).

There have been many reports that the pathogenic potential of silica dusts depends on particle size, with the finest materials being the most harmful However, much of the experimental work has been done using the extremely artificial technique of intravenous injection (Gardner and Cummings, 1933; Tebbens et al., 1945; Swensson et al., 1956). It has been suggested that this is an effect of particle surface area, but following intratracheal injection into rats of a variety of quartz preparations with a constant surface area, maximum fibrosis was still found with the finest material (King et al., 1953b). The most fibrogenic size of quartz dust particles in lung tissue appears to be 1-2 μm in diameter (King and Nagelschmidt, 1960; Goldstein and Webster, 1966).

Particle size alone, however, is not the only factor in determining the harmful effects of polymineralic assemblages containing quartz. This has been found particularly in the coal mining industry where quartz levels in mine dusts can range between 1 and 25% or more. In some cases, the inhalation of dust with high quartz content can produce multiple round fibrotic nodules very similar to silicosis, but in other mining areas, the same proportion of quartz in mine dust causes little disease (Jacobsen et al., 1971; Davis et al., 1991). It would appear that in some circumstances quartz can be more toxic than in others. It has been suggested that this could be due to the sorption of protective substances onto the reactive surfaces of quartz particles. Reissner (1983) has shown that there is better correlation between coal mine dust toxicity and *uncontaminated quartz surface area* (measured by thermoluminescence) than between toxicity and the *total quartz mass content* of the dusts. Le Bouffant et al. (1983) has suggested that this protective effect may be due to aluminum leached from clay minerals (e.g., illite or mica), and they have demonstrated experimentally that such inhibition may not be permanent. It is suggested that continued exposure may maintain protection whereas removing a man or an experimental animal from exposure to coalmine dust may result in the termination of the inhibition of quartz toxicity. Experimental studies have demonstrated that, within any one colliery, the proportion of quartz in inhaled dust determines the level of pneumoconiosis (Robertson et al., 1984), but more needs to be determined about the conditions under which quartz surface protection occurs. Davis et al. (1991) treated rats with dusts from two collieries, both with very high quartz levels. One colliery had a high recorded incidence of pneumoconiosis, the other a very low incidence. Certainly the low incidence colliery had the highest levels of illite in the dust yet the high incidence colliery also had a significant amount of illite (17 wt %). In spite of this multiple silicotic nodules developed in rats treated with this dust, whereas no pneumoconiotic nodules at all were found in the other treatment group.

Although experimental studies on the *fibrotic potential* of silica have in general produced similar findings to those produced from humans, rather different results have been reported on the *carcinogenic effects* of silica. There have been suggestions that populations of workers in a number of mining industries have an increased incidence of lung cancer, but, in spite of the high numbers of people involved, this has been difficult to substantiate (IARC, 1987). Experimental quartz inhalation studies in rats, particularly female rats, have however produced significant numbers of pulmonary carcinomas (Dagle et al., 1986; Holland et al., 1986). Similar results have been obtained when quartz has been mixed with other dusts, particularly coal (Martin et al., 1977), but both Robertson et al. (1984) and Davis et al. (1991b) found no evidence of carcinogenicity in rats treated with coalmine dusts containing up to 20% quartz. No evidence of carcinogenicity was found in mice treated with quartz by inhalation (Wilson et al., 1986). Quartz has been found to be carcinogenic to rats both by intratracheal injection (Holland et al., 1983; Groth et al., 1986) and intrapleural injection (Wagner, 1976), but studies using hamsters treated by intratracheal injection have been negative (Holland et al., 1983; Rennie et al., 1985).

The finding that quartz can be carcinogenic in rats but not in other experimental animals is part of a pattern which indicates that the rat lung responds to widespread chronic damage and fibrosis with tumor production much more readily than other species. This makes the prediction of carcinogenic hazard to humans very difficult, because results from rats tend to exaggerate danger levels, particularly when extremely high doses are used.

REFERENCES

Attygalle, D., Harrison, C.V., King, E.J. and Mohanty, G.P. (1955). Infective pneumoconiosis. The influence of dead tubercule bacilli (BCG) on the dust lesions produced by anthracite coalmine dust and kaolin in the lungs of rats and guinea pigs. Brit. J. Indus. Med. 11, 245-254.

Bischoff, F. and Bryson, G. (1976). Talc at the rodent intrathoracic, intraperitoneal and subcutaneous sites. Proc. Amer. Assoc. Cancer Res. 17, 1.

Cauer, H. and Neymann, N. (1953). New studies on electro-aerosol inhalation fields. Staub-Reinhalt Luft. 33, 293-307.

Churg, A. (1988) Chrysotile, tremolite and malignant mesothelioma in man. Chest 93, 621-628

Condie, L.W. (1983). Review of published studies of orally administered asbestos. Environ. Health Perspec. 53, 3-11.

Corrin, B. and King, E. (1970). Pathogenesis of experimental pulmonary alveolar proteinosis. Thorax 25, 230-236.
Dagle, G.E., Wehner, A.P., Clark, M.L. and Buschbom, R.L. (1986). Chronic inhalation exposure of rats to quartz. In: Silica, Silicosis and Cancer. Controversy in Occupational Medicine. Eds. D.F. Goldsmith, D.M. Winn and C.M. Shy. Praeger, New York. p. 255-266.
Davis, J.M.G., Beckett, S.T., Bolton, R.E., Collings, P. and Middleton, A.P. (1978). Mass and number of fibres in the pathogenesis of asbestos-related lung disease in rats. Brit. J. Cancer 37, 673-688
Davis, J.M.G., Addison, J., Bolton, R.E., Donaldson, K., Jones, A.D. and Wright, A. (1984). The pathogenic effects of fibrous ceramic aluminium silicate glass administered to rats by inhalation or peritoneal injection. In: Biological Effects of Man-made Mineral Fibres. Proc. Symp. 1982. World Health Organisation; Copenhagen, p. 303-322
Davis, J.M.G., Bolton, R.E., Donaldson, K., Jones, A.D. and Miller, B. (1985). Inhalation studies on the effects of tremolite and brucite dust in rats. Carcinogenesis 6, 667-674
Davis, J.M.G., Addison, J., Bolton, R.E., Donaldson, K. and Jones, A.D. (1986a). Inhalation and injection studies in rats using dust samples from chrysotile asbestos prepared by a wet dispersion process. Brit. J. Exper. Path. 67, 113-129
Davis, J.M.G., Addison, J., Bolton, R.E., Donaldson, K., Jones, A.D. and Smith, T. (1986b). The pathogenicity of long versus short fibre samples of amosite asbestos administered to rats by inhalation and intraperitoneal injection. Brit. J. Exper. Path. 67, 415-430
Davis, J.M.G. and Jones, A.D. (1988). Comparisons of the pathogenicity of long and short fibres of chrysotile asbestos in rats. Brit. J. Exper. Path. 69, 717-737
Davis, J.M.G. (1989). Mineral fibre carcinogenesis: experimental data relating to the importance of fibre type, size, deposition, dissolution and migration. In: Non-occupational exposure to mineral fibres, Eds. J. Bignon, J. Peto, and R. Saracci. IARC Publ. No. 90, 33-46. Int'l Agency for Cancer Research, Lyon.
Davis, J.M.G. and Cowie, H.A. (1990). The relationship between fibrosis and cancer in experimental animals exposed to asbestos and other fibres. Environ. Health Perspec. 88, 305-309.
Davis, J.M.G., Addison, J., McIntosh, C., Miller, B. and Niven, K. (1991a). Variations in the carcinogenicity of tremolite dust samples of differing morphology. Annals New York Acad. Sci. 643, 473-491.
Davis, J.M.G., Addison, J., Brown, G.M., Jones, A.D., McIntosh, C., Miller, B.G. and Whittington, M. (1991b). Further studies on the importance of quartz in the development of coalworkers' pneumoconiosis. Report Inst. Occupational Medicine, Edinburgh, No. TM/91/05.
Davis, J.M.G. (1993). Information from experimental studies relating to the effects of ingestion of mineral fibres. Paper presented at Int'l Symp. on the Health Effects of Low Exposure to Fibrous Materials, Kitakyushu, Japan. (Conference report in press).
Denny, J.J., Robson, W.D. and Irwin, D.A. (1939). The prevention of silicosis by metallic aluminium. Canadian Med. Assoc. J. 40, 213-228.
Gardner, L.U. and Cummings, D.E. (1933). The reaction to fine and medium-sized quartz amd aluminium oxide particles. Amer. J. Path. 9 (Suppl.), 751-769.
Gardner, L.U. (1935). The experimental production of silicosis. U.S. Public Health Report, 50, 695-702.
Gibbs, A.R. (1990). Human pathology of kaolin and mica pneumoconiosis. In: Health Effects of Phylbosilicates, Ed. J. Bignon. NATO ASI Series G. Ecological Sci. Vol G21, 217-227. Springer-Verlag, Heidelberg.
Goldstein, B. and Webster, I. (1966). Intratracheal injection into rats of size-graded silica particles. Brit. J. Indus. Med. 23, 71-74.
Gross, P., Westrick, M.L. and McNerney, J.M. (1960). Glass dust: a study of its biological effects. Arch. Indus. Health 21, 10-23.
Gross, P., De Treville, R.T.P., Tolker, E.B., Kaschak, B.S. and Babyak, M.A. (1967). Experimental asbestosis: the development of lung cancer in rats with pulmonary deposits of chrysotile asbestos dust. Arch. Environ. Health 15, 343-355
Gross, P. and De Treville, R.T.P. (1968). Alveolar proteinosis. Its experimental production in rodents. Arch. Path. 86, 255-261.
Gross, P., Kaschak, M., Tolker, E.B., Babyak, M.A. and De Treville, R.T.P. (1970). The pulmonary reaction to high concentrations of fibrous glass dust. Arch. Environ. Health, 20, 696-704.
Groth, D.H., Stettler, L.E., Platek, S.F., Lal, J.B. and Burg, J.R. (1986). In: Silica, Silicosis and Cancer. Controversy in Occupational Medicine. Eds. D.F. Goldsmith, D.M. Winn, and C.M. Shy. Praeger, New York. p. 243-253.
Heppleston, A.G. (1962). The disposal of dust in the lungs of silicotic rats. Amer. J. Path. 40, 493-506.
Heppleston, A.G. (1967). Atypical reaction to inhaled silica. Nature 213, 199.
Heppleston, A.G., Wright, N.A. and Stewart, J.A. (1970). Experimental alveolar lipoproteinosis following the inhalation of silica. J. Path. 101, 293-307.

Hesterberg, T.W., Mast, R., McConnell, E.E., Chevalier, J., Bernstein, D.M., Burn, W.B. and Anderson, R. (1992). Chronic inhalation toxicity of refractory ceramic fibers in Syrian hamsters. In: Mechanisms of Fibre Carcinogenesis. Eds. R.C. Brown, J.A. Hoskins and N.F. Johnson. NATO ASI Series A. Life Sci. Vol. 223, 519-539. Plenum, New York.

Holland, L.M. Gonzales, M., Wilson, J.S. and Tillery, M. I. (1983). Pulmonary effects of shale dusts in experimental animals. In: Health Issues Related to Metal and Nonmetallic Mining. Eds. W.L. Wagner, W.N. Rom and J.A. Merchant. Butterworths, Boston. p. 485-496.

Holland, L.M., Wilson, J.S., Tillery, M.I. and Smith, D.M. (1986). Lung cancer in rats exposed to fibrogenic dusts. In: Silica, Silicosis and Cancer. Controversy in Occupational Medicine. Eds. D.F. Goldsmith, D.M. Winn and C.M. Shy. Praeger, New York. p. 267-279

Hunter, B. and Thomson, C. (1973). Evaluation of the tumorigenic potential of vermiculite by intrapleural injection. Brit. J. Indus. Med. 30, 167-173.

IARC: International Agency for Research on Cancer (1987). Silica and some silicates. IARC Monographs on the evaluation of the carcinogenic risk of chemicals to humans. Vol. 42. Int'l Agency for Research on Cancer, Lyon.

Jacobsen, M., Rae, S., Walton, W.H. and Rogan, J.M. (1971). The relation between pneumoconiosis and dust exposure in British coalmines. In: Inhaled particles III, Ed. W.H. Walton. Unwin Bros., Boston, p. 903-917.

King, E.J., Clegg, J.W. and Rae, V.M. (1946). Effect of asbestos and asbestos and aluminium on the lungs of rabbits. Thorax 1, 118-124.

King, E.J., Gilchrist, M. and Rae, M.V. (1947). Tissue reaction to sericite and shale dusts treated with hydrochloric acid: experimental investigations on lungs of rats. J. Path. Bacteriol. 59, 324-327.

King, E.J. Wright, B.M., Ray, S.C. and Harrison, C.V. (1950). Effect of aluminium on the silicosis-producing action of inhaled quartz. Brit. J. Indus. Med. 7, 27-36.

King, E.J., Mohanty, G.P., Harrison, C.V. and Nagelschmidt, G. (1953a). The action of different forms of pure silica on the lungs of rats. Brit. J. Indus. Med. 10, 9-17.

King, E.J., Mohanty, G.P., Harrison, C.V. and Nagelschmidt, G. (1953b). The action of flint of variable size injected at constant weight and constant surface into the lungs of rats. Brit. J. Indus. Med. 10, 76-92.

King, E.J. and Nagelschmidt, G. (1960). The physical and chemical paroperties of silica, silicates and modified forms of these in reaction to pathogenic effects. In: Proc. Pneumoconiosis Conf., Johannesburg. Ed. A.J. Orenstein. Churchill, London. p. 78-83.

Klosterkötter, W. and Jötten, K.W. (1953). The action of different forms of silica in animal experiments. Archiv. Hygiene und Bakteriol. 137, 625-636.

Le Bouffant, L., Daniel, H., Martin, J.C., Aubin, C. and Le Heuede, P. (1983). Recherche communautaire sur le role de quartz dan la pneumoconiose des mineurs de charbon et sur l'influence des mineraux d'accompagnement. Verneuil-en-Halatte, Cerchar.

Le Bouffant, L., Daniel, H., Henin, J.P., Martin, J.C., Normand, C., Tichoux, G. and Trolard, F. (1987). Experimental study on long-term effects of inhaled MMMF on the lungs of rats. Annals Occup. Hygiene 31, 765-791

Lee, K.P., Barras, C.E., Griffith, F.D. and Waritz, R.S. (1979). Pulmonary response to glass fiber by inhalation exposure. Laboratory Invest. 40, 123-133

Liu, Z., Fang, Y., Yie, F. and Xiao, X. (1990). Study of fibrogenic effect of vermiculite dust on rat lung. Proc. VIIth Int'l Pneumoconiosis Conf. U.S. Dept. Health and Human Services, DHHS (NIOSH) Publ. No. 90-108, Part II. p. 1290-1292.

Maltoni, C., Minardi, F. and Morisi, L. (1982). The relevance of the experimental approach in the assessment of the oncogenic risks from fibrous and non-fibrous particles. The ongoing project of the Bologna Inst. of Oncology. Medicina del Lavoro 4, 394-407.

Marenghi, B. and Rota, L. (1953). Effect of cortisone on experimental silicosis in rats. Medicina del Lavoro,44, 383-397.

Martin J.C., Daniel, H. and Le Bouffant, L. (1977). Short and long-term experimental study of the toxicity of coalmine dust and some of its constituents. Inhaled particles IV, Ed. W.H. Walton. Pergamon Press, Oxford. p. 361-370.

McConnell, E.E., Wagner, J.C., Skidmore, J.W. and Moore J.A. (1984). A comparative study of the fibrogenic and carcinogenic effects of UICC Canadian chrysotile B asbestos and glass microfibre (JM100). In: Biological Effects of Man-made Mineral Fibres. Report of WHO/IARC meeting, World Health Organization, Copenhagen, 234-252.,

Miller, A., Teirstein, A.S., Bader, M.E., Bader, R.A. and Selikoff I.J. (1971). Talc pneumoconiosis. Significance of sublight microscopic mineral particles. Amer. J. Med. 50, 395-402.

Mühle, H., Pott, F., Bellman, B., Takenaka, S. and Ziem, V. (1987) Inhalation and injection experiments in rats to test the carcinogenicity of MMMF. Annals Occup. Hygiene 31, 755-765

National Toxicology Program Technical Report (1992). Toxicology and carcinogenesis studies of talc in F344 rats and B6C3F mice. U.S. Dept. Health and Human Services (National Institutes of Health) NIH Publ. No. 92-3152.

Oldham, P.D. (1983). Pneumoconiosis in Cornish china clay workers. Brit. J. Indus. Med. 40, 131-137.

Ozesmi, M., Patiroglu, T.E., Hillerdal, G. and Ozesmi, C. (1985). Peritoneal mesothelioma and malignant lymphoma in mice caused by fibrous zeolite. Brit. J. Indus. Med. 42, 746-749.

Policard, A. and Collet, A. (1957). Experimental study on pathological reactions following the introduction of inframicroscopic particles of amorphous silica into the organism. Arch. maladies professionelles 18, 508-510.

Pooley, F.D. (1976). An examination of the fibrous mineral content of asbestos lung tissue from the Canadian chrysotile mining industry. Environ. Res. 12, 281-298

Pott, F. and Friedrichs, K.H. (1972). Tumours in rats after intraperitoneal injection of asbestos dusts. Naturwissenschaften 59, 318-332

Pott, F., Friedrichs, K.H. and Huth, F. (1976). Ergebnisse aus Tierversuchen zur kanzerogenen Wirkung faserformiger Staube und ihre Dentung im Hinblick auf die Tumorenstehung beim Menschen. Zentralblatt Bakteriol. Parasitkenkunde. Infectionskrankheiten und Hygiene Abt. I Originale 162, 467-505.

Pott, F., Bellmann, B., Mühle, H., Rodelsperger, K., Rippe, R.M., Roller, M. and Rosenbruch, M. (1990). Intraperitoneal injection studies for the evaluation of the carcinogenicity of fibrous phyllosilicates. In: Health Related Effects of Phyllosilicates, Ed. J. Bignon. NATO ASI Series G. Ecological Sci. Vol. G21, 319-331. Springer-Verlag, Heidelberg.

Pylev, L.N., Kulagina, T.F., Vasilyeva, L.A., Chelischev, N.F. and Berenstein, B.G. (1986). Blastomogenic activity of erionite (Nidale erionite). Gigiena Truda I Professional 'Nye Zabolevaniia 161, 33-37.

Reeves, A.L., Puro, H.E. and Smith, R.G. (1974) Inhalation carcinogenesis from various forms of asbestos. Environ. Res. 8, 178-202.

Reisner, M.T.R. (1983). Untersuchungen der Beziehuwgen zwischen den epidemiologische Ergebnissen und der Schädlichkeit der Feinstaube im Kohlenbergbau. Bergbau-Forschung GmbH, Essen, Germany.

Rennie, R.A., Eldridge, S.R., Lewis, T.R. and Stevens, D.L. (1985). Fibrogenic potential of intratracheally instilled quartz, ferric oxide, fibrous glass and hydrated alumina in hamsters. Toxicological Path. 13, 306-314.

Robertson, A., Bolton, R.E., Chapman, J., Davis, J.M.G., Dodgson, J., Gormley, I.P., Jones, A.D. and Miller, B. (1984). Animal inhalation experiments to investigate the significance of high and low percentage concentrations of quartz in coalmine dusts in relation to epidemiology and other biological tests. Report No. TM/84/2, Inst. Occupational Medicine, Edinburgh.

Rosmanith, J., Hilscher, W. and Schyma, S.B. (1990). The effect of the surface quality on the fibrogenicity of the phyllosilicates muscovite and kaolinite. In: Health Related Effects of Phyllosilicates. Ed. J. Bignon. NATO ASI Series G. Ecological Sci. Vol G21, 123-129. Springer-Verlag, Heidelberg.

Saffiotti, U. (1962). The histogenesis of experimental silicosis. III. Early cellular reactions and the role of necrosis. Medicina del Lavoro 53, 5-18.

Sahu, A.P., Shanker, R. and Zaidi, S.H. (1978). Pulmonary response to kaolin, mica and talc in mice. Experimental Path. 16, 276-282.

Sahu, A.P. (1990). Biological effects of mica dust in experimental animals. In: Health Effects of Phyllosilicates. Ed. J. Bignon. NATO ASI Series G. Ecological Sci. Vol G21, 387-395. Springer-Verlag, Heidelberg.

Schepers, G.W.H. and Delahant, A.B. (1955) An experimental study of the effects of glass wool on animal lungs. AMA Arch. Indus. Health 12, 276-279

Schmidt, K.D. and Lüchtrath, H. (1955). Comparison of the effects of quartz, cristobalite and tridymite dust in intratracheal studies in rats. Beitrage Silikose-Forschung 37, 1-43.

Selikoff, I.J., Churg, J. and Hammond, E.C. (1964). Asbestos exposure and neoplasia. J. Amer. Med. Assoc. 188, 22-26.

Silverman, L. and Moritz, A.R. (1950). Peritoneal reaction to injected fused (spherical) and unfused (spiculate) quartz. Arch. Indus. Hygiene Occup. Med. 1, 499-505.

Sluis-Cremer, G.K., Hessel, P.A. (1990). Palabora Vermiculite. In: Health Related Effects of Phyllosilicates. Ed. J. Bignon. NATO A5I Series G. Ecological Sci. Vol. G21, 227-234. Springer-Verlag, Heidelberg.

Smith, D.M., Ortiz, L.W., Archuleta, R.F. and Johnson, N.F. (1987). Long-term health effects in hamsters and rats exposed chronically to man-made vitreous fibers. Annals Occup. Hygiene 31, 731-743

Stanton, M.F. and Wrench, C. (1972). Mechanisms of mesothelioma induction with asbestos and fibrous glass. J. National Cancer Inst. 48, 797-821

Stanton, M.F., Layard, M., Tegeris, A., Miller, M. and Kent, E. (1977). Carcinogenicity of fibrous glass: pleural response in the rat in relation to fibre dimension. J. National Cancer Inst. 58, 587-603

Stanton, M.F., Layard, M., Tegeris, A., Miller, E., May, M., Morgan, E. and Smith, A. (1981). Relation of particle dimension to carcinogenicity in amphibole asbestoses and other fibrous minerals. J. National Cancer Inst. 67, 965-975.
Stenbäck, F. and Rowland, J. (1978). Role of talc and benzo(a)pyrene in respiratory turnover formation. An experimental study. Scandinavian J. Respiratory Dis. 59, 130-140.
Suzuki, Y. (1982). Carcinogenic and fibrogenic effects of zeolites. Environ. Res. 27, 433-445.
Suzuki, Y. and Kohyama, N. (1984). Malignant mesothelioma induced by asbestos and zeolite in the mouse peritoneal cavity. Environ. Res. 35, 277-292.
Swensson, A., Glomme, J. and Bloom, G. (1956). On the toxicity of silica particles. Arch. Indus. Health 14, 482-486.
Teblens, B.D., Schulz, R.Z., Drinker, P. (1945). The potency of silica particles of different size. J. Indus. Hygiene Toxicol. 27, 199-200.
Vallyathan, N.V. and Craighead, J.E. (1981). Pulmonary pathology in workers exposed to nonasbestiform talc. Human Path. 12, 28-35.
Vallyathan, N.V. (1990). Pulmonary response to inhaled fibrogenic minerals. Div. of Respiratory Disease Studies, NIOSH, U.S. Dept Health and Human Services, Morgantown, West Virginia.
Vorwald, A.J., Durkan, T.M. and Pratt, P.C. (1951). Experimental studies of asbestosis. AMA Arch. Indus. Hygiene Occup. Med. 3, 1-43
Wagner, J.C., Sleggs, C.A. and Marchand, P. (1960). Diffuse pleural mesotheliomata and asbestos exposure in North Western Cape Province. Brit. J. Indus. Med. 17, 260-271
Wagner, J.C. (1962). Experimental production of mesothelial tumours of the pleura by implantation of dusts in laboratory animals. Nature 196, 180-181
Wagner, J.C., Berry, G., Skidmore, J.W. and Timbrell, V. (1974). The effects of the inhalation of asbestos in rats. Brit. J. Cancer 29, 252-269
Wagner, J.C., Berry, G., Cooke, T.J., Hill, R.J., Pooley, F.D. and Skidmore, J.W. (1977). Animal experiments with talc. In: Inhaled Particles IV, Ed. W.H. Walton. Pergamon Press, Oxford. p. 647-654.
Wagner, J.C., Berry, G., Skidmore, J.W. and Pooley, F.D. (1980). The comparative effects of three chrysotiles by injection and inhalation in rats. In: J.C. Wagner, ed. Biological Effects of Mineral Fibres. Int'l Agency Research on Cancer, Lyon. IARC Publ. No. 30, p. 363-373.
Wagner, J.C., Berry, G., Hill, R.J., Munday, D.E. and Skidmore, J.W. (1984). Animal experiments with MMM(V)F. Effects of inhalation and intraperitoneal inoculation in rats. In: Biological Effects of Man-made Mineral Fibres. Report of a WHO/IARC meeting. World Health Organization, Copenhagen, 207-233.
Wagner, J.C., Skidmore, J.W., Hill, R.J. and Griffiths, D.M. (1985). Erionite exposure and mesothelioma in rats. Brit. J. Cancer 51, 727-730.
Wagner, J.C., Griffiths, D.M. and Munday, D.E. (1987). Experimental studies with palygorskite dusts. Brit. J. Indus. Med. 44, 749-763.
Waxweiler, R., Zumwalde, R.D., Ness, G.O. and Brown, D.P. (1988) A retrospective cohort mortality study of males mining and milling attapulgite clay. Am. J. Indus. Med. 13, 305–315.
Wilson, T., Scheuchenzuber, W.J., Eskew, M.L. and Zankower, A. (1986). Comparative pathological aspects of chronic olivine and silica inhalation in mice. Environ. Res. 39, 331-344.
Wright, G.W. and Kuschner, M. (1977). The influence of varying lengths of glass and asbestos fibres on tissue response in guinea pigs. In: Inhaled Particles IV, Ed. W.H. Walton. Pergamon Press, Oxford. p. 455-472.

CHAPTER 16

IN VITRO EVALUATION OF MINERAL CYTOTOXICITY AND INFLAMMATORY ACTIVITY

Kevin E. Driscoll

*The Procter & Gamble Company
Miami Valley Laboratories
Cincinnati, Ohio 45239 U.S.A.*

INTRODUCTION

Occupational and environmental exposure to certain mineral dusts can cause chronic interstitial lung diseases, such as asbestosis, silicosis and coal worker pneumoconiosis (Seaton, 1984). In addition, exposure to fibrous minerals (e.g., crocidolite and erionite) increase the risk of bronchogenic carcinoma and malignant mesothelioma (Mossman et al., 1990). Health risks associated with inhalation of mineral dusts have been identified to a great extent through epidemiology studies but have also been assessed by conducting subchronic and chronic inhalation studies (typically using rats, mice, or hamsters). Controlled inhalation studies provide a means for hazard identification and defining no-effect exposure levels in animals, the latter being used to extrapolate to safe human exposures using appropriate risk assessment models. However, *in vivo* studies are costly and time-consuming, and inhalation exposure requires complex technologies for aerosol generation and characterization, which further increase costs and limit the number of laboratories with adequate capabilities. In addition, whole animal studies alone are limited in that they may not yield information on cellular and molecular mechanisms of response, which is critical for appropriate extrapolation of results, development of new therapeutic approaches for treating disease, or guiding development of alternative technologies (e.g., asbestos substitutes). In this respect, the use of *in vitro* cell culture systems has played an important role in studying the toxicology of mineral dusts and providing insight into mechanism of action.

The use of cells or tissues in culture offers a number of advantages for investigating chemical toxicity not found in other models used for toxicological research. *In vitro*, cell populations can be manipulated and monitored more readily under rigorously controlled conditions. In addition, when functional properties (such as cell metabolism, permeability, or secretory activity) are retained *in vitro*, they can be subjected to detailed investigation. Furthermore, the use of isolated cells in culture allows identification and characterization of cell-type-specific toxicities as well as investigation of cell–cell interactions important in response to chemical exposure. Also, using *in vitro* techniques, human cells may be exposed to chemicals that could not be investigated *in vivo* for ethical concerns (e.g., exposure to carcinogens). Lastly, the cost of time, research dollars and potentially the number of experimental animals needed for toxicological research can be reduced when *in vitro* models rather than whole animals are used. Thus, when cells or tissues retain function in culture, *in vitro* techniques represent a powerful adjunct to animal or clinical studies for investigating human health effects resulting from chemical exposures.

IN VITRO APPROACHES TO STUDY MINERAL DUST TOXICITY

Several different systems have been implemented to investigate the toxicity of mineral dusts including: erythrocytes to study particle–cell membrane interactions (Harley and Margolis, 1961); macrophages as key phagocytic targets and effector cells in the lung (Allison et al., 1966); lung epithelial cells, fibroblasts, and mesothelial cells as critical components of carcinogenic and fibrogenic responses (Woodworth et al., 1982; Shatos et al., 1987; Paterour et al., 1985); and explanted lung tissue systems which preserve the cell–cell interactions present *in vivo* (Placke and Fischer, 1987). Early investigations focused on the use of cell cultures for characterizing cytotoxic effects of various minerals in an effort to identify key chemical and physical properties that contribute to toxicity. In addition, a considerable effort has been made using *in vitro* models to investigate the genotoxic potential of mineral dusts using both gene mutation and cell transformation assays (Huang et al., 1978; Paterour et al., 1985). More recently *in vitro* studies have focused on (1) the ability of mineral dusts to activate lung cells with respect to proliferative responses and production of inflammatory mediators and (2) deciphering the specific transmembrane signalling pathways and intracellular regulatory mechanisms underlying mineral dust-induced changes in cell function (Marsh and Mossman, 1988; Roney and Holian, 1989; Driscoll et al., 1990a).

Overall, *in vitro* studies have provided useful information to make preliminary assessments of a mineral's potential toxicity *in vivo* and have provided key insight into basic biochemical and molecular mechanisms of how mineral dusts interact with cells to elicit inflammation, fibrosis, and cancer. The following chapter will review selected *in vitro* approaches that are used to investigate the cytotoxic and inflammatory properties of mineral particles and fibers. This chapter will also present some key findings as they relate to understanding the toxicology of mineral dusts.

CYTOTOXICITY ASSAYS

Erythrocyte hemolysis

One of the first *in vitro* systems used to investigate the toxicity of mineral dusts was the erythrocyte (or red blood cell) hemolysis assay (Harley and Margolis, 1961). Although hemolysis of erythrocytes does not play a role in the pathogenesis of mineral dust-induced lung disease, it provides a simple and rapid approach for studying the effects of mineral particles on cell membranes, a potentially important target site of toxicity. In addition, numerous studies have suggested a relationship between the hemolytic activity of various minerals and their cytotoxicity (Marks, 1957; Allison, 1966, 1971; Stadler and Stöber, 1965; Hefner and Gehring, 1975). In the hemolysis assay, red blood cells (RBCs) isolated from human or animal donors are placed in suspension culture with the test material (i.e., mineral dust); the cells and particles are incubated (typically for 1–4 hours); and cell lysis (or, more explicitly, cell membrane permeability/integrity) is determined by release of hemoglobin into the cell culture medium.

There appears to be at least two types of mineral-dust-induced erythrocyte lysis: one involves direct hemolysis resulting from contact of the particles with the erythrocyte membrane, and the other involves sensitization of the erythrocyte to complement-induced cell lysis (Harington et al., 1975). The latter mechanism has been observed for silicic acid and amphibole asbestos (i.e., crocidolite and tremolite), but not for chrysotile asbestos or quartz (Harington, 1971a,b). Regarding the mechanism of direct hemolysis by silica early studies indicated that proton donating silanol groups on the surface of hydrated silica are important, likely due to their ability to denature proteins and induce cell membrane damage (Nash et al., 1966). The importance of silanol groups is supported by the protective effect of proton accepting chemicals, such as polyvinylpyridine-N-oxide (PVPNO), which has been shown to inhibit silica-induced hemolysis (Stadler and Stöber, 1965; Nash et al., 1966). Subsequent studies demonstrated correlation between negative charge on silica, measured as zeta potential, and hemolytic activity (Little and Wei, 1977). Nolan et al., (1981) reported that both functionalities—the proton donor silanol and the ionized silanol group (SiO^-)—are critical to silica's hemolytic activity. These investigators showed that hydrogen bonding of PVPNO with silanol groups as well as the bonding of metal cations (e.g., Al^{3+}) to ionized silanols dramatically reduces hemolytic activity (Nolan et al., 1981). Langer and Nolan (1985) proposed the proton-donating silanol facilitates attachment of the mineral species to the cell membrane and the negative charge supplied by the ionized silanol is critical for cell lysis. More recently, Razzaboni and Bolsaitis (1990) have shown that silica hemolysis is attenuated by catalase suggesting that hydrogen peroxide (possibly formed through reactions between silanols and carbonyl oxygens in the cell membrane) is an active intermediate in the hemolytic response. Along these lines, Shatos et al. (1987) and Kennedy et al. (1989) reported that several mineral dusts (e.g., chrysotile, amosite, crocidolite, and quartz) may act as Fenton catalysts, causing cytotoxicity through conversion of superoxide to hydrogen peroxide. This ability to catalyze Fenton reactions relates to iron associated with the minerals. Because quartz and chrysotile should contain relatively small amounts of iron (at most), it has been postulated that some minerals sorb iron on their surfaces, either during sample processing or *in vivo*.

The mechanism by which chrysotile asbestos causes hemolysis appears to differ from that of silica polymorphs, since the activity of the former is not blocked by PVPNO (Macnab and Harington, 1967). As discussed by Harington (1975), hemolytic activity of chrysotile asbestos is blocked by chelators, particularly those with an affinity for magnesium. In addition, fiber treatments which leach or remove magnesium reduce chrysotile's hemolytic activity. These findings indicate the magnesium ion is critical to the hemolytic action of chrysotile and, in fact, hemolytic activity is positively correlated with magnesium-silicon ratios for different asbestos fiber types (Harrington et al., 1971b). The dependence hemolytic activity on magnesium would explain why chrysotile is more active than other forms of asbestos that are lower in magnesium (e.g., crocidolite and amosite, which are only weakly hemolytic). However, the numerous other structural and compositional differences between chrysotile and the amphibole asbestoses may also be important in determining the differences in hemolytic activity for these minerals. As with silica, the generation of active oxygen species appears to play a role in asbestos-induced cytotoxicity (Shatos et al., 1987; Kennedy et al., 1989).

The hemolysis assay has been useful in defining the physical and chemical features of mineral dusts that influence their interaction with cell membranes. Other factors shown to be important in mineral-induced hemolysis include: particle surface area, particle or fiber number, size, shape, and the sorptive properties of the mineral with respect to lipids and proteins. The latter may be important, because the adsoprtion of lipids and proteins can attenuate hemolytic activity of both fibrous and nonfibrous minerals (Kozin et al., 1982). Table 1 summarizes some key observations derived from hemolysis assays and macrophage cytotoxicity assays (which are discussed in the next section).

Macrophage toxicity

The alveolar macrophage and other lung macrophages (e.g., interstitial macrophages, airway macrophages) are critical components of the lung's cellular defenses (Murray and Driscoll, 1992). One primary function of these cells is to clear particles from the lung via phagocytosis (see Chapter 14). However, macrophages participate in lung defense in numerous other ways, including the mediation of the immune response by the presentation of antigen; and the release of various lipid and protein mediators, which activate immunocompetent cells (e.g., lymphocytes, mast cells, and eosinophils). Macrophages also play an important role in maintaining the homeostasis of the respiratory region as well as initiating lung inflammation and repair responses. In this respect, alveolar macrophages represent important targets as well as potential effectors of respiratory tract toxicity.

Numerous *in vitro* studies on mineral dust cytotoxicity have utilized macrophages as the cellular targets. The emphasis on the macrophage has been due to the general acceptance that interactions with lung macrophages is a first critical step in biological effects of mineral dusts. Many studies on macrophage cytotoxicity have utilized peritoneal macrophages, because these cells can be obtained in greater numbers than alveolar macrophages or other lung macrophage populations. The cytotoxic effects of various mineral dust species on peritoneal macrophages appear to be similar to those on alveolar macrophages (Dogra and Kaw, 1988). However, peritoneal and alveolar macrophage populations do differ in a number of important functional activities, such as oxidative metabolism (Lasser, 1983), therefore the suitability of peritoneal cells as a model to evaluate effects of minerals on lung macrophages depends on the specific function/response being investigated. In addition to mechanistic considerations, another reason for the widespread use of macrophages for *in vitro* assays is the ease with which these cells can be obtained from the peritoneum or lung in high yield and purity. The

Table 1

Key observations from *in vitro* hemolysis and macrophage-cytotoxicity studies

Observation	References
Mineral particles can cause hemolysis	Dognon and Simonot, 1951 Harley and Margolis, 1961 Stadler and Stober, 1965 Macnab and Harington, 1967
Sorption of proteins and lipids to minerals attenuates hemolytic and cytotoxic activities	Marks 1957 Jaurand et al., 1979 Stadler and Stober, 1965 Kozin et al., 1982 Wallace et al., 1985, 1992
There are parallels between hemolytic activity and/or cytotoxicity of nonfibrous minerals and their fibrogenicity, however, some exceptions exist (i.e., bentonite, kaolin amorphous silica)	Marks and Nagelschmidt, 1959 Stadler and Stober, 1965 Allison et al., 1966, 1971 Styles and Wilson, 1973 Hefner and Gehring, 1975 Davies et al., 1984
Hemolytic activity of asbestos parallels *in vitro* cytotoxicity, but these *in vitro* responses do not correlate with fibrogenicity or carcinogenicity	Harington et al., 1971
Hemolysis and/or cytotoxicity by silica polymorphs but not chrysotile asbestos are blocked by proton acceptor compounds such as PVPNO	Stadler and Stober, 1965 Nash et al., 1966 Macnab and Harington, 1967 Kaw et al., 1975
Hemolytic and cytotoxic activity of silica results from proton donating silanol groups	Nash et al., 1966
Hemolytic activity of minerals is related to charge in the mineral particle determined as zeta potential	Light and Wei, 1977
Hemolytic activity of chrysotile involves, at least in part, magnesium; there is good correlation between magnesium content and hemolytic activity of asbestos	Macnab and Harington, 1967 Harington et al., 1971 Schnitzer et al., 1971
Silica hemolysis involves both silanol groups and ionized silanols; hemolysis blocked by both proton acceptors and cations such as aluminum	Nolan et al., 1981 Langer and Nolan, 1985
Generation of active oxygen species catalyzed by mineral surfaces and/or associated metals contributes to hemolysis and cytotoxicity	Shatos et al., 1987 Kennedy et al., 1989 Iguchi and Kojo, 1989 Razzaboni and Bolsaitis, 1990

most frequently used method for harvesting macrophages from these anatomic sites is through the use of saline lavage. Lavage involves "washing" the target organ with saline and can be performed *in vivo* or on lungs removed by excision. The lavage technique and factors that influence its efficiency are discussed for the peritoneal and alveolar spaces in detail elsewhere (Meltzer, 1984; Brain et al., 1970; Reynolds, 1987).

Much of the early research on interactions between macrophages and mineral particles focused on decrements in cell viability as a means to screen or rank relative toxicity. Cell viability is generally determined by exclusion of vital dyes or release of cytoplasmic enzymes, i.e., it is a measure of the permeability or integrity of the cell membrane. Other responses examined included monitoring the release of specific lysosomal enzymes, measurement of functional activities (such as phagocytosis), and characterization of oxidative metabolism. More recent studies have examined the potential of mineral particles to activate other macrophage functions, including the secretion of inflammatory mediators such as cytokines (small proteins with hormone-like actions), metabolites of arachidonic acid, or active oxygen species; factors important to the pathogenesis of mineral-induced lung disease. The cytotoxic response of macrophages to nonfibrous mineral dusts appears to parallel that characterized for hemolysis; including the protective effect of PVPNO (Davies and Preece, 1983). Thus, macrophage cytotoxicity likely involves, at least partly, mechanisms of particle–cell-membrane interaction similar to those active in hemolysis: hydrogen bonding by silanol groups and generation of active oxygen, the latter potentially involving Fenton-type reactions catalyzed by iron present on mineral surfaces (Shatos et al., 1987; Brown et al., 1987).

There are at least two potential distinctions between mineral hemolysis and cytotoxicity to macrophages. First, the source of oxidants in macrophage toxicity may not be limited to the reactive surface of the mineral. As discussed below, macrophages can generate active oxygen species (e.g., superoxide anion or O_2^-) through the action of membrane-bound enzymes. Several studies have suggested that mineral particles can activate macrophage oxidant synthesis, and this may contribute to cytotoxic responses in phagocytic cells as well as airway epithelial cells (Mossman and Landesman, 1983; Shatos et al., 1987; Goodlick and Kane, 1986). In this respect, the mineral likely acts both as a stimulus for cellular oxidant production and as a source of iron for catalyzing the generation of oxidant species with greater toxicity (Goodlick and Kane, 1986; Shatos et al., 1987). Second, the site of membranolytic activity in macrophages may be the lysosomal membrane and not the outer cell membrane. Because macrophages are avidly phagocytic, mineral particles are rapidly internalized, resulting in the formation an intracellular vacuole (called a *phagosome*) that fuses with enzyme-containing cellular structures (called *lysosomes*) to form a *phagolysosome*.[1] Allison et al. (1971) characterized the intracellular distribution of lysosomal enzymes and the exogenous enzymes taken up into phagosomes concurrent with particles. Their results indicate that the site of silica membranolytic activity is the phagolysosome, which results in leakage of lysosomal enzymes into the cell and subsequent lysis. Allison et al. (1971) suggested that the phagolysosome is a primary target for mineral–membrane interactions, because sorption of proteins and lipids onto mineral particles protects the outer cell membrane from reactive sites on the particle's surface. However, this protective coating is removed by lysosomal enzymes, thereby exposing the phagolysosomal membrane to the reactive surface of the mineral.

---

[1] This process is normally used by phagocytes to destroy foreign matter (e.g., bacteria).

Hemolytic activity and cytotoxicity: Their relationship to *in vivo* toxicity

Early studies by Stadler and Stöber (1965) compared the hemolytic activities of numerous mineral dusts and reported that several silica polymorphs (quartz, tridymite, cristobalite, and coesite) are more hemolytic than relatively inert minerals (e.g., stishovite, anatase, corundum [i.e., $\alpha-Al_2O_3$], and $\gamma-Al_2O_3$). The relative biological inactivity of stishovite (also reported by Brieger and Gross, 1967) may relate to the octahedral coordination of silicon (e.g., Wiessner et al., 1988), which is different from the tetrahedral coordination more commonly exhibited by Si (see Chapter 5). A correlation between hemolytic activity for an extensive series of mineral dusts and *in vivo* fibrogenicity was also reported by Hefner and Gehring (1975). Similar to findings for hemolysis initial studies on macrophage responses to silica polymorphs demonstrated parallels between fibrogenicity and cytotoxicity for macrophages (Marks et. al., 1956; Marks and Nagelschmidt, 1959; Allison et al., 1966; Styles and Wilson, 1973). Quartz, cristobalite, tridymite, and coesite were shown in several studies to be cytotoxic to macrophages, and are all fibrogenic; however, coesite is less hemolytic than quartz, cristobalite, and tridymite (Stadler and Stöber, 1965). Stishovite and several other relatively innocuous minerals—titanium dioxide (including anatase and rutile), aluminum oxide (including corundum and $\gamma-Al_2O_3$), feldspar, and magnetite—have been shown to possess low cytotoxicities (Marks et. al., 1956; Marks and Nagelschmidt, 1959; Davies et al., 1983, 1984; Lock et al., 1987; Driscoll et al., 1990a). Thus, studies on several mineral dusts have demonstrated a positive correlation among *in vitro* hemolytic activity, cytotoxicity, and *in vivo* toxicity (i.e, fibrogenicity), providing some enthusiasm for *in vitro* assays as screens for mineral toxicity. However, some minerals known to be relatively innocuous (although *not* inactive) *in vivo* (e.g., montmorillonite, kaolinite, and amorphous silica) have been shown to possess hemolytic activity or cytotoxicity similar to quartz, cristobalite, or chrysotile (Kozin et al., 1982; Woodworth et al., 1982; Davies et al., 1984; Wallace et al., 1985; Lock et al., 1987). In addition, crocidolite and other amphibole asbestoses (e.g., amosite) are only very weakly hemolytic and cytotoxic but are well documented fibrogenic and carcinogenic agents in man or experimental animals. In this respect, although overall there is some agreement between the *in vitro* hemolytic/cytotoxic activity of mineral dusts and fibrogenic activity, clear exceptions exist.

Recent studies suggest that differences in the sorptive capacities of minerals for lung surfactant may explain some discontinuities between *in vitro* membranolytic activity and *in vivo* fibrogenicity. Sorption of surfactant phospholipid has been shown *in vitro* to attenuate the membranolytic activity of mineral dusts, including quartz and chrysotile (Marks, 1957; Wallace et al., 1985, 1992). Wallace et al. (1992) demonstrated that when coated with dipalmitoyl phosphotidylcholine, a phospholipid component of pulmonary surfactant, quartz and kaolin exhibit significantly reduced membranolytic ability. However, in the presence of phospholipase enzymes, such as those present in alveolar macrophages, the phospholipid was more readily removed from quartz compared to kaolin, resulting in a more rapid restoration of the membranolytic activity of quartz. These investigators suggested differential susceptibility to the removal of protective lipids on silicates may explain, at least in part, discontinuities between *in vitro* and *in vivo* effects.

Overall, *in vitro* hemolysis and cytotoxicity assays represent useful models for studying potential mechanisms of mineral–cell-membrane interactions. In addition, these assays have some utility as rapid screening techniques to provide preliminary information on potential *in vivo* toxicity and to guide more definitive assays. As suggested by the studies of Wallace et al. (1992), as our understanding of factors that modify the reactivity of minerals *in vivo* increases, we will be in a better position to develop *in vitro* cytotoxicity assays (or batteries of assays) that better predict *in vivo* responses.

MACROPHAGE ACTIVATION AND INFLAMMATORY ACTIVITY

Macrophages can release a diverse array of products into their extracellular environment (Nathan, 1987), ranging from low molecular weight substances (e.g., arachidonic acid metabolites [eicosanoids] and cyclic nucleotides) to much larger molecules, including components of the complement system and a variety of proteolytic and hydrolytic enzymes. Since many macrophage-derived products play important roles in host defense and contribute to the pathogenesis of disease, the secretory activity of these cells represents a critical component of their overall function.

Early investigations on the pathogenesis of mineral-dust-induced fibrosis suggested a critical role for macrophages, in which the macrophages release cellular constituents (e.g, proteolytic and hydrolytic enzymes) upon succumbing to the cytotoxic effects of the inhaled dust. Clearly cytotoxicity and functional impairment of macrophages are important effects contributing to the pulmonary toxicity of an inhaled material, however, there is increasing evidence that activation of macrophages to release mediators (chemical signals sent to other cells) also plays a key role in determining the degree and nature of response to an inhaled agent. For example, activation of macrophages to release various cytokines and growth factors plays an important role in the pathogenesis of mineral-dust-induced fibrosis (Bitterman et al., 1981; Rom et al., 1991; Mossman et al., 1990). The importance of macrophage activation in the disease process has led to an increased number of studies utilizing *in vitro* models to investigate the direct effects of inhaled materials on macrophage secretory activity. Below is provided an overview of some key inflammatory mediators released by macrophages, their potential contribution to mineral-dust-induced lung disease, and a discussion regarding how mineral dust exposure *in vitro* influences their release. Although a wide range of macrophage secretory products have been examined *in vitro*, the following discussion focuses on the production of active oxygen species and two inflammatory cytokines (tumor necrosis factor α, or TNF-α, and interleukin-1, or IL-1).

Active oxygen species

Phagocytic cells such as macrophages and neutrophils can be stimulated by a variety of soluble and particulate agents (e.g., n-formyl tripeptides, phorphol esters, zymosan) to undergo a "respiratory burst," which is characterized by a 2 to 20-fold increase in cellular oxygen consumption, increased glucose metabolism, and production of active oxygen species (including O_2^- and H_2O_2) (see also Chapter 14). The enzyme responsible for this increased oxygen consumption and O_2^- generation is the membrane bound NADPH oxidase. Formation of other oxidant species is dependent on enzymes (such as superoxide dismutase, which catalyzes $O_2^- \rightarrow H_2O_2$) and myeloperoxidase (which, in the presence of halide, converts H_2O_2 into hypohalous ion). Toxic oxygen radicals can also be formed through a Fenton-like reaction, in which O_2^- reacts with the oxidized form of a trace metal such as iron, causing reduction of the metal and generation of O_2. The reduced metal can then react with H_2O_2, regenerating the initial oxidized metal and forming OH^- and •OH (the highly toxic hydroxyl radical). These phagocyte-derived oxidants play an essential role in defending the lung against micro-organisms. However, they also have been implicated in lung injury caused by a variety of agents, including paraquat, bleomycin, hyperoxia, and mineral dusts. In this role, these oxidants can cause peroxidation in cell membranes, inactivate extracellular proteins such as alpha-1-antitrypsin, and damage DNA (Carp and Janoff, 1979; Freeman and Crapo, 1982).

Table 2 summarizes findings from studies investigating *in vitro* effects of mineral dusts on macrophage oxidant production. These studies demonstrate that mineral dust particles (including quartz, aluminum oxide, titanium dioxide) can stimulate oxidant production by rabbit, rat, hamster, and/or human alveolar macrophages (Vilim et al., 1987; Wilhelm et al., 1987; Perkins, 1991). With respect to fibrous minerals, both chrysotile and crocidolite have been shown to induce the production of O_2^- by alveolar macrophages derived from rats, hamsters, mice, and humans. In contrast, guinea pig macrophages will release of O_2^- in response to chrysotile but not to various amphibole asbestoses (i.e., crocidolite, anthophylite, or amosite), suggesting species differences may exist for asbestos-induced macrophage respiratory burst.

The shape and size of mineral particles influences their ability to stimulate macrophage oxidant production. Hansen and Mossman (1987) compared the *in vitro* oxidant responses of hamster and rat alveolar macrophages to a variety of fibrous minerals, including crocidolite, erionite, and glass fibers and their nonfibrous analogs riebeckite, mordenite, and glass beads.[†] Although all samples elicited some increase in O_2^- production, the nonfibrous samples were less active than the fibrous ones, indicating that the fibrous geometry influences the generation of reactive oxygen by phagocytic cells. Fiber length also affects the release of O_2^- in that exposure to long (>10 μm) fibers of chrysotile asbestos activates a greater increase in superoxide release by rat alveolar macrophages than

---

[†] It should be noted that mordenite is *not* a nonfibrous analog for erionite, as it was used in the original studies. Mordenite and erionite have different structures and compositions (see Chapter 4), and, in fact, most natural mordenite is fibrous. The original mordenite-bearing material for the Hansen and Mossman (1987) study was contaminated with nonfibrous minerals, such as clinoptilolite (Guthrie and Bish, 1991; Guthrie, 1992; Guthrie, Bish, and Mossman, unpublished data). Hence, the mordenite results from this study apply to a complex mixture of minerals and not pure mordenite. Other *in vivo* studies of mordenite (e.g., Suzuki and Kohyama, 1984) similarly used impure mordenite samples. In fact, as for most natural zeolites, pure samples of native mordenite are uncommon.—*Eds.*

exposure to short (<2 μm) fibers (Mossman et al., 1989); the greater effect of long thin fibers parallels data on the carcinogenic potency of fibers after intraperitoneal or intrapleural injection (see Chapter 15).

Studies by Holian and co-workers have provided insight into the mechanisms by which mineral dusts, specifically chrysotile, stimulate macrophage O_2^- production. Alveolar macrophage oxidant generation can be activated by several agonists through a pertussis-toxin-sensitive pathway involving activation of phospholipase C (Holian et al., 1982, 1986). Phospholipase^2 C hydrolyses phosphatidylinositol in cellular membranes to generate inositol 1,4,5 triphosphate and diacylglycerol. Inositol triphosphate stimulates mobilization of intracellular calcium and diacylglycerol activates protein kinase C; and the latter activates the NADPH oxidase enzyme, which catalyzes reduction of oxygen to O_2^-. Studies on chrysotile demonstrated activation of alveolar macrophage O_2^- production is associated with increased phosphatidylinositol turnover and mobilization of intracellular calcium (Roney and Holian, 1989). In addition, asbestos-induced oxidant production by guinea pig alveolar macrophages can be attenuated using three putative protein kinase C inhibitors (i.e., staurosporine, sphingosine, and fluphenazine) and can be partially blocked by pertussis toxin. Overall these findings indicate chrysotile may stimulate macrophage oxidant production, at least in part, via activation of phospholipase C. Other studies have shown that chrysotile-induced O_2^- production can be prolonged by increasing extracellular calcium and is attenuated by the calcium channel blocker verapamil (Kalla et al., 1990). This effect of asbestos on calcium was not affected by protein kinase C inhibitors, indicating that the calcium effects are at least partly independent of this enzyme and that asbestos exposure may perturb calcium channels. Verapamil also inhibits chrysotile-induced increases in ornithine decarboxylase activity (a marker of cell proliferation) in tracheal epithelial cells (Marsh and Mossman, 1988), indicating asbestos effects on calcium mobilization are not unique to macrophages. Recently, silica treatment of rat alveolar macrophages was shown to increase intracellular calcium levels (Chen et al., 1991). Importantly, calcium can influence several key cellular processes in addition to O_2^- production, such as activation of phospholipase A_2 with subsequent synthesis of inflammatory lipids (Van Kuijk et al., 1987). In summary, there is compelling evidence that chrysotile and potentially other mineral particles can activate macrophage oxidant production by a phospholipase-C-dependent pathway and potentially by influencing cellular calcium channels.

As occurs for macrophage cytotoxicity, coating or opsonization of minerals clearly effects the oxidant response of macrophages. Specifically, sorption of IgG was shown to increase the ability of chrysotile to stimulate O_2^- release by guinea pig alveolar macrophages (Scheule and Holian, 1990). The effect of IgG in this study was unique to chrysotile as sorption to several other minerals had no effect on macrophage oxidant release. In human macrophages, the oxidant stimulating activity of both chrysotile and crocidolite asbestos was shown to be increased by IgG, with the greatest effect demonstrated for the IgG_1 isotype; IgA has also been

---

[2] Phospholipase C is one of a series of enzymes that catalyze the hydrolysis of phospholipids (i.e., the material that constitutes cell membranes).

Table 2
In vitro effect of mineral dusts on macrophage production of active oxygen species.

Mineral	Species: Cell type	Response	Reference
crocidolite, chrysotile	hamster: alveolar macrophage	Chrysotile and crocidolite stimulate production of superoxide	Case et al. (1986)
crocidolite, titanium dioxide	mouse: peritoneal P388D cell line	Crocidolite toxicity to macrophages is attenuated by oxygen radical scavengers. TiO_2 is only toxic in the presence of iron, and this toxicity is blocked by oxygen scavengers.	Goodlick and Kane (1986)
quartz, aluminum oxide, anatase, chrysotile	rabbit: alveolar macrophage	Mineral particles increase chemiluminescence, which was used as an indicator of oxidant production. No differences in chemiluminescence based on type of mineral.	Vilim et al. (1987)
crocidolite, erionite, Code 100 glass, sepiolite, riebeckite, mordenite, glass	hamster & rat: alveolar macrophage	Fibrous minerals are more effective at activating superoxide compared to their nonfibrous analogs. Rat cells are more responsive than hamster cells to the minerals.	Hansen and Mossman (1987)
quartz	rabbit: alveolar macrophage	Quartz-induced chemiluminescence is inhibited by superoxide dismutase, indicating the cells produce superoxide.	Wilhelm et al. (1987)
chrysotile	rat: alveolar macrophage	Long fibers (>10 μm) are more effective than short fibers (<2 μm) in stimulating dose-related increases in superoxide production.	Mossman et al. (1987)
chrysotile, crocidolite, anthophyllite, amosite	guinea pig: alveolar macrophage	Chrysotile stimulates superoxide production, whereas amphibole does not. Oxidant production is blocked by protein kinase C inhibitors. Chrysotile increases phosphatidyl inositol turnover, and calcium mobilization. Asbestos-induced oxidant production may occur via a phospholipase C pathway.	Roney and Holian (1989)
chrysotile, crocidolite, quartz, aluminum beads	guinea pig: alveolar macrophage	Oxidant production was stimulated by chrysotile but not by other minerals. In contrast to other minerals, IgG increases chrysotile's ability to stimulate oxidant production.	Scheule and Holian (1989)
chrysotile, crocidolite	guinea pig: alveolar macrophage	Sorption of proteins onto asbestos is influenced by mineral surface charge, protein charge and the presence of other proteins. When macrophages are exposed to chrysotile or crocidolite to which IgG is bound, superoxide production is stimulated.	Scheule and Holian (1990)
chrysotile	human: monocyte	IgA and IgG increase chrysotile-stimulated oxidant production by human monocytes.	Nyberg and Klockers (1990)
chrysotile	guinea pig: alveolar macrophage	Chrysotile-activated superoxide production is associated with increased intracellular calcium. Increasing extracellular Ca^{2+} levels prolongs asbestos-induced oxidant production. Verapamil (a Ca^{2+} channel blocker) attenuates superoxide production.	Kalla et al. (1990)
chrysotile, quartz	human: monocyte macrophage	γ-interferon increases quartz- and chrysotile-stimulated oxidant production. Combining γ-interferon with IgG increases chrysotile oxidant production ~20 fold.	Nyberg and Klockers (1991)
chrysotile, crocidolite, quartz	human: alveolar macrophage	Chrysotile and crocidolite stimulate oxidant production, whereas silica and aluminum do not. IgG coating increases oxidant production after asbestos and silica. Lipid blocks the effect of IgG on silica but not on asbestos.	Perkins et al. (1991)

shown to enhance asbestos-induced macrophage oxidant production (Perkins et al., 1991; Nyberg and Klockers, 1990). Interestingly, in contrast to guinea pig cells, immunoglobulin treatment in human cells increases macrophage oxidant production in response to amosite, quartz, chrysotile, and aluminum beads (Perkins et al., 1991). In contrast to IgG, sorption of lung surfactant lipids—e.g., phosphatidylinositol, phosphatidylserine, and phosphatidylglycerol—onto asbestos inhibits O_2^- production by exposed guinea pig macrophages (Jabbour et al., 1991). Lung lipids can attenuate the effect of IgG on silica and aluminum bead stimulated macrophage O_2^- synthesis; however, this effect is selective since lipids do not effect the enhanced O_2^- response observed with IgG coated chrysotile or crocidolite. These studies suggest components of lung lining fluid (e.g., phospholids or immunoglobulins) may significantly influence macrophage oxidant production *in vivo*. However, the precise effect may depend on both the coating materials present and the type of mineral particle.

Overall, the data on *in vitro* mineral exposure and macrophage reactive oxygen production indicates a pattern of response different from that seen for hemolysis and cytotoxicity. It appears that phagocytic stimulation of macrophages, in general, is sufficient to activate an oxidative burst, however, there exist mineral specific differences related to particle shape, size, and mineral species that influence the magnitude of oxidant production. Regarding the disparate toxicity of some minerals, the differential effects of lung lining materials on mineral-induced oxidant production may be a contributing factor.

In vitro effect of minerals on macrophage cytokine production

Although a wide range of macrophage secretory products have been examined *in vitro*, the following discussion focuses on recent studies examining the pro-inflammatory cytokines, TNF-α and IL-1. TNF-α is a 17-kilodalton (kd) protein produced by a variety of phagocytic and nonphagocytic cell types including: macrophages, monocytes, polymorphonuclear leukocytes, lymphocytes, smooth muscle cells, and mast cells (Kawakami and Cerami, 1981; Beutler et al., 1985; Dubravec et al., 1990; Young et al., 1987; Warner et al., 1989). IL-1 exists

Table 3

Inflammatory and immune processes stimulated by IL-1 or TNF-α

T- and B-lymphocyte proliferation and activation
Acute phase protein response
Arachidonic acid metabolism (i.e., prostaglandin E2, prostacyclin)
Inflammatory cell oxidative burst and degranulation
Expression of adhesion molecules (e.g., ELAM, ICAM, VCAM)
Chemotactic protein (chemokine) expression (e.g., IL-8, MCP-1, MIP-2)
Cytokine expression (e.g., IL-1, TNF-α, IL-6, GM-CSF)

in two structurally related but biochemically distinct forms called IL-1α and IL-1β; both these proteins are ~17 kd in size (Cameron et al., 1986). Although in early studies IL-1 production was primarily associated with mononuclear phagocytes other cell types now known to produce this cytokine include: polymorphonuclear leukocytes, keratinocytes, Langerhans cells, endothelial cells, smooth muscle cells, fibroblasts, and natural killer cells (Tiku et al., 1986; Sauder et al., 1985; Libby et al., 1986a, b; Akahosi et al., 1988). TNF-α and IL-1 are multifunctional cytokines and share a wide range of biologic activities that have been the subject of several recent detailed reviews (Jaattela, 1991; Le and Vilcek, 1987; Driscoll, 1993a). Table 3 summarizes some of the processes influenced by these cytokines that are of particular relevance to inflammatory and immune responses. Briefly, TNF-α and IL-1 can influence the infiltration of inflammatory cells by stimulating expression of adhesion proteins on vascular endothelium and the release of specific inflammatory cell chemotactic proteins called chemokines from a number of cell types including macrophages, epithelial cells, and fibroblasts (Gamble et al., 1985; Bevilaqua et al., 1985; Driscoll et al., 1993b). Additionally, TNF-α and IL-1 can stimulate secretion of tissue damaging active oxygen species and proteolytic enzymes by phagocytic cells (Klebanoff et al., 1986; Klempner et al., 1978; Warren et al., 1988). Furthermore, recent *in vivo* studies have demonstrated an association between local release of IL-1 and/or TNF-α and increased cellular proliferation and connective tissue accumulation (Dunn et al., 1989; Piquet et al., 1990a).

Given the bioactivities of IL-1 and TNF-α, it is clear these cytokines have the potential to play a role in the types of responses associated with toxic mineral dusts. As summarized in Table 4 several studies have demonstrated that mineral dust particles activate macrophages *in vitro* to release IL-1 and TNF-α, and this response appears to correspond to *in vivo* inflammatory and fibrogenic activity. For example, a study by Driscoll et al. (1990a) compared the effect of two fibrogenic and nonfibrogenic mineral dusts on TNF-α release by rat alveolar macrophages. As shown in Table 5 these investigators found exposure to quartz (Min-U-Sil) or crocidolite stimulated dose-related increases macrophage TNF-α production. In contrast, treatment with anatase or aluminum oxide had no significant effect on macrophage TNF-α production. It is noteworthy that the effects of crocidolite on TNF-α were detected in the absence of any detectable cytotoxic effects (Table 5, see LDH release).

Human alveolar macrophages also demonstrate a differential cytokine response to mineral dusts. Gosset et al. (1991) reported that coal dust and quartz increased steady-state levels of TNF-α mRNA and release of TNF-α protein, with titanium dioxide having only a minimal effect on macrophage TNF-α. Similar studies have demonstrated that chrysotile asbestos stimulates TNF-α production by human alveolar macrophages *in vitro* (Perkins et al., 1993). Preliminary findings indicate that in addition to crocidolite and silica, refractory ceramic fibers are potent stimulators of macrophage TNF-α release *in vitro* (Leikauf et al., 1993).

As for TNF-α, the differential effects of mineral dusts on macrophage IL-1 production appear to parallel their *in vivo* inflammatory and fibrogenic activity. *In vitro* exposure of macrophages to noncytotoxic levels of crystalline silica

(DQ-12 quartz and Min-U-Sil), coal mine dust (anthracite), and chrysotile or crocidolite asbestos stimulates monocytes, peritoneal macrophages, and (in some studies) alveolar macrophages to produce IL-1, whereas minerals of low inherent toxicity (e.g., titanium dioxide or diamond dust) have minimal effect on IL-1 production (Gery et al., 1981; Lepe-Zuniga and Gery, 1984; Schmidt et al., 1984; Kampschmidt et al., 1986).

Table 4

In vitro effect of mineral dusts on monocyte and macrophage production of IL-1 or TNΦ-α[a]

Mineral	Species: Cell type	Response	Reference
quartz	mouse: peritoneal macrophage	Silica increases extracellular IL-1 and to a lesser extent intracellular IL-1. Latex beads increase both intra- and extracellular IL-1 activity. A correlation was observed between cell damage and IL-1 release.	Gery et al. (1981)
quartz	human: monocyte	Silica stimulates production and release of IL-1.	Lepe-Zuniga and Gery (1984)
quartz, diamond dust	human: monocyte	Silica but not diamond particles stimulates dose-related increases in IL-1 production. A factor released by monocytes stimulates fibroblast growth and appears identical to IL-1.	Schmidt et al. (1984)
silica	rabbit: peritoneal macrophage	Combined LPS and silica exposure stimulates oil elicited macrophages to release increased amounts of IL-1-like activity.	Kampschmidt et al. (1986)
silica, chrysotile, crocidolite	rat: alveolar macrophage	Silica, chrysotile and crocidolite increase production of IL-1 and fibroblasts' proliferation activity. Combination of LPS and mineral dust synergistically increase IL-1.	Oghiso and Kubota (1986)
silica (Sigma), α-quartz, chrysotile, crocidolite, titanium dioxide	rat: alveolar macrophage	The silica polymorphs and both asbestos types, but not titanium dioxide, increase IL-1 release by macrophages.	Oghiso (1987)
chrysotile, quartz (DQ-12), rutile	mouse: peritoneal macrophage	Chrysotile and silica, but not titanium dioxide, stimulate release of IL-1. Increased IL-1 release was not due to leakage of preformed IL-1.	Godelaine and Beaufay (1989)
silica (Min-U-Sil 5), chrysotile	rat: alveolar macrophage	TNF and leukotriene B_4 (LTB_4) production is increased by silica and chrysotile; latex beads do not stimulate this response. Lipoxygenase inhibitors attenuated mineral dust-induced TNF release and reconstituting LTB_4 restores the TNF response.	DuBois et al. (1989)
silica (Min-U-Sil), crocidolite, anatase, corundum	rat: alveolar macrophage	TNF and LTB_4 production are increased by silica and crocidolite, but not by titanium dioxide or aluminum oxide. None of the mineral dusts increases IL-1 release.	Driscoll et al. (1990)
coal dust, quartz, titanium dioxide	human: alveolar macrophage	Coal dust stimulates release of TNF but not IL-1. Silica stimulates TNF release, but is less potent that coal. Titanium dioxide has a minimal effect on TNF or IL-1 release.	Gosset et al. (1991)
silica	rat: alveolar macrophage	Silica exposure increases production of IL-1. This response os blocked by tetrandrine, an antifibrotic drug.	Kang et al. (1992)
silica	human: THP-1 (myelomonocytic cell line)	Silica stimulates TNF gene transcription and release of TNF protein.	Savici et al. (1992)
chysotile	human: alveolar macrophage	Chrysotile stimulates macrophage production of TNF but not IL-1, IL-6, GM-CSF or PGE_2.	Perkins et al. (1993)

Table 5

Cytotoxicity of mineral dusts to rat alveolar macrophages [a]

		Measure of Cytotoxicity	
		LDH Release (U/mL)	TNF Production (units/10^6 cells)
quartz			
	10 µg/ml	34 ± 1[b]	11 ± 2[b]
	30 µg/ml	40 ± 2[b]	18 ± 2[b]
	100 µg/ml	50 ± 1[b]	41 ± 7[b]
	300 µg/ml	47 ± 1[b]	55 ± 6[b]
crocidolite			
	10 µg/ml	17 ± 2[b]	8 ± 2[b]
	30 µg/ml	19 ± 1[b]	17 ± 2[b]
	100 µg/ml	19 ± 5[b]	20 ± 1[b]
	300 µg/ml	16 ± 1[b]	38 ± 2[b]
anatase			
	10 µg/ml	25 ± 1	3 ± 1
	30 µg/ml	25 ± 1	3 ± 1
	100 µg/ml	28 ± 2	4 ± 1
	300 µg/ml	35 ± 4[c]	3 ± 1
corundum			
	10 µg/ml	25 ± 1	4 ± 1
	30 µg/ml	24 ± 2	3 ± 1
	100 µg/ml	24 ± 1	3 ± 1
	300 µg/ml	33 ± 1[c]	4 ± 1

[a] N = 3 experiments
[b] different from respective titanium dioxide or aluminum oxide mean; $p < 0.05$
[c] different from respective asbestos mean; $p < 0.05$

It is noteworthy that several recent studies have shown that, in addition to TNF-α release, inflammatory mineral dusts (i.e., quartz, crocidolite, and refractory ceramic fibers) activate alveolar macrophages to release leukotriene B_4 (LTB_4), a lipoxygenase metabolite of arachidonic acid (Dubois et al., 1989; Driscoll et al., 1990a; Leikauf et al., 1993). While LTB_4 itself has inflammatory activity, it also may influence macrophage release of TNF-α and other cytokines. Dubios et al. (1989) reported that treatment of macrophages with the lipoxygenase inhibitors, nordihydroguaiaretic acid (NDGA) and AA861, blocked LTB_4 release and attenuated quartz or chrysotile-induced TNF-α release. Reconstituting LTB_4 to inhibitor and mineral dust treated cells partially restored dust-induced TNF-α production. These results indicate that LTB_4 can act to amplify TNF-α release and suggests a common LTB_4 dependent mechanism for silica and chrysotile stimulated TNF-α production. In this respect, LTB_4 has been previously shown to enhance the production of other inflammatory cytokines including IL-1, IL-2, and γ interferon (Rola-Pleszczynski et al., 1985, 1987). It is tempting to speculate that the marked inflammatory activity of crystalline silica, asbestos, and refractory ceramic fibers may result from their ability to stimulate both LTB_4 and TNF-α production by macrophages.

The significance of correlations between *in vitro* mineral effects on macrophage-derived cytokines (particularly TNF-α) and *in vivo* inflammatory and fibrogenic activity is underscored by recent studies demonstrating a critical role for macrophage-derived cytokines in mineral-dust-induced lung disease. Alveolar macrophages obtained from crystalline-silica-exposed rats produce increased levels of TNF-α and IL-1 (Driscoll et al., 1990b; 1993c; Oghiso and Kubota, 1986; Struhar et al., 1989). Likewise, macrophages obtained from asbestos-exposed humans release increased levels of TNF-α and IL-1 as well as IL-6 and prostaglandin E_2 (Lassalle et al., 1990; Zhang et al., 1992; Perkins et al., 1993). These studies clearly show an association between mineral-dust-induced lung disease and macrophage cytokines. Work reported by Piguet et al. (1990b) has provided compelling evidence for a causal relationship between TNF-α and pathogenesis of pulmonary fibrosis, including silicosis. These investigators reported that passive immunization of mice with an anti-TNF-α antibody prior to silica instillation blocks the development of fibrosis. Similarly, passive immunization of mice with anti-TNF-α antibody was shown to block pulmonary fibrosis resulting from exposure to the pneumotoxic anti-cancer drug, bleomycin (Piguet et al., 1989).

Studies by Driscoll and coworkers (1990b; 1993a,d, Driscoll and Maurer, 1991) suggest that a mechanism by which TNF-α may contribute to mineral-dust toxicity is involves facilitating the recruitment and activation of inflammatory cells. These investigators demonstrated a significant positive correlation between mineral-dust-induced neutrophilic inflammation in rat lungs and activation of macrophages to produce TNF-α. Subsequent studies have shown that, as with silica-induced fibrosis, the recruitment of neutrophils after high doses of titanium dioxide can be attenuated by pretreatment of rats with anti-TNF-α antibody (Driscoll et al., 1993a,d). Although TNF-α is not directly chemotactic for inflammatory cells, it can activate release of chemotactic proteins (called *chemokines*) by a number of cell types including macrophages, fibroblasts, and epithelial cells (Driscoll et al., 1993a,b). For example, TNF-α and IL-1 can induce lung cells to produce IL-8 and macrophage inflammatory protein-2, proteins that are highly chemotactic for neutrophils. In addition, production of MCP-1 a potent monocyte chemotactic factor is upregulated by TNF-α and IL-1 (Driscoll et al., 1993a). Thus, as illustrated in Figure 1, mineral dusts appear to elicit recruitment of inflammatory cells through a network of cytokine and cell–cell interactions involving, at least in part, activation of macrophages to produce TNF-α. TNF-α then acts via autocrine and paracrine pathways to stimulate lung cells such as macrophages, fibroblasts, and epithelial cells to release chemokines and, in conjunction with increased endothelial adhesion molecule expression, this results in an infiltration of inflammatory cells to the lung.

SUMMARY

In vitro cell culture methods provide the ability to investigate the toxicity of mineral particles. Using hemolysis and macrophage-cytotoxicity assays, investigators have identified some of the important physical and chemical characteristics of minerals that influence their interactions with biological membranes. For

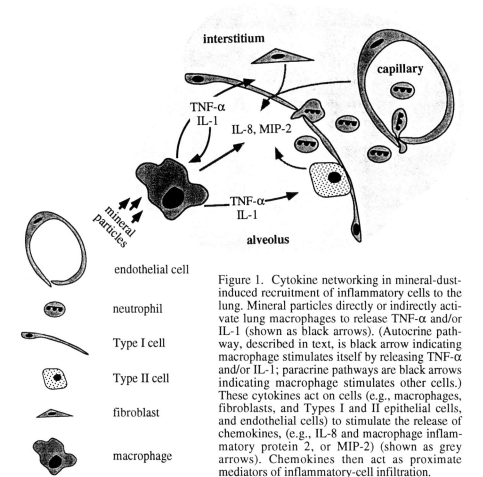

Figure 1. Cytokine networking in mineral-dust-induced recruitment of inflammatory cells to the lung. Mineral particles directly or indirectly activate lung macrophages to release TNF-α and/or IL-1 (shown as black arrows). (Autocrine pathway, described in text, is black arrow indicating macrophage stimulates itself by releasing TNF-α and/or IL-1; paracrine pathways are black arrows indicating macrophage stimulates other cells.) These cytokines act on cells (e.g., macrophages, fibroblasts, and Types I and II epithelial cells, and endothelial cells) to stimulate the release of chemokines, (e.g., IL-8 and macrophage inflammatory protein 2, or MIP-2) (shown as grey arrows). Chemokines then act as proximate mediators of inflammatory-cell infiltration.

example, studies have demonstrated the importance of reactive silanol groups on silica polymorphs and of particle surface charge, the potential role of magnesium in the membranolytic ability of chrysotile, and the significance of metals (e.g., iron) associated with mineral particles. In addition, although these relatively simple assay systems are not consistently predictive of *in vivo* toxicity, they possess sufficient positive correlation with *in vivo* toxicity that they may be useful as preliminary screens to guide decision-making on the need for further more definitive toxicity testing.

More recent studies are focusing on substances that stimulate macrophages (i.e., that promote activation) rather than substances that lyse macrophages (i.e., that are cytotoxic). This emphasis on macrophage activation and the associated secretory activity stems from the rapidly growing data base demonstrating factors secreted by macrophages play an important role in the pathogenesis of lung disease. Studies on macrophage oxidant production suggest that particle phagocytosis represents a nonspecific stimulus for oxidant release. However, the magnitude of the oxidant response is influenced by the chemical and physical characteristics of the mineral particle as well as molecules that may be sorbed on particles. Regarding the latter, it is likely the lung environment plays an important role in modifying mineral surfaces and influencing their ability to elicit inflammatory mediator release from macrophages or other lung cells. Developing better understanding of how constituents in lung fluid modifies a mineral's bioactivity should guide development of better *in vitro* systems to predict *in vivo* toxicity.

There is increasing evidence that cytokines (e.g., TNF-α and IL-1) play key roles in the inflammatory and fibrotic responses to a variety of pneumotoxic agents. Several studies have characterized release of potent inflammatory cytokines by macrophages in response to mineral particles. These studies demonstrate a good correlation between the ability of mineral dusts to stimulate macrophage cytokine release (particularly TNF-α) and their inflammatory and fibrogenic activity *in vivo*. Clearly work is needed in this area to expand the variety of minerals tested and, more importantly, understand the molecular mechanisms by which minerals exert their apparently disparate effects on activation of macrophages and other lung cells. Research in this area will assist in develop-ment and validation of mechanistically based *in vitro* models which focus on aspects of particle–cell interactions critical to their *in vivo* bioactivity .

REFERENCES

Akahosi, T., Oppenheim, J.J. and Matsushima, J. (1988) Interleukin-1 stimulates its own receptor expression on human fibroblasts through the endogenous production of prostaglandin(s). J. Clin. Invest. 82, 1219-1224

Allison, A.C., Harrington, J.S. and Birbeck, M. (1966) An examination of the cytotoxic effects of silica on macrophages. J. Exp. Med. 124, 141-153.

Allison, A.C. (1971) Lysosomes and the toxicity of particulate pollutants. Arch. Intern. Med. 128, 131-139.

Allison, A.C. (1973) Experimental methods-cell and tissue culture: Effects of asbestos particles on macrophages, mesothellal cells and fibroblasts. In: Biological Effects of Asbestos. IARC Conference. P. Bogovski, J.C. Gilson, V. Timbrell and J.C. Wagner, Eds. Vol. 8, 89-93.

Bevilacqua, M.P., Pober, J.S., Wheeler, M.E., Cotran, R.S. and Gimbrone, M.A., Jr. (1985) Interleukin-1 acts on cultured human vascular endothelium to increase the adhesion of polymorphonuclear leukocytes, monocytes and related leukocytic cell lines. J. Clin. Invest. 76, 2003-2011.

Beutler, B., Mahoney, J., LeTrang, N., Pekala, P. and Cerami, A. (1985) Purification of cachectin a lipoprotein lipase-suppressing hormone secreted by endotoxinn-induced raw-264.7 cells. J. Exp. Med. 161, 984-995

Bignon, J. and Jaurand, M.C. (1983) Biological *in vitro* and *in vivo* responses of chrysotile versus amphiboles. Environ. Health Perspect. 51, 73-80.

Bitterman, P.R., Rennard, S.I. and Crystal, R.G. (1981). Environmental lung disease and the interstitium. Clinics in Chest Medicine. Eds. S.M. Brooks, J.E. Lockey and P. Harper, Eds. Occup. Lung Dis. II, 2,393-412.

Brain, J.D. (1970) Free cells in the lungs. Arch. Intern. Med. 126, 477-487.

Brieger, H. and Gross, P. (1967) On the theory of silicosis: III. Stishovite. Arch. Environ. Health 15, 751–757.

Brown, R.C., Poole, A., Turver, C.J. and Vann, C. (1987) Role of iron-mediated free-radical generation in asbestos-induced cytotoxicity. Toxic. *In vitro* 1, 67-70.

Cameron, P.M., Limjuco, G.A., Chin, J., Silberstein, L. and Schmidt, J.A. (1986) Purification to homogeneity and amino acid sequence analysis of two anionic species of human interleukin 1. J. Exp. Med. 164, 237-250.

Carp, H. and Janoff, A. (1979) *In vitro* suppression of serum elastase inhibitory capacity by reactive oxygen species generated by phagocytosing polymorphonuclear leukocytes. J. Clin. Invest. 63, 793-797.

Case, B.W. Ip, M.P.C., Padilla, M. and Kleinerman, J. (1986) Asbestos effects on superoxide production: An *in vitro* study of hamster alveolar macrophages. Environ. Res. 39, 299-306.

Chamberlain, M. and Brown, R.C. (1977) The cytotoxic effects of asbestos and other mineral dust in tissue culture cell lines. Brit. J. Exp. Path. 59, 183-189.

Chen, J., Armstrong, L.C., Liu, S., Gerriets, J.E. and Last, J.A. (1991) Silica increases cytosolic free calcium ion concentration of alveolar macrophages in vitro. Toxicol. Appl. Pharm. 111, 211-220.

Davies, R. and Preece, A.W. (1983) The electrophoretic mobilities of minerals determined by laser Doppler velocimetry and their relationship with the biological effect of dusts towards macrophages. Clin. Phys. Physiol. Meas. 4, 129-140.

Davies, R., Skidmore, J.W., Griffiths, D.M. and Moncrieff, C.B. (1983) Cytotoxicity of talc for macrophages in vitro. Fd. Chem. Toxic. 21, 201-207.

Davies, R., Griffiths, D.M., Johnson, N.F., Preece, A.W. and Livingston, D.C. (1984) The cytotoxicity of kaolin towards macrophages in vitro. Brit. J. Exp. Path. 65, 453-466.

Dogra, S. and Lai Kaw, J. (1988) Changes in some histochemically demonstrable enzymes in macrophages exposed to quartz dust in vitro. J. Appl. Toxicol. 8, 23-27.

Driscoll, K.E., Higgins, J.M., Leytart, M.J. and Crosby, L.L. (1990a) Differential effects of mineral dusts on the in vitro activation of alveolar macrophage eicosanoid and cytokine release. Toxic in vitro 4, 284-288.

Driscoll, K.E., Lindenschmidt, R.C., Maurer, J. K., Higgins, J.M. and Ridder, G. (1990b) Pulmonary response to silica or titanium dioxide: inflammatory cells, alveolar macrophage-derived cytokines and histopathology. Amer. J. Respir. Cell Mol. Biol. 2, 381-390.

Driscoll, K.E. and Maurer, J.K. (1991) Cytokine and growth factor release alveolar macrophages: potential biomarkers of pulmonary toxicity. Toxicol Pathol. 19, 398-405.

Driscoll, K.E. (1993a) The role of interleukin-1 and tumor necrosis factor in the lung's response to silica. In: Silica and Silica-Induced Lung Disease: Current Concepts. V. Castranova, V. Vallyathan and W. Wallace, Eds. CRC Press, Boca Raton, Florida (in press).

Driscoll, K.E., Hassenbein, D., Carter, J., Poynter, J. Asquith, T., Grant, R.A., Whitten, J., Purdon, M. and Takigiku, R. (1993b) Macrophage Inflammatory Proteins 1 and 2: Expression by rat alveolar macrophages, fibroblasts and epithelial cells and in rat lung tissue after mineral dust exposure. Amer. J. Resp. Cell Mol. Biol. 8, 311-318.

Driscoll, K.E., Strzelecki, J., Hassenbein, D., Janssen, Y.M.W., March, J.K., Oberdoerster, G. and Mossman, B.T. (1993c) Tumor Necrosis Factor (TNF): Evidence for the role of TNF in increased expression of manganese superoxide dismutase after inhalation of mineral dusts. Ann. Occup. Hygiene (in press).

Driscoll, K.E., Maurer, J.K., Hassenbein, D., Carter, J., Janssen, Y.M.W., Mossman, B.T.,· Osier, M. and Oberdorster, G. (1993d) Contribution of macrophage-derived cytokines and cytokine networks to mineral dust-induced lung inflammation. Proc. 4th Int'l Symp. on Toxic and Carcinogenic Effects of Particles, ILSI Press (in press).

Dubois, C.M., Bissonnette, E. and Rola-Pleszczynski, M. (1989) Asbestos fibers and silica particles stimulate rat alveolar macrophages to release tumor necrosis factor. Amer. Rev. Respir. Dis. 139, 1257-1264.

Dubravec, D.B., Spriggs, D.R., Mannick, J.A. and Rodrick, M.L. (1990) Circulating human peripheral blood granulocytes synthesize and secrete tumor necrosis factor. Proc. Nat'l Acad. Sci. U.S.A. 87, 6758

Dunn, C.J., Hardee, M.M., Gibbons, A.J., Staite, N.D. and Richard, K.A. (1989) Local pathological responses to slow-release recombinant interleukin-1, interleukin-2 and γ-interferon in the mouse and their relevance to chronic inflammatory disease. Clin. Sci. 76, 261.

Freeman, B.A. and Crapo, J.D. (1982) Free radicals and tissue injury. Lab. Invest. 47, 412-426.

Gamble, J.R., Harlan, J.M., Klebanoff, S.J. and Vadas, M.A. (1985) Stimulation of the adherence of neutrophils to umbilical vein endothelium by human recombinant tumor necrosis factor. Proc. Nat'l Acad. Sci. U.S.A. 82, 8667-8671.

Gery, I., Davies, P., Derr, I., Krett, N. and Barranger, J.A., (1981) Relationship between production and release of lymphocyte-activating factor (interleukin 1) by murine macrophages. Effects of various agents. Cell Immunol. 64, 293-303.

Godelaine, D. and Beaufay, H. (1989) Comparative study of the effect of chrysotile, quartz and rutile on the release of lymphocyte-activating factor (interleukin 1) by murine peritoneal macrophages in vitro. IARC Sci. Publ. 90, 149-155.

Goodglick, L.A. and Kane, A.B. (1986) Role of reactive oxygen metabolites in crocidolite asbestos toxicity to mouse macrophages. Cancer Res. 46, 5558-5566.

Gosset, P., Lassalle, P., Vanhee, D., Wallaert, B., Aerts, C., Voisin, C. and Tonnet A.-B. (1991) Production of tumor necrosis factor- and interleukin-6 by human alveolar macrophages exposed in vitro to coal mine dust. Amer. J. Resp. Cell Mol. Biol. 5, 431-436.

Guthrie, G.D., Jr. (1992) Biological effects of inhaled minerals. Amer. Mineral. 77, 225–243.

Guthrie, G.D., Jr. and Bish, D.L. (1991) Quantitative phase analysis of mordenite samples using the Rietveld method, Amer. College of Chest Physicians, 4th Int'l Conf. on Environmental Lung Disease (abstr.).

Hansen, K. and Mossman, B.T. (1987) Generation of superoxide (O_2) from alveolar macrophages exposed to asbestiform and nonfibrous particles. Cancer Res. 47, 1681-1686.

Harington, J.S., Macnab, G., Miller, J. and King, P.C. (1971a) Enhancement of haemolytic activity of asbestos by heat-labile factors in fresh serum. Med. Lav. 62, 171-176.

Harington, J.S., Miller, K. and MacNab, G. (1971b) Hemolysis by asbestos. Environ. Res. 4, 95-117.

Harington, J.S., Allison, A.C. and Badami, D.V. (1975) Mineral fibers: Chemical, physiochemical, and biological properties. Adv. Pharm. Chemo. 12, 291-402.

Harley, J.D.and Margolis, J. (1961) Haemolytic activity of colloidal silica. Nature 189, 1010-1011.

Heflin, A.C. and Brigham, K.L. (1981) Prevention by granulocyte depletion of increased vascular permeability of sheep lung following endotoxemia. J. Clin. Invest. 68, 1253-1260.

Hefner, R.E. and Gehring, P.J. (1975). A comparison of the relative rates of hemolysis induced by various fibrogenic and non-fibrogenic particles with washed erthrocytes *in vitro*. Amer. Indus. Hygiene Assoc. J. 36, 734-740.

Holian, A. and Daniele, R.P. (1982) The role of calcium in the initiation of superoxide release from alveolar macrophages. J. Cell. Physiol. 113, 87-93.

Holian, A. (1986) Leukotriene B4 stimulation of phosphatidylinositol turnover in macrophages and inhibition by pertussis toxin. FEBS Letters 201, 15-19.

Huang, S.L., Saggioro, D., Michelmann, H. and Malling, H.V. (1978) Genetic effects of crocidolite asbestos in Chinese hamster lung cells. Mutation Res. 57, 225-232.

Iguchi, H. and Kojo, S. (1989) Possible generation of hydrogen peroxide and lipid peroxidation of erythrocyte membrane by asbestos: cytotoxic mechanism of asbestos. Biochem. Int'l 5, 981-990.

Jaattela, M. (1991) Biology of Disease: Biologic activities and mechanisms of action of tumor necrosis factor-/cachectin. Lab. Invest. 64, 724.

Jabbour, A.J., Holian, A. and Scheule, R.K. (1991) Lung lining fluid modification of asbestos bioactivity for the alveolar macrophage. Toxicol. Appl. Pharm. 110, 283-294.

Jaurand, M.C., Kaplan, H., Thiollet, J., Pinchon, M.C., Bernaudin, J.F. and Bignon, J. (1979). Phagocytosis of chrysotile fibers by pleural mesothelial cells in culture. Amer. J. Pathol. 94, 529-538.

Kalla, B., Hamilton, R.F., Scheule, R.K. and Holian, A. (1990) Role of extracellular calcium in chrysoltile asbestos stimulation of alveolar macrophages. Toxicol. Appl. Pharm. 104, 130-183.

Kampschmidt, R.F., Worthington, M.L. and Mesecher, M.I. (1986) Release of interleukin-1 (IL-1) and IL-1-like factors from rabbit macrophages with silica. J. Leukocyte Biol. 39, 123-132.

Kaw, J.L., Beck, E.G. and Bruch J. (1975) Studies of quartz cytotoxicity on peritoneal macrophages of guinea pigs pretreated with polyvinylpyridine N-Oxide. Environ. Res. 9, 313-320.

Kawakami, M. and Cerami, A. (1981) Studies of endotoxin-induced decrease in lipoprotein lipase activity. J. Exp. Med. 154, 631-639.

Kennedy, T.P., Dodson, R., Rao, N.V., Ky, H., Hopkins, C., Baser, M., Tolley, E. and Hoidal, J.R. (1989) Dusts causing pneumoconiosis generate OH and produce hemolysis by acting as fenton catalysts. Arch. Biochem. Biophys. 269, 359-364.

Kessel, R.W.I., Monaco, L. and Marchisio, M.A. (1962) The specificity of the cytotoxic action of silica: A study *in vitro*. University of Milan, Milano, Italy, p. 351-364.

Kozin, F., Millstein, B., Mandel, G. and Mandel, N. (1982) Silica-induced membranolysis: A Study of different structural forms of crystalline and amorphous silica and the effects of protein adsorption. J. Colloid Interface Sci. 88, 326-523.

Klebanoff, S.J., Vadas, M.A., Harlan, J.M., Sparts, L.H., Gamble, J.R., Agosti, J.M. and Waltersdorph, A.M. (1986) Stimulation of neutrophils by tumor necrosis factor. J. Immunol. 136, 4220-4225

Klempner, M.S., Dinarello, C.A. and Gallin, J.I. (1978) Human leukocytic pyrogen induces release of specific granule contents from human neutrophils. J. Clin. Invest. 61, 1330-1336

Langer, A.M. and Nolan, R.P. (1985) Physiochemical properties of minerals relevant to biological activities: State of the art. In *In vitro* Effects of Mineral Dusts. D.C. Beck and J. Bignon, Eds. Springer-Verlag, Berlin. 9-24.

Lasser, A. (1983) The mononuclear phagocytes system: A review. Human Pathol. 14, 108-126.

Le, J. and Vilcek, J. (1987) Biology of Disease: Tumor necrosis factor and interleukin 1: cytokines with multiple overlapping biological activities. Lab. Invest. 56, 234-248.

Leikauf, G.D., Fink, S.P., Miller, M.L., Lockey, J.E. and Driscoll, K.E. (1993) Refractory ceramic fibers activate alveolar macrophage eicosanoid and cytokine release. Toxicologist 13, 266 (abstract).

Lepe-Zuniga, J.L. and Gery, I. (1984) Production of intra-and extracellular interleukin-1 (IL-1) by human monocytes. Clin. Immunol. Immunopathol. 31, 222-230.

Libby, R., Ordovas, J.M., Auger, K.R., Robbins, A.H., Birinyl, L.K. and Dinarello, C.A. (1986a) Inducible interleukin-1 gene expression in vascular smooth muscle cells. J. Clin. Invest. 78, 1432-1438.

Libby, R., Ordovas, J.M., Auger, K.R., Robbins, A.H., Birinyl, L.K. and Dinarello, C.A. (1986b) Endotoxin and tumor necrosis factor induce interleukin-1 expression in adult human vascular endothelial cells. Amer. J. Pathol. 124, 179-185.

Light, W.G. and Wei, E.T. (1977) Surface charge and asbestos toxicity. Nature 265, 537-539.

Lock, S.O., Jones, P.A., Friend, J.V. and Parish, W.E. (1987) Extracellular release of enzymes from macrophages *in vitro* for measuring cellular Interaction with particulate and non-particulate materials. Toxic *In vitro* 1, 77-83.

Macnab, G. and Harington, J.S. (1967) Hemolytic activity of asbestos and other mineral dusts. Nature. 214, 522-523.

Marks, J. (1957) The neutralization of silica toxicity *in vitro*. Brit. J. Indus. Med. 14, 81-84.

Marks, J. and Nagelschmidt, G. (1959) Study of the toxicity of dust with use of the *in vitro* dehydrogenase technique. A.M.A. Arch. Indus. Hlth. 20, 382.

Marsh, J.P. and Mossman, B.T. (1988) Mechanisms of induction of ornithine decarboxylase activity in tracheal epithelial cells by asbestiform minerals. Cancer Res. 48, 709-714.

Meltzer, M.S. (1981) Peritoneal mononuclear phagocytes from small animals. Methods for Studying Mononuclear Phagocytes. Adams, D.O., Edelson, P.J. and Koren, H., Eds. Academic Press, New York, p. 63-67.

Mossman, B.T. and Landesman, J.M. (1983) Importance of oxygen free radicals in asbestos-induced injury to airway epithelial cells. Chest 83, 50S-51S.

Mossman, B.T., Marsh, J.P. and Shatos, M.A. (1986) Alteration of superoxide dismutase activity in tracheal epithelial cells by asbestos and inhibition of cytotoxicity by antioxidants. Lab. Invest. 54, 204-212.

Mossman, B.T., Marsh, J.P., Shatos, M.A., Doherty, J., Gilbert, R. and Hill, S. (1987) Implication of oxygen species as second messengers of asbestos toxicity. Drug Chem. Toxicol. 10, 157-180.

Mossman, B.T., Marsh, J.P., Gilbert, R., Hardwick, D., Sesko, A., Hill, S., Shatos, M.A., Doherty, J., Bergeron, M., Adler, K.B., Hemenway, D., Mickey, R., Vacek, P. and Kagan, E. (1990) Inhibition of lung injury inflammation and interstitial pulmonary fibrosis by polyethylene glycol-conjugated catalase in rats exposed by inhalation to asbestos. Amer. Rev. Respir. Dis. 141, 1266-1271.

Murray, M.J. and Driscoll, K.E. (1992) Immunology of the respiratory system. In: Comparative Biology of the Normal Lung. Vol 1, R. Parent, Ed., CRC Press, Boca Raton, Florida p. 725-746.

Nash, T, Allison, A.C. and Harington, J.S. (1966) Physico-chemical properties of silica in relation to its toxicity. Nature 210, 259-261.

Nathan, C.F. (1987) Secretory products of macrophages. J. Clin. Invest. 79, 319-326.

Nolan, R.P., Langer, A.M., Harington, J.S., Oster, G. and Selikoff, I.J. (1981) Quartz hemolysis as related to its surface functionalities. Environ. Res. 26, 503-520.

Nyberg, P. and Klockars, M. (1990) Effect of immunoglobulins on mineral dust-induced production of reactive oxygen metabolites by human macrophages. Inflammation 4, 621-629.

Nyberg, P. and Klockars, M. (1991) Interferon- and immunoglobulin enhance mineral dust-induced production of reactive oxygen metabolites by human macrophages. Clin. Immunol. Immunopathol. 60, 128-136.

Oghiso, Y. and Kubota, Y. (1986) Interleukin-1-like thymocyte and fibroblast activating factors from rat alveolar macrophages exposed to silica and asbestos particles. Japan. J. Vet. Sci. 48:461-471.

Oghiso, Y. (1987) Heterogeneity in immunologic functions of rat alveolar macrophages-their accessory cell function and IL-1 production. Microbiol. Immun. 31, 247-260.

Paterour, M.J., Bignon J. and Jaurand, M.C. (1985) *In vitro* transformation of rat pleural mesothelial cells by chrysotile fibres and/or benzo(a)pyrene. Carcinogenesis 6, 523-529.

Perkins, R.C., Scheule, R.K. and Holian, A. (1991) *In vitro* bioactivity of asbestos for the human alveolar macrophage and its modificaiton by IgG. Amer. J. Respir. Cell Mol. Biol. 4, 532-537.

Perkins, R.C., Scheule, R.K., Hamilton, R., Gomes, G., Freidman, G. and Holian, A. (1993) Human alveolar macrophage cytokine release in response to *in vitro* and *in vivo* asbestos exposure. Exp. Lung Res. 1, 55-65.

Piguet, P.F., Collart, M.A., Grau, G.E., Kapanci, Y. and Vassalli, P. (1989) Tumor necrosis factor/cachectin plays a key role in bleomycin-induced pneumopathy and fibrosis. J. Exp. Med. 170, 655-663.

Piguet, P.F., Grau, G.E. and Vassalli, P. (1990a) Subcutaneous perfusion of tumor necrosis factor induces local proliferation of fibroblasts, capillaries, and epidermal cells, or massive tissue necrosis. Amer. J. Pathol. 136, 1990-1998.

Piguet, P.F., Collart, M.A., Grau, G.E., Sappino, A.P. and Vassalli, P. (1990b) Requirement for tumour necrosis factor for development of silica-induced pulmonary fibrosis. Nature 344, 245-247.

Placke, M.E. and Fischer, G.L. (1987) Adult peripheral lung organ culture-a model for respiratory tract toxicology. Toxicology 90, 284-298.

Razzaboni, B.L. and Bolsaitis, P. (1990) Evidence of an oxidative mechanism for the hemolytic activity of silica particles. Environ. Health Perspectives. 87, 337-341.

Reynolds, H.Y. (1987) Bronchoalveolar lavage. Amer. Rev. Respir. Dis. 135, 250-263.
Rola-Pleszczynski and M., Lemaire, I. (1985) Leukotrienes augment interleukin-1production by human mono-cytes. J. Immunol. 135, 3958-3961.
Rola-Pleszcyzynski, M., Bouvrette, L., Gingras, D. and Girard, M. Identification of interferon as the lympho-kine that mediates leukotriene B4 induced immunoregulation. (1987) J. Immunol. 139, 513-517.
Rom, W N., Travis, W.D. and Brody, A.J. (1991) Cellular and molecular basis of the asbestos-related diseases. Amer. Rev. Respir. Dis. 143, 408-412.
Roney, P.L. and Holian, A. (1989) Possible mechanism of chrysotile asbestos-stimulated superoxide anion production in guinea pig alveolar macrophages. Toxicol. Appl. Pharm. 100, 132-144.
Sauder, D.N. (1985) Epidermal-derived cytokines: Properties of epidermal-derived thymocyte activating factor. Lymphokine Res. 3, 145-151.
Scala, G., Allavena, P. and Djeu, et al. (1984) Human large granular lymphocytes are potent producers of interleukin-1. Nature 309, 56-59.
Scheule, R.K. and Holian, A. (1989) IgG specifically enhances chrysotile asbestos-stimulated superoxide anion production by the alveolar macrophage. Amer. J. Respir. Cell Mol. Biol. 1, 313-318.
Scheule, R.K. and Holian, A. (1990) Modification of asbestos bioactivity for the alveolar macrophage by selective protein adsorption. Amer. J. Respir. Cell Mol. Biol. 2, 441-448.
Schmidt, J.A., Oliver, C.N., Lepe-Zuniga, J.L., Green, I. and Gery, I. (1984) Silica-stimulated monocytes release fibroblast proliferation factors identical to interleukin 1: A potential role for interleukin 1 in the pathogenesis of sillicosis. J. Clin. Invest. 73, 1462-1472.
Schnitzer, R.J., Bunescu, G. and Boden, V. (1971) Interactions of mineral fiber surfaces with cells *in vitro*. Ann. New York Acad. Sci. 172, 759-772.
Seaton, A. (1984) Silicosis. Occupational Lung Diseases 2, 250-294.
Shatos, M.A., Doherty, J.M., Marsh, J.P. and Mossman B.T. (1987) Prevention of asbestos-induced cell death in rat lung fibroblasts and alveolar macrophages by scavengers of active oxygen species. Environ. Res. 44, 103-116.
Sherry, B. and Cerami, A. (1988) Cachectin/Tumor necrosis factor exerts endocrine, paracrine, and autocrine control of inflammatory responses. J. Cell Biol. 107, 1269-1277.
Stalder, K. and Stöber, W. (1965) Haemolytic activity of suspensions of different silica modifications and inert dusts. Nature 207, 874-875.
Stuhar, D.J., Harbeck, R.J., Gegen, N., Kawada, H. and Mason, R.J. (1989) Increased expression of class II antigens of the major histocompatibility complex on alveolar macrophages and alveolar type II cells and interleukin-1 (IL-1) secretion from alveolar macrophages in an animal model of silicosis. Clin. Exp. Immunol. 77, 281-284.
Styles, J.A. and Wilson, J. (1973) Comparison between *in vitro* toxicity of polymer and mineral dusts and their fibrogenicity. Ann. Occup. Hygiene 16, 241-250.
Suzuki, Y. and Kohyama, N. (1984) Malignant mesothelioma induced by asbestos and zeolite in the mouse peritoneal cavity. Environ. Res. 35, 277–292.
Tiku, K., Tiku, M.L. and Skosey, J.L. (1986) Interleukin-1 production by human polymorphonuclear neutrophils. J. Immunol. 136, 3677-3685.
Van Kuijk, F.J.G.M.M., Sevaniam, A., Handelman, G.J. and Dratz, E.A. (1987) A new role for phospholipase A2: Protection of membrane from lipid peroxidation damage. Trends Biochem. Sci. 12, 31-38.
Vilím, V., Wilhelm, J., Brzák, P. and Hurych, J. (1987) Stimulation of alveolar macrophages by mineral dusts *in vitro*: Luminol-dependent chemiluminescence study. Environ. Res. 42, 246-256.
Wallace, W.E., Jr., Vallyathan, V., Keane, M.J. and Robinson, V. (1985) *In vitro* biologic toxicity of native and surface-modified silica and kaolin. J. Toxicol. Environ. Health. 16, 415-424.
Wallace, W.E., Keane, M.J., Mike, P.S., Hill, C.A., Vallyathan, V. and Regad, E.E. (1992) Contrasting respirable quartz and kaolin retention of lecithin surfactant and expression of membranolytic activity following phospholipase A_2 digestion. J. Toxicol. Environ. Health 37, 391-409.
Warner, S.J.C. and Liby, P. (1989) Human vascular smooth muscle cells. Target for and source of tumor necrois factor. J. Immunol. 142, 100-104.
Warren, J.S., Kunkel, S. L., Cunningham, T.W., Johnson, K.J. and Ward, P.A. (1988) Macrophage-derived cytokines amplify immune complex-triggered O_2-responses by rat alveolar macrophages. Amer. J. Pathol. 130, 489-495.
Wiessner, J.H., Henderson, Jr., J.D. Sohnle, P.G., Mandel, N.S. and Mandel, G.S. (1988) The effect of crystal structure on mouse lung inflammation and fibrosis. Amer. Rev. Respir. Dis. 138, 445-450.
Wilhelm, J., Vilím, V. and Brzák, P. (1987) Participation of superoxide in luminol-dependent chemiluminescence triggered by mineral dust in rabbit alveolar macrophages. Immun. Letters 15, 329-334.
Wolpe, S.D. and Cerami, A. Macrophage inflammatory proteins 1 and 2: members of a novel superfamily of cytokines. FASEB J. 3, 2565-2572.

Woodworth, C.D., Mossman, B.T. and Craighead, J.E. (1982) Comparative effects of fibrous and nonfibrous mineral on cells and liposomes. Environ. Res. 27, 190-205.

Young, J.D., Liu, C., Butler, G., Chon, Z.A. and Galli, S..J. (1987) Identification, purification, and characterization of a mast cell-associated cytolytic factor related to tumor necrosis factor. Proc. Nat'l Acad. Sci. U.S.A. 84, 9175-9189.

CHAPTER 17

CELLULAR AND MOLECULAR MECHANISMS OF DISEASE

Brooke T. Mossman

Department of Pathology
University of Vermont College of Medicine
Burlington, Vermont 05405 U.S.A.

INTRODUCTION

Previous chapters have addressed the epidemiology and pathology of asbestiform and non-asbestiform mineral-related diseases and the results of inhalation and injection studies in rodent models. A synthesis of this information reveals a spectrum of biologic events that occur primarily in the respiratory tract and pleura after exposure to minerals. With transient or low-dose exposures, changes may be reversible, and cells may adapt to initial injury. However, prolonged exposure to many minerals (particularly the asbestos minerals and the silica-group minerals) at high airborne concentrations results in disease.

Minerals exhibit a range in biological activity. Some minerals (such as hematite or rutile) are conventionally regarded as "nuisance dusts" in that they are not associated with disease in animal models or cytotoxic effects on cells. Yet other minerals (such as crocidolite) cause both malignant and non-malignant pulmonary and pleural disease in man and rodents. Because interactions between minerals and lung tissue have been described previously in this volume, this chapter will focus on cellular and molecular events observed primarily in cells and organ cultures after introduction of asbestos and silica, the mineral types most widely examined by *in vitro* studies. These studies have provided much insight into the mechanisms which may be intrinsic to the initiation of mineral-induced disease.

THE DISEASE PROCESS

Before considering the general phenomenon of disease and the results of mechanistic studies, it is important to realize that the lung consists of over 40 different cell types, each with its own specialized function. Epithelial cells line the airways and air spaces of the lung. Some types of epithelial cells (i.e., ciliated cells and mucin-secreting cells) may act to defend the lung from particulates and other foreign material. Other types of epithelial cells in the peripheral lung (i.e., type I and type II epithelial cells) function in gas exchange and make surfactant to enable more efficient respiration. Epithelial cells of the trachea and bronchi can be likened to tube-like structures supported by other cell types, i.e., fibroblasts, which make collagen and other cell products comprising the matrix (or interstitium) of the lung (Fig. 1). Both epithelial cells and fibroblasts are target cells of mineral-induced disease, giving rise to bronchogenic carcinomas (or other lung tumor types) and fibrosis (asbestosis, silicosis), respectively.

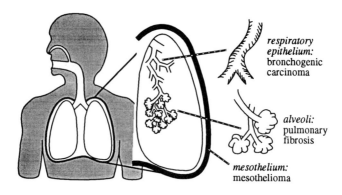

Figure 1. A schematic drawing of the respiratory tract and pleural mesothelium indicating how asbestos fibers are inhaled and the sites of development of disease. Kindly supplied by Dr. Craig Woodworth, a former student in B.T. Mossman's laboratory.

Cells of the immune system [macrophages, neutrophils (also called polymorphonuclear leukocytes or PMNs), lymphocytes, etc.] occur both in the airways and airspaces and in the interstitium of the lung. Multiple subpopulations of these cell types may occur and be recruited to the lung in response to inhaled materials. These events are broadly termed "inflammation" or an "inflammatory response." The cells comprising the inflammatory response are called "effector" cells in that they may contribute to the disease process and/or clearance of inhaled matter from the lung but do not give rise themselves to disease.

Another cell type affected and/or targeted in pleural disease associated with amphibole asbestos and erionite fibers is the mesothelial cell, a cell type lining the thin body sac enclosing the chest (pleural mesothelial cell) or abdominal organs (peritoneal mesothelial cell). Although it is unclear how inhaled asbestos reaches these cells or whether cell contact is necessary to initiate disease, human mesothelial cells appear to be exquisitely sensitive to asbestos fibers.

Figure 2 shows a schematic drawing of the cellular events occurring in the lung and pleura after exposure to asbestos and their relationship to the disease process. As seen in the diagram, inflammation occurs at the site of injection or deposition of inhaled fibers when cells such as alveolar macrophages (AM) try to engulf asbestos fibers and clear them from the lung. Macrophages also release a number of chemical substances, i.e., chemoattractants, which recruit other inflammatory cell types to this site. Alternatively, macrophages and neutrophils produce a number of enzymes, oxidants, chemoattractants, and growth factors (cytokines) which may either (1) "deactivate" minerals, (2) activate cell defense mechanisms, or (3) cause cell injury and abnormal cell function of epithelial cells, fibroblasts, and mesothelial cells. They can therefore initiate or potentiate the disease process.

It is unclear whether asbestos directly affects cells or does so indirectly through the generation of active oxygen species (AOS). Figure 2 also indicates mechanisms for the production of various AOS, including the superoxide anion

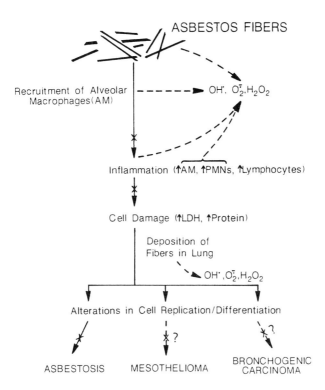

Figure 2. Cellular responses to asbestos after inhalation indicating possible pathways of generation of active oxygen species (AOS). The Xs indicate points in the pathway which can be blocked using administration of the antioxidant enzyme, catalase, to rats. Reproduced with permission from J. Kehrer, B.T. Mossman, A. Sevanian, M. Trush and M. Smith (1988) in Toxic. Appl. Pharm. 95, 349-362.

(O_2^-), the hydroxyl radical (·OH), and hydrogen peroxide (H_2O_2). On the one hand, asbestos fibers may generate AOS by redox reactions, involving iron in a Fenton-like reaction (Weitzman and Graceffa, 1984; Gulumian and Van Wyk, 1987). Alternatively, phagocytosis of asbestos fibers by AM, PMN, or target cells of disease results in AOS liberated by an oxidative burst (Hansen and Mossman, 1987), which is sometimes called the respiratory burst (see Chapter 14). The evidence for a cause-and-effect relationship between AOS and asbestos-induced lung damage is strengthened by studies showing amelioration of many of the events depicted in Figure 2 by the antioxidant enzyme, catalase (Mossman et al., 1990b).

As indicated in Figure 2, the processes of altered cell division (i.e., replication, proliferation, and mitogenesis) and abnormal differentiation, defined broadly as when a cell loses its normal cell function, are intrinsic to the development of disease. With respect to the asbestos-associated diseases, the latency period from the time of first exposure to the manifestation of clinical disease is extremely long and may average 35 or more years in mesothelioma (see Chapters 11 & 13; also reviewed in Mossman and Gee, 1989; Mossman et al.,

1990a). A causal link between inflammation and the development of fibrosis may also exist, as fibrogenic agents in general (e.g., bleomycin, phorbol esters, fibrogenic minerals) induce a potent and sustained inflammatory response.

CARCINOGENESIS AND FIBROSIS: GENERAL PRINCIPLES

Carcinogenesis, the process of cancer development, is a complex, multistage process occurring over long latency periods. The process is thought to begin with a genetic "hit" on a cell. This genetic hit results in a mutagenic change in the cell's genome that can then be inherited by daughter cells. This process is called "initiation." An initiated cell can sit dormant for many years, or it can be induced to divide by a substance known as a tumor "promoter." (Tumor promoters are generally nongenotoxic substances which cause cell replication or division.) Tumor promotion can be reversible in early stages, but it eventually becomes irreversible as tumor cells become autonomous and acquire subsequent genetic changes which render them fully malignant.

The sequence of events described above has been largely deciphered using models of chemical carcinogenesis. In such models, the "initiators" are soluble chemicals that are genotoxic either after metabolism by cells or by their direct interaction with DNA. However, carcinogenesis by relatively insoluble fibers and particles may be more complex and governed by physical and chemical properties of the fibers or particles. These properties may include properties that affect the material's durability (or persistence) in the lung or pleura and surface properties that affect the material's interaction with the cells or fluid. For example, as discussed in many preceding chapters, amphibole asbestos may be more tumorigenic in man than chrysotile, because amphibole asbestos is apparently more durable than chrysotile and it may persist indefinitely in the lung or pleura.

In contrast to true initiators described above, some agents appear to be carcinogenic but not mutagenic. An intriguing theory put forth recently to explain this apparent lack of mutagenicity of many carcinogens is that chronic proliferation can cause carcinogenesis (Preston-Martin et al., 1990; Ames and Gold, 1990). In this scenario, agents which cause chronic inflammation and cell proliferation render cells more susceptible to oxidative stress and DNA damage, thus causing them to accumulate multiple hits to their DNA. The importance of this concept as related to asbestos-induced carcinogenesis can be seen by the classical experiments of Brand (1975) who studied "foreign body carcinogenesis" after implantation of plastic films and synthetic materials under the skin of rodents. If implants were removed prior to a critical period of time, tumors did not develop.

Fibrogenesis, or the development of pulmonary fibrosis, is also a disease of altered cell replication and differentiation, and it may involve similar mechanisms of cell proliferation by minerals. However, the cell type affected here is the fibroblast, as opposed to the epithelial cell and mesothelial cell which are the target cells for the cancers described above.

MECHANISMS OF LUNG CANCER BY ASBESTOS AND SILICA

Increased risk of lung cancer has been observed in a number of asbestos-exposed occupational cohorts (see Chapter 11), but the incidence of tumors varies with fiber type and usage, exposure regimen and co-factors such as smoking history and exposure to other carcinogens in the work place (Mossman and Gee, 1989; Mossman et al., 1990a). Experimental evidence in a number of bioassays suggests that asbestos is primarily a tumor promoter in the development of lung cancers, but other interactions between asbestos and components of cigarette smoke may also be important (reviewed in Wehner and Felton, 1989; see Fig. 3). For example, asbestos fibers and particulates in urban air appear to adsorb chemical carcinogens, such as polycyclic aromatic hydrocarbons (PAH), in smoke and act as condensation vehicles for delivery of these agents to cells of the respiratory tract after inhalation. Furthermore, if the potent carcinogen benzo[a]pyrene (BaP) is introduced into cultures of epithelial cells on fibers, as opposed to directly in culture medium, more BaP enters the cell and remains associated with the DNA for prolonged time periods (Eastman et al., 1983). This maximizes the time available for initiation of cells. Under these circumstances, no DNA breakage is observed with asbestos alone, but fibers with adsorbed BaP cause DNA breaks (Mossman et al., 1983). Asbestos also causes cells to metabolize BaP, thereby forming more carcinogenic products. Pathways activated by asbestos include the mixed microsomal enzyme system used for metabolism of a variety of chemicals and pollutants, i.e. the aryl hydrocarbon hydroxylase (AHH) enzymes and redox reactive pathways (Graceffa and Weitzman, 1987). Lastly, smoking inhibits the normal removal of asbestos fibers from the lung, thus increasing retention of fibers and their uptake by tracheobronchial epithelial cells (McFadden et al., 1986a,b). Asbestos may cause cell damage directly to epithelial cells, i.e., ciliostasis, or deplete pools of antioxidants which may normally protect cells from oxidant injury by asbestos. Enzymes such as lactate dehydrogenase (LDH) and protein released by cells in response to asbestos are used as markers of cell injury in bronchoalveolar lavage or cell culture (Fig. 2).

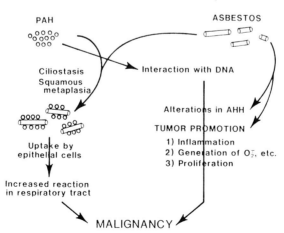

Figure 3. Mechanisms of interaction between asbestos fibers and PAH in cigarette smoke in the development of lung cancer.

If one compares asbestos fibers to classical tumor promoters such as the phorbol ester compounds, there are a constellation of similar biological effects which may contribute to the process of tumor promotion (reviewed in Mossman et al., 1985). These include stimulation of cell proliferation through common cell signaling pathways initiated at the cell surface, generation of AOS, and induction of polyamine synthesis, a necessary event in cell division. Work from our laboratory shows that asbestos causes increased expression of genes (ornithine decarboxylase, c-*fos*, c-*jun*) which regulate the initiation of DNA synthesis and related events in both tracheal epithelial and mesothelial cells (Heintz et al., 1993; Marsh and Mossman, 1991).

It is controversial whether asbestos, particularly chrysotile, can act directly with the DNA of cells to cause a mutagenic event (Shelby, 1988), but chrysotile fails to induce significant numerical chromosomal changes in human bronchial epithelial cells (Kodama et al., 1993), whereas amosite asbestos does so in human mesothelial cells (Lechner et al., 1982). These observations indicate that amphibole asbestos may be a complete carcinogen possessing properties of both an initiator and tumor promoter in the development of human mesotheliomas. The high spontaneous rate of cell transformation in rodent cells and the often increased incidence of asbestos-induced chromosomal changes and lung tumors reported in some rodent bioassays after exposure to asbestos suggest that important cell-type- and species-specific differences exist in response to this mineral (reviewed in Health Effects Asbestos Research, 1991).

Whether silica plays a role in the development of lung cancer in man is also a subject of debate, but several scenarios may be possible. First, inhalation of silica by rodents at high concentrations results in type II epithelial cell proliferation and increases in antioxidant enzymes, an event triggered by oxidant stress (Janssen et al., 1992). Thus, assuming that these cell types are initiated by chemical carcinogens in cigarette smoke, silica might act as a tumor promoter by giving them a selective advantage over normal cells. Secondly, a recent report indicates that intratracheal injection of silica increases the activity of certain microsomal enzymes that may metabolize chemical carcinogens to forms which may interact with the DNA of cells (Miles et al., 1993). Lastly, some preparations of silica have redox activity as determined by electron spin resonance (ESR) which may result in oxidative damage to isolated DNA (Daniel et al., 1993). Cell transformation by silica also was reported in one rodent cell bioassay, but silica was much less active than asbestos, and non-carcinogenic control dusts were not examined (Hesterberg and Barrett, 1984).

MECHANISMS OF MESOTHELIOMA BY ASBESTOS

In contrast to the situation with epithelial cells, rodent mesothelial cells show chromosomal aberrations after exposure to chrysotile asbestos (reviewed in Walker et al., 1992b), and human mesothelial cells acquire both chromosomal changes and altered growth characteristics after multiple exposures to amosite asbestos (Lechner et al., 1982). Moreover, many non-random cytogenetic changes have been reported in human mesotheliomas that may be important in the initiation and/or development of human tumors (Walker et al., 1992b).

Searches for conventional oncogenes which are activated in carcinogenesis or tumor suppressor genes which may be inactivated in this process have not been fruitful (Metcalf et al., 1992), and it is likely that asbestos might induce unique genes which have yet to be identified.

Several laboratories have also focused on genes which may encode growth factors and be differentially expressed in mesotheliomas and normal mesothelial cells. For example, c-*sis*, an oncogene encoding platelet derived growth factor (PDGF), a factor which induces proliferation of fibroblasts and transformation of an immortalized human mesothelial cell line (Van der Meeren et al., 1993), is overexpressed in human mesotheliomas in comparison to normal human mesothelial cells (Gerwin et al., 1987). However PDGF and its receptors are not increased in rodent mesotheliomas, indicating that it may not play a causal role in development of tumors (Walker et al., 1992a). Additional evidence that growth factors may act in a peripheral capacity in the induction of cell proliferation and tumor development is indicated from recent studies by Adamson and Bowden (1990) in which both long (carcinogenic and fibrogenic) and short (noncarcinogenic and nonfibrogenic) crocidolite fibers were introduced into cultures of alveolar macrophages which indiscriminately produced cytokines stimulating proliferation of lung fibroblasts *in vitro*. However, cytokine secretion induced by short fibers *in vivo* did not result in pulmonary fibrosis. A recent review summarizes the spectrum of growth factors which appear to be produced in mesothelial cells and other cell types after exposure to asbestos, but points out that a causal relationship between cytokines and mineral-induced diseases has yet to be solidified (Walker et al., 1992b).

MECHANISMS OF PULMONARY FIBROSIS (ASBESTOSIS AND SILICOSIS)

As emphasized previously, pulmonary fibrosis is also a disease of unregulated cell division and altered differentiation in which fibroblasts of the lung produce abnormal and excessive amounts of collagen. This extracellular material then fills the interstitium of lung making it stiff and less compliant. Using rat lung fibroblasts in culture, we have shown that increased amounts of cell-associated collagen occur in cultures after exposure to crocidolite asbestos in the absence of changes in expression of collagen genes (Mossman et al., 1986). However, asbestos-induced synthesis of collagen may involve additional molecular mechanisms such as partitioning of mRNAs into polysomes, a subject of current research. It will also be critical to determine whether or not the growth regulatory genes (c-*fos*, c-*jun*, ODC) activated by asbestos in mesothelial and tracheal epithelial cells are also involved in cell proliferation in fibroblasts. Chapter 16 (Driscoll) describes the cytokine networks activated by asbestos and silica in fibroblasts and their possible relationship to the development of fibrotic lung disease. Whether certain of these cytokines, such as TGF-β (Williams et al., 1993), are important in epithelial cell and fibroblast proliferation in asbestosis and silicosis will be critical to understanding cell-cell interactions in disease and possible relationships between the development of asbestosis, silicosis and lung cancer.

ACKNOWLEDGMENTS

This research is supported by the Environmental Protection Agency and the National Institutes of Health (NIHLBI and NIEHS). Yvonne M.W. Janssen, Barbara Cady and Judith Kessler provided valuable technical assistance in the preparation of this manuscript.

REFERENCES

Adamson, I.Y.R. and Bowden, D.H. (1990) Pulmonary reaction to long and short asbestos fibers is independent of fibroblast growth factor production by alveolar macrophages. Amer. J. Path. 137, 523-529.
Ames, B.N. and Gold L.S. (1990) Mitogenesis increases mutagenesis. Science 249, 970-971.
Brand, K.G. (1975) Foreign body induced sarcomas. In Becker, F.F., ed. , Cancer: A Comprehensive Treatise. Etiology: Chemical and Physical Carcinogenesis. New York, Plenum Press, p. 485-511.
Daniel, L.N., Mao, Y. and Saffiotti, U. (1993) Oxidative DNA damage by crystalline silica. Free Radical Biol. & Med. 14, 463-472.
Eastman, A., Mossman, B.T. and Bresnick, E. (1983) Influence of asbestos on the uptake of benzy{a}pyrene and DNA alkylation on hamster tracheal epithelial cells. Cancer Res. 43, 1251-1255.
Gerwin, B.I., Lechner, J.F., Reddel, R.R., Roberts, B.B., Robbin, K.C., Gabrielson, E.W. and Harris, C.C. (1987) Comparison of production of transforming growth factor-β and platelet-derived growth factor by normal human mesothelial cells and mesothelial cells and mesothelioma cell lines. Cancer Res. 47, 6180-6184.
Graceffa, P. and Weitzman, S.A. (1987) Asbestos catalyzes the formation of the 6-oxobenzo{a}pyrene radical from 6-hydroxybenzo{a}pyrene. Arch. Biochem. Biophys. 258, 481-484.
Gulumian, M. and Van Wyk, J.A. (1987) Hydroxyl radical production in the presence of fibers by a Fenton-type reaction. Chem. Biol. Interactions 62, 89-97.
Hansen, K. and Mossman, B.T. (1987) Generation of superoxide (O_2^-) from alveolar macrophages exposed to asbestiform and nonfibrous particles. Cancer Res. 47, 1681-1686.
Health Effects Institute - Asbestos Research (1991) Asbestos in Public and Commercial Buildings: A Literature Review and Synthesis of Current Knowledge, Cambridge, Massachusetts, p. 1-A2-18.
Heintz, N.H., Janssen, Y.M. and Mossman, B.T. (1993) Persistent induction of c-*fos* and c-*jun* expression by asbestos. Proc. Nat'l Acad. Sci. USA 90, 3299-3303.
Hesterberg, T.W. and Barrett, J.C. (1984) Dependence of asbestos and mineral dust-induced transformation of mammalian cells in culture on fiber dimension. Cancer Res. 44, 2170-2180.
Janssen, Y.M.W., Marsh, J.P., Absher, M.P., Hemenway, D., Vacek, P.M., Leslie, K.O., Borm, P.J.A. and Mossman, B.T. (1992) Expression of antioxidant enzymes in rat lungs after inhalation of asbestos or silica. J. Biol. Chem. 267, 10625-10630.
Kodama, Y., Boreiko, Boreiko, C.J., Maness, S.C. and Hesterberg, T.W. (1993) Cytotoxic and cytogenetic effects of asbestos on human bronchial epithelial cells in culture. Carcinogenesis 14, 691-697.
Marsh, J.P. and Mossman, B.T. (1991) Role of asbestos and active oxygen species in activation and expression of ornithine decarboxylase in hamster tracheal epithelial cells. Cancer Res. 51, 167-173.
McFadden, D., Wright, J.L., Wiggs, B. and Churg, A. (1986a) Smoking inhibits asbestos clearance. Amer. Rev. Respiratory Dis. 133, 372-374.
McFadden, D., Wright, J.L., Wiggs, B. and Churg, A. (1986b) Cigarette smoke increases the penetration of asbestos fibers into airway walls. Amer. J. Path. 123, 95-99.
Metcalf, R.A., Welsh, J.A., Bennett, W.P., Seddon, M.B., Lehman, T.A., Pelin, K., Linnainmaa, K., Tammilehto, L., Mattson, K., Gerwin, B.I., et al. (1992) *p53* and Dirsten-*ras* mutations in human mesothelioma cell lines. Cancer Res. 52, 2610-2615.
Miles, P.R., Bowman, L. and Miller, M.R. (1993) Alterations in the pulmonary microsomal cytochrome P-450 system after exposure of rats to silica. Amer. J. Resp. Cell. Mol. Biol. 8, 597-604.
Mossman, B.T., Bignon, J., Corn, M., Seaton, A. and Gee, J.B.L. (1990a) Asbestos: scientific developments and implications for public policy. Science 247, 294-301.
Mossman, B.T., Cameron, G.S. and Yotti, L.P. (1985) Cocarcinogenic and tumor promoting properties of asbestos and other minerals in tracheobronchial epithelium. In, Mass, M.J., Kaufman, D.G., Siegfried, J.M., Steele, V.E., Nesnow, S., eds., Carcinogenesis: a Comprehensive Survey. Cancer of the Respiratory Tract: Predisposing Factors (Vol. 3), New York, Raven Press, p. 217-238.
Mossman, B.T., Eastman, A., Landesman, J.M. and Bresnick, E. (1983) Effects of crocidolite and chrysotile asbestos on cellular uptake, metabolism and DNA after exposure of hamster tracheal epithelial cells to benzo{a}pyrene. Environ. Health Persp. 51, 331-338.

Mossman, B.T. and Gee, J.B.L. (1989) Asbestos-related diseases. New England J. Med. 320, 1721-1730.
Mossman, B.T., Gilbert R., Doherty, J., Shatos, M.A., Marsh, J., and Cutroneo, K. (1986) Cellular and molecular mechanisms of asbestosis. Chest 89, 160s-161s.
Mossman, B.T., Marsh, J.P., Sesko, A., Hill, S., Shatos, M.A., Doherty, J., Petruska, J., Adler, K.B., Hemenway, D., Mickey, R., Vacek, P. and Kagan, E. (1990b) Inhibition of lung injury, inflammation, and interstitial pulmonary fibrosis by polyethylene glycol-conjugated catalase in a rapid inhalation model of asbestosis. Amer. Rev. Respiratory Dis. 141, 1266-1271.
Preston-Martin, S., Pike, M.C., Ross, R.K., Jones, P.A. and Henderson, B.E. (1990) Increased cell division as a cause of human cancer. Cancer Res. 50, 7415-7421.
Shelby, M.D. (1988) The genetic toxicity of human carcinogens and its implications. Mutation Res. 204, 3-15.
Van der Meeren, A., Seddon, M.B., Betsholtz, C.A., Lechner, J.F. and Gerwin, B.I. (1993) Tumorigenic conversion of human mesothelial cells as a consequence of platelet-derived growth factor-A chain overexpression. Amer. J. Respiratory Cell Mol. Biol. 8, 214-221.
Walker, C., Bermudez, E., Stewart, W., Bonner, J., Molloy, C.J. and Everitt, J. (1992a) Characterization of platelet-derived growth factor and platelet-derived growth factor receptor expression in asbestos-induced rat mesothelioma. Cancer Res. 52, 301-306.
Walker, C., Everitt, J. and Barrett, J.C. (1992b) Possible cellular and molecular mechanisms of asbestos carcinogenicity. Amer. J. Indus. Med. 21, 253-273.
Wehner, A.P., Felton, D-L., eds. (1989) Biological Interaction of Inhaled Mineral Fibers and Cigarette Smoke. Columbus, Ohio, Battelle Press, 611 p.
Weitzman, S.A. and Graceffa, P. (1984) Asbestos catalyzes hydroxyl and superoxide radical generation from hydrogen peroxide. Arch. Biochem. Biophys. 228, 373-376.
Weitzman, S.A. and Weitberg, A.B. (1985) Asbestos-catalyzed lipid peroxidation and its inhibition by desferrioxamine. Biochem. J. 225, 259-262.
Williams, A.O., Flanders, K.C. and Saffiotti, U. (1993) Immunohistochemical localization of transforming growth factor-β_1 in rats with experimental silicosis, alveolar type II hyperplasia, and lung cancer. Amer. J. Path. 142, 1831-1840.

CHAPTER 18

BIOLOGICAL STUDIES ON THE CARCINOGENIC MECHANISMS OF QUARTZ

Umberto Saffiotti, Lambert N. Daniel, Yan Mao, A. Olufemi Williams,
M. Edward Kaighn, Nadera Ahmed, and Alan D. Knapton

Laboratory of Experimental Pathology
National Cancer Institute
Bethesda, Maryland 20892 U.S.A.

COORDINATED *IN VIVO*, CELLULAR, AND MOLECULAR APPROACHES

This chapter discusses experimental research approaches to the identification of carcinogenic effects of quartz, other crystalline silica polymorphs, and other mineral particles of respirable size. The epidemiological and experimental evidence for the carcinogenic effect of crystalline silica emerged in the early 1980s and was brought together for the first time at the symposium on "Silica, silicosis and cancer" held in 1984 (Goldsmith et al., 1986). The experimental evidence of silica-induced carcinogenesis was based on the results of long-term tests in rats exposed to quartz, either by inhalation or by intratracheal instillation, and included tests of different quartz samples. The rats exposed to quartz developed both fibrosis and epithelial tumors of the lung, most frequently malignant. Hamsters, however, developed neither lung fibrosis nor tumors, indicating a host-dependent effect of silica. These results were reviewed in 1986 by the International Agency for Research on Cancer (IARC), which is an agency of the World Health Organization. The IARC publishes a series of monographs that assemble all published data on the carcinogenicity of environmental chemicals, and these data are reviewed by working groups of experts. The review on silica (IARC, 1987a) concluded that there was sufficient evidence for the carcinogenicity of crystalline silica to experimental animals and limited evidence for its carcinogenicity to humans.[1] Since then, silica-induced carcinogenesis mechanisms have been investigated by several laboratories. However, most research has focused on quartz and to a lesser extent cristobalite and tridymite. Few data are available on other crystalline silica polymorphs.

In our laboratory, we have developed a multidisciplinary approach, in order to correlate biological mechanisms of silica at the tissue, cellular, and molecular levels. In summary, we have studied the effects of silica in animal models exposed to quartz, cristobalite, and tridymite *in vivo*. Our work has included the identification of patterns of tumor induction (for quartz) and the determination of early histogenetic pathways of tumor development. In the treated lung tissues,

---

[1] In a subsequent classification, the IARC designated crystalline silica as "probably carcinogenic to humans" (Group 2A), whereas it designated amorphous silica as "not classifiable as to its carcinogenicity to humans" Group 3), because of inadequate data (IARC, 1987b).

we investigated the localization of molecular markers for cellular mediators (growth factors and cytokines) and altered gene expression. The cellular localization of these markers in the lung (observed at progressive time intervals from a single exposure of the animals to crystalline silica) provides a picture of the cellular interaction mechanisms involved in the pathogenesis of silicosis, of the adjacent proliferation of the lining epithelial cells of the alveoli, and of the formation of alveolar-cell-derived tumors, both benign and malignant. Studies on silica uptake and localization and on cytotoxicity and transformation are conducted using two cell lines. Other studies investigate the direct effects of crystalline silica on the induction of DNA damage *in vitro* and in cells, the production of active oxygen species by the crystalline silica surface, the specific binding of silica to DNA, and the characterization of surface properties (surface area and charge) related to toxic effects. Additional studies are planned to detect the induction of mutations in cells, indicative of genetic damage. For this purpose, we will use a method suitable for the detection of large DNA deletions, which has proven adequate for mutagenesis assays of chrysotile and crocidolite asbestos (Hei et al., 1992).

All of these studies are conducted with a series of samples of quartz and several other minerals, in order to correlate biological effects with mineralogical properties. Our research approach is aimed at identifying basic mechanisms of silica-induced carcinogenesis, and their eventual relevance to specific mineral properties and to specific biological host factors. The new studies on silica-induced carcinogenesis complement an extensive body of knowledge (IARC, 1987a) that had previously been developed on silica toxicity and silicosis. New methods in cellular and molecular biology now offer a unique opportunity to investigate the mechanisms of the biological effects of crystalline silica and other minerals, not only for fibrogenesis, but also for carcinogenesis.

CRYSTALLINE SILICA USED BY BIOLOGICAL STUDIES

The biological effects induced by minerals need to be related to the mineralogical properties of the tested materials, in order to clarify the mechanisms by which interaction of mineral surfaces with biological targets leads to specific effects. As it was pointed out in a critical review of the literature on the biological effects of inhaled minerals (Guthrie, 1992), the minerals tested in biological studies have often been inadequately defined, even to the point of being named inappropriately. As biologists addressing the study of minerals, we experienced difficulties in obtaining mineral samples with the desired detailed characterization, including origin of the sample, mineralogical classification, nature and level of impurities, particle morphology, and particle size distribution. For example, only a few large samples of quartz have been prepared, characterized, and made available to biological investigators, such as the German (DQ12: Robock, 1973) and Chinese (Fu et al., 1984) standards. In the United States, Min-U-Sil[2] has been

---

[2] Originally supplied by Pennsylvania Glass Sand Co., now US Silica. Various samples of Min-U-Sil with different particle sizes are available commercially. They may vary in their mineral content. The sample of Min-U-Sil 5 used in all our experiments was analyzed at IITRI, Chicago, Illinois. An X-ray diffraction pattern showed no extraneous, non-quartz peaks. Amorphous material, if present, represents less than 1%. X-ray fluorescence analysis showed a small peak for iron (0.1%).

widely used as a "standard" for biological studies, and various size fractions have been available. Various analyses of Min-U-Sil samples have been reported by different investigators for the samples they used, but a comprehensive report has not been published on the analytical characteristics of a standard sample, and the reproducibility of its aliquots.

Unfortunately, standards represented by large-quantity samples of crystalline-silica-dusts have not been generally available. Such standards are crucial to furthering our understanding of the relationships between mineralogical properties and pathogenesis. Several standards need to be developed in order to test biological effects due to mineralogical properties. These include: (1) various size fractions; (2) samples with different physical and chemical properties (e.g., different impurities, different surface area, and different surface charge); and (3) samples of various mineral and mineraloid species (e.g., samples of quartz, tridymite, cristobalite, coesite, stishovite, opal, amorphous silica). Samples with very small particle size (<0.5 µm) would be particularly interesting, because these would permit the investigation of the role of such fine particles in nuclear localization, DNA damage, and carcinogenesis. Recent electron microscopy studies in our laboratory revealed intranuclear localization of such small particles in cells exposed to quartz in culture (Williams et al., unpublished observation).

For biological studies of a mineral, e.g., quartz, inclusion of a series of different samples, representative of different properties, provides a basis for interpretation of the results in relation to those properties. An example is offered by the studies on freshly ground versus aged quartz (Vallyathan et al., 1988).

ANIMAL MODELS AND SPECIES SUSCEPTIBILITY

The biological response occurring in humans from exposure to inhaled silica is a function of numerous complex factors, including deposition and clearance of particles in the lungs, cellular uptake and localization, interaction of mineral particles with the cell membrane, with cytoplasmic organelles, and with the nucleus (e.g., DNA or genetic material). The biological response also depends on the reactions of various types of cells, giving rise to nodular aggregates of inflammatory cells with progressive fibrosis (silicotic granulomas) and possibly to the proliferation of epithelial cells into malignant lung tumors (carcinomas).

The ability of a chemical or physical agent to induce cancer in humans can be estimated by four complementary approaches:

- Evaluation of the epidemiological evidence of excess cancers, usually of a given type, in human populations that have excess exposure to the agent (including occupational exposures).

- Evaluation of the agent's ability to induce cancer in experimental animals under controlled conditions of exposure.

- Evaluation of the agent's activity from measured responses in tests for genetic damage (e.g., chromosomal damage, mutagenesis, neoplastic cell transformation).

- Prediction of the agent's activity based on its chemical composition or structure.

Each of these approaches requires careful evaluation of the methods used and of the possible role of confounding factors and competing risks. Exposure to crystalline silica is ubiquitous, and human lungs from the general population contain relatively numerous crystalline silica particles (Churg et al., 1992). Although occupationally-exposed subjects have much higher amounts of crystalline silica in their lungs, the general population is not a completely unexposed control.

The experimental studies on cancer induction (carcinogenesis) by crystalline silica include animal bioassays protracted for most of the lifespan of the laboratory animals (under adequately controlled protocols) to allow for the long latent period for the development of detectable cancer. The guidelines for the conduct of carcinogenesis bioassays in laboratory rodents (usually rats, mice, and hamsters) have been formulated on the basis of experience. For example, the National Cancer Institute initiated a systematic program for carcinogenesis testing (Sontag et al., 1976), which was subsequently greatly extended by the National Toxicology Program. For the evaluation of the carcinogenic activity of an agent, in addition to the results of adequate long-term animal bioassays, much can be learned from studies of toxicity, tissue localization, metabolism, species susceptibility, development of precursor lesions in target animal tissues (histogenesis), and induction of mutation and neoplastic cell transformation.

The main basis of evidence for carcinogenesis by crystalline silica derives from experimental groups of rats in which exposure to crystalline silica by inhalation or by intratracheal instillation caused lung cancer. Occurrence of lung carcinomas in untreated rats is very rare. The available data on silica-induced carcinogenesis in rats are summarized in Tables 1 and 2.

The cells that give rise to lung tumors induced by crystalline silica in the rat lung model are the epithelial cells of the lining of the alveoli. Normal lung alveolar lining cells, which are flat and thin, permit gaseous exchanges with the adjacent blood capillaries and are called alveolar type I cells. A few cells in the normal alveolar lining are cuboidal and contain lamellar bodies (round cytoplasmic organelles, which produce the pulmonary surfactant). These cells are called alveolar type II cells and are the stem cells (or progenitor cells) that divide and differentiate into type I cells. When exposed to crystalline silica, alveolar type II cells become larger and proliferate (hyperplasia) (Miller et al., 1986). In crystalline silica-treated rats, hyperplasia of the alveolar type II cells develops adjacent to the silicotic granulomas. This hyperplasia gives rise to (1) adenomatoid (*i.e.*, "glandular-like") lesions, showing many contiguous alveoli lined by cuboidal cells, and eventually to (2) tumors, including benign adenomas and malignant carcinomas[3] (Saffiotti, 1986, 1992; Williams et al., 1993). We examined

---

[3] Epithelial tumors of the lung can originate from (1) the cells of the mucosa lining the bronchi or (2) the cells lining the distal airways (bronchioles and alveoli). They can be benign (e.g., adenomas) or malignant (e.g., adenocarcinomas). Depending on the main types of cell differentiation, the

Table 1
Summary of data on lung tumors induced in rats by crystalline silica

Sample	Exposure conditions	Rat strain	Sex	Incidence of lung tumors[*] Treated	Incidence of lung tumors[*] Controls	Reference
quartz (Min-U-Sil 5)	Intratracheal instillation of 7 mg weekly for 10 weeks	Sprague-Dawley	?	6/36[a]	0/58	Holland et al. (1983)
quartz (Min-U-Sil 5)	Inhalation[†] of 12±5 mg/m^3 for up to 2 years	Fischer 344	F	20/60[b]	0/54	Holland et al. (1986)
quartz (Min-U-Sil 5)	Inhalation exposure of 51.6 mg/m^3 for various durations; sacrificed at 24 months	Fischer 344	F	10/53[c]	0/47	Dagle et al. (1986)
quartz (Min-U-Sil 5)		Fischer 344	M	1/47[d]	0/42	Dagle et al. (1986)
quartz (Min-U-Sil 5)	Intratracheal instillation of 20 mg in left lung[††]	Fischer 344	M	30/67[e]	1/75[f]	Groth et al. (1986)
novaculite	Intratracheal instillation of 20 mg in left lung[††]	Fischer 344	M	21/72[g]	same	Groth et al. (1986)
raw shale dust	Inhalation of 152±51 mg/m^3 dust with an average quartz content of 8–12%	Fischer 344	F	17/59[h]	0/54 1/15[i]	Holland et al. (1986)
spent shale dust	Inhalation of 176±75 mg/m^3 dust with an average quartz content of 8–12%	Fischer 344	F	11/59[j]	same as above	Holland et al. (1986)
quartz (DQ12)	Inhalation of 1 mg/m^3 for 24 months	Fischer 344	F	12/50[k]	3/100[n]	Muhle et al. (1989)
quartz (DQ12)			M	6/50[l]		
quartz (DQ12)	Inhalation[†] of 6-mg/m^3 for 29 days followed by lifetime observation.	Wistar	F	62/82[n]	0/85	Spiethoff et al. (1992)
quartz (DQ12)	Inhalation[†] of 30-mg/m^3 for 29 days followed by lifetime observation.	Wistar	F	69/82[o]	0/85	Spiethoff et al. (1992)

[*] Number of lung tumors per number of rats observed.
[†] Nose only exposure.
[††] Sacrificed at 12, 18, and 22 months or found dead.

[a] 1 adenoma and 5 carcinomas; [b] 6 adenomas, 11 adenocarcinomas, and 3 epidermoid carcinomas; [c] all epidermoid carcinomas; [d] epidermoid carcinoma; [e] all adenocarcinomas; [f] 1 adenocarcinoma; [g] 20 adenomas, 8 adenocarcinomas and 1 epidermoid carcinoma; [h] 2 adenomas, 8 adenocarcinomas, and 7 epidermoid carcinomas; [i] 1 adenoma; [j] 2 adenomas, 8 adenocarcinomas, 1 adenosquamous carcinoma, 1 epidermoid carcinoma, and 2 keratinizing cystic squamous cell tumors; [m] 2 adenomas and 1 adenocarcinoma; [n] 8 adenomas, 17 bronciololaveolar carcinomas, and 37 epidermoid carcinomas; [o] 13 adenomas, 26 bronchioloalveolar carcinomas, and 20 epidermoid carcinomas.

carcinomas can be classified as adenocarcinomas (or glandular-like, made of secretory cells), epidermoid carcinomas (or epidermis-like, also called squamous cell carcinomas), undifferentiated carcinomas (with small cell or large cell types), and mixed types. The histological diagnosis of the tumor types is an important parameter for the evaluation of experimental carcinogenesis.

Table 2

Summary of data on lung tumors induced in rats by crystalline silica
(Laboratory of Experimental Pathology, National Cancer Institute)

Sample	Exposure conditions*	Rat strain	Sex	Comments	Incidence of tumors**	Total number of lung tumors
untreated	none	Fischer 344	M	died after 17 months	0/32	
			F	died after 17 months	1/20	1[a]
quartz (Min-U-Sil 5)	12 mg in 0.3 ml saline	Fischer 344	M	sacrificed, 11 months	3/18 (17%)	
				sacrificed, 17 months	6/19 (32%)	
				died after 17 months	12/14 (86%)	37[b]
quartz (Min-U-Sil 5)	12 mg in 0.3 ml saline	Fischer 344	F	sacrificed, 11 months	8/19 (42%)	
				sacrificed, 17 months	10/17 (59%)	
				died after 17 months	8/9 (89%)	59[c]
quartz (Min-U-Sil 5)	20 mg in 0.3 ml saline	Fischer 344	F	died after 17 months	6/8 (75%)	13[d]
quartz (Min-U-Sil 5 etched in HF)	12 mg in 0.3 ml saline	Fischer 344	M	sacrificed, 11 months	2/18 (11%)	
				sacrificed, 17 months	7/19 (37%)	
				died after 17 months	7/9 (78%)	20[e]
quartz (Min-U-Sil 5 etched in HF)	12 mg in 0.3 ml saline	Fischer 344	F	sacrificed, 11 months	7/18 (39%)	
				sacrificed, 17 months	13/16 (81%)	
				died after 17 months	8/8 (100%)	45[f]

Sources: Saffiotti et al. (1990, 1992) and Williams et al. (1993) and unpublished data.
* Single intratracheal instillation.
** Number of lung tumors per number of rats observed.
a 1 adenoma; *b* 6 adenomas, 25 adenocarcinomas, 1 undifferentiated carcinoma, 2 mixed carcinomas, 3 epidermoid carcinomas; *c* 2 adenomas, 46 adenocarcinomas, 3 undifferentiated carcinomas, 5 mixed carcinomas, 3 epidermoid carcinomas; *d* 1 adenoma, 10 adenocarcinomas, 1 mixed carcinomas, 1 epidermoid carcinoma; *e* 5 adenomas, 14 adenocarcinomas, 1 mixed carcinoma; *f* 1 adenoma, 36 adenocarcinomas, 3 mixed carcinomas, 5 epidermoid carcinomas.

the patterns of tissue reaction to crystalline silica in rats sacrificed at serial intervals after a single intratracheal dose of 12- or 20-mg quartz. The temporal sequence of development of alveolar hyperplasia and lung tumors (adenomas and carcinomas of alveolar cell origin) indicates that silica has the effect of stimulating alveolar cells to proliferate and eventually to undergo malignant transformation. For comparison, a single intratracheal dose of 50 mg of hematite does not induce fibrotic lesions, alveolar hyperplasia or tumors: these findings indicate that the reactions observed after treatment with quartz are not the result of a non-specific effect due to interaction with a particle but, instead, result specifically from interaction with quartz. The early reactions induced by crystalline silica in the rat model are schematically illustrated in Figure 1.

The results of long term bioassays of crystalline silica in rats, from different laboratories, presently comprising 17 individual experimental groups, provide strong evidence of the carcinogenic effect of crystalline silica in the induction of peripheral lung tumors in rats, with a predominance of the adenocarcinoma type. The current experimental evidence for lung carcinogenesis by crystalline silica in rats can be summarized by stating that lung cancer was induced in rats treated by several different samples of quartz (Min-U-Sil, HF-etched Min-U-Sil, DQ12), by novaculite,[4] and by samples of quartz-bearing shale. Furthermore, lung cancer was induced by quartz in rats of three strains (Fischer 344, Sprague-Dawley, and Wistar), in both sexes, by either long-term or short-term inhalation or by single or repeated intratracheal instillation. The prevalent histologic types of lung cancer reported in rat experiments with crystalline silica are carcinomas (predominantly adenocarcinomas) associated with fibrosis and originating from peripheral lung epithelium.[5]

Additional evidence of carcinogenesis by crystalline silica in rats is provided by the results of other experiments, in which crystalline silica was injected into the pleural cavity and produced malignant hystiocytic lymphomas; these results were confirmed with four different samples of quartz, with tridymite, and with cristobalite (Wagner et al., 1980; IARC, 1987).

In contrast with the results in rats, five tests of silica in hamsters failed to induce fibrosis or any tumors (IARC, 1987). Serial sacrifice experiments conducted in our laboratory, following single intratracheal instillation of quartz or tridymite in hamsters, showed that the lung reaction in hamsters is limited to storage of crystalline silica in macrophages, with no fibrosis and no epithelial proliferative lesions. Niemeier et al. (1986) reported a series of long-term bioassays in hamsters by repeated intratracheal instillation. The tested dusts, of respirable size, included Ottawa silica sand (Sil-Co-Sil), Min-U-Sil 5, and a sample of ferric oxide. Each sample was tested alone (suspended in saline) or by mixing

---

[4] Novaculite is a microcrystalline variety of quartz similar to chert.
[5] Epidermoid carcinomas were reported at relatively low frequenceies in most of the experiments, except two (Dagle et al., 1986; Spiethoff, 1992). Strain differences, dietary factors (e.g., vitamin A levels, not reported in these studies), or other factors may have caused the higher incidence of epidermoid carcinomas in only two of the rat experiments with quartz.

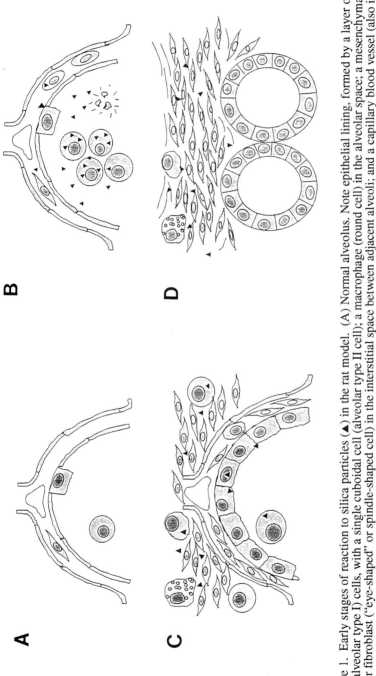

Figure 1. Early stages of reaction to silica particles (▲) in the rat model. (A) Normal alveolus. Note epithelial lining, formed by a layer of flat (alveolar type I) cells, with a single cuboidal cell (alveolar type II cell); a macrophage (round cell) in the alveolar space; a mesenchymal cell or fibroblast ("eye-shaped" or spindle-shaped cell) in the interstitial space between adjacent alveoli; and a capillary blood vessel (also in the interstitial space). (B) Alveolus after exposure to crystalline silica particles. Note macrophages with internalized (or phagocytized) silica particles; a lysed macrophage (cell fragments on right side of alveolus); and macrophages and silica particles in the interstitial space. Note also macrophages recruited to the interstitium (ovals, one of which contains a silica particle). (C) Early silicotic granuloma with alveolar type II hyperplasia. The alveolus is lined with type II cells (one of which is shown with a silica particle in its nucleus); the number of fibroblasts has increased dramatically; and a mast cell is present in the upper left. (D) Alveolar type II hyperplasia with adenomatoid pattern (formation of multiple spaces lined by cuboidal type II cells) adjacent to silicotic granuloma.

with a suspension of finely ground benzo[a]pyrene (BP).[6] The incidence of respiratory tract tumors in hamsters was not significant for treatment with the dusts alone. On the other hand, it was 47% in hamsters treated with BP alone, 76% in hamsters treated with ferric oxide + BP, 77% in hamsters treated with Ottawa silica sand + BP, 90% in hamsters treated with Min-U-Sil + BP, and 81% in hamsters treated with Min-U-Sil + ferric oxide + BP.[7] This study confirms that the hamster is not susceptible to carcinogenesis by silica alone, whereas it is highly susceptible to the induction of lung tumors by the carcinogen, BP, administered together with a mineral dust. This effect was shown for many different types of dust (Saffiotti et al., 1985). The multifactorial mechanisms involved in this animal model of lung carcinogenesis (Keenan et al., 1989) suggest a co-carcinogenic or synergistic role for the mineral dust, not specific for crystalline silica.

In mice, following single intratracheal administration of quartz or tridymite, we found that silicosis was readily induced (although with less fibrosis then in rats), but that neither alveolar epithelial hyperplasia nor tumor induction were observed, even in mice of strain A, which are highly susceptible to induction of lung tumors by many carcinogens (Saffiotti, 1992).

In conclusion, the animal experiments offer three distinct animal models for the study of the effects of crystalline silica:
- *rats* develop pulmonary fibrosis, alveolar hyperplasia, and lung tumors
- *mice* develop pulmonary fibrosis, no persistent epithelial hyperplasia, and no tumors
- *hamsters* develop neither fibrosis nor tumors

Therefore, these animal models allow comparative investigations on the mechanisms responsible for susceptibility to lung cancer induced by crystalline silica.

What is the response to silica in the human species? We know that humans are susceptible to silicosis and to silicosis-associated lung cancer. Individuals in the human population, however, may have different levels of susceptibility to either one of these effects. Saffiotti (1992) suggested the hypothesis that the three pathways observed in rodent species may be representative of host susceptibility differences in different subsets of the human population. In other words, some humans may respond to silica in a manner similar to rats, i.e., develop fibrosis and lung cancer; others may respond as mice, i.e., develop moderate fibrosis but no lung cancer; and others may respond as hamsters, i.e., be resistant to both silica-induced fibrosis and lung cancer. By identifying the critical factors responsible for the differences in susceptibility exhibited by rats, mice, and

---

[6] Benzo[a]pyrene is a well known carcinogen for the hamster lung (Saffiotti et al., 1968). It is present in cigarette smoke and other combustion products.

[7] Most tumors observed in the hamsters treated with dusts + BP were adenocarcinomas, adenomas, and adenosquamous tumors of the bronchi and lungs, with a smaller number of squamous cell carcinomas. Adenomas were more frequently induced by silica sand + BP and by Min-U-Sil + BP than by ferric oxide + BP (Niemeier et al., 1986).

hamsters, it may be possible to understand the different susceptibilities reflected in the human population.

Cytokines and gene expression in the pathogenesis of silicosis, alveolar type II hyperplasia and lung carcinogenesis

The starting point of these studies was the observation that, in the rat (a species susceptible to silica-induced fibrogenesis and carcinogenesis), the development of silicotic granulomas (composed of macrophages, fibroblasts, and other cell types, typical of chronic inflammation with progressive fibrosis) is accompanied by adjacent hyperplasia of epithelial cells lining the alveoli (Saffiotti, 1986). The close association of epithelial reactions with adjacent silicotic granulomas suggested a role for cell–cell interactions between the cells of mesenchymal origin (macrophages, monocytes, fibroblasts, mast cells) and the adjacent epithelial cells (alveolar type II cells).

Cell–cell interactions can be mediated by proteins or peptides (called cytokines or growth factors) produced by a cell type, that stimulate adjacent cells (paracrine factors) or even the same cell that produces them (autocrine factors). Several mediators bind to specific receptors on cell membranes. Mediators can therefore regulate the functions of other cells. Their effect is modulated by mechanisms that control their synthesis, secretion, diffusion, and interaction with target cells and their receptors. A great deal of new knowledge has been acquired, especially in the last decade, on these complex signals that represent major mechanisms for the maintenance of normal tissue functions and for the cellular responses to pathogenic stimuli. Much of the early evidence on the role of cytokines came from studies on interactions of various cell types in the immune system, but more recent evidence has demonstrated their importance also for epithelial cells.

A cell type that may be further investigated for its possible role in the pathogenesis of silicosis and associated carcinogenesis is the mast cell. Mast cells produce secretory granules.[8] In old experiments on silicosis in rats, a marked increase in the number of mast cells and in the proportion of degranulating mast cells was observed during the development of silicotic granulomas (Saffiotti et al., 1960; Saffiotti, 1962). Preliminary studies in rats, mice, and hamsters (Saffiotti, 1992) showed that mast cells are frequent in normal rat lung tissues, but rare in mice and hamsters, even after treatment with crystalline silica. The analogy with species susceptibility to crystalline-silica-induced lung cancer suggests that the role of these cells in carcinogenesis mechanisms should be further investigated.

Active oxygen species, produced by macrophages in response to stimulation by crystalline silica, are other mediators that may reach adjacent cells, including alveolar epithelial cells, and contribute to stimulate their proliferation.

---

[8] Membrane-bound round bodies in the cytoplasm of mast cells, containing histamine, serotonin and other products. The granules are released into the intercellular spaces under various stimuli (e.g., in allergic reactions) during a process called mast cell degranulation.

A working hypothesis for carcinogenesis by crystalline silica, first proposed at the 1984 meeting on "Silica, Silicosis and Cancer" (Saffiotti, 1986), and modified by recent updates, has been a useful framework for our investigations. It is illustrated in Figure 2, and can be stated as follows:

(a) Cell mediators, released by silicotic granulomas (from macrophages and other cells), include cytokines, such as interleukin-1 (IL-1), tumor necrosis factor-α (TNF-α) and transforming growth factor-β (TGF-β), mast cell products and oxygen radicals; some of these mediators act upon the adjacent epithelial cells of the distal airways and induce cell injury and/or cell proliferation, on a continuous basis.

(b) Crystalline silica-induced hyperplastic epithelial cells also produce mediators (e.g., TGF-β), that stimulate fibrogenesis (feedback effect).

(c) Crystalline silica penetrates into alveolar cells and causes DNA damage and/or chromosomal alterations, directly and/or through oxygen radicals.

(d) The combined effects of direct genetic damage to target epithelia and their chronic stimulation by cell mediators produced during fibrogenesis can account for crystalline-silica-induced carcinogenesis in pulmonary epithelia of susceptible hosts.

Let us consider the evidence so far available on the production of cytokines in silicotic lesions. IL-1 was found to be released *in vitro* by silica-stimulated human macrophages stimulated by crystalline silica, resulting in fibroblast proliferation (Schmidt et al., 1984). Crystalline silica induced the release of increased levels of IL-1 from alveolar macrophages in rats, following inhalation exposure (Oghiso and Kubota, 1986) or intratracheal instillation (Driscoll et al., 1990). Rabbit macrophages stimulated by endotoxin were found to release increased amounts of IL-1 when incubated with silica (Kampschmidt et al., 1986). In contrast, studies by Piguet et al. (1990) showed that IL-1 was not overproduced in the lung tissues of mice treated with an intratracheal instillation of 2 mg of a fibrogenic crystalline silica sample in 0.1 ml saline. In the same study, IL-6 was reported to be detectable in most cells of the silicotic nodule, and the expression of TNF-α mRNA was increased in silicotic mouse lungs and, in addition, TNF-α administration for 15 days through an osmotic minipump significantly increased collagen content of the lungs in crystalline silica-treated but not in control mice. In this study, TNF-α was detected in cells interspersed in the silicotic nodules by *in-situ* hybridization and by immunohistochemical staining,[9] but TGF-β mRNA levels were not significantly altered in mice after crystalline silica treatment (Piguet et al., 1990). In rats, after intratracheal instillation of crystalline silica, TNF-α was found to be released in increased amounts from alveolar macrophages (Driscoll et al., 1990).

---

[9] Immunohistochemical methods are now widely used to localize molecular markers (e.g., proteins or sequences of amino acids) in animal or human tissues. They use antibodies to the target protein coupled with chemical reagents that react by staining tissue or cellular sites where the target molecule is localized. *In situ* hybridization is a technique using radiolabelled antibodies to the messenger RNA for a given gene product.

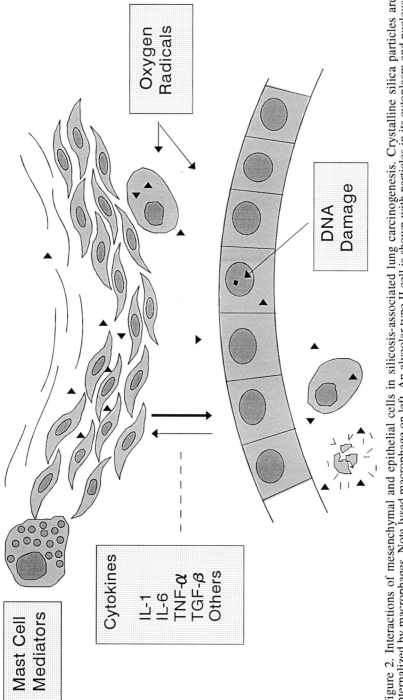

Figure 2. Interactions of mesenchymal and epithelial cells in silicosis-associated lung carcinogenesis. Crystalline silica particles are internalized by macrophages. Note lysed macrophage on left. An alveolar type II cell is shown with particles in its cytoplasm and nucleus. Particles are present in the adjacent granuloma. Shown are fibroblasts, a macrophage, a granular mast cell, and strands of collagen fibers. Mediators from the granuloma include oxygen radicals, cytokines, and mast cell products. Some factors (e.g., TGF-β) can be produced by epithelial cells and stimulate granuloma cells.

Recently developed methods allow the localization of specific markers for cell mediators and for altered gene expression in animal (or human) tissues, that were fixed in formalin and embedded in paraffin. Such archival tissue sources can thus be used to investigate specific factors of interest. In our laboratory, we have used this approach to study the pathogenesis of crystalline silica-induced fibrogenesis and associated carcinogenesis. Williams et al. (1993) used antibodies specific for mature TGF-β1 or for its precursor form to identify TGF-β1 in the lungs of rats that had received a single intratracheal instillation of quartz. Preliminary results in our laboratory indicate that TGF-β1 localization is minimal in the lungs of quartz-treated mice and not detected in quartz-treated hamsters, the two species that were found to be resistant to quartz-induced pulmonary carcinogenesis (Williams et al., unpublished observation). These results indicate that TGF-β1 is produced by hyperplastic alveolar type II cells, and suggest its role in crystalline silica-induced mesenchymal and epithelial lesions. We are currently studying the localization of several mediators in the lungs of crystalline silica-treated animals that were serially sacrificed, in order to evaluate the temporal sequence of induction of these cell mediators in the pathogenesis of the granulomatous and epithelial lesions.

The identification of biologically active cell mediators is likely to provide evidence for the interpretation of pathogenic mechanisms in different conditions, including the reasons for marked host differences in the reaction to crystalline silica, now known from species differences in the response to crystalline silica among rodents. It is hoped that similar differences can also be identified as individual markers of susceptibility in the human population.

Another critical mechanism, that needs to be explored for the crystalline silica carcinogenesis model, is the alteration of specific genes in the target tissues. These alterations can be due to altered levels of expression of certain genes or to their mutation. The genes that are of major current interest for the study of mechanisms of carcinogenesis are oncogenes and tumor suppressor genes. Altered expression of the *ras* oncogenes and of the *p53* suppressor gene (chosen because of their frequent activation in lung cancer) was studied using antibodies to gene products. In the lungs of rats treated with crystalline silica, the gene product of the *ras* family of genes (pan-reactive ras p21 protein) showed increased expression in hyperplastic alveolar type II cells adjacent to silicotic granulomas, and in the cells forming adenomatoid patterns, but not in adenomas and carcinomas. Control rat lungs showed no increased expression of this *ras* gene product. The *p53* protein showed no reactivity in hyperplastic alveolar type II cells and adenomas, but 25% of the carcinomas showed reactivity. A proposed model for the relationships between TGF-β on one hand and the *p53* and *ras* genes on the other is being investigated in silicosis and associated lung cancer. Similar studies on gene expression are under way in mice and hamsters, species that were found resistant to silica-induced lung carcinogenesis, to compare them with susceptible rats (Williams et al., unpublished observations).

It is important to relate the study of cellular mechanisms to the animal species that show susceptibility or resistance to different biological effects of crystalline silica (summarized above for responses in rats, mice, and hamsters). The observations on cellular mediators, cited above, indicate that patterns for fibrogenesis may differ from those for epithelial proliferation and carcinogenesis. After crystalline silica treatment, both IL-1 and TGF-β1 were found to be increased in rats but not in mice (in analogy with alveolar hyperplasia and carcinogenesis), whereas TNF-α was increased in both rats and mice treated with crystalline silica. IL-6 was increased in crystalline silica-treated mice, but it was not studied in rats. The study of cytokines and gene expression in the pathogenesis of silicosis and associated lung carcinogenesis is still in its beginnings. It is expected to provide valuable further insights into the mechanisms involved in the reactions of lung tissues to crystalline silica.

CELLULAR MODELS FOR TOXICITY AND NEOPLASTIC TRANSFORMATION

The effects of crystalline silica and other minerals have been studied in two cell lines. Cell lines, expanded from a single clone of cells and preserved by freezing, are advantageous, because the cells have defined desired properties and can be used reproducibly for many studies. In a sense, they represent cell "standards." We worked with (a) a cell line of mesenchymal origin (BALB/3T3/A31-1-1), derived from mouse embryo cells, that was characterized for use in neoplastic transformation assays, and (b) a cell line of fetal rat lung epithelial origin (FRLE), chosen because it is derived from alveolar type II cells, which are the cells of origin for silica-induced rat lung carcinomas. The cells attach to the surface of the culture dish and are grown in a nutrient medium. Exposure of the cells to dusts is obtained by replacing the culture medium with an aliquot of the same medium in which the dust has been suspended at the desired concentration.[10]

Exposure of cells in culture to minerals can be used to measure different biological effects, including: (1) uptake of particles by cells; (2) cell toxicity as measured by cell survival and colony forming efficiency (CFE); (3) cell growth rate (population doublings per day); and (4) induction of neoplastic transformation (identified by criteria discussed below). There is an extensive literature on cellular effects of crystalline silica, especially on macrophages in culture and on cells obtained by bronchial lavage. The recent evidence of marked species differences in their reaction to silica *in vivo*, discussed above, suggests that particular attention should be given to the source of cells used for tests in cell cultures.

The ability of a substance to induce neoplastic transformation in target cells can be tested, outside living animals, by exposing cells in culture. The advantages of such methods are that the cells are exposed under conditions that can be well controlled. Thus, the chosen cells are cultured and exposed in isolation from other cell types and from organismic factors, the cell milieu (nutrient medium, serum, and/or growth factors) can be directly controlled and/or experimentally

---

[10] In order to obtain good dispersion of the dust in the medium, the suspension can be sonicated in a bath for 2–3 minutes. The dust suspension should be prepared immediately before use. After the dust-containing medium is placed in the culture dishes, the dust rapidly settles on the surface of the dish and on the previously attached cells. The exposure is usually expressed in µg dust per cm^2 of culture surface.

manipulated, and the assays are rapid (they can be completed in a few weeks). The cell populations, in which neoplastic transformation was induced in culture, are usually recognizable by morphological changes. Their neoplastic properties can be confirmed by isolating and subculturing the morphologically altered cell colonies and testing them for their ability to grow as a neoplasm when injected in appropriate host animals.[11] The induction of neoplastic transformation in cultured cells by chemical and physical agents, in a quantitative dose-dependent manner, however, requires special test conditions that have been established, so far, only for a small number of cell systems. The most frequently used assay systems, which were well described (Kakunaga and Yamasaki , 1985) are the following three:

(a) The Syrian hamster embryo (SHE) system, which uses primary or secondary cultures of hamster embryo cells, plated at low density, that respond to carcinogens by developing morphologically altered colonies (a colony assay).

(b) The $C3H/10T^1/2$ system, using a cell line derived from mouse embryo cells with fibroblastic morphology. These cells are grown to form a confluent monolayer; when the cells are initially treated with a carcinogen, they develop morphologically altered foci that grow over the monolayer and are scored for frequency (a focus assay).

(c) The BALB/3T3-A31 system, using one of several cell lines derived from the BALB/3T3/A31 line of mouse embryo cells, isolated for their susceptibility to transformation by carcinogens (also a focus assay).

We used the BALB/3T3/A31-1-1 cell line (Kakunaga and Crow, 1980), which was characterized in our laboratory and used for transformation studies with different classes of carcinogens, including soluble metals (Saffiotti et al., 1985; Bertolero et al., 1987). With this assay,[12] we tested 5 samples of quartz, including Min-U-Sil 5 (MQZ), hydrofluoric-acid-etched MQZ (HFMQZ), DQ12, F600 and Chinese standard (CSQZ), each at doses from 6.25 to 100 µg/cm^2. All tested quartz samples showed a dose-dependent induction of transformed foci at lower doses, followed by a plateau at higher doses. Three repeated tests of MQZ showed a reproducible pattern, with maximal transformation reached at 25 µg/cm^2

---

[11] "Tumorigenicity assay," usually performed in "nude mice" that are genetically iummunosuppressed and therefore unable to reject implanted foreign tissues.

[12] For the transformation assay, cells are plated at the density of $10^4/50$ mm dish (20 cm^2) in MEM medium with 10% fetal bovine serum (FBS). The test agent is applied within 24 hr of plating. The cells grow to form a continuous monolayer of cells; if transformation is induced, foci of morphologically altered cells appear (usually in 3–5 weeks). The morphology of transformed foci is characterized by a criss cross growth pattern of spindle shaped cells (Type III foci), evident on staining. The frequency of transformed foci is calculated over the number of surviving cells, determined from concurrent assays for colony-forming efficiency. The transformation frequency of unteated control cells should be very low (usually below $1/10^5$ surviving cells), whereas the transformation frequency of carcinogen-treated cells should be significantly higher, with dose-dependent increases. The neoplastic nature of the transformed foci is verified by their subculture and subcutaneous injection in nude mice, where they give rise to sarcomas (tumorigenicity assay). The transformation assay requires carefully controlled culture conditions. The cells should be used within passage 15 (we used passage 12), and the serum batch should be selected after appropriate assays to yield low spontaneous and high induced rates of transformation.

(53 transformed foci/10^5 surviving cells), followed by a response plateau at doses of 50 and 100 µg/cm^2. Untreated controls showed transformation frequencies between 0.16 and 0.6 foci/10^5 surviving cells. The findings are compatible with a range of transformation responses common to all five tested quartz samples.[13] Quartz-transformed foci from all these assays were tested in nude mice and found to be rapidly tumorigenic. Examination of the chromosomes from these quartz-transformed cell lines showed that all had one or more altered chromosomes not seen in the untreated cell line (Ahmed and Saffiotti, 1992, and unpublished observations).

The only previous study on cell transformation by crystalline silica was done in the Syrian hamster embryo cell system and showed transforming activity of two samples of quartz, one of low toxicity and low transforming activity and the other (Min-U-Sil) with higher activity. The activity of both quartz samples was lower than that of two samples of chrysotile and crocidolite at comparable doses in terms of µg/cm^2 (Hesterberg and Barrett, 1984).

Another endpoint needs to be studied as a cellular effect of crystalline silica: the induction of mutations. Several assays that are widely used to evaluate mutagenesis by chemical agents, including bacterial mutagenesis test systems (e.g., the Ames assay), are likely to be inadequate for the study of the mutagenesis by materials, possibly because the particles do not gain the necessary access to target DNA. A few assays for silica were reported by the IARC (1987a). Pairon et al. (1990) reported the induction of sister chromatid exchanges in human lymphocytes cocultured with monocytes, after treatment with tridymite of fine particle size (87.9% <1 µm), but not with Min-U-Sil quartz of larger average particle size; this effect may have been due to mediators released by the monocytes (cytokines, reactive oxygen), because it was not observed in cultures of purified lymphocytes. Quartz-induced damage was demonstrated by the formation of binucleated cells and micronuclei in Syrian hamster embryo cells exposed to 20 µg/cm^2 of Min-U-Sil 5, a dose that induced neoplastic transformation in the same cell type (Hesterberg et al., 1986). A cell model that appears suitable for the study of mutagenesis by silica has been described by Hei et al. (1992). They used the human-hamster hybrid cell line A_L, which contains a single copy of human chromosome 11, in addition to normal Chinese hamster chromosomes. This cell line allows detection of both intragenic and multilocus mutations and even large deletions, using monoclonal antibodies to surface antigenic markers mapped on the human chromosome. Even extensive deletions on the hybrid human chromosome are not lethal in this cell line. Marked mutagenic activity has been demonstrated in this cell system for crocidolite and chrysotile asbestos (Hei et al., 1992). We plan to investigate the mutagenic activity of silica in this system in collaboration with Dr. Hei.

The target cells for silica-induced neoplasia in the rat lung model are epithelial cells, namely the alveolar type II cells. Therefore, a cell line of alveolar

---

[13] The present results are too limited to demonstrate reproducible differences in transforming activity among the tested quartz samples. Several repeated tests of each sample would be needed to identify whether the differences observed among different samples are reproducible and significant. Tests of several other minerals (e.g., hematite, anatase) were consistently negative.

type II rat cells would be the most relevant culture model for the study of transformation mechanisms by crystalline silica. Quantitative transformation assay systems for chemical and physical carcinogens were described for epithelial cells of the upper respiratory tract (trachea and bronchi), derived from animal models and from human sources (Harris, 1987). For rat tracheal epithelial cells, methods have been developed for culture and transformation (Nettesheim and Barrett, 1984), even in serum-free medium (Thomassen et al., 1986). However, until recently, no methods have been reported for the establishment of rat alveolar type II cell lines and for their transformation. After the first cell line of rat embryo alveolar type II origin (FRLE) was established (Leheup et al., 1989), we undertook, in our laboratory, to develop a transformation model in these target cells, adequate for studies on silica-induced transformation (Saffiotti et al., 1993). This cell line maintains differentiated characteristics of alveolar type II cells, such as the surfactant-producing lamellar bodies. The cell line was found to grow in several different culture media, even with low serum levels. FRLE cells exposed to quartz (Min-U-Sil 5) showed rapid and marked uptake of particles in their cytoplasm. Interestingly, small quartz particles (<0.4 µm) were found by transmission electron microscopy to be localized in the nucleus of these cells, and were confirmed by energy dispersive X-ray spectroscopy. The possible significance of intranuclear particles in the induction of DNA damage and transformation is the subject of current research. We recently obtained neoplastic transformation of this cell line by introducing into it a mutated K-*ras* oncogene (a mutated gene that has been frequently found in lung adenocarcinomas in humans and experimental animals) (Saffiotti et al., 1993; unpublished observations). We are using these *ras*-transformed cell lines as prototypes in the development of assays in FRLE cells for the detection of neoplastic transformation induced by other agents, including crystalline silica.

DNA BINDING AND DNA DAMAGE BY CRYSTALLINE SILICA

We have investigated the ability of crystalline silica to cause direct DNA damage *in vitro* using the five quartz samples listed above: MQZ, HFMQZ, DQ12, F600, and CSQZ. In addition, a series of particle size fractions, derived from Min-U-Sil 10 were used, as well as samples of cristobalite and tridymite. These studies were designed to identify and to quantify surface characteristics of crystalline silica, which could be involved in its carcinogenic effects. These include surface area, surface charge, and the production of oxygen free radicals, as well as the ability of crystalline silica to bind to DNA and to produce strand breakage and altered bases in DNA. The surface of crystalline silica is capable of interacting with biological systems in a way that leads to specific biological effects distinct from those of many other minerals. Elucidating the properties that confer such biological activities to the crystalline silica surface is a key to understanding its mechanisms of action. Because DNA is a critical target for carcinogenesis, our approach has been to study the effects of the crystalline silica surface directly on DNA, initially studied in vitro, as a basis for understanding and possibly modifying crystalline silica carcinogenesis mechanisms in the cell.

Reactive groups on the crystalline silica surface that can be important in direct interactions are probably either hydrated silanol groups (Si–OH) or ionized groups (Si–O⁻). We recently developed a simple dye binding assay (Daniel et al., in press), which may provide an easy measure of surface charge on crystalline silica in aqueous medium. A solution containing a positively charged dye, Janus Green B (25 µg/ml), is incubated in 10 mM sodium phosphate buffer containing crystalline silica in suspension (500 µg/ml). The amount of dye bound by the crystalline silica is obtained by spectrophotometric determination of the Janus Green B concentration remaining in the supernatant following removal of the suspended silica. Janus Green B binding correlated strongly with crystalline silica surface area, as measured by the BET nitrogen adsorption technique (Brunauer et al., 1938). We also determined the binding of crystalline silica to poly(2-vinylpyridine-N-oxide) (PVPNO), which is believed to bind to silanol groups on the silica surface (Nolan et al., 1981). PVPNO binding also strongly correlated with crystalline silica surface area, but PVPNO binding did not interfere with Janus Green B binding (Daniel et al., in press). It is likely that PVPNO binding measures surface silanol binding whereas Janus Green B binding measures surface binding to ionized groups. We have not yet tested the correlation of the binding of the cationic dye with the surface charge determined by zeta potential, which measures mobility of suspended particles subjected to an electric field (Adamson, 1990).

Associated with the negatively charged silica surface are traces of metal impurities which can undergo redox cycling because of the reducing surface microenvironment. We have observed rapid consumption of dissolved oxygen by silica immediately following exposure to an aqueous environment. This consumption of oxygen is associated with the appearance of the one electron reduction product of oxygen, the superoxide free radical. By spectrophotometric determination of Fe^{2+} and Fe^{3+} chelates from the surface of crystalline silica, it was also shown that oxidation of iron takes place at the same time as the superoxide production and oxygen consumption (Daniel et al., 1993). In addition to superoxide, crystalline silica surfaces produce hydrogen peroxide (H_2O_2) and the highly damaging hydroxyl radical (•OH). Disruption of the crystal structure by crushing of the particles also produces silicon based radicals (Si• and Si–O•). These radicals can be detected by electron spin resonance trapping (Vallyathan et al., 1988).

Effects of crystalline silica-produced oxygen-free radicals on DNA can be determined directly. By gas chromatography/mass spectrometry, we have measured the formation of the oxidized DNA base, thymine glycol, in incubations of calf thymus DNA with crystalline silica and found it to be elevated (Markey et al., unpublished observation). Oxidized DNA bases lead to DNA point mutations by inaccurate replication of the damaged DNA strand, unless corrected by DNA repair.

Another effect of oxygen free radicals on DNA is strand breakage. We measured DNA strand breaks produced by crystalline silica using a simple electrophoresis assay. Crystalline silica preparations produced direct damage to DNA, recognized by the appearance of DNA with lower molecular weight below the bands of control undamaged DNA (indicating breakage of longer DNA strands into shorter strands). The effects of modifiers of oxygen free radicals (*e.g.*, metal chelators, antioxidants and cellular enzymes) on DNA strand breakage were consistent with a mechanism of crystalline silica-induced DNA damage by hydroxyl radical (Daniel et al., 1993).

Damage to DNA by free radicals is dependent on the distance between DNA target sites and the site of radical formation. For example, the •OH radical, which is the most damaging radical, has a reaction distance of only 15 Å, about $2/3$ the diameter of a DNA double helix. Chemical bonding between DNA and silica increases the likelihood that silica-derived free radicals would hit a DNA target (assuming the free radicals are produced at the silica surface). Therefore we examined the possible binding of crystalline silica to DNA *in vitro*, using Fourier transform infrared spectroscopy techniques for characterization of the DNA-silica interaction. These techniques are similar to those used recently to study binding of drugs (Theophanides, 1981) or protein (Dev and Walters, 1990) to DNA. Representative reference spectra of DNA and silica alone were obtained in a zinc selenite attenuated total reflectance (ATR) cell. Coincubation of DNA with silica produced a complex composite spectrum from which the individual reference spectra for DNA and silica could be subtracted. The results showed that chemical bonding does in fact occur between silica and DNA (Mao et al., unpublished observation). Significant changes occurred in both the DNA and silica spectra during coincubation. In the DNA spectrum, an increase in intensity was observed for the PO_2^- asymmetric stretch at 1225 cm^{-1}, and this band was also shifted to a slightly lower frequency. At the same time, the symmetric stretch at 1086 cm^{-1} was markedly increased in intensity and the band at 1053 cm^{-1}, representing either the phosphodiester or the C–O stretch of the DNA backbone, was significantly reduced in intensity. Changes in the silica spectrum included marked narrowing of the Si–O stretching band at 1080 cm^{-1}. Taken together, these FT-IR findings indicate a bonding interaction between the phosphate backbone of DNA and the surface silanol groups of silica. This is the first evidence of a specific binding of silica to DNA. The finding that silica–DNA binding occurs at the phosphate backbone suggests that closeness of the DNA phosphate backbone to the silica surface may be important in the induction of DNA strand breaks by silica through hydroxyl radicals. The binding of silica surface with DNA may also interfere with DNA replication during mitosis, causing chromosomal breaks, or affect DNA repair by limiting access of the DNA to repair enzymes. Further work is needed to identify and measure silica-induced DNA damage in cells.

CONCLUSIONS

Mechanism studies conducted so far have provided a broader basis of understanding for the carcinogenic activity of silica. They have demonstrated that silica can bind directly to DNA and cause DNA damage and neoplastic transformation. They have also shown marked host differences in the susceptibility to silica-induced alveolar cell hyperplasia and carcinogenesis. The basis for such differences is being investigated at the cellular level in order to identify markers of susceptibility that may identify high risk individuals in the human population.

REFERENCES

Adamson, A.W. (1980) Physical Chemistry of Surfaces. Wiley, New York, p. 421-721.
Ahmed, N. and Saffiotti, U. (1992) Crystalline silica-induced cytotoxicity, transformation, tumorigenicity, chromosomal translocations and oncogene expression. Proc. Amer. Assoc. Cancer Res. 33, 119.
Bertolero, F., Pozzi, G., Sabbioni, E. and Saffiotti, U. (1987) Cellular uptake and metabolic reduction of pentavalent to trivalent arsenic as determinants of cytotoxicity and morphological transformation. Carcinogenesis 8, 803-808.
Brunauer, S., Emmett, P.H. and Teller, E. (1938) Adsorption of gases in multimolecular layers. J. Amer. Chem. Soc. 60, 309-319.
Churg, A., Wright, J.T., Stevens, B. and Wiggs, B. (1992) Mineral particles in the human bronchial mucosa and lung parenchyma. II. Cigarette smokers without emphysema. Exp. Lung Res. 18, 687-714.
Dagle, G.E., Wehner, A.P., Clark, M.L. and Buschbom, R.L. (1986) Chronic inhalation exposure of rats to quartz. In: Silica, Silicosis and Cancer. D.F. Goldsmith, D.M. Winn and C.M. Shy, Eds. Preager, New York, p. 255-266.
Daniel, L.N., Mao, Y. and Saffiotti, U. (1993) Oxidative DNA damage by crystalline silica. Free Radic. Biol. Med. 14, 463-472.
Daniel, L.N., Mao, Y., Vallyathan, V.,and Saffiotti, U. (in press) Binding of the cationic dye, Janus Green B, as a measure of the specific surface area of crystalline silica in aqueous suspension. Toxicol. Appl. Pharmacol.
Dev, S.B. and Walters, L. (1990) Fourier transform infrared spectroscopy for the characterization of a model peptide. Biopolymers 29, 289-299.
Driscoll, K.E., Lindenschmidt, R.C., Maurer, J.K., Higgins, J.M. and Ridder, G. (1990) Pulmonary response to silica or titanium dioxide: Inflammatory cells, alveolar macrophage-derived cytokines and histopathology. Amer. J. Respir. Cell Mol. Biol. 2, 381-390.
Fu, S.C., Yang, G.C., Shong, M.Z. and Du, Q.Z. (1984) Characterization of a new standard quartz and its effects in animals (in Chinese). Chinese J. Indust. Hygiene Occup. Dis. 2, 134-137.
Goldsmith, D.F., Winn, D.M. and Shy, C.M., Eds. (1986) Silica, Silicosis, and Cancer. Praeger, New York.
Groth, D.H., Stettler, L.E., Platek, S.F., Lal, J.B. and Burg, J.R. (1986) Lung tumors in rats treated with quartz by intratracheal instillation. In: Silica, Silicosis, and Cancer. D.F. Goldsmith, D.M. Winn and C.M. Shy, Eds. Praeger, New York, p. 243-253.
Guthrie, G.D. (1992) Biological effects of inhaled minerals. Amer. Mineral. 77, 225-243.
Harris, C.C. (1987) Human tissues and cells in carcinogenesis research. Cancer Res. 47, 1-10.
Hei, T.K., Piao, C.Q., He, Z.Y., Vannais, D. and Waldren, C.A. (1992) Chrysotile fiber is a strong mutagen in mammalian cells. Cancer Res. 52, 6305-6309.
Hesterberg, T.W. and Barrett, J.C. (1984) Dependence of asbestos- and mineral dust-induced transformation of mammalian cells in culture on fiber dimension. Cancer Res. 44, 2170-2180.
Hesterberg, T.W., Oshimura, M., Brody, A.R. and Barrett, J.C. (1986) Asbestos and silica induce morphological transformation of mammalian cells in culture: A possible mechanism. In: Silica, Silicosis, and Cancer. D.F. Goldsmith, D.M. Winn and C.M. Shy, Eds.Preager, New York, p. 177-190.
Holland, L.M., Gonzales, M., Wilson, J.S. and Tillery, M.I. (1983) Pulmonary effects of shale dusts in experimental animals. In: Health Issues Related to Metal and Nonmetallic Mining. W.L. Wagner, W.N. Rom and J.A. Merchant, Eds. Butterworth, Boston, p. 485-496.
Holland, L.M., Wilson, J.S., Tillery, M.I. and Smith, D.M. (1986) Lung cancer in rats exposed to fibrogenic dusts. In: Silica, Silicosis, and Cancer. D.F. Goldsmith, D.M. Winn and C.M. Shy, Eds.Preager, New York, p. 267-279.
International Agency for Research on Cancer (1987a) Silica. In: IARC Monographs on the evaluation of the carcinogenic risk of chemicals to humans. International Agency for Research on Cancer, Lyon, France, Vol. 42, 39-143.
International Agency for Research on Cancer (1987b) Overall evaluations of carcinogenicity: an updating of IARC Monographs, Volumes 1 to 42. In: IARC Monographs on the evaluation of the carcinogenic risk of chemicals to humans. International Agency for Research on Cancer, Lyon, France, Suppl. 7, 341-343.
Kakunaga, T. and Crow, J.D. (1980) Cell variants showing different susceptibility to ultraviolet light-induced transformation. Science 209, 505-507.
Kakunaga, T. and Yamasaki, H., Eds. (1985) Transformation assay of established cell lines: Mechanisms and applications. IARC Sci. Publ. No. 67. International Agency for Research on Cancer, Lyon, France.
Kampschmidt, R.F., Worthington, M.L. and Mesecher, M.I. (1986) Release of interleukin 1 (IL-1) and IL-1-like factors from rabbit macrophages with silica. J. Leucocyte Biol. 2, 39-123.

Keenan, K.P., Saffiotti, U., Stinson, S.F., Riggs, C.W. and McDowell, E.M. (1989) Mulyifasctorial hamster respiratory carcinogenesis with interdependent effects of cannula-induced mucosal wounding, saline, ferric oxide, benzo[a]pyrene and N-methyl-N-nitrosourea. Cancer Res. 49, 1528-1540.

Leheup, B.P., Federspiel, S.J., Guerry-Force, M.L., Wetherall, N.T., Commers, P.A., DiMari, S.J. and Haralson, M.A. (1989) Extracellular matrix biosynthesis by cultured fetal rat epithelial cells. I. Characterization of the clone and the major genetic types of collagen produced. Lab. Invest. 60, 791-807.

Miller, B.E., Dethloff, L.A. and Hook. G.E.R. (1986) Silica-induced hypertrophy of type II cells in the lungs of rats. Lab. Invest. 55, 153-163.

Muhle, H., Takenaka, S., Mohr, U., Dasenbrock, C. and Mermelstein, R. (1989) Lung tumor induction upon long- term low-level inhalation of crystalline silica. Amer. J. Indust. Med. 15, 343-346.

Nettesheim, P. and Barrett, J.C. (1984) Tracheal epithelial cell transformation: a model system for studies on neoplastic progression. CRC Crit. Rev. Toxicol. 12, 215-239.

Niemeier, R.W., Mulligan, L.T. and Rowland, J. (1986) Cocarcinogenicity of foundry silica sand in hamsters. In: Silica, Silicosis, and Cancer. D.F. Goldsmith, D.M. Winn and C.M. Shy, Eds. Preager, New York, p. 215-227.

Nolan, R.P., Langer, A.M., Harington, J.S., Oster, G. and Selikoff, I.J. (1981) Quartz hemolysis as related to its surface functionalities. Environ. Res. 26, 503-520.

Oghiso, Y. and Kubota, Y. (1986) Enhanced interleukin 1 production by alveolar macrophages and increase in IA-positive lung cells in silica-exposed rats. Microbiol. Immunol. 30, 1189-1198.

Pairon, J.C., Jaurand, M.C., Kheuang, L., Janson, X., Brochard, P. and Bignon, J. (1990) Sister chromatid exchanges in human lymphocytes treated with silica. Brit. J. Indust. Med. 47, 110-115.

Piguet, P.F., Collart, M.A., Grau, G.E., Sappino, A-P and Vassalli, P. (1990) Requirement of tumor necrosis factor for development of silica-induced pulmonary fibrosis. Nature 344, 245-247.

Robock, K. (1973) Standard quartz DQ-12 <12 μm for experimental pneumoconiosis research projects in the Federal Republic of Germany. Ann. Occup. Hygiene 16, 63-66.

Saffiotti, U. (1962) The histogenesis of experimental silicosis. III. Early cellular reactions and the role of necrosis. Med. Lavoro 53, 5-18.

Saffiotti, U. (1986) The pathology induced by silica in relation to fibrogenesis and carcinogenesis. In: Silica, Silicosis, and Cancer. D.F. Goldsmith, D.M. Winn and C.M. Shy, Eds.Preager, New York, p. 287-307.

Saffiotti, U. (1990) Lung cancer induction by silica in rats, but not in mice and hamsters: Species differences in epithelial and granulomatous reactions. In: Environmental Hygiene II. N.H. Seemayer and W. Hadnagy, Eds. Springer-Verlag, New York, p. 235-238.

Saffiotti, U. (1992) Lung cancer induction by crystalline silica. In: Relevance of Animal Studies to the Evaluation of Human Cancer Risk. R. D'Amato, T.J. Slaga, W.H. Farland and C. Henry, Eds. Wiley-Liss, New York, p. 51-69.

Saffiotti, U., Bignami, M. and Kaighn, M.E. (1985) Parameters affecting the relationships among cytotoxic, genotoxic, mutational and transformational responses in BALB/3T3 cells. In: Mammalian Cell Transformation. J.C. Barrett and R.W. Tennant, Eds. Raven Press, New York, p. 139-151.

Saffiotti, U., Kaighn, M.E., Knapton, A.D. and Williams, A.O. (1993) Transformation of the fetal rat alveolar type II cell line, FRLE, by lipofection with a mutated K-ras gene. Proc. Amer. Assoc. Cancer Res. 34, 102.

Saffiotti, U., Tommasini, Degna, A. and Mayer, L. (1960) The histogenesis of experimental silicosis. II. Cellular and tissue reactions in the histogenesis of pulmonary lesions. Med. Lavoro 51, 518-552.

Saffiotti, U., Cefis, F. and Kolb, L. (1968) A method for the experimental induction of bronchogenic carcinoma. Cancer Res. 28, 104-124.

Saffiotti, U., Stinson, S.F., Keenan, K.P. and McDowell, E.M. (1985) Tumor enhancement factors and mechanisms in the hamster respiratory tract carcinogenesis model. In: Cancer of the Respiratory Tract, Predisposing Factors. M.J. Mass, D.G. Kaufman, J.M. Siegfried, V.E. Steele and S. Nesnow, Eds. Carcinogenesis, a Comprehensive Survey, vol. 8, Raven Press, New York, p. 63-92.

Schmidt, J.A., Oliver, C.N., Lepe-Zuniga, J.L., Green, I. and Grey, I. (1984) Silica-stimulated monocytes release fibroblast proliferation factors identical to interleukin-1. A potential role for interleukin-1 in the pathogenesis of silicosis. J. Clin. Invest. 73, 1462-1472.

Sontag, J.M., Page, N.P. and Saffiotti, U. (1976) Guidelines for carcinogen bioassay in small rodents. NCI Carcinogenesis Tech. Report Series No. 1. DHEW Publ. No. (NIH)76-801. U.S. Gov't Printing Office, Washington, D.C.

Spiethoff, A., Wesch, H., Wegener, K. and Klimisch, H-J. (1992) The effects of thorotrast and quartz on the induction of lung tumors in rats. Health Phys. 63, 101-110.

Theophanides, T. (1981) Fourier transform infrared spectra of calf thymus DNA and its reactions with the anticancer drug cysplatin. Applied Spectros. 35, 461-465

Thomassen, D.G., Kaighn, M.E. and Saffiotti, U. (1986) Clonal proliferation of rat tracheal epithelial cells in serum-free medium and their responses to hormones, growth factors and carcinogens. Carcinogenesis 7, 2033-2039.

Vallyathan, V., Shi, X., Dalal, N.S., Irr, W. and Castranova, V. (1988) Generation of free radicals from freshly fractured silica dust. Amer. Rev. Resp. Dis. 138, 1213-1219.

Wagner M.M.F., Wagner J.C., Davies R. and Griffiths D.M. (1980) Silica-induced malignant hystiocytic lymphoma: incidence linked with strain of rat and type of silica. Brit. J. Cancer 41, 908-917.

Williams, A.O., Flanders, K.C. and Saffiotti, U. (1993) Immunohistochemical localization of transforming growth factor-β1 in rats with experimental silicosis, alveolar type II hyperplasia, and lung cancer. Amer. J. Pathol. 142, 1831-1840.

CHAPTER 19

REGULATORY APPROACHES TO REDUCE HUMAN HEALTH RISKS ASSOCIATED WITH EXPOSURES TO MINERAL FIBERS

Vanessa T. Vu

Office of Pollution and Prevention and Toxics
U.S. Environmental Protection Agency
401 M Street, S.W.
Washington, District of Columbia 20460 U.S.A.

INTRODUCTION

An important task for environmental protection is to identify, and subsequently to prevent the hazards to human health posed by toxic substances. Asbestos and related mineral fibers are one group of substances that have been identified as priority substances for risk reduction and pollution prevention. Because of the known health effects associated with past occupational exposures to elevated levels of asbestos, and because of the widespread use of asbestos in commerce, there has been considerable concern that exposures to asbestos may present a health hazard to workers and the general public. All major types of asbestos are associated with pulmonary fibrosis (asbestosis), lung cancer, mesotheliomas of the pleura and peritoneum in a dose-related manner.[†] Cancer at other sites (e.g., gastrointestinal cancer, laryngeal cancer) has also been shown to be associated with asbestos exposure, but the degree of excess risk and the strength of association are considerably less than for lung cancer and mesothelioma (IPCS, 1986; USEPA, 1986; ATSDR, 1990).

There is also a health concern for many other types of natural and synthetically made fibers whose commercial uses have been growing in recent years as replacement materials for asbestos-containing products. Yet, only limited information is available concerning their potential health effects and the exposure levels to workers, consumers, and the general public.

Studies conducted to date suggest that occupational exposures to rock wool and slag wool have produced an increased incidence of lung cancer in humans. Whether this increase is actually due to mineral wool exposure, to other contaminants, or to other factors remains to be determined (IARC, 1988; USEPA, 1988; HEI, 1991). In experimental studies, man-made mineral fibers and a variety of synthetic organic and inorganic fibers cause pulmonary fibrosis, lung cancer, and/or mesotheliomas in rats and hamsters under certain exposure conditions (IARC, 1988; USEPA, 1988; Vu and Dearfield, 1993). However, to date, only refractory ceramic fibers (RCF) have been shown conclusively to induce lung fibrosis, lung cancer and/or mesotheliomas in exposed animals by inhalation (IRIS, 1992; Vu, 1992; Vu, 1993).

[†] The relationship between chrysotile and mesothelioma is currently hotly debated (e.g., see Chapters 11 and 13).—*Eds.*

Erionite is the only natural fiber other than asbestos for which a high incidence of mesothelioma resulting from environmental exposures has been documented. Erionite has also been found to be extremely carcinogenic in rats following inhalation (IPCS, 1986; IARC, 1987; USEPA, 1988a). Erionite, however, is not known to be available in commerce at this time.

This chapter provides an overview of past and current regulatory activities relating to mineral fibers. Various approaches have been utilized by the federal agencies in the U.S. to reduce health risks associated with exposures to asbestos and other mineral fibers. These approaches are generally in the form of regulations, enforceable consent orders, negotiated voluntary actions, advisories, hazard communication, and guidance documents.

MAJOR REGULATIONS AND GUIDELINES ON ASBESTOS

There are many sources of exposures to asbestos. In addition to exposures from natural sources (e.g., see Chapter 2), humans are exposed to asbestos fibers during activities such as mining, milling, manufacturing, use, demolition, and disposal. There can be exposure to asbestos from other sources including schools, public and private buildings that have asbestos-containing materials, ambient air and water, and drinking water. Regulations and guidelines have been established by the various regulatory authorities in the U.S. (1) to limit exposure to asbestos in the workplace; (2) to minimize emissions of asbestos into the atmosphere from activities involving the milling, manufacturing, and processing of asbestos, demolition and renovation of asbestos-containing buildings, and the handling and disposal of asbestos-containing waste materials; (3) to control asbestos-containing materials in schools and in buildings; (4) to limit the level of asbestos in ambient water and drinking water; and (5) to restrict or to prohibit the use of asbestos in certain products and applications.

Occupational exposure limits and work practices

Asbestos was the first group of substances for which a comprehensive standard was issued in 1972 by the Occupational Safety and Health Administration (OSHA) under section 6(b) of the Occupational Safety and Health Act (OSH Act). The OSH Act of 1970 established OSHA to provide working conditions that are safe for employees, and it empowers the agency to prescribe mandatory occupational safety and health standards "which most adequately assures, to the extent feasible, on the basis of best available evidence, that no employee will suffer material impairment of health or physical capacity even if such employee has regular exposure for the period of his working life."

The 1972 asbestos standard established a Permissible Exposure Limit (PEL) for asbestos of 2.0 fibers per cubic centimeter (or f/ml) as an 8-hour time-weighted average (TWA). The standard also prescribed methods of compliance, personal protective equipment, employee monitoring, medical surveillance, hazard communication to employees, housekeeping procedures, and record keeping (OSHA, 1972). The standard of 1972 was intended primarily to protect workers against asbestosis and thereby to provide some protection from asbestos-associated cancer. In 1986, OSHA revised the asbestos standard based on the sufficient evidence that asbestos is a human carcinogen, and that the 1972 standard does not adequately protect workers from asbestos-related hazards. The 1986 asbestos standards reduced the PEL from 2.0 f/ml to 0.2 f/ml and updated other requirements. These standards, which remain in effect at present, apply to all industries including the construction and maritime industries and general industry (OSHA, 1986). As pointed out by OSHA, the current exposure limits do not represent "safe" levels of exposure, but are the lowest levels that industry can feasibly achieve using current control technologies.

Regulations to limit asbestos exposure during mining and milling activities have been issued by the Mine and Safety and Health Administration (MSHA) under the Federal Mine Safety and Health Act (Mine Act). The Mine Act of 1977 established MSHA to control the hazards of exposure to potentially harmful substances generated by mining activity or used in the mining or milling process. The Mine Act requires that MSHA, in promulgating a standard, attain the highest degree of health and safety protection for the miner, with feasibility of engineering controls and cost of compliance as additional considerations. The current health standard for asbestos specifies an 8-hour TWA exposure limit of 2 f/ml and provisions for labeling, use of protective equipment, engineering controls, and monitoring miners' exposures (MSHA, 1977). Consistent with OSHA's asbestos standard, MSHA recently proposed to lower the asbestos exposure limit to 0.2 f/ml (MSHA, 1989).

Since OSHA's asbestos health standards only apply to worker exposures in the private sector, the Environmental Protection Agency (EPA) has used its legal authority under Title II of the Toxic Substances Control Act (TSCA) to issue a regulation known as EPA Asbestos Worker Protection Rule (USEPA, 1987a). This rule requires comprehensive work practices as provided under the OSHA asbestos standard to protect employees in the public sector (state and local government employees) who are engaged in asbestos abatement work. The EPA rule also contains a provision not included in the OSHA rule, i.e., notification to EPA generally 10 days before an asbestos abatement project is begun when public employees are doing the work.

Air emissions control and waste disposal

Emissions of asbestos to the ambient air are regulated under the Clean Air Act (CAA). EPA, under the CAA of 1971, is required to develop and to enforce regulations necessary to protect the general public from exposure to air pollutants that are known to be hazardous to human health. EPA designated asbestos as a hazardous air pollutant and issued a National Emission Standards for Hazardous Air Pollutant (NESHAP) rule for asbestos in 1973 under section 112 of the CAA. The Asbestos NESHAP has been amended several times; the last revision was promulgated in 1990 to enhance enforcement and to promote compliance (USEPA, 1990a).

The Asbestos NESHAP requires specific emission control requirements for the milling, manufacturing, and fabricating of asbestos, for activities associated with the demolition and renovation of asbestos containing buildings. The Asbestos NESHAP does not set a quantitative fiber release level but requires work practices at demolition or renovation sites, and no "visible emissions" from any asbestos milling, manufacturing, fabricating, demolition, or renovation operation. This regulation also requires a facility survey for asbestos prior to the commencement of a demolition or renovation activity that is subject to the NESHAP.

Asbestos-containing waste is generally deposited in landfills. Asbestos is regulated as a solid waste for land disposal under the Resource Conservation and Recovery Act (RCRA) of 1976. EPA does not consider asbestos a hazardous waste under RCRA because asbestos does not pose a potential risk of leaching into groundwater. However, under expanded authority of RCRA, a few states have classified asbestos-containing waste as a hazardous waste, and these states require stringent handling and disposal procedures. The Asbestos NESHAP regulates emissions of asbestos from landfills. The rule prohibits visible emissions to the ambient air by requiring emission control procedures and appropriate work practices during collection, packaging, transportation, and disposal of friable asbestos-containing waste materials.

Asbestos is also subject to public reporting requirements for releases of hazardous substances under the Emergency Planning and Community Right-to-Know Act (EPCRA) of 1986 and the Comprehensive Emergency Response, Compensation and Liability Act (CERCLA) of 1980. EPCRA requires emergency notification to appropriate state and local authorities of any release of asbestos, and the submission of annual Toxic Release Inventory (TRI) reports to EPA and designated officials. The TRI reports include the amount of asbestos released into each environmental medium including air, water, and land (USEPA, 1988b).

Control of asbestos exposure in schools and buildings

Because the health risks of school children being exposed to low levels of asbestos is a concern, Congress passed the Asbestos Hazard Emergency Response Act (AHERA) in 1986 as a Subchapter II of TSCA to protect school children and employees from exposure to asbestos in school buildings. The Act required EPA to develop regulations creating a comprehensive framework for dealing with asbestos in public and nonprofit private elementary and secondary schools. To implement AHERA, EPA issued the Asbestos-Containing Materials in School Rule in 1987 (USEPA, 1987b). The AHERA school rule requires local education agencies to identify asbestos-containing materials in school buildings and take appropriate action to control release of asbestos, including inspections for asbestos, development of management plans, and to carry out the plan in a timely fashion. The school rule also requires the development of an asbestos operations and maintenance plan for schools where asbestos materials remain in place. The AHREA school regulations do not require schools to remove asbestos-containing materials.

AHERA also requires that EPA conduct a study to determine (1) the extent and condition of asbestos in public and commercial buildings; and (2) whether public and commercial buildings should be subject to the same inspection and response action requirements that apply to school buildings under the AHERA school rule. In response to Congressional mandate, in February 1988, EPA completed a study known as "EPA Study of Asbestos-Containing Materials in Public Buildings—A Report to Congress" (USEPA, 1988c).

EPA's study determined that friable asbestos-containing materials can be found in about one-fifth of the public and commercial buildings in the U.S. Two-thirds of these asbestos-containing buildings have at least some asbestos that is already damaged. Although EPA believed that asbestos in commercial buildings represents a potential health hazard that deserves attention, EPA did not recommend a comprehensive regulatory inspection and abatement program such as was implemented for school buildings. This was because there is only a limited supply of the accredited professionals and laboratories that are needed for the implementation of AHERA school rule, which has priority attention. Rather, EPA recommended to Congress that the Agency work during the next three years to enhance the nation's technical capability in asbestos by helping building owners better select and apply appropriate asbestos control and abatement actions in their buildings. To carry out that recommendation, EPA published a comprehensive asbestos guide known as "Managing Asbestos in Place" in July 1990 (USEPA, 1990b). This publication provides detailed and up-to-date instruction to building owners to help them successfully manage asbestos-containing materials in place.

On March 6, 1991, EPA published "An Advisory to the Public on Asbestos in Buildings" to provide guidance to the public for reducing asbestos exposure in buildings and to clarify EPA's policies regarding asbestos in schools and buildings (USEPA, 1991a). The advisory is in the form of five major facts that the Agency presented in congressional testimony. EPA concluded that on the basis of limited data, prevailing asbestos levels in buildings with asbestos management programs were very low. Although the data are not conclusive, available information suggests that health risks to building occupants are likely to be low when their buildings have active asbestos management programs. EPA recommended in-place management to control fiber release when the asbestos-containing materials are not significantly damaged. EPA also pointed out that removal of asbestos is not always the best alternative from a public health perspective. Improperly performed removal of asbestos can result in a very high level of exposure for building occupants. When removal is deemed necessary, i.e., when asbestos containing materials are damaged beyond repair, careful procedures to prevent exposure to the public both and during and after the removal are mandated.

EPA's findings concerning health risks to building occupants are consistent with conclusions reached by the Health Effects Institute—Asbestos Research (HEI, 1991). EPA and HEI recognized that building workers (i.e., service and custodial workers) may face greater health risks than building occupants, if they are not properly trained and protected, since they are more likely to be transiently exposed to higher levels of asbestos. OSHA and EPA have agreed that OSHA will take the lead in pursuing regulation to address these potential risks, and both agencies will work cooperatively to this end.

Health standards for drinking water and effluent guidelines

Under the Safe Drinking Water Act of 1972 (SDWA), EPA is required to regulate drinking water contaminants which "may have an adverse effect on human health." Drinking water in the U.S. is known to be contaminated with asbestos fibers resulting from mining operation, geologic erosion, the disintegration of asbestos cement pipe, and atmospheric sources. The 1986 SDWA amendments subsequently direct EPA to regulate asbestos in public water supplies. A Maximum Contaminant Level Goal (MCLG) of 7 millions fibers exceeding 10 microns in length per liter of drinking water was promulgated in 1991 (USEPA, 1991b).

EPA recognizes that there is insufficient evidence to demonstrate that asbestos in drinking water is associated with organ-specific cancer. However, EPA believes that there is a sufficient basis to regulate asbestos as a possible human carcinogen in drinking water (Regulatory Category II). The MCLG for asbestos is primarily based on the evidence that asbestos may be associated with an increase risk of gastrointestinal cancer through occupational exposure, and animal data showing that chrysotile asbestos fibers greater than 10 microns in length may be carcinogenic by ingestion.

Asbestos is also regulated under the Federal Water Pollutants Control Act of 1972 (amended by the Clean Water Act of 1977). Under this regulation, effluent limitations and technology performance standards have been established for eleven asbestos manufacturing point sources subcategories using the best available control technology that is economically achievable (USEPA, 1974).

Restriction or prohibition of the use of asbestos in certain products and applications

Release of asbestos fibers occurs not only in the manufacture and processing of asbestos, but also in their use and maintenance. Several regulatory actions have been taken by federal agencies to reduce asbestos exposure from certain uses or applications of asbestos-containing products or materials.

In 1973, EPA prohibited the spraying of asbestos-containing materials on buildings and structures for fireproofing and insulation purposes under the Clean Air Act (Asbestos NESHAP). The ban of the use of spray-on asbestos was later expanded to cover applications of asbestos-containing materials for decorative purposes (USEPA, 1990). In addition, the Consumer Product Safety Commission (CPSC) has banned use of asbestos-containing patching compounds (mostly for dry wall use) and artificial fireplace emberizing materials containing respirable free-form asbestos under the Consumer Product Safety Act (CPSC, 1977). In 1979, CPSC developed voluntary agreements under which hair dryer manufacturers stopped the use of asbestos heat shields.

EPA is empowered by section 6 of TSCA to ban or to restrict the manufacture, processing, distribution, use, or disposal of a chemical substance when there is a "reasonable basis" to conclude any such activity poses an "unreasonable risk of injury to health or environment," while taking into consideration the benefits of the chemical substance for various uses and the availability of substitutes, along with economic consequences of the regulation. In 1989, EPA issued a rule, known as the Asbestos Ban and Phase Out Rule (ABPO), under the authority of TSCA, to prohibit the manufacture, importation, processing and distribution in commerce of asbestos and most asbestos-containing products in the U.S. in three stages over seven years beginning in 1990 and ending in 1996. The regulation was intended to further reduce health risks to workers and the general public from many sources of asbestos releases.

The ABPO rule, however, was challenged in the U.S. court by the asbestos industry. In October 1991, the U.S. 5th Circuit Court of Appeals vacated and remanded most of the rule. The Court's decision did not question EPA's findings on the health effects associated with asbestos exposure; rather, the decision was based on differences in legal interpretation of TSCA, the authority under which the rule was issued. The rule is still in effect for those products which were no longer in commerce when the rule was issued on July, 1989. EPA is presently considering a number of regulatory and non-regulatory actions on asbestos in response to the Court's decision.

REGULATORY ACTIVITIES ON OTHER MINERAL FIBERS

Few actions have been taken by the U.S. regulatory authorities to prevent or limit exposures to other mineral fibers. This is primarily due to the lack of hazard and exposure information which serves as the basis for any risk reduction measures. EPA has recently identified a "respirable fibers" category as priority substances for hazard and exposure testing (USEPA, 1992). EPA is presently considering various approaches to obtain such information so that fibers of high concern can be identified for further regulatory investigation. Additionally, the following steps have been taken to address the potential risk posed by a number of specific non-asbestos fibers.

Erionite

EPA has promulgated a significant new use rule (SNUR) under section 5(e) of TSCA for erionite fiber. Because of the known health effects of erionite, EPA believes that any use may result in significant human exposure. This rule requires persons who intend to manufacture, import, or process any article containing erionite fiber to submit a significant use notice to EPA at least 90 days before any manufacturing, importation, or processing. The required notice will provide EPA with the opportunity to evaluate the intended use and, if necessary, to prohibit or to limit that activity before it occurs (USEPA, 1991c).

Refractory ceramic fibers

Based on animal inhalation data of RCFs submitted under section 8(e) of TSCA, EPA concluded in November 1991, that RCF may present an unreasonable risk of cancer to human health (USEPA 1991d). After conducting an accelerated review of RCF under section 4(f) of TSCA, EPA concluded that although there is sufficient evidence to classify RCF as a probable human carcinogen, exposure data are inadequate to determine whether or not RCFs pose an unreasonable health risk to workers. However, there was sufficient basis to support a concern for RCF and to initiate a regulatory investigation of RCF. Since there is a need to develop additional worker exposure data, EPA considered requiring the testing by promulgating test rules or by adopting enforceable consent agreements under section 4 of TSCA. In light of the manufacturers' willingness to work with EPA on the development of an exposure testing program to monitor workplace exposures (i.e., manufacturing, fabrication, processing, installation, and removal), EPA signed an enforceable testing consent order with the Refractory Ceramic Fibers Coalition (RCFC) in May 1993 (USEPA, 1993).

In addition to developing the exposure monitoring consent order with EPA, RCFC has developed and implemented a Product Stewardship Program which includes an implementation of workplace exposure control measures and a 1 f/ml industry recommended exposure guideline. Results from the exposure testing consent order should help determine the effectiveness of industry's stewardship of RCF.

OSHA has also proposed a 1 f/ml 8-hour TWA limit for respirable RCF for the construction, maritime, agriculture, and general industry. The proposed exposure limit is based on non-malignant respiratory disease, although OSHA has pointed out that the proposed limit will also increase the protection of workers from the potential carcinogenic effects (OSHA, 1992).

Glass fiber and mineral wool

Title III of the 1990 Clean Air Act amendments establishes a control technology-based program to reduce stationary source emissions of hazardous air pollutants. Man-made mineral fibers (including glass fibers, rock wool, and slag wool fibers) have been designated as hazardous air pollutants under section 112 (b) of the 1990 CAA amendments (CAA, 1990). EPA is in the process of establishing emissions standards for this group of substances.

OSHA has also proposed, under section 6(a) of OSH Act, a 1 f/ml 8-hour TWA limit for the respirable fibers of fibrous glass, rock wool, and slag wool for the construction, maritime, agriculture, and general industry. OSHA believes that this limit will protect worker from the risk of nonmalignant respiratory disease (OSHA, 1992).

CONCLUSIONS

Extensive regulations and guidelines have been established by several U.S. federal agencies to control or to limit the exposure of asbestos to humans. In contrast, only limited activities have been focused on other mineral fibers. However, it is generally recognized that there is an adequate basis to support a concern for respirable fibers, particularly those which are durable. Hence, there is a need to develop a comprehensive strategy for reducing risks from exposures to all respirable fibers. Components of such a strategy should include the practice of pollution prevention, development and implementation of product stewardship program, design of safer products (e.g., development of non-respirable fibrous products), the conduct of health effects research and testing, and exposure monitoring. Cooperative efforts among the federal agencies, industrial sectors, and public interest groups are necessary to achieve this goal, which is aimed at protecting the public from an unreasonable risk of injury.

REFERENCES

Agency for Toxic Substances and Disease Registry (ATSDR) (1990) Toxicological profile for asbestos. U.S. Department of Human and Health & Human Services. TP-90-04.
Clean Air Act Amendments (CAA) (1990) Public Law 101–549; 101st Congress, November 15, 1990.
Consumer Product Safety Commission (CPSC) (1977) Consumer patching compounds and artificial emberizing materials (embers and ash) containing respirable free-form asbestos; Final rule. Fed. Register 42, 63354–63365.
Health Effects Institute (HEI) (1991) Asbestos in public and commercial buildings: A literature review and synthesis of current knowledge. Health Effects Institute-Asbestos Research, Cambridge, Mass.
IRIS (1992) Integrated Risk Information System. Refractory ceramic fibers; Carcinogenicity assessment 09/01/92. U.S. Environmental Protection Agency, Washington, D.C.
International Agency for Research on Cancer (IARC) (1987) Silica and some silicates. Monographs on the evaluation of carcinogenic risk of chemicals to humans, Vol. 42, Lyon, France.
International Agency for Research on Cancer (IARC) (1988) Man-made mineral fibres and radon. Monographs on the evaluation of carcinogenic risks to humans, Vol. 43, Lyon, France.
International Programme on Chemical Safety (IPCS) (1986) Asbestos and other natural mineral fibres. Environmental Health Criteria 53. World Health Organization, Geneva.
International Programme on Chemical Safety (IPCS) (1988) Man-made mineral fibres. Environmental Health Criteria 77. World Health Organization, Geneva.
Mine Safety and Health Administration (MSHA) (1977) 30 CFR Part 56 et al.
Mine Safety and Health Administration (MSHA) (1989) Air quality, chemical substances, respiratory protection standards; proposed rule. Fed. Register 54, 35760–35852.
Occupational Safety and Health Administration (OSHA) (1972) Standard for exposure to asbestos dust. Fed. Register 37:11318–11322.
Occupational Safety and Health Administration (OSHA) (1986) Occupational exposure to asbestos, tremolite, anthophyllite, and actinolite; final rules. Fed. Register 51, 22612–22790.
Occupational Safety and Health Administration (OSHA) (1992) Air contaminants; Proposed rule. Fed Register 57, 26002–26601.
U.S. Environmental Protection Agency (USEPA) (1974) Effluent Guidelines and Standards; Asbestos manufacturing point source category. CFR Part 427.
U.S. Environmental Protection Agency (USEPA) (1986) Airborne asbestos health assessment update. EPA/600/8–84/003F.
U.S. Environmental Protection Agency (USEPA) (1987a) EPA worker protection rule. 40 CFR 763 Subpart G.
U.S. Environmental Protection Agency (USEPA) (1987b) EPA Asbestos Hazard Emergency Response Act (AHERA) Regulations. 40 CFR 763 Subpart E.
U.S. Environmental Protection Agency (USEPA) (1988a) Health hazard assessment of nonasbestos fibers. OPTS-62036. By V.T. Vu, Office of Toxic Substances, U.S. Environmental Protection Agency, Washington, D.C.
U.S. Environmental Protection Agency (USEPA) (1988b) Toxic chemical release reporting. Fed. Register 53, 4500–4539.
U.S.Environmental Protection Agency (USEPA) (1988c) EPA study of asbestos-containing materials in public buildings: A report to Congress. USEPA, Washington, D.C.
U.S. Environmental Protection Agency (USEPA) (1989) Asbestos: Manufacture, importation, processing, and distribution in commerce prohibitions; Final rule. Fed Register 54, 29460–29513.
U.S.Environmental Protection Agency (USEPA) (1990a) National Emission Standards for Hazardous Air Pollutants (NESHAP). 40 CFR 61 Subpart M.
U.S. Environmental Protection Agency (USEPA) (1990b) Managing asbestos in place- A building owner's guide to operations and maintenance programs for asbestos-containing materials. 20T–2003.
U.S. Environmental Protection Agency (USEPA) (1991a) An advisory to the public on asbestos in buildings: The facts about asbestos in buildings. Washington, D.C.
U.S. Environmental Protection Agency (USEPA) (1991b) National primary drinking water regulations; Final rule. Fed. Register 56, 3526–3597.

U.S. Environmental Protection Agency (USEPA) (1991c) Erionite fiber; Significant new use of a chemical substance. Fed. Register 56, 5647056472.

U.S. Environmental Protection Agency (USEPA) (1991d) Refractory Ceramic Fibers; Initiation of priority review. Fed. Register 56, 58693–58694.

U.S. Environmental Protection Agency (USEPA) (1992) Master Testing List. Office of Prevention, Pesticides and Toxics, USEPA, Washington, D.C.

U.S. Environmental Protection Agency (USEPA) (1993) Testing consent order for refractory ceramic fibers. Fed. Register 58, 28517–28520.

Vu, V.T. (1992) Refractory ceramic fibers: EPA's risk assessment and regulatory status. Proceedings of the Toxicology Forum 1992 Annual Winter Meeting. Toxicology Forum Inc., Washington, D.C., p. 400–416.

Vu, V.T., and Dearfield, K.L. (1993) Biological effects of fibrous materials in experimental studies and related regulatory aspects. In Warheit DB (ed.) Fiber Toxicology. Academic Press, New York, p. 449–492.

Vu, V.T. (1993) Assessment of the potential health effects of natural and man-made fibers and their testing needs: perspectives of the U.S. Environmental Protection Agency. Proc. 4th Int'l Inhalation Symp., in press.

GLOSSARY

The following glossary contains several biological and geological terms that may be unfamiliar to some readers. A more thorough listing of terms can be found in a general scientific dictionary, such as *Dictionary of Scientific and Technical Terms* (1989, 4th edition, S.P. Parker, Editor, McGraw Hill, New York, 2138 pp.), or in dictionaries specific to each of the disciplines (e.g., *Glossary of Geology*, 1980, 2nd edition, R.L. Bates and J.A. Jackson, Editors, American Geological Institute, Falls Church, Virginia, 751 pp.; *Stedman's Medical Dictionary*, 1990, 25th edition, W.R. Hensyl, Editor, Williams and Wilkns, Baltimore, Maryland, 1784 pp.). Many of the definitions below are modified from these sources. Accepted mineral species names and formulae can be found in the *Mineral Reference Manual* (1991, E.H. Nickel and M.C. Nichols, van Nostrand Reinhold, New York, 250 pp.), and we have generally followed their usage. However, in some cases, errors in the *Mineral Reference Manual* have been corrected here. The editors assume ultimate responsibility for the correctness of the following definitions. However, we acknowledge the assistance of many in developing this glossary, including the authors of chapters in this book and H.C.W. Skinner. Many of the definitions below use words that are also defined in the glossary, and these words are generally italicized.

a-axis: One of the three principle axes used to describe the coordinate system of a crystal structure. See *crystallographic axes*.

accessory mineral: Any mineral that is present in a rock but is not essential to classifying the rock. Generally accessory minerals are present in minor quantities.

acicular: Said of a crystal that is needlelike in form. A high aspect ratio mineral particle formed during growth or crushing. See *asbestiform, fibrous, prismatic, equant, tabular*.

actinolite: An amphibole with the ideal composition $Ca_2(Mg,Fe^{2+})_5Si_8O_{22}(OH)_2$. Actinolite is a species in the $Mg-Fe^{2+}$ series, tremolite–ferro-actinolite, with $0.9 > Mg/(Mg+Fe^{2+}) > 0.5$. See *amphibole, ferro-actinolite, tremolite*.

activation energy: The additional energy required to allow a system to proceed from one energy state to another, e.g., to make a reaction proceed.

active oxygen species: Oxygen free radicals. Reactive metabolites or reduced species of oxygen that can react with cellular targets, including DNA. These species possess a non-equilibrium number of electrons (i.e., they possess an unpaired electron), such that the species is unstable and can function as either an electron donor/acceptor or a proton donor/acceptor. See *hydroxyl radical, superoxide*.

additive: The condition when two or more agents induce a biological response that is the sum of the weighted biological responses of each agent individually.

aeolian: See *eolian*.

AEM: Analytical electron microscopy. This is typically done using a transmission electron microscope equipped with a capability such as energy-dispersive spectrometry. By performing AEM with a transmission electron microscope, addition important mineralogical information may be obtained, such as electron diffraction information, particle morphology, microstructures, etc.

AES: Auger electron spectroscopy.

AFM: Atomic force microscopy or atomic force microscope.

agate: A type of microcrystalline quartz.

akaganeite: β-FeOOH. See *lepidocrocite* and *goethite*.

alkali feldspar: A feldspar with an ideal composition of $(K,Na)AlSi_3O_8$.

allophane: A amorphous clay with a composition approximating that of the kaolin group.

alteration: In mineralogy, any change in the chemical or mineralogical composition of a rock or mineral, typically resulting from the interaction between a mineral and an aqueous fluid. Often, the alteration process involves the introduction of some constituents (e.g., H_2O, cations, or anions) and the release of other constituents of the mineral. Weathering reactions are one example of mineral alteration.

alumina: A chemical term for aluminum oxide, Al_2O_3. See *corundum*.

alveoli: Small sacs or compartments located at the terminus of the respiratory tract. Alveoli are the sites at which gaseous exchange with the blood occurs.

Ames assay: A bacterial assay used to assess the mutagenic potential of an agent.

amesite: A 1:1 layer silicate with an ideal composition of $Mg_2Al[SiAl]_2O_5(OH)_4$.

amosite: A varietal name for brown (or sometimes gray) asbestos. Amosite was derived from the acronym for *A*sbestos *M*ines *o*f *S*outh *A*frica (AMOSA), and it is a commercial term (i.e., it refers to a product and not a mineral). Amosite generally implies asbestiform cummingtonite–grunerite; however, some amosite samples have been reported to contain other asbestiform amphiboles as well, including anthophyllite, actinolite, and riebeckite. The mineral content may vary from sample to sample, because different source materials may have been used. If the mineral species in a sample is known to be cummingtonite or grunerite, the terms asbestiform cummingtonite or asbestiform grunerite can be used to provide more detailed information. See *amphibole, cummingtonite, grunerite*.

amorphous: Lacking long-range, periodic atomic order or translational symmetry. Non-crystalline.

amphibole: A mineral group of chain silicates possessing a double-chain structure with an ideal composition of $AB_2C_5T_8O_{22}(OH,F,Cl)_2$. The A-site is 6- to 12-coordinated and can be unoccupied or can contain monovalent cations (e.g., K or Na); the B site is represented by the M4 crystallographic site, which is 6-, 7-, or 8-coordinated and can contain Mg, Fe^{2+}, Ca, or Na; the C site is represented by the M1, M2, and M3 crystallographic sites, which are octahedrally coordinated and can contain Mg, Fe^{2+}, Al, Fe^{3+}, Ti, Mn, and Li; and the T-site is tetrahedrally coordinated and generally contains Si^{4+} but can also accommodate Al^{3+}.

anatase: A mineral species that is a polymorph of TiO_2. See *brookite* and *rutile*.

anionic group: A molecular unit possessing a net negative charge. SiO_4^{4-} and CO_3^{2-} are examples of anionic groups.

antagonistic: The condition when two or more agents induce a biological response that is less than the sum of the weighted biological responses of each agent alone, e.g., one agent inhibits the response of another.

anthophyllite: A species of amphibole with the ideal composition of $(Mg,Fe^{2+})_7Si_8O_{22}(OH)_2$. The iron content of anthophyllite is typically small but not zero. Anthophyllite differs from *magnesio-cummingtonite* in that it possess an orthorhombic structure.

anthropogenic: Relating to the impact of man on nature. That which is found in the environment but which was created by man.

antigorite: A serpentine mineral with corrugated layers and a composition that differs slightly from the ideal composition for chrysotile and lizardite.

antibody: An immune or protective protein that reacts with a specific antigen on foreign material (e.g., bacteria or viruses), i.e., the antibody possess structural components that complement structural components on the antigen. The binding of the antibody to the antigen aids the immune response by processes such as targeting foreign bodies for phagocytosis or neutralizing infectious agents. See *immunoglobulin*.

antigen: A molecule that induces the production of an antibody.

antioxidants: Naturally occurring or synthetic scavengers of active oxygen species.

AOS: See *active oxygen species*.

apical oxygen: In polymerized sheets of tetrahedra, the apical oxygens are those not shared between tetrahedra. See *basal oxygen*.

arachidonic acid: An unsaturated fatty acid essential in nutrition. A precursor of the eicosanoids, a group of physiologically active substances (e.g., leukotrienes, which mediate inflammation and are involved in alergic reactions, and prostaglandins, which affect vasodilation/constriction and stimulate bronchial smooth muscle, as well as other functions).

asbestiform: An adjective describing inorganic materials that possess the form and appearance of asbestos. Asbestiform is a subset of fibrous, where asbestiform implies relatively small fiber thickness and large fiber length, flexibility, easy separability, and a parallel arrangement of the fibers in native (unprocessed) samples. Often, asbestos fibers occur in bundles, i.e., they are often polyfilamentous. See *acicular, fibrous, prismatic, equant, tabular*.

asbestos: A term applied to asbestiform varieties of serpentine and amphibole, particularly chrysotile, "crocidolite," "amosite," asbestiform tremolite, asbestiform actinolite, and asbestiform anthophyllite. The asbestos minerals possess *asbestiform* characteristics.

asbestosis: A fibrotic lung disease associated with inhalation of asbestos. The disease is characterized by the inability of the lung to oxygenate blood or to eliminate carbon dioxide and a decrease in the ability to expand or to respond to the action of the diaphragm.

aspect ratio: The ratio of length to width.

atomic positions: The position of atoms within a crystal structure. See *fractional coordinates*.

attapulgite: A varietal term for palygorskite. Attapulgite is not a mineral species name. Often the distinction between attapulgite and palygorskite has been that one is fibrous and the other is not. However, industrial usage often assumes attapulgite is non-fibrous, whereas biological usage often assumes attapulgite is fibrous. The term attapulgite should not be used; rather palygorskite should be used in conjunction with modifiers such as fibrous or non-fibrous. See *palygorskite*.

authigenic: Said of minerals or rock constituents formed or generated in place.

autocrine effect: See *cytokine*.

b-axis: One of the three principle axes used to describe the coordinate system of a crystal structure.

basal oxygen: In polymerized sheets of tetrahedra, the basal oxygens are those shared between tetrahedra. See *apical oxygen*.

bentonite: A soft, plastic, porous rock that contains major amounts of clay minerals, particularly montmorillonite.

berthierine: A 1:1 layer silicate with an ideal composition of $(Fe,Al)_3(Si,Al)_2O_5(OH)_4$. Berthierine-1M and berthierine-1H are the two ordered *polytypes* observed for berthierine.

biopyribole: See *pyribole*.

biotite: A series of 2:1 layer silicates of ideal composition $K(Mg,Fe)_3Si_4O_{10}(OH)_2$. Phlogopite is the magnesium end member of the series, and annite is the iron end member.

blue asbestos: See *crocidolite*.

boehmite: γ-AlOOH. See *diaspore*.

Brazil twin: A common type of twin in quartz, resulting from regions with different chirality or handedness. See *Dauphiné twin*.

bridging oxygen: An oxygen atom shared between two coordination polyhedra, particularly tetrahedra.

bright-field TEM image: An image formed using the central ("undiffracted") electron beam. Regions that scatter electrons strongly appear dark, and regions that do not scatter electrons appear bright. The scattering responsible for the variation in image contrast can result from structural differences (e.g., diffraction related to the orientation of a crystal) or from compositional differences (e.g., diffuse scattering from large atoms). See *dark-field TEM image*.

brindleyite: A 1:1 layer silicate with an ideal composition of $(Ni,Al)_3(Si,Al)_2O_5(OH)_4$.

bronchus: One of two primary branches of the respiratory tract below the trachea.

bronchiole: Small, thin-walled branch of a bronchus. Bronchioles usually terminate in alveoli.

brookite: A mineral species that is a polymorph of TiO_2.

brown asbestos: See *amosite*.

brucite: Magnesium hydroxide [$Mg(OH)_2$] that generally exhibits a planar morphology. However, a fibrous variety (nemalite) also occurs.

byssolite: A fibrous variety of amphibole. Dana restricted the term to the stiff, fibrous variety of actinolite (see Ch. 3 under "Byssolite and nephrite"). Also, a variety of quartz containing fibrous inclusions of actinolite or asbestos.

c-axis: One of the three principle axes used to describe the coordinate system of a crystal structure.

calcified: In biology, a process whereby tissue or noncellular material in the body is hardened as a result of the precipitation of calcium and magnesium salts, particularly calcium carbonate and phosphate.

carcinogen: A cancer-causing agent.

carcinogenic: Possessing the ability to induce cancer.

carcinogenesis: The process of cancer development. Carcinogenesis can be divided into the initiation phase (implying interaction with DNA) and the promotion or progression phase (implying alterations in replication affecting cell division and conversion to malignancy).

chabazite: A species of zeolite with ideal composition $Ca(Al_2Si_4)O_{12} \cdot 6H_2O$, where the "·" indicates that the water molecules are not integral to the mordenite structure. See *zeolite*.

chalcedony: A type of microcrystalline quartz.

chain silicate: A mineral class consisting of species with SiO_4^{4-} tetrahedra polymerized in one dimension. Included in this class are the amphiboles, pyroxenes, and pyroxenoids.

chemography: In mineralogy, the graphical representation of mineral compositions in terms of end-member components.

Chemokines: A family of 8 to 10-kd proteins that possess chemoattractant activity for a variety of cell types (e.g., monocytes, neutrophils, lymphocytes, and fibroblasts). Chemokines are important mediators of inflammation and tissue repair processes.

chemotaxis: Movement of cells or organisms in response to chemicals. *chemotactic,* adj.

chert: A type of microcrystalline quartz.

chlorite: A layer silicate characterized by 2:1 layers alternating with hydroxide sheets. Compositionally, chlorites can be similar to 1:1 layer silicates such as the serpentine- and kaolin-group minerals.

chrysotile: A subgroup name for tubular serpentine minerals of ideal composition $Mg_3Si_2O_5(OH)_4$. The term chrysotile can be modified by the prefixes ortho-, clino-, and para- to identify the crystallographic arrangement of the sheets in the structure. The resulting terms ("orthochrysotile," "clinochrysotile," and "parachrysotile") are mineral-species names. Chrysotile is frequently (but not always) asbestiform. See *serpentine*.

clay: In geology, this term is applied to particles <4 μm, because this size is approximately the upper size limit for particles exhibiting colloidal properties. However, in soil science, the term applies to particles <2 μm, and in engineering, the term applies to particles ≤74 μm. Clay also refers to any fine-grained layer silicate, such as chrysotile, kaolinite, smectite (chief constituent of bentonite), and illite.

cleavage: The property of an individual crystal to fracture or break along crystallographically defined planes determined by the structure of the material. Cleavage can be pronounced (as in the cleavage that produces the sheets in micas), to weak, to absent (as in materials such as glasses). Minerals can possess more than one cleavage direction. C.f., *parting*.

clino–: A prefix used to connote a monoclinic structure, e.g., clinoptilolite is the monoclinic variety of the mineral ptilolite (a formerly used term for mordenite).

clinoptilolite: A zeolite with the ideal composition $(Na,K)_6(Al_6Si_{30})O_{72} \cdot 20\ H_2O$, where the "•" indicates that the water molecules are not integral to the clinoptilolite structure. Natural clinoptilolite is generally non-fibrous. See *zeolite*.

cocarcinogen: An agent that is generally not carcinogenic by itself but may increase the tumor yield of a known carcinogen.

coesite: A high pressure polymorph of SiO_2.

cohort: A group of individuals with similar backgrounds or characteristics. Cohorts are used in epidemiological studies to determine the biological effects related to a specific activity or agent.

collagen: A fibrous protein found in the connective tissue. Also, the principle component of scar tissue that forms during fibrosis of the lung.

comminution: Pulverization or reduction to minute particles.

complement: A protein complex, normally present in the serum. The activation of complement is part the immune response and involves a sequence of events leading to the destruction of certain bacteria and other cells.

confluence: The condition when cells in culture have reached a monolayer coverage and growth rate diminishes.

congruent: See *dissolution*.

connective tissue: A collection of similar cells and intercellular substances that for the framework of the animal body.

coordination number: The number of an ion's nearest neighboring ions.

coordination polyhedron: A volume described by an ion's nearest neighboring ions. Typically, minerals are represented by coordination polyhedra consisting of a cation in the interior of the polyhedron and anions at the apices of the polyhedron. *Tetrahedra* and *octahedra* are examples of coordination polyhedra.

coordination structure: A crystal structure that can be represented with coordination polyhedra.

corundum: A mineral species that is a polymorph of $Al_2O_3 = \alpha\text{-}Al_2O_3$. Corundum is isostructural with hematite ($\alpha\text{-}Fe_2O_3$).

cristobalite: A polymorph of SiO_2. The field of stability for cristobalite occurs at high temperature; however, cristobalite can exist or even grown metastably at low temperature.

crocidolite: A varietal name for blue asbestos, usually asbestiform riebeckite. If the mineral species in a sample is known to be riebeckite, the term asbestiform riebeckite can be used to provide greater detail. See *amphibole, riebeckite*.

cronstedtite: A 1:1 layer silicate with the ideal composition $Fe_2^{2+}Fe^{3+}(Si,Fe)_2O_5(OH)_4$. There are a number of polytypes with specific mineral species names.

crystal: A homogeneous, solid body of a chemical element, compound, or solid solution having a regularly repeating atomic arrangement.

crystal structure: The structure of a crystalline material, i.e., a material possessing long-range periodic order and translational symmetry. Mineral crystal structures are typically represented using a unit cell, which defines a subset of the entire structure. The entire structure can then be produced by translating the unit cell along the cell edges. The shape of the unit cell is described with axes (**a**, **b**, and **c**) and their angular relationships ($\alpha = \mathbf{b}\angle\mathbf{c}$, $\beta = \mathbf{a}\angle\mathbf{c}$, and $\gamma = \mathbf{a}\angle\mathbf{b}$), and these also define the coordinate system used to describe atomic positions. The magnitudes of these axes and the angles between them (i.e., the lattice parameters) are denoted by a, b, c, α, β, and γ. Atomic positions are generally described in terms of fractional coordinates, which give the positions of the atom relative to the origin of the unit cell. The origin has the coordinates (x,y,z) of (0,0,0); the corner of the cell at the end of the **a**-axis has the coordinates (1,0,0); the **b**-axis has the coordinates (0,1,0); and so on. An atom half-way along the **b**- and **c**-axes has the coordinates (0,0.5,0.5). This notation for coordinates should not be confused with the notation for *Miller indices*, which are used to describe the orientation of a crystal face. Another similar notation is used to describe the orientation of a vector, e.g., [100] describes a vector of unit length along the direction of the **a**-axis. In general, these notations for vectors and planes differ from the notation for atomic coordinates by the absence of commas separating the values. The above description provides information on the ideal structure of the crystal, as determined by techniques that measure the average properties of a crystal (e.g., X-ray diffraction). Most natural materials, however, deviate from the ideal structure. These deviations include compositional variations, slight displacements in atom positions, and other defects. These properties can be measured with techniques that examine local environments, such as IR spectroscopy or transmission electron microscopy.

crystalline: An adjective to describe any material possessing the properties of a crystal, i.e., long-range translational periodicity.

crystallographic axis: One of the three vectors (**a**, **b**, and **c**) that define the unit cell of a crystal. The lattice parameters (a, b, and c) refer to the lengths (magnitutdes) of the vectors **a**, **b**, and **c**.

crystallographic plane: See *Miller indices*.

culture: The propagation of microorganisms or cells on or in media.

cummingtonite: A series of amphiboles with the generalized composition $(Mg,Fe^{2+})_7Si_8O_{22}(OH)_2$. Three mineral species consitute this series: magnesio-cummingtonite with $Mg/(Mg+Fe^{2+}) > 0.7$; cummingtonite with $0.7 > Mg/(Mg+Fe^{2+}) > 0.3$; and grunerite with $0.3 > Mg/(Mg+Fe^{2+})$. Asbestiform varieties of these amphiboles are often referred to as "amosite." Cummingtonite differs from *anthophyllite* in that it possess a monoclinic structure.

cytokine: A protein released from one cell type and that modifies the biological responses of the producing cell (autocrine effect) or those of other cell types (paracrine effect). Cytokines include tumor necrosis factor (TNF), interleukins (e.g., IL-1, IL-2), and transofmring growth factors (e.g., TGF-α and TGF-β).

cytotoxic: An adjective to describe any substance that possesses the ability to kill (or to lyse) cells.

dark-field TEM image: An image formed by using an aperture to exclude the central ("undiffracted") electron beam. Regions that are diffracting into the aperture appear bright, and regions that are not diffracting into the aperture appear dark. See *bright-field TEM image*.

Dauphiné twin: A common type of twin in quartz, resulting from the two possible orientations assumable by tetrahedra during the transition from β-quartz to α-quartz. Parts of the crystal will assume one orientation and other parts will assume the other orientation. See *Brazil twin*.

defect structure: The structures that define the departure of a crystal from its ideal structure. Defect structures do not occur in a long-range ordered fashion. Point defects (e.g., misplaced atoms) and planar defects (e.g., stacking faults) are examples of defect structures.

d_{hkl}: The interplanar spacing (i.e., the length of the vector normal to the planes) for the plane (hkl).

diagenesis: Alteration of a rock or mineral resulting from conditions <1 kbar and <100 to 300°C. Diagenesis is analogous to mild metamorphism.

diaspore: α-AlOOH. See *boehmite*.

dickite: A 1:1 layer silicate with an ideal composition of $Al_2Si_2O_5(OH)_4$. Dickite is a member of the kaolin group.

dioctahedral sheet: An octahedral sheet in which two thirds of the octahedral sites (cf. trioctahedral sheet) are occupied by a cation.

dioctahedral substitution: The substitution of vacancies in a trioctahedral sheet.

diopside: A pyroxene mineral with an ideal composition of $CaMgSi_2O_6$. It is the pyroxene analog of tremolite.

displacive transformation: A phase transition characterized by the bending of chemical bonds but for which no chemical bonds are broken. These transitions occur rapidly and are reversible. See *reconstructive transformation*.

dissolution: The process of releasing solid constituents to a fluid. A mineral is said to dissolve congruently when the release of material to the fluid coincides with the stoichiometry of the dissolving mineral. During incongruent dissolution, the stoichiometry of the material released to the fluid differs from that of the mineral, so a solid with a different composition (e.g., a leached layer) remains behind.

ditrigonal: A form consisting of six sides with alternate angles of equal magnitude.

dolomite: A carbonate (structurally somewhat similar to calcite) with an ideal composition of $CaMg(CO_3)_2$.

dyspnea: Difficult or labored breathing.

EDS: Energy-dispersive spectrometry. The intensity of X-rays emanating from a sample is measured by a detector as a funtion of energy (i.e., wavelength). All energies of interest are measured simulataneously.

effector cell: A cell that may function by elaborating growth factors or other factors contributing to inflammation of disease. Generally refers to cells of the immune system. *C.f.*, *target cell*.

eicosanoid: See *arachidonic acid*.

enantiomorphic twin: A twin in which one domain is related to the other by a change in handedness, i.e., by reflection across a mirror or inversion through a point.

endothelium: The layer of cells that line the blood and lymph vessels, the heart, and other cavities of the body.

end member: A mineral species that defines one end point of a mineral solid-solution series. Typically, end member refers to a compositional end point. For example, tremolite [$Ca_2Mg_5Si_8O_{22}(OH)_2$] and ferro-actinolite [$Ca_2Fe_5Si_8O_{22}(OH)_2$] are end members of a mineral series with a varying Fe/Mg ratio.

enzyme: An organic catalyst, often a protein secreted by cells. Enzymes catalyze chemical reactions.

eolian: Derived by a process involving the wind, e.g., involving transportation by the wind.

eosinophil: A type of *leukocyte*.

epidemiology: The study of disease in human populations.

epitaxy: The condition where one crystal serves as a template during the growth of another crystal and the resulting mineral pair exhibits specific crystallographic relationships and the boundary is often periodic.

epithelium: The layers of cells and connecting tissues that form the covering of internal and external body surfaces, including the lining of the respiratory tract.

epithelial cell: A type of cell that occurs in an external surface or in the lining of an internal surface. The skin and the surface of the respiratory tract consist of epithelial cells.

equant: Said of a crystal having the same or nearly the same dimensions in all directions. See *asbestiform, fibrous, prismatic, acicular, tabular*.

erionite: A fibrous zeolite of ideal composition $K_2NaCa_{1.5}Mg(Al_8Si_{28})O_{72} \cdot 28H_2O$, where the "·" indicates that the water molecules are not integral to the erionite structure. Erionite has been implicated as the causative agent in mesotheliomas in non-occupationally-exposed individuals in Turkey, and it has been shown to be cytotoxic, genotoxic, fibrogenic, and carcinogenic in a variety of *in vitro* and *in vivo* assays. See *zeolite*.

erythrocyte: Red blood cell.

etiology: The science dealing with the causes of disease.

euhedral: Said of a mineral grain possessing a morphology defined by its own rational crystal faces.

exsolution: The process whereby a crystal separates into two or more crystalline phases without addition or removal of material.

feldspar: A mineral group with the general formula $K_xNa_yCa_{1-(x+y)}Al_{2-(x+y)}Si_{2+(x+y)}O_8$. Feldspar is the most common mineral group at the earth's surface.

Fenton reaction: A generic oxidation/reduction reaction based on the reaction $Fe^{2+} \rightarrow Fe^{3+} + e^-$. The reaction was first proposed by H.J.H. Fenton in 1894 (*J.*

Chemical Society 65, 899–910) as a mechanism to oxidize tartaric acid. A similar mechanism was proposed by F. Haber and J. Weiss in 1934 (*Proc. Royal Society London A* 147, 332–351) in the breakdown of hydrogen peroxide. In research on mineral-induced disease, *Fenton reaction* is often used to refer to the Fe-catalyzed oxidation/reduction process resulting in the formation of active oxygen species, and *Haber-Weiss reaction* is sometimes used to refer specifically to the reaction involving the the Fe-catalyzed breakdown of H_2O_2.

ferritin: The iron-containing protein associated with ferruginous bodies. Originally thought to be a product of red blood cell lysis, ferritin occurs in several different forms, originates from different biological processes, and is generated in several different organs.

ferro-actinolite: An amphibole with ideal composition $Ca_2(Fe^{2+},Mg)_5Si_8O_{22}(OH)_2$. Ferro-actinolite is the iron-rich end member of the $Mg-Fe^{2+}$ series tremolite–ferro-actinolite, with $0.5 > Mg/(Mg+Fe^{2+})$. See *actinolite, amphibole,* and *tremolite.*

ferruginous body: Mineral fibers coated with an iron-rich material believed to be derived from ferritin or hemosiderin. Sometimes, ferruginous bodies are described in terms of the mineral forming the core, e.g., *asbestos body* or *zeolite body*.

fiber: Mineralogists have generally applied the term fiber to minerals with a highly elongate morphology developed during growth. However, numerous other definitions have been assumed for the term fiber, including: (1) a 3:1 aspect ratio; (2) being respirable (e.g., less than a critical size so that it can pass through the respiratory tract to the bronchioles and alveolar ducts); (3) being an individual particle of asbestos; and (4) being less than a specific diameter but greater than a specific length [e.g., Stanton et al. (1981, *J. National Cancer Institute* 67, 965) proposed that most carcinogenic fibers are ≤0.25 µm in diameter and >8 µm in length, whereas other workers have suggested other size criteria]. Clearly, the definition for the term fiber is not rigorous, rather it is generally defined operationally.

fibril: An individual fiber of asbestos, generally a single crystal. Often, an asbestos particle consists of numerous fibrils bound together. Chrysotile fibrils are generally ~25 nm in diameter with a central tube of ~7 nm, but there is much variation in these dimensions.

fibroblast: The primary cell of connective tissue, including in the lung. Fibroblasts secrete molecular collagen that is polymerized to form connective-tissue fibers.

fibrogenic: Possessing the ability to induce fibrosis.

fibrosis: See *pulmonary fibrosis*.

fibrous: In mineralogy, according to Zoltai (1981, *Reviews in Mineralogy* 9A, 237–278), a mineral is said to be fibrous if it "gives the appearance of being composed of fibers, whether the mineral actually contains separable fibers or not." See *asbestiform, acicular, prismatic, equant, tabular*.

flint: A type of microcrystalline quartz.

foci: A concentration of cells that may pile up on one another.

fractional coordinates: The coordinates (designated x, y, and z) that define the position of an atom within the unit cell. The coordinates are defined in terms of fractions of each axis of the coordinate system (i.e., each side of the unit cell).

framework silicate: A mineral class consisting of species with SiO_4^{4-} tetrahedra polymerized in three dimensions, along with other cations. Included in this class are the feldspar minerals and the zeolites. The silica minerals (e.g., quartz, tridymite, cristobalite) are commonly grouped among the silicates for convenience, though they technically are oxides, not silicates.

fraipontite: A 1:1 layer silicate with an ideal composition of $(Zn,Al)_3(Si,Al)_2O_5(OH)_4$.

fuller's earth: A naturally occurring, fine-grained, earthy material made up of various silicates, chiefly the clay minerals montmorillonite and palygorskite.

gangue mineral: The valueless minerals of an ore, where an ore is a rock containing a mineable metal or mineral.

garnet: A mineral group with the general formula $A_3B_2Si_4O_{12}$, where A is typically Ca^{2+}, Mg^{2+}, Fe^{2+}, or Mn^{2+}, and B is typically Al^{3+}, Fe^{3+}, or Cr^{3+}.

genotoxic: Possessing the ability to induce damage to the genetic material (i.e., DNA) of a cell.

gibbsite: $Al(OH)_3$.

goethite: α-FeOOH. See *lepidocrocite* and *akaganeite*.

granite: An igneous rock containing major amounts of quartz and feldspar and lesser amounts of other minerals.

grossular: A garnet mineral with an ideal composition of $Ca_3Al_2Si_4O_{12}$.

growth factor: Highly specific proteins in serum that are generally present in very low concentrations but that are necessary for cell growth by stimulating cell division. See *cytokine*.

grunerite: See *cummingtonite*.

habit: The shape or morphology that a crystal or aggregate of crystals assumes during crystallization.

Haber-Weiss reaction: See *Fenton reaction*.

halloysite: A member of the kaolin group of minerals. Halloysite can exhibit tubular, fibrous, and spherical habits.

hematite: A mineral species that is a polymorph of $Fe_2O_3 = \alpha\text{-}Fe_2O_3$. See *maghemite*.

hemolysis: The destruction, alteration, dissolution, or lysis of red blood cells. This results in bursting of the red blood cell to release hemoglobin.

hexagonal symmetry: Sixfold rotational symmetry.

hexagonal system: One of the six crystal systems, characterized by either a threefold or sixfold axis that is perpendicular to three identical axes that intersect at angles of 120°.

histology: Microscopic anatomy, i.e., the study of the structure and chemistry of tissues as related to their functions.

homeostasis: Equilibrium in the body with respect to various functions and the chemical compositions of fluids and tissues.

HRTEM: High-resolution transmission electron microscopy. Typically, modern 100 to 300-keV TEMs have point-to-point resolutions in the range of 1.6 to 3.0 Å, i.e., approximately equivalent to most nearest-neighbor distances in minerals. Hence, these TEMs can produce images of the atomic structure of a material under fortuitous conditions.

hydrologic cycle: The cycling of water between the atmosphere and the earth's surface, either via transpiration or evaporation. The cycle includes the movement of water to the seas.

hydrothermal activity: A metamorphic or igneous process involving heated (up to hundreds of degrees Celsius) water. Typically, this heated fluid interacts with the rocks through which it flows, forming new (often hydrous) minerals.

hydroxyl group: In mineralogy and geochemistry, the term applies to the negatively charged OH (or hydroxide ion), which may be present as an anionic group in a mineral structure or as a dissolved species in a fluid (sometimes functioning as complexing agent). This is distinct from the "hydroxyl radical."

hydroxyl radical: In biology, the term generally applies to the neutrally charged OH•, which is a highly active free radical capable of damaging other hydrogen-bearing molecules, such as DNA. This is distinct from the "hydroxyl group." The hydroxyl radical is sometimes written as •OH, to emphasize that the unpaired electron is associated with the oxygen atom.

hyperplasia: An increase in the normal numbers of cells. Hyperplasia often causes an increase in the size of an organ.

IARC: International Agency for Research on Cancer.

ideal composition: The composition typically given for a mineral species. Many minerals, however, exhibit solid solution. So the actual composition of a mineral sample may differ from the ideal composition.

Ig-: A prefix indicating immunoglobulin, e.g., IgA and IgG. See *immunoglobulin*.

IL-: A prefix indicating interleukin-, e.g., IL-1, IL-2, etc. See *interleukin*.

illite: A 2:1 layer silicate with the ideal composition $(K,H_3O)Al_2(Si_3Al)O_{10}(OH)_2$ (as given by Nickel and Nichols, 1991). However, most clay mineralogists agree that illite generally contains a partially unoccupied interlayer site, and a more generalized formula for illite is $A_x^+Al_2(Si_{4-x}Al_x)O_{10}(OH)_2$, where $x \approx 0.7$ p.f.u. and A^+ is generally K.

immune response: The host response to pathogens. This response may include the activation of cells, an increase in the numbers of specific cells, and a release of substances that can neutralize the pathogen (e.g., active oxygen species) or mediate other aspects of the immune response (e.g., cytokines).

immunoglobulin: One of a class of structurally related proteins. Antibodies are immunoglobulins, and immunoglobulins probably all function as antibodies. Abbreviated *Ig*.

incongruent: See *dissolution*.

index of refraction: The ratio of the velocity of light *in vacuo* to the velocity of light in the material (e.g., crystal).

inflammation: A process characterized by the influx of cells of the immune system into a tissue. These cells release substances that may either exacerbate or mitigate disease. Inflammation is generally a localized response to injury to cells or tissues. The inflammatory process may be acute (ending after a period of time) or chronic (which may lead to the formation of damaged or scarred tissue).

initiation: A heritable change or modification of the DNA of a cell that is an early step in carcinogenesis. See *carcinogenesis*.

inosilicate: See *chain silicate*.

interlayer region: The region between the 2:1 or 1:1 layers in layer silicates. This region is typically either unoccupied or occupied with cations, water, or hydroxide sheets.

interleukin: A group of lymphokines and polypeptide hormones. Lymphokines are released by lymphocytes following contact with a specific antigen, and they help effect cellular immunity by stimulating monocytes and macrophages. See *cytokine*.

Interleukin-1 (IL-1): A protein that can modulate immune and inflammatory responses. IL-1 exists in two forms (α and β), both having a molecular weight of ~17 kd.

Interleukin-8 (IL-8): An ~6–8-kd protein that is a potent chemoattractant for neutrophilic leukocytes. IL-8 is an important mediator of inflammatory-cell recruitment to sites of tissue injury or infection and is a memeber of the *Chemokine* cytokine family.

interstitium: A small area or space within an organ or tissue (e.g., the lung).

in vitro: In an artificial environment, e.g., in a test tube or media.

in vivo: In the living body.

ionic strength: A measure of the salinity of an aqueous solution, given by the equation $I = ½ \Sigma m_i z_i^2$, where I is the ionic strength, m_i and z_i are the molality and charge of each species (i) in the solution.

isomorphous substitutions: Chemical substitutions that occur in a mineral series possessing one structure across the series.

isostructural: Possessing the same structure.

isotropic: In optical microscopy, said of a mineral or material through which light travels the same speed in any direction.

jimthompsonite: A triple-chain silicate with a structure similar to an amphibole but with wider chains. The ideal formula for jimthompsonite is $(Mg,Fe^{2+})_{10}Si_{12}O_{32}(OH)_4$. Jimthompsonite often occurs as defects within magnesium-rich amphiboles, such as anthophyllite or cummingtonite. The monoclinic variety is named clinojimthompsonite.

kaolin: A soft earthy rock that contains major amounts of kaolin group minerals, particularly kaolinite.

kaolin group: A group of 1:1 layer silicates with the general formula $Al_2Si_2O_5(OH)_4$. Minerals in this group include kaolinite, dickite, nacrite, and halloysite.

kaolinite: A member of the kaolin group.

kd: Kilodalton. A dalton is equivalent to an atomic mass unit, which is defined as $1/12$ the atomic weight of ^{12}C.

kellyite: A 1:1 layer silicate with an ideal composition of $(Mn,Mg,Al)_3(Si,Al)_2O_5(OH)_4$.

lavage: A procedure used to harvest cells (and any other easily mobilized material such as particles or fluids) from a specific organ. Lavage can be performed *in situ* (i.e., on a living organism) or on a specific organ removed by excision.

layer silicate: A mineral class consisting of species with SiO_4^{4-} tetrahedra polymerized in two dimensions to form tetrahedral sheets. Included in this class are 1:1 layer silicates (e.g., serpentine, kaolinite), 2:1 layer silicates (e.g., talc, mica), the 2:1:1 layer silicates (e.g., chlorite), and modulated layer silicates (e.g., palygorskite, sepiolite). 1:1 layer silicates have one tetrahedral sheet bonded to one sheet of octahedrally coordinated cations (termed octahedral sheet), and these 1:1 units are stacked atop one another. 2:1 layer silicates have a tetrahedral sheet bonded to each side of the octahedral sheet to form 2:1 units that are stacked atop one another. Modulated layer silicates generally consist of 1:1 or 2:1 units that are discontinuous, forming islands or ribbons of 1:1 or 2:1 structure that are interrupted by regions with complex structures. Examples of modulated layer silicates include *antigorite* and greenalite (modulated 1:1 layer silicates) and *palygorskite* and *sepiolite* (modulated 2:1 layer silicates).

leach: To remove constituents selectively from a mineral or rock.

lepidocrocite: γ-FeOOH. See *goethite* and *akaganeite*.

lesion: An injury or other change of an organ or tissue of the body which leads to impairment or loss of function.

leukocyte: White blood cell.

leukotriene: See *arachidonic acid*.

lithic: Pertaining to or made of stone or rock.

lizardite: A subgroup name for planar serpentine minerals of ideal composition $Mg_3Si_2O_5(OH)_4$. The term lizardite can be modified by the suffixes *-1T* and *-2H₁* to identify the crystallographic arrangement of the sheets in the structure. The resulting terms ("lizardite-*1T*" and "lizardite-*2H₁*") are mineral-species names. See *serpentine*.

lumen: The interior space within a tubular structure, such as a bronchus.

lymphocyte: Lymph cell.

lyse: To kill a cell.

lysosome: An internal vacuole containing enzymes used by a cell to break down a foreign body, such as a bacterium.

macrophage: A cell type that is part of the immune system. Macrophages are types of phagocytes, i.e., they are capable of engulfing (and subsequently transporting or destroying) foreign bodies.

macrophage inflammatory protein 2 (MIP-2): An ~6-kd protein that is chemotactic for neutrophilic leukocytes. MIP-2 is an important mediator of inflammatory-cell recruitment.

maghemite: A mineral species that is a polymorph of Fe_2O_3. γ-Fe_2O_3. See *hematite*.

magnesio-cummingtonite: See *cummingtonite*.

magnesio-riebeckite: Magnesium end member of the riebeckite series of amphiboles.

magnesite: A carbonate (similar to calcite) with an ideal composition of $MgCO_3$.

magnetite: A mineral species in the spinel group with the ideal composition of $(Fe^{2+},Mg)Fe_2^{3+}O_4$.

major element: Any element present in major amounts, generally $>\sim 0.5$ wt % for the element's oxide. See *minor element* and *trace element*.

massive: Said of a mineral that is physically isotropic, e.g., lacking a platy, fibrous, asbestiform, or acicular morphology. Massive minerals are commonly polycrystalline.

mast cell: A cell type of the connective tissue.

medium: The solution used to grow cells *in vitro*. Media generally consist of both nutrients and antibiotics to protect the cells from contamination.

mesothelioma: A type of malignant tumor arising from mesothelial cells.

mesothelium: The lining of the lung (*pleural mesothelium*), digestive organs (*peritoneal mesothelium*), or heart (*pericardial mesothelium*). A single layer of mesothelial cells constitutes the mesothelium.

metamorphism: The mineralogical, chemical, and structural adjustment of rocks in response to changes in the original physical and chemical conditions to which the rock equilibrated. Metamorphism is restricted to changes that occur at high pressure and temperature. Lower temperature processes similar to metamorphism include *diagenesis* and *weathering*.

metastable: Said of a mineral or phase that is under conditions outside of its thermodynamic stability field or of a mineral or phase that has no field of thermodynamic stability.

mica: A group of 2:1 layer silicates with an ideal charge of -1 p.f.u. on the 2:1 layers. This charge is introduced by the substitution of Al for Si in the tetrahedral sheets, and it is compensated by univalent cations (e.g., K and Na) in the 12-coordinated sites within the interlayer region.

microtopography: The morphology of the surface on the scale of the chemical interactions that take place there, usually on the order of 10^{-1} to 10^0 nm.

Miller indices: The indexes h, k, and l used to describe a crystal face. The values for Miller indices are determined by the reciprocals of the intercepts made by the plane on the **a**-, **b**-, and **c**-axes, respectively. Hence, the (100) plane intersects the **a**-, **b**-, and **c**-axes at

1, ∞, and ∞. The use of brackets "{h k l}" indicates a class of faces or planes that are crystallographically equivalent, i.e., they are related by a symmetry operation.

mineral: A naturally occurring inorganic substance possessing a composition that is fixed or that varies within well defined limits and a periodic structure with translational symmetry. A substance that does not possess translational symmetry (i.e., is not crystalline) is sometimes referred to as a mineraloid.

mineral family: The broadest division within the classification scheme for minerals. *Silicates* comprise a mineral family.

mineral group: A sub-division of a mineral family consisting of mineral species with similar structures. *Amphiboles* comprise a mineral group.

mineral series: Two or more mineral species with the same structure but different compositions, such that intermediate compositions can occur. For example, tremolite and ferro-actinolite form a mineral series with a varying Fe:Mg ratio.

mineral species: Analogous to animal and plant species. The most specific distinct division within the classification scheme for minerals. A mineral species name defines a specific structure and specific composition or compositional range. Mineral species are sometimes subdivided into varieties (i.e., crocidolite is a varietal name for asbestiform riebeckite).

minor element: Any element present in minor amounts, generally ~0.05 to 0.5 wt % for the element's oxide. See *major element* and *trace element*.

MIP-2: See *macrophage inflammatory protein 2*.

mitosis: Asexual reproduction. Nuclear division resulting in exact duplicates of a cell.

modulated layer silicate: A group of minerals with structure based on a layer-silicate-like structure, where the layers are no longer continuous in two dimensions but form strips or islands. The silicate sheets remain continuous but have nonbridging oxygens pointing in two different directions.

moganite: A metastable polymorph of SiO_2.

Mohs hardness scale: Hardnesses of minerals are typically reported using the Mohs hardness scale, which ranges from 1 to 10 (softest to hardest). The scale is defined by ten minerals that have been assigned integral hardnesses: talc (with a hardness of 1), gypsum, calcite, fluorite, apatite, orthoclase, quartz, topaz, corundum, and diamond (with a hardness of 10). Other common minerals are used to define the remaining integral hardness values.

monoclinic system: One of the six crystal systems. Monoclinic minerals have unit cells defined by a coordinate system in which one of the axes (the unique axis) is perpendicular to the other two but the remaining two axes need not be orthogonal. Specifically, the monoclinic system is characterized by only a single twofold axis, a single plane of symmetry, or a combination of the two.

monocyte: A mononuclear leukocyte (9 to 12 µm in diameter) that normally contitutes 3 to 8% of the leukocytes in the circulating blood. Monocytes can transform to macrophages.

montmorillonite: A mineral species of the smectite group with the ideal composition $(Na,Ca)_{0.3}(Al,Mg)_2Si_4O_{10}(OH)_2 \cdot nH_2O$, where the "•" indicates that the water molecules are not integral to the montmorillonite structure. Montmorillonite is a member of the *smectite* group.

mordenite: Species of zeolite with ideal composition $K_{2.8}Na_{1.5}Ca_2(Al_9Si_{39})O_{96} \cdot 29H_2O$, where the "•" indicates that the water molecules are not integral to the mordenite structure. Thus, this zeolite generally has an Al:Si ratio of ~1:4, although the exact ratio varies between samples and can be chemically manipulated. See *zeolite*. Most natural mordenites are fibrous.

mucociliary escalator: The apparatus responsible for clearing foreign agents from the respiratory tract. Particles, other foreign agents, and cells (e.g., macrophages) can be entrapped in the mucus lining the respiratory tract. This mucus is propelled upward and out of the respiratory tract by the beating action of epithelial cells' cilia. This mucus can then be swallowed or expectorated. For a more detailed discussion, see Chapter 14.

muscovite: A 2:1 layer silicate with an ideal composition of $KAl_2(Si_3Al)O_{10}(OH)_2$. Muscovite is a member of the mica group.

nacrite: A 1:1 layer silicate with an ideal composition of $Al_2Si_2O_5(OH)_4$. Nacrite is a member of the kaolin group.

népouite: A 1:1 layer silicate with an ideal composition of $Ni_3Si_2O_5(OH)_4$.

neoplasm: New and abnormal growth.

neutrophil: A mature white blood cell normally constituting 54 to 65% of the total number of leukocytes.

NMR: Nuclear magnetic resonance.

O_2^-: See *superoxide*.

octahedral sheet: A polymerized sheet of edge-sharing octahedra. A structural unit common to many hydroxides and layer silicates.

octahedron: A coordination polyhedron characterized by 6 apices and 8 triangular sides. (octahedra, *pl.*)

odinite: A 1:1 layer silicate with an ideal composition of $(Fe,Al)_3(Si,Al)_2O_5(OH)_4$. Odinite-1M and odinite-1T are the species names.

OH$^\bullet$: See *hydroxyl radical*.

OM: Optical microscopy.

opal: An amorphous or poorly crystalline form of SiO_2 (i.e., a mineraloid). Often opal contains a significant amount of water, so its formula is written as $SiO_2 \cdot nH_2O$, where n is generally <0.5. There are several varieties of opal (e.g., opal-A and opal-CT) that differ in structure. For example, opal-CT has cristobalite- and tridymite-like aspects to its structure.

opsonin: A substance that enhances phagocytosis. A specific or immune opsonin is an antibody formed in response to a specific antigen, i.e., it has structural components that recognize structural components on the antigen. See *immunoglobulin* and *antibody*.

organelles: Structural components of a cell that are analogous to organs in an animal.

ortho–: A prefix used to connote an orthorhombic structure.

orthorhombic system: One of the six crystal systems. Orthorhombic minerals have unit cells defined by an orthogonal coordinate system but with axes of different lengths.

OSHA: Occupational Safety and Health Administration.

oxide: In mineralogy, the oxides are a family of minerals consisting of oxygen+cations, where the cations are typically metals or alkaline earth cations. Examples include magnetite (Fe_3O_4), hematite (α-Fe_2O_3), rutile (TiO_2), and corundum (α-Al_2O_3).

oxygen radicals: See *active oxygen species*.

palygorskite: A mineral species name for a modulated layer silicate with the ideal composition $(Mg,Al)_2Si_4O_{10}(OH) \cdot 4H_2O$, where the "·" indicates that the water molecules are not integral to the palygorskite structure. Although many non-fibrous palygorskites have been described, Jones and Galan (1988, *Reviews in Mineralogy* 19, 631) suggest that these occurrences are better described as other clays (e.g., illite) and that all palygorskites are fibrous. Other names that have been used to describe palygorskite include attapulgite, pilolite, and lassalite. The terms mountain leather and mountain wood have also been used to describe palygorskite, although some of these occurrences may refer to sepiolite, chrysotile, or other minerals. See *attapulgite, sepiolite*.

parachrysotile: A serpentine polymorph with a structure similar to chrysotile.

paracrine effect: See *cytokine*.

paragonite: A 2:1 layer silicate with an ideal composition of $NaAl_2(Si_3Al)O_{10}(OH)_2$. Paragonite is a member of the mica group.

particle: An individual unit constituting a dust or a separable distinct unit in a rock. Particle implies no restrictions with respect to composition, internal structure, shape, or mineral content. However, in a dust, individual particles are frequently monomineralic.

parting: A weak tendency for minerals to fracture along planes. Parting differs from *cleavage* in that it is generally related to the presence of defects (e.g., exsolution lamellae) within the mineral rather than the crystal structure of the mineral. Parting varies from sample to sample.

pathogen: An agent capable of producing disease.

pathogenesis: The origin and development of disease.

pathology: The study of the nature, causes, and effects of disease.

pecoraite: A 1:1 layer silicate with an ideal composition of $Ni_3Si_2O_5(OH)_4$.

pericardium: The membrane lining the cavity containing the heart. Sometimes called the pericardial sac. The pericardium consist of a layer of connective tissue covered with a layer of mesothelium. The part adjacent to the heart is the epicardium or visceral pericardium.

peritoneum: The membrane lining the cavities containing the stomach, intestines, *etc.* The peritoneum consists of a layer of connective tissue covered with a layer of mesothelium. The part adjacent to the organs is the visceral peritoneum, and the part lining the cavity is the parietal peritoneum.

p.f.u.: per formula unit.

phagocyte: A type of cell capable of internalizing a foreign body via the process called phagocytosis.

phagocytosis: The process of engulfing a foreign body. Typically, a cell (phagocyte) will extend pseudopodia around the foreign body and internalize it in a cavity (phagosome). Once internalized, the cell can secrete various agents in an attempt to destroy the foreign body.

phagosome: The internal vacuole formed during phagocytosis. This vacuole will contain the foreign body targeted by the phagocyte.

phagolysosome: The internal vacuole formed by the merging of a phagosome and a lysosome.

phillipsite: A species of zeolite with ideal composition $K(Ca_{0.5},Na)_2(Al_3Si_5)O_{16} \cdot 6H_2O$, where the "•" indicates that the water molecules are not integral to the mordenite structure.. See *zeolite*. Most natural phillipsites are fibrous.

phlogopite: A 2:1 layer silicate with an ideal composition of $KMg_3(Si_3Al)O_{10}(OH)_2$. Phlogopite is the magnesium end member of biotite series.

phosphate: A family of minerals with structures containing phosphorus–oxygen tetrahedra.

phospholipase: An enzyme that catalyzes the hydrolysis of a phospholipid.

phospholipid: A lipid containing phosphorus. Phospholipids consitute cell membranes.

phyllosilicate: See *layer silicate*.

plane: See *Miller indices*.

plaque: A deposit of material on a flat surface, such as the epithelium of the lung or pleura.

pleura: The membrane lining the cavities containing the lungs. The pleura consists of a layer of connective tissue covered with a layer of mesothelium. The part covering the lungs is the visceral pleura; the part lining the cavity is the parietal pleura; and the potential space between the visceral and parietal pleura is the pleural cavity or space.

pneumoconiosis: In pathology, non-neoplastic reaction of the lung due to the accumulation of inhaled dusts. Some uses of the term include all lung diseases caused by the accumulation of inhaled dusts.

point of zero charge: Points of zero charge are pH values corresponding to specific states of surface charge. The conventional PZC is the pH value at which the net particle charge is zero. The point of zero net proton charge (PZNPC) is the pH value at which the net proton surface charge is zero, and this can be measured by potentiometric titration. A thorough discussion of the distinctions between the various points of zero charge is given by Sposito (1984, *The Surface Chemistry of Soils*, Oxford University Press, New York, 234 p.).

polyanions: Polymerized anionic groups.

polygonal serpentine: A chrysotile-like serpentine mineral consisting of a tube-like crystal with planar serpentine (lizardite) sides.

polyhedron: See *coordination polyhedron*.

polymorph: Polymorph is a term applied to materials with the same composition but with different structures.

polymerized: Said of molecular groups (e.g., coordination polyhedra) that share common apices, edges, or faces with adjacent molecular groups. For example, a SiO_4^{4-} tetrahedron can share three of its oxygens with adjacent SiO_4^{4-} tetrahedra (and so on) to form a polymerized sheet.

polysomatic series: A series of structures made from differing proportions or arrangements of a pair of polysomatic slabs or modules. For example, pyroxene–amphibole–triple-chain-silicate (e.g., jimthompsonite)–mica form a polysomatic series of P (pyroxene) and M (mica) modules: (P)–(MP)–(MMP)–(M).

polysome: In cell biology, a cluster of ribosomes active in protein synthesis.

polysome: In mineralogy, modular structures constructed from slabs of less complex structures. For example, the amphibole structure can be represented by slabs of pyroxene structure (P slabs) alternating with slabs of mica structure (M slabs). Thus, the amphibole structure is a polysome that can be denoted by ...MPMPMP... or (MP).

polytype: A structure formed by ordered stacking of essentially identical layers in a specific sequence. Polytype is a subset of polymorph.

prismatic: A term used to describe crystals exhibiting aspect ratios >1 and having parallel sides. *Glossary of Geology* offers no restrictions on the aspect ratio; however, Skinner et al. (1988, *Asbestos and Other Fibrous Materials*, Oxford Univ. Press, New York) suggest that prismatic crystals have aspect ratios in the range 1–3. See *asbestiform, fibrous, acicular, equant, tabular*.

prograde metamorphism: Metamorphism that results from increasing the pressure and temperature such that the minerals which crystallized under lower pressures and temperatures are no longer stable, and the rock recrystallizes to other minerals.

progression: See *carcinogenesis*.

proliferation: The sequence of events leading to DNA synthesis and mitosis; the process of cell replication or division.

promotion: A necessary series of steps in tumor development characterized by increased proliferation (cell division) of initiated cells. See *carcinogenesis*.

prostaglandin: See *arachidonic acid*.

protein: Macromolecules consisting of long sequences of amino acids.

protolith: The original rock from which a metamorphic rock developed during metamorphism.

pulmonary fibrosis: A lung disease characterized by increased deposition of collagen and other proteins in the lung. The fibroblast is the primary cell type affected.

PVPNO: polyvinyl-2-pyridine *N*-oxide. A polymer capable of bonding to negatively charged sites. PVPNO has been shown to decrease the bioactivity of negatively charged surfaces, such as quartz and the edges of kaolinite crystals, ostensibly by blocking the "dangling" Si–O bonds at the surface. A detailed discussion of PVPNO is presented in Nolan et al. (1981, *Environmental Research* 26, 503–520.).

pyriboles: A term used to describe any *pyr*oxene or amph*ibole*. These minerals can be difficult to distinguish in hand sample, so pyribole was a term introduced by field geologists to categorize these occurrences until identification could be made in the laboratory. Biotite (a 2:1 layer silicate) can sometimes be confused with pyriboles in hand sample, so the term *biopyribole* is sometimes used to indicate the pyriboles plus micas and talc.

pyrophyllite: A mineral species that is a 2:1 layer silicate with the ideal composition $Al_2Si_4O_{10}(OH)_2$.

pyroxenes: A mineral group of chain silicates possessing a single-chain structure with an ideal composition of BCT_2O_6. The pyroxene M1 site (designated C in the formula) is octahedrally coordinated; commonly contains Mg and/or Fe^{2+} (but can contain elements such as Al, Fe^{3+}, and Mn); and is analogous to the amphibole M2 site. The pyroxene M2 site (designated B in the formula) is 6-, 7-, or 8-coordinated; can contain Mg, Fe^{2+}, Ca, or Na; and is analogous to the amphibole M4 site. The T-site is tetrahedrally coordinated and generally contains Si but can also accommodate Al.

pyroxenoids: A mineral group of chain silicates possessing a single-chain structure. The pyroxenoids differ from the pyroxenes in that their silicate chains possess a different topology. Generally, the chains in pyroxenoids are kinked relative to the linear pyroxene chains. Pyroxenes and pyroxenoids can be treated as a polysomatic series between true pyroxene and wollastonite (a pyroxenoid with an ideal composition of $CaSiO_3$). The topologies of the remaining pyroxenoids can be constructed by assembling P and W slabs.

PZC: Point of zero charge. See *point of zero charge*.

PZNPC: Point of zero net proton charge. See *point of zero charge*.

quartz: A member of the silica group. At temperatures below ~573°C, all "quartz" is low quartz or α-quartz.

RBC: Red blood cell.

reconstructive transformation: A phase transformation characterized by the breaking of chemical bonds. These transitions occur slowly and are simply irreversible. Often, the activation energy associated with breaking bonds and reorganizing the structure is so large that a mineral may exist outside of its stability field, i.e., it may exist metastably. See *displacive transition*.

refractive index: See *index of refraction*.

relative risk: The ratio of the rate of disease in one population compared to a reference population.

respirable: Said of any agent able to pass through the respiratory tract. Generally, in man particles must be smaller than about 10 μm in length and about 3 μm in diameter to be respirable. However, in other species, these figures are different. For example, in rats, the particles must be less than about 5 μm in length and about 1 μm in diameter.

retrograde metamorphism: Metamorphism that results from decreasing the pressure and temperature such that the minerals which crystallized under higher pressures and temperatures are no longer stable and the rock recrystallizes to other minerals.

richterite: A species of amphibole with ideal composition $Na_2Ca(Fe,Mg)_5Si_8O_{22}(OH)_2$.

riebeckite: An amphibole with ideal composition $Na_2(Fe^{2+},Mg)_3Fe^{3+}_2Si_8O_{22}(OH)_2$. The asbestiform variety of this amphibole is often referred to as "crocidolite."

Rietveld analysis: A method for analyzing powder diffraction data to extract various parameters, including abundances of minerals, mineral lattice parameters, atomic coordinates, and other structural information about the minerals present in a sample. Typically, Rietveld analysis is applied to X-ray or neutron powder diffraction data.

rock: Generally an aggregate of one or more minerals, e.g., granite, limestone, sandstone.

rotational symmetry: The property possessed by structures that are identical when rotated through an angle about an axis. The degree of the rotational symmetry describes the number of such operations possible in 360°. For example, the letter "N" has 2-fold rotational symmetry, since a 180° rotation reproduces the letter and two 180° rotations are possible in 360°.

rutile: A mineral species that is a polymorph of TiO_2.

SAED: See *selected-area electron diffraction.*

SAM: Scanning Auger electron microscopy.

SEM: Scanning electron microscopy or scanning electron microscope. Sometimes SEM is used to refer to secondary electron microscopy, one of the techniques done with a scanning electron microscope.

selected-area electron diffraction: An electron diffraction technique that utilizes an aperture on the image plane of the objective lens to select a specific area of the image from which to form a diffraction pattern. The minimum unique area that can be selected for diffraction is ~0.5 μm, because, even if smaller apertures are used, this represents the approximate minimum area from which diffracted beams will originate.

sepiolite: A mineral species name for a modulated layer silicate with the ideal composition $Mg_4Si_6O_{15}(OH)_2·6H_2O$, where the "·" indicates that the water molecules are not integral to the sepiolite structure. Although many non-fibrous sepiolites have been described, Jones and Galan (1988, *Reviews in Mineralogy* 19, 631) suggest that these occurrences are better described as other clays (e.g., illite) and that all sepiolites are fibrous. Other names that have been used to describe sepiolite include parasepiolite, gunnbjarnite, xylotile, falcondoite, Meerschaum, Myrsen, and Ecume de Mer. See *palygorskite.*

sericite: A petrological term generally applied to fine-grained muscovite or paragonite.

serpentine: A mineral group of 1:1 layer silicates with a general formula of $(Mg,Al,Fe,Mn,Ni,Zn)_{2-3}(Si,Al,Fe)_2O_5(OH)_4$. Also used to denote the subroup consisting of the Mg 1:1 layer silicates lizardite, chrysotile, and antigorite.

serpentinite: A green rock consisting dominantly of serpentine minerals. Serpentinites generally form by the alteration of pyroxene- and olivine-bearing rocks.

sheet silicate: See layer silicate.

silanol group: The functional group SiOH.

silica: A chemical term for silicon dioxide, SiO_2. Also a term for material consisting of SiO_2, e.g., quartz and silica glass.

silicate: A family of minerals with structures containing silicon–oxygen polyhedra plus other cations (e.g., Mg, Ca, and Na). Typically, these polyhedra are tetrahedra, and they can be isolated or polymerized to one, two, or three other tetrahedra by sharing of their oxygen atoms.

silicosis: A nonmalignant disease of the lung caused by inhalation of dust containing silica minerals. Not all silica minerals may cause silicosis.

smectite: A mineral group of 2:1 layer silicates with the general formula $(Ca,Na,Li)_{0-1}(Mg,Fe,Al,Li,Ni,Cr,Zn)_{2-3}(Si,Al)_4O_{10}(OH)_2 \cdot nH_2O$. Smectites possess interlayer cations that can be easily exchanged with cations in a solution.

solid solution: substitutional, interstitial, and omission...

space group: The group of symmetry operations (e.g., rotation, reflection, translation) that can be performed on a crystalline structure and leave the arrangement of atoms in the structure unchanged. There are 230 possible space groups.

specific gravity: Specific gravity is a unitless measure of density and is equivalent to the ratio between the density of a substance relative to the density of water at $4°C$ (i.e., the maximum density of water). Specific gravity is often determined by comparing the weight of a material in air with the weight of a material in water; hence, it is sometimes defined as the ratio of a material's weight to the weight of an equivalent volume of water.

squamous: Scaly or scale-like. A layer of flattened cells lining or covering the surface of organs such as the skin or esophagus. The lining of the bronchus may change its state of differentiation from epithelial to squamous cells in response to irritants such as cigarette.

stacking fault: A type of planar defect in crystalline materials in which two adjacent planes are stacked differently relative to each other than the other planes in the material are stacked.

stishovite: A very high-pressure polymorph of SiO_2, containing 6-coordinated Si.

STM: Scanning tunneling microscopy or scanning tunneling microscope.

stoichiometry: The numerical relationship between elements in a reaction or formula.

stuffed derivative: A group of minerals with frameworks isostructural with one of the silica polymorphs but with Al and cations substituted for Si and vacancies. For example, eucryptite ($LiAlSiO_4$) is a stuffed derivative of quartz (SiO_2 or $\square SiSiO_4$); the two minerals have frameworks with the same structure and are related by the substitution $\square + Si \rightarrow Li + Al$.

substitution: The exchange a different atom or vacancy for an atom or vacancy in a crystal structure (e.g., the exchange of Al for Si in the tetrahedral sites of a silicate).

superoxide: An active oxygen species with the general formula of O_2^-.

superstructure: A structure with a longer-range periodicity than the dominant periodicity, or substructure. For example, the dominant periodicity in the 2:1 layer silicates is ~10 Å normal to the sheets; however, some 2:1 layer silicates have an additional periodicity that repeats every two 2:1 layers, i.e., they have a superstructure normal to the 2:1 layers. When the superstructure repeats at an integral multiple of the dominant periodicity, the superstructure is said to be commensurate. Incommensurate superstructures do not repeat at an integral frequency of the dominant periodicity. Diffraction maxima arising from a superstructure are generally weaker than those arising from the substructure.

synergistic: The characteristic of two or more substances acting together to produce an effect greater than the additive effects of the individual substances.

tabular: Said of a crystal form that has one dimension markedly smaller than the other two. See *asbestiform, fibrous, prismatic, equant, acicular.*

talc: A mineral species that is a 2:1 layer silicate with the ideal composition $Mg_3Si_4O_{10}(OH)_2$.

target cell: A cell that may be affected in disease.

tectosilicate: See framework silicate.

TEM: Transmission electron microscopy or transmission electron microscope.

tetrahedral sheet: A polymerized sheet of tetrahedra. A structural unit common to the layer silicates.

tetrahedron: A coordination polyhedron characterized by four apices and four triangular sides. (tetrahedra, *pl.*)

TGF-β: Transforming growth factor β. TBF-β is a *cytokine*, and it potentiates or inhibits (depending on the cell type) the response of most cells to other growth factors. TBF-β also regulates the differentiation of some cell types.

titania: A chemical term for titanium dioxide, TiO_2.

TNF-α: Tumor necrosis factor α or cachectin. A 17-kilodalton (kd) protein (a cytokine) produced by a variety of phagocytic and nonphagocytic cell types including: macrophages, monocytes, polymorphonuclear leukocytes, lymphocytes, smooth muscle cells, and mast cells. TNF-α has the ability to lyse tumor cells *in vitro*.

torr: A measure of pressure (or vacuum), where 1 torr = $1/760$ atmosphere ≈ 1 mmHg.

toxicology: The study of toxic substances (or poisons).

trace element: Any element present in minute amounts, generally <~0.05 wt % for the element's oxide. See *major element*

translocation: The process that transports a particle from its site of deposition (e.g., in the lung) to another site, such as the pleura.

tremolite: A species of amphibole with the ideal composition $Ca_2(Mg,Fe^{2+})_5Si_8O_{22}(OH)_2$. Tremolite is the magnesium-rich end member of the Mg–Fe^{2+} series tremolite–ferro-actinolite, with $Mg/(Mg+Fe^{2+}) \geq 0.9$. See *actinolite, ferro-actinolite* and *amphibole*.

triclinic: One of the six crystal systems. Triclinic minerals have unit cells that have a onefold axis of symmetry.

tridymite: A high-temperature polymorph of SiO_2.

trigonal symmetry: Threefold rotational symmetry.

trioctahedral sheet: An octahedral sheet in which all unique (i.e., 3 out of 3) octahedral sites are occupied by a cation.

trioctahedral substitution: The substitution of cations into the otherwise unoccupied sites in a dioctahedral sheet.

tumor: The pathologic term for a mass, usually used to describe a neoplasm. New and abnormal growth. Tumors may be benign (i.e., will not metastasize or invade adjacent tissue) or malignant.

tumorigenic: Possessing the ability to induce a tumor.

twinning: The property of minerals to intergrow rationally with one or more units of the same mineral and where the two units are related in a mathematically described manner (i.e., by a symmetry operator).

unit cell: A subunit of a structure possessing translational periodicity. The entire crystal structure can be described by repeating the unit cell along the translation directions.

vacancy: An unoccupied crystallographic site.

vacuole: A membrane-bound cavity within a cell.

vermiculite: A mineral group of 2:1 layer silicates with the general formula $(Mg,Fe,Al)_3(Si,Al)_4O_{10}(OH)_2 \cdot 4H_2O$.

viability: Capability of living.

vug: A small cavity in a rock, usually lined with crystals.

weathering: The mineralogical and chemical adjustment of a rock or mineral in response to interactions with atmospheric agents. Most weathering occurs at or near the earth's surface and includes processes such as the hydration of minerals stable at higher temperatures and pressures to form clay minerals and other soil constituents.

winchite: An amphibole with the ideal composition $NaCa(Mg,Fe^{2+})_5(Si,Al)_8O_{22}(OH)_2$.

XAS: X-ray absorption spectroscopy.

XPS: X-ray photoelectron spectroscopy.

XRD: X-ray diffraction.

X-ray diffraction: A series of techniques that exploit the ability of crystalline substances to diffract X-rays. X-ray diffraction (XRD) techniques include methods for analyzing individual crystals or powders. The diffracted X-rays contain information about the types and arrangements of atoms with each crystal structure and the abundances of each mineral species present. See *Rietveld Analysis*.

zeolite: A mineral group.

zeta potential: A measure of the potential energy difference between a bulk solution and the boundary between the free solvent and the solvent adhering to the fiber surface. Zeta (ζ) potential is proportional to the total surface potential.

ZPC: Zero-point-of-charge. Equivalent ot point of zero charge, or PZC. See *point of zero charge*.

INDEX

Absorption. *See* Sorption
Actinolite. *See* Amphibole, actinolite–tremolite
Active oxygen species, 287, 356, 494, 514–515, 518.
  hydrogen peroxide 285, 299, 355, 446, 515, 540.
  hydroxyl radical, 285, 299, 355, 446, 515, 540–541.
  superoxide, 285, 299, 355, 446, 494–500, 514, 540.
Adenosine triphosphate (ATP), 205.
Adsorption. *See* Sorption
Adventitious molecules, 277.
Agate. *See* Silica
Albumin, 342–343, 443.
Alumina, γ-Al_2O_3, 495. *See also* Corundum
Alveoli, 439.
Amorphous calcium phosphate, 225.
Amorphous silica. *See* Silica
Amosite. *See* Asbestos
Amphibole, 8–17, 26, 42–45, 53–54, 67–71, 101–132, 351, 362, 412, 413, 414, 472. *See also* Asbestos.
  actinolite–tremolite, 10, 12–14, 17, 43–45, 69–75, 103–104, 113–114, 120–122, 130, 285, 296, 316, 350, 353, 366, 374, 391–396, 413–423, 472–480, 491. *See also* Amphibole, byssolite
  anthophyllite, 13–14, 17, 43, 69–70, 105–106, 108–117, 120–127, 296, 316, 411, 472–475, 497.
  byssolite, 110, 114, 117.
  cummingtonite–grunerite, 13–14, 16, 69–70, 102–105, 112–114, 120–121, 285, 296, 366–367, 372–375. *See also* Asbestos, amosite
  hornblende, 297, 366.
  richterite, 75, 122, 391.

Amphibole (*continued*)
  riebeckite, 14–16, 50– 53, 69–70, 102–105, 111–119, 293, 497. *See also* Asbestos, crocidolite
  winchite, 70, 75, 125.
Analytical electron microscopy, (AEM) 45–46, 114, 266–267.
  energy dispersive spectroscopy (EDS), 10, 13, 42–46, 49, 126–131, 266–267.
  electron-energy loss spectroscopy (EELS), 266–267.
Anatase, 206–216, 495, 501. *See also* Brookite; Rutile
Antigen, 492.
Antigorite. *See* Serpentine
Apatite, 224–225.
Arachidonic acid, 494, 496, 503.
Asbestos, 16–17, 19, 43, 46–50, 52, 57, 61–64, 69–71, 74–75, 78, 81, 92–93, 95, 100–103, 106, 108–131, 293, 299, 309–311, 337, 347–351, 435, 437, 442.
  amosite 16–17, 43, 76, 101–102, 110–114, 120, 127–130, 285, 296, 301–302, 311, 316, 321, 350, 369, 409, 414–423, 472–475, 481, 491–492, 495, 497, 500. *See also* Amphibole, cummingonite–grunerite
  chrysotile 10, 17, 19, 43, 45–55, 61, 69–87, 90–100, 111, 114, 118, 120–124, 127–131, 284–285, 287, 289–293, 297–298, 300, 310–313, 316, 321, 350–352, 395–396, 409, 412–423, 435, 472–476, 481, 491, 495–502, 518. *See also* Serpentine
  crocidolite, 10, 16–17, 43, 50, 53, 69, 101–102, 110–114, 120–121, 127–130, 286, 293, 296–302, 311, 313, 316, 321, 324, 349, 350, 369, 398, 409, 414–420, 472–475, 481, 491–497, 500–503. *See also* Amphibole, riebeckite

Asbestos bodies, 352, 414.
Asbestosis. See Fibrosis
Aspect ratio, 10, 17, 113, 120, 127, 197, 212, 409, 423.
Atomic force microscopy (AFM), 96, 269, 280, 290, 293.
 crocidolite, 294–295.
 lizardite, 96, 290.
 roggianite, 29.
Attapulgite. See Palygorskite
Bentonite, 388.
Benzo[a]pyrene, 517, 531.
Berthierine, 76, 91.
BET, surface area, 286, 540.
Biotite, 11, 23, 76, 106, 114, 390. See also Phlogopite
Black lung, 384–385.
Brazil twins, 193–196, 203. See also Twinning
Brindleyite, 76, 91.
Bronzite, 297.
Brookite, 207–216. See also Anatase; Rutile
Brucite, 74, 77, 95, 139, 142, 143, 144, 146, 155. See also Nemalite
Byssolite. See Amphibole
c-fos, 518–519.
c-jun, 518–519.
c-sis, 519.
Calcification, 353, 396.
Cancers
 adenocarcinoma, 365, 529.
 adenoma, 526, 529.
 carinoma, 348, 365, 414, 473, 478, 513, 526, 529.
 colon, 354.
 gastrointestinal, 474.
 larynx, 354.
 lung, 351–353, 361, 366–389, 392–394, 415, 423, 475.
 lymphoma, 354, 365, 529.
 mesothelioma 309–312, 348–356, 368, 373, 395–397, 414–423, 473–480.

Cancers (continued)
 prostate, 385.
 stomach, 354, 370, 385.
Cappadocian region, Turkey, 396–398.
Carboxymethylcellulose, 285.
Carcinogenesis, 191, 207, 516.
Case studies, 349.
Catalysis, 75, 97, 161–162, 169–170, 179–181, 197, 204, 206–209, 212–215, 285–286, 292, 299–300, 304.
Cation exchange, 167, 170, 179–181.
Cadmium compounds, 442.
Cell types
 eosinophils, 492.
 epithelial cells, 353, 355, 432–4357, 451–452, 490, 501, 504, 513, 526, 532.
 erythrocytes (red blood cells), 286, 490–491.
 fibroblasts, 355, 363, 422, 453, 490, 501, 504, 513–514, 532.
 lymphocytes, 439, 443, 492, 500, 514.
 macrophages, 352, 355, 363–364, 384, 422, 433, 439–459, 490–505, 514, 532.
 mast cells, 492, 500, 532.
 mesothelial cells, 490, 514, 519.
 monocytes, 441, 500–502, 532.
 neutrophils, 514.
 pneumocytes, 447, 450–451.
 polymorphonuclear leukocytes (PMNs), 434–449, 500–501, 514.
 type I epithelial cell, 447–451, 455–456, 513, 526.
 type II epithelial cell, 365, 450, 456–459, 482, 513, 518, 526, 532–539.
Chabazite. See Zeolite
Chalcedony. See Silica
Chemisorption, 329.
Chemoattractants, 434, 440, 448, 514.
Chemokines, 501.

Chert. *See* Silica
Chlorite, 10, 19, 24–26, 45.
Chrysotile. *See* Asbestos
Cigarette smoke. *See* Smoking
Cilia, 431–432, 438.
Clearance 356, 383, 412–414, 423, 437–439, 447–455, 477.
Cleavage 14, 17, 19, 23, 26, 71, 92, 113–114, 117–120, 123–124, 146–147, 158, 162, 179, 202.
Clinoptilolite. *See* Zeolite
Coal, 383–386.
Coesite. *See* Silica
Cohort studies, 348, 351.
Collagen, 363, 355, 384.
Comminution, 7, 33, 117.
Complement, 440, 443, 491, 496.
Contact angles, 331, 333.
Corundum, 220–221, 256, 495.
Coughing, 438–439.
Cristobalite. *See* Silica
Crocidolite. *See* Asbestos
Cronstedtite, 76, 92, 123.
Cummingtonite. *See* Amphibole, cummingtonite–grunerite
Cytokine, 363, 494, 496, 500, 514, 519, 532.
Cytotoxicity, 223, 355, 494–495.
Dahllite, 224.
Dauphiné twins, 193. *See also* Twinning
Defects, 61, 108–110, 113–114, 118, 160, 165–166, 185, 206, 211, 220.
Deferoxamine, 285–286, 301–303.
Deposition, particles in lung, 427–430, 446.
Diagenesis, 170, 187, 202–203.
Diatomaceous earth, 363, 378.
Dickite. *See* Kaolin minerals
Dioctahedral sheet, 22, 141. *See also* Octahedral sheet
Diopside, 12, 74.
Energy-dispersive analysis (EDS). *See* Analytical electron microscopy

Displacive transitions, 193, 199.
Dissolution, 61, 95, 97, 99–100, 118, 120–121, 164, 203–205, 222, 292, 297, 412–413, 449–450, 476–477.
DNA, 516, 518, 539–540, 541.
Dolomite, 23.
DQ12. *See* Silica
Dust, 7–10, 13–14, 17–19, 26, 33–57.
  eolian transport of, 26, 39–40.
Electron-energy loss spectroscopy (EELS). *See* Analytical electron microscopy.
Effusions, pleural, 354, 396.
Eicosanoids, 496.
Electrical double layer, 283.
Electron diffraction. *See* Transmission electron microscopy
Electron probe microanalysis. *See* Scanning electron microscopy
Enantiomorphic twins, 196, 221. *See also* Twinning
Endocytosis 436, 445.
Erionite. *See* Zeolite
Eucryptite, 204.
Exocytosis, 447, 449.
Exposure techniques, animals
  inhalation 471, 477–483, 526, 529.
  intraperitoneal injection, 472, 474, 478–481.
  intrapleural injection, 472, 474, 479, 483.
  intratracheal injection, 471, 476, 478–483, 526, 529, 535.
Exsolution, 62, 71, 110, 218, 222.
Feldspar, 8, 10–14, 36, 40–45, 200, 208, 238, 362, 375.
Fenton reaction, 269, 300, 491, 494, 497, 515.
Ferruginous bodies, 352, 414.
Fibronectin, 443.
Fibrosis, 348, 351–352, 355, 396, 422, 437, 473, 478–482, 516.

Fibrosis (*continued*)
    asbestosis, 348, 351–352, 355–356, 383, 414, 417, 419, 422–423, 473, 475, 513
    kaolinosis, 388.
    silicosis, 363–364, 373–383, 481, 513.
Flint. *See* Silica
Fraipontite, 91.
Francolite, 223.
Free-cell response, 441.
Free energy of cohesion, 328–329.
Free radical. *See* Active oxygen species
Goethite, 139, 142–150, 215, 222, 369.
Gold miners, 371, 375.
Granite workers, 362, 375.
Growth factors, 356, 363, 422, 496, 514.
Grunerite. *See* Amphibole, cummingtonite–grunerite
Guinea pigs, 479, 481, 497–498, 500.
Gypsum, 320.
Haber–Weiss reaction, 300.
Habit 10, 14, 19, 26, 30, 64, 67–71, 84, 90–93, 102–103, 110–119, 124, 129. *See also* Morphology
Halloysite. *See* Kaolin minerals
Hamsters, 478, 483, 497, 523, 529, 531, 532, 535.
Hectorite. *See* Smectite
Hematite, 53, 142, 211, 217–222, 281, 366, 369, 377, 529.
Hemolysis, 287–288, 490–495.
Hornblende. *See* Amphibole
Humoral immunity, 443.
Hydrogen peroxide. *See* Active oxygen species
Hydrologic cycle, 7–8, 33–34, 38.
Hydrolytic enzymes, 496.
Hydroxyl radical. *See* Active oxygen species
Hyperplasia, 526, 529–532.
Illite, 139, 150, 154, 162, 339, 482, 483.
Immunoglobulins, 440.
    IgA, 443, 498.
    IgE, 443.
    IgG, 440, 443–446, 498, 500.
    IgM, 440, 443.
Index of refraction, 121–122.
Inflammation, 285–286, 352–354, 355, 478, 481, 490–492, 496, 514–516.
Interfacial tension, 329.
Interleukin, 496.
    IL-1, 500–504, 533, 536.
    IL-2, 503.
    IL-6, 536.
    IL-8, 504.
Interstitium, 513.
Ionic strength, 335.
Iron content, importance in disease, 355.
Iron ore miners, 366, 369–370.
Janus Green B, 540.
Jimthompsonite, 70, 84, 107–108, 117, 122–124, 131.
Kaolin minerals, 10, 21, 62–64, 71–76, 85–91, 121–127, 139, 142, 477–478, 495.
    dickite 75–76, 85–88, 121–123, 126.
    halloysite, 71, 75–76, 87, 90, 94, 121–124, 127.
    kaolinite 19–22, 45, 48, 61, 71–76, 85–90, 94, 121–127, 142, 164, 339, 370, 388.
Kaolinosis. *See* Fibrosis
Karain, Turkey, 397.
Keatite. *See* Silica
Kellyite, 91, 123.
Lactate dehydrogenase, 517.
Latent period, 349–354, 368–369, 397.
Lavage, 439, 493.
Layer silicate, 8, 13–14, 17–26, 36, 45, 48, 57, 74, 91–94, 102, 108, 139, 141–144, 147, 150–165, 171, 176, 183.
Lepidocrocite, 139, 142–143, 146–147, 150.

Leukotriene B$_4$, 440–441, 503.
Lipid peroxidation, 285.
Lizardite. *See* Serpentine
Lung
    digestion of, 411.
    mineral burden in, 43–45, 293, 298, 350, 409–424.
    fiber concentration in, 411–424.
    fiber distribution in, 421–422.
    fiber sizes in, 422–424.
    particle deposition in, 441.
    particle overload in, 455–456.
Lysosomes, 494.
Macrophage inflammatory protein 2, 440, 441, 504.
Maghemite, 217–222.
Magnetite, 216–222, 366, 377.
Man-made mineral fibers, 357, 476.
Mazzite. *See* Zeolite
MCP-1, 504.
Mica, 8–10, 19–23, 42–48, 76, 106, 114, 362, 479, 482. *See also* Layer silicate
Mice, 478, 479, 481, 483, 497, 531, 532, 535.
Microtopography, 276–281, 287, 303.
Miller indices, 111, 121.
Mimetite, 223.
Min-U-Sil. *See* Silica
Mineral species, importance of, 254.
Mineral surfaces, 276–277.
Manganese compounds, 442.
Moganite. *See* Silica
Monkeys, 481.
Montmorillonite. *See* Smectite.
Mordenite. *See* Zeolite
Morphology, 9, 21, 46, 49, 67, 70–71, 79, 92–94, 109–110, 115, 124–127, 131, 165–166, 178. *See also* Habit
    acicular, 14, 17, 68, 147.

Morphology (*continued*)
    asbestiform, 14, 16f, 53, 61, 64, 67–71, 92f, 102, 103, 104, 111, 112, 113, 114, 117, 119, 120, 121, 122, 125, 129.
    equant, 14, 30, 68, 181, 222.
    euhedral, 213.
    fibrous, 14, 19, 26, 30, 62, 67–71, 92–95, 103, 108–117, 122, 123–125, 146–147, 162, 180–181, 186, 195–197, 212–214, 296, 337.
    prismatic 14, 17, 68, 114, 117, 124, 181, 195–196.
Mucociliary escalator, 431, 438, 448.
Mucous blanket, 433.
Muscovite, 10, 19–24, 331, 339–343, 390, 479. *See also* Mica
NADPH oxidase, 497–498.
Nemalite, 350. *See also* Brucite
Nepouite, 76, 92, 95.
Nuclear magnetic resonance (NMR), 201, 206.
Octadecyldimethylchlorosilane, 285.
Octahedral sheet, 19, 21, 24, 26, 66, 71–72, 77–78, 82–84, 88, 91, 97, 141–144, 147, 152–157, 161, 165–167, 290, 291.
    dioctahedral sheet, 22, 141.
    trioctahedral sheet, 141.
Octahedra, 65, 140.
Octyldimethylchlorosilane, 285.
Odinite, 76, 92, 123.
Opsonin, 443–445.
Oxidative burst, 515.
p53 suppressor gene, 535.
Palygorskite, 19, 26, 139, 141, 164–165, 167–170, 240, 338, 388–390, 479–480. *See also* Sepiolite
Particle analysis, 266.
Phagocytosis, 355, 440, 443–445–447, 450.
Phagosome, 494.
Phlogopite, 10, 22, 75, 85, 390.
Phosphate, 223–224.

Phospholipase C, 498.
Physisorption, 329.
Proton induced X-ray emission (PIXE), 267–268.
Platelet derived growth factor, 519.
Pleural effusions, 354, 396.
Pleural plaques, 348–349, 353, 387, 395–396, 409, 419, 423.
Pleural space, 353.
Poly-D-alanine, 205.
Poly-L-alanine, 205.
Polyhedra
    octahedron, 65, 140.
    tetrahedra, 65–66, 140.
Polysomes, 62, 82–83, 106–109.
Polyvinylpyridine-N-oxide (PVPNO), 321.
Probe molecule, pyridine, 291, 296.
Proliferation, 356.
Proteolytic enzymes, 496.
PVPNO 206, 285, 288, 491, 494, 540.
Pyribole 62, 106–131.
Pyridine, 291, 296.
Pyromorphite, 222.
Pyrophyllite, 139, 150–161, 340, 387–388.
Pyroxenes, 297, 362.
    diopside, 12, 74.
Point of zero charge (PZC), 148, 204, 283.
Point of zero net proton charge (PZNPC), 283–284, 288, 296.
Rabbits, 481–482, 497.
Radon, 366–367, 370–375, 383.
ras oncogenes, 535.
Rats, 473–483, 497, 523–532.
Reconstructive transition, 199.
Refractory ceramic fibers 503, 552.
Regulations, 351, 356, 545–553.
    asbestos, 546–551.
    erionite, 351.
    glass fiber, 552.
    mineral wool, 552.
    phosphates, 223.

Regulations (*continued*)
    refractory ceramic fibers, 552.
    silica, 190, 365.
Respirable particles, sizes of, 472.
Respiratory burst, 446, 497, 515.
Richterite. *See* Amphibole
Riebeckite. *See* Amphibole
Rietveld method. *See* X-ray diffraction
Roggianite. *See* Zeolite
Rutile, 10, 45, 200–216, 377, 495. *See also* Anatase; Brookite
Sample preparation/purification 235–249.
    density separation, 238.
    disaggregation, 236.
    field-flow fractionation, 240.
    heavy liquids, 238.
    magnetic separation, 237.
    settling velocity, 239, 243.
    sieving, 237.
    size fractions, preparation of, 241.
Sandblasting, 381.
Sarihidir, Turkey, 397.
Scanning electron microscopy (SEM), 10, 14, 42, 45, 49, 269, 293.
    crocidolite, 294.
    electron probe microanalysis 261–266.
    erionite, 31.
    mazzite, 30.
    mordenite, 32.
    quartz, 29.
    roggianite, 29.
    wavelength dispersive spectroscopy, 266.
    winchite, 125.
Scanning tunneling microscopy (STM), 269, 280–281.
    hematite, 281.
Sepiolite, 19, 26, 139, 141, 164, 165, 167, 168, 169, 170, 338, 388, 479. *See also* Palygorskite
Sericite, 19, 479. *See also* Illite; Layer silicate; Mica; Muscovite

Index 583

Serpentine, 8, 17–19, 52, 61–97, 100, 106, 114, 121–127, 414–415, 472.
    antigorite 17, 74–85, 92–93, 106, 114, 121, 124, 127, 395.
    chrysotile. *See* Asbestos
    lizardite, 17, 22, 66, 71–86, 91–96, 106, 114, 122–124, 127, 290, 395.
    polygonal serpentine, 74, 84–85, 90, 93, 124.
Serpentinite, 19, 50–52, 74, 76, 84, 92–93, 415.
Silanol, 162, 203–205, 282, 288, 491, 494, 539, 540.
Silica 26, 42, 45, 55, 185–206, 221, 286–288, 340, 362–366, 375, 442, 481–483, 491, 523–541.
    agate, 195.
    amorphous silica, 189–206, 286, 363, 378, 388, 482.
    chalcedony, 196–197, 202–203.
    chert, 26, 56, 196, 217, 378.
    coesite, 187, 200, 286, 495.
    cristobalite, 197–206, 286, 363, 378–379, 388, 482, 495, 529, 539.
    DQ12 (quartz), 502, 524, 537, 539.
    flint 196, 202.
    keatite 42, 190.
    Min-U-Sil, 502, 524, 537–539.
    moganite, 196, 202.
    opal, 200–203, 363, 378.
    quartz, 185–206, 240, 286–288, 340, 362–381, 388, 479–483, 491, 495, 497, 500–503, 529–531, 535–537.
    stishovite, 189, 200, 209, 286, 495.
    tridymite, 197–206, 238, 286, 482, 495, 529–531, 539.
Silicic acid, 491.
Silicosis. *See* Fibrosis
Secondary ion mass spectrometry (SIMS), 267–270.
Slate, 362, 377–378.

Smectite, 22–23, 139, 150–164, 168–171, 240, 333, 337, 339, 343.
    hectorite, 337.
    montmorillonite, 19, 22, 23, 150, 153, 159, 333, 388, 397.
Smoking, 348, 353–356, 370, 381–382, 387, 391, 423, 434, 517–518, 350–352.
Sodium polytungstate, 238.
Solid solution, 10–14, 24, 78, 81, 91, 110–111, 202, 217–223.
Sorption, 148–149, 160–161, 166–167, 276, 277, 278, 279, 280, 283, 284, 288, 291.
Specific gravity, 238.
Stacking faults, 73, 83, 89, 109, 113, 117, 178, 198.
Stanton hypothesis, 2, 4, 304, 309–325, 355.
Stishovite. *See* Silica
Stokes' law, 239.
Stuffed derivative, 202.
Substitution, compositional, 10–16, 24, 68, 73–101, 111, 118, 141–161, 168–173, 176, 180, 195, 211–212, 220–224.
Superoxide. *See* Active oxygen species
Superoxide dismutase, 497.
Surface
    area, BET, 286, 540.
    atomic structure, 278.
        low-energy electron diffraction (LEED), 269.
        scanning tunneling microscopy, 269, 280–281.
        atomic force microscopy, 96, 269, 280, 290, 293.
    charge, 282, 288, 290.
    composition, 277.
    roughness, 287, 288.
    tension, 330, 331, 337.
    thermodynamics, 339.
Survival, following diagnosis, 353–354.
Taconite, 366, 367, 368, 369.

Talc, 74–77, 86, 94, 103, 106, 108, 114, 122, 139, 150–152, 155–158, 160–165, 333, 340, 387, 411, 478–479.
Tetrahedra, 65–66, 140.
Thin layer wicking, 333.
Tracheobronchial tree, 431.
Transforming growth factor β (TGF-β), 519, 533, 535–536.
Translocation of particles, 453, 472.
Transmission electron microscopy (TEM), 10, 17, 42, 56, 64, 70, 78–86, 93, 95, 108–131, 195, 200, 214, 260, 270, 316, 410, 411. *See also* Analytical electron microscopy
   electron diffraction, 64, 83, 90–94, 121, 127–131.
   amphibole, 86.
   anthophyllite, 116.
   chlorite, 27.
   chrysotile, 18, 79–80, 86, 116, 289.
   crocidolite, 294, 298–299.
   kaolinite, 23.
   lizardite, 86.
   montmorillonite, 25.
   palygorskite, 28.
   sepiolite, 27.
   smectite, 25.
   talc, 86.
   vermiculite, 24.
Tremolite. *See* Amphibole
Tridymite. *See* Amphibole
Trioctahedral sheet, 141. *See also* Octahedral sheet
Tuberculosis, 364, 372–385, 396.
Tumor necrosis factor α (TNF-α), 496, 500–504, 533, 536.
Twinning, minerals, 73, 109–114, 117–129, 193–200, 212, 221.
   Brazil, 193–196, 203.
   Dauphiné, 193.
   enantiomorphic, 196, 221.
Vanadinite, 223.

Vermiculite, 19, 22, 45, 390–392, 480.
Volcanic ash, 340.
Winchite. *See* Amphibole
Wollastonite, 394.
X-ray diffraction (XRD), 82–84, 87–93, 121–124, 130, 255.
   detection limits, 256.
   determination of mineral content, 255.
   microdiffraction, 261.
   reference-intensity-ratio method, 256.
   Rietveld method, 257–261.
X-ray photoelectron spectroscopy (XPS), 284, 297–302.
ζ-potential (zeta-potential), 97, 120, 283, 288–290, 335.
Zeolite, 10, 28, 30, 32, 45, 57, 139, 141–142, 163–164, 168–181, 394, 480.
   chabazite, 397.
   clinoptilolite, 139, 170, 174–183, 240, 258, 397.
   erionite, 30–33, 139, 170, 173–183, 257, 259, 286, 353, 397–398, 480–481, 497, 546, 551.
   mazzite, 30–31.
   mordenite, 30–33, 139, 170–183, 240, 258, 481, 497.
   roggianite, 30.
   synthetic zeolite 4A, 481.